Perspectives on the History of Chemistry

Series Editor

Seth C. Rasmussen, Department of Chemistry and Biochemistry, North Dakota State University, Fargo, ND, USA

Commonly described as the "central science", chemistry and the chemical arts have an extremely long history that is deeply intertwined with a wide variety of other historical subjects. Perspectives on the History of Chemistry is a book series that presents historical subjects covering all aspects of chemistry, alchemy, and chemical technology.
Potential topics might include:

- An updated account or review of an important historical topic of broad interest
- Biographies of prominent scientists, alchemists, or chemical practitioners
- Translations and/or analysis of foundational works in the development of chemical thought

The series aims to provide volumes that advance the historical knowledge of chemistry and its practice, while also remaining accessible to both scientists and formal historians of science. Volumes should thus be of broad interest to the greater chemical community, while still retaining a high level of historical scholarship. All titles should be presented with the aim of reaching a wide audience consisting of scientists, chemists, chemist-historians, and science historians.

All titles in the book series will be peer reviewed. Titles will be published as both printed books and as eBooks. Both solicited and unsolicited manuscripts are considered for publication in this series.

More information about this series at http://www.springer.com/series/16421

Anders Lennartson

Carl Wilhelm Scheele and Torbern Bergman

The Science, Lives and Friendship of Two Pioneers in Chemistry

Springer

Anders Lennartson
Gothenburg, Sweden

ISSN 2662-4591 ISSN 2662-4605 (electronic)
Perspectives on the History of Chemistry
ISBN 978-3-030-49193-2 ISBN 978-3-030-49194-9 (eBook)
https://doi.org/10.1007/978-3-030-49194-9

This Springer imprint is published by the registered company Springer Nature Switzerland AG
The registered company address is: Gewerbestrasse 11, 6330 Cham, Switzerland

Map over southern Sweden with a few places of interest for the reader of this book. Provinces (landskap) are shown in grey. *Image*: Wikimedia Commons/Anders Lennartson

Nature may be compared not improperly to an immense book, written in an unknown language.

Torbern Bergman 1779

Preface

In order to really understand Scheele, one has to be an experimental chemist. If one can read about a chemical substance, without feeling an irresistible desire to explore its properties in the laboratory, one can never fully understand Scheele's motivations and driving forces. Just like Scheele, I came across out-dated chemistry books as a boy. I had many magic moments when I managed to get hold of the chemicals and equipment needed to repeat the experiments form these books. Next, I learned to design my own simple experiments and to predict their outcome and finally, many years later, I was in the position to perform experiments not previously described in the literature. Thus, I was introduced to chemistry in a similar fashion as Scheele and I believe I am in a good position to understand him. Unfortunately, the image of Scheele in the literature is quite often misleading, as authors have tried to fit Scheele into contexts where he does not properly belong. In the late nineteenth century, for instance, Scheele was turned into a national hero. To be a hero implies that someone makes a large sacrifice or takes risks without personal benefits. That would be a false picture of Scheele. Scheele was an outsider in the scientific community; he did not have a plan for a brilliant career, but was completely satisfied as long as he had free access to a laboratory.

Bergman is easier to grasp as a person since he shares many features with modern successful scientists. He was very good at selling his ideas, excellent at securing support from the right people, and a master in presenting his results. If he lived today, he would probably have been very successful in writing proposals for research grants. Unlike Scheele, he carefully planned the steps in his career. The study of Bergman still presents some problem: while all of Scheele's work is in the field of chemistry, Bergman's work ranges from celestial bodies to atoms, from caterpillars to volcanoes. No modern scientist can cover such a wide field. As a chemist, I have focused on Bergman's chemical research.

In 2015, I published the first extensive biography of Scheele that has been written since the 1930s. The book, *Ett kemiskt äventyr—Carl Wilhelm Scheele och hans värld* (A Chemical Adventure—Carl Wilhelm Scheele and his World) was written in Swedish and contained chapters on pharmaceutical history written by Professor Emeritus Björn Lindeke and Bo Ohlson. I divided the book into two parts; the first part was a biographical part, while the second part discussed each of Scheele's publications, explaining his results in modern terms. This second part was adapted to English and published by Springer in 2017 as *The Chemical Works of Carl Wilhelm Scheele*. For a more thorough discussion of the scientific aspects of Scheele's work, the reader is referred to that book. The present book is based on material from my Swedish biography of Scheele, combined with new material on Bergman. Scrutinising every detail of Bergman's and Scheele's lives and research could easily have taken decades and filled several volumes. Instead, I have tried to keep the page numbers down in order to create a book that can be read rather used as a reference volume only.

As a chemist, I have approached Bergman and Scheele in a different way than a historian would have done. I have spent much more time in the laboratory repeating crucial experiments than I have spent in archives. As a chemist, it is the published results and their interpretation that is most interesting for me, and I assume, most other chemists. The aim of this book is to explain the history of Bergman and Scheele to scientists, but I also hope the book will be usefull for historians who want to understand the science of Bergman and Scheele. No more than basic chemical knowledge is required in order to understand the text, and no previous knowledge of eighteenth-century chemistry or Swedish history is assumed.

In order to make the text understandable to a modern reader, modern chemical nomenclature and terminology has been applied throughout the text. I clearly remember my struggle when I first read a book on the history of chemistry and had to try to figure out the meaning of terms such as "muriatic acid" from the context. The reader must however keep in mind that terms such as oxidation were unknown in the eighteenth century. The original works by Bergman and Scheele used a mixture of old traditional Swedish (or German) and Latin nomenclature. To give the reader a flavour of the nomenclature they used, I have often added the original Latin names, or the closest possible English eighteenth-century equivalents of the Swedish and German names, in parenthesis. Quotations have, as long as possible, been taken from eighteenth-century translations, where such exist. Explanations of terms and chemical names within quotations have been added in brackets [ ].

Although there are many excellent texts (see Appendix C) on many different aspects of Bergman's and Scheele's lives and scientific contributions, most of the literature has been available in Swedish and German only. It is my hope that I have filled a gap in the literature by writing this book.

Last, but not least, I would like to express my gratitude to Dr. Andreas Furängen, former CEO of the Swedish Pharmaceutical Society for support and interesting discussions, Prof. Em. Björn Lindeke and Bo Ohlson for their extensive knowledge of pharmaceutical history, Prof. Seth Rasmussen for many valuable comments, Dr. Petra Rönnholm for beautiful photos, and my wife for endless patience.

Göteborg, Sweden Anders Lennartson

Contents

Introduction 1

Torbern Bergman and Carl Wilhelm Scheele were among the most successful chemists of their time—a short period of about 15 years in the 1770s and 1780s. They both died young at the peak of their careers. Their work was closely connected, yet their collaboration and friendship was very unlikely. Scheele was the enthusiast who became absorbed by chemistry as a young boy and devoted his life to chemistry entirely for the pleasure. He had no academic education and seems to have had little interest in other subjects than chemistry. Bergman, on the other hand, had a broad academic background; initially inspired by the biological work of Linnaeus he studied insects, then he turned to astronomy and the rapidly evolving field of electricity. Without any prospects of becoming professor in physics, he finally turned to chemistry simply because a professorship became available. He was soon knighted and was well integrated in the network of European scientists. A friendship and collaboration between an apothecary assistant like Scheele and a professor like Bergman was indeed uncommon in the hierarchic eighteenth century.

Although Scheele has been underestimated as a theoretician, his strength was definitely in experimental chemistry. As an experimenter he was curious, honest and very observant. For Scheele, it was the work in the laboratory that was important. It was certainly in the lab with his crucibles and flasks that he was most happy. For Scheele, writing publications was not a means of achieving fame and high social status. Putting so much devotion into his experiments he could not, however, let anyone else take the credit. Anyone who has published a scientific paper will probably agree that it is a pleasant feeling to see the final publication after a lot of hard work. For Bergman, who had become a chemist for entirely different reasons, it was the search for the larger contexts that was important. Whether this was in the field of chemistry, physics, or biology was perhaps not crucial. His skill was to organise information and to rationalise results, using Newton and Linnaeus as inspiration. Thus, for Bergman, it was probably not so important to carry out the actual experiments, and he probably entrusted his students to carry out a large portion of the laboratory work. Scheele would only ask for

A. Lennartson, *Carl Wilhelm Scheele and Torbern Bergman*, Perspectives on the History of Chemistry, https://doi.org/10.1007/978-3-030-49194-9_1

assistance in a few rare cases, when his simple apothecary laboratory prevented him from performing the experiments he envisioned, such as the reduction of manganese and molybdenum oxide and the decomposition of silver carbonate with a burning glass. In Scheele's publications, the experiments are extremely well described and it is typically not difficult to reproduce his experiments.[1] In contrast, Bergman's experiments are not so well described and in order to reproduce his work, more chemical experience and improvisation is needed. While Scheele's research was entirely devoted to chemistry, Bergman had a much broader interest. During his career, he also wrote auroras, lightning conductors, leeches, caterpillars, geology, bee keeping, mineralogy, crystallography and meteorology. One of the more bizarre research projects that Bergman embarked on was to prepare ethanol from faeces. The idea came from his former student Johan Gottlieb Gahn (1745–1818; Sect. 13.1), who told Bergman in May 1772 that peasants in Falun had prepared vodka from human faeces due to lack of grain [1]. Bergman took the information seriously and apparently asked Scheele to purify sodium carbonate for the project.

1.1 Sweden in the Eighteenth Century

As a protestant country, Sweden entered the Thirty Years War in 1630. Troops lead by Swedish King Gustavus Adolphus (Gustav II Adolf; 1594–1632, King from 1611) were so successful that large areas of present day Germany became Swedish provinces after the Peace of Westphalia in 1648. This is important for the present storey, as Scheele's ancestors now became Swedish citizens. Swedish Pomerania (Vorpommern), where Scheele was born, remained Swedish until 1815, two decades after Scheele's death.

Throughout the seventeenth and early eighteenth centuries, Sweden was involved in a number of wars with archenemies Denmark, Poland, and Russia, and given Sweden's small population (approximately 1,4 million in 1700), it became increasingly difficult to protect the overseas provinces. While Gustavus Adolphus' troops were largely composed by German soldiers under Swedish command, King Charles XII (Karl XII; king 1697–1718) had to fight his wars in Russia largely with Swedish men. The large costs and the high death rates lead to an increased opposition against the sovereign King. The King died in 1718 during the siege of the fortress of Fredriksten in Norway.[2] As Charles XII had made the controversial decision that he would only marry for love and not as a part of a political scheme,

[1]Provided, of course, one can access the correct starting materials. Another complicating factor is that Scheele had no means to measure high temperatures, and for some experiments a charcoal fire would be needed. A word of caution could also be in place: some of the experiments are very dangerous to perform and can, if proper safety measures are not taken, result in serious injuries or death.

[2]The circumstances of his death are unclear, he may either have been killed by an enemy bullet, or assassinated.

and that he could not marry during the war, he was still unmarried as he died. Thus, his sister Ulrika Eleonora became Queen. However, she was forced to sign a document that essentially transferred all political power to the Swedish Diet[3] (Riksdagen), thus putting an end to the absolute monarchy that had been established in 1680. The following year, she abdicated in favour for her husband, Frederick I (Fredrik I; Fig. 1.1). Frederick was a German prince and general who had married Ulrika Eleonora speculating he could become sovereign King of Sweden in the—not very unlikely—event that Charles XII would not return from the battlefields. Although he actually became King of Sweden, he found himself deprived of all political power and spent his days eating, drinking, and involving himself in amorous affairs that gave him a solidly bad reputation in Swedish history. When Torbern Bergman graduated from the Gymnasium of Skara, he did so with an obituary of Frederick I, who had recently died. The reign of Frederick I and his successor, another German prince, Adolph Frederick (Adolf Fredrik) is called the Age of Liberty. Neither of the two kings had any political power, and Sweden was ruled by the Diet. In fact, when Adolph Frederick refused to sign documents, a stamp with his signature was made.

From 1738, two political parties, called the Hats and the Caps, fought for power. It was a politically turbulent period with economic corruption on a massive scale. The Hat party, which had assumed power in 1738, was sponsored by France. They favoured grand new ideas and a strong military. The Hat party started two wars, against Russia (1741–1743) and against Prussia (1757–1762). Lacking the military competence of Gustavus Adolphus or Charles XII, Sweden lost both wars. During the Prussian war, troops surrounded Scheele's hometown Stralsund. The Hat party supported mercantilism and tried to minimise import and reduce consumption of luxury items. Coffee, for example, was banned in 1756. Of importance for this storey is the great interest the government took in new technical and scientific ideas that could improve Swedish economy. The Hat party supported early industrialism, and the Swedish economy improved. In 1765, the Hat party lost power to the Cap party that was sponsored by England, Russia, and Denmark. The Cap party supported a more cautious economic policy and tried to reduce governmental spending. The Cap party's seize of power led to an economic crisis, but the party also introduced one of the world's first freedom of press acts in 1766. The Hat party reassumed power in 1769. A third party, the Court party, struggled in vain to increase the power of the King.

In April 1772, the Cap party once again rose to power, but their position did not last long. The Age of Liberty, with all its political conflicts, was put to an abrupt end on August 19, 1772, when King Gustav III (Fig. 1.2), who had succeeded his father Adolph Frederick in February 1771, performed a coup d'état and turned Sweden into an almost absolute monarchy. As Crown Prince, Gustav had played an important role in Bergman's promotion to professor, and to some extent, he regarded Bergman as his client. With a new, more stable government, corruption

[3]The Swedish Diet would eventually develop into the Swedish Parlament. Before the nineteenth century, it only assembled on orders from the King.

Fig. 1.1 King Frederick I and Queen Ulrika Eleonora. Georg Engelhard Schröder, 1733. *Photo* Nationalmuseum, Stockholm

Fig. 1.2 King Gustav III. The white ribbon around his arm was the sign used by those involved in the 1772 coup. Alexander Roslin, undated. *Photo* Nationalmuseum, Stockholm

decreased and economy improved, although Gustav's interests were more devoted to culture than to science, and the freedom of press was once again suppressed. Although Scheele and Bergman grew up and lived during a rather turbulent time in Swedish history, neither of them seems to have had any interest in political matters.

1.2 Science: Linnaeus and Klingenstierna

The Age of Liberty coincided with the enlightenment movement, which was welcomed by Sweden, with its strong connections to France during the power of the Hat party. Swedish science soon reached a very high place of eminence. Science was seen by the leaders as a way to improve the society by, for instance, increasing the quality and yield of iron and copper, improve agriculture and reduce import. Among Swedish scientists, botanist Carl Linnaeus (1707−1778) and physicist Samuel Klingenstierna (1698−1765) were the most influential, and their names will appear repeatedly throughout this book.

Linnaeus (Fig. 1.3) was born in May of 1707 in Råshult, a vicarage in Stenbrohult, Småland. It was his father, Nils Ingemarsson, who took the name Linnæus. Nils became vicar in Stenbrohult, but was also a devoted amateur botanist. Carl Linnaeus attended grammar school in Växsjö before he enrolled at Lund University in 1727. He received private tutoring from Killian Stobæus (1690–1742), professor of natural history, in whose house Linnaeus also lived. In 1728, Linnaeus moved to Uppsala, where he studied medicine under professors Lars Roberg (1664–1742) and Olof Rudbeck the younger (1660–1740).[4] He was also appointed by botanist, linguist, and theology professor Olof Celsius the elder (1670–1756) as private tutor for his children. After Linnaeus had presented a thesis on sexual reproduction in plants in 1729, Rudbeck allowed him to give lectures on botany. On recommendation by Rudbeck and Celsius, he was commissioned by the Royal Society of Sciences at Uppsala to travel through Lappland, the northernmost province of Sweden, and northern Finland in 1732. During his travel, he studied botany, zoology, geology, as well as local history and ethnology. Latter, Linnaeus undertook similar trips to Dalarna in 1743 (where he met his wife to be), Öland and Gotland in 1741, Västergötland in 1746, and Skåne in 1749. The books that he wrote after returning from his trips have been reprinted many times through the years and are still appreciated accounts of the life in Sweden in the eighteenth century.

In February 1735, Linnaeus travelled to Hardwijk in the Netherlands to obtain his M.D. degree. The University of Hardwijk was well known for issuing a doctoral degree within a week. Linnaeus had written his thesis in Uppsala, and he also brought manuscripts for a number of scientific publications, most notably his *Systema naturæ*, which would revolutionise biology by introducing the system of classification still used today. After receiving his M.D. degree, Linnaeus went to

[4]Olof Rudbäck the younger was a son of Uppsala professor Olof Rudbäck the Elder (1630–1702), who discovered the lymphatic system.

Fig. 1.3 Carl Linnaeus painted in 1775 by Alexander Roslin. *Photo* Nationalmuseum, Stockholm

Leiden to study with the celebrated physician and chemist Herman Boerhaave (1668–1738). It was in Leiden he published his *Systema naturæ*, and rapidly gained a reputation. Boerhaave introduced Linnaeus to the wealthy banker George Clifford (1685–1760), director of the Dutch East Indian Company, who hired Linnaeus to catalogue his botanical garden and extensive natural history collections. Linnaeus was also sent to England and France on Clifford's expense. In 1738, Linnaeus returned to Sweden, and with an improved economy and increasing reputation, he was finally able to marry Sara Moræus, whom he had met during his travel in Dalarna. His appointment as professor of medicine in Uppsala is discussed in Chap. 9.1. After his ennoblement in 1761, he took the name von Linné, and in Sweden is invariably known as Carl von Linné. He died in 1778 after a series of strokes and years of declining health.

The second most important scientist in Sweden during the early eighteenth century was probably Samuel Klingenstierna (Fig. 1.4). Klingenstierna [2] was born in 1698 in Linköping and he enrolled at Uppsala University in 1717. Initially, he studied law, but when he had to consult Euclid to find the answer to a problem, he was struck by the beauty of mathematics. In 1720, he obtained a civil servant position in Stockholm but used his spare time to study mathematics. After five years, he was back in Uppsala, now giving private tutoring in mathematics and he also published his first mathematical works in *Acta Societatis Regiae Scientiarum Upsaliensis*. In 1727, he embarked on a journey through Europe: Germany,

Fig. 1.4 Swedish physicist and mathematician Samuel Klingenstierna. Drawing by Jean Eric Rehn 1747. Nationalmuseum, Stockholm

Switzerland, France, and England. On recommendation by mathematician Christian von Wolff (1679–1754) in Marburg, he was appointed professor in mathematics in Uppsala in 1728. He did not, however, return to Sweden and Uppsala until 1731. His appearance in Uppsala would become very important for the development of mathematics and physics at the university: with Klingenstierna, the theories of Newton received a definite breakthrough in Uppsala on the expence of the old Aristotelean system. From England, Klingenstierna bought physical instruments, including a static electricity generator, a hydrostatic balance, and a pressure boiler, and thereby introducing experimental physics in Uppsala. When, in 1750, a professorship in physics[5] was established, Klingenstierna was the obvious choice. Due to poor health, he had to apply for a leave of absence after just two years and devoted his time to research in ballistics for the army. In 1756, the royal couple, Adolph Frederick and Lovisa Ulrika, attempted a coup d'état, which failed. In the aftermaths of the coup, the Crown Prince Gustav's teacher, Carl Gustaf Tessin (1695–1770; Sect. 9.3), was fired. Count Carl Fredrik Scheffer took over Tessin's position and Klingenstierna became the Crown Prince's teacher in geometry and surveying [3]. He now finally resigned from his professorship in Uppsala. As the

[5]A chair in chemistry was established at the same time (Chap. 9).

sensitive Prince had developed strong bonds to Tessin, this was initially a very unfortunate solution. The situation was further complicated by conflicts between Scheffer and the Queen, Lovisa Ulrika [4]. Still, Klingenstierna remained in the court until 1764, the year before his death. Having an accomplished physicist as teacher does not seem to have evoked any scientific interest in the young Prince, however, who never paid much attention to science.

Klingenstierna was a perfectionist, and only by way of exception did he publish his discoveries: after his death, 200 unpublished papers were found. Thus, many of his most important discoveries were independently discovered and published later by others, in some cases, as late as in the nineteenth century. His contemporaries in Sweden regarded Klingenstierna and Linnaeus as Sweden's most important scientists, but while Linnaeus is still well known, few remember Klingenstierna today. Although Klingenstierna left Uppsala University the same year that Bergman enrolled, Bergman's teachers, such as Mårten Strömer and Bengt Ferner, were students of Klingenstierna and it was in Klingenstierna's steps that Bergman followed as a young lecturer in physics.

1.3 The Phlogiston Theory

In order to understand the chemistry in this book, a brief knowledge of the phlogiston theory is needed. The phlogiston theory was introduced by German chemist and physician Georg Ernst Stahl (1659–1734) in the early eighteenth century [5] and was based on ideas of Johann Joachim Becher (1635–1682), who in turn based his theories on the old alchemical principles mercury, sulphur and salt which were believed to comprise matter at the time. The three principles were strongly advocated by Paracelsus, but their origin goes back to old Arabic alchemy. The phlogiston theory taught that combustible bodies contained an inflammable principle called *phlogiston* that escaped upon combustion. This caused chemists to believe that our modern elements were compounds. Metals were thought to consist of a metal calx (metal oxide in modern terminology) and phlogiston, while sulphur was composed of sulphuric acid and phlogiston. On burning sulphur, the phlogiston escaped leaving sulphuric acid (actually sulphur dioxide). Different chemists had different conceptions of phlogiston. Some, e.g. Stahl, regarded it as a *principle* that could not be isolated in free form. Others believed phlogiston to be a *material substance* and both charcoal and hydrogen were claimed to be almost pure phlogiston. On watching the flames of burning matter, it was quite natural to regard combustion as a process where something escaped. Also, when reducing lead(II) oxide (calx of lead) with charcoal, the calx and the charcoal seemed to disappear leaving metallic lead; the gaseous carbon dioxide escaped the early chemist's attention. It thus seamed logical to regard lead as a compound of lead calx and charcoal (phlogiston) and to regard metals as compounds was not as far-fetched as could be imagined. The phlogiston theory was the most widely accepted chemical theory in the mid-eighteenth century, and for the first time, chemists could explain what is now known as redox-reactions using a single theory. By the end of the century, Lavoisier's[6] theory of combustion (*i.e.* the absorption of oxygen by a burning body) had largely replaced the phlogiston theory.[7] In some sense phlogiston would be revived in the twentieth century: in the reduction of copper(II) sulphate to metallic copper by zinc in aqueous solution, phlogistonists like Bergman and Scheele believed that zinc gave off phlogiston to the copper calx which was present in the solution united with sulphuric acid, forming copper and zinc sulphate. In this process, phlogiston is nothing but valence electrons. This reaction was not easily explained by Lavoisier and his supporters but was studied in detail by Bergman.

References

1. Trofast J (1994) Johan Gottlieb Gahn: brev, vol 2, Lund, p 49
2. Lindroth S (1975–1977) Samuel Klingenstierna, Svenskt biografiskt lexikon, vol 21, Stockholm, p 319
3. Landen L (2004) Gustaf III—en biografi. Stockholm, Wahlström & Widstrand, p 31
4. Landen L (2004) Gustaf III—en biografi. Stockholm, Wahlström & Widstrand, p 32
5. Stahl GE (1703) Speciminis Becheriani Principia Mixtionis Subterraneae demonstrandi. Pars prima. In: Becher J (1703) Physica Subterranea… editio novisima…& Specimen Becceranum, Joh. Ludov. Gleditschium, Leipzig, p 39

[6] Antoine Lavoisier (1743–1794), French chemist. See Chap. 30.8.

[7] A few chemist, including Joseph Priestley and Henry Cavendish, refused to abandon the phlogiston theory.

Bergman and Scheele: Childhoods

2

Torbern Olof Bergman[1] was born on March 20, 1735. As Sweden used the older Julian calendar until February 1753, Bergman's birthday was March 9 by that calendar. He was born on the Katrineberg royal residence (kungsgård; Fig. 2.1), which Bergman's father had rented at the time. Katrineberg is located in the Låstad hamlet between Mariestad and Skövde in the Västergötland province of Sweden. Carl Linnaeus travelled through Västergötland in the summer of 1746, and described Mariestad as "small, but pleasant, lying on the eastern side of great lake Vänern, built with small, but pleasant wooden houses. Streets are straight and light" [1]. Mariestad has a cathedral, but no bishop since 1647. As Linnaeus noted, it is located on the shore of Vänern, the third-largest lake in Europe (after Ladoga and Onega in Russia). Linnaeus never passed Bergman's childhood home, but later arrived in Skövde, "a very small spot located on the eastern side of [table hill] Billingen without lake or any particular situation, houses were small and the streets irregular and the churchyard surrounded by beautiful ash trees" [2].

The Bergman family has been traced to Michel Esbjörnsson, a district scribe (häradsskrivare) in the Kålland district of Västergötland, who lived in the mid-seventeenth century [3]. He had four children, Esbjörn, Jonas, Torbjörn (or Torbern), and Lars. Little is known about Lars, but the other three sons followed in their father's steps and became district scribes; Esbjörn finally became Mayor in Borås. It was these three sons who started to use the name Bergman around 1700; the first documented usage of the name is by Jonas in 1694. There are still descendants of Michael Esbjörnsson named Bergman in Sweden, but it should be

[1]Pronounced ["bærjman"]. The correct spelling of his family name is "Bergman", but he was frequently referred to as "Bergmann" in German literature, and several translations were published under that name. On his first publications, he spelt his first name "Thorbern" rather than "Torbern". His middle name, Olof, was only used on his first dissertation in 1755.

A. Lennartson, *Carl Wilhelm Scheele and Torbern Bergman*, Perspectives on the History of Chemistry, https://doi.org/10.1007/978-3-030-49194-9_2

Fig. 2.1 Katrineberg in Låstad, where Bergman was born. The yellow wooden main building, partly obscured by trees, with its two wings may originate from Bergman's time. *Photo* Anders Lennartson, June 2018

noted that Bergman is a common name in Sweden, used by several unrelated families. Torbjörn Michelsson, who used the name Bergman from 1703, was district scribe in Läckö district and bailiff in Höjentorp. He had two children, Barthold Bergman (Torbern Bergman's father) and a daughter, Brita, who married a priest Erik Afzelius in Mariestad. Thus, Bergman was probably distantly related to Johan Afzelius (Chap. 11), his student and successor as professor in Uppsala.

Bergman's father, Barthold Bergman, was the royal bailiff (kronobefallningsman) in the Vadsbo district and was born in 1704. Thomson wrote in his *History of Chemistry* about Bergman's father that "A receiver of the revenues was at the time, in Sweden, a post both disagreeable and hazardous" [4]. Bergman's mother, Sara Hägg, was a merchant daughter born in Gothenburg in 1697 and a cousin of the wealthy and successful merchant Niclas Sahlgren (1701–1776), director and co-founder of the Swedish East Indian Company. Sara Hägg married for the first time in the village of Björsäter, Västergötland, in 1721, with district official Olof Myra, born in 1689. The couple had three children: Margaretha (1721–1763), Petrus (1726–1741), and Jonas (1730–1737). Of these, only Margaretha could have made any lasting impression on Torbern. Olof Myra died in Mariestad in early 1732. On November 9, 1733, Sara married Bergman's father in Mariestad. Torbern was the couple's first born child. In 1737, Carl Fredric Bergman was born; he would become an official in the Swedish Navy [5] and died in Stockholm in 1789; he married in 1770 and had six children. In 1738, Torbern's sister, Maria Regina Bergman, was born. She married a customs inspector, Andreas Bark from Kristinehamn, in 1761 and had six children. One of them was called Torbern and was recorded as a silk weaver apprentice. At some point in the 1770s, the family moved from Kristinehamn to Stockholm. Bergman's father died in Mariestad in 1770, aged 66, while Bergman's mother died in 1776, aged 80. She spent her last years in Uppsala, living with Torbern.

Bergman only gives a small clue about his childhood in his short autobiography, written in 1782: "In my early years as a child, I am supposed to have been exceptionally lively, but that has largely disappeared, as soon as I started to study more seriously. If one can judge from a child's inclinations, what will become their main occupation in the future, if they are allowed to counsel themselves, one would already then have considered me as meant to become a chemist, since nothing amused me more than being able to throw different available substances in the fire and see their changes there" [6]. This fascination for fire indicates that Bergman was a curious boy, but should perhaps not be taken too seriously. It may have been included merely for stylistic reasons. In another text, Bergman concluded that impressions could have strong effect on children's future interests, as their minds were still plastic [7]. All children did not have the same inclinations and talents; this was according to Bergman an act of God to ensure that all different tasks in society could be performed.

2.1 Grammar School in Skara

As a child, Bergman was taught by army preacher Johan Ödman, and by the teachers at the school in Mariestad, Lars Otter and Bengt Ljungblad [8]. Aged 11, he had learned to read well enough to be admitted at the grammar school in Skara, a city of about 800 inhabitants. He was registered on November 13, 1746 along with his younger brother [9]. On arrival, the children took a test to determine in which class they should begin. The age of the pupils in a class therefore varied [10]. Bergman entered the third class, while his brother entered the second class.

Carl Linnaeus had visited Skara just a few months earlier, arriving in Skara on June 25. Skara "is now a small city, although in the past it was the seat of several kings and the capital of the Göta Kingdom. The houses were wooden except for the bishop's residence, the school, the Gymnasium and the library. The streets were irregular, swampy and crooked" [11]. The church, originally built in the eleventh century, but expanded to a gothic cathedral in the fourteenth century, did impress Linnaeus, and most likely also on Bergman. Bergman had of course never seen such a large building before. A modern visitor meets a very different city compared to what Bergman saw. The cathedral remains, although its exterior was altered in the nineteenth century.

The school building was a two-storey building of stone, with small windows. The roof was wooden and painted with tar [12]. At the grammar school, Bergman had two pupils from the Gymnasium supervising him: Lechart Bodell (who would become customs official in Uddevalla) and Joseph Billgren (who would become teacher of theology at the Gymnasium of Skara and member of the Swedish Diet) [8]. This was a common way for pupils at the Gymnasium to earn extra money [9].

In the third class of grammar school,[2] where Bergman entered, all education was in Latin [13]. A pupil who talked Swedish during a lesson was punished by having to sit at the "donkey bench" or carry a picture of a donkey. The education involved the complete Latin grammar, and the pupils began studying Greek. In music, they practised choral and learned the keys.

In the fourth and final class of the grammar school, the pupils continued practising speaking and writing Latin and Greek. The pupils started studying logics and theory of music. Neither history nor geography appears to have been taught in the grammar school in Skara at Bergman's time. Also, the pupils did not get any teaching in Swedish, the only practise in writing Swedish came from translation of Latin texts to Swedish.

After three years at grammar school, Bergman was transferred to the Gymnasium. The Gymnasium, founded in 1641, was divided into two classes, and each class was further dived into two subclasses. Typically, a pupil would spend one year in each subclass and thus four years at the Gymnasium. Talented pupils could be offered to skip one subclass, while pupils making less progress would have to remain an additional year. Bergman was offered to move directly from the first to the second class at the Gymnasium, but he declined the honour [6]. Bergman's class had 33 pupils, the majority being sons of priests [14]. In the catalogue, the pupils were listed based on their skill. The best pupil, the Primus, was listed first and so on. In 1748, Bergman's last year at the grammar school, he was listed number six from the end, but the following year he advanced to a second place from the top [14]. He never managed to pass the Primus, Christian Söderqvist.

At the Gymnasium, Bergman studied Latin, Greek, Hebrew, theology, logics, rhetoric, arithmetic, geometry, geography, astronomy and history. As a part of the education, the pupils had to preach in the grammar school on Saturdays. The pupils also had to practise by either defending theses written by the teachers or acting as opponents against their classmates, a very good training for someone aiming for university. During the holidays, Bergman got private teaching in theology, Hebrew, and Greek from Knös, the rector of the school in Mariestad [15].

Of Bergman's teachers [16], one is of particular interest: Sven Hof (1703–1786). Hof was born in Skara and after several years at the Uppsala University, he obtained a master's degree in 1731. He was interested in science, collected plants and minerals, but failed to obtain any academic position in Uppsala or Åbo. Instead,

[2]Rules from 1649, 1693 and 1724. According to Warne, these probably applied to Bergman's education.

he had to return to Skara with bitterness. He was stubborn and hot tempered [17], but Linnaeus, who met Hof in Skara, was impressed by his botanical knowledge [18]. His most important works include a book on the proper style of writing Swedish, published in 1753, and a book on the Västergötland dialect of Swedish, *Dialectus vestrogothica*, including a dictionary of 3,000 words, published in 1772. Although there was no education in natural science in Skara, specially gifted pupils could get private teaching in botany from Hof. Bergman was one of these pupils, and he got private education in both Latin and botany [19]. Most probably, Hof's interest in science strongly influenced Bergman's choice of career.

Bergman graduated from the Gymnasium in 1752 with a speech expressing the great sorrow caused by the death of King Frederick I in April 1751. In fact, as indicated in the introduction, few actually mourned Frederick I. Of course, we have not much more information on Bergman's childhood (Fig. 2.2).

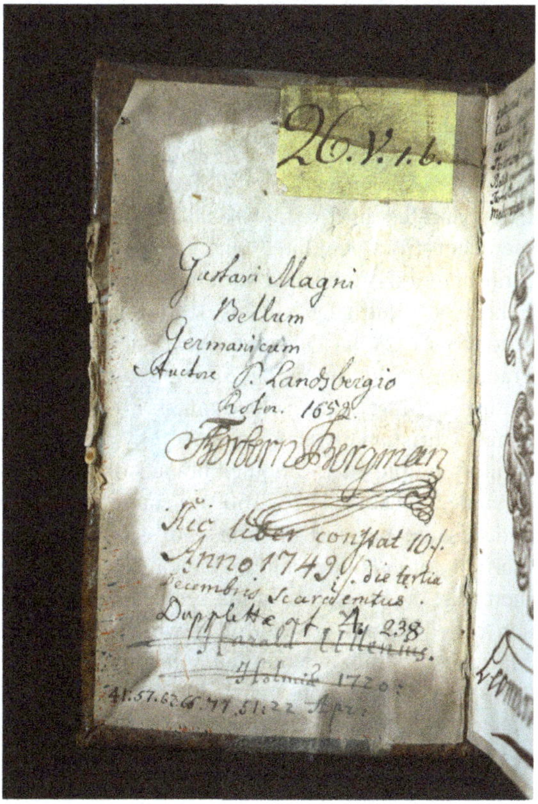

Fig. 2.2 A copy of P Lansbergius' book *Gustavi Magni Bellum Germanicum* (Gustavus Adolphus German War) published in Rotterdam in 1652. The book was bought on auction by 14 year old Torbern Bergman on December 4, 1749, his first year at the Gymnasium. The auctioned books came from the library of Skara vicar (domprost) Harald Ullenius, who collected and wrote down local history. *Photo* Anders Lennartson

2.2 Scheele's Childhood

Carl Wilhelm Scheele[3] was born on December 9, 1742 in Stralsund (Fig. 2.3), the capital of Swedish Pomerania (Vorpommern). Since the peace of Westphalia in 1648, Vorpommern was a Swedish province, and Scheele was born as a Swedish citizen. Occasionally, Scheele's birthday is stated to be December 19, but this is wrong [20]. This date appears to originate from Crell's biography [21] and may simply be a printing error. Carl Wilhelm was baptised in St. Nicholas church (Nikolaikirche; Fig. 2.4) on December 21 and had three godparents: Michael Christian Küdelback, Karl Heinrich Hagemeister, and a Mrs. Warnekros. Among the family, he was called Wilhelm [22].

Zekert has traced the origins of Scheele's family [23], but some aspects of his genealogical study have been criticised [20]. The oldest undisputed ancestor of Scheele was Johannes Scheele, born in Tribsees, Vorpommern (Fig. 2.5), in 1526. After studies at the University of Greifswald, he became vicar in Wiek, a small hamlet on Rügen, where he died in 1600. With his wife Margarethe von Zuhmen, he is reported to have had 16 children, 9 of which are known. The oldest son, Martin (1554–1620), became merchant in Stralsund, where he married Katarina Bünsow. Their son, also called Martin (1593–1652 or 1662), studied in Greifswald and became priest in Pasewalk in 1630. Martin Scheele and his third wife, Elisabeth Dewitz, the daughter of a priest, had a son called Benjamin (1642–1694) who became merchant and established a brewery in Anklam. Benjamin married three times, and his third wife was called Elisabeth Suter (1649–1700), the daughter of a merchant from Anklam. One of their sons, Benjamin (1677–1721), moved to Stralsund and established a brewery there in 1701. He would become Carl Wilhelm Scheele's grandfather. In his 1931 book [23], Zekert named Benjamin Scheele's second wife, Anna Dorothea Wulffrath, born 1682 in Stralsund, as Carl Wilhelm's grandmother, but in his 1963 biography [24], he changed this to merchant daughter Katharina Margaretha Scharow (*c.*1680–1716). On March 1, 1703, Scheele's father, Johan Christian, was born. In 1733, he married Margaretha Warnecross, who actually belonged to another branch of the Scheele family. The youngest son of Johannes Scheele (1526–1600) mentioned above was called Johannes as his father, and became burgher in Stralsund. The younger Johannes Scheele (*c.*1565–1640) married Anna Nidder and had a son called Bartholomäus (born 1588) who became merchant in Stralsund. In his second marriage to Margarethe Seelmann, he had a son called Viktor (1631–1681) who became Mayor in Stralsund. Victor married Margarethe Klinkow and their son Johan Friedrich (1661–1710) became lawyer. In 1683 Johan Friedrich married Anna Schütte and two years later, Ilsabe Maria Scheele (1685–1716) was born. She married Christoph Warnecros, the director of the Stralsund brewer community, and on July 11, 1713, Scheele's mother, Margaretha Eleonora Warnecross, was born. Thus, the Scheele family had predominantly been burghers and priests in northern Germany for at least two hundred years before Scheele's parents were born (Fig. 2.6).

[3]Pronounced [ʻʃe:lə], but in Sweden the pronunciation [ɧe:lɛ] is actually more common.

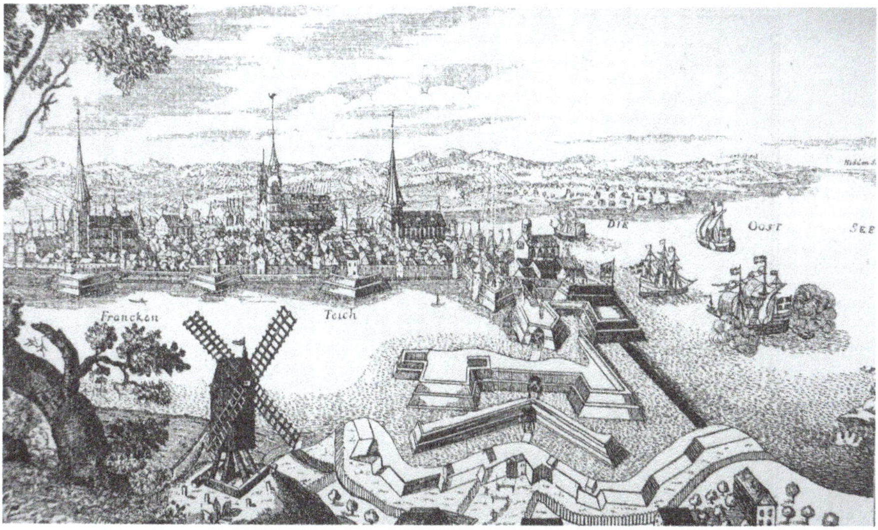

Fig. 2.3 A wood cut of Stralsund in 1715. Reproduced from *Danmarks Riges Historia* by Edvard Holm

Fig. 2.4 St. Nicholas church in Stralsund and the old City Hall. The church was damaged by American bombing in Second World War and is still missing the top of one tower. *Photo* Anders Lennartson, August 2018

Fig. 2.5 Map of northern Germany, with some of the places related to the Scheele family. *Image* Anders Lennartson

Scheele's father obtained a brewer license in 1733, the same year as he married Margaretha Eleonora. The couple settled at Fährstraße 23[4] (Fig. 2.7) in Stralsund, with the brewery located in the adjacent building (Fig. 2.8). It was the family of Mrs. Scheele who provided the house. The house was built in the fourteenth century and has survived to the present day; the facade was altered in the nineteenth century but restored in the 1980s.

In the 1740s, Stralsund's brewing industry faced serious problems due to increased competition [25], and in 1744 the Scheele brewery went bankrupt and the family was forced to sell the house at Fährstraße. Their next address, where Scheele spent most of his childhood, is not known. With assistance from friends, Scheele's father could continue his business until 1746, when he was forced to leave the brewer guild to become a broker. One can suspect that he was dealing with barley or other brewing-related products.

[4]23 is the modern address, in Scheele's days the number was 65.

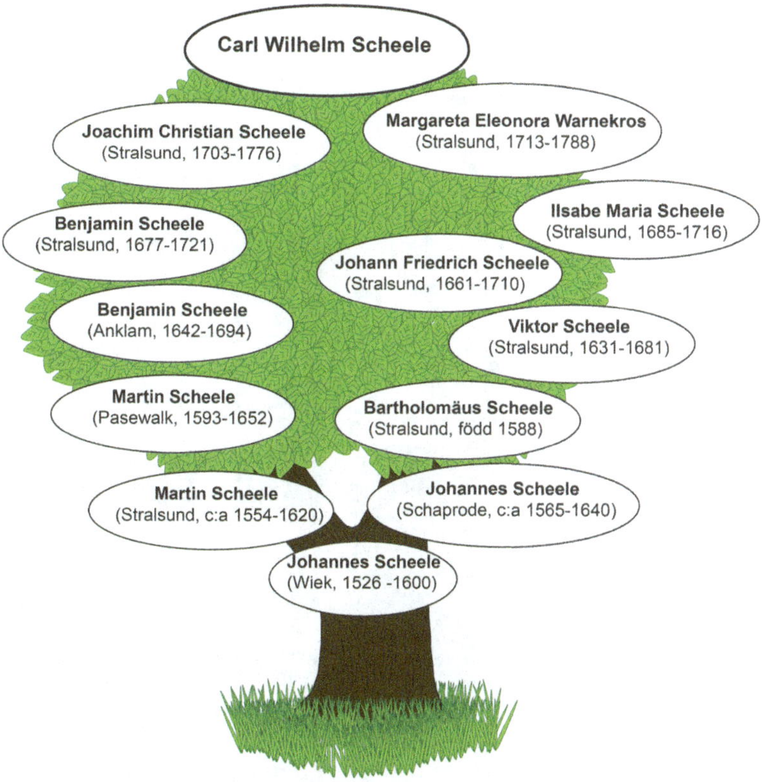

Fig. 2.6 Family tree of Scheele, based on the research of Zekert. *Image* Anders Lennartson

As a consequence of the bankruptcy, we have a good insight into the home of the Scheele family around the time of Scheele's birth [26]. A visitor entering the front door had Joachim Christian's office to the left. The furniture included a desk, a table with six chairs and an oak casket. On the walls were family portraits in golden frames. To the right was the living room with a large table, twelve English chairs, a wing chair, a large mirror in golden frame and eight sconces. The family had two maids, a servant girl, a wet nurse, a shop assistant, and a worker.

Our knowledge of Scheele's childhood is, of course, very limited. According to Scheele's biographer and friend Carl Johan Wilcke (1732–1796; Sect. 6.2) [27], he had been a quiet and brooding child. He preferred to be on his own, rather than playing with his brothers and sisters. He played with silkworms, grew plants, did carpentry work, turning, painting, and made models from cardboard. Aged six, he went to a private school with a teacher called Schmidt, a popular school among Stralsund's burgher children. Sjöstén claimed that Scheele studied at the Stralsund Gymnasium [28], which does not seem to be true; rather it was his younger brother who studied at the Gymnasium [29].

Fig. 2.7 The house in Stralsund where Scheele was born. *Photo* Anders Lennartson, August 2018

2.3 Scheele's Brothers and Sisters

Johan Marti(e)n Scheele (February 14, 1734–January 15, 1754). Carl Wilhelm had probably only vague memories of his oldest brother, who was sent to Gothenburg in Sweden at a young age to become an apothecary apprentice. It has been assumed that apothecary Andreas Bauch (Chap. 9) at the Unicorn pharmacy in Gothenburg was a friend of the Scheele family, as he also had his roots in northern Germany. Johan Martin arrived in Gothenburg around 1748; there is a preserved apothecary manual written by him in Gothenburg that year. The manual, written in Latin, is an alphabetical list of different medicines accompanied by a list of medicaments that should be kept on board the Swedish East Indian Company's ships bound for China. This manual, now preserved at the Swedish Pharmaceutical Society, has been transcribed and published by Zekert as the third volume of his Scheele biography [30]. After only a few years, Johan Martin died from typhus at the age of 19.

Fig. 2.8 The house where the Scheele family had their brewery. *Photo* Anders Lennartson, August 2018

Anna Margaretha Scheele (January 4 or 5, 1736–*c.* 1782). Scheele's oldest sister died unmarried in Stralsund. Her younger brother Friedrich Christoph, who was providing for her, attempted to persuade her to move to Carl Wilhelm in Sweden. As will be seen in Chap. 19, she remained in Stralsund.

David Benjamin Scheele (January 7, 1737–?) David Benjamin was the black sheep of the family. He moved to Saint Petersburg where he worked for Count Orlov, and then turned to the tobacco trade. For several years, the family seems to have lost contact with him. However, he was mentioned in a letter to Carl Wilhelm from his younger brother Friedrich Christoph, dated January 1783. He was alive, but lived under poor circumstances—it appears he had been involved in some kind of fraud and thus he had avoided seeking contact with his family. He appears to have returned to Stralsund, since he is listed as a godfather there in 1788.

Catharina Juliana Scheele (February 17 or 18, 1738–May 6, 1739). Catharina Juliana died in her infancy.

Christian Heinrich Scheele (February 14, 1740–1766). Christian Heinrich moved to Batavia (present day Jakarta in Indonesia), where he became a merchant, but died young.

Friedrich Christoph Scheele (June 14, 1741–1817). When his father, Joachim Christian, died in 1776, Freiedrich Christoph took over his father's business, but he also had to support his mother and unmarried sisters Anna and Maria. He also supported his younger brother Paul's medical studies, which contributed to his financial problems [31]. Soon after his father's death, he had to give up the business for some time [32]. After marrying the daughter of wealthy Stralsund apothecary Frederici in 1779, his economy improved [33]. Carl Wilhelm corresponded with his brother for his entire life, and Friedrich Christoph is listed as a subscriber of Carl Wilhelm's collected works [34], as was father-in-law, Apothecary Frederici.

Catharina Ilsabetha Scheele (April 14, 1744–April 15, 1766). Catharina Ilsabetha died young, and no details of her life are known.

Christiana Rosina Scheele (December 22, 1745–1782). Christina Rosina was the only of Scheele's sisters who married, which she did on October 11, 1775, with clergyman Gabriel Daniel Groth of Altenkirchen on the Wittow peninsula of Rügen.

Maria Juliana Scheele (June 6, 1748–May 19, 1780). In the summer of 1779, when Maria was 31 years old, she moved to her brother Wilhelm in Köping (Chap. 19). Unfortunately she died the following year, probably from tuberculosis.

Paul Joachim Scheele (December 6, 1749–November 12, 1825). Paul was not born when Scheele left Stralsund, and they can only have met once, when Paul was *c.* 16–18 years old. Paul was the most well-educated of the Scheele-children: he studied at Gymnasium Sundensis in Stralsund 1760–1765 and was registered at the University of Halle on April 28, 1773 where he studied medicine. He graduated on January 16, 1775 and in the introduction to his thesis [35], he wrote that he initially had intended to write a thesis on a chemical subject [29]. First, he practised in Dresden, but in autumn 1775, he moved to Stralsund, where he stayed until 1778, when he moved to Neiße (present day Nysa in Poland). The same year he mentioned plans to move to Breslau (present day Wrocław in Poland) as a physician in the Prussian army. In 1781, he moved to Berlin, where he lived with an uncle on his father's side. He appears to have planned to move to Sweden, but instead moved to Köslin (present day Koszalin in Poland), where he married Anna Zarnin on December 19, 1783. He died in Köslin in 1825. He kept himself informed about his brother Wilhelm's research and showed interest in the physiological effects of oxygen [36].

Carl Wilhelm was 11 years old when his older brother Johan Martin died in Gothenburg. Wilcke/Sjöstén claims that young Scheele had expressed a wish to take his brother's position at the pharmacy in Gothenburg. A physician called Schütte and an apothecary called Cornelius were friends of the Scheele family [37] and may have inspired Scheele. At any rate, it was decided that Scheele should go to apothecary Bauch in Gothenburg after receiving elemental education from Schütte in recipe-writing and chemical symbols [37]. Scheele left Stralsund in 1758, when he was 15 years old. This was during the Prussian war, and the same year Stralsund was sieged by Swedish troops. As we do not know exactly the date when Scheele left Stralsund, we cannot know whether the decision to send Scheele to Gothenburg had anything to do with the situation in Stralsund.

References

1. Linnæus C (1747) Wästgöta Resa, Lars Salvius, Stocholm p 18
2. Linnæus C (1747) Wästgöta Resa, Lars Salvius, Stocholm p 75
3. Bergman P-O (1998) Bergman från Västgöta-Dal, Eskekärr-Tomasbolssläktens på Dal släktförening, Sturefors
4. Thomson T (1831) The history of chemistry, vol II. Henry Colburn & Richard Bentley, London, p 27
5. Hjelm PJ (1786) Åminnelse-tal öfver...Herr Torbern Bergman, Stockholm p 7
6. Schück H (1916) Torbern Bergmans självbiografi, äldre svenska biografier 3–4, Uppsala, p 85
7. Bergman T (1779) Åminnelse-tal öfver...Högvälborne Friherren Herr Carl de Geer, Kongl. Vetenskapakademien, Stockholm, p 7
8. Hjelm PJ (1786) Åminnelse-tal öfver...Herr Torbern Bergman, Stockholm, p 8
9. Warne K (1966) Skollivet i Skara under Torbern Bergmans djäknetid omkring år 1750. Skaradjäknarnas förening, Borås p 11
10. Warne K (1966) Skollivet i Skara under Torbern Bergmans djäknetid omkring år 1750. Skaradjäknarnas förening, Borås, p 13
11. Linnæus C (1747) Wästgöta Resa, Lars Salvius, Stocholm p 56
12. Warne K (1966) Skollivet i Skara under Torbern Bergmans djäknetid omkring år 1750. Skaradjäknarnas förening, Borås, p 137
13. Warne K (1966) Skollivet i Skara under Torbern Bergmans djäknetid omkring år 1750. Skaradjäknarnas förening, Borås, p 19
14. Warne K (1966) Skollivet i Skara under Torbern Bergmans djäknetid omkring år 1750. Skaradjäknarnas förening, Borås, p 15
15. Hjelm PJ (1786) Åminnelse-tal öfver...Herr Torbern Bergman, Stockholm p 9
16. Warne K (1966) Skollivet i Skara under Torbern Bergmans djäknetid omkring år 1750. Skaradjäknarnas förening, Borås, p 30
17. Lindroth S (1978) Svensk lärdomshistoria. Frihetstiden., Norstedts, Stockholm p 599
18. Linnæus C (1747) Wästgöta Resa, Lars Salvius, Stocholm p 57
19. Schück H (1916) Torbern Bergmans självbiografi, äldre svenska biografier 3–4, Uppsala, p 86
20. Hildebrand B (1936) Scheeleforskning och Scheelelitteratur, Lychnos, p 76–102
21. Crell L (1787) Einige Nachrichten von den Lebensumständen Carl Wilhelm Scheeléns, Chem Ann 1:175–192
22. Gentz L (1958) Hur såg Scheele ut? Sv Farm Tidskr 17:405–421
23. Zekert O (1931) Carl Wilhelm Scheele Sein Leben und seine Werke, 1. Theil, Gesellschaft für Geschichte der Pharmazie, Mittelwald
24. Zekert O (1963) Carl Wilhelm Scheele, Wissenschaftliger Verlagsgesellschaft, Stuttgart
25. Friedrich C (1992) Carl Wilhelm Scheele, Greifswald, p 10
26. Friedrich C (1992) Carl Wilhelm Scheele, Greifswald, p 9
27. Sjöstén CG (1801) Åminnelse-tal, hållit för Kongl. Vetenskaps Academien öfver dess framledne ledamot herr Carl Vilhelm Scheele, den 14 oct. 1799. Stockholm
28. Sjöstén CG (1801) Åminnelse-tal, hållit för Kongl. Vetenskaps Academien öfver dess framledne ledamot herr Carl Vilhelm Scheele, den 14 oct. 1799. Stockholm, p 6

29. Zekert O (1931) Carl Wilhelm Scheele. Sein Leben und seine Werke. Vol 1. Mittenwald, p 28
30. Zekert O (1933) Carl Wilhelm Scheele. Sein Leben und seine Werke. Vol 3. Mittenwald
31. Lindhagen A (1922) Brev till Carl Wilhelm Scheele från hans fader och bröder, Stockholm, p 14
32. Lindhagen A (1922) Brev till Carl Wilhelm Scheele från hans fader och bröder, Stockholm, p 15
33. Lindhagen A (1922) Brev till Carl Wilhelm Scheele från hans fader och bröder, Stockholm, p 18
34. Scheele CW (1793) Sämmtliche physische und chemische Werke. Vol 1, Berlin, p XI
35. Scheele PJ (1775) Dissertatio Inauguralis medica de difficili saepe causarum scrutinio in morbis, exemplo icteri in puero verminoso observati illustrato, Halle
36. Lindhagen A (1922) Brev till Carl Wilhelm Scheele från hans fader och bröder, Stockholm, p 32
37. Sjöstén CG (1801) Åminnelse-tal, hållit för Kongl. Vetenskaps Academien öfver dess framledne ledamot herr Carl Vilhelm Scheele, den 14 oct. 1799, Stockholm, p 7

The Two Men

3

This chapter attempts to paint the picture of the two main characters of the book—Carl Wilhelm Scheele and Torbern Bergman. From the scattered information available via letters and early biographies, we can get at least some idea on their personal characters.

3.1 Wilhelm

Scheele is described as a man of average height, sturdily built with round face and red complexion [1]. His friend Wilcke gave the following characteristics [1]:

> The clear expression of his face did not immediately reveal anything extraordinarily remarkable, but the eyes revealed a fire, and the speech a steadiness that arouse attention and confidence, the latter, however, not easy to detect without closer relations. One noticed it especially when he, in discussions with others, with great curiosity listened to what he himself did not know, and with confidence talked about his own discoveries, explained his experiments and often ended up in instructive lectures.

In scientific discussions, he listened to opposing opinions and could always back up his claims with experimental evidence. In speech, he expressed himself briefly and was usually seriously minded, thoughtful, and unaffected. He was unpretentious, lived a simple life, and for instance, never consumed alcohol [2]. Except for science, he rarely engaged in longer discussions. Anders Jahan Retzius (1842–1821), a friend from his youth (Sect. 8.1), wrote to Wilcke after Scheele's death: "His genius was completely intended for physical sciences, others he had absolutely no inclination for. The result is, without doubt, that he could appear slow when it came to other things" [3]. Retzius also remarked that Scheele had an excellent memory. Most biographers have indicated that Scheele throughout his life preferred to express himself in German, but it should be noted that he, at least later in life, wrote letters in flawless Swedish. It is generally believed that all his manuscripts

A. Lennartson, *Carl Wilhelm Scheele and Torbern Bergman*, Perspectives on the History of Chemistry, https://doi.org/10.1007/978-3-030-49194-9_3

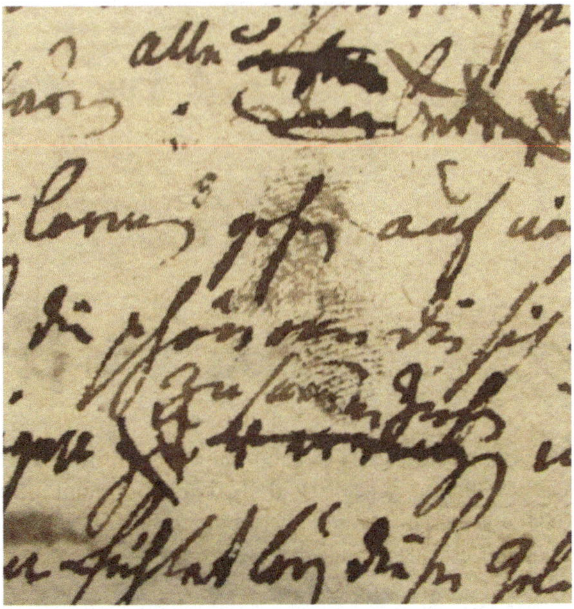

Fig. 3.1 Inspection of the manuscripts in the Brown Book (Sect. 28.7) revealed what is probably Scheele's fingerprint. Note that the finger print is in ink of the same colour as the text, and not created by a wet finger smearing out the original text at a later date. Carl Wilhelm Scheele's Archive, Centre for History of Science, Royal Swedish Academy of Sciences. *Photo* Anders Lennartson

were originally written in German and later translated, but Swedish concepts for a few of his later papers are preserved in the archive of the Royal Swedish Academy of Sciences (Fig. 3.1).

Scheele was always friendly and helpful. For example, he lent his brother Friedrich Christoph 100 riksdaler [4],[1] a large sum of money. He also relieved the economical strain on his brother by accepting one of his sisters, who moved to Köping in 1779. He was always reliable when it came to business transactions.

In his letters (Fig. 3.2), Scheele never discussed religion or politics, and his letters generally contain few details about his private life. In a letter to his friend Johan Gottlieb Gahn (1745–1818), dated April 1772, Scheele apologised that he had not written for a long time, but other occupations had prevented him from experimenting and without chemical news, he was not inclined to waste his time on writing. He appeared to be almost embarrassed when praised and wrote for example to Wilcke that men like Bergman and Pierre Macquer (1718–1784) had more knowledge in their fingers than he had in his whole head. At Scheele's funeral, his friend vicar Ahlström gave the following picture of Scheele [5]:

> When you first met him, little or none of all the good that was inside was revealed. As little as one could notice, during the first or second encounter with him, his remarkable intellect and thinking soul, as little could one in the first encounter get an impression of the humanity that lived in his chest. While other people often lose much of their value, the more one get to know them, Scheele always gained esteem, since, when others appear better than they are, he was actually better than he appeared, and he did not know the bad art of combining sweet words with an ugly inside.

[1]For an explanation of the Swedish currency, see Appendix C.

Fig. 3.2 Scheele's signature from six different letters. As can be seen, he did not have a distinct autograph. Carl Wilhelm Scheele's Archive, Centre for History of Science, Royal Swedish Academy of Sciences. *Photo* Anders Lennartson

3.2 Torbern

Unfortunately, Bergman's private life and personality is even less known than Scheele's. This may be surprising, as Bergman lived a far more public life than Scheele, but it should be remembered that much of our knowledge of Scheele's personality comes from the work of Wilcke, while Bergman's early biographers, Peter Jacob Hjelm (1746–1813) and Pehr Fabian Aurivillius (1756–1829) did not mention much about Bergman's private life. This is not surprising, as they (unlike Wilcke) delivered their memorial lectures soon after Bergman's death, and the majority of their audiences probably knew Bergman in person. In the case of Scheele, the situation was different; he had probably only visited Stockholm once since 1770 and had already become somewhat of a myth by the time Wilcke was collecting his material. Thus, we have to search for more indirect details in letters and other sources in order to get a picture of Bergman.

Hjelm described Bergman as tall and slim [6]. According to Aurivillius, Bergman tended to avoid social life when it interfered with his work, but he enjoyed spending time with a few friends [7]. Some authors claim that Bergman's only pleasure was his work, [8] while others are better informed. When he had a spare moment he would, at least in the summer, go out in his garden; look after his bees (Sect. 12.2) or his birds, which he in later years breed in large number in order to observe their habits [9]. As everyone in Sweden at the time, Bergman was a believing protestant (anything else was illegal). He attended the morning service in the church every Sunday, as long as his health permitted, and would otherwise read Christian texts to compensate [10].

Aurivillius said that if Bergman "had found good reasons to make a certain decision, then he was relentless" [11]. Aurivillius also said in his memorial lecture that:

> among You, who had gained access to his closest confidence, know that his friendship was noble and trustworthy. If he promised something, one could be confident [that he would keep his promise]. If advice was needed, one could be certain to receive, if not always the most pleasant advice, but usually the advice that turned out to be the best in the end. [12]

A comment in a letter from Scheele to Gahn, dated December 1771, gives a slightly different picture: "I have noted, that when I am not taking measures myself, Bergman does not fulfil his promises".

The picture of Bergman that emerges from letters and his writings is largely positive and Bergman appears to have been a friendly, respected, and agreeable man. Unlike his predecessor as professor in chemistry, Johan Gottschalk Wallerius (1709–1785; Sect. 9.1), he did not seek open confrontation, and when he criticised the works of others in his papers, he was always polite.[2] His friendship with Scheele, who had a far lower social status, is also notable. In a letter to French chemist Guyton de Morveau (1737–1816), Bergman even wrote that his greatest discovery was that of Scheele's brilliant mind [8]. Perhaps Hjelm was not far from the truth, when he wrote that Bergman did not care much about attention, but that it was important for him to be remembered after his death [13]. Bergman's largest shortcoming was, according to Hjelm, that he exhausted himself and ruined his health [14].

Much of Bergman's success was probably due to his skills as an author. Whatever the topic, his writing is always captivating and inspiring. He was well acquainted with the classical authors. Interestingly, Bergman seldom sat down as he wrote; instead he put the paper on a book and wrote as he was walking around in the room [9]. Occasionally he would rest a knee on a chair. Still, his handwriting is very easy to read (Fig. 3.3). While other authors have to edit their texts multiple times, Bergman wrote efficiently and it has been said that he rarely edited his texts. Bergman was also good at expressing himself orally; he was happy and polite towards everyone [15].

Schufle came to the conclusion that Bergman was very well off, [16] but his calculations are based on a misconception regarding the value of the Swedish currency (Appendix 3). In fact, Bergman's economy was rather strained (Sect. 9.3). It seems that Scheele, who could afford to buy and renovate a house in Köping, had a more solid economy.

[2]For the single exception, see Sect. 9.6.

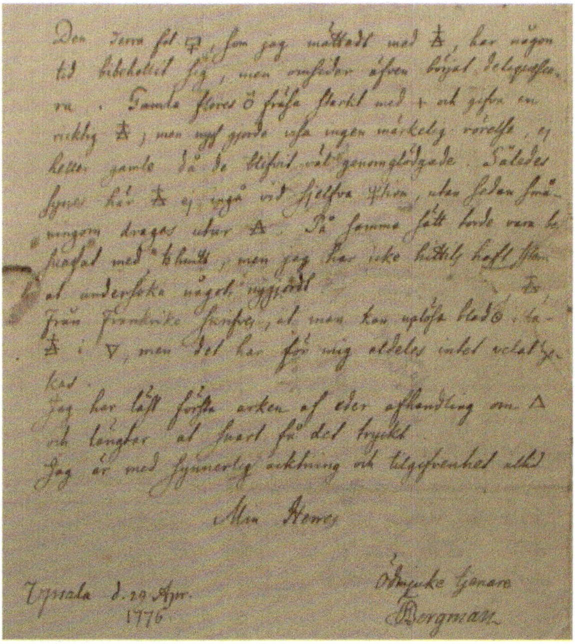
Min Herres
Upsala d. 29 Apr. 1776
ödmjuke tjenare
Bergman

Fig. 3.3 The last page of a letter from Bergman to Scheele, written in 1776. Unlike Scheele, Bergman had a characteristic signature which remained virtually the same over the years. See also Fig. 2.2. Carl Wilhelm Scheele's Archive, Centre for History of Science, Royal Swedish Academy of Sciences. *Photo* Anders Lennartson

3.3 Portraits of Bergman

Although we now slightly more about Scheele's private life compared with Bergman's, we actually do not know what Scheele looked like. Scheele's interest was chemistry, and he was rather uninterested in fame and public recognition. For Bergman, on the other hand, such things were important. Bergman wanted a place in the history books and it is not surprising that we have several portraits of Bergman by some of the most celebrated artists of his time.

There are three contemporary paintings of Bergman; one from 1778 (Fig. 3.4) at Uppsala University painted by Lorens Pasch the younger (1733–1805), one of the most appreciated Swedish portrait painters of the time. Viewing the painting, is it clear that every detail had been planned by Bergman. In this painting, Bergman wears the so-called national costume, introduced by King Gustav III in 1778 as a standard costume for the nobility and burghers to limit the import of luxury items. The costume was mandatory in his court, and by wearing the costume in the painting, Bergman probably wanted to express his loyalty to the King, who had helped him to the chair in chemistry (Sect. 9.3). Bergman is wearing the Order of Wasa in its green silk ribbon around the neck and is holding a scroll in his hand. The scroll is an attraction table (Chap. 20), which Bergman considered his most important contribution to science. Interestingly enough, the painter did not fully

Fig. 3.4 Torbern Bergman at an age of about 43, painted by Pasch in 1778. *Photo* Mikael Wallerstedt, Uppsala University

appreciate Bergman's work, as the chemical symbols in the table appear in a random order, and not according to Bergman's theory. Bergman must have noted this.

There is a second painting in the national Swedish portrait gallery at Nationalmuseum in Stockholm (Fig. 3.5), painted by Pasch's sister, Ulrika Pasch (1735–1796) in 1779. This painting was acquired by Nationalmuseum in 1989.

A third, undated, portrait was painted by the Per Krafft the younger (1724–1793), also one of the most appreciated Swedish portrait painters of the late eighteenth century. This painting was donated to the Royal Swedish Academy of Sciences in 1789 by Bergman's brother, Carl Fredric, after his death. It seems therefore that the painting had likely been private property in Bergman's family. Perhaps Carl Fredric inherited the painting after Bergman's death? The painting, which perhaps is the most aesthetic of the Bergman portraits, is hanging in the Session Hall of the Academy, the hall where the Nobel Prizes in chemistry and physics are announced every year.

There is also a contemporary plaster medallion (Fig. 3.6) sculptured by the Swedish neoclassical sculptor Tobias Sergel (1740–1814). It belongs to the collections at the national Swedish portrait gallery at Nationalmuseum in Stockholm and is a number in a series of similar medallions made by Sergel. In 1782,

Fig. 3.5 Torbern Bergman aged about 44. This painting by Ulrika Pasch shows less likeliness with Bergman and lacks, for example, Bergman's characteristic chin. *Photo* Nationalmuseum, Stockholm

Bergman's friend Schwediauer in London asked Bergman for a portrait, as he wished to order a medallion from the Wedgwood firm [17], famous for their porcelain (Jaspar ware) medallions in white relief against blue background. Not until January 1784, Bergman sent a copy of the Sergel medallion to Schwediauer, and the Wedgewood medallion was not finished until after Bergman's death that summer (Fig. 3.7). Copies of this medallion are now presented to the winners of the Torbern Bergman Medal in analytical chemistry by the Swedish Chemical Society (Sect. 28.8).

There are two contemporary engravings depicting Bergman, one (12.1 × 8.1 cm; Fig. 7.5) made by engraver Johan Fredrik Martin (1755–1816) (Fig. 3.8). Variants of this engraving were used by Gustav Georg Endner (1754–1824) to create the frontispiece for the first volume of *Allgemines Journal der Chemie* in 1798 and by K. Mackenzie in 1801 to illustrate a biographical paper on Bergman in *Philosophical Magazine* [18].

Another engraving (Fig. 3.9), enface, was made by engraver Fredrik Akrel (1748–1804), who worked for, e.g., the Royal Academy of Sciences and probably made illustrations for some of Bergman's and Scheele's works. Several latter engravings from the nineteenth century were used to illustrate, e.g., encyclopedias and lithographic magazines.

Fig. 3.6 Sergel's plaster medallion, 66 cm in diameter. *Photo* Nationalmuseum, Stockholm

Fig. 3.7 Wedgewood medallion presented to the receivers of the Torbern Bergman Medal in analytical chemistry. *Photo* Swedish Chemical Society

Fig. 3.8 Torbern Bergman, engraving by Johan Fredrik Martin. Nationalmuseum, Stockholm

Shortly before his death in 1784, the Finnish Student nation in Uppsala commissioned a silver medallion (Fig. 3.10), engraved by Swedish medal engraver Gustaf Ljungberger (1733–1787). Bergman was presented with a copy in gold at a ceremony in Stockholm on his last trip to Medevi (Sect. 28.2) [19].

The Royal Swedish Academy of Sciences in Stockholm issued a silver medallion to commemorate Bergman in 1785, also engraved by Ljungberger (Fig. 3.11). A copy in gold of this medal was presented to Bergman's widow [20].

In 1826, French sculptor, engraver and medal producer Pierre Amédée Durand (1789–1873) issued a bronze medal over Bergman in his series of medals over historic celebrities (Fig. 3.12). The medal was engraved by Swedish engraver Johan (born Isaac) Salmson (1789–1859) who was active in Paris from 1820. The medal has the word Monachii (Latin for "from Munich") stamped on the edge, as a way to circumvent the French Mint's monopoly on striking medals.

The Swedish Academy issued a silver medallion (Fig. 3.13) of Bergman engraved by Lea Ahlborn (1826–1897) in 1859. Ahlborn was a niece and student of Salmson.

Fig. 3.9 Torbern Bergman, engraving by Fredrik Akrel

In 1957, the Royal Academy of Sciences in Stockholm issued a second medallion to commemorate Bergman engraved by Erik Lindberg (Fig. 3.14) [21]. It shows Bergman in profile on one side, and one of his "reaction formula" (Sect. 20.2) on the opposite side.

3.4 Portraits of Scheele

While there are several contemporary images of Bergman, there is no known portrait of Scheele made during his lifetime. On April 22, 1788, 2 years after Scheele's death, the Royal Swedish Academy of Sciences decided to commemorate Scheele with a medallion. The task was assigned to engraver Johan Gabriel Wikman (1753–1821), but the problem was that there was no authentic picture of Scheele. Instead, an image was created based on the accounts of people who had met Scheele. Unfortunately, 10 years had passed since Scheele last visited

Fig. 3.10 Silver medallion by Ljungberger. *Photo* Uppsala University

Fig. 3.11 A 1984 bronze re-issue of Ljungberger's 1785 medal. *Photo* Anders Lennartson

Stockholm, so the task was not easy. The medallion (Fig. 3.15) was struck the following year, and it is the only image of Scheele that can be expected to show any resemblance to the great chemist. However, in the minutes from an Academy meeting on February 2, 1791, there are complaints about the image, which was said not to show a good enough likeliness of the deceased chemist [22]. The original medallion was made from silver, but bronze copies have been struck from the original matrix in 1942 (200 years after Scheele's birth) and in 1986 (200 years after Scheele's death). A variant (in bronze) was issued at the International congress

Fig. 3.12 Medallion over Bergman engraved by Salmson in 1826. *Photo* Anders Lennartson

Fig. 3.13 Medallion issued by the Swedish Academy in 1859. *Photo* Anders Lennartson

of physiology in Stockholm in 1926. It used the original Wikman matrix for one side, and the inscription "conventus physiologorum internationalis duodecimus Holmiae MCMXXVI" on the other side.

In 1821, the Swedish Academy suggested honouring Scheele with a medallion, but it would take until 1827 before it was realised (Fig. 3.16). The image was based on Wikman's medallion. Wikman's image was also used to create an epitaph in the church of Köping at the 40th anniversary of Scheele's death in 1826.

Fig. 3.14 The 1957 medallion issued by the Royal Swedish Academy of Sciences. Compared to the previous medallions, Bergman looks much older. The other side of the medallion depicts Bergman's reaction scheme for the reduction of silver nitrate with metallic copper in water. *Photo* Anders Lennartson

Fig. 3.15 Wikman's Scheele medallion, the only known portrait of Scheele produced in the eighteenth century. How closely the medallion resembles Scheele is impossible to say. *Photo* Anders Lennartson

There are a large number of different engravings depicting Scheele, most of which are based on Wikman's medallion [23]. One of the most frequently reproduced images (Fig. 3.17), a wood cut made by block cutter Amanda Maria Falander (1842–1927), was originally used to illustrate Cleve's Scheele biography in 1886 [24]. Although the artist probably used Wikman's medallion as a source of inspiration, it is not a reliable image of Scheele.

Fig. 3.16 Silver medallion issued by the Swedish Academy in 1826. The engraver has signed it with an "E". *Photo* Anders Lennartson

Fig. 3.17 Nineteenth century woodcut by Falander depicting Scheele. Although aesthetically appealing, it is unlikely to show any close likeliness with Scheele

Another classic depiction of Scheele is John Börjesson's sculpture (Fig. 3.18) in Humlegården in Stockholm, unveiled on the 150th anniversary of Scheele's birth in 1892. In 1958, apothecary Ebba Hugosson said that Börjesson had told a colleague of her that it was actually his own son, Caspar, who was the model. Caspar would later become vicar in Saltsjöbaden. In February 1992, the Scheele statue was

Fig. 3.18 Statue of Scheele on Floras kulle behind the Royal Library in Humlegården, Stockholm. Note that the rod Scheele is using to stir the contents in his crucible is missing, probably since 1992 bombing. *Photo* Anders Lennartson, October 2014

severely damaged by a homemade bomb constructed by some teenage boys. It was probably just by chance that the Scheele statue became the target. Fortunately, the statue could be reassembled. The original plaster concept for the statue is preserved in the entrance hall of the Swedish Pharmaceutical Society.

To celebrate the unveiling of the statue, the Pharmaceutical Society in Stockholm issued a medal depicting the statue on one side, and Scheele's recently demolished pharmacy in Köping on the other side (Fig. 3.19). This medal, designed by Adolf Lindberg (1839–1916), was made from aluminium, which was still an exotic and expensive metal in 1892.

Another statue (Fig. 3.20) was erected 1912 in Köping, this time the internationally recognised sculptor Carl Milles (1875–1955) was contracted. This sculpture shows no resemblance with Wikman's portrait or any other Scheele depiction, and thus only represent Milles' own mental image of Scheele. A concept for the statue, one metre high, was put on display in February 2006 at Waldemarsudde, Stockholm, but was stolen the same weekend [25].

Fig. 3.19 Aluminium medallion to celebrate the unveiling of the Scheele statue in Humlegården. The medallion is rather large, with a diameter of 43 mm. *Photo* Anders Lennartson

Fig. 3.20 Milles' Scheele statue in Köping, where Scheele lived the last 16 years of his life. *Photo* Anders Lennartson, June 2014

Fig. 3.21 The "Scheele medallion", depicting an unknown man from the eighteenth century. It is now in the possession of the Swedish pharmaceutical society. *Photo* Bo Ohlson

In the late 1920s, the president of the Swedish Pharmaceutical Society, Stellan Gullström, was told that in the late nineteenth century, singer Angelica Brüggemann, a grandchild of one of Scheele's brothers, had possessed a painted medallion depicting Carl Wilhelm Scheele as a young man [26]. The medallion had last been seen in the 1880s, some 50 years and one world war earlier, but Gullström did not hesitate. He sent young pharmacist John Quist to Germany. Quist's mission was certainly not easy. Brüggemann was dead since several years, but she had been survived by a brother, who had died in Berlin under poor circumstances. His widow had died soon after, but the widower of one of her nieces was found to have the medallion in his possession. The medallion (Fig. 3.21) was purchased by Gullström and presented to the Swedish Pharmaceutical Society. The joy would not last, however.

In 1958, Lauritz Gentz analysed the image, and showed that the clothes worn by the man on the image were typical for the 1820s, and could therefore not be a genuine image of Scheele [23]. Fortunately, Gullström passed away in 1951 and did not have to face the disappointing news. By then, the image had unfortunately already been used for two Swedish stamps, commemorating the 200th anniversary of Scheele's birth in 1942 (Fig. 3.22). About 75 million of these false Scheele images were produced.

Fig. 3.22 Two Swedish stamps, 5 and 60 öre, issued for the 200th anniversary of Scheele's birth in 1942. As they are based on Gullström's medallion, no one knows who the man on the stamp actually is. *Photo* Anders Lennartson

References

1. Sjöstén CG [aut] (1801) Åminnelse-tal, hållit för Kongl. Vetenskaps Academien öfver dess framledne ledamot herr Carl Vilhelm Scheele, den 14 oct. 1799. Stockholm, p 77
2. Boklund U [aut] (1961) Carl Wilhelm Scheele bruna boken del 2, Stockholm, p 402
3. Boklund U [aut] (1961) Carl Wilhelm Scheele bruna boken del 2, Stockholm, p 390
4. Lindhagen A (1922) Brev till Carl Wilhelm Scheele från hans fader och bröder, Stockholm, p 22
5. Ahlström CJ (1786) Tal vid Herr Carl Wilhelm Scheeles graf. Reprint 1936, Stockholm, p 10
6. Hjelm PJ [aut] (1786) Åminnelse-tal öfver...Herr Torbern Bergman, Stockholm, p 77
7. Aurillius PF (1785) Åminnelse-tal, öfver chemiæ professoren i Upsala...Herr magister Torbern Bergman...Uppsala, p 47
8. von Beskow B (1860) Minne öfver kemie professoren Torbern Bergman. Svenska Akademiens Handlingar 32:23–70
9. Hjelm PJ [aut] (1786) Åminnelse-tal öfver...Herr Torbern Bergman, Stockholm, p 83
10. Aurillius PF (1785) Åminnelse-tal, öfver chemiæ professoren i Upsala...Herr magister Torbern Bergman...Uppsala, p 51
11. Aurillius PF (1785) Åminnelse-tal, öfver chemiæ professoren i Upsala...Herr magister Torbern Bergman...Uppsala, p 48
12. Aurillius PF (1785) Åminnelse-tal, öfver chemiæ professoren i Upsala...Herr magister Torbern Bergman...Uppsala, p 49
13. Hjelm PJ [aut] (1786) Åminnelse-tal öfver...Herr Torbern Bergman, Stockholm, p 6
14. Hjelm PJ [aut] (1786) Åminnelse-tal öfver...Herr Torbern Bergman, Stockholm, p 92
15. Hjelm PJ (1786) Åminnelse-tal öfver...Herr Torbern Bergman, Stockholm, p 85
16. Schufle JA (1985) Torbern Bergman A Man Before His Time. Cornado Press, Lawrence, Kansas, p 134

17. Carelid G, Nordström J (1965) Torbern Bergman's Foreign Correspondence, Almqvist & Wiksell, Uppsala, p XLVI
18. Anonymous (1801) Account of the life and writings of Olof Torbern Bergman, professor of chemistry at Upsal, Phil Mag 9:193–200
19. Aurillius PF (1785) Åminnelse-tal, öfver chemiæ professoren i Upsala...Herr magister Torbern Bergman...Uppsala, p 44
20. Tiselius A (1958) Torbern Olof Bergman, Levnadsteckningar över K. Svenska Vetenskapsakademiens ledamöter, vol 9, p 140
21. Tiselius A (1958) Torbern Olof Bergman, Levnadsteckningar över K. Svenska Vetenskapsakademiens ledamöter, vol 9, p 127
22. Lindroth S (1967) Kungl. Svenska vetenskapsakademiens historia, vol. 1:1, KVA, Stockholm, p 551
23. Gentz L [aut] (1958) Hur såg Scheele ut? Sv Farm Tidskr 17:373–394; 17:405–421
24. Cleve PT [aut] (1886) Carl Wilhelm Scheele. Ett minnesblad på årsdagen af hans död. M. Barkéns förlagsbokhandel, Köpin
25. Konststöld på Waldemars udde, Svenska dagbladet, 2006 02 06
26. Gullström S [aut] (1932) Carl Wilhelm Scheeles Porträtt in Zekert O[aut] Carl Wilhelm Scheele. Sein Leben und seine Werke. vol. 2, Mittenwald, p 31–35

4 Bergman's and Scheele's Education

The educations of the two chemists Bergman and Scheele were very different: while Bergman had a formal university education, Scheele was accepted as an apprentice in a pharmacy.

4.1 Bergman's Education in Uppsala

In the fall of 1752, Bergman, aged 17, arrived at Uppsala University (Fig. 4.1) [1]. Here, he was supposed to study philosophy and language, subjects needed for a career as a civil servant. However, Bergman's interests were already completely devoted to science and mathematics. Bergman used the following words to described the delight of scientific investigation in a memorial lecture over entomologist de Geer: "Let us consider our own natural instinct. Is it not like this, that we like changes in our pleasures? Do we not, with surprise, delight and rapture view what is unusual and different from what we are familiar to?" [2] A cousin of Bergman, Jonas Victorin, latter vicar in Fägred, east of Mariestad, was living next door to Bergman and was ordered by the family to keep an eye on Torbern and his studies [3]. He repeatedly tried to convince Bergman that the future prospects for a mathematician or a scientist were not very promising, and Bergman could not find any convincing arguments [1]. Thus, Bergman put a small bookshelf under his table and covered the table with a large table cloth. When his cousin came for inspection, there were only philosophical books on the table; all scientific works were properly hidden under the table. Bergman claims that he rarely visited lectures, but spent

A. Lennartson, *Carl Wilhelm Scheele and Torbern Bergman*, Perspectives on the History of Chemistry, https://doi.org/10.1007/978-3-030-49194-9_4

his time reading Euclid's geometry,[1] Keill's *Physica et Astronomia*,[2] Wolff's *Logica*[3] and Wallerius' *Systema metaphysicum*.[4], [1] He got up at 4 a.m. every morning and went to bed at 11 p.m. He ate in his room, which he often did not leave for an entire week. It seems quite safe to assume that Bergman, with his deep interest in biology, also approached Linnaeus, and probably joined him and his students on excursions in the area. Not surprisingly, this double life and hard work exhausted Bergman, and his studies in Uppsala thus ended at midsummer the following year, when he went back to his parents in Mariestad.

Bergman spent 15 months recovering at Tjos farmstead (Fig. 4.2), 17 km north-east of Mariestad, where his family now lived. He used this time for studies rather than recreation. Bergman expanded his herbarium, which he had compiled in Skara, comparing the plants he collected with Linnaeus' descriptions from the books he had brought with him from Uppsala. Most interest, however, he found in collecting insects. Those that he could not identify using Linnaeus' *Fauna suecia*, were sent with a traveller to Linnaeus that Christmas [4]. Most of these were, according to Bergman, unknown also to Linnaeus, who sent Bergman a letter acknowledging the discoveries. As Linnaeus was one of Europe's most celebrated scientists, this must have been very inspiring for young Bergman. Apart from this, he also found time to expand his knowledge in mathematics: he studied Fredrik Palmquist's book *Inledning til algebra* (Introduction to Algebra) but was annoyed by the frequent printing errors. Without assistance, a few lines could keep him busy for an entire week, but as he finally resolved a problem, the pleasure was indescribable [4].

Bergman returned to Uppsala and the University in the fall of 1754, now with his parents' permission to study mathematics and physics. He continued his old habit, largely ignoring the lectures and spending the time reading books that he bought. In his autobiography, he gives the impressions of being an autodidact, which is actually not true.

Bergman spend a lot of time in the astronomical observatory (Fig. 4.3), and in the absence of the observatory assistants, he often acted as manager [5]. On the initiative of Anders Celsius,[5] an old, possibly mediaeval, two-storey stone building had been supplied with an observation tower on the roof. The observatory was less than 10 years old when Bergman arrived, but had unfortunately turned out to be quite unsuitable for astronomical observations. Since it was built in the middle of Uppsala, the view towards the horizon was obscured by roofs and smoke from

[1] *Elementa* by Greek mathematician Euclid (around 300 BC). Possibly he used Mårten Strömer's Swedish translation from 1744: *Euclidis Elementa eller grundeliga inledning til geometrien*.

[2] *Introductiones ad veram Physicam et veram Astronomiam*, the posthumous collected works of Scottish mathematician John Keill (1671–1721), published in 1725.

[3] *Vernünftige Gedanken von den Kräften des menschlichen Verstandes* (1712) by German philosopher and mathematician Christian von Wolff (1679–1754). Possibly in a Latin translation.

[4] *Systema metaphysicum* (3 vols. 1750–1752) by Swedish philosopher Nils Wallerius (1706–1764), professor in Uppsala.

[5] Anders Celsius (1701–1744); Swedish astronomer, well known for the temperature scale he invented.

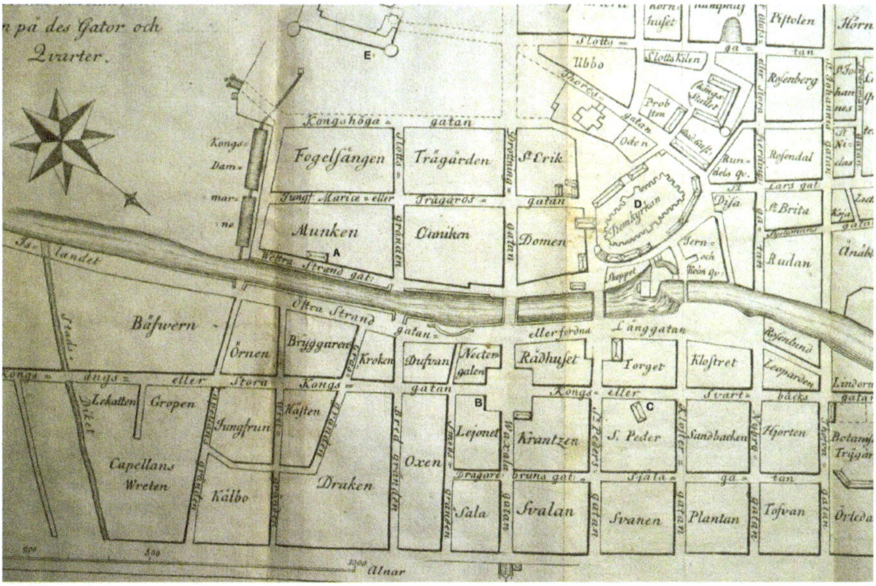

Fig. 4.1 Map over Uppsala showing the places of interest in this book; A: The chemical laboratory where Bergman worked and lived as professor, B: The Arms of Upland pharmacy where Scheele worked and lived 1770–1775, C: The astronomical observatory, D: the Cathedral, E: Uppsala Castle. Engraving from Busser's book *Utkast till beskrifning om Upsala* from 1773

chimneys. Frequently, the traffic interfered with the delicate instruments [6]. Still, it served as observatory until 1853. Although the tower was removed in 1857, the building still stands today.

On March 19, 1755, Bergman defended a 16-page thesis titled *Dissertatio de crepusculis* (Dissertation on the twilights) under supervision of Mårten Strömer. It was common practice to defend two theses, one for practice, and one to obtain the degree. This was the training (*pro exercitio*) thesis, and since Bergman refers to it as "my thesis" in his autobiography it is possible that it was written by himself rather than by Strömer.

Mårten Strömer (Fig. 4.4) was born in Örebro in 1707 and had studied under Anders Celsius. When Celsius died, aged only 42, in 1744, he was succeeded by Strömer as professor in astronomy. Strömer was more oriented towards mathematics, and although he was a good teacher, he made no significant original contributions to astronomy. He is most famous for his Swedish translation of the works of Euclid [6]. In 1757, while Bergman was a student, Strömer left for Karlskrona, the site of Sweden's primary naval base, where he helped to establish the new Naval Academy.

The new Naval Academy was established in 1756, and when Strömer learned of the plans, he went into action [7]. During his presidency in the Royal Swedish Academy of Sciences, Bengt Ferner (1724–1802; Fig. 4.5) was inaugurated.

Fig. 4.2 Tjos, where Bergman spent 15 months with his parents in 1753–1754. *Photo* Anders Lennartson, July 2019

Ferner, or Ferrner as he was called after his ennoblement, was born in the Värmland province and enrolled at Uppsala University in 1743. He was a student of Klingenstierna and became observatory assistant (observator) at the astronomical observatory in 1751. When Strömer concluded his term as president in the Academy, he did so with a lecture on astronomy and navigation, and the Secretary of the Academy, Wargentin (Sect. 6.2), assured in his answering address that Strömer always had been interested in navigation. When the professorship in astronomy at the Naval Academy was to be filled, Ferner was appointed on the condition that Strömer first organised the education. Thus, both Strömer and Ferner applied for a leave of absence, and swopped positions. Through this process, Strömer got the position in Karlskrona keeping his salary from Uppsala and in addition he obtained a sum of 6,000 silver dalers. Strömer extended his contract in Karlskrona and remained there for the rest of his career. Ferner, however, left Uppsala the following year for an extensive trip through Europe as a companion of the son of wealthy merchant Jean Henri Lefebure. While travelling he studied the construction of astronomical instruments and met many prominent scientists. The trip was an offer that Ferner could not resist, and it has been suggested that the offer went to Ferner on suggestion by mathematician and astronomer Daniel Melander (Melanderhjelm after his ennoblement) who was eager to take over Ferner's position in Uppsala.

Fig. 4.3 The old Uppsala observatory. Engraving from Busser's *Utkast till bskrifning af Upsala*, part 2 from 1769

Bergman, who had so far been supported by his parents, was now forced to take a job to get some income but was fortunately allowed to continue his studies at the University. Bergman was hired by General and Count Wolter Reinhold Stackelberg to privately teach his son, Adolph Fredric, who would follow in his father's steps and become a colonel and knight [8].

After passing compulsory exams in theology, Bergman passed an *examen rigorosum* on May 24, 1756, now being only one step away from the desired master's degree, the highest degree at the faculty of philosophy, equivalent to a Ph.D. degree in modern terminology. He now only had to successfully defend his inaugural thesis.

Bergman was, as always, hardworking and besides his studies in astronomy and mathematics, and his job as a private tutor, he also found time for his passion for insects. He was surprised to find that *Coccus aquaticus* described by Linnaeus as an insect in his *Fauna Suecica*, actually was the larvae of a leech. According to Linnaeus the females of the genus *Coccus* attached themselves to plants or trees and

Fig. 4.4 Bergman's teacher Mårten Strömer. Oil painting by Olof Arenius after an original by Johan Henrik Scheffel. *Photo* Uppsala University

transformed into something similar to an oak apple. However, no one had actually seen the insects attaching to the vegetation. Bergman collected these hard shells and found that they developed into leeches which he identified as the species *Hirudo depressa fusca*. He also witnessed one of the leeches laying eggs. Although the leeches are hermaphrodites, the lonely leech's eggs did not develop. At first, Linnaeus refused to believe Bergman but was finally convinced when Bergman had shown him the hatching. Linnaeus then asked Bergman to write a paper [9] for the *Transactions of the Royal Academy of Sciences*. Bergman did so, and Linnaeus sent the manuscript to Stockholm after signing it with the words "Vidi & obstupui" [10],[6] (I have seen and been surprised) thus ensuring a swift publication. Surprisingly enough, this is not mentioned by Bergman in his autobiography. The *Transactions* (Sect. 6.1) was the leading scientific journal in Sweden and also received considerable international interest, so this publication was a major merit for Bergman, now aged 21.

Inspired by this success, Bergman undertook an extensive study of leeches (Fig. 4.6) resulting in another paper in the *Transactions* [11]. Here Bergman gave a general description of the Swedish species of leeches, describing six new species.

[6]The original manuscript has not survived.

Fig. 4.5 Bergman's teacher Bengt Ferner. *Photo* Nationalmuseum, Stockholm

Bergman's paper describes both the anatomy and ethology of the leeches and although the title of the paper suggests that it is the first part, no further publications on the subject followed. Bergman had made such an impression on Linnaeus that he named a moth, *Tortrix bergmanniana* (modern name *Acleris bergmanniana* or yellow rose button moth) after Bergman in 1758. Bergman now wrote a paper in Latin on the classification of larvae, which he sent to The Royal Society of Sciences at Uppsala in 1757, [12] but which was not printed until 1773 [13]. Swedish entomologist Charles de Geer (1720–1778) borrowed the manuscript from the society and wrote in a letter to Linnaeus (April 1757) that "Bergman must have large knowledge in entomology and must have read and observed much. Such things always pleases me much". A few years later Bergman, anonymously, reviewed a work on insects by de Geer, [14] and he also gave the memorial lecture over de Geer in the Royal Swedish Academy of Sciences [15].

On Mars 15, 1758, Bergman defended his thesis on correcting errors in astronomical observations, *de interpolatione astronomica* (astronomical interpolations), now under supervision of Bengt Ferner. By midsummer, he graduated as master of philosophy, and when the faculty members voted who would be the primus, the highest ranked student at the ceremony, Bergman lost with only one vote [8]. The statement by Schufle [16] that Bergman immediately joined the faculty staff and

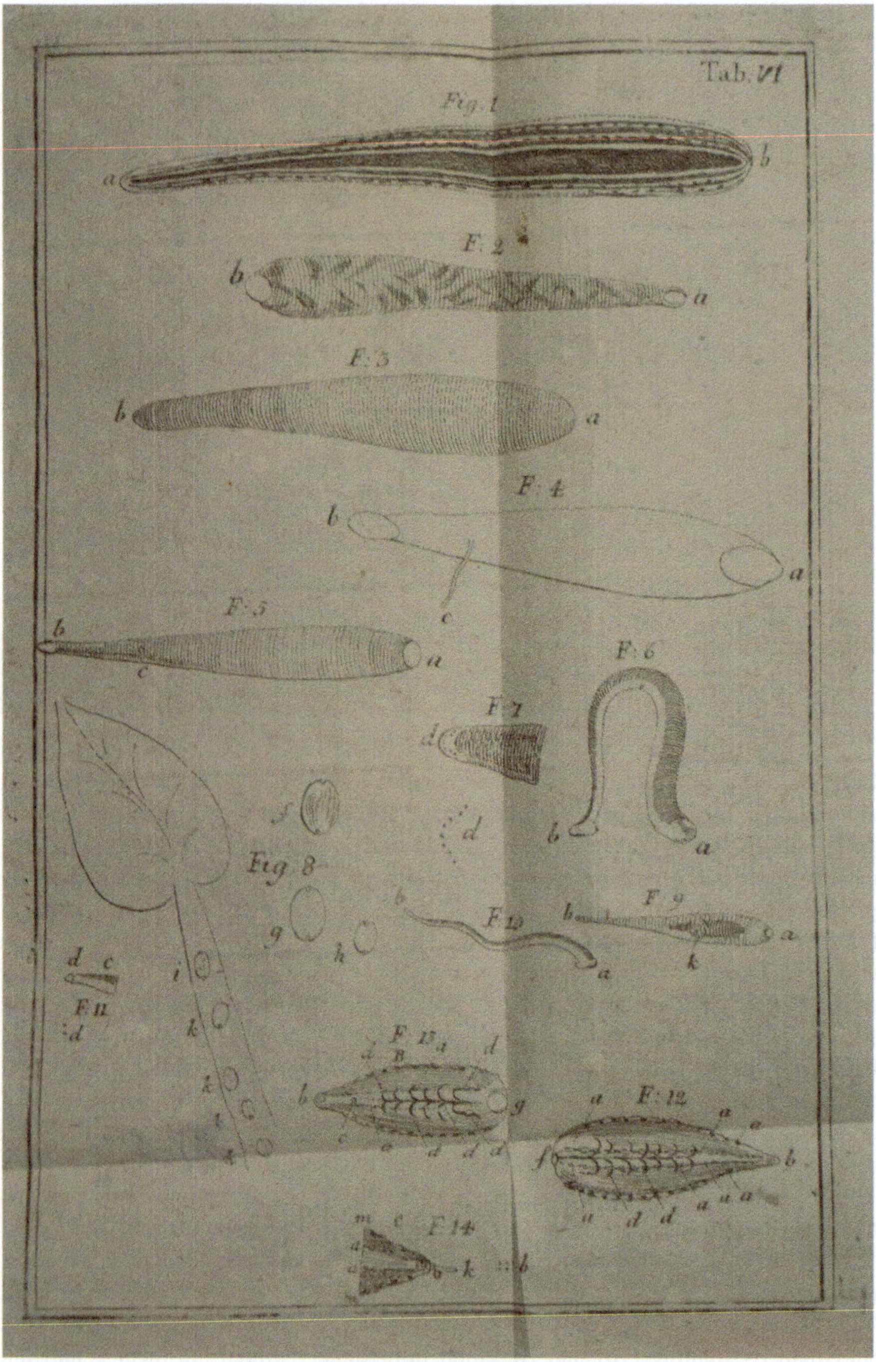

Fig. 4.6 Illustration of leeches from Bergman's 1757 paper

took part in a voting on self-government of the University is based on a translation error by Schufle; no such voting took place and Bergman did not get a permanent position.[7]

4.2 Scheele's Education in Gothenburg

Aged 15, Scheele arrived in Gothenburg in 1757, i.e. around the time when Bergman was about to finish his studies in Uppsala. Gothenburg was the second largest city of Sweden, having about 10, 000 citizens. It was still a young city, founded only 100 years earlier at the strategic point where the river Göta älv meets the sea. The city was built by Dutch experts, and during the seventeenth century the majority of the citizens were Dutch. Linnaeus, who visited Gothenburg in 1746, described a city of round layout surrounded by embankments. The streets were straight with canals cutting through the city. Trees were planted along the canals and most of the houses were wooden, built in two storeys, painted red or yellow, with white or blue window frames and corners. The roofs were covered with clay tiles [17]. Only a few seventeenth and eighteenth century buildings remain in Gothenburg, and only fragments of the embankments have survived.

In Scheele's time, Gothenburg was the home of the Swedish East Indian Company, which has been described as the most successful company in Swedish history. Founded in 1731, the company imported luxury items such as porcelain, tea, medicinal plants, and spices from Guangzhou (Canton) in China. The headquarters were located just opposite the canal from where Scheele worked and lived. The building of the East Indian Company still remains today and houses the City Museum of Gothenburg. The director of the East Indian Company, the wealthy merchant Niklas Sahlgen, lived in the adjacent building. Slightly further away was the German Christinæ church, which had burned in 1746 and was under reconstruction when Scheele arrived.

The Unicorn pharmacy (Apoteket Enhörningen), established in 1641 by Kilian Treutiger from East Frisia, was the fifth pharmacy in Sweden, and the first in Gothenburg. The original pharmacy building at Södra Hamngatan burned down in 1721. Apothecary Ermersch, the third owner of the pharmacy and like his predecessors born in Germany, could not afford to rebuild it, but sold the lot to his brother in law, Martin Andreas Bauch (1693–1766) who moved to Gothenburg from Güstrow in Mecklenburg. In the eighteenth century, many Swedish pharmacies were run by German immigrants, so this was no coincidence. Bauch built a two-storey building in wood, where the pharmacy reopened in 1724. This building, where Scheele worked, burned down in 1803. It was now rebuilt in stone, and the pharmacy remained there until 1915 (Fig. 4.7). Like so many other old buildings in Swedish cities, it was demolished by city planners during a wave of destruction in the 1960s and 1970s.

[7]The voting that Schufle referred to was actually the voting of who would be the *primus*.

Fig. 4.7 Södra Hamngatan, where the Unicorn pharmacy was once located. The current building (behind the red car) was built in 1970 and bears no memories from Scheele's days. The ground floor houses a gym and a hairdresser. *Photo* Anders Lennartson, July 2014

Typically, an apothecary's education in Sweden at Scheele's time would take between 3 and 5 years. It was not a formal education, instead the apprentice's parents would write a contract with an apothecary. The apothecary would provide food and sleeping quarters, while the family had to provide clothes. It was hard work from the very first day. Scheele was, however, very ambitious and spent the nights reading the books from Bauch's book collection. Scheele's main teacher in chemistry and pharmacy appears to have been an older apprentice called Grünberg, who later became apothecary in Stralsund. Grünberg has later told Scheele's friends that Scheele was very quiet and serious. Apart from the books in Bauch's library, Scheele borrowed books from Grünberg. Two of Scheele's favourite books were *Laboratorium Chymicum* by Johannes Kunckel (1630–1703) and *Praelectiones chemiae* by Caspar Neumann (1683–1737) [18]. Not only did Scheele read, he also repeated the experiments described in the books and often found deviations. This may be the origin of Scheele's habit to always question published results and repeat every published experiment that appeared to contradict his knowledge or chemical intuition. Zekert has published an inventory list of Bauch's shop from 1755, [19] and among the other chemistry books Scheele had access to were Lémery's *Cours de chymie*, Rothe's *Gründliche Anleitung zur Chymie* and Boerhaave's *Elementa chemiæ*. It is said that Bauch wrote back to Scheele's parents in Stralsund, worried that his young apprentice would ruin his health by his frequent reading of books which Bauch thought were too difficult for a 15-year-old boy [20].

In 1784, when Scheele was a famous chemist, he was contacted by Grünberg. In his answer, Scheele wrote:

> The first reason for that [i.e. his success] is you, my dear friend, as you already in the very beginning of my education encouraged me to read Neumann's Chemistry. Through the reading I got a desire to do my own experiments; and I still remember so well how I mixed oil of clove with fuming nitric acid in a small jar, and it immediately ignited;[8] I never told anyone about it, but I can still see the unfortunate experiment for my eyes. [21]

At one point, his colleagues joked with him and placed fulminating powder[9] in one of his bottles causing a violent explosion in the night [21]. It is quite remarkable that Bauch tolerated Scheele's experiments.

According to C. G. Helling, later apothecary in Lidköping, Scheele, already as an apprentice, had accuired a remarkable knowledge in chemistry [22]. After finishing the education it was customary for the apprentice to stay some time as an assistant at the pharmacy as a tribute to his master. When Bauch, aged 72, sold the Unicorn pharmacy in 1765, Scheele took the opportunity to move on after 8 years in Gothenburg. Bauch died from a stroke on August 18, 1766 and never experienced the fame of his apprentice.

References

1. Schück H (1916) Torbern Bergmans självbiografi, äldre svenska biografier 3–4, Uppsala, p 86
2. Bergman T (1779) Åminnelse-tal öfver...Högvälborne Friherren Herr Carl de Geer, Kongl. Vetenskapakademien, Stockholm, p 27
3. Hjelm PJ (1786) Åminnelse-tal öfver...Herr Torbern Bergman, Stockholm p 10
4. Schück H (1916) Torbern Bergmans självbiografi, äldre svenska biografier 3–4, Uppsala, p 87
5. Schück H (1916) Torbern Bergmans självbiografi, äldre svenska biografier 3–4, Uppsala, p 88
6. Lindroth S (1978) Svensk lärdomshistoria. Frihetstiden. Norstedts, Stockholm, p 311
7. Lindberg SG (1956) Bengt Ferner, Svenskt biografiskt lexicon, vol 15, p 635
8. Hjelm PJ (1786) Åminnelse-tal öfver...Herr Torbern Bergman, Stockholm p 17
9. Bergman T (1756) Rön om Coccus aquaticus. Linn Faun. Suec. n. 725. KVA Handl 17: 199–204
10. Hjelm PJ (1786) Åminnelse-tal öfver...Herr Torbern Bergman, Stockholm p 15
11. Bergman T (1757) Afhandling om iglar. KVA handl 18:304–314
12. Schück H (1916) Torbern Bergmans självbiografi, äldre svenska biografier 3–4, Uppsala, p 97

[8]When a drop of eugenol, the main constituent of oil of clove, is mixed with a drop of fuming nitric acid, it ignites with a bright flash. Scheele's experiment was probably carried out on a much larger scale.

[9]Most likely a mixture of potassium nitrate, sulphur and sodium or potassium carbonate. On heating, the carbonate yields a polysulphide with the sulphur, which in turn is ignited by potassium nitrate.

13. Bergman T (1773) Classes larvarum, definitæ Nova Acta Regiæ Societatis Scientiarum Upsaliensis, 1:58–65
14. [Bergman T] (1771) Mémoires pour servir à l'historie des insectes par Charles de Geer, Lärda Tidningar, 350–352
15. Bergman T (1779) Åminnelse-tal öfver…Högvälborne Friherren Herr Carl de Geer, Kongl. Vetenskapakademien, Stockholm
16. Schufle JA (1985) Torbern Bergman A Man Before His Time. Cornado Press, Lawrence, Kansas,p 33
17. Linnæus C (1747) Wästgöta Resa, Lars Salvius, Stocholm p 135
18. Leventin A, Schimmelpfennig CFV & Ahlberg KA (1910–1918) Sveriges apotekarhistoria, vol 1, Stockholm, p 9–202
19. Zekert O (1933) Carl Wilhelm Scheele. Sein Leben und seine Werke. Vol 3. Mittenwald, p 85–192
20. Sjöstén CG (1801) Åminnelse-tal, hållit för Kongl. Vetenskaps Academien öfver dess framledne ledamot herr Carl Vilhelm Scheele, den 14 oct. 1799. Stockholm, p 9
21. Crell L (1787) Einige Nachrichten von den Lebensumständen Carl Wilhelm Scheeléns, Chem Ann 1:175–192
22. Sjöstén CG (1801) Åminnelse-tal, hållit för Kongl. Vetenskaps Academien öfver dess framledne ledamot herr Carl Vilhelm Scheele, den 14 oct. 1799. Stockholm, p 10

5 Bergman's Early Scientific Career

On November 29, 1758, less than half a year after receiving his master's degree, Bergman presided over his first own dissertation. Matthias Rydell, born in the Västergötland province just as Bergman, defended a thesis titled *On universal attractions* [1]. A few days later Bergman was promoted to *docent*[1] in physics on recommendation by Ferner [2]. Bergman was now entitled to lecture, and his lectures on experimental physics were popular despite his limited excess to scientific instruments [3].

In 1761, the adjunct (approximately equivalent to an assistant professor) in physics, Daniel Melander (Melanderhjelm) was appointed professor in physics. Essentially, he had to buy the title, as he agreed to pay 15,000 copper dalers, a very large sum, to a fund established to raise the salary of the observatory assistants [4]. As Melander was appointed, the position as adjunct in mathematics and physics became vacant. By the end of the year, the University Chancellor, Anders Johan von Höpken, appointed Bergman. With a position at the University, Bergman could finally give up his job as private teacher. The following spring, Bergman gave lectures on algebra, replacing Professor Jonas Meldercreutz (1715–1785), who took part in the Diet meetings (Riksdagen) in Stockholm 1760–1762.

As mentioned earlier, Bergman's teacher Ferner left Uppsala for a 5-year journey through Europe in 1758, shortly after Bergman had defended his thesis. During this time, they corresponded regularly by exchanging letters. Ferner's trip proved to be very important for Bergman's career, as Ferner actively sought for suitable European correspondents for Bergman [5]. International contacts were as important then for a successful scientist, as they are today. The first contact was

[1]Docent is an academic title in Sweden. In the present day, a docent title implies that a person has proved capable of carrying out independent research and shown pedagogic skills. Typically, this is equivalent to be appointed to associate professor.

A. Lennartson, *Carl Wilhelm Scheele and Torbern Bergman*, Perspectives on the History of Chemistry, https://doi.org/10.1007/978-3-030-49194-9_5

established in London, where Benjamin Wilson (1721–1788) agreed to become Bergman's correspondent. Göte Carelid, who collected and transcribed the letters to Bergman from his foreign correspondents, lists 84 scientists that Bergman corresponded with during his career [6]. Wilson was not only an accomplished physicist working in the field of electricity and a correspondent of Benjamin Franklin[2] in Philadelphia, but also a skilled artist. As Bergman had started to study electrical phenomena (Sect. 5.2), Wilson was a perfect contact, and of Bergman's letters to Wilson, five where published in *Philosophical Transactions*. One of Bergman's experiments was demonstrated in the Royal Society and was communicated by Wilson to Franklin [7]. Bergman also demonstrated experiments by Wilson in the Royal Swedish Academy of Sciences in Stockholm and published an excerpt from one of Wilson letters describing some electrical experiments [8]. Next, Ferner travelled to Paris, where he continued to search for possible contacts for Bergman.

5.1 Astronomical Research

As a student, Bergman had spent much time in the Observatory in Uppsala. Bergman's first original paper on astronomy, *Remarks on silent fires*, was published in 1760 [9]. In this paper, Bergman relates his own, and others, observations of lightning phenomena without any audible thunder during the last winter. The nature of the lightning was still not known with certainty, although its connection with electricity had recently been established. There was also confusion with other atmospheric phenomena such as auroras and falling stars. Bergman speculated on a possible connection with several earthquakes during the same period. An earthquake could perhaps release flammable particles from the underground which would collect at high altitude and ignite. Bergman also wrote two, what we today would call review-papers, summarising published research. The first of these papers appeared in 1759 and dealt with the rainbow, [10] the second appeared in 1760; here Bergman returned to the topic of his 1755 dissertation, *On the twilights* [11]. Bergman realised that the twilights were a very fortunate phenomenon:

> To be thrown from the brightest day to a dark night, and to see the sun from the deepest darkness, would certainly bring about the greatest inconvenience. So rapid and strong changes would soon destroy the animals' eyes, many wanderers and travellers would be lost and birds would fall to their destruction. All of this has the wise Creator prevented by adding an atmosphere or circle of air to our Earth, so that we do not immediately lose all light although the sun has went under the horizon, and also not kept in complete darkness until she appear again, but the change between day and night occurs gradually through the twilights.

[2]Benjamin Franklin was one of the foremost authorities on electricity in the mid-eighteenth century.

In his paper, Bergman summarised the observations and explanations of twilights back to the days of Johannes Kepler and Tycho Brahe and provided a geometrical description accompanied by figures.

A long paper on the altitude of aurora borealis, based on his own and other's observations was published in two parts in 1764 [12] with an addition in 1766 [13]. During the period 1645–1715, the so-called Maunder minimum, a period of anomalously low solar activity, auroras had been very rare phenomena. In the eighteenth century the solar activity increased, and the solar maximum in 1761 gave Bergman good opportunities to study auroras. In order to understand the nature of auroras, their altitude was of great importance. If they appeared at very high altitudes, it was unlikely that they could be attributed to combustion of vapours from the ground. Bergman described a method reported by a F C Mayer which could be used to calculate the altitude of auroras under certain circumstances. Bergman then described a simpler method which was based on two observations of the same aurora. If the angle from the ground to the aurora was measured at two locations, and the distance between the two observations was known, the altitude of the aurora could be calculated through geometry.[3] The great draw-back was of course the need of two measurements, but by comparing his diary with the diary of Nils Gissler (1715–1771) in Härnösand,[4] he found that they both had observed the same auroras and made the necessary measurements. Other pairs of measurements were taken from literature sources. Bergman concluded that the auroras always appeared over the clouds, usually at altitudes between 530 and 1,070 km (50–100 Swedish *mil*). Auroras typically appear at 90–150 km altitude, but may occasionally stretch up to 1,000 km. After becoming professor in chemistry, it is not likely that Bergman had much time for astronomical observations and he published no more astronomical papers.

On June 6, 1761, planet Venus was scheduled to transit the sun (Fig. 5.1), a very rare event. Transits of Venus occur in pairs separated by 8 years followed by a 100-year long gap to the next transit. The first transit to be observed was in 1639, some 30 years after the invention of the telescope, followed by transits in 1761, 1769, 1874, 1882, 2004, and 2012. The transit gives a unique opportunity to study planet Venus, and one of the objectives was to determine whether Venus had an atmosphere or not.

Thus, astronomers had waited a long time for this opportunity and it was perhaps the first international scientific collaboration on a large scale, with astronomers distributed over the area were the transit would be visible. The Royal Swedish Academy of Sciences considered sending Bergman to Torneå in northern Finland (at the time still a Swedish province) as an observer [3]. After some consideration, it was decided that Kajaneborg would be a better location for observations, and Anders Planman (1724–1803), who was born in Hattula in Finland and spoke

[3]Bergman does not give a reference and seems to have developed this method by himself. Whether he was the first astronomer to use the method is unknown to me.

[4]Gissler was physician and lecturer in Härnösand. He had studied physics for Anders Celsius and Klingnestierna, chemistry for Wallerius and finally medicine for Linneaus and Rosén.

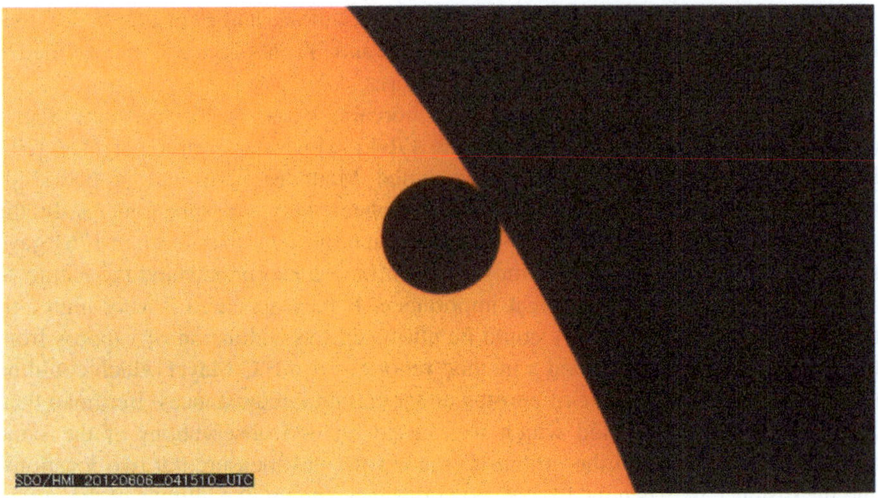

Fig. 5.1 The June 6, 2012 transit of Venus imaged from the Solar Dynamics Observatory (SDO) in geosynchronous orbit. *Photo* NASA

Finnish, was sent instead of Bergman. It would actually turn out to be fortunate for Bergman, since his colleague Fredric Mallet (1728–1797), who later became professor of mathematics, actually travelled to Torneå, but could not make any observations due to the cloudy weather [14]. Bergman instead assisted Strömer in Uppsala, where the weather was more favourable. Bergman described his observations in a letter to Benjamin Wilson, who published the letter, written in Latin, in *Philosophical Transactions* [15]. The observation of an atmosphere around Venus by the Russian scientist Mikhail Lomonosov during the 1761 transit was later found to be an error. Not until the nineteenth century was the atmosphere Venus correctly observed and not until the space age did we obtain any data about its remarkable nature.

In 1765, Bergman published a paper in *Economic Newspapers* on the importance of weather forecasts for agriculture [16]. Bergman noted that within the tropics, the weather was easily predicted, while weather was more unpredictable in Sweden. Bergman suggested that this was due to lack of reliable data, and encouraged people to use thermometers, barometers, and hygrometers to collect data and send the observations to the Royal Swedish Academy of Sciences for evaluation. Bergman stressed to the reader that observed connections could be coincidental. For instance, it could be observed that the lightning of a blast furnace diverted a raincloud, but based on a single observation this was not a reliable conclusion.

5.2 Bergman's Research on Electricity

Of the equipment that Bergman managed to acquire as a newly promoted *docent*, the most important piece was an electricity machine (Fig. 5.2), a device that would be more than useful. Electricity was one of the hottest topics in physics at the time, so it was probably a good career choice for a young physicist. In Bergman's time, only static electricity was known; the first galvanic cell, the voltaic pile, was not invented until 1799, so a source for continuous electric current was not available. In an electricity machine, a glass sphere or glass disc was rotated on a shaft against a pad creating a charge separation.

Static electricity had been known for centuries and was studied by the ancient Greeks. In 1733, French physicist Charles François de Cisternay du Fay (1698–1739) had made a surprising discovery: static electricity could give rise not only to attraction, but also to repulsion. There appeared to be two forms of electricity: glass acquired one form of electricity (vitreous electricity) on rubbing, while lacquer acquired the opposite form (resinous electricity). du Fay proposed that both forms of electricity were present, in the same amounts, in uncharged objects. In 1745, Ewald Georg von Kleist (1700–1748) in Kammin, Pommerania, invented the Leyden jar, a simple capacitor which greatly simplified electrical research. In 1747, Benjamin Franklin and Scottish physician William Watson (1715–1787) independently proposed that there was only one form of electricity and that an object could have an excess or deficiency of electricity. Franklin was the first to use the terms positive and negative charge. In 1750, Franklin proposed charging a Leyden jar by sending a kite connected via a metal wire into a thunderstorm, an experiment carried out 2 years later proving the electrical nature of lightings.

It was Klingenstierna who introduced research on electricity in Sweden. In 1740 and 1742, his student Jah. Morthensson defended two theses on electricity. Following its invention by Kleist, the Leyden jar was also described by Dutch mathematician Professor Pieter Musschenbroek (1692–1761) in Leiden (hence the name); whether he was an independent inventor of the Leyden jar is unknown. Klingenstierna translated Musschenbroek's text book, *Beginsel der Naturunde* (*Introduction to Natural History*; 1737), to Swedish with numerous additions. In this translation, [17] published in 1747, Klingenstierna described some electrical experiments performed in Uppsala. The same year Strömer published a paper on electrical phenomena, [18] and professor Pehr Elvius (1710–1749), another student of Klingenstierna, published a paper on the history of electricity [19]. A few years later, Strömer published a paper on the effects of electricity on the human body [20]. After Klingenstierna gave his lecture *On the Latest Discoveries in Electricity*, [21] when he concluded his term as president in the Royal Swedish Academy of Sciences in October 1755, the electrical research in Uppsala seems to have declined [22]. The research on electricity was, however, picked up by Klingenstierna's and Strömer's former student Carl Johan Wilcke in Stockholm.

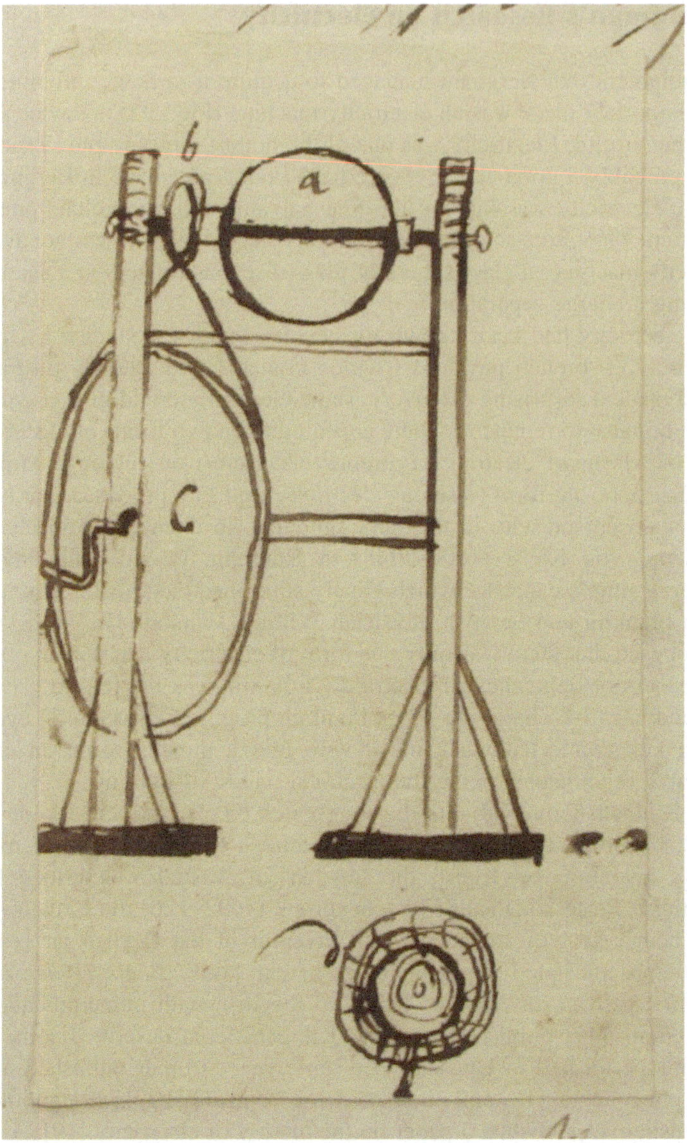

Fig. 5.2 A 1777 drawing of an electricity machine by Scheele in a letter to his friend Hjelm. On turning the crank *C* the glass ball *A* rotates and is charged by friction against a pad. Carl Wilhelm Scheele's Archive, Centre for History of Science, Royal Swedish Academy of Sciences. *Photo* Anders Lennartson

Wilcke,[5] [23] was born on September 6, 1732 in Wismar, Swedish Pommerania, where his father, Samuel Wilcke, was a priest. His mother was Anna Scheele, a distant relative to Carl Wilhelm—she was a half second cousin to Carl Wilhelm's father. In 1739, his father was appointed vicar in the German Church in Stockholm, and Johan Carl attended the German School. In 1750, he enrolled at Uppsala University, where he was supposed to study theology, but devoted his time to mathematics and physics—i.e. just like Bergman would do a few years later. His principle teachers were Klingenstierna and Strömer. In 1751, he moved to Rostock, where his father had friends: as young, Samuel Wilcke had been private teacher to Angelius Iohannes Aepinus, who since had become professor in rhetoric. Wilcke also befriended Aepinus brother, natural scientist Franz Ulrich Theodorius Aepinus (1724–1802). In 1753, he continued his studies in Göttingen and in 1755 he moved to Berlin, where he lived in the same house as Franz Aepinus and befriended the famous mathematician Leonard Euler (1707–1783). Wilcke presented his inaugural thesis [24] in Rostock in October 1757, a thesis concerning electrical experiments. He returned to Uppsala soon after in order to become *docent*. Unfortunately, there was only one professorship in physics in Uppsala, and with a young professor—Samuel Duræus (1718–1798)—recently appointed, he could not hope for a professorship. Another possibility would however arise. In 1727, Samuel Tham had donated money to Collegium illustre, a society which had been established by the House of Nobility (Riddarhuset) in 1625. The money was later transferred to the Royal Swedish Academy of Sciences and its secretary, Wargentin (Sect. 6.2), suggested that money from the fund should be used to appoint a Thamian lecturer, who would serve as vice secretary in the Academy and who would give lectures on experimental physics. The Academy turned to Klingenstierna for advice, and Klingenstierna recommended Wilcke, who was appointed in 1759. Wilcke was an active researcher publishing numerous papers in the *Transactions of the Royal Academy of Sciences*. The papers published during the period 1758–1772 mainly concerns electricity and magnetism, where after his interest shifted to heat. When Wargentin died in 1784, Wilcke succeeded him as secretary in the Academy.

When Bergman started to study electrical phenomena in the early 1760s, Wilcke was the main authority on electricity in Sweden. It does not seem like Wilcke and Bergman knew each other in Uppsala, but Bergman wrote to Wilcke in Stockholm in spring 1763, asking Wilcke to become his correspondent. Wilcke accepted the invitation.

In 1762, Bergman published a paper on the electrical properties of calcite (Iceland spar) [25]. Benjamin Wilson had found that calcite crystals would only be electrically charged on rubbing at low temperature, and he therefore asked Bergman to test this in Sweden, as the temperatures in Uppsala typically are much lower than in London in the winter. Bergman only noted weak charge at −3 °C. He expected this charge to disappear on heating, but, to his surprice, he found that the charge increased. Ten crystals displayed the same effect.[6]

[5]There are no contemporary portraits of Wilcke, except a silhouette.

[6]I have found no reliable source referring to any pyroelectric properties of calcite.

His next paper on electricity concerned the rubbing of silk ribbons of different colours [26]. The idea of using ribbons of different colours may sound naïve today, but originated from Robert Symmer (1707–1763) who had reported that when white and black silk stockings were rubbed against each other, the former became positively charged and the latter negatively charged. Bergman built a devise where a silk ribbon was attached and stretched. The ribbon was then rubbed with another silk ribbon, either along or against its direction. The sign of the charge was determined with a piece of gold leaf attached to a charged silk thread. The most important conclusion was that if a ribbon was heated before rubbing, it would acquire a negative charge on rubbing. As expected, the colour of the ribbon was found to have no effect on the charge. In a paper on the rubbing of glass plates, published in 1765 [27], Bergman concluded that (1) there are two types of electricity, having the same effects but annihilating each other and (2) both types of electricity consists of an excess of a particular substance. The particles of the two substances can saturate each other giving rise to an inactive compound. In the natural (uncharged) state, any substance is saturated with this inactive compound. Thus, Bergman did not agree with Franklin and proposed a chemical view of electricity.

Bergman's final paper on electricity appeared in 1766 and concerned the electrical properties of tourmalines [28], a study which British chemist Thomas Thomson (1773–1854), the author of the influential book *History of Chemistry* regarded as important [29]. Tourmalines were mined in Sri Lanka and brought to Europe by the Dutch East Indian Company. They were very expensive and had attracted much attention, also among scientists. Axel Fredrik Cronstedt (the discoverer of nickel) had asked mineralogist Bengt Qvist (1733–1790), who was in Amsterdam at the time, to acquire tourmalines on the behalf of the Royal Swedish Academy of Sciences in Stockholm. These stones, there were five of them (three from Sri Lanka and two from Brazil), were investigated by mineralogist Sven Rinman and by Bergman. Rinman analysed the composition of small fragments, which had come loose during the shipping to Sweden [30]. Wilcke, finally, wrote a review article summarising the known literature on tourmalines, including the works by Rinman and Bergman [31] and a fourth paper on tourmalines from Brazil was published later the same year by Rinman [32]. Rinman (Fig. 5.3), who would become an important friend for Bergman, was born in Uppsala in 1720. He studied chemistry for Wallerius and became an auscultator (a civil servant position) at the Board of Mines (Bergskollegium) in 1740. In 1746, he left Sweden to undertake a study trip through Europe: the Netherlands, Germany and France. In 1762, he left Board of Mines to take up a position as inspector for the Swedish Steel Association (Jernkontoret). Rinman was not particularly happy with this job [33], since the position meant that he had to constantly travel across the country to inspect the different iron works and in addition he had to deal with problems of little scientific challenge. Largely disconnected from libraries and scientific news, Bergman became an important link to the scientific community for Rinman. In 1773, he could

Fig. 5.3 Bergman's friend Sven Rinman. Plaster medallion by Tobias Sergel. *Photo* Nationalmuseum, Stockholm

finally settle down in Eskilstuna, where he also set up his own laboratory. He is today best known for his textbooks on mining and metallurgy. He was elected a member of the Royal Swedish Academy of Sciences in 1751 and died in Eskilstuna in 1792.

Franz Aepinus had studied tourmalines in Berlin during the time when Wilcke visited him. In 1757, Aepinius reported that the tourmaline crystals had two poles, and upon heating, one pole became positively charged while the other pole became negatively charged. In 1759, Benjamin Wilson had reported contradictory results: when a large crystal was heated on only one side, both sides would acquire the same charge. Bergman showed that the effect was due to thermal expansion (Fig. 5.4). When one pole was heated the crystal expanded and the opposite pole acquired the opposite charge, but as the heated pole cooled, the polarisation switched. Bergman showed that the crystal would not acquire any net charge: there were always equal amounts of positive and negative charge. The original stones that

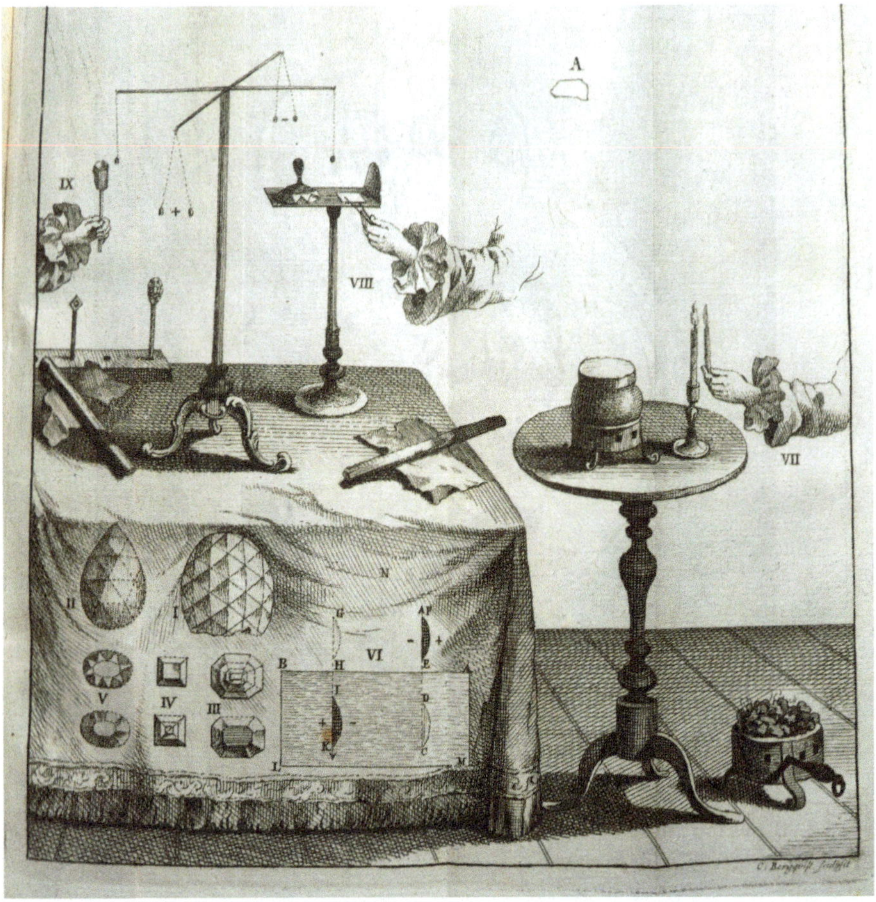

Fig. 5.4 Copper plate from Bergman's paper on the electric properties of tourmalines, including images of the now lost stones

Bergman studied seem to be lost.[7] Interestingly enough, the cause of the pyroelectricity of tourmalines is still, as of 2019, a matter of discussion [34]. In short, the pyroelectricity can be correlated to the change in dipole moment of the coordination polyhedra along the crystallographic axes as a function of temperature.

Bergman did not only study the theoretical aspects of electricity, he also engaged himself in a more practical aspect of the field: the design of lightning conductors. This research is discussed in Sect. 6.3.

[7]In 1849, the Royal Academy transferred its mineral collection to the Museum of Natural History in Stockholm; no stones corresponding to Bergman's and Rinmans' descriptions are found there.

5.3 Bergman's Research in Entomology

Bergman never gave up his interest in insects. In 1762, he published a paper describing an oak apple, which he could not find in the works of de Réaumur,[8] and which he had shown to Linnaeus 2 years earlier [35]. It was known that oak apples were formed by insects, but much of the details of their formation were still unknown at Bergman's time.

Bergman's next contribution was a long paper on sawflies (genus *Tenthredinidae*) and their larvae [36]. In this paper, which must have taken several years of studies to prepare, Bergman gave an extensive description of the different sawflies, their ethology and life cycles.

In 1762, Bergman had an excellent chance to apply his knowledge about insects. The Royal Swedish Academy of Sciences had received a donation from Count Fredrik Sparre, who had died in 1747. From 1762, the fund was used to finance yearly competitions [37]. The members of the Academy were entitled to suggest topics for the competitions and the Academy then selected topics on their meetings, encouraging the public to send in their solutions. Usually, the topics were related to the improvement of agriculture and other questions of economic interest. From the answers, the Academy selected a winner, who was awarded a gold medal.

In 1762, The Academy published the question "How worms [caterpillars], that make damage on fruit trees by eating flowers and leaves, can be circumvented or expelled in the best way". The Academy received eleven answers, of which answer number six was found most promising; after evaluation it was awarded the gold medal [38]. Four additional answers were found suitable for publication (Fig. 5.5). It turned out that the winning contribution came from Bergman. The other published contributions came from Professor Johan Leche (1704–1764) in Åbo, Professor Erik Gustaf Liedbeck (1724–1803; one of Linnaeus' apostles), Roland Schröder, a merchant from Stockholm, and finally a Mr C. N. Nelin. The later never revealed himself, but C. N. Nelin is an anagram of C. N. Linné (e.g. Carl Linnaeus, who changed his name to Carl von Linné after his ennoblement in 1761. The N. would stand for "Nilsson"). Linnaeus was very sensitive and believed himself to be commissioned directly by God to reveal the wonders of nature, and it has been suggested that the fact that Bergman won would have turned Linnaeus against him. If so, Linnaeus did not show it. When Bergman's suggestions were attacked by an anonymous writer, Linnaeus sent Bergman a polite letter[9] were he gave Bergman his support. The anonymous author had also criticised Linnaeus for publishing his books in Latin rather than Swedish. Linnaeus was annoyed and wrote to Bergman that by writing in Latin he could serve the whole world, and if he would also publish in Swedish, he would have worked himself to death.

Bergman's essay stresses that a deep knowledge is important in order to find a solution to the problem. "He who believes that a worm [*i.e.* a caterpillar] can arise from putrefaction (although it is more complicatedly built than an elephant and, as

[8]René Antoine Fercault de Réaumur (1683–1757), French physicist and zoologist.
[9]Signed June 25, 1764.

Fig. 5.5 The printed answers to the question "How worms, that make damage on fruit trees by eating flowers and leaves, can be circumvented or expelled in the best way". *Photo* Anders Lennartson

little as he, arise without a father and mother) will not easily find a reliable means against them" [39]. He concluded that "An injurious bear is easily caught or killed, but to catch or kill the same weight, if it could be transformed into flees, would be impossible". In another text, Bergman concluded that caterpillars could eat twice their body weight a day, and continued: "If horses or cattle, which can weigh from 700 to over 1,000 *skålpund* [290–420 kg], could eat their double weight in a day, these for humans so useful animals would since long been completely exctinct by hunger" [40]. Bergman then described the most notorious species of butterflies that attack apple- and pear trees.

Bergman had studied the subject thoroughly and described many methods that had been suggested to him, and explained why they would not work. Fortunately, he could also deliver several suggestions on how to get rid of the butterflies and their caterpillars. First, the species had to be identified as it determined the remedy. Some species could be caught by fires by night: they were attracted by the light but had to pay with their lives. Others could be killed with swatters, a proper work for children according to Bergman. Other species could be caught by catching the females and attaching them to the trees with needles. The males could then be caught when they came for mating with the immobilised females.

To kill the caterpillars, Bergman had two solutions depending on the species. Those who spent the winter nestled in the trees could be picked manually. Another solution was to release beetles of the gender *Carabus*, especially the blue ground beetle, which eat caterpillars. The beetles themselves were harmless since "as little as a wolf can learn to eat apples, as impossible is it for these small predators to eat vegetables". After criticism from an anonymous author, claiming that the caterpillars of the butterfly *Aporia crataegi* were the main problem, Bergman had to deliver a defence of his ideas in a public letter to the Secretary of the Academy [41].

Apparently, the solutions Bergman provided were not effective enough, and caterpillars continued to eat the leaves from the Swedish fruit trees. Therefore, the Academy renewed the question in 1768. This time four answers were delivered, and after opening the sealed author tickets it turned out that Bergman, now professor of chemistry, once again was the winner. The second prize went to Carl Fredric Lund (1716–1776), the Mayor of Linköping, who suggested planting spruce between the fruit trees and Adolph Modéer (1739–1799), who frequently delivered answers to the Academy's competitions. The fourth answer came from an unidentified person with initials C. F. H. [42].

Since 1762, Bergman had gained much additional knowledge, and in his second essay he focused on two insects. The first, the winter moth, he had found responsible for most of the damaged during the past years. He had collected the caterpillars and placed them in a bell jar with suitable food and soil. After some time, the caterpillars buried themselves in the soil, and Bergman buried the bell jar outside. In October, moths appeared, some with wings (males) and some without (females). The wingless females climbed the tree in order to mate with the males. Bergman could observe the mating and how the small eggs were laid close to the buds of the fruit trees. With this knowledge the solution was simple: in September 1767 he attached birch bark around the trunk of the apple trees in a small garden and painted it with sticky tar. In October the female moths, attempting to climb the trees, got stuck in the tar. The other insect was a weevil *Curculio pomorum*, for which there was no better treatment than to cut off and burn infected twigs. After a few years, the problems would be under control, Bergman promised. Bergman's method of capturing the climbing insects was successfully tried by count Cronstedt and Mine Councillor (bergsrådet) Adlerheim, who caught 28,716 females the following year [43].

References

1. Bergman T, Rydell M (1758) Dissertatio physico-mathematica de attractione universali cujus partem priorem, cum consens. ampliss. facult. philos. in regia academia Upsaliensi, publico bonorum examini submittunt stipendiarius Thorbernus Bergman ... et Matthias Rydell, V. Gothi, In audit. Carol. maj. die XXIX. Nov. anni MDCCLVIII. H. a. m. s. Uppsala
2. Hjelm PJ (1786) Åminnelse-tal öfver...Herr Torbern Bergman, Stockholm, p 18
3. Schück H (1916) Torbern Bergmans självbiografi, äldre svenska biografier 3–4, Uppsala, p 88
4. Lindroth S (1978) Svensk lärdomshistoria. Norstedts, Stockholm, Frihetstiden, p 20
5. Carelid G, Nordström J (1965) Torbern Bergman's foreign correspondence. Almqvist & Wiksell, Uppsala, p XXII
6. Carelid G, Nordström J (1965) Torbern Bergman's foreign correspondence. Almqvist & Wiksell, Uppsala
7. Carelid G, Nordström J (1965) Torbern Bergman's foreign correspondence. Almqvist & Wiksell, Uppsala, p XXIII
8. Wilson B (1761) Utrag af et bref från Herr Benjamin Wilson...til Thorbern Bergman... Några nya rön vid electriciteten, KVA Handl, 22:318–320
9. Bergman T (1760) Anmärkningar om tysta eld-sken, KVA Handl 21:63–69
10. Bergman T (1759) Vetenskaps historien om rägnbågens förklaring KVA Handl 20:239–251
11. Bergman T (1760) Vetenskaps historien om skymningarna KVA Handl 21:239–251
12. Bergman T (1764) Afhandling om nordskenens högd. Förra stycket KVA Handl 25:193–210; Bergman T (1764) Afhandling om nordskenens högd. Sednare stycket KVA Handl 25:249–261
13. Bergman T (1766) Tilläggning om nordskenens högd KVA Handl 27:224–227
14. Lindroth s (1978) Svensk lärdomshistoria. Frihetstiden. Stockholm: Norstedts. P 319
15. Bergman T (1761) An account of the observations made on the same transit at Upsal in Sweden: In a letter to Benjamin Wilson. Phil Trans 52:227–230
16. Bergman T (1765) Landthushållningens nytta af noggranna meteorologiska observationer, Oeconomiska tidningar, 1765, N:o 23
17. van Musschenbroeck P (1747) Inledning til Naturkunnigheten, Gottfried Kiesewtter, Uppsala and Stockholm
18. Strömer M (1747) Rön Angående electriciteten, KVA Handl 8:138–142
19. Elvius P (1747) Vetenskapernas historia. Om electriciteten, KVA Handl 8:161–167
20. Strömer M (1752) ...Om electicitetens verkan på människans kropp, KVA Handl 13:193–203
21. Klingenstierna S (1755) Tal om de nyaste rön vid electriciteten, KVA, Stockholm
22. Oseen CW (1939) Johan Carl Wilcke – Experimentalfysiker. Svenska Vetenskapsakademien, Uppsala, Kungl, p 12
23. Oseen CW (1939) Johan Carl Wilcke – Experimentalfysiker. Kungl, Svenska Vetenskaps-akademien, Uppsala
24. Wilcke JC (1757) Disputatio Physica Experimentalis de Electricitatibus contrariis, Rostock
25. Bergman T (1762) Anmärkning om Islands krystalls electricitet. KVA Handl 23:62–65
26. Bergman T (1763) Electriska försök med siden-band af åtskillig färg. KVA Handl 24:323–330
27. Bergman T (1765) Electriska försök med sammangnidna glas-skifvor. KVA Handl 26:127–142
28. Bergman T (1766) Afhandling om tourmalinens electriska egenskaper KVA Handl 27:57–68
29. Thomson T (1831) The history of chemistry, vol II. Henry Colburn & Richard Bentley, London, p 32
30. Rinman S (1766) Mineralogiska rön om tourmalinen eller askblåsare-stenen, KVA Handl 27: 45–57
31. Wilcke JC (1766) Historien om tourmalin KVA Handl 27:90–108
32. Rinman S (1766) Ytterligare mineralogiska rön om brasiliansk tourmalin KVA Handl 27:109–116

33. Fors H (2003) Mutual favours (diss.). Depatment of History of Science and Ideas, Uppsala University, Uppsala, p 107
34. Zhou G, Liu H, Chen K, Gai X, Zhao C, Liao L, Shen K, Shan Fan Z Y (2018) The origin of pyroelectricity in tourmaline at varying temperature. J Alloys Coumpd 744:328–336
35. Bergman T (1762) Et sällsamt galläple, beskrifvit KVA Handl 23:139–143
36. Bergman T (1763) Anmärkningar om vild-skråpukar och såg-flugor KVA Handl 24:154–175
37. Lindroth S (1967) Kungl. svenska vetenskapsakademiens historia 1739–1818, vol 1.1, Stockholm, KVA, p 142
38. Svar på frågan Huru kunna maskar, som göra skada på frukt-träd, medelst blommornas och löfvens affrätande, bäst förekommas och fördrifvas? Hvilken fråga, af Kongl. Vetensk. Academien blef upgifven, år 1762. Stockholm, KVA 1763 p i
39. Several authors (1763) Svar på frågan Huru kunna maskar, som göra skada på frukt-träd, medelst blommornas och löfvens affrätande, bäst förekommas och fördrifvas? Hvilken fråga, af Kongl. Vetensk. Academien blef upgifven, år 1762. Stockholm, KVA p 3
40. Bergman T (1779) Åminnelse-tal öfver...Högvälborne Friherren Herr Carl de Geer, Kongl. Vetenskapakademien, Stockholm, p 33
41. Bergman T (1764) Bref til Kongl. Vetenskaps Academiens secereterare, angående anmärkningarna, som utkommit öfver det svar på frågan om skadeliga frukt-träd-maskar, hvilket vunnit den af Kongl. Academien utlåfvda belöningen. Stockholm
42. Several authors (1769) Svar på den af Kongl. Vetenskaps Academien andra gången framställda frågan, huru maskar, som göra skada på frukt-träd, medelst blommornas och bladens förtärande, bäst kunna förekommas och fördrifvas, Stockholm KVA
43. Hjelm PJ (1786) Åminnelse-tal öfver...Herr Torbern Bergman, Stockholm, p 30

Bergman, Scheele and the Royal Academy of Sciences

6

An important step in Bergman's career was his election as member of the Royal Swedish Academy of Sciences (Kungl. Vetenskapsakademien) in Stockholm. Scheele was eventually also made a member, but as an apothecary, the membership had no direct bearing on his career. Through the membership he did, however, receive financial support, important contacts and a higher international status. In addition to the Academy, Scheele also developed strong bonds to *Collegium medicum*.

6.1 Royal Swedish Academy of Sciences

Mårten Triewald (1691–1747) was a Swedish engineer who had spent ten years in England studying engineering and who was especially interested in the first steam engines that had found use in British mines. After returning to Sweden in 1726, he gave lectures in mechanics in Stockholm and was the first to introduce the steam engine in Sweden. During his stay in London, he had been inspired by the Royal Society, and he now dreamed of establishing a learned society in Sweden. The Royal Society of Sciences at Uppsala (Kungliga Vetenskapssocieteten i Uppsala) had been established in 1710 but focused on humanities, and was far from the society that Triewald had in mind [1].

During the winter of 1738/39, Triewald had befriended Carl Linnaeus, who had returned to Sweden from the Netherlands. He convinced Linnaeus of the need for a society for applied science in Stockholm. A young politician, Anders Johan von Höpken (1712–1789),[1] who had made his way to the inner political circles in

[1]It may be recalled that it was von Höpken, as University Chancellor, who appointed Bergman to adjunct.

A. Lennartson, *Carl Wilhelm Scheele and Torbern Bergman*, Perspectives on the History of Chemistry, https://doi.org/10.1007/978-3-030-49194-9_6

Fig. 6.1 Riddarhuset, or the House of Nobility in Stockholm, where the Royal Swedish Academy of Sciences had their early meetings. *Photo* Anders Lennartson, April 2019

Stockholm, joined the two, and they met several times during spring 1739 in the home of Jonas Alströmer (1685–1761), one of the main figures in the industrial revolution in Sweden. Alströmer had also been to England, been elected a fellow of Royal Society and with two of the founding members strongly oriented towards British traditions, it was natural to use the independent Royal Society as a model rather than the more state-controlled Académie des Sciences in Paris [1]. It became von Höpken's role to write down the by-laws for the new society. Two other men joined, Carl Wilhelm Cederhielm and Sten Carl Bielke, but they played minor roles. The Academy of Sciences was founded on Saturday June 2, 1739 in Auditorium illustre in the House of Nobility (Riddarhuset) in Stockholm (Fig. 6.1). It was in this building that the Academy had their meetings during the first years.[2] In 1740, von Höpken managed to get the Academy's by-laws approved by the King, thus raising the status to a Royal Academy. This was merely a name, as the Academy had no formal connections to the state.

In contrast to the Royal Society of Sciences at Uppsala, the goal for the Royal Swedish Academy of Sciences in Stockholm was to reform the Swedish society through modern science and technology [2]. This was in line with the Enlightment ideas of the time. One of the main instruments of the Academy was the journal *Transactions of Royal Academy of Sciences* (Kongl. Vetenskapsakademiens Handlingar; Fig. 6.2), the first issue appearing in 1739. The journal was issued four times a year, and in contrast to the journals published by Académie des Sciences, and the Uppsala Society, the Stockholm *Transactions* were always published on

[2]In 1779 the Academy got its first own house.

Fig. 6.2 Title page of volume 32 of the *Transactions of the Royal Swedish Academy of Sciences*, published in 1771. The volume contains Scheele's first paper (Sects. 13.4 and 14.1), Bergman's paper on the manufacture of bricks (Sect. 9.5), as well as papers by Bergius (Chap. 10), Hermelin (Sect. 13.1), Mallet (Sect. 7.1), Melander (Chap. 5), Planman (Sect. 5.1), Wargentin (Sect. 6.2), Wilcke (Sect. 6.2) and Wåhlin (Sect. 28.1). The old man planting palm trees for future generations was used, in slightly different designs, well into the nineteenth century. *Photo* Anders Lennartson

time, thus assuring rapid publication of new results. The printing of some of Bergman's publications in the *Transactions of the Royal Society of Science in Uppsala*, such as the paper on classification of larvae (Sect. 5.3) and his last paper on classification of minerals (Sect. 25.8) were delayed by several years. The high publication rate of the Stockholm Academy was partly due to the fact that the president of the Academy was elected for a three-month period, and his main task was to deliver an issue of the *Transactions* by the end of his term. In order to reach a broader audience and not only the professors in Uppsala, the *Transactions* were published in Swedish, although some professors would still have preferred Latin in order to more easily attract an international audience. The *Transactions* were translated to German, but with a few years delay. Some of Scheele's and Bergman's papers were, however, published in German translation in Crell's *Chemische Annalen*. It was Wilcke who provided Crell with translations, at least in 1784 [3].

The choice of typeface for the journal resulted in considerable debate. von Höpken advocated the old-fashioned black letter typeface,[3] which was used until 1743, when a switch was made to a modern Roman typeface [4]. Manuscripts were submitted by both members of the Academy and by the public. Manuscripts that appeared to meet the standard were read in the Academy and then sent to referees [5]. The accepted manuscripts were finally edited by the secretary and longer essays, such as Scheele's paper on manganese (Sect. 15.1), were frequently divided in parts. The contents of the *Transactions* were very varied and included, e.g. medicine, technology, agriculture, geography, physics, chemistry, mathematics botany, zoology and mineralogy. There was a deep focus on new useful inventions. Some papers may seem rather naive, ridiculous or extremely out-dated today, while other papers report important discoveries, such as the discovery of several chemical elements, e.g. nickel, tantalum, tungsten, molybdenum, manganese, barium, chlorine, silicon and selenium.

6.2 The Academy Secretary

The work in the Academy was organised by the Secretary. The first two years, von Höpken served as secretary, then the position was briefly held by Jacob Faggot (1699–1777) and Pehr Elvius. Upon Elvius' death in 1749, Pehr Wargentin was appointed secretary, and he would hold the position for 34 years, longer than any Secretary in the history of the Academy.[4]

[3]Sweden had by tradition used black letter typefaces, like in Germany, but in the Age of Enlightment the Roman typefaces became increasingly popular, especially in science. In more traditional fields such as theology and law, the black letter typefaces survived into the nineteenth century.

[4]Chemist Jacob Berzelius came close, by serving 30 years as Secretary (1818–1848).

Fig. 6.3 Physicist Wargentin, secretary of the Royal Swedish Academy of Sciences 1749–1784. Engraving by Johan Elias Cardon. Nationalmuseum, Stockholm

Wargentin (Fig. 6.3) was the son of a priest and was born in Sunne, the province of Jämtland in 1717. He enrolled at Uppsala University in 1735, and studied physics and astronomy under Anders Celsius and Samuel Klingenstierna. He received his master degree in 1743. Wargentin played a very important role in the Swedish scientific community, both as Secretary of the Academy in Stockholm and as a researcher. He was instrumental in the building of the new astronomical observatory in Stockholm and published a large number of publications in the *Transactions of the Royal Swedish Academy of Sciences*. Among these were a long series of review papers on astronomical, biological and mathematical subjects. Today, Wargentin is best remembered as the father of Swedish vital statistics. He died in 1784 and was succeeded as Secretary by Wilcke (Sect. 6.2), who served until his death in 1796.

6.3 Bergman Becomes a Member of the Royal Academy of Sciences

Initially, the Royal Academy had no limit of the number of members in contrast to most other scientific societies, and by the late 1740s the number of members approached one hundred [6]. This evoked the idea to try to limit the number of members to 100, and in the 1750s, the Academy became more restrictive when it came to electing new members. In 1762, it was finally decided to limit the number of members to 100 [7]. After the death of three members, industrialist Jonas Alströmer (1685–1761), officer and politician Mattias Alexander von Ungern Sternberg (1689–1763), and officer and physicist Sven Ljungenstierna (1717–1763), there was an opening for election of a new member and on May 4, 1763, Wargentin suggested that the vacancy should be filled by electing Bergman as a member of the Academy. By suggesting Bergman, Wargentin strengthened the group of young experimental physicist who were fighting for influence. In 1764, Bergman was elected a member of the Royal Academy of Sciences and took his seat on May 23, aged 29. On this occasion, he gave an inaugural lecture *On the Possibility to Circumvent the Damaging Effects of Lightning*. The lecture was subsequently printed as a 103 pages long booklet including a copper plate (Fig. 6.4) [8]. Of these 103 pages, Bergman's lecture makes up the first 27 pages. It is followed by a two pages long answer from the Secretary of the Academy, Wargentin, and a 73 pages long appendix. As electricity was his main topic of research at the time (Sect. 5.2), and research on lightings had been conducted at the observatory of Uppsala at least since 1753, [9] the choice of topic was by no means accidental. It proved Bergman's role as an individual thinker and showed that he could turn physics into practical and useful ideas from which the Academy and the society could benefit. Lightings had been the topic of several papers in the Academy's *Transactions* and a dissertation by the professor of chemistry, Wallerius titled "*Chemical Remarks on the Lightning Strike at the Royal Castle of Uppsala*", [10] so Bergman could expect considerable interest from the audience.

Bergman's work on lightning conductors was ahead of its time in Sweden. When Bergman held his lecture, there was not a single lightning conductor installed in Scandinavia, Germany or France and possibly two or three in England [11]. They were more common in America, the home of their inventor, Benjamin Franklin. In Europe, two major factors contributed to the scepticism towards lightning conductors [12]. A misconception, possibly fuelled by the death in 1753 of Richman, a scientist in St Petersburg, lead people to believe that a lightning conductor could attract the lightning and thus actually possess a danger. Another factor was the strong opposition from French physicist Jean-Antoine Nollet (1700–1770) who had a very strong influence on electrical research in Europe. He was a strong advocate of the two-fluid theory of electricity and a strong opponent to the ideas of Franklin. Nollet said that a metal rod could provide as small means of conducting away the electricity from a cloud covering a whole city, as a thin pipe could provide protection from a flooding river. Bergman said in his lecture that this argument was

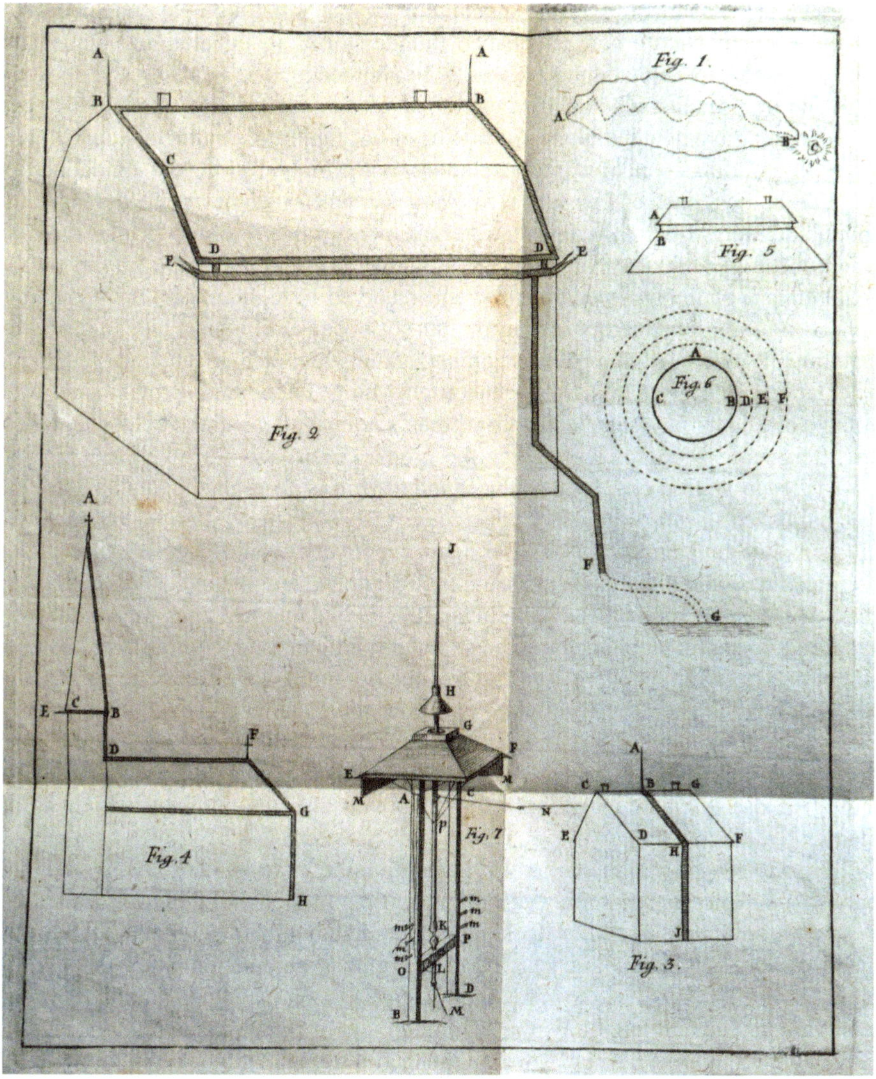

Fig. 6.4 Illustration for Bergman's lecture on lightning conductors

improper, as the velocity of electricity in a metal conductor is very high, and if a pipe could provide a comparable water flow, it would indeed be enough to divert a flood [13].

An important question to ask in the context of lightning conductors was whether humans should try to protect themselves from the lightning at all. If the lightning was a means of God to punish the Earthlings, then precautions against the effects of lightning was an act against the will of God. Bergman did not agree with this view and made a comparison with the inoculation against smallpox. He concluded that "If the Lord of Nature had established thunderstorms to threaten and punish the Earth's disobedient inhabitants, it would be another matter; but now she does not flash and thunder just when and where the greatest sins are being committed" [14].

Bergman gave detailed advice of how to instal lightning conductors on different types of buildings, and his proposal does not simply follow the guidelines of previous authors such as Franklin [15]. Actually, when it comes to the height of the conductor, Bergman's recommendations are more in line with modern views, than where those of Franklin. Nevertheless, Bergman's lecture had little impact on the development of lightning conductors, and received little attention [16]. Construction of lightning conductors in Europe did not occur until after 1770, [17] and by that time Bergman's interest was almost entirely devoted to chemistry. On the bicentennial of Bergman's lecture, Dietrich Müller-Hillebrand, professor of high voltage physics in Uppsala, studied Bergman's contributions to the field, [12] and at last, Bergman's thoughts were given the proper credit.

A letter to Wilson on a spectacular thunderstorm in Uppsala on August 24, 1760 was published in *Philosophical Transactions* in 1764 [18]. Wallerius, the professor of chemistry in Uppsala whom Bergman eventually would succeed (Sect. 9.1), presented a dissertation about the same thunderstorm, and the effects of a lightning striking Uppsala Castle [19]. Wallerius tried to prove that lightings were caused by a special form of sulphur formed by water particles and combustible material combining in the atmosphere. The fact that the lightning was attracted to metals was compared to the chemical reactivity of sulphur towards metals. Another proof was the presence of sulphuric acid in the atmosphere, which Bergman would later disprove (Chap. 18). Wallerius also claimed that this special form of sulphur did not only give rise to electrical phenomena, but it was also the ultimate nutrient of all plants.

Bergman returned to the topic of preventing lighting strikes in 1770. Wilcke had published a paper on a lightning striking a building in Stockholm in May 1769 [20]. Wilcke referred to Bergman's work and concluded that the discharge had followed metal plates on the building which had prevented more serious damage. On the other hand, he was sceptic towards the prospects of protecting a building by a tall metal rod, as the building hit by the lightning by no means was the highest building in the neighbourhood. The tall tower of the German church was not hit, for instance. Wilcke's paper was followed a comment by Bergman, where he maintained his idea that a city could be protected from lightning by an array of tall metal poles, provided they were correctly designed [21].

References

1. Lindroth S (1967) Kungl. Svenska Vetenskapsakademiens Historia 1739–1818, vol 1:1, Kungl. Vetenskapsakademien, Stockholm, p 2
2. Lindroth S (1967) Kungl. Svenska Vetenskapsakademiens Historia 1739–1818, vol 1:1, Kungl. Vetenskapsakademien, Stockholm, p 111

3. Boklund U (1961) Carl Wilhelm Scheele bruna boken, Stockholm, p 19
4. Lindroth S (1967) Kungl. Svenska Vetenskapsakademiens Historia 1739–1818, vol 1:1, Kungl. Vetenskapsakademien, Stockholm, p 122
5. Lindroth S (1967) Kungl. Svenska Vetenskapsakademiens Historia 1739–1818, vol 1:1, Kungl. Vetenskapsakademien, Stockholm, p 116
6. Lindroth S (1967) Kungl. Svenska Vetenskapsakademiens Historia 1739–1818, vol 1:1, Kungl. Vetenskapsakademien, Stockholm, p 14
7. Lindroth S (1967) Kungl. Svenska Vetenskapsakademiens Historia 1739–1818, vol 1:1, Kungl. Vetenskapsakademien, Stockholm, p 15
8. Bergman (1764) Inträdes-tal, om möjeligheten at förekomma åskans skadeliga verkningar; hållit för Kongl. Vetenskaps Academien den 23 maji 1764. Stockholm
9. Müller-Hillebrand D (1963) Torbern Bergman as a lightning scientist. Stockholm, p 2
10. Wallerius JG, Wibon CP (1761) Animadversiones chemicæ, ad ictum fulminis in arce regia Upsaliensi d. XXIV. Aug. MDCCLX/Chemiska anmärkningar öfver åskeslaget på kongl. slottet i Upsala d. 24 aug. 1760. Uppsala
11. Müller-Hillebrand D (1963) Torbern Bergman as a lightning scientist. Stockholm, p 18
12. Müller-Hillebrand D (1963) Torbern Bergman as a lightning scientist. Stockholm
13. Bergman (1764) Inträdes-tal, om möjeligheten at förekomma åskans skadeliga verkningar; hållit för Kongl. Vetenskaps Academien den 23 maji 1764. Stockholm, p 13
14. Bergman (1764) Inträdes-tal, om möjeligheten at förekomma åskans skadeliga verkningar; hållit för Kongl. Vetenskaps Academien den 23 maji 1764. Stockholm, p 19
15. Müller-Hillebrand D (1963) Torbern Bergman as a lightning scientist. Stockholm, p 26
16. Müller-Hillebrand D (1963) Torbern Bergman as a lightning scientist. Stockholm, p 4
17. Müller-Hillebrand D (1963) Torbern Bergman as a lightning scientist. Stockholm, p 13
18. Bergman T (1764) Observations in electricity and on a thunderstorm: In a letter to Benjamin Wilson Phil Trans 53:97–100
19. Wallerius JG, Wibom CP (1761) Animadversiones chemicae, ad ictum fulminis in arce regia Upsaliensi,..., Uppsala
20. Wilcke JC (1770) Anmärkningar vid et, d. 30 maij 1769, här I staden timadt åskeslag, KVA Handl 31:112–124
21. Bergman T (1770) Tilläggning i föregående ämne, KVA Handl, 31:125–129

7 Bergman's Geological Work

Aged 30, Bergman found a new interest in geology and the development of the Earth. Bergman's role as a geologist was first examined in 1900 by John Lindquist in his Ph.D. thesis [1] and later by Hedberg [2] and Frängsmyr [3]. The subject is also discussed in a chapter in Schufle's Bergman biography [4]. Bergman's main contribution to the field of geology is the 1766 book *Physisk beskrifning öfver jord-klotet* (Physical Description of the Earth; Sect. 7.2) but also includes a paper published in 1768 on the mountains of Västergötland [5] and an essay on geysers and volcanoes on Iceland (Sect. 7.3).

Early thinkers within the field of geology included da Vinci, who, for example, recognised that fossilised seashells found on land proved that these areas had once been covered with water [6], and Robert Hooke who suggested that some fossils corresponded to extinct life forms. Many of the ideas put forward by Bergman in his work had been formulated in 1669 by Danish scientist Nicolaus Steno (Niels Steensen; 1638–1686) [7]. Steno stated that (1) rocks had been deposited in water, (2) rocks had originally been deposited in horizontal layers, (3) the order of these layers indicated their relative age, (4) fossils are remains of buried life forms, (5) the nature of the fossils reveals whether they lived in fresh or salt water, (6) older rocks contain no fossils and (7) thought layers of rock were originally deposited horizontally, they have latter occasionally been tilted by the Earth's movements [8]. Oddly enough, Bergman never referred to Steno. Whether Bergman was aware of Stenos work is not known, [9] but it appears likely when considering Bergman's extensive knowledge.

Geology emerged as a more mature science in eighteenth century, but it was sensitive subject, as it could possibly contradict the Bible. The more conservative minds still held that Earth was created in six days about 6,000 years ago and that all substantial change had been caused by the Deluge. Two rivalling theories eventually emerged: plutonism, which taught that rocks were formed by fire or volcanic activity and neptunism which taught that rocks had formed by sedimentation from water. Neptunism is said to have been founded by German geologist Abraham

A. Lennartson, *Carl Wilhelm Scheele and Torbern Bergman*, Perspectives on the History of Chemistry, https://doi.org/10.1007/978-3-030-49194-9_7

Gottlob Werner (1749–1817), who has been called the father of geology. Werner and Bergman exchanged both letters and mineral samples [10].

On the other side was James Hutton (1726–1797), one of the important figures in the development of the plutonist school. Although Bergman is rarely mentioned in the history of geology, Werner was clearly inspired by Bergman, but also Hutton derived some of his ideas from Bergman [11]. In turn it has been pointed out [12] that Bergman probably derived some of his ideas on the formation of layered rocks from Johan Gottlob Lehman (1719–1767) [13]. Bergman was also clearly inspired by Linnaeus' geological writings, especially when it came to the question of water decrease [14].

In principle, Bergman thought that rocks had formed by sedimentation of particles suspended in water and by crystallisation of dissolved matter from water. The crystallisation hypothesis was introduced in the second edition of his *Physical Description of the Earth*, published after he became professor of chemistry [15]. Other rocks had formed by volcanic activity, and Bergman clearly suggested that Earth had undergone much change since its first formation and not only through the action of water. He also clearly recognised that the lower layers in mountains were older than the upper layers.

The mountains in Bergman's home province Västergötland, e.g. Billingen (Fig. 7.1) and Kinnekulle, caused considerable problems for Bergman. They are mainly sedimentary, consisting of sandstone resting on the bedrock, followed by alum shale, lime stone and shale deposited from Cambrian to early Silurian but covered with a layer of volcanic basalt from Permian. It is this basalt cap that has protected the underlying layers from erosion and the destructive effects of the glacial periods. The lower layers contain fossils and Bergman correctly recognised them as sedimentary. Bergman, however, did not believe in widespread volcanism and thus had to propose a sedimentary origin for the basalt as well [16]. The origin of volcanism was unknown, but was believed to be caused by chemical reactions or combustion of coal deep underground, thus it was unlikely to occur other than in special locations such as Iceland and Italy. In Sweden, there were absolutely no signs of volcanic activity. The igneous nature of basalt was not perhaps first recognised by Beddoes in 1791, a few years after Bergman's death [17].

In the fall of 1768, a year after becoming professor of chemistry, Bergman published a paper on the mountains of Västergötland [18]. Bergman had travelled around exploring the mountains over the years and noted that they all consisted of the same layers in the same order. It is interesting to note that Bergman's discussions on the chemical composition of the different layers are strictly qualitative, and he was far from the expert in chemical analysis he would be a decade later. He had learned to use the blow-pipe (Sect. 23.5) and treated the minerals with acids, but did not report more thorough analyses. For example, when discussing lime stone, he wrote that "It holds something ferrous,[1] as it turns yellow or rusts with time, and is drawn to the Magnet after burning". In his later writing, he would never use such a vague description, but would more probably report the percentage of iron oxide in

[1]"järnaktigt" in the original text, meaning something iron-like.

Fig. 7.1 Basalt formations at Ryds grottor, Billingen. *Photo* Anders Lennartson, March 2019

the mineral. Of interest is also the statement that alum shale "is so penetrated with phlogiston, that it is completely black and often burn in fire". It seems that Bergman had rather vague ideas of phlogiston at this point. It is almost as if he regards it as a material substance, perhaps even synonymous with the oil in the shale. A footnote in the paper, discussing the similarity between quartz and flint would be attacked by his predecessor Wallerius (Chap. 9.1).

7.1 The Cosmographic Society

In 1758, the Cosmographic Society was formed in Uppsala. Bergman was among the founding members, the others were Bergman's supervisor Bengt Ferner, Per Arrhenius, who would become vicar in Funbo, Johan Gustaf Zegollström, who became a teacher in mathematics at the Naval Academy in Karlskrona, Stefan Insulin, later bishop in Strängnäs and engraver Anders Åkerman [19]. Soon they were joined by, e.g. Mårten Strömer, astronomical observer Fredric Mallet in Uppsala and Wargentin. The first undertaking of the society was the production of two globes, one of Earth and one of the stars, and an extensive treatise about the world in three volumes. The globes were produced by Åkerman and Fredric Mallet undertook the writing of the first volume or the treaties, *Mathematical Description of the Earth*. The second volume, *Physisk beskrifning öfver jord-klotet* (Physical Description of the Earth), became Bergman's task. The third and final volume, dedicated to the ethnology of human races, was first commissioned to an unnamed author,[2] but was finally written by Stefan Insulin. Bergman's book was the first to appear in 1766 while the other two volumes appeared in 1772. Of the three volumes, Bergman's book was by far the most popular, and the only to appear in a second edition.

7.2 Physical Description of the Earth

It was *Physical Description of the Earth* that was Bergman's major scientific breakthrough, and the book received considerable interest internationally. From now on, Bergman did not have to search for correspondents abroad, but would instead receive letters from European scientists [20]. These new contacts did not only lead to exchange of knowledge, an important part was also the exchange of mineral specimens. This exchange would continue throughout his career, and analysis of such samples would be the topic of his last original paper. The first edition of Bergman's book, 431 pages long, sold out in six months, [21] and is a very rare book today. The length of the book had to be limited to a certain number of sheets, and Hjelm claims that each page Bergman wrote was immediately printed, [22] which might explain why Bergman's volume was the first to appear.

A second extended and more widespread edition appeared in two volumes (470 + 535 pages) in 1773–1774 (Fig. 7.2). In the preface of the second edition, Bergman wrote that "If health and other circumstances do not put insuperable obstacles in the way, I hope I can one day publish this work in a reasonably complete form" [23]. This third edition, presumably planned to be written in Latin to attract an international readership, [24] would unfortunately never be realised. Bergman's work still reached European readers [25]. The first edition was translated to German by German mathematician and astronomer Lambert Hindrich Röhl (1724–1790) and published in 1769. It is a book of 487 pages, containing additions by Bergman. Apparently, Bergman was not satisfied with this translation, since he made the Danish translator, Brandt, alerted to its shortcomings [26]. Brandt's Danish edition (648 pages) appeared in 1771 with additions by Bergman. The second Swedish edition also appeared in a German translation by Röhl in 1780 (two volumes, 387 + 426 pages). An abbreviated German edition (454 pages) appeared in 1781 and a reprint of the second German edition appeared in 1791. A Russian translation (possibly by Karamychev) [27] was published in 1791–1794 (267 + 311 pages). From the correspondence between Bergman and Kirwan, [28] its seems like Kirwan was encouraging British physician and scientist Charles Blagden (1748–1820) to translate Bergman's book to English, but he then learned that Franz Xaver

[2]Ferner and Ziervogel have been suggested.

Fig. 7.2 Frontispieces from the second volume of the second edition of Bergman's Physical Description of the Earth. *Photo* Anders Lennartson

Schwediauer (1748–1824) was actually working on such a translation in 1784 [29]. Unfortunately, this English translation was never published, nor was Bergman's book published in French. Still, *Physical Description of the Earth* reached a wide audience and had a significant impact. Its main importance is as a critical review collecting and discussing previous knowledge; it contains little original research [30]. In the second edition, written after Bergman had become professor in chemistry, the chemical sections are greatly expanded and contain references to work by Scheele that was yet to be published.

The introduction to the book is a good example of Bergman's writing talent: "If Earth is viewed from a long distance and compared with other celestial bodies, its size virtually disappears: it will look like an ant-hill against a mountain" [31]. After discussing the size of Earth compared to the Sun and the planets he concluded: "Such contemplations should diminish our pride and exterminate such presumptuous illusions, that all of this vast and amazing creation has been begun and completed for man's sake alone". This view differentiated Bergman from an older school of scientists, who regarded the whole world as created for us by God. Later in the introduction he wrote that man has come to a point where the size and trajectories of the planets can be determined, the orbits of their moons calculated, the heights of the mountains on the Moon measured and "we even take the time to speculate over the shapes and moods of the inhabitants of other celestial bodies, but of our own home, we are still in such ignorance, that we do not even know if there is solid ground or sea under the poles" [32]. Bergman assumed without discussion that the stars were surrounded by planets, (of which there were no proofs until 1988), and taking into account the vast number of stars in the universe, extra-terrestrial life was a natural thought for Bergman [33].

The book is divided into six parts, and the parts are divided into chapters. The second edition has the same disposition as the first, with no new parts or chapters. The first part is called "on Earth's surface in general" and is a purely geographical overview of Earth. The second part "on land" is divided into seven chapters. The first chapter is the geography and history of exploration of the known world. Through the works of Snorri Sturlason (*c*.1179–1241), Bergman knew of Leif Ericsson, the Viking who visited North America 500 years before Columbus [34]. Chapter 2 deals with islands and Chapter 3 with "lesser known lands", a hundred pages about places which were not sufficiently studied and the places which were completely unknown, such as the Polar Regions. Chapter 4 deals with mountains and describes different types of mountains as well as a geographical overview.

It is Chapter 5 "on the soil layers" that Bergman developed his ideas of stratigraphy that has attracted most interest from later scientists and historians. The discussion is continued in the chapter "on the creation of Earth" in the fifth part. Bergman started by describing the different accessible soil layers at different geographical locations. "Thus the Earth crust, down to a considerable depth, appears to consist of spherical shells that differ from each other with regard to size, composition and thickness. Such layers really arise when a water several times are mixed with different materials and between each time is allowed the stillness, so the intermixed particles are allowed to settle at the bottom" [35]. Just three pages latter, however, he described lava fields created by volcanoes, so Bergman was certainly not purely a neptunist. Bergman continued by discussing layered mountains, the occurrence of metals, ores and precious stones.

Chapter 6 is devoted to petrifications, or fossils as we say today; in Bergman's vocabulary the word "fossil" meant mineral in general. Bergman was apparently slightly disturbed by the fact that elephant fossils had been found in Siberia, where there should be no elephants. Of course, he was actually referring to mammoths. The final chapter of the second part deals with caves.

The third part concerns water. The first chapter deals with springs and the origin of the water from springs. Rainwater was a frequent source, but a spring may also be connected to an underground cavity, a lake or the sea. Of special interest to Bergman were hot springs, especially the geysers on Iceland, which Bergman claimed were powered by underground fire [36]. Not surprisingly, the sections dealing with the chemical compositions of mineral waters were expanded in the second edition, written after he became a professor of chemistry (Sect. 9.3). The remaining chapters of part 3 deal with rivers, swamps, lakes and seas, respectively. In the chapter of lakes, Bergman claimed that some lakes were stirred up by activity on the bottom as waves form in calm weather without the aid of wind. Bergman attributed this to chemical reactions in underground caves: "Nature's workshop is constantly in action, new bodies are formed, old are destroyed, and during all this an air-like spirit is released from the compound where it used to be a part" [37]. Regarding the sea, Bergman remarked that the sea floor is not different from dry land, it just happens to be covered with water. Bergman wrote that the salt in the sea serves the purpose of facilitating fast degradation of dead plants and animals and that the Lord had added just the perfect amount of salt to assure that dead bodies rapidly dissolve [38].

The topic for the fourth part is the atmosphere. The most striking thing is that there is no mention of Scheele's discovery of oxygen (Chapter 21) in the second edition, and it is clear that this important information was still unknown to Bergman in 1774. Instead, Bergman appears to have regarded air as a single substance: "The main substance [in the atmosphere] is air" [39]. This air was, according to Bergman, composed of particles. The atmosphere also contained many foreign substances such as water vapour and combustible dust, which gave rise to "air fires", e.g. auroras. Bergman mentioned carbon dioxide (aerial acid) and the electrical substance that gives rise to lightings.

Bergman discussed the expansion of air by heat, and the much lower density of steam compared to liquid water. He speculated that steam is a compound between fire and water, but he did not believe that steam was air, since air cannot be condensed by compression. Through his work on the twilights, [40] he knew that the atmosphere only refracts light to a certain altitude and that the atmosphere does not extend infinitely into space.

Chapter 4 deals with "air fires", a category including such different phenomena as lightning, auroras and meteors. One explanation was volatile flammable substances that evaporated from plants and busted into flame at high altitude. The true nature of meteors was not recognised until 1794 [41] and received more attention from the works of Davy [42]. In many cases, however, electricity was involved, but Bergman could not explain how electrical charges arise in the atmosphere. Rubbing of air particles against each other would, according to Bergman, give equal amounts of positive and negative electricity that would soon cancel each other out. Since lightings are rarely seen in the winter, temperature differences might play a crucial role according to Bergman. Meteorites were difficult to explain, [43] and Bergman discussed the different theories. Meteors may be aggregates of particles formed within the atmosphere, although Bergman found this hard to believe. Meteors may also be bodies traversing Universe or possibly bodies orbiting the sun like comets. A less likely explanation was that they were ejected from volcanoes. Bergman was not aware of any chemical investigations of meteorites.

The fifth part deals with changes occurring on Earth. First, Bergman discussed day and night, different climates and seasons. It is clear that there was a reservoir of heat at the core of the Earth, but Bergman could not imagine any other source for that heat than the Sun: Bergman believed that Earth had accumulated heat from sunshine since the creation [44]. Of course, Bergman and his contemporaries could not have anticipated the true source of this energy over a century before the discovery of radioactivity. In the next chapter, Bergman discussed changes of geological origin. For instance, fossilised trees had been found deep down in the ground, a clear proof that Earth is changing. Subterranean fire [45] could, according to Bergman, cause both land rise and land sinking. Volcanoes, Bergman wrote, are chimneys where vapours, smoke and ashes can escape from these subterranean fires; without means of releasing this pressure, earthquakes arise. After a long description of volcanoes and volcanic materials, Bergman turned to the explanation of these supposed subterranean fires [46]. He found a constant fire within Earth unlikely and was more inclined towards local chemical reactions. One possibility, according to Bergman, was the reaction between iron and sulphur. When a moist mixture of powdered iron and sulphur is exposed to air, it turns red-hot. This reaction was also investigated by Scheele [47]. As the Earth was unlikely to contain large amounts of elemental sulphur and iron, Bergman instead suggested that pyrites (iron sulphide) may ignite layers of coal or alum shale. In 1780, Bergman returned to volcanoes in a long paper reporting chemical analysis of volcanic materials [48].

Chapter 3 deals with an important issue for the geologists of Bergman's time: water decrease. Sedimentary rocks and giant's kettles, for instance, indicated that dry land had once been covered with water. So, where is this water today? First of all, rather than a decrease in water, it could be a land rise, although Bergman found water decrease quite likely [49]. A common explanation was the conversion of water to earth, but Bergman was not convinced and referred to Lavoisier's famous experiment where it was found, by weighing, that the earth deposited on refluxing water in a glass vessel originated from the glass. He was apparently unaware of Scheele's similar results that are found in Gahn's notes from 1770 (Sect. 13.2).

Chapter 4, 120 pages long in the second edition, deals with the creation of the Earth. Bergman's opinion was that God had established the natural laws that regulated the changes on Earth. Bergman rejected the idea that the interior of Earth was filled with fire, instead he agreed with those suggesting a homogeneous magnetic core [50]. The core was surrounded with bed rock that Bergman classified as ancient (uråldig), as it was probably as old as Earth itself.[3] This type of rock formed many of the mountains and contained no petrifications (fossils). The second type of rocks was far less voluminous, and Bergman called them layered (flolägrige) rocks.

[3]This is not true. There are no remaining traces of Earth's original crust.

The third type was composed of irregular aggregates of sand and rock. An addition to the second edition is a long description of the different inorganic bodies in the Earth crust: salts, earths, combustible bodies and metals. The work with Gahn in the late 1760s on crystals (Sect. 25.9) also inspired him to write a paragraph on crystallography and crystallisation.

The formation of the magnetic core was difficult for Bergman to discuss, as no knowledge of its nature was available. It was clear, however, that the surrounding crust once had been liquid or at least soft enough for the centrifugal force to flatten the Earth at the poles. It appeared unlikely to Bergman that God had designed the Earth that way. The mechanism behind the creation of the chemical elements was completely unknown, but it was clear that they initially had been very finely divided so that they could be dissolved or suspended in water. At this point Bergman does not seem to differentiate between the concept of element and particle. Bergman regarded the mountains as the skeleton of the Earth, which prevented the surface from collapsing and flattening. Valleys had once been carved out by water perhaps fluctuations of the location of the Earths centre of gravity had been a driving force.

Bergman continued the discussion of petrifications from part 1, and the questions why fossilised rhinoceroses were found in Siberia. Some authors had argued that the climate once had been much warmer in Siberia, which Bergman concluded was astronomically impossible. This was long before there was any evidence suggesting the continental drift. The only possibility, according to Bergman, was that their dead bodies were brought there by the Deluge. When discussing the formation of minerals, Bergman concluded that there did not appear to be any formation of new minerals in the present day. He also briefly mentioned alchemy and the possibility of preparing gold. Bergman argued that it would be ridiculous to establish rules for what nature can or cannot do; there were, however, no evidence for transmutation [51]. Bergman concluded that "Thus, water has been the most prominent reason to the state, which the Earth's surface has adopted" [52]. The Deluge, which other more traditional authors attributed a great importance, is only discussed briefly on the last page of the chapter.

The last, sixth, part of the book is devoted to organic bodies: plants and animals. Of special interest is the discussion of the nutrients of plants. Bergman discussed the famous willow tree experiment of van Helmont, where a willow tree was found to increase in weight although only water was added and without reducing the weight of the soil. Bergman rejected the idea that the soil merely supports the roots, as he had found lime, magnesia, clay and silica in the ashes from plants [53]. Although it was clear that water was the main nutrient for plants, it was also clear that they took up air and light. Fertilisers caused plants to grow faster, as the plant did not have to prepare vital substances by themselves. In the second edition, Bergman also referred to Priestley's finding that green plants can refurbish foul air (i.e. converts carbon dioxide to oxygen).

7.3 Geysers and Volcanoes on Iceland

Uno von Troil (1746–1803), who would later become archbishop of Sweden, travelled Europe after his graduation in 1770. While in London in 1772, he decided to make a trip to Iceland together with English naturalist Sir Joseph Banks (1743–1820), Scottish physician James Lind (1736–1812) and Swedish botanist Daniel Solander (1733–1782), one of Linnaeus' apostles.

After returning home, Troil wrote several letters about his experiences to publicist Christoffer Gjörwell (1731–1811), linguist and professor Johan Ihre (1707–1780), physician Abraham Bäck (Chap. 10), a Mrs. S. Carlsson in Gothenburg and Axel Lejonhufvud (1717–1789), cavalier of King Gustav III. Letters concerning mineralogical and geological matters were sent to Bergman. Troil collected and published his letters including the answers he received from Bäck, Ihre and Bergman in a book (Fig. 7.3) called *Letters, concerning a journey to Iceland MDCCLXXII* (*Bref, rörande en resa til Island MDCCLXXII*) in 1777. His book was later translated to English and French.

Six letters in total were sent to Bergman in 1773. The contents of the first letter are largely geographic, while the remaining letters mainly concern volcanoes and hot springs; basalt pillars are also described at some length. Troil had also collected mineral samples that he presented to Bergman, with the request that Bergman gave his opinion on the geology of Iceland. As this was a very rare and valuable

Fig. 7.3 Engraving from the title page of Troils book about Iceland, depicting a volcano, a geyser and hexagonal basalt pillars. *Photo* Anders Lennartson

Fig. 7.4 The basin of the great Geysir, which unfortunately does not erupt any more. The orifice is surrounded by a crust that has been deposited by the hot water over the past centuries. *Photo* Anders Lennartson, 2013

collection, it was a request that Bergman could not turn down: "It would be very ungrateful if I hesitated to comply with this request, as you presented me with the intire collection you made there, that I might chemically examine the nature of each" [54]. Bergman rejected the idea of working out theories about the origin of Earth behind the writing-desk; theories had to be based on experiments [55]. Thus, Troil had to wait until June 1776 for an answer, but then he received a long[4] essay from Bergman, which he included in his book.

The first section of Bergman's essay dealt with geysers, and Bergman examined several mineral samples, e.g. the solid crust formed in the basin of the great geyser Geysir (Fig. 7.4). The crust was found to be composed of silica, which confirmed Bergman's suspicion that silica could be made slightly water soluble at high temperatures. Bergman did not supply a theory for the mechanism of the eruptions of geysers.

Bergman's theories on volcanoes from his *Physical Description of the Earth* were also supported by the samples from Troil. Pyrite had been found at all examined volcanoes, and Troil had also brought several samples of oil-containing alum shale.

[4]The printed version is 49 pages long.

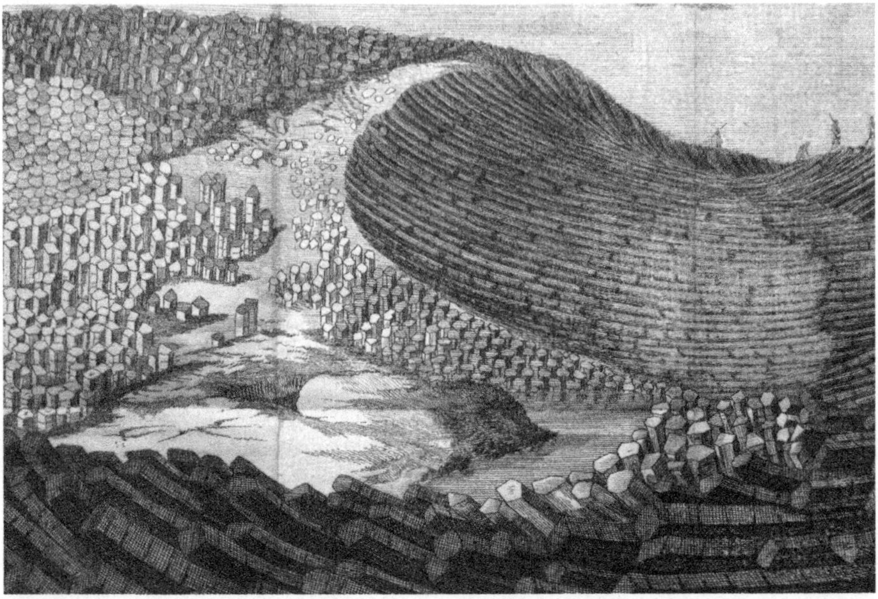

Fig. 7.5 Illustration of hexagonal basalt pillars from Troil's book. The image has been croped

Bergman could not conclude whether Iceland had formed solely by volcanic activity. To make such a conclusion, an extensive survey of the whole island had to be performed in order to rule out the occurrence of granite and other forms of ancient rock, which Bergman believed had formed in water. Bergman did not believe it likely for granite to form from a melt, since he thought that the crystals of quartz and feldspar would crack and become opaque due to the heat [56]. Bergman also discussed the formation of basalt pillars (Fig. 7.5). Since a fresh basalt surface examined by microscope (Fig. 7.6) was found to be inhomogeneous, Bergman found it unlikely that basalt pillars had formed by crystallisation [57]. Thus, two possible mechanisms remained: cooling of a melt or drying of wet mud. As Bergman found no glassy appearance of basalt, he rejected the idea of basalt being solidified lava, and favoured the hypothesis of formation from a wet mixture cracking into pillars as it slowly dried. This was of course wrong: the hexagonal basalt pillars arise when thick layers of lava solidify. As the lava cools, it contracts which causes it to crack. The most efficient way of cracking is the formation of hexagonal pillars.

Fig. 7.6 Bergman's microscope, manufactured by F. J. Stegman in Kassel. Museum Gustavianum. *Photo* Anders Lennartson

References

1. Lindquist J (1900) Framställning af Torbern Bergmans fysiska geografi. I. Stockholm
2. Hedberg HD (1969) Influence of Torbern Bergman (1735-1784) on stratigraphy. Stockh Contrib Geol 20:19–47

3. Frängsmyr T (1969) Geologi och Skapelsetro. Stockholm
4. Schufle JA (1985) Torbern Bergman a man before his time. Cornado Press, Lawrence, Kansas, p 141
5. Bergman T (1768) Anmärkningar om Vestgötha bergen KVA Handl 29:324–336
6. Hedberg HD (1969) Influence of Torbern Bergman (1735-1784) on stratigraphy. Stockholm contributions in geology 20:19–47
7. Stenonis N (1669) De solido intra solidum naturaliter contento dissertationis prodromus. Florence
8. Hedberg HD (1969) Influence of Torbern Bergman (1735-1784) on stratigraphy. Stockholm contributions in geology 20:19–47
9. Hedberg HD (1969) Influence of Torbern Bergman (1735-1784) on stratigraphy. Stockholm contributions in geology 20:19–47
10. Carelid G, Nordström J (1965) Torbern Bergman's foreign correspondence. Almqvist & Wiksell, Uppsala, p XXX
11. Schufle JA (1985) Torbern Bergman a man before his time. Cornado Press, Lawrence, Kansas, p 142
12. Lindquist J (1900) Framställning af Torbern Bergmans fysiska geografi. I. Stockholm
13. Lehman JG (1756) Versuch einer Geschichte von Flötz-Gebürgen. Klüterschen Buchhandlung, Berlin
14. Hedberg HD (1969) Influence of Torbern Bergman (1735-1784) on stratigraphy. Stockholm contributions in geology 20:19–47
15. Schufle JA (1985) Torbern Bergman a man before his time. Cornado Press, Lawrence, Kansas, p p145
16. Bergman T (1768) Anmärkningar om Vestgötha bergen KVA Handl 29:324–336
17. Beddoes T (1791) Observations on the affinity between basaltes and granite. Phil Trans 81:48–70
18. Bergman T (1768) Anmärkningar om Vestgötha bergen KVA Handl 29:324–336
19. Rosenhane S (1811) Anteckningar hörande till Kongl. Vetensk. Akademiens historia, KVA Stockholm p, p 139
20. Carelid G, Nordström J (1965) Torbern Bergman's foreign correspondence. Almqvist & Wiksell, Uppsala, p XXVI
21. Carelid G, Nordström J (1965) Torbern Bergman's foreign correspondence. Almqvist &Wiksell, Uppsala, p xxvi
22. Hjelm PJ (1786) Åminnelse-tal öfver...Herr Torbern Bergman. Stockholm, p 36
23. Bergman T (1773) Physisk beskrifning öfver jord-klotet, på cosmographiska sällskapets vägnar författad. Second edition, vol 1. Uppsala, p xvi
24. Hjelm PJ (1786) Åminnelse-tal öfver...Herr Torbern Bergman, Stockholm, p 36
25. Moström B (1957) Torbern Bergman a bibliography of his works. Almqvist & Wiksell, Stockholm
26. Carelid G, Nordström J (1965) Torbern Bergman's foreign correspondence, Almqvist &Wiksell, Uppsala, p xxvii
27. Moström B (1957) Torbern Bergman a bibliography of his works. Almqvist & Wiksell, Stockholm, p 93
28. Carelid G, Nordström J (1965) Torbern Bergman's foreign correspondence. Almqvist & Wiksell, Uppsala
29. Hedberg HD (1969) Influence of Torbern Bergman (1735-1784) on stratigraphy. Stockholm contributions in geology 20:19–47
30. Hedberg HD (1969) Influence of Torbern Bergman (1735-1784) on stratigraphy. Stockholm contributions in geology 20:19–47
31. Bergman T (1773) Physisk beskrifning öfver jord-klotet, på cosmographiska sällskapets vägnar författad. Second edition, vol 1. Uppsala, p i
32. Bergman T (1773) Physisk beskrifning öfver jord-klotet, på cosmographiska sällskapets vägnar författad. Second edition, vol 1. Uppsala, p iv

33. Bergman T (1773) Physisk beskrifning öfver jord-klotet, på cosmographiska sällskapets vägnar författad. Second edition, vol 1. Uppsala, p xv
34. Bergman T (1773) Physisk beskrifning öfver jord-klotet, på cosmographiska sällskapets vägnar författad. Second edition, vol 1. Uppsala, p 23
35. Bergman T (1773) Physisk beskrifning öfver jord-klotet, på cosmographiska sällskapets vägnar författad. Second edition, vol 1. Uppsala, p 202
36. Bergman T (1773) Physisk beskrifning öfver jord-klotet, på cosmographiska sällskapets vägnar författad. Second edition, vol 1. Uppsala, p 341
37. Bergman T (1773) Physisk beskrifning öfver jord-klotet, på cosmographiska sällskapets vägnar författad. Second edition, vol 1. Uppsala, p 413
38. Bergman T (1773) Physisk beskrifning öfver jord-klotet, på cosmographiska sällskapets vägnar författad. Second edition, vol 1. Uppsala, p 442
39. Bergman T (1774) Physisk beskrifning öfver jord-klotet, på cosmographiska sällskapets vägnar författad. Second edition, vol 2. Uppsala, p 1
40. Strömer M, Bergman TO (1755) Dissertatio de crepusculis, Uppsala
41. Chladni EFF (1794) Ueber den ursprung der von Pallas gefundenen und anderen ihr ähnlicher Eisenmassen, Riga
42. Davy H (1825) On the Safety Lamp for Coal Miners with some researches on flame. London, p 102
43. Bergman T (1774) Physisk beskrifning öfver jord-klotet, på cosmographiska sällskapets vägnar författad. Second 44.edition, vol 2. Uppsala, p 94
44. Bergman T (1774) Physisk beskrifning öfver jord-klotet, på cosmographiska sällskapets vägnar författad. Second edition, vol 2. Uppsala, p 141
45. Bergman T (1774) Physisk beskrifning öfver jord-klotet, på cosmographiska sällskapets vägnar författad. Second edition, vol 2. Uppsala, p 173
46. Bergman T (1774) Physisk beskrifning öfver jord-klotet, på cosmographiska sällskapets vägnar författad. Second edition, vol 2. Uppsala, p. 173
47. Scheele CW (1777) Chemische Abhandlung von der Luft und dem Feuer, M Swederus & S. L Crusius, Uppsala & Leipzig, p 52
48. Bergman T (1780) Producta ignis subterranei chemice considerate, Nova Acta Regiæ Societatis Scientiarum Upsaliensis, 3:59–136
49. Bergman T (1774) Physisk beskrifning öfver jord-klotet, på cosmographiska sällskapets vägnar författad. Second edition, vol 2. Uppsala, p 268
50. Bergman T (1774) Physisk beskrifning öfver jord-klotet, på cosmographiska sällskapets vägnar författad. Second edition, vol 2. Uppsala, p 293
51. Bergman T (1774) Physisk beskrifning öfver jord-klotet, på cosmographiska sällskapets vägnar författad. Second edition, vol 2. Uppsala, p 375
52. Bergman T (1774) Physisk beskrifning öfver jord-klotet, på cosmographiska sällskapets vägnar författad. Second edition, vol 2. Uppsala, p 385
53. Bergman T (1774) Physisk beskrifning öfver jord-klotet, på cosmographiska sällskapets vägnar författad. Second edition, vol 2. Uppsala, p 404
54. Troil U (1780) Letters on Iceland J. Robson, London, p 338
55. Troil U (1777) Bref, rörande en resa til Island MDCCLXXII Magnus Swederus,Uppsala, p 328
56. Troil U (1777) Bref, rörande en resa til Island MDCCLXXII Magnus Swederus,Uppsala, p 350
57. Troil U (1777) Bref, rörande en resa til Island MDCCLXXII Magnus Swederus,Uppsala, p 363

8 Scheele in Malmö

After leaving Gothenburg in 1765, Scheele, now aged 22, was hired by Peter Magnus Kjellström (1725–1803) at the Split Eagle (Fläkta Örn) pharmacy in Malmö (Fig. 8.1). The exact date of his arrival in Malmö is not known, but as he is listed in Malmö's register for 1766, he must have arrived in Malmö at the latest in November 1765. In the tax register, his salary is reported to be 100 silver dalers a year. At the time, Malmö had approximately 3,000 citizens, so it was, by modern standards, a rather small city. Linnaeus visited Malmö in 1749, and gave the following description:

> [Malmö] has large houses, mainly of half-timber with tiled roofs, except 40 [stone] buildings, and broad streets. The square [*i.e.* where Scheele worked] is one of the largest in the country, 200 steps long and equally broad, on all sides surrounded by tall trees, lime, horse-chestnut and walnut trees. On the square a [water pump] is built opposite Corps de Guardie. Hereto water is directed through underground gutters from the Pil dam, [...] south of the city and from [the pump] the water is redirected through [wooden tubes] to most blocks in the city. [1]

Scheele's new master was born in Karlshamn and received his pharmaceutical training from Bauch in Gothenburg 1740–1746, so one can assume that it was through Bauch that Scheele came in contact with Kjellström. The Split Eagle pharmacy had been founded by Thomas Arendt in 1731, but the name Split Eagle was not documented until 1782. Arendt's son, Fredrik Christoffer, took his apothecary exam in 1742, but later went bankrupt, and Kjellström came into possession of the pharmacy by marrying Fredrik Christoffer's sister. Kjellström grew medicinal plants on a large scale, had a large natural historical collection and is said to have been interested in chemistry [2], so the Split Eagle was a good and natural choice for Scheele. Kjellström seem to have encouraged Scheele in his, probably costly, chemical studies and to the best of our knowledge, Scheele was never accused of neglecting his work. Kjellström, who outlived Scheele, wrote to Wilcke many years later that Scheele never trusted a piece of information in a book until he had time to test it himself:

A. Lennartson, *Carl Wilhelm Scheele and Torbern Bergman*, Perspectives on the History of Chemistry, https://doi.org/10.1007/978-3-030-49194-9_8

Fig. 8.1 The Split Eagle pharmacy in Malmö, where Scheele made his first documented discoveries, is gone since centuries. A cinema (the grey building at the centre of the picture) was erected at the site in 1934. *Photo* Anders Lennartson, July 2016

> during the whole of the three year long stay at my pharmacy, Scheele was occupied by chemical investigations; acquired several chemical books, saying while reading the copies through: That could be. That was not true. This I will investigate. [3]

During his stay in Malmö, Scheele appears to have made his first and last visit to his family in Stralsund since he moved to Gothenburg [4].

8.1 Anders Jahan Retzius

In Malmö, Scheele befriended Anders Jahan Retzius (Fig. 8.2), born on October 3, 1742, the son of a provincial physician. Born the same year as Scheele, Retzius had made a much faster career than Scheele. He took his apothecary exam in 1761, the youngest apothecary ever in Sweden. When Scheele met him, he had also received a degree in chemistry [5] and worked as a lecturer in chemistry at Lund University, the first at that university to pay any serious attention to chemistry. As will be seen, he lived in Stockholm 1768–1772, where he was an auscultator at the Swedish Board of Mines. He also lectured on chemistry at *Collegium Illustre* and worked as an assistant to Abraham Bäck, president of *Collegium medicum*. After returning to Lund in 1772, he mainly worked in botany. He died on October 6, 1821.

Fig. 8.2 Anders Jahan Retzius on a commemorative coin issued by the Royal Swedish Academy of Sciences. *Photo* Linköpings mynt och antikhandel

8.2 Scheele's First Discoveries

According to Retzius, Scheele worked without a plan with whatever he found interesting. Retzius believed that this gave Scheele an advantage over more traditional chemists, as he lacked preconceived opinions and undertook investigations that educated chemists would find pointless [6]. Retzius persuaded Scheele to write down the results of his experiments. As we will see later, some of these notes still exist today and contain a blend of laboratory notes, excerpts from books and other notes. Scheele never dated any of his notes, the only dated documents are the persevered letters written by Scheele to his friends. "A few times a year", Retzius wrote to Wilcke, "he visited me in Lund for a day, and then we rarely had time to eat dinner, since he had performed a long row of experiments that we had to discuss" [6]. Retzius also recalled that Scheele used most of his income to buy chemistry books:

> During his stay in Malmö, he bought from Copenhagen as many books as his salary admitted. Those he read one or two times, then he remembered all what he wanted to remember from them, and never asked for that book again.

Scheele's first documented discovery was that of nitrous acid (HNO_2) in 1767. It is first described in a letter to Retzius, dated December 1, 1767, the oldest preserved letter from Scheele (Fig. 8.3). He asked Retzius whether he was familiar with a volatile nitric acid. If not, Scheele offered to write a report to him. Retzius answered that no such acid was known, and on December 11, Scheele sent a detailed report on the new acid to Retzius. This report shows that Scheele already was a very experienced chemist, although it lacks the structure and order of the papers he eventually would publish.

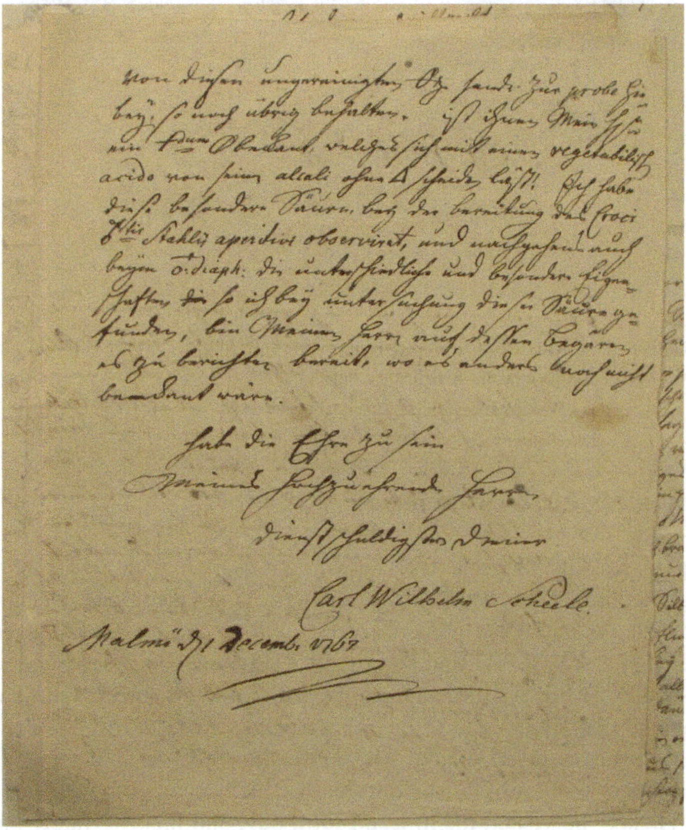
Carl Wilhelm Scheele

Malmö d. 1 Decemb. 1767

Fig. 8.3 The last page from a letter from Scheele to Retzius written in December 1767. This is the oldest dated document by Scheele. Carl Wilhelm Scheele's Archive, Centre for History of Science, Royal Swedish Academy of Sciences. *Photo* Anders Lennartson

While working in the pharmacy, Scheele sometimes prepared a medicine called *Crocus Martis Stahlii aperitivi* by melting a mixture of iron flings, antimony(III) sulphide and potassium nitrate (saltpetre). He also prepared another medicine called *Antimonium diaphoreticum*, which was prepared by igniting a mixture of one part antimony(III) sulphide and three parts potassium nitrate, after which the residue was washed with water [7]. Per standard methods, the water used for the washing the *Crocus Martis* or *Antimonium diaphoreticum* (in both cases containing, e.g. potassium nitrite, KNO_2) was of course thrown away, but Scheele, always curious, decided to analyse it. He found that it gave off red fumes (NO_2) when treated with acids (Fig. 8.4); such a weak acid as acetic acid was sufficient.[1] This is due to a rapid disproportion of the nitrous acid formed:

$$KNO_2 + CH_3COOH \rightleftarrows HNO_2 + CH_3COOK.$$

$$2HNO_2 \rightleftarrows H_2O + NO + NO_2.$$

The fumes could be dissolved in water to a previously unknown acid, similar to nitric acid but weaker and more volatile.

Scheele found that the new acid, now known as nitrous acid (HNO_2), could be prepared more easily by heating potassium nitrate alone, and adding acetic acid to the residue after cooling. Evolution of oxygen upon heating potassium nitrate was not mentioned by Scheele at this point:

$$2KNO_3 \rightarrow 2KNO_2 + O_2.$$

Scheele found that the red fumes (NO_2) obtained by dissolving iron or copper in nitric acid could be dissolved in potassium hydroxide solution to give the potassium salt of the new acid. He also obtained a white crystalline substance (now known as nitrosyl sulphuric acid, $HOSO_3NO$) by treating nitrous acid with sulphuric acid.

Nitrous acid had actually been observed by Willis,[2] did not realise that it was an acid, but merely concluded that the aqueous solution was yellow and had a repulsive taste. The fact that potassium nitrate gave red fumes with acids after heating had been observed by Boerhaave, but this was unknown to Scheele and Retzius.

Scheele had to wait a long time for a reply from Retzius. The reason for the delay is not known, but it may perhaps have taken Retzius quite some time to digest Scheele's discoveries, which were quite exceptional. In February 1768, Scheele wrote back to Retzius, asking whether he had received the report. Retzius reply is not preserved, but by end of March, he had replied, although Scheele seems not to have been too impressed by Retzius' explanations. Apparently, Retzius had suggested that potassium nitrate had reacted with a part of the phlogiston on heating since Scheele stresses in his answer to Retzius that phlogiston is a principle without

[1]Nitrous acid is a stronger acid than acetic acid (pK_a 3.39 and 4.75, respectively), but the decomposition of nitrous acid to gaseous nitrogen oxides drives the reaction.

[2]Thomas Willis (1621–1675), an English scientist and co-founder of the Royal Society.

Fig. 8.4 Formation of nitrogen oxides by addition of acetic acid to potassium nitrite. *Photo* Petra Rönnholm

constituents. Scheele had now noted that silver nitrate (*Lapis infernalis*) was reduced to phlogiston-rich silver upon heating, even in the absence of phlogiston-containing materials such as charcoal. Thus, reductions could be performed without the addition of phlogiston-rich materials. In reality, the silver ions are reduced by the nitrate ions, giving rise to oxygen gas.

The most important point is not that Scheele had isolated nitrous acid, but that he had found phenomena that were incompatible with the traditional phlogiston theory the way it was taught by Stahl and the way that Scheele had learned it from the books in Bauch's shop [8]. Not only had he found that potassium nitrate and silver nitrate could (in his view) extract phlogiston from heat, he also made another curious discovery. When he mixed potassium nitrite (obtained by heating potassium nitrate) with an acid, vegetable or mineral, in a distillation apparatus, he found that red fumes formed without the need to apply heat. Still, the alembic (Fig. 8.5) became hot. This heat was generated by the exothermic reaction between colourless nitrogen oxide and oxygen to give red nitrogen dioxide:

$$2NO + O_2 \rightarrow 2NO_2.$$

Scheele noted that when the reaction had proceeded for half an hour (which had depleted the oxygen in the apparatus), the contents of the flask were colourless, but when the alembic was removed (and oxygen entered the system), the flask was once again filled with red fumes. Probably these findings inspired Scheele to study fire and combustion. As will be seen in Chap. 21, it soon led him to the discovery of oxygen.

In Malmö, Scheele also began his difficult studies of Prussian blue (Sect. 14.3), which he mentioned for the first time in a letter to Retzius dated February 1768, although he had apparently already studied the topic for 3 years. In the same letter,

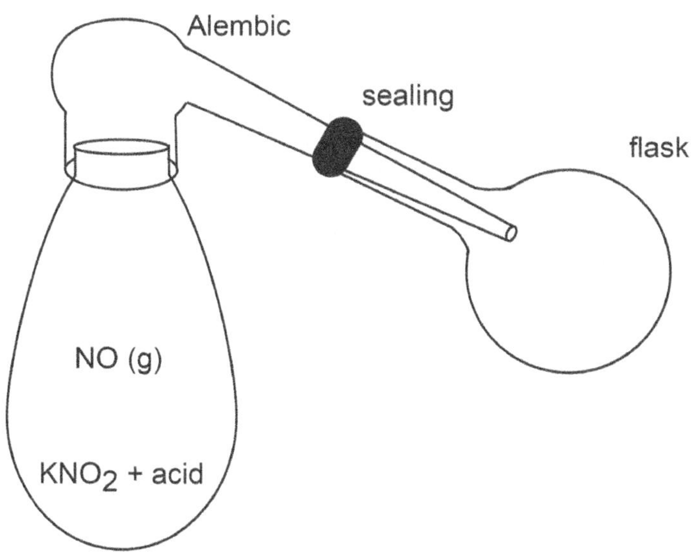

Fig. 8.5 Distillation apparatus with alembic, used by chemical practitioners since the early days of alchemy in Alexandria. See also a genuine piece of glassware in the background of Fig. 7.6. *Image* Anders Lennartson

he also reports on an investigation of borax (sodium tetraborate, $Na_2B_4O_7$). Scheele isolated boric acid, which was known and used in medicine under the name *Sal sedativum*, but it had previously not been noticed that it was, in fact, an acid. Since borax was imported from India and China, and its origin was virtually unknown and surrounded by myths, Scheele regarded boric acid as an organic acid. Baumé's report that borax could be prepared from lard was disproved by Scheele in 1773.

The studies on nitrous acid and borax were never published. Scheele's first attempt to publish his discoveries, probably encouraged by Retzius, was in 1767 or 1768. Scheele wrote a paper on *Globuli martiales*, which were prepared by boiling a mixture of iron flings with water and potassium hydrogen tartrate (cream of tartar) and forming the product into balls, which were dried. These balls were used to prepare medicines or were added to bath water (iron baths). Retzius claims that the manuscript was sent to the Royal Swedish Academy of Sciences, although it was actually never recorded in Stockholm [9]. It is possible that Scheele sent the manuscript directly to one of the members, most probably to Bergman. Retzius wrote that the paper contained a large number of experiments, but

> without the order, that is present in his latter works, and the concluded theory was so elaborate, that nothing could be extracted from it without great effort. I know nothing more than that it came into the hands of Bergman and disappeared. [10]

Bergman may have kept the manuscript (or received a latter version after he befriended Scheele, as he wrote to his friend Gahn in December 1771:

> I now send you some other experiments [by Scheele]. The beginning is the end of a description of the preparation of Globuli ♂les, which is neither so interesting, nor so short, that it needs to be attached.

Although the manuscript is lost, a summary has survived among Gahn's notes (Sect. 20.2) [11]. This proves that Gahn did not share his notes from his meetings with Scheele; why would Bergman otherwise send an excerpt to Gahn? This may be of some importance, as it seems that Bergman was the last of Scheele's friends to learn about the discovery of oxygen. If Gahn knew of Scheele's discovery of oxygen, he would not necessarily have told Bergman. Among the experiments on *Globuli maritiales*, Scheele showed for the first time that hydrogen is evolved when iron was treated with organic acids.

Nordenskiöld assumed that the paper on *Globuli martiales* was written before the report on nitrous acid [12], but Boklund was of another opinion [13]. In his report on nitrous acid and subsequent letters to Retzius, Scheele stressed that phlogiston is a principle, and still shares the old views of Stahl. The study on *Globuli martiales*, however, forced Scheele to change his view. When he observed that hydrogen was evolved from iron flings and vegetable acids, he came to the conclusion, independently of Cavendish, that the evolved gas was phlogiston in an elastic state. This was a very strong deviation from the view he held in late 1767. Boklund has also shown how Scheele changed his view once again. In the version of the *Globuli martiales* manuscript copied by Gahn in 1770, Scheele describes an experiment where water acts on iron flings for several weeks. He found that iron slowly reacted with water, and that the water contained no dissolved carbon dioxide and that the evolved gas was combustible (i.e. hydrogen):

$$2Fe(s) + 2H_2O(l) + O_2(g) \rightarrow 2FeO(OH)(s) + H_2(g).$$

There is a later manuscript in the so-called Brown Book (a collection of Scheele's notes, Sect. 28.7), which according to Boklund was written by Scheele no later than 177 [1, 14] and here all mentions of elastic phlogiston have been removed. The changed view is most likely due to Scheele's advances in his studies of combustion that would eventually result in his book on air and fire, *Chemische Abhandlung von der Luft und dem Feuer*, a few years later.

References

1. Linnaeus C (1751) Skånska resa, på Höga Öfwerhetens befallning förrättad år 1749, Lars Salvius, Stockholm, p 180
2. Åberg G (1982) Apotekarna på Fläkta Örn och deras medicinalväxtodlingar. Sv Farm Tidskr 86:34–48
3. Nordenskiöld A (1892) Carl Wilhelm Scheele. Stockholm, p XIII, Efterlämnade bref och anteckningar
4. Sjöstén CG (1801) Åminnelse-tal, hållit för Kongl. Vetenskaps Academien öfver dess framledne ledamot herr Carl Vilhelm Scheele, den 14 oct. 1799. Stockholm, p 12
5. Wollin C, Retzius A (1764) Dissertatio academica de natura ac indole chemiæ puræ cujus parten priorem auxiliante Deo … præside doct. Christiano Wollin … publico examini submittit Andreas Jahannes Retzius scanus ad diem XI febr. MDCCLXIV. Lund
6. Boklund U (1961) Carl Wilhelm Scheele bruna boken, Stockholm, p 399
7. von Hellwig C (1718) Lexicon medico chymicum, Frankfurt and Leipzig. p 8
8. Boklund U (1961) Carl Wilhelm Scheele, bruna boken de, Stockholm, p 43ff
9. Fredga A (1943) Carl Wilhelm Scheele. Stockholm, KVA, p 3
10. Boklund U (1961) Carl Wilhelm Scheele bruna boken, Stockholm, p 400
11. Nordenskiöld AE (1892) Carl Wilhelm Scheele. Stockholm, Efterlämnade bref och anteckningar, p 49
12. Nordenskiöld AE (1892) Carl Wilhelm Scheele. Stockholm, Efterlämnade bref och anteckningar, p 53
13. Boklund U (1961) Carl Wilhelm Scheele, bruna boken, Stockholm, p 62
14. Boklund U (1961) Carl Wilhelm Scheele, bruna boken del 2, Stockholm, p 349

9 Bergman Becomes a Chemist

Sweden has a long tradition of mining, and copper mined in the Falun copper mine was vital for financing Sweden's extensive participation in, for example, the Thirty Years' War. As much as two-thirds of Europe's copper consumption originated from Falun during its peak in the mid-seventeenth century; 3,000 tons of copper was produced in 1650. Sweden was also one of the major European iron producers in the early eighteenth century, with bar-iron being Sweden's most important export product. With falling iron prices in the 1730s, Sweden limited its iron production, allowing Russia to take an increasing market share, but the Swedish iron export still amounted to 46,700 tons in 1748 [1]. "Iron is Sweden's most significant commodity, but [Sweden] has now strong competitors, especially since one in foreign countries has started to content with Russian [iron] for inferior needs," as Bergman wrote [2]. In 1637, a special authority, the Board of Mines (Bergskollegium), was established in order to control the mining. The following year, the Board of Mines established a small laboratory for assaying (determining the metal content of ores and analysing the produced metals) in Stockholm.

By the mid-seventeenth century, Sweden had four universities (called Academies): Uppsala (founded in 1477), Dorpat (present-day Tartu in Estonia; founded in 1632), Åbo (in present-day Finland; founded 1640) and in Lund (founded in 1666). In the seventeenth and early eighteenth centuries, there was no chemical research at the Swedish universities and hardly any chemistry teaching [3]. Instead, the first Swedish chemist was a multitalented physician, Urban Hiärne (Fig. 9.1), who was instrumental in the establishment of a chemical research institute, *Laboratorium chymicum*, in Stockholm in 1683. Hiärne was born in Svorits (present-day Kolppana in Russia) in Swedish Ingermanland, where his father was vicar. He attended school in nearby Nyen (present-day St Petersburg), but due to war, he ended up in Stockholm in 1657, at the age of 16. Throughout his life, Hiärne had the ability to secure support from influential persons and, without money, he could still attend the Gymnasium in Strängnäs and Uppsala University, where he studied botany and medicine. He earned money as a portrait painter and, in his spare time, he was a playwright and poet. His tragedy Rosimunda was played at the Uppsala castle with

A. Lennartson, *Carl Wilhelm Scheele and Torbern Bergman*, Perspectives on the History of Chemistry, https://doi.org/10.1007/978-3-030-49194-9_9

Fig. 9.1 Sculpture of Hiärne at Medevi *Photo* Anders Lennartson, 2018

King Charles XI (Karl XI) in the audience. As a private physician for Officer Jacob Staël von Holstein, he travelled to the hot spas of Aachen in Germany, where he learned the art of analysing mineral waters (Chap. 17). He continued his trip through Leiden, Riga, London (where he became a fellow of the Royal Society) and Paris, where he studied chemistry for Christopher Glaser (born 1615) at Jardin du Roi. In Copenhagen, he deepened his chemical knowledge under Professor Ole Borch (1626–1690), the first chemist in Scandinavia. Hiärne became a convinced paracelsian, and an opponent to Robert Boyle (1627–1691). Back in Stockholm, he practised as a physician and operated a private laboratory before he became the director of *Laboratorium chymicum*. One of the main responsibilities of *Laboratorium chymicum* was to provide chemical (paracelcian) medicine for the army, but from the end of the century, Hiärne could devote his time to basic chemical research. Among Hiärne's, more important works were his analyses of mineral waters and a study of formic acid, which he found to be different from acetic acid.[1]

[1]This discovery was not, however, published until after his death.

It was Hiärne's water analysis that established Medevi as the first Swedish spa (Sect. 17.1). Bergman would refine the analysis of the water (Sect. 17.2), and it was to Medevi that Bergman made his last trip in hope of curing his health. Hiärne died in 1722.

Laboratorium chymicum was reopened by Georg Brandt (1694–1768), who was awarded a scholarship from the Board of Mines in 1721 to study chemistry in Leiden with the celebrated Herman Boerhaave. Under Brandt, *Laboratorium chymicum* became the place of choice for mining students who wanted some knowledge of chemistry. Bergman, who held the memorial lecture over Brandt in the Royal Swedish Academy of Sciences, described Brandt as a very hard working man who lived a very quiet and simple life [4]. As a chemist, Brandt strongly believed that chemistry should be based on mathematics [5], an idea that was hard to realise given the state of chemical knowledge in the early eighteenth century, but which Bergman would do his best to put into practice. Today, Brandt is best remembered for his 1735 discovery of cobalt [6], the first new metal discovered for over a thousand years. Brandt was succeeded by Gustaf von Engeström (1738–1813; Sect. 9.6), who would become one of Bergman's main opponents.

The main chemical competence in Sweden during the first half of the eighteenth century was connected to the Board of Mines. Anton von Swab (1702–1768; Fig. 9.2) was born in the mining city of Falun and thus came in early contact with mining as did his half-brother Anders Swab (1681–1731). Anton von Swab published a number of papers on mineralogy and mining in the *Transactions of the Royal Swedish Academy of Sciences*. Daniel Tilas (1712–1772) was a grand-child of Urban Hiärne. His main interests were in geology and mineralogy and his views are best summarised in two lectures given in the Royal Swedish Academy of Sciences in 1742 [7] and 1765 [8]. His view in 1742 was that all rocks had formed during the third day of the Creation, but that a redistribution had occurred during the Deluge. He also supported the alchemical idea that noble metals gradually formed in minerals.

Sven Rinman (1720–1792) not only published several important handbooks on mining and metallurgy but also wrote papers on chemical analysis of minerals. Two important figures in the early history of Swedish chemistry had recently died when Bergman entered the scene. Henrik Theophil Scheffer (1710–1759) belonged to a noble family, members of which were influential politicians in the eighteenth century. From 1748, he ran a private laboratory in Stockholm where he gave private lectures and performed chemical investigations for his customers. His investigation on platinum [9] established platinum as a new metal rather than a form of gold or silver, as was commonly believed. Additional investigations of platinum were performed by Bergman (Sect. 15.6). Bergman also published Scheffer's lecture notes and used them in his own teaching (Sect. 11.1). In the preface of this book, Bergman, who never met Scheffer, relates a story that von Swab had told him several times. Apparently, Scheffer was rather hot tempered, and when von Swab

Fig. 9.2 Mineralogist Anton von Swab. Plaster medallion by Tobias Sergel. *Photo* Nationalmuseum, Stockholm

faced a problem that he did not have time to solve, he turned to Scheffer and expressed a preconceived opinion. This would immediately anger Scheffer, who worked day and night in his laboratory to find the truth. Fortunately, he would always admit if von Swab actually had been correct. Scheffer's habit of getting into disputes was challenging for his friends, Bergman wrote, but the reason was that he was so used to criticise his own ideas, and his main focus was always on finding the truth.

Finally, Axel Fredrik Cronstedt (1722–1765; Fig. 9.3) came from a Swedish-Finnish noble family, several members of which had high positions in the Swedish army. His main contributions to science were his discovery of nickel in 1751 [10] and his 1758 book on mineralogy, which introduced a purely chemical system for classifying minerals (Sect. 25.7). Both Bergman and Scheele regarded Cronstedt's book as the standard work on mineralogy to which they always referred.

Fig. 9.3 Axel Fredrik Cronstedt, the discoverer of nickel and reformer of mineralogy. Memorial coin issued by the Royal Swedish Academy of Sciences in 1882. *Photo* Anders Lennartson

9.1 Wallerius—The First Chemistry Professor in Sweden

Johan Gotschalk[2] Wallerius (Fig. 9.4) was born in 1709 as the son of a vicar. After studying in Uppsala, he took a master degree in April 1731 under the guidance of Klingenstierna [11] and, only a few months later, he took a degree in medicine under the guidance of Nils Rosén (1706–1773)[3] [12]. His older brother, Nils Wallerius (1706–1764), also studied in Uppsala and eventually become a professor of theology, but he also studied physics and philosophy. It was Bergman who held the memorial lecture over Nils in the Royal Swedish Academy of Sciences [13]. It could also be recalled that Nils Wallerius' *Systema metaphysicum* was one of the books that Bergman kept hidden under his table during his first year at the university.

[2]Often spelt Gottschalk, but Wallerius used the spelling Gotschalk on most of his publications.
[3]Rosén von Rosenstein after his enoblement.

Fig. 9.4 Johan Gotschalk Wallerius. Engraving by J. Gillberg, 1772

Wallerius first held a position in Lund, but the medicine students were few and Wallerius' income was consequently low. He then returned to Uppsala in hope of a better position. When Professor Olof Rudbäck the younger died in March 1740, Wallerius applied for the chair in medicine, but so did his former teacher Rosén, as did Carl Linnaeus, who had returned to Sweden 2 years earlier. Rosén, who had worked 13 years as *adjunct* and 9 years as acting professor was the most qualified candidate and was appointed on July 10, 1740 [14]. In connection to the appointment, another possibility arose, as Lars Roberg (1664–1742), a professor in medicine and botany, decided to retire. There were now two candidates left, Linnaeus and Wallerius. Linnaeus main merits were in botany and the university initially required Linnaeus to undergo an examination by presenting a thesis, a decision which was cancelled by the University Chancellor. Wallerius, whose main competence was in chemistry, still had to present a thesis for evaluation. Wallerius was very upset over this injustice. The thesis that Wallerius presented [15] was a detailed attack on Linnaeus' main works, trying to discredit Linnaeus. The thesis (which had been approved by Rosén, now Dean at the Faculty of Medicine) was defended on February 25, 1741, a day that few would forget. The nature of the thesis attracted a large audience and the defense act was the noisiest in the history of the university [16]. The first hour was calm, but as the second opponent, theologian and later Archbishop Magnus Beronius (1692–1775), entered the stage, the scene changed. Beronius attacked Wallerius and was often far from the subject. He also refused to let Wallerius answer, and finally, the Dean, Rosén, had to intervene. This resulted in a quarrel between Rosén and Beronius. When Wallerius finally was allowed to speak, he was so angry that he was stomping his feet and beating the table with his fists. The quarrel ended in an uproar where Linnaeus' supporters in the audience ripped the copies of Wallerius dissertation into pieces. Needless to say, it was Linnaeus who was appointed professor. Wallerius' thesis also criticised a paper on *Sal natron* by apothecary Julius Salberg (1680–1753), an apostle of Hiärne [17], which was the start of a 6-year-long debate in a series of papers and pamphlets between the two men. Wallerius' dissertation was officially defended by Johan Anders Darelius (1718–1780) who, interestingly enough, later studied under Linnaeus, although it appears that Linnaeus was rather cold against his student [18]. Darelius became a physician and was later ennobled and took the name af Darelli. When Scheele took his apothecary exam (Sect. 19.4), af Darelli was on the examination board.

In 1747 and 1748, Wallerius published two books, *Mineralogia* and *Hydrologia*. Mineralogia was an attempt to organise the mineral kingdom in a more scientific way than Linnaeus had done in his *Systema naturae*. It was an improvement but was inferior to the work presented by Cronstedt a decade later (Sect. 25.7). In the book *Hydrologia*, Wallerius took his organising skills to the extreme. Here he attempted to find a system of classification for different types of water, such as seawater, rainwater, lake water, etc., all based on superficial observations.

In the late 1740s, the university started to realise the importance of experimental physics, and as both the universities in Lund and Åbo had chairs in physics, it was found unacceptable that Uppsala had no professor in physics [19]. The idea was to withdraw another professorship to free resources for the new chair. There were several alternatives, and while trying to avoid the merge of the chairs in Greek and oriental languages, bishop Samuel Troilius (1706–1764; father to Uno von Troil; Sect. 7.3) suggested, on February 18, 1749, that a chair in chemistry also should be established. The university was positive, the arguments being that the Faculty of Medicine did not take the chemistry teaching seriously and that mining students had to travel abroad to obtain the important knowledge in chemistry. To free resources, the chairs in poetry and oriental languages (which in reality only included Hebrew) were to be withdrawn. The Royal Council approved the plans on February 23, 1750. The appointment of Klingenstierna (Sect. 1.2) as a professor in physics was, of course, the natural choice.

Wallerius was the only applicant for the chair in chemistry. Attempts were made to recruit Georg Brandt or Anton von Swab, but they did not apply [20]. Thus, on July 17, 1750, Wallerius was appointed professor of chemistry. Wallerius took the appointment seriously, and his work would set the stage for Bergman and Scheele.

As a chemistry professor, Wallerius needed a laboratory. Initially, he wanted a building in the city centre, close to the observatory and the botanical garden [21], but a building site was finally bought at Västra Ågatan,[4] close to the river, in 1751. The location was not ideal, as it was in the periphery of Uppsala and frequently flooded by the river in spring. The laboratory was inaugurated in 1752 or 1754 [22]. Wallerius had planned a two-storey building but only one floor was built during his time as a professor (Fig. 9.5). To the left of the entrance hall was Wallerius private apartment, and to the right were two laboratories, one larger for public demonstrations and one smaller for private lessons.

In 1761, chairs in chemistry were established in Lund and Åbo (Finland remained a Swedish province until 1809), but hardly any chemical research was performed there during Scheele's and Bergman's lifetimes. The professor in Lund, Christian Wollin (1731–1798), was a former student of Wallerius but had little interest in chemistry. In 1775, he became a professor of medicine and it was not until 1812 that the professorship in chemistry in Lund was re-established. The professor in Åbo, Per Adrian Gadd (1727–1797), was also a professor of economy, and his main interest was in agriculture and he did not conduct any significant original research in chemistry. Gadd visited Bergman in Uppsala in 1772, as mentioned in a letter to Gahn [23]. It must have been a short meeting since Bergman had to leave to Mariestad the following day to take care of his mother. Gadd was succeeded by Bergman's student Gadolin in 1789.

In 1759, Wallerius published the first part of his textbook *Chemia Physica*, which (despite its Latin title) was the first textbook on chemistry in Swedish (the first chemistry book in Swedish being published by Hiärne in 1706). In addition, a steady flow of chemical dissertations was being defended by his students and chemical papers by his hand appeared in the *Transactions of the Royal Swedish Academy of Sciences*. Wallerius also published some of the unpublished works of Hiärne. Wallerius is not remembered for any chemical discoveries, but as an author who gained international recognition and his books were translated to Latin and German. The fact that he did not make any important discoveries was not because he was a poor chemist, but rather because he belonged to the old school of professors who saw as their main task to study literature rather than performing new experiments. It had long been believed that the world was a message from God, a message that could be read—and was intended for us to read—by observing nature [24]. Wallerius fiercely objected to anything contradicting this view. This does not mean that he did not perform experiments at all; he claimed to have worked in his laboratory every day for over 30 years [25]. Although he hardly made any impact on the history of chemistry, he was very important for chemistry in Sweden, as he built up a chemical institution and developed chemistry as an academic discipline in contrast to the practical chemistry carried out at the Board of Mines in Stockholm. Wallerius wanted his laboratory placed under the Medical Faculty, but after

[4]Known as Westra Strandgatan at the time.

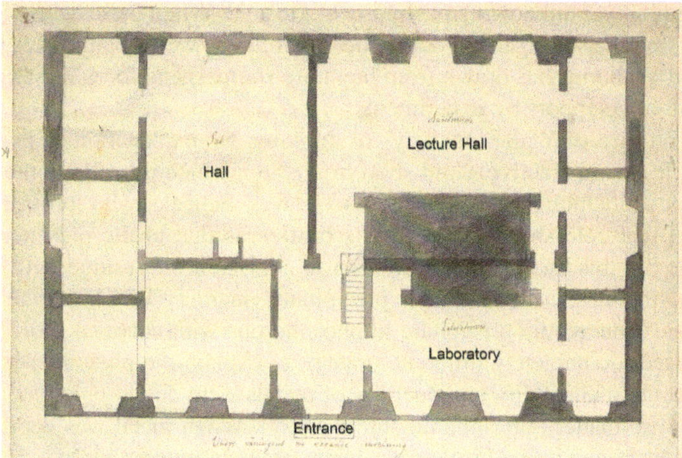

Fig. 9.5 Layout of Wallerius' laboratory. The image is cropped and annotated. Uppsala University Library

opposition from older professors, it was instead placed under the Philosophical Faculty. Although not intended by Wallerius, this was important as it helped to differentiate chemistry from medicine making chemistry a more independent discipline.

On January 14, 1767, Wallerius had to retire due to poor health. He had more and more hard of hearing, suffered from general weakness and vertigo [26]. In addition, his private apartment in the laboratory building was not very comfortable [26]. After his retirement, he became largely detached from the scientific community [27], but continued to perform agricultural experiments and served as president of the Royal Swedish Academy of Sciences for three months in 1783.

As a person, Wallerius was not always very diplomatic, to say the least, and his autobiography [26] gives the impression of a rather bitter man. He had a tendency to make enemies, although it seems that he and Linnaeus respected each other after their first clash (they had, for example, their strong religiosity in common). Unfortunately, however, he would come to regard Bergman as one of his main enemies and he seems to have largely ignored Scheele. In the beginning of his career, he had a few innovative ideas, such as his book Hydrologia, but towards the end of his life, he became more conservative, as can be seen, for example, in his attacks on Bergman and Scheele.

Wallerius' chemical theories were very different from those that Bergman eventually would develop. For instance, Wallerius believed the salts to be composed of water and earth, possibly in combination with a salty principle, *Sal principiale* [28]. From earth and water, the general salt, sulphuric acid, was generated. Nitric acid was formed from sulphuric acid in combination with something combustible, preferably from the vegetable kingdom. Sulphuric acid in combination with something yet unknown gave hydrochloric acid. When these acids were mixed with earth, middle salts (salts in our modern definition) were obtained [29]. This is a system that Wallerius has drawn from literature sources, and for which he could not have found any support by experiments.

To get an idea of Wallerius' ways of thinking, his explanation of the so-called cold-short iron is instructive. Cold-short iron is iron which is brittle in the cold state due to contamination with phosphorus. It was Scheele who proved in 1785 (Chap. 15) that cold-short iron owed its brittleness due to the presence of phosphorus. He did this by dissolving the iron in acid, boil the residue with alkali and dissolve it in nitric acid. He then precipitated mercury(II) phosphate from the solution and reduced the precipitate to phosphorous with charcoal [30]. Wallerius, in 1750, used a completely different strategy to explain the phenomenon [31]. He argued that the malleability and ductility of metals were due to phlogiston (he used the terms "fire matter" or "inflammable matter"). When metal was deprived of its phlogiston, it turned into a brittle calx (metal oxide in modern words). He had also performed an experiment where he heated potassium carbonate (*Sal tartari*) with charcoal (rich in phlogiston) and found the residue[5] so soft that it could be beaten with a hammer without cracking. From these two phenomena, he drew the, in our eyes far-fetched, conclusion that cold-shortness in iron was caused by too low phlogiston content. Red-short iron (iron which is brittle due to a high sulphur content) was, according to Wallerius, brittle due to high phlogiston content. Since Wallerius believed in the transmutation of metals in the Earth's crust, he finally came to the conclusion that red-short iron came from ore which was young and not ripe, while the cold-short iron came from ores which were too old. Neither Scheele nor Bergman would ever have used such analogies to draw conclusions about a completely unrelated chemical system.

9.2 Bergman's First Chemical Study

Bergman had probably never paid much attention to chemistry, but the illness of Wallerius gave him a reason to do so. As he saw no openings for a professorship in physics,[6] the chair in chemistry was a possibility. Probably, he initially saw it as a step towards a future professorship in physics. Although the branches of natural sciences were not as differentiated as they are today and the step from physics to chemistry not as great as it would be today, Bergman definitely needed more chemical experience in order to be appointed professor. He had studied chemistry under Wallerius but had only received the second-lowest grade on a four-step scale (admittitur cum approbatione) [32]. Of all his publications, only *Physical Description of the Earth* approached the field of chemistry. Most probably, the

[5]No chemical reaction is expected in this case.

[6]The chair in physics was held by Samuel Duræus (1718–1789). As a former student of Klingenstierna, he became acting professor when Klingestierna applied for leave of absence in 1752 and finally succeeded Klingenstierna in 1757. As it happened, he would outlive Bergman.

rumours of Wallerius retirement had reached Bergman in advance, but he still had very little time to acquire the required chemical experience. It was probably to Rinman, whom he had likely befriended during the studies of tourmalines (Sect. 5.2), that he turned for advice [33]. Rinman was at the time developing an alum extraction plant in Garphyttan, and improved methods for the production of alum became Bergman's introduction to chemistry. The results of his studies were published in a paper in the *Transactions of the Royal Academy of Sciences* [34]. The topic was well-chosen, as it was of great practical importance, well in line with what the government expected from a chemistry professor in eighteenth-century Sweden.

Alum (aluminium potassium sulphate dodecahydrate, $KAl(SO_4)_2 \cdot 12H_2O$; Fig. 9.6) was an important chemical product used in large quantities, for example, not only as a fixative in dyeing textiles but also found use in medicine. In Sweden, it was produced by burning alum shale and extracting the residue with water. Unfortunately, the quality of the alum was often poor, and the product contaminated with iron and a fatty substance (oily organic residues from the shale). It had been suggested that iron could be precipitated with alkali, but Bergman showed that alkali actually had a negative effect, decomposing alum and increasing the solubility of the fatty substances. Bergman found that, in modern terms, aluminium actually precipitates before iron(II) on adding alkali to a solution of alum and iron(II) sulphate (green vitriol).[7] Bergman instead suggested fractional crystallisation. He also found that the mother liquor contained an excess of sulphuric acid (probably originating from the combustion of sulphurous compounds in the shale), and this acid prevented crystallisation of alum.[8] Bergman's solution to this problem was to add lime-free clay, which neutralised the acid giving alum, and absorbing the fatty substance. Bergman's opinion that alum consists of pure clay and sulphuric acid (vitriolic acid) is not correct and suggests that Bergman did not discriminate between alum and aluminium sulphate. In fact, neutralisation of any excess sulphuric acid with potassium carbonate generates potassium sulphate which gives alum with aluminium sulphate. It is a fine balance, however, since a too high pH leads to precipitation of aluminium hydroxide and aluminium carbonate. Bergman's paper in the *Transactions* was followed by papers by Jacob Faggot and Anton von Swab. Faggot, who was more oriented towards economics, was pleased that Bergman had paid attention to a problem of economic importance, but was sceptic to Bergman's proposed method of purifying alum by adding clay. The criticism was probably justified, as the outcome of Bergman's method would have been highly dependent on the quality of the clay. Whether it would have worked or not is not easy to determine, since it would be virtually impossible to recreate the conditions of Bergman's experiments. The author of this book only noticed a small degree of absorption of Fe^{2+} by a native Swedish clay, and no absorption at all by pure

[7]The situation is rather complicated with several competing acid–base and solubility equilibria, but upon adding sodium carbonate solution to a solution containing of aluminium sulphate laced with iron(II) sulphate, I found Fe^{2+} in both the precipitate and the remaining solution, thus proving Bergman's criticism to be justified.

[8]I found no significant difference in solubility of alum in water compared to 0.1 M sulphuric acid.

Fig. 9.6 A crystal of alum. *Photo* Anders Lennartson

montmorillonite clay.[9] Absorption of Fe^{3+} was even smaller.[10] Bergman was not happy with Faggot's paper since a good reception of this work was essential if he should have any chance of becoming a professor. He contacted Wargentin, and von Swab's addendum was written on Wargentin's request [35]. Swab wrote that Bergman's study was so thorough, that he had little to add.

Bergman adapted his work on alum to a dissertation in Latin, *Disquisitio chemica de confectione aluminis* which was defended by Gustaf Swedelius on April 1, 1767 [36]. This was Bergman's first dissertation in chemistry; opponent was Carl Peter Wibom (1736–1801), Wallerius assistant [37]. Although Bergman's work on alum gained general recognition and served its primary purpose of establishing Bergman as a competent candidate for the chair in chemistry, there is no evidence that Bergman's theories were actually put in practice in Garphyttan [38]. It is, after all, far from certain that it would have worked.

In 1774, Gustaf von Engeström (one of Bergman's opponents; Sect. 9.6) published a long paper on alum production. Bergman is not mentioned, but it is clear that Bergman's study is his target [39]. von Engeström reported that sulphuric acid did not prevent the crystallisation of alum (here, von Engeström seems to be correct), and he had never observed any fatty residue in the alum lye. Bergman

[9]A 0.01 M Fe^{2+} solution was divided into three parts. One part was stirred over night with suspended clay, the second part was treated in the same way with montmorillonite and the third part was left as a reference. The samples were centrifuged and titrated with $KMnO_4$.

[10]A 0.01 M Fe^{3+} solution was treated in the same manner was the Fe^{2+} solution. The concentration was determined by addition of excess SCN^- and measurement of the absorbance at 490 nm.

continued his work on alum and published a second paper[11] in 1776 presenting a systematic study of the crystallisation of alum from solutions containing excess sulphuric acid [40]. Here, he made the interesting discovery that "common alum always holds a small excess of acid". As a trivalent ion with small ionic radius, hydrated $[Al(H_2O)_6]^{3+}$ is acidic, and it was this hydrolysis Bergman had noted:

$$[Al(H_2O)_6]^{3+}(aq) + H_2O(l) \rightleftarrows [Al(OH^-)(H_2O)_5]^{3+}(aq) + H_3O^+(aq)$$

9.3 Bergman is Appointed Professor

Bergman was not the only candidate for the position as Wallerius' successor. Four candidates applied and the other three candidates were chemists who had worked as assistants to Wallerius.

Anders Tidström (1723–1779) was born in Mariestad and was the son of a potter. Like Bergman, he attended the school in Mariestad and the Gymnasium in Skara. He enrolled at Uppsala University in 1744 and received private lessons from Linnaeus. He defended two theses about his home town, Mariestad, under the supervision of history professor and later Bishop of Uppsala, Olof Celsius the younger (1716–1794)[12] in 1748 [41] and 1752; [42] he received his master degree in 1752 [43]. He then turned to chemistry and became Wallerius' assistant in the late 1750s. In 1765, his own student, Daniel Herlenius, defend a thesis on vegetable acids [44]. This is Tidström's only publication in chemistry. Tidström undertook scientific journeys to Dalarna in 1754, Halland and Skåne in 1756 and Västergötland in 1756 and 1760. Bergman and his friends did not seem to have been too impressed by Tidström. In 1774, Bergman's student Gahn wrote to Bergman complaining over French chemist Balthazar Georges Sage (1740–1824) and his sloppy work saying that "I have difficult to imagine him being anything else but another Tidström" [45].

Mathias Kewenter (1735–1805) was born in Karlshamn and began his studies at Lund University in 1750 [46]. He presented a thesis under the supervision of history professor Sven Lagerbring (1707–1787)[13] in 1752 [47] and received his master degree in 1757. He moved to Uppsala and enrolled at the university in 1759. He first studied under Linnaeus before he turned to Wallerius and became his assistant. Kewenter's first own thesis and only chemical publication was defended in 1762 [48]. He became *docent* in 1763 but, in 1765, he left his academic career to take over his father in law's successful restaurant business. As a burgher, he was very

[11]Bergman did not actually mention von Engeström's study in his paper.

[12]Olof Celsius the younger was the son of Olof Celsius the elder, Linnaeus mentor, and cousin of astronomer Anders Celsius.

[13]Sven Lagerbring is regarded as the first "modern" historian in Sweden, as he was the first to critically review his sources. His daughter was married to chemist Gustaf von Engeström (Sect. 9.6).

active and was appointed a member of the Diet in 1786 and joined the opposition against the King's restrictions to the freedom of press.

Lars Hiortzberg (or Hjortsberg; 1727–1789) had defended a thesis on the connection between chemistry and society under the supervision of Wallerius in 1751 [49]. He received his master's degree in 1752 and, the same year, he was promoted to *docent* after his student, Gustaf Elfsten, presented his thesis [50]. This thesis is actually a continuation of Hiortzberg's own inaugural thesis. In 1754, Hiortzberg obtained his M.D. degree under the supervision of Linnaeus [51]. Hiortzberg presided over two more theses, in 1756 [52] and 1757 [53], defended by Erik Holmén and Peter Boling, respectively. In 1762, he left Uppsala for Karlskrona, where he became navy physician (amiralitetsmedicus) [54]. Thus, by the time of the application, both Kewenter and Hiortzberg had left the university, and Kewenter's change of career was used as an argument against his application. Hiotzberg made a few more appearances at the university, and served as an opponent on a thesis defence of one of Bergman's students, Plomgren, in 1770 [55].

The first step in the process of appointing a professor was for a board consisting of the professors at the university to set up a ranking list of the three top candidates that was presented to the University Chancellor, at the time Crown Prince Gustav. The Chancellor would pass his opinion to the King, or in reality the Royal Council, for the formal decision [56]. In the appointment of the chemistry professor, the opinion of the Board of Mines would also be taken into account [57]. Thus, there were many factors that would affect the outcome of the process, and in order to be appointed, an influential network was of vital importance [57]. The intrigues leading to the appointment of Bergman have been studied in detail in a doctoral thesis by Fors [58], and most of the following discussion is based on his work.

Tidström's position as Wallerius' assistant for 8 years does not seem to have been as meriting as one could have suspected. Rumours at the university said that Wallerius actually tried to prevent his appointment by means of an elaborate plot, where Wallerius would have switched position with another professor before his retirement. It still appears that he preferred Tidström over the other three candidates. The highly religious and conservative Wallerius was certainly not impressed by Bergman's *Physical Description of the Earth* and definitely did not want to see a Newtonian physicist as his successor. When Wallerius wrote a book on the creation of Earth in 1776, he found Bergman's book hardly worth mentioning [59].

Although Tidström did not have Wallerius' full support, he did have an influential supporter in Linnaeus. Linnaeus generally supported his apostles, and although Linnaeus had encouraged Bergman in his early biological work, and acknowledged Bergman as a skilled physicist, he did not think Bergman could replace Wallerius. In a letter, Linnaeus wrote that he had "never heard, before the chemical vacancy, that he [Bergman] had ever thought of chemistry". He was probably correct. Through Linnaeus, Tidström also had the support of Carl Gustaf Tessin (Fig. 9.7), who had once supported Linnaeus in his campaign against Wallerius in 1742. Tessin was a highly influential diplomat, politician and a patron for culture and science. Tidström had for example, been appointed by Tessin to

Fig. 9.7 Carl Gustaf Tessin, influential politician and diplomat and former teacher of Crown Prince Gustav. Painting by Jakob Björk. *Photo* Nationalmuseum, Stockholm

organise his mineral collection. Linnaeus also engaged politician Carl Johan Gyllenborg, son of Linnaeus' deceased patron Carl Gyllenborg. Unfortunately for Tidström, the Hat party, to which Tessin and Gyllenborg belonged, had just lost power to the Cap party (Chap. 1), and Tessin and Gyllenborg had lost some of their former power.

Through his membership in the Cosmographic Society (Sect. 7.1) and the Royal Swedish Academy of Sciences, Bergman was well acquainted with Wargentin and it should be recalled that Bergman was elected a member of the Academy on the recommendation by Wargentin. None of the other applicants were members of the Academy. As a physicist, Wargentin could fully appreciate Bergman's talent, and could probably see merit in a more physical and mathematical approach to chemistry rather than the purely descriptive approach of Wallerius.

Wallerius' former teacher, professor Rosén von Rosenstein, also supported Bergman, and according to a preserved letter to Wargentin, he attempted to get the Queen's support for Bergman. The Queen had a genuine interest in science, and although the royal couple had no political power, it was strategically important.

While Wallerius' connections to the Board of Mines were rather weak, Bergman, with his connections to the Royal Swedish Academy of Sciences and Wargentin, approached two of its more influential members, Tilas and von Swab, and sent them copies of his *Physical Description of the Earth* and his dissertation on alum. Tilas wrote to Bergman and thanked him for the book and gave several comments. Swab had already read the manuscript of Bergman's paper on alum and was thus already familiar with Bergman's work. He wrote to Bergman that he could count on the support from the Board of Mines and that he had discussed the matter with the president.[14] While Tilas wrote to Bergman on a more equal level, the older von Swab used a more formal language.

When the board at the university voted on April 25, Bergman was ranked number one with 10 of 22 votes [60]. Five delegates (e.g. Linnaeus) ranked Bergman second, and seven put Bergman on the third place, so there was clearly some opposition. The fact that Bergman only had been employed at the university for about six years was the main argument against him. Tidström and Hjortsberg shared second place with six votes each [61]. Kewenter received no votes and was very upset. His letter to the university was dismissed as unworthy. The matter was now presented to the Chancellor, Crown Prince Gustav, who passed his decision to the Royal Council on May 14 [60]. The Prince seems to have put considerable effort into the question. According to Bergman, Gustav consulted Tilas and von Swab, who both supported Bergman [62]. It should be recalled that Tidström's patron Tessin was Gustav's former teacher, and had been almost as a father for the young prince (Chap. 2). With the support of the Chancellor, the Royal Council appointed Bergman on July 7, 1767. As Fors put it: "Tidström didn't lose because Bergman was a better chemist. He lost because he had put the faith in the wrong patrons." [63] Had the process occurred a few years earlier, the situation may have been quite different. Linneaus expressed his disappointment over the decision in a letter to Gyllenborg two weeks later [64]. Crown Prince Gustav, who became Gustav III a few years later, had actually little interest in science, and as King, he had little contact with the professors in Uppsala. Bergman and historian Johan Henric Lidén (1741–1793) were the exceptions [65]. The King took no interest in Scheele. When Claudine Picardet (1735–1820), wife of French chemist Louis-Bernard Guyton de Morveau (1737–1816; Sect. 26.7) translated Scheele's works to French,[15] she sent a letter to the Swedish King, a letter which the King did not respond to [65]. A nineteenth-century anecdote tells that Gustav III learned of Scheele while on a trip to Italy, and immediately decided to award him the Order of Vasa. In Stockholm, no one had heard of Scheele and thus gave the Order to Scheele's cousin, Christian Benjamin von Schéele, a highly ranked civil servant, instead [66]. In reality, Scheele was never considered as a receiver of the Order of Vasa, and von Schéele was awarded the Order by Gustav III's son, Gustav IV Adolf in 1796, 10 years after Scheele's death.

Now Bergman was finally a professor, but the first years would be challenging. For example, he had to wait 5 years for his salary. The problem was that Wallerius retained his salary on retirement, and not until Bergman's former teacher, Mårten Strömer, died in 1770 could Wallerius take Strömer's emeritus salary, and Bergman was finally paid [67]. In 1777, Bergman's salary was 58 riksdalers and 16 skillings

[14]At the time, politician Jean Georg Lillienberg (1713–1782) was president at the board of Mines.
[15]*Mémoires de Chymie*, 2 vols, Dijon 1785.

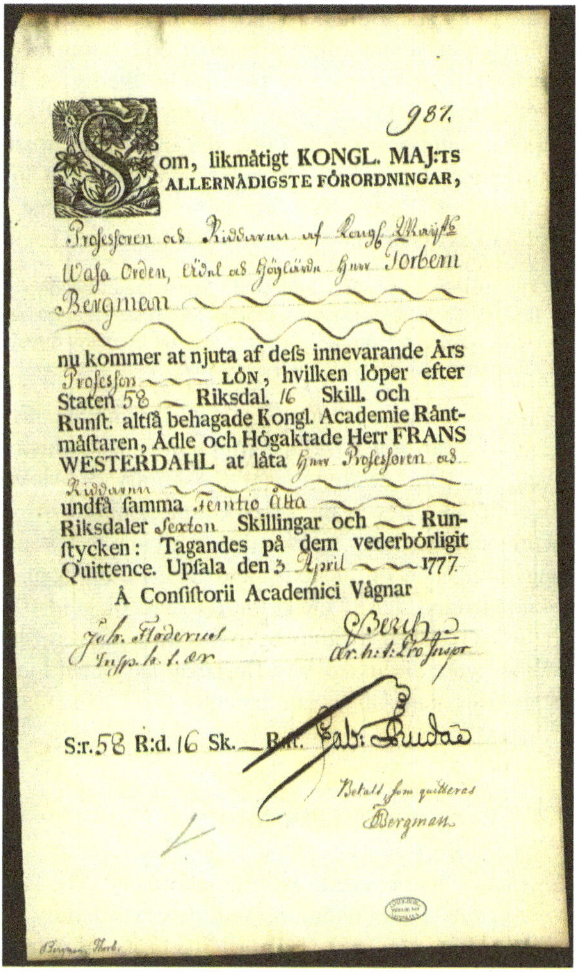

987.

Som, likmåtigt KONGL. MAJ:TS ALLERNÅDIGSTE FÖRORDNINGAR,

Professoren och Riddaren af Kongl. Maj:ts Wasa Orden, Edel och Höglärde Herr Torbern Bergman

nu kommer at njuta af deſs innevarande Års Professions LÖN, hvilken löper efter Staten 58 Riksdal. 16 Skill. och Runſt. altſå behagade Kongl. Academie Råntmåſtaren, Ädle och Högaktade Herr FRANS WESTERDAHL at låta Herr Professoren och Riddaren undfå ſamma Femtio åtta Riksdaler Sexton Skillingar och Runſtycken: Tagandes på dem vederbörligit Quittence. Upſala den 3 April 1777.

Å Conſiſtorii Academici Vågnar

S:r. 58 R:d. 16 Sk.

Betalt som quitteras
Bergman

Fig. 9.8 A receipt of the payment of Bergman's salary for 1777. *Photo* Uppsala University Library

(Fig. 9.8). Apart from smaller funds from the university to cover the running costs of the laboratory, Bergman had to rely on the fees from students for his income. In addition, Bergman had to compete for students and money with the Board of Mine's laboratory in Stockholm and it's director Gustaf von Engeström, and also with his former competitor Tidström, who remained as an assistant after Bergman's appointment and gave his own competing lectures until his death in 1779 [68]. Bergman also inherited Wallerius' amanuensis Wibom [69]. As will be seen (Sect. 9.6), Bergman took the competition with von Engeström very seriously.

Bergman was inaugurated as a professor at the university with a lecture on the history of chemistry [70]. With little chemical experience but a talent for literature studies, it was certainly a suitable topic for Bergman. Ten years later, Bergman would reuse the text in a dissertation, *On the Origin of Chemistry*, defended by one of his students, Bergman and Paulin [71]. This thesis mainly covers Egyptian alchemy (an "extraordinary rage of converting every thing into gold" [72]) and Greek philosophy, such as the atomic theory of Democritus. Bergman not only based much of his account on biblical sources but also made extensive use of classical authors such as Pliny the elder. Bergman acknowledged that the great knowledge of practical chemical processes in China had been developed independently of Western science. Bergman wrote that it was very natural that chemistry had evolved via theories which had latter been disproved:

> The history of Natural Philosophy must therefore in a great measure consist of errors, falsehoods, and conjectures: For in all cross ways we seldom arrive at the truth by the shortest path; nor do we reach it at last but by many circuitous wanderings, and after every other road has been tried unsuccessfully. [73]

The thesis was followed in 1782 by another historical dissertation, the master thesis of Pehr von Afzelius, which covers the history of chemistry in the Middle Ages, from the destruction of the library of Alexandria to the establishment of scientific societies in the seventeenth century [74]. The main topics discussed are medicine, metallurgy, glass-making, dying and preparation of mortar followed by a long discussion on alchemy and transmutation of base metals to gold. Bergman acknowledged that if the primitive acid of gold (Chap. 15) could be obtained, it could be reduced to gold by phlogiston. Several historical cases of alleged gold making were discussed, and Bergman attributed the vast majority to contamination by gold in materials or equipment.

9.4 Bergman's Laboratory

Bergman's first concern was the laboratory. The laboratory and the living quarters were in very poor conditions and there was hardly any equipment apart from a few mortars (Fig. 9.9) [62]. On April 30, 1766, a fire had raided Uppsala burning down a hundred houses. The fire, which spread rapidly due to high winds had burned down some wooden buildings on the yard, and severely damaged the laboratory building [70]. Among the other houses destroyed was the home of Bergman's friend Rinman, who lost all his books, manuscripts and his mineral collection [75]. Wallerius had been quite successful in raising funds for his institution, but of the planned two-storey building, only the ground floor had been built. Thus, there was no room to display the mineral collection that the university had bought from von Swab in 1751 [76]. Bergman made up ambitious plans for his new institution but ran almost immediately into problems. In early 1768, it was revealed that the university accountant, Petter Julinsköld, had embezzled a sum corresponding to twice the annual budget of the university in order to cover losses he had made on financial speculations [77]. Meanwhile, Bergman gave his lectures in the Theatrum Oeconomicum (Fig. 9.10), where he also displayed Swab's mineral collection [78].

Fig. 9.9 A mortar from Wallerius laboratory, one of a few pieces of equipment that survived the fire in 1766. Museum Gustavianum, Uppsala. *Photo* Anders Lennartson

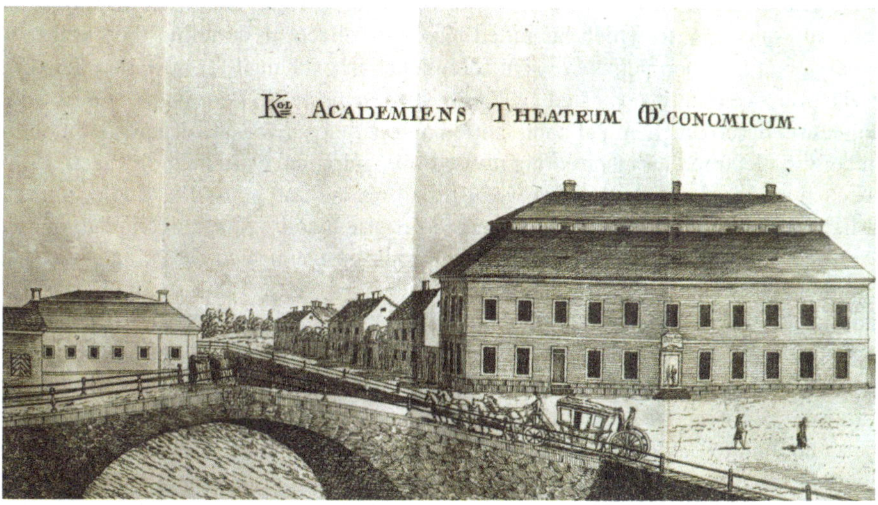

Fig. 9.10 Theatrum Oeconomicum, where Bergman lectured as a newly appointed professor. From Busser—*Utkast til beskrifning om Uppsala*, 1770

By taking over the property from Julinsköld, the university could recover some of the lost money, and the renovation of the laboratory building could start. The architect Carl Johan Cronstedt (1709–1777; a student of the influential engineer Christopher Pohlhem and renowned architect Carl Hårleman, a pioneer of Swedish rococo) was consulted and the layout was approved by the university on June 11,

1768. The plan included the addition of a second floor, which Wallerius had only dreamt of. The laboratory, located on the ground floor, was the first part of the building to be finished. The most striking difference between Bergman's laboratory compared to that of Wallerius (Fig. 9.5), was that the lecture hall was removed and lectures were instead given in the laboratory.

Bergman's chemical laboratory (Figs. 9.11, 9.12, 9.13 and 9.14), *Laboratorium chemicum*, was a white stone building surrounded by gardens. On entering the building through the main entrance, there was an entrance hall with stairs to the second floor, where Bergman lived (Fig. 9.13). To the left (Wallerius' former apartment) was a kitchen and to the right the combined laboratory and lecture hall. The laboratory part of the room had a stone floor [79] and was equipped with different furnaces and heating baths [80]. Next to the laboratory, there was a small room for glassware and instruments. The acute lack of instruments was solved by purchasing the collection of the late professor of medicine, Samuel Aurivillius (1721–1767) [81]. From Bergman's correspondence with his former student, Johan Gottlieb Gahn in 1770, it is apparent that he had asked Gahn to make inquiries on glassware, porcelain and thermometers in Stockholm. Another door in the laboratory led to a small room for tools and materials. This room also had a back door. Near this backdoor was the storage for charcoal.[16] A gallery stretched along almost the entire backside of the building towards the river. Here, Bergman displayed the mineral collection, to which he added his own collection, doubling the number of specimens [76]. Doublets were sent to colleges in exchange for new samples. The collection eventually consisted of 7,888 specimens [82]. Bergman organised the collection based on chemical composition and perhaps it was the organisation of the mineral collection that inspired his revised mineralogical classification (Sect. 25.7). At each end of the gallery, there was a small room. One housed a smaller mineral collection with Swedish minerals arranged geographically; this collection amounted to 3,420 specimens [82]. Here, a student could see which minerals and rocks he could expect to find in different Swedish mines [76]. In the other room on the other side of the gallery, there was a collection of 35 painted wood models [82]. Here, students could study scale models of gold refining in Ädelfors, blast furnaces for iron production, the brass works at Skultuna, the facility for copper refining in Falun, and so on [76]. All models were made in the same scale and painted; Bergman had to pay for these models with his private money [83]. Here was also a cupboard with "remarkable" products prepared in the laboratory and various crystals. The new laboratory left the retired Wallerius unimpressed [84]. He accused Bergman of prioritising his own comfort rather than science [85]. This was because Bergman used the entire upper floor as his private apartment. Wallerius feared that the mineral collection would be damaged by water from the river which frequently flooded in spring. This was actually a problem; in March 1780, Bergman wrote to Scheele: "Last night the river rose up so that I am now completely locked in. The laboratory building stands like in a lake".

[16]This shed is apparently omitted on the engravings in Figs. 9.11 and 9.14.

Fig. 9.11 Bergman's chemical laboratory. From Busser—*Utkast til beskrifning om Upsala*, 1770

Fig. 9.12 The building where Bergman had his laboratory has fortunately survived. Note that the photo was taken from the opposite side of the building compared to Fig. 9.11. More modern buildings have been erected in the former gardens. *Photo* Andreas Furängen 2013

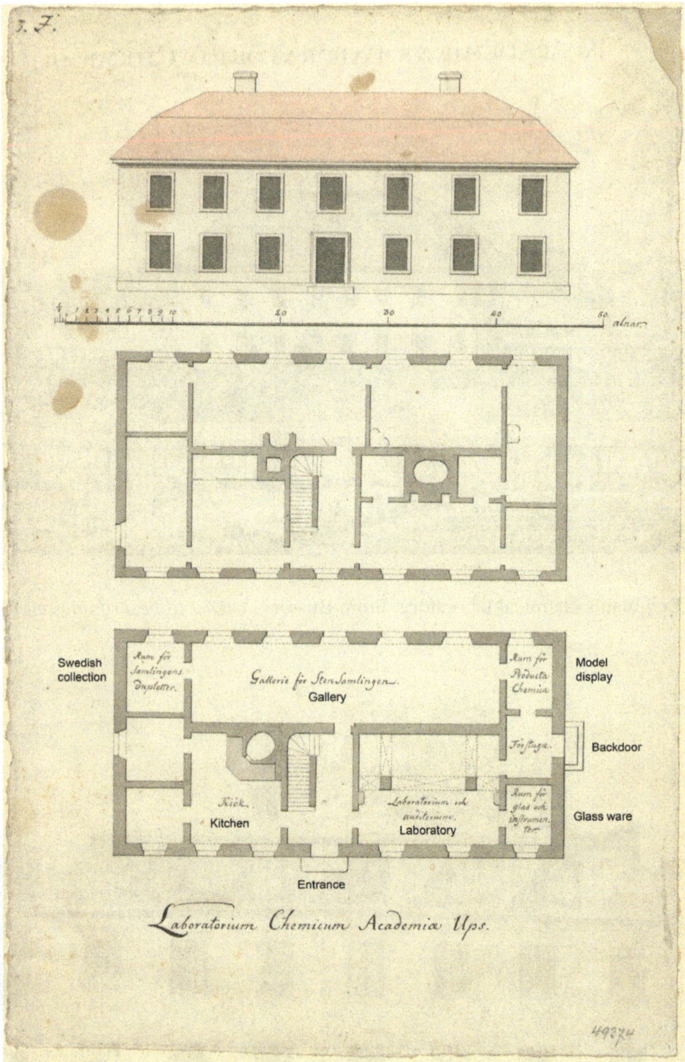

Fig. 9.13 Original drawings of Bergman's laboratory building. Uppsala University Library

When the chemist and historian of chemistry Thomas Thomson visited Uppsala in 1812, the geographical mineral collection was gone, and he thought that it might have been transferred to the Board of Mines in Stockholm [86]. In the nineteenth century, the collections from the Board of Mines were transferred to the Museum of Natural History in Stockholm, where there is no trace of such a collection. Perhaps it was kept in Uppsala but merged with the larger collection. Bergman's collection of models also seems to have disappeared, possibly being out-dated by 1812. The

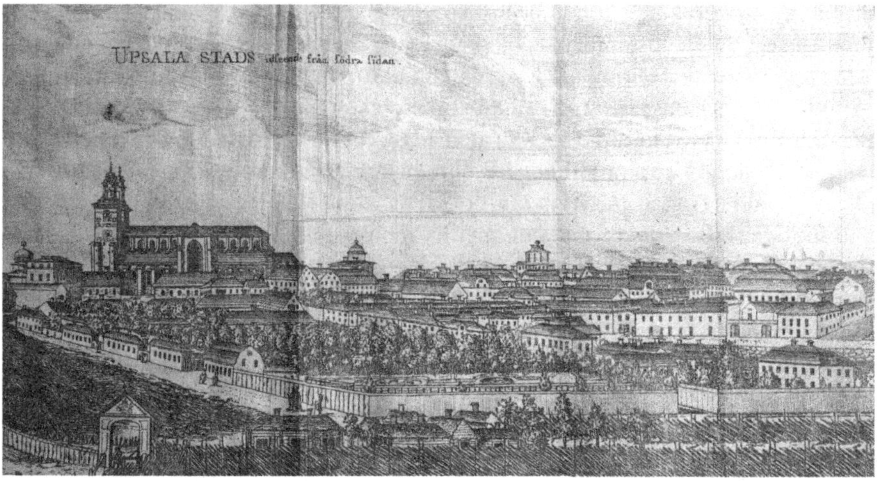

Fig. 9.14 View of Uppsala with Bergman's laboratory surrounded by gardens by the river. The castle is on the hill to the left and the cathedral at the centre. Bergman's laboratory is on the lower right. From Busser—*Utkast til beskrifning om Upsala*, part 1, 1773

laboratory itself had become seriously out-dated in the nineteenth century, and after years of discussions, a new building was erected on the initiative of chemistry professor Lars Fredrik Svanberg (1805–1878), a student of Berzelius, who moved from Bergman's laboratory in 1857 [87]. Luckily, the old laboratory was not demolished and has survived to the present day. When the chemists left the building, it was converted to a printing house. In the mid-1930s, it was the home for Statens institut för rasbiologi (the Institute of Racial Biology), which became the Department of Medical Genetics and incorporated into the university in 1958. Still, in the 1980s, it was used as a hydrological laboratory. At present (2019), the newly renovated building is used by the Department of Law. Bergman's layout is still recognisable, but some of the larger rooms have been split into smaller rooms with new passages between the rooms. Bergman's former laboratory, with its preserved stone vaults, is a meeting room for students, and the main part of the gallery has been turned into a conference room.

9.5 Bergman's Early Chemical Work

Following his appointment as a professor, it would take some time for Bergman to build up his new career. In 1768, he published only a single paper, the paper on the mountains of Västergötland discussed in Chap. 7. In 1769, he presented his first dissertation as a professor. It was a study on fulminating gold, a shock-sensitive compound of the ill-defined composition prepared by adding ammonia to a solution of gold in *Aqua regia*.[17] It had been known since the days of the alchemists in the middle ages, but its instability still today makes chemical studies difficult. Bergman correctly found that ammonia was essential for the explosive properties of the precipitate, other bases formed non-explosive precipitates with gold. As a noble metal, Bergman wrote, gold in its calcined state (i.e. an oxidation number higher than 0) has a very strong affinity for phlogiston. Thus, the gold calx (gold oxide) attracted phlogiston from ammonia liberating large volumes of elastic fluid (gaseous nitrogen). Bergman could not identify the gas but found that it was at least not carbon dioxide (fixed air). Bergman determined that fulminating gold evolved four times more gas per weight than black powder, and wrote that an explosion is nothing else but a rapid evolution of gas. Fulminating gold was also studied by Scheele [88], who also found it to be composed of gold calx and ammonia (alkali volatile). The gas evolved on detonation, he found to be identical to the gas (i.e. nitrogen) evolved by decomposition of ammonium chloride (salmiack) with MnO_2. As will be seen in the next section, Bergman's thesis led to an infected dispute with Wallerius.

In 1770, Bergman started his important studies of water analysis (Chap. 17) with a thesis by Bergman and Dubb [89]. The same year he also published the first part of a long review on the chlorides of mercury [90], the third and last part of which appeared in 1772. This work is essentially a literature study but Bergman has made occasional control experiments. It is also here that we find the first seed to his chemical nomenclature (Chap. 26) in the form of the name *Mercurius Nitratus* for mercury nitrate.

Of more interest is his paper on the preparation of more durable bricks and roof tiles [91]. Bergman complained that the quality of bricks had decreased in recent years, so that it almost seemed like the knowledge of preparing bricks had been lost. Brick walls could be protected by cement rendering, but roof tiles made from clay was a more serious issue. Clay tiles were the perfect material for roofs, Bergman wrote, as wooden boards, wood shingles or sod used up much more valuable resources. The tiles that Bergman saw around him were too porous, which caused the roofs to leak and the water-soaked tiles cracked by freezing in the winter. A few years earlier, Bergman had visited a tile factory belonging to the university and he had found the manufacturing process unsatisfactory.

By reading the works of Macquer,[18] Bergman had found out that pure clay mixed with lime does not melt on heating. On addition of silica, however, the mixture melts on heating, but as the silica content was increased still further, the melting point increased. Bergman had made experiments with both lime-free clays and lime rich clays (marl) from the Uppsala area and mixed them with sand in different proportions. He found that bricks made from pure clay shrunk and cracked on burning, so the addition of sand was necessary. Too much sand, however, prevents proper sintering. The degree of heating was also critical. The tiles must be

[17]Fulminating gold is a mixture of different species but appears to be square planar Au(III) complexes with chloride and ammonia ligands.
[18]Pierre Macquer (1718 – 1784). See Sect. 14.7.

burned enough for the surface to melt in order for the tiles to be waterproof. Also, clay contains sulphates which are reduced to gaseous sulphur dioxide on burning. With insufficient heating, Bergman noted that the sulphate remaining tended to be leached away by water as alum, leaving pores in the material. Bergman's samples had been exposed to the open air for 3 years when he published his paper, and while the improperly burnt specimens were almost entirely destroyed, the properly burnt specimens were unaffected. The paper also includes a discussion on the chemical analysis of clays. Bergman identified calcium carbonate (lime) by its effervescence with nitric acid. The lime was reprecipitated from the acidic solution with "stinking spiritus", a sulphide solution. Later in his career, he did not precipitate metals as sulphides, which is common in modern procedures. By repeated washing with water, clay was separated from sand.

In this paper, as in the case of the work on alum refining, Bergman addressed a practical problem and showed that he could offer a simple solution based on chemistry. This was the kind of paper that the people who had helped him to the professorship wanted to see. As will be seen, however, as his position became more secure and his chemical knowledge increased, his work would become less applied. Bergman complained that many people had little or no interest in things that could not be used as food or for clothes [92].

There are probably no surviving artifacts from Bergman's early career, but the museum of Uppsala University, Museum Gustavianum, has some of Bergman's chemical equipment on display (Fig. 9.15).

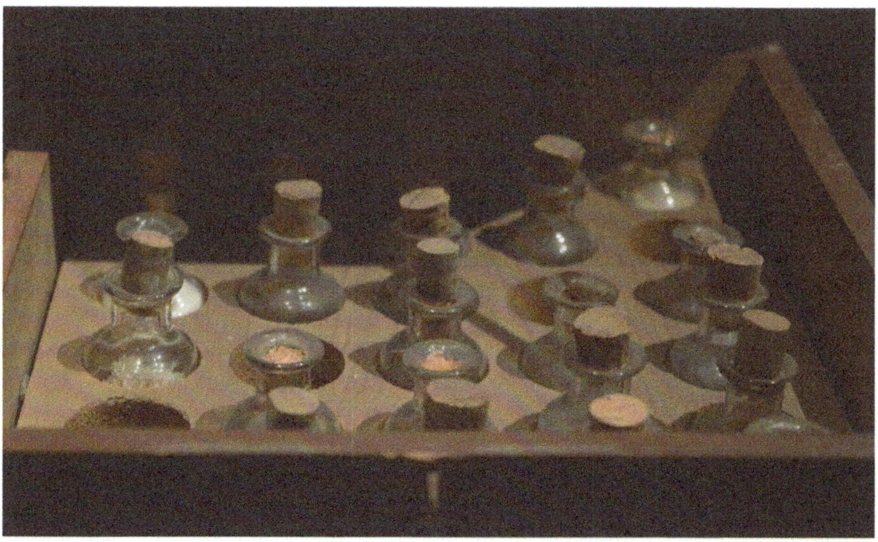

Fig. 9.15 Sample bottles from Bergman's collections. Museum Gustavianum. *Photo* Anders Lennartson

9.6 Bergman's Relations with Wallerius and von Engeström

Wallerius (Fig. 9.16) had spent years building up the chemical institution at Uppsala. He had started from scratch and both established chemistry as an independent discipline at the university and himself as a respected scientist well known in Europe. He must have been quite proud of what he had achieved and, of course, he had hoped for a successor who continued in his footsteps. The appointment of Bergman as his successor must have come as a chock, and soon Wallerius would see his worst fears come true. Bergman had a completely different philosophy compared to Wallerius and he rapidly and systematically erased every trace of Wallerius' work, starting with the layout of the laboratory.

After the infamous dissertation in 1741, Wallerius and Linnaeus appears to have accepted each other as they had much in common and worked along similar lines of thought. As a botanist, Linnaeus did not have to conduct experiments. He believed that God had created the World and given him the mission to explain the creation to mankind. Thus, Linnaeus observed nature and tried, very successfully, to develop a system for classification. By only observing God's creation, he could classify nature based on the principles used by God in the creation. Essentially, Wallerius believed he could approach chemistry in a similar manner. His main focus was not experimentation, but he attempted to create a chemical system largely based on known facts [93]. As he rejected most new discoveries in the rapidly developing field of chemistry,[19] this strategy was of course doomed to fail, but one must still acknowledge his serious attempts. When reading his textbook, *Physica Chemica*, his generous use of literature references is rather striking, not least considering that references to primary literature are quite uncommon in modern chemistry textbooks on the undergraduate level. Unfortunately, Wallerius often relied on old and out-dated sources.

While Wallerius mainly relied on older chemical authorities and the Bible, Bergman based much of his philosophy on Newton. In his thesis on the history of chemistry (Sect. 9.3), for example, he wrote "...Newton, the great Newton, the glory and ornament of the human understanding..." [94] Chemistry based on the theories of Newton became of course rather different from the qualitative chemistry of Wallerius. While Wallerius simply did not believe that the recent developments in chemistry had revealed any significant facts unknown to earlier researchers [95], Bergman was of a completely different opinion and introduced modern experimental chemistry in Swedish academic chemistry.

Wallerius, who suddenly found himself isolated and without authority after Bergman's appointment, did not accept his new situation and launched a campaign against Bergman. On September 25, 1769, an anonymous paper appeared in *Inrikes Tidningar* (Domestic Newspapers) [96]. The author, which Bergman recognised as Wallerius, criticised a footnote in his paper on the Mountains of Västergötland (Chap. 7). In this note, Bergman wrote: "That clay on drying can gain the hardness

[19] At least those contradicting his own views.

Fig. 9.16 Wallerius and Bergman starring at each other at the façade of the former office of the Swedish Steel Association at Kungsträdgårdsgatan, Stockholm, built in 1875. Scheele, the outsider in the background, is locking in a completely different direction. The artist, Frithiof Kjellberg (1836–1885), was of course unaware of Wallerius and Bergman's frosty realtions. *Photo* Anders Lennartson, April 2019

required to give fire against steel, is a reason to believe it to be one of the elements in siliceous stones, which is fully confirmed by the fact that the earth precipitated from Liquor Silicium give alum with vitriolic acid" [97]. Liquor Silicium was a solution of sodium silicates, and the earth (hydrated silicon dioxide) precipitated by acids should not have reacted with sulphuric acid under any reaction conditions, and certainly not to give alum (aluminium potassium sulphate). Bergman probably made the mistake to prepare his Liquor Silicium by fusing sodium carbonate with silica in a clay crucible which, as Scheele later showed (Sect. 22.4), dissolved some aluminium-rich clay from the crucible. Wallerius did not question that alum was formed by treating the precipitate with sulphuric acid, this observation was correct according to Wallerius, but he questioned the nature of the earth. Wallerius' point was that the formation of alum said nothing about the nature of the Earth since earths could transform into each other in solution, in precipitates as well as in nature. Wallerius claimed that alum could be obtained from, for example, pyrites (FeS_2) and iron(II) sulphate (green vitriol). This paper was written in an objective manner, and Bergman did not respond although he certainly did not agree. This was probably a disappointment for Wallerius, who probably had hoped for a long debate with Bergman, where he could expose Bergman's lack of chemical knowledge.[20]

On March 1, 1770, Wallerius published another anonymous article entitled *Necessary, but well-meant reminders for young chemists*, in one of the most widely read Swedish newspapers of the time, *Lärda Tidningar* (Learned newspapers) [98]. This paper was written in a very different style, possibly since his previous paper had little effect. Wallerius paper does not mention Bergman, but the true target was easily realised by the readers since Wallerius illustrated his criticism by examples

[20]From that point of view, his attack on Salberg (Sect. 9.1) was more successful.

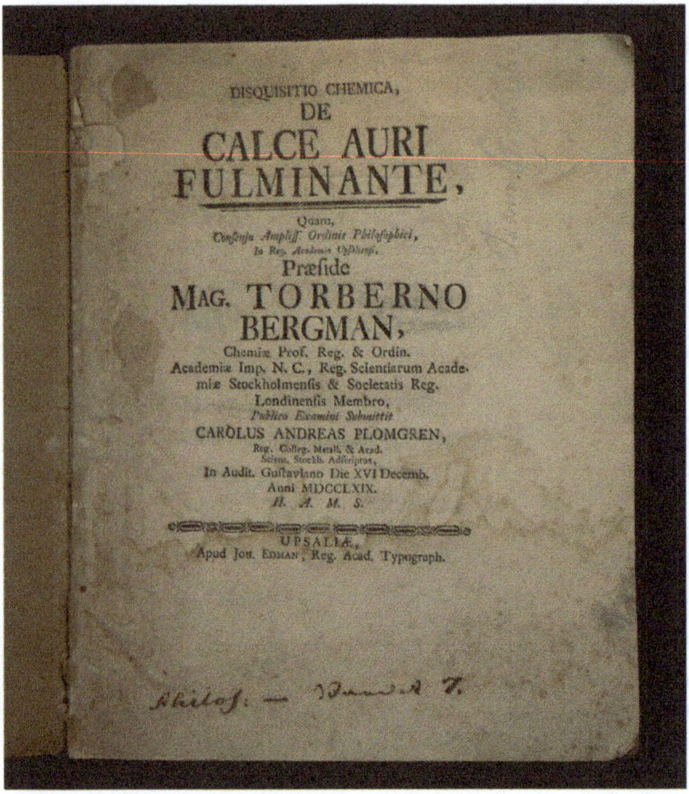

Fig. 9.17 Bergman's dissertation on fulminating gold, defended by Plomgren (Sect. 11.4). This was one of Bergman's early chemical studies and used by Wallerius to illustrate the unreliability of young chemists in general and Bergman in particular. *Photo* Anders Lennartson

from Bergman's dissertation on fulminating gold (Fig. 9.17). The introduction to the article reads:

> It is often noted among those who have interest for chemical studies, that as soon as they can perform one or another experiment, they believe themselves able to perform all the other [experiments], and if the results from their experiments are different from what an experienced chemist had said, they imagine that the error is to be sought in [the writings of] the latter rather than in their own.

Wallerius' advice to young chemists was delivered in ten points, some more general and some specific criticism of Bergman's thesis. First, he advised that before trying to study chemistry in-depth, one has to learn to prepare his own reagents. The second point was more serious and more typical for Wallerius' way of thinking. He cautioned young chemists from challenging the theories of older more experienced chemists. Experiments should be repeated before making any conclusions. Wallerius listed a number of chemists who had made trustworthy investigations of fulminating gold: Kunckel, Neumann, Zimmermann, Pott, Lewis,

Brandt, Becher and Stahl. It is quite ironic that Wallerius included Becher, who had a very liberating view of the truth, to say the least. It is important, Wallerius continued, to mix the ingredients in the correct proportions as described in the literature. This is an interesting remark, as Bergman became one of the early proponents of a stoichiometric view of chemistry, in contrast to Wallerius. Wallerius also stressed accuracy in order to obtain reproducible results. Wallerius attacked Bergman's definition of salts and while he agreed that salts are soluble in water, he had noted that also substances that separate (precipitates) from water may have saline properties. As an example, he gave silver chloride (*Luna cornua*) which contained hydrochloric acid (*Spiritu salis*). Here, Wallerius indicated that a salt should contain an acid. This view had been proposed earlier, i.e. by seventeenth-century chemist Glauber and although Bergman simply defined salts by their solubility in water, he was clearly aware of the saline properties of, for example, calcium carbonate [99]. Wallerius went as far in his criticism as to question all of Bergman's work: "May not such experiments put an attentive [person] in the thought, that other experiments, originating from the same hand, should not be of considerable value, no matter how expensive they could have been." The tone is rather sarcastic throughout the paper.

Wallerius' paper does not seem to have gained much support for his behaviour. Gahn supported Bergman fully [100], as did Rinman, who was a former student of Wallerius, and who wondered whether Wallerius had gone mad [101]. After Bengt Bergius had confirmed to Bergman that Wallerius really was the author of the paper, Bergman wrote a reply. Bergman's article was divided into four parts and published in four subsequent issues (March 21–March 27) of another newspaper of the time, *Allmänna tidningar* (*Public Newspapers*) [102–105].

The style of Bergman's reply is very different from everything else Bergman has written, and it is clear that Bergman had finally had enough of his opponent. He opened his paper with an excuse:

> It is reluctantly that I this time grab the pen. He how devote his time to experimental science has his hands full with more urgent work than discussing unfounded opinions, especially when they are accompanied by bitterness and having completely different purposes than revealing the truth.

Bergman assured that this was not the first time he had been attacked by Wallerius, he had previously been attacked both verbally and in print, and Bergman's silence had only increased his opponent's anger. Bergman then proceeded to demolish Wallerius ten points: "The first one is a blind shot, before he dares to go into deepness". He invited Wallerius, whom he referred to as the Old Chemist or simply "O.C" (den Gamle Chemisten, G.C.) to join him and see how the experiments were carried out. Of course, he did not actually expect his opponent to show up. Wallerius criticism that experiments had to be repeated exactly as described was dismissed by Bergman. It is important, Bergman wrote, to test different experimental approaches, if one not only wish to confirm if a written statement is correct or not. In one of his ten points, Wallerius had claimed that heat itself, when applied to a substance immersed in a liquid had no effect, but that the heat acted via the liquid which constituted the reagent. Bergman reminded Wallerius that an egg is cooked after being immersed in boiling water, although the shell prevents direct contact with the water. He concluded by writing that

> Thus, not a single circumstance of the O.C.'s accusations holds, and never the less must solid rules, unclear concepts, false conclusions, serious mistakes about clear opinions, yes untruth serve him when he everywhere vents his spite. [—] I happily leave to every impartial reader to discern whether they [Wallerius' reminders] are necessary and whether they are well-meant. When his heart, his Theoria and Praxis of this specimen [his paper], that now to his commemoration lays in front of the eyes of the public, is to be judged, I fear that the verdict will not please him. I, by the way, know that man. The number of his important discoveries, the accuracy in his results and the swiftness of his operations are also quite well known. Some people are born to worry others.

Bergman then turned to Wallerius' criticism of the paper on the mountains of Västergötland and disproved all Wallerius' points. "It is", Bergman wrote, "human to fail, but he who with the utmost rigour examines others works, and who cleaves hairs in order to, with utmost bitterness impose them with errors, which they are not guilty of, should one not of him expect perfection in every way" In fact, Wallerius' short paper still casts long shadows over his memory, two centuries later.

In an introductory essay, called *Investigation of the Truth*, in the first volume of his collected works, *Opuscula Chemica et Physica* published in 1779, Bergman described his own ideas of how research should be conducted. He criticised Descartes and priced the works of Newton, who first collected facts and then analysed them thoroughly. Bergman then introduced a set of guidelines for a scientific approach to chemistry. (1) On investigating a substance, conclusions should not be drawn from slight similarities with other substances. Each substance has to be analysed and the analysis confirmed with synthesis. (2) Chemical analysis should preferably be conducted in a solution (Chap. 23). (3) Experiments should be designed in order to reveal the truth: it is not the number of experiments that is important, but their quality. (4) Experiments should be made with the greatest possible accuracy. (5) Experiments by others, "particularly the more remarkable ones." should be reviewed critically. (6) Investigations should start from known facts and proceed to more complex questions. (7) A discovered phenomenon should not be trusted until it is confirmed by other experiments. (8) The relation between the cause and the effect should be determined quantitatively. As most chemists still worked qualitatively, this was rather controversial. When discussing the accuracy of experiments, he wrote: "I am almost ashamed to relate, that I knew a chemist who considered thermometers, and such instruments, as physical subtleties, superfluous and unnecessary in a laboratory" [106]. It is almost certainly Wallerius that Bergman is referring to [107].

Apart from Wallerius, there was one more important chemist in Sweden at the time, Gustaf von Engeström,[21] who was born in Lund 1738, and was thus about three years younger than Bergman. He began his studies in Lund, but continued in Stockholm, studying under Scheffer, Cronstedt and von Swab. After spending some time as auscultator and extraordinary assayer at the Board of Mines in Stockholm,

[21]There are no known portraits of von Engeström.

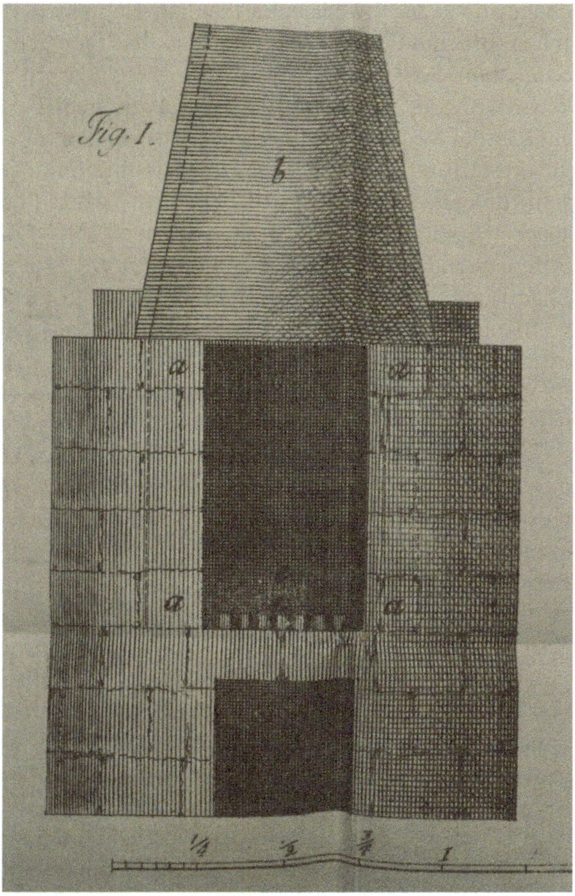

Fig. 9.18 A furnace from von Engeström's laboratory. Engraving from his book Laboratorium Chymicum, 1781

he undertook a journey in 1764–1765, travelling first to England and then to Germany, where he studied under Andreas Sigismund Marggraf (1709–1782). After the death of Georg Brandt, von Engeström succeeded him as director of *Laboratorium chymicum*, where he remained until his retirement in 1794. As director of *Laboratorium chymicum*, he was succeeded by Bergman's student Peter Jacob Hjelm; he died in 1813. Gahn had visited von Engeström in Stockholm, and gave the following account to Bergman in November 1770:

> I have once been to Mr v. Engeström. His Laboratory is now almost finished. The room is rather cramped [...] He has a glass furnace that he brags a lot about, [...] I am not sure how cautious he is with his knowledge, and whether one can be allowed to copy his furnaces (Fig. 9.18) and learn the art. In addition he spoke a lot that I do not believe, and that I do not even remember. Otherwise, he was so courteous and showed me some stone boxes and asked me to visit him again. [108]

von Engeström's most important contribution to science was his English translation of Cronstedt's mineralogy with an addition on the use of the blow-pipe. In the *Transactions of the Royal Swedish Academy of Sciences*, he published twelve papers, among them papers on the analysis of materials from China which had obtained via the Swedish East Indian Company. One of his contacts was Peter Johan Bladh (1746–1816) who also had contacts with Scheele's employer in Stockholm, Scharenberg (Chap. 10), where he lived in 1775. von Engeström's most extensive work, *Laboratorium Chymicum*, published in three parts 1781–1784, is entirely devoted to the refinement and separation of gold and silver, especially from the waste from goldsmiths. Despite the first line of the preface "After I have, for some time, been occupied with chymical[22] work, I have thought it to be my duty to share with the public what I have achieved" [109] the book contains little of what Bergman and Scheele would call chemistry, and it seems like "chymistry" in von Engeström's eyes would be better translated to metallurgy. The only chemistry to be found in the book is associated with the separation of gold and silver with, for example, nitric acid and antimony(III) sulphide.

Wallerius and von Engeström had in practice divided the education between them: a student would be taught basic chemistry by Wallerius in Uppsala, and von Engeström taught them the art of assaying in Stockholm [110]. Unfortunately for the students, von Engeström had a strained economy, and thus charged the students very high fees. The fee for a chemistry course was 3,000 copper dalers, for a course in mineralogy it was 2,000 and a course in assaying was 1,000 copper dalers [111]. Bergman, also in need of students and money, broke this silent agreement and started to compete with von Engeström. With lower fees, he was able to keep at least some of the students in Uppsala, certainly annoying von Engeström.

Two particular incidents increased the strain between the two men. Bergman and von Engeström had agreed to exchange mineral samples, a common practice among mineralogists and a way to establish bonds. They had agreed that von Engeström should give Bergman two Swedish or one foreign sample for each foreign sample Bergman sent him. Apparently, von Engeström only sent one Swedish mineral for each sample he received from Bergman [111]. Bergman wrote a letter to Wargentin in 1778, where he expressed his feelings about his competitor [110]. He complained that von Engeström was arrogant and imagined that he knew more than anyone else.

The next episode was more serious. von Engeström had a copy of the lecture notes from the chemical lectures of Henrik Theofil Scheffer (Chap. 9), which he allowed the students to copy once they had paid the fee. von Engeström regarded this text as a secret which he shared with his students [111]. Thus, this text was very precious to von Engeström. One can assume that the students had to promise not to distribute copies. von Engeström's copy was, however, not the only copy of the text. Bergman was acquainted with industrialist Patrick Alströmer (1733–1804), who was the son of Jonas Alströmer, an important figure in Swedish industrialisation. Patrick Alströmer had a documented interest in chemistry and had attended Scheffer's lectures in chemistry 1749–1751. Alströmer had kept notes from the lectures, notes which had been read and corrected by Scheffer himself [112]. In a letter to Wargentin, Bergman wrote that he learned of this copy by chance and that Alströmer asked him if it deserved to be published [111]. Bergman needed a book for his teaching, and thus published Scheffer's lectures with his own remarks in

[22]von Engeström consequently used the old spelling "chymistry" (chymie).

1775 (Sect. 11.1). In the preface, Bergman gives the impression that Alströmer had turned to him with the manuscript, but one cannot avoid suspecting that Bergman played a more active role and, at least partly, published the text in order to deprive von Engeström of his secret. In December 1770, Gahn wrote to Bergman: "Baron Hermelin have given me a rejection regarding Dir. Scheffer's papers. If I wanted to use them these days in Stockholm, it was fine, but then he wanted to take them with him..." [113]. It seems that Bergman was actively searching for copies of Scheffer's lectures.

Bergman had to spend a lot of time adding comments to Scheffer's text, and the question is how much time Bergman actually saved compared to writing his own book. According to Bergman, von Engeström saw the publication as a declaration of war [111]. It was probably not only economic reasons and the competition for students that drove Bergman to the publication of Scheffer's lectures. As an enlightenment era scientist, Bergman also found it unacceptable to keep knowledge secret [114]. For the chemistry students, Bergman's action must have been more than welcome. Not only did they no longer have to hand-copy the manuscript, the price also dropped from 3,000 copper dalers to 13 copper dalers and 16 öre [115].

According to Bergman, von Engeström constantly tried to discredit him; for instance, he told the students that came from Uppsala to *Laboratorium chymicum* that they had to forget everything that Bergman had taught them [114]. Wargentin wished to put an end to the fight by asking Rinman to talk to the two, but Rinman was wise enough not to get involved [114]. Bergman thought less of von Engeström and Wallerius, as they had no interest in theoretical chemistry. In 1777, for example, he wrote a preface to Scheele's book on air and fire (Sect. 21.3), and he wrote to Gahn:

> I have on request written a small preface to the work [i.e. Scheele's book], dealing primarily with chemistry's, in particular the finer chemistry's [i.e. theoretical and analytical chemistry in contrast to applied chemistry] usefulness. The latter is despised by a W., a v. E. [i.e. Wallerius and von Engeström] and others who do not understand it. [116]

The same year, 1777, Bergman gave a lecture in front of the Royal Swedish Academy of Sciences (see Sect. 19.5), as he handed over the presidency that he had upheld for three months. The topic of the lecture was "*On the Latest Progress of Chemistry*". The topic was very well chosen by Bergman. The King was in the audience, and in his lecture, Bergman argued for the importance of basic chemical research and talked about the amazing discoveries made by chemists in recent years. Thus, Bergman assured that invested tax money was used to for the benefit of Sweden and Swedish people. The topic also distanced Bergman from Wallerius' school, who did not believe in new discoveries.

von Engeström and Wallerius also held the president position (preses) of the Academy; von Engeström in 1774 and 1782 and Wallerius in 1783. The lectures they delivered are very different from Bergman's lecture of 1777. When von Engeström handed over the presidency in 1774, he spoke of *The Obstacles and Progress of Mineralogy During the Last Years* [117]. A large portion of the lecture was used to establish Cronstedt's priority for the development of the chemical mineralogy system (Sect. 25.7). He said that chemistry is the only correct ground for mineralogy, but also said that two other sciences were necessary in the study of mineralogy: metallurgy and experimental physics [118]. Bergman certainly agreed on this point. When, however, von Engeström claimed that the amount of sulphuric acid in calcium sulphate (gypsum) varied between different samples, Bergman did not agree.

In 1782, von Engeström once again held the president position in the Academy. The lecture he delivered, *Speech, on some Difficulties, and other Circumstances, Faced on Practicing Chymistry* [119] was very different from Bergman's lecture in 1777. When Bergman painted a very bright picture of the progress of chemistry and its usefulness for society, von Engeström's lecture focused on problems encountered in the study of chemistry. As a chemist, it would be in von Engeström's interest to keep the interest in chemistry high, so the choice of topic is difficult to understand unless one assumes that von Engeström's intention was to point out that the discoveries made by academic chemists were unreliable in contrast to his own, practical work. von Engeström's mistrust was not, however new. He had expressed himself in a similar manner in a paper on alum (which also criticised Bergman, Sect. 9.2) in 1774: "As is often found in Chymistry, that different operations in the same type of Experiment, give changes in the effect [...]" [39].

von Engeström started by saying "The chymical science could be regarded as a game, if it was not for several problems that are often encountered in its practice, and which reveal much uncertainty in its very foundations." As the first example, he spoke of Scheele's discovery of hydrofluoric acid (Sect. 14.1) and several competing interpretations of Scheele's results by other chemists. After discussing various ideas of the composition of fluorspar (fluorite), he said:

> it is fortunate, after all, when such conflicts concern a subject, which is of less importance and use in everyday life. When chymistry is viewed only from the useful side, it almost does not matter if fluorspar contains a unique acid, or one previously known, [...], but is still disturbing to find from such incidents that chemical experiments are subjected to such uncertainty. Here is consequently revealed a large deficiency in this science, that is, reliable grounds for performing chymical experiments and the conclusions therefrom drawn. [120]

It is quite clear that von Engeström did not care much about theoretical chemistry. He admitted that some discrepancy could arise from different reaction conditions and impure reagents, but he said that the largest problem was the limits of the human mind. Perhaps later generations would be able to resolve some problems, but only to evoke new problems in their place. He discussed the different competing theories of carbon dioxide (Aër Fixus; Chap. 18) but without mentioning the work of Bergman. His next point of attack was the scientific journals, pointing out that two French journals had published very different reviews of a book published a few years earlier. Interestingly, he then entered a long discussion on alchemy, or "the higher chemistry" as he called it. von Engeström was convinced that there was much sound knowledge in at least some of the alchemical works and that there would be much gain if chemists and physicists would study alchemy. It should be noted that Bergman with "higher chemistry" meant theoretical or physical chemistry.

As customary, the lecture was followed by a short answer by the Secretary of the Academy, and it is not difficult to imagine that Wargentin was disturbed by the fact that a member of the Academy attacked the works of other academy members in a presidency lecture. He pointed out in a polite way that "one should however not conclude from this, that it is the science itself and its foundations that are uncertain". Of course, said Wargentin, scientists will have different opinions in such a rapidly evolving science as chemistry.

The following year, Wallerius delivered a lecture in front of the Academy as he ended his term as president [121]. This lecture can be regarded as a direct attack on Bergman, Scheele and the circles around them. Wallerius said that he had performed experiments on a daily basis during his 30 years in Uppsala, and had learned not to accept other chemists' opinions prematurely, especially not the newer ideas which were "often only guesses". Wallerius sought confirmation in experiments, or even better, by observing nature. This was a very important point in Wallerius' philosophy.

Wallerius was a strong opponent to analytical chemistry, and he found support in the works of Boerhaave, Pott and a doctoral dissertation by Daniel Triller published in Wittenberg in 1767 [122]. The important point in Wallerius' philosophy was that analysis did not necessarily reveal the actual composition of a body. He gave two examples. The precipitation of iron oxides (ochre) and calcium sulphate (gypsum) upon evaporating mineral water did not mean that these substances were present as such in the water. The other example was the distillation of ethanol (Spritus vini) with sulphuric acid to give "a penetrating Liquor mineralis", "a fine aether" (diethyl ether), a "sulphur spiritus" (SO_2), a "clear heavy oil" (diethyl sulphate), "something tar-like" and a siliceous earth (SiO_2). This could, according to Wallerius only happen through a complete decomposition and recombination. The siliceous earth that Wallerius referred to must, of course, have originated from the glassware. "Here against are a large number of chemists, who live in the conviction within themselves, that everything that a body, by the action of more or less artificial solutions, can yield, is a significant part of the body." These are quite harsh words from a former professor. The best way, according to Wallerius, was to "ask nature". The natural processes occurring within Earth are slow and by studying these processes one could make "absolutely certain conclusions". Instead of the direct quantitative measurements of Bergman and others, Wallerius suggested qualitative conclusions made through analogies with natural processes. First, he discussed carbon dioxide (fixed air), a subject where he and Bergman had very different ideas. Bergman (Chap. 18) had shown that carbon dioxide is an acid present in the air, while Wallerius believed that air contained sulphuric acid. Wallerius said that some (i.e. Bergman) had suggested that sulphuric acid in the atmosphere would react with potassium carbonate to give potassium sulphate (which of course is correct). According to Wallerius, however, one could not expect that such a "fine air-like" sulphuric acid would behave as ordinary sulphuric acid. Wallerius gave no reasons for that assumption. He discussed different theories of carbon dioxide, without naming Bergman, and finally came to the conclusion that carbon dioxide is not an independent acid. On which grounds he favoured this view is not told, but Wallerius often had a tendency to select examples that fitted his ideas. Then he attacked Priestley and his use of a eudiometer, a graded glass tube for measuring the changes in gas volume. Wallerius had misunderstood Priestley's experiment, but the important thing is his conclusion: Priestley had

> not needed an oxymeter or eudiometer to measure how much acid there was in the atmosphere, much less of which kind, but needed only to listen to nature speaking with horrifying voice, lightning and thunder and unpleasant sulphur smell [...] He had been able to find, with his nose, mouth and ears, of smell, smoke, and flowers of sulphur, that there is no other acid in our air, but a quite subtle, volatile, elastic and active sulphuric spirit.

Thus, the smell after a lighting (which is actually due to ozone and very different from the smell of sulphur dioxide) was enough for Wallerius to conclude the presence of sulphuric acid in the atmosphere. Wallerius concluded that this acid was the same that Urban Hiärne (Sect. 17.1) had found in mineral waters and that Seip had found in the Pyrmont water [123]. These studies that Wallerius referred to were purely quantitative, and the next sentence is nothing but an insult on Bergman and his thorough studies of mineral waters, including the Pyrmont water (Sect. 17.4): "No one has investigated this water [the water from Bad Pyrmont] more thoroughly and extensively [then Hiärne and Seip]". As there was no carbonic acid (aerial acid) according to Wallerius, the gas (carbon dioxide) evolved when adding sulphuric acid (vitriolic acid) to calcium carbonate (lime) clearly originated from the acid, not from the carbonate.

Next, Wallerius attacked the views of metal salts held by Bergman and Scheele. They both correctly argued that metal was calcined (i.e. oxidised) when dissolved in acid, and not present in its metallic state in solution. According to Wallerius, however, phlogiston did not separate from the metal unless it was precipitated (by adding alkali) to form metal calx. Wallerius made this conclusion since elemental copper had been found in nature, apparently crystallised from aqueous solution. He did not mention that, for example, copper sulphide and copper carbonate (i.e. copper in the oxidised form) are far more common minerals than native copper.

Bergman and Scheele had put much effort in trying to establish the number of elemental earths (Sect. 22.4), and this work was also attacked by Wallerius. Without mentioning Scheele, he mentioned Scheele's work on siliceous earth (SiO_2; Sect. 22.4). Wallerius did not believe in the primitive nature of earths, but was convinced that earths could transmute to each other. As a piece of evidence, he took fossilised sea shells that had turned into flint and pieces of wood turned into agate. Not only could lime turn to flint, the opposite process could also occur. For example, cavities in flintstone could be found to be filled with lime stone.

Once again, the Secretary of the Academy had to come with some objections. After acknowledging that even great scientist could have different opinions, he concluded that by the hard work of chemists many certain and useful facts would be revealed. He also reminded that chemistry had undergone such a development that "the great Boerhaave himself, if he awoke, would hardly recognise it".

9.7 Bergman's European Contacts

When appointed professor, Bergman already had a fairly strong network in Sweden, at least in Stockholm within the Board of Mines and the Royal Swedish Academy of Sciences. He also had contacts with European mineralogists and geologists, but some of his old correspondents within the field of physics, such as Wilson, were now of less importance. One of Bergman's first attempts to make contact with European chemists was with Pierre Macquer in Paris, whom Bergman contacted in January 1768 [124]. Macquer accepted Bergman's invitation and the two started to exchange chemical news. For instance, Bergman mentioned his crystallographic theories (Sect. 25.9) to Macquer in 1769, years before the paper was published [124]. Among Bergman's other chemical contacts were Lorenz Crell (1744–1816), Louis Bernard Guyton de Mourveau (1737–1816), Richard Kirwan (1833–1812), and Joseph Priestley (1733–1804) [125]. His friendship with Kirwan is noteworthy, as Kirwan was very hostile towards Scheele, which will be seen later. Bergman would, however, apparently never have any contact with Sigismund Marggraf (1709–1782) in Berlin [126].

In order to have his works more widely spread, Bergman began publishing his collected works in Latin translation in 1779 as *Opuscula physica et chemica* (*Physical and Chemical Essays*). It is very important to note that the versions that appear in *Opuscula* were usually extended and reworked rather than just being translations or reproductions. It was, for example, the updated *Opuscula* version of Bergman's paper on chemical attractions (Chap. 20) that was translated into English. Bergman's first intention was to publish the work in Germany and negotiated with a book dealer in Göttingen, Johan Christian Dietrich, but without result [127]. Thus, Bergman settled with Magnus Swederus (1748–1836), a book dealer and publisher in Uppsala, who had previously published Bergman's edition of *The Chemical Lectures of H.T. Scheffer* (Sect. 11.1). Swederus was actually an ambitious student who took a chance as there were much complaint with the university book dealer, Kristian Eberhart Steinert [128]. Swederus thus tried to open a competing book store. The university tried to prevent this, but Swederus registered as a burgher and got a licence from the city council in 1771. When Steinert died in 1776, Swederius had no competitor and the univeristy finally accepted Swederus as university book dealer in 1780. In 1789, however, Swederus left the publishing business to become a priest in Västerås.

The first volume of *Opuscula* appeared in 1779, but Bergman was (just like Scheele, see Sect. 21.3) unsatisfied with Swederus and once again tried to find a German publisher for the next volume. None of the publishers he approached was, however, interested in taking up the publication of a work with the second volume. Bergman was thus forced to publish the second (1780) and third (1783) volume on his own expense and had large trouble with the European distribution [127]. Volumes four to six were published posthumously in Leipzig in 1787–1790 [129].

Bergman rapidly rose to an important position in European chemistry, and an indication of this is his membership in European scientific academies. As early as 1764, the same year he was elected a member of the Royal Swedish Academy of Sciences in Stockholm, he was elected a member of Academia Casarea Naturae Curiosorum in Halle [130]. The following year, he was elected a member of the prestigious Royal Society in London. After his appointment as professor, he was elected a member of the American Philosophical Society, presided by Benjamin Franklin, but apparently, the notification letter did not reach Bergman [130]. The membership is, for example, not mentioned in his autobiography [131]. In 1776, Bergman was elected a corresponding member of Académie des Sciences in Paris. The same year he was elected a member of Gesellschaft naturforschender Freunde in Berlin. In 1778, he became a member of Könglishe Gesellschaft der Wissenschaften in Göttingen, in 1779 corresponding member of Société Royal de Médecine in Paris, the letter being signed in Versailles by Amelot de Chaillou, a minister of Louis XVI [132]. In 1781, he became a member of the Academy in Dijon, the hometown of Bergman's correspondent de Morveau (Sect. 26.7), and in 1782 he became the corresponding member of Société Royal des Sciences in Montpellier. In 1782, Bergman was also honoured with an election as one of only eight associés étrangers of Académie des Sciences in Paris [133]. This was regarded as the highest scientific honour in Europe and only one Swede, Linnaeus, had received the honour before Bergman [132]. In 1783, the year before his death, he was appointed one of 20 foreign members of Reale Accademia delle Scienze in Turin by King Victor Amadeus III [132]. In addition, he was a member of a number of Swedish societies: the Royal Society of Arts and Sciences in Gothenburg, The Physiographic Society in Lund (1776) and the Patriotic Society in Stockholm.

Scheele's most important European contacts were German apothecarys Lorenz Crell (1744–1816) and Johan Christian Friedrich Meyer (1739–1811). By the time of his death, Scheele was a member of Gesellschaft Naturforschender Freunde zu Berlin (1778), Regia Scientiarum Taurinensis Academia (Turin, 1784), Societa Italiana (Verona, 1785) and Churmainzische Gesellschaft der Wissenschaften (Ehrfurt, 1785). In August 1785, he had been elected a member of Société Royale de Médecine in Paris, but unfortunately the letter did not arrive in Köping until five days after Scheele's death.

References

1. Boethius B (1958) Swedish iron and steel. Scand Econ Hist Rev 6:144–175
2. Bergman T (1779) Åminnelse-tal öfver...Högvälborne Friherren Herr Carl de Geer, Kongl. Vetenskapakademien, Stockholm, p 21
3. Olsson H (1971) Kemiens historia i Sverige intill år 1800. Almqvist & Wiksel, Stockholm
4. Bergman T (1769) Åminnelse-tal öfver...Herr Georg Brandt. Lars Salvius, Stockholm
5. Fors H (2015) The limits of matter. The University of Chicago Press, Chicago, p 90
6. Brandt G (1735) Dissertatio de semimetallis. Acta Literaria et Scientiarum Sveciae 4:1–10
7. Tilas D (1742) Stenrikets historia utförd i det wid præsidii afläggande håldne talet in för kongl. swenska wetenskaps-academien den 14 Apr 1742. Lorentz Ludwig Grefing, Stockholm
8. Tilas D (1765) Utkast til Sveriges mineral-historia. Framlagdt uti det tal, som vid præsidii afläggande hölts in för kongl. svenska vetenskaps academien den 6 februarii 1765. Lars Salvius, Stockholm
9. Scheffer HT (1752) Det hvita gullet, eller den sjunde metallen...beskrifvit til sin natur. KVA Handl 13:269–275
10. Cronstedt (1751) Rön och försök gjorde med en ny malm-art, från Los kobolt grufvor, i Färila socken och Helsingeland. KVA Handl 12:287–292
11. Klingenstierna S, Wallerius JG (1731) Articuli generales de natura et utilitate methodi scientificæ, quos, ... præside ... D:no Samuele Klingenstierna, ... examini publico pro gradu

sistit Joh. Gottzk. Wallerius Er. fil. Nericius, ad diem [3] Aprilis anni 1731. In Aud. Gustav. majori H A M S, Uppsala
12. Rosén N, Wallerius JG (1731) Tentamen anthropologiæ experimentalis, quo demonstratur existentia vasorum absorbentium in intestinis, partem chyli ad venas mesentericas immediate deferentium, quod, consensu ampliss. senat. med. in reg. academ. Upsaliensi praeside [...] Nicolao Rosén [...] examini publico sistit Johannes Gotzkalk Wallerius [...] ad diem 18. Decembris anni 1731. In audit. gust. majori. H A M S, Uppsala
13. Bergman T (1765) Åminnelse-tal öfver framledne theol. professoren i Upsala ... Nils Wallerius, på kongl. academiens befallning hållet uti stora riddarhus-salen den 2. februarii 1765, af Torbern Bergman ... Stockholm, tryckt hos direct. Lars Salvius, Stockholm
14. Annerstedt C (1913) Uppsala universitets historia, vol 3:1. Almqvist & Wiksell, Uppsala, p 108
15. Wallerius JG, Darelius A (1741) Decades binae thesium medicarum, quas, jussu max. ven. senatus acad. et suffragio amplissimae facultat. medicae in reg. Upsal. athenaeo, publico bonorum examini modeste subjiciunt auctor Johan. Gotschalk Wallerius ... et respondens Johan. A. Darelius, W-Gothus. In audit. Carol. maj. ad diem 25. febr. anni MDCCXLI., Uppsala
16. Annerstedt C (1913) Uppsala universitets historia, vol 3:1. Almqvist & Wiksell, Uppsala, p 111
17. Salberg JJ (1739), Beskrifning på et Sal Natron, funnit i Swerike. KVA Handl 1:245–248
18. Neander G (1931) Johan Anders af Darelli. Svenskt biografiskt Lexikon 10:295
19. Annerstedt C (1913) Uppsala universitets historia, vol 3:1. Almqvist & Wiksell, Uppsala, p 265ff
20. Fors H (2003) Mutual favours (diss). Depatment of history of science and ideas. Uppsala University, Uppsala, p 45
21. Fors H (2003) Mutual favours (diss). Depatment of history of science and ideas. Uppsala University, Uppsala, p 91
22. Olsson H (1971) Kemiens historia i Sverige intill år 1800. Stockholm, Almqvist & Wiksel, p 110
23. Trofast J (1994) Johan Gottlieb gahn brev, vol 2, Lund, p 54
24. Principe LM (2011) The scientific revolution. Oxford University Press, Oxford, p 37
25. Wallerius JG (1783) Tal, om nödig jämförelse emellan de chemiska undersökningar, och naturens verkningar, hållet i Kongl. Vetensk. Academien vid præsidii nedläggande, den 9 julii 1783, Stockholm, p 4
26. Wallerius JG (1953) självbiografi, Lychnos, pp 235–259
27. Wallerius JG (1783) Tal, om nödig jämförelse emellan de chemiska undersökningar, och naturens verkningar, hållet i Kongl. Vetensk. Academien vid præsidii nedläggande, den 9 julii 1783, Stockholm, p 3
28. Wallerius JG (1750) Inträdes-tal om salternas ursprung och anledning, at utleta orsaken til kallbräckt järn. Hållit i kongl. vetenskaps academien af ... Johan Gottschalk Wallerius, då han första gången der intog sit säte den 2 junii 1750, Stockholm, p 4
29. Wallerius JG (1750) Inträdes-tal om salternas ursprung och anledning, at utleta orsaken til kallbräckt järn. Hållit i kongl. vetenskaps academien af ... Johan Gottschalk Wallerius, då han första gången der intog sit säte den 2 junii 1750. Stockholm, p 8
30. Lennartson A (2017) The chemical works of Carl Wilhelm Scheele. Springer, Cham, p 87
31. Wallerius JG (1750) Inträdes-tal om salternas ursprung och anledning, at utleta orsaken til kallbräckt järn. Hållit i kongl. vetenskaps academien af ... Johan Gottschalk Wallerius, då han första gången der intog sit säte den 2 junii 1750. Stockholm
32. Annerstedt C (1913) Uppsala universitets historia, vol 3:1. Almqvist & Wiksell, Uppsala, p 413
33. Fors H (2008) Matematiker mot Linneaner. In: Widmalm S (ed) Vetenskapens sociala strukturer. Nordic Academic Press, Lund, p 34
34. Bergman T (1767) Förslag, at förbättra alun-luttringen. KVA Handl 28:73–80

35. Fors H (2003) Mutual favours (diss). Depatment of history of science and ideas, Uppsala University, Uppsala, p 126
36. Bergman T, Suedelius G (1767) Disquisitio chemica de confectione aluminis, quam, consensu amplis. ordinis philosophici in regia academia Upsaliensi publico examini submittunt … Torbernus Bergman … et … stipendiarius regius Gustavus Suedelius, Westmannus, in auditorio Carolino majori, die I Aprilis, anni MDCCLXII. Horis ante meridiem solitis. Uppsala
37. Fors H (2003) Mutual favours (diss). Depatment of history of science and ideas. Uppsala University, Uppsala, p 130
38. Fors H (2003) Mutual favours (diss). Depatment of history of science and ideas. Uppsala University, Uppsala, p 125
39. von Engström G (1774) Anmärkningar vid alun tilverkningen. KVA Handl 35:273–297
40. Bergman T (1776) Ytterligare anmärkningar om aluntilverkningen. KVA Handl 37:177–189
41. Celsius O & Tidström A (1748) Dissertationis academicæ, de Mariæstadio, Vestrogothiæ urbe, cum vicina Mariæholmia dynastica sede provinciæ Skaraborgensis pars prior. Quam, consensu ampl. facult. philos. acad. Ups. præside … Olavo Celsio … publicæ bonorum disquisitioni modeste submittit Andreas Ph. Tidstroem Mariæstadio-V-Gothus, in aud. Carol. min. die XXIII Decembr. an. MDCCXLVIII, horis ante meridiem solitis, Uppsala
42. Celsius O & Tidström A (1752) Dissertationis academicæ, de Mariæstadio Vestrogothiæ urbe, partis prioris continuatio, quam consensu ampl. facult. philos. acad. Ups. præside … Olavo O. Celsio … pro gradv publico examini committit Andreas Ph. Tidstroem Mariæstadio-V-Gothus, in audit. Carol. maj. d. 28 Apr an. MDCCLII, Uppsala
43. Silvius S (1752) Då Professoren vid Kongl. Academien i Upsala, Högädle och Vidtberömde Herr Mag. Johan Ihre promoverade til Philosophiæ Magistrar följande Herrar af Vestgötha Nation Mag. Anders Boström, Mag. Nils Bredberg, Mag. Anders Tidström, Mag. Nils Themptander, Mag. Petter Tegnæus, Mag. Jonas Victorin, Mag. Petter Tenggren. d. 15 June 1752, Uppsala
44. Tidström A, Herlenius D (1765) Dissertatio chemica acidum vegetabile sistens, cujus part. I. Uppsala
45. Trofast J (1994) Johan Gottlieb Gahn brev, vol 2, Lund, p 75
46. Söderberg T (1975–1977) Mathias Kewenter, in Svenskt Biografiskt lexikon, Stockholm, p 74
47. Lagerbring S & Kewenter M (1752) Dissertatio historica de commerciis longinquis, cujus partem posteriorem, divina annuente gratia, suffragante amplis. ord. philos. in alma Goth. Carolina, moderante d:no Sven Bring, hist. prof. reg. et ord. publicæ eruditorum censuræ modeste subjicit Matthias Keventer, Blekingus ad diem XIX Decembr. anni MDCCLII, Uppsala
48. Kewenter M & Ehrengren S (1762) Chemisk-metallurgisk avhandling om koppar-gärningen uti Avesta. Med ampliss. facult. philos. samtycke vid kongl. academien i Upsala til försvarande utgifven av auctor och præses Matthias Keventer … och Samuel Ehrengren, gestrike, uti gustavianska lärosalen f. m. den 22 decemb. 1762, Uppsala
49. Wallerius JG, Hiorzberg L (1751) Dissertatio chemica, de nexu chemiæ cum utilitate reipublicae, cujus partem primam, Uppsala
50. Hiotzberg, L Elfsten G (1752) Dissertatio chemica, de nexu chemiæ cum utilitate reipublicæ, cujus partem secundam, Uppsala
51. Linnaeus C, Hiortzberg L (1754) Dissertatio chemico medica de methodo investigandi vires medicamentorum chemica, Uppsala
52. Hiortzberg, L, Holmén E (1756) Dissertatio chemica, fundamentum Halurgiæ systematicæ sistens, cujus partem priorem, Uppsala
53. Hiortzberg L, Boling A (1757) Dissertatio physica de caussa maxime probabili attractionis corporum, Uppsala
54. Anonymous (1971–1973) Svenskt Biografiskt Lexicon, vol 19, Stockholm, p 132
55. Trofast J (1994) Johan Gottlieb Gahn: brev, vol 2, Lund, p 32

56. Lindroth S (1978) Svensk lärdomshistoria. Frihetstiden. Norstedts, Stockholm p 19
57. Fors H (2008) Matematiker mot Linneaner. In: Widmalm S (ed) Vetenskapens sociala strukturer. Nordic Academic Press, Lund, p 37
58. Fors H (2008) Matematiker mot Linneaner. In: Widmalm S (ed) Vetenskapens sociala strukturer. Nordic Academic Press, Lund, pp 25–53
59. Wallerius JG (1776) Tankar om verldenes, i synnerhet jordenes, danande och ändring, Stockholm, p 1
60. Annerstedt C (1913) Uppsala universitets historia, vol 3:1. Almqvist & Wiksell, Uppsala, p 414
61. Olsson H (1971) Kemiens historia i Sverige intill år 1800. Stockholm, Almqvist & Wiksel, p 122
62. Schück H (1916) Torbern Bergmans självbiografi, äldre svenska biografier 3–4, Uppsala, p 89
63. Fors H (2003) Mutual favours (diss). Depatment of history of science and ideas. Uppsala University, Uppsala, p 65
64. Fors H (2003) Mutual favours (diss). Depatment of history of science and ideas. Uppsala University, Uppsala, p 64
65. Landen L (2004) Gustaf III—en biografi. Stockholm, Wahlström & Widstrand, p 222
66. Franzén FM (1829) Minne af Chemisten Carl Wilhelm Scheele. Svenska Akademiens Handlingar 12:533–544
67. Annerstedt C (1913) Uppsala universitets historia, vol 3:1. Almqvist & Wiksell, Uppsala, p 463
68. Annerstedt C (1914) Uppsala universitets historia, vol 3:2. Almqvist & Wiksell, Uppsala, p 447
69. Fors H (2003) Mutual favours (diss). Depatment of history of science and ideas, Uppsala University, Uppsala, p 48
70. Hjelm PJ (1786) Åminnelse-tal öfver...Herr Torbern Bergman, Stockholm, p 40
71. Bergman T, Paulin J (1779) Dissertatio gradualis de primordiis chemiæ, Uppsala
72. Bergman T (1791) Physical and chemical essays, vol III. G, Mudie, J & J Fairbairn and J Evans, Edinburgh, p 39
73. Bergman T (1791) Physical and chemical essays, vol III. G, Mudie, J & J Fairbairn and J Evans, Edinburgh, p 3
74. Bergman T Afzelius P (1782) Dissertatio gradualis sistens chemiæ progressus a medio sæc. VII ad medium sæc. XVII, Uppsala
75. Fors H (2003) Mutual Favours (diss.). Depatment of History of Science and Ideas, Uppsala University, Uppsala, p 110
76. Schück H (1916) Torbern Bergmans självbiografi, äldre svenska biografier 3–4, Uppsala, p 90
77. Annerstedt C (1913) Uppsala universitets historia, vol 3:1. Almqvist & Wiksell, Uppsala, p 422
78. Hjelm PJ (1786) Åminnelse-tal öfver...Herr Torbern Bergman, Stockholm, p 41
79. Schück H (1916) Torbern Bergmans självbiografi, äldre svenska biografier 3–4, Uppsala, p 102
80. Schück H (1916) Torbern Bergmans självbiografi, äldre svenska biografier 3–4, Uppsala, p 91
81. Schück H (1916) Torbern Bergmans självbiografi, äldre svenska biografier 3–4, Uppsala, p 92
82. Hjelm PJ (1786) Åminnelse-tal öfver...Herr Torbern Bergman, Stockholm, p 45
83. Aurillius PF (1785) Åminnelse-tal, öfver chemiæ professoren i Upsala...Herr magister Torbern Bergman...Uppsala, p 23
84. Fors H (2003) Mutual favours (diss). Depatment of history of science and ideas. Uppsala University, Uppsala, p 95
85. Wallerius JG (1953) Självbiografi, Lychnos, 235–259

86. Thomson T (1831) The history of chemistry, vol II. Henry Colburn & Richard Bentley, London, p 34
87. Söderbaum HG (1913) Lars Fredrik Svanberg. Almqvist & Wiksell, Minnesteckning, Stockholm & Uppsala, p 77
88. Scheele CW (1777) Chemische Abhandlung von der Luft und dem Feuer, M Swederus & S L Crusius, Uppsala & Leipzig, p 110
89. Bergman T, Dubb P (1770) Chemisk undersökning om källe-vatnen uti och närmast kring Upsala, Uppsala
90. Bergman T (1770) Historien om qvicksilvers föreningar med koksalts-syra. KVA Handl 31:79–102
91. Bergman T (1771) Anledningar, at tilverka varaktigt tegel. KVA Handl 32:211–220
92. Bergman T (1779) Åminnelse-tal öfver...Högvälborne Friherren Herr Carl de Geer, Kongl. Vetenskapakademien, Stockholm, p 28
93. Fors H (2003) Mutual favours (diss). Depatment of history of science and ideas. Uppsala University, Uppsala, p 78
94. Bergman T (1791) Physical and chemical essays, vol III. G, Mudie, J & J Fairbairn and J. Evans, Edinburgh, p 6
95. Fors H (2003) Mutual favours (diss). Depatment of history of science and ideas. Uppsala University, Uppsala, p 80
96. [Wallerius JG] (1769) Tilfällige Tanckar om de Kieselartade, eller så kallade Glasartige Stenars generation, Inrikes Tidningar, No 75
97. Bergman T (1768) Anmärkningar om Vestgötha bergen KVA Handl 29:324–336
98. [Wallerius JG] (1770) Nödige, men wälmente påminnelser för unga chemister, Lärda Tidningar, Nr 18, pp 68–72
99. Bergman T (1773) Om Luftsyra KVA Handl 34:170–186
100. Trofast J (1994) Johan Gottlieb Gahn: brev, vol 2, Lund, p 18
101. Fors H (2003) Mutual favours (diss). Depatment of history of science and ideas. Uppsala University, Uppsala, p 74
102. [Bergman T] (1770) Swar, på de så kallade nödige och wälmente påminnelser, som finnas i N. 18 af Lärda Tidningarne, för innew. År, Allmänna tidningar, No 39 (Mar 21), p 156
103. [Bergman T] (1770) Swar, på de så kallade nödige och wälmente påminnelser, som finnas i N 18 af Lärda Tidningarne, för innew. År, Allmänna tidningar, No 40 (Mar 24), pp 159–160
104. [Bergman T] (1770) Swar, på de så kallade nödige och wälmente påminnelser, som finnas i N 18 af Lärda Tidningarne, för innew. År, Allmänna tidningar, No 41 (Mar 26), pp 162–163
105. [Bergman T] (1770) Swar, på de så kallade nödige och wälmente påminnelser, som finnas i N 18 af Lärda Tidningarne, för innew. År, Allmänna tidningar, No 42 (Mar 27), pp 166–168
106. Bergman T (1784) Physical and chemical essays, vol 1. J Murray, London, p XXX
107. Fors H (2003) Mutual favours (diss). Depatment of history of science and ideas. Uppsala University, Uppsala, p 88
108. Trofast J (1994) Johan Gottlieb Gahn: brev, vol 2, Lund, p 35
109. von Engeström (1781) Laboratorium Chymicum Första delen om gull och silver fineraren. P A Brodin, Stockholm, first page of the preface
110. Fors H (2003) Mutual favours (diss.). Depatment of history of science and ideas. Uppsala University, Uppsala, p 96
111. Fors H (2003) Mutual favours (diss.) Depatment of history of science and ideas. Uppsala University, Uppsala, p 97
112. Bergman T (1775) H T Scheffers Chemiske föreläsningar, rörande salter, jordarter, vatten, fetmor, metaller och färgning, samlade, i ordning stälde och med anmärkningar utgifne af T B, M Swederus, Uppsla, p I
113. Trofast J (1994) Johan Gottlieb Gahn: brev, vol 2, Lund, p 36
114. Fors H (2003) Mutual favours (diss). Depatment of history of science and ideas. Uppsala University, Uppsala, p 98

115. Bergman T (1775) H T Scheffers Chemiske föreläsningar, rörande salter, jordarter, vatten, fetmor, metaller och färgning, samlade, i ordning stälde och med anmärkningar utgifne af T B, M Swederus, Uppsla, p 448
116. Oseen CW (1940) Torbern Bergman och Carl Wilhelm Scheele. KVA, Stockholm, p 15
117. von Engeström G (1774) Tal om mineralogiens hinder och framsteg i senare åren; hållet för kongl. vetenskaps academien, vid præsidii nedläggande, den 4 maji 1774, Stockholm
118. von Engeström G (1774) Tal om mineralogiens hinder och framsteg i senare åren; hållet för kongl. vetenskaps academien, vid præsidii nedläggande, den 4 maji 1774. Stockholm, p 9
119. von Engeström G (1782) Tal, om vissa svårigheter och andra omständigheter, som möta vid utöfvandet af chymien; hållet för kongl. vetensk. academien, vid præsidii nedläggande, den 6 Nov 1782, Stockholm
120. von Engeström G (1774) Tal om mineralogiens hinder och framsteg i senare åren; hållet för kongl. vetenskaps academien, vid præsidii nedläggande, den 4 maji 1774, Stockholm, p 5
121. Wallerius JG (1783) Tal, om nödig jämförelse emellan de chemiska undersökningar, och naturens verkningar, hållet i kongl. vetensk. academien vid præsidii nedläggande, den 9 julii 1783, Stockholm
122. Michaelis JF, Triller, DW (1767) Dissertatio medica inauguralis de fallacia examinis chemici in exploranda intima thermarum natura, Wittenberg
123. Seip JP (1738) Relatio de Caverna vaporifera Sulphurea in Lapicidina Pyrmontana, quæsimilis est foveæ neapolitanæ grotta del cane dictæ, à dno misson, & aliis descriptæ, Regali Societati communicata. Phil Trans 40:266–277
124. Carelid G, Nordström J (1965) Torbern Bergman's foreign correspondence. Almqvist & Wiksell, Uppsala, p XXXIV
125. Carelid G, Nordström J (1965) Torbern Bergman's foreign correspondence. Almqvist & Wiksell, Uppsala
126. Carelid G, Nordström J (1965) Torbern Bergman's foreign correspondence. Almqvist & Wiksell, Uppsala, p XL
127. Carelid G, Nordström J (1965) Torbern Bergman's foreign correspondence. Almqvist & Wiksell, Uppsala, p LII
128. Annerstedt C (1914) Uppsala universitets historia, vol 3:2. Almqvist & Wiksell, Uppsala, p 532
129. Moström B (1957) Torbern Bergman a bibliography of his works. Almqvist & Wiksell, Stockholm, p 48
130. Carelid G, Nordström J (1965) Torbern Bergman's foreign correspondence. Almqvist & Wiksell, Uppsala, p LIII
131. Schück H (1916) Torbern Bergmans självbiografi, äldre svenska biografier 3–4, Uppsala, p 94
132. Carelid G, Nordström J (1965) Torbern Bergman's foreign correspondence. Almqvist & Wiksell, Uppsala, p LIV
133. Tiselius A (1958) Torbern Olof Bergman. Levnadsteckningar över K. Svenska Vetenskapsakademiens ledamöter 9:141

10 Scheele Moves to Stockholm

In 1768, after 2 years in Malmö, Scheele, now aged 25, moved to Stockholm and the Raven pharmacy (Apoteket Korpen; Fig. 10.1) [1]. The pharmacy had been founded in 1674 by Georg Brandt, who sold it in 1692 to buy Hed iron works outside Skinskatteberg together with his cousin. His son, also called Georg, became a chemist (Chap. 9) and is best remembered for the discovery of cobalt. The pharmacy was initially located at Södermalm and called The Eagle (Apoteket Örnen), but after a few years, it was moved to Stortorget in Old Town. In 1721, the pharmacy was bought by the German apothecary Ludwig Hindrich Dowe, who renamed it The Raven (Apoteket Korpen) and later changed the name to The Gilded Raven (Förgyllda Korpen). He also moved the pharmacy to an adjacent building, where it remained until 1924. Since then, it has been located at nearby Väster Långgatan. A large brass mortar from 1694 can be viewed in the store and was probably used by Scheele.

During Scheele's time at the Raven, it was run by apothecary Johan Scharenberg (1714–1778). Scharenberg was born in Stockholm, but the family originated from Germany and had emigrated from Lübeck in the mid-seventeenth century. Scharenberg studied in Stockholm and in Germany before he took over the pharmacy in 1743. Not only did he run his pharmacy and grew medicinal plants but it is also known that he invested 9,000 riksdaler specie in the Swedish East Indian Company [2]. He also received warnings after involving himself with charlatans and for illegally selling opium. There is no indication that Scharenberg had any interest in chemistry, but his son studied under Bergman (Sect. 11.4).

Scheele made the journey to Stockholm by boat sometime between April 26 and April 30, 1769. In the custom's register, it has been found that shipper Jöns Börgesson sailed from Malmö to Stockholm on April 26, carrying, *for example,* apothecary equipment [3]. This may very well be Scheele's belongings. Stockholm had, at the time, 68,700 citizens, and Sweden's population had just reached two million.

A. Lennartson, *Carl Wilhelm Scheele and Torbern Bergman*, Perspectives on the History of Chemistry, https://doi.org/10.1007/978-3-030-49194-9_10

Fig. 10.1 The building where Scheele worked and lived in Stockholm has changed little since Scheele's days. It is the only of Scheele's former working places that have survived *Photo* Anders Lennartson, June 2016

Scheele's friend Retzius moved to Stockholm the same year as Scheele, and although Scheele got company, it means that there are no letters that can inform us about Scheele's life and research in Stockholm. It was probably Retzius who introduced Scheele to the learned circles in Stockholm and it is quite remarkable that the young apothecary apprentice made several important and influential friends during his short time in Stockholm. One of them was archiater Abraham Bäck, the president (*preses*) in *Collegium medicum*, the medical authority of Sweden and one of the main figures in eighteenth-century Swedish medicine. Bäck [4] was born in Söderhamn in 1713. After the Russian raids of the Swedish coast in 1720, Bäck grew up with his uncle. His initial intention was to become a priest, but after witnessing an anatomical demonstration, he turned to medicine, becoming one of Rosén's students. He defended his inaugural thesis on tuberculosis in 1740, after which he undertook a European journey 1741–1745, visiting the Netherlands, England, France and Germany. With recommendation letters from Linnaeus, all doors were open. After returning to Sweden, he unsuccessfully applied for several professorships but became auscultator at *Collegium medicum*, court physician to Adolph Frederick[1] in 1749 and received a Professor's title. In 1752, he became the President of *Collegium medicum*, a position he held for over 40 years. He died in 1795.

Scheele also became acquainted with botanist and physician Peter Johan Bergius, [5] who was born in the province of Småland in 1730. An orphan at a young age, he still had the opportunity to study at Lund University (from 1746) and Uppsala University (from 1749). He studied under Linnaeus and Rosén, obtaining his M.D. degree in 1754. Bergius became one of the most popular physicians in Stockholm, being appointed professor of natural history and pharmacy in 1761 and becoming auscultator at *Collegium medicum* in 1766. He published a large number of papers in the *Transactions of the Royal Swedish Academy of Sciences*, being a member since 1758. He died in 1790. His older brother, Bengt Bergius (1723–1784), was an accomplished historian, also a member of the Royal Swedish Academy of Sciences. Among Scheele's other friends from Stockholm was physician David von Schulzenheim, born in the province of Dalarna in 1732 [6]. Before his ennoblement in 1770, he was called Schultz, and was an uncle of Scheele's friends Johan Gottlieb Gahn (Sect. 13.1) and Henric Gahn. David von Schulzenheim began his studies at the University of Königsberg in 1747, but moved to Uppsala in 1751, where he studied under Rosén von Rosenstein and Linnaeus. He obtained his doctoral degree in 1754 and undertook a European journey, during which he studied the new technique of inoculation against smallpox in England, a knowledge he brought to Sweden. He returned to Sweden in 1756, becoming auscultator in *Collegium medicum* in 1765. In 1778, he unexpectedly retired from all his positions and became a farmer. His growing dissatisfaction with the politics of Gustav III's Sweden was most likely the reason. He died in 1823.

Scheele's new friends were all connected to the Royal Swedish Academy of Sciences and *Collegium medicum*, while there is no evidence of any connections between Scheele and the chemists at the Board of Mines, *for example,* von Engeström, during his years in Stockholm [7].

[1]The heir of the Swedish throne.

10.1 Scheele's Research in Stockholm

It has been said that the time in Stockholm was a disappointment for Scheele since he had to work in the shop, rather than the laboratory [8]. It is quite clear, however, that Scheele carried out important research in Stockholm, but perhaps he had less time for experiments in Stockholm than in Malmö. It is also quite probable that Scharenberg was less forgiving and supporting compared to Kjellström. According to tradition, it was in the window of the Raven Pharmacy that Scheele studied the action of solar light on silver chloride, although the results were not published until 1777 [9]. This was the first photochemical experiment involving a pure compound, and the first step towards photography (Fig. 10.2).

Scheele seems to have collaborated with Retzius, who many years later wrote to Scheele's biographer Wilcke that "none of us made any experiments without communicating them with each other" [10]. It was in Stockholm that Scheele discovered the acidic properties of carbon dioxide (Chap. 18) and according to Oseen, it may have been in Stockholm rather than in Uppsala that Scheele discovered oxygen (Sect. 21.1).

The isolation of tartaric acid became the first of Scheele's discoveries to be published, but it is impossible to determine to what extent Retzius contributed to the discovery. It is believed that Scheele, probably already in Malmö, had found a way to isolate tartaric acid from potassium hydrogen tartrate (cream of tartar). Retzius was now granted 300 copper dalers for experiments on cream of tartar [11] and the results of his experiments were published in the July to September 1770 issue of the *Transactions, i.e.* after Scheele's move to Uppsala. This paper, *On cream of tartar and its acid* [12] describes the isolation of tartaric acid and is, probably to a large extent, based on experiments by Scheele. After Scheele's death, Retzius wrote to Wilcke

> "The discovery of tartaric acid is the fruit of our common labour, although I, by agreement, undertook the experiments with it, while Scheele performed a succession of experiments that formed the base of his beautiful book on air and fire".

Although the paper on tartaric acid is written by Retzius[2], he did not take sole credit for the discovery and it is in this paper that Scheele's name first appears in print: "These experiments I related to Mr Carl Wilhelm Scheele (a fast and curious pharmaciæ studiosus)". Potassium hydrogen tartrate is only sparingly soluble in water (0.37 g per 100 ml at 20 °C) and is deposited in wine barrels on storage. Scheele found that the addition of calcium carbonate gave calcium tartrate, with even lower solubility. Treating calcium tartrate with sulphuric acid gave calcium sulphate (gypsum), which could be filtered off leaving tartaric acid in solution.

[2]The eighteenth-century papers in the *Transactions* typically had a single author.

Fig. 10.2 Silver chloride exposed to sunlight. The originally white powder darkens by metallic silver particles *Photo* Anders Lennartson

While in Stockholm, Scheele made a new attempt to publish a paper in the *Transactions of Royal Swedish Academy of Sciences*. A paper called *"Chemical findings on Sal acetocellae etc."* was read in the Academy on August 17, 1768, but was rejected on November 9 on the recommendation by Bergman, who found no novelty in the paper [13]. The manuscript is not preserved, and its contents are not known. *Sal acetocellae* is potassium hydrogen oxalate, but the paper cannot have described the isolation of oxalic acid, since Scheele wrote in a letter to Hjelm dated November 1780 that he had yet not prepared that acid. It has been indicated by Gahn, Bäck and Ehrhart that Scheele made a third unsuccessful attempt to send a manuscript to the *Transactions of the Royal Swedish Academy of Sciences*, and this paper is supposed to have reported the isolation of tartaric acid. If so, the paper on tartaric acid by Retzius, who had a far stronger position than Scheele, may have been a way to help Scheele to establish himself as a chemist, rather than a way to take credit of Scheele's research, which has occasionally been suggested.

References

1. Nordholm U, Härdelius M (1974) Apoteket Korpen Gamla stan 300 år, Stockholm
2. Lindeke B, Yue Q-Y (2000) Ett kinesiskt apotek in: Ostindiska kompaniet – Affärer och föremål, Gothenburg, p 217
3. Ersgård H (1982) Apoteket Fläkta Örn i Malmö tiden 1731–1820. Sv Farm Tidskr 86:28
4. Boëthius B (1927), Abraham Bäck, Svenskt biografiskt lexikon vol 7, Stockholm, p 71
5. Fries RE, Hult OT (1922), Peter Jonas Bergius, Svenskt biografiskt lexikon vol 3, Stockholm, p 565
6. Åman A (2000–2002) David von Schultzenheim, Svenskt biografiskt lexikon 31:665
7. Fors H (2003) Mutual favours (diss). Depatment of history of science and ideas. Uppsala University, Uppsala, p 174
8. Sjöstén CG (1801) Åminnelse-tal, hållit för Kongl. Vetenskaps Academien öfver dess framledne ledamot herr Carl Vilhelm Scheele, den 14 Oct 1799, Stockholm, p 12

9. Scheele CW (1777) Chemische Abhandlung von der Luft und dem Feuer. M Swederus & SL Crusius, Uppsala & Leipzig, p III
10. Boklund U (1961) Carl Wilhelm Scheele bruna boken, Stockholm, p 400
11. Lindroth S (1967) Kungl. Vetenskapsakademiens historia 1739–1818, vol 1:1. KVA, Stockholm, p 554
12. Retzius A (1770) Försök med vinsten och dess syra. KVAH 31:207–226
13. Fredga A (1943) Carl Wilhelm Scheele. Stockholm, KVA, p 19

11 Bergman as a Teacher

A large part of Bergman's time was taken up by teaching (Fig. 11.1), and it appears that he was a devoted and appreciated teacher. In early 1769, he was teaching assaying, the art of analysing minerals, to a class of 15 students [1]. The classes took place at 2 p.m. on Wednesdays and Saturdays, but were usually not finished until 11 or 12 p.m. This was a test for Bergman, whose health was not strong:

> I had to be constantly present, since so many youngsters left by themselves, would soon have taken up other amusements than those intended. On top of this, half of the laboratory had stone floor, since several youngsters should learn how to use the power of fire, and the house would otherwise soon have been on fire: we know that caution is not the character of that [young] age. In addition, there had to be a strong draught in the room in order to maintain strong and brisk fire. On top of this an extreme cold set in February and March. All these circumstances meant that after each lesson, I was so chilled that after getting to bed I could not get warm and fall to sleep until 4 a.m. in the morning.

As there is a continuous series of temperature measurements from Uppsala dating back to 1722, Bergman's claims are easily verified. In late January to early March 1769, there were several cold days. On Wednesday February 1, the temperature was −18 °C, and the following Saturday the temperature was −11 °C. Next Wednesday, the temperature had risen to +1 °C, but was back down at −15 °C the following Saturday.

11.1 Bergman's Edition of the Chemical Lectures of H.T. Scheffer

As already mentioned (Sect. 9.6), Bergman needed a textbook for his lectures, and he was not inclined to use the material that Wallerius had written. In 1751, Wallerius published a 32-page long introduction to chemistry called *Letter, on the true nature, benefit and value of chemistry*, [2] with a second extended edition in 1767. Wallerius also wrote an extensive textbook in Swedish, *Chemia Physica*,

A. Lennartson, *Carl Wilhelm Scheele and Torbern Bergman*, Perspectives on the History of Chemistry, https://doi.org/10.1007/978-3-030-49194-9_11

which was planned to consists of three volumes. The first volume appeared in 1759 [3] the second volume appeared in two parts in 1765 and 1768. The third volume was never finished [4]. Wallerius' textbook was translated to Latin and German.

As described in Sect. 9.6, Bergman got hold of a copy of the lecture notes from Henrik Theofil Scheffer's chemical lectures of 1749–1751 from Patrick Alströmer. Bergman found that Scheffer had discussed the same topics that he himself usually included in his lectures, and came to the conclusion that an abbreviated form of Scheffer's lectures supplemented with notes to bring the text up to date would be a suitable textbook for his students [5]. In addition to the manuscript written by Alströmer, Bergman had access to additional lecture notes written by other students [6]. The book (Fig. 11.2) was published by Magnus Swederus in 1775; the preface was signed by Bergman in November 1774, and by August 1775, the printing was nearly finished, except for the copper plate with the attraction table (Chap. 20) [7]. A German translation by Professor Weigel in Greifswald was published in 1779. Schufle published an English translation in 1992.[1] Bergman organised Scheffer's text in sections, chapters and paragraphs, with his own notes and comments printed in a smaller font and thus easily recognised. The extensive preface gives a first account of Bergman's theory of affinities (Chap. 20) and explains the two copper plates: an attraction table and Bergman's revised set of chemical symbols. Bergman also gave some information on the personality of Scheffer [8]. From von Swab, Bergman knew that accuracy was very important to Scheffer (Chap. 9), so Bergman probably felt quite confident that the experiments presented in the lectures were performed thoroughly.

Bergman published a revised edition of *The Chemical Lectures of H.T. Scheffer* in 1779 with a slightly altered title (Fig. 11.3). This edition has a 63-page appendix (Sect. 11.2), *Instructions to the lectures concerning the nature and benefits of chemistry and the general differences between natural bodies* [9]. This appendix was also published separately. A third edition (but presented as the second edition) was published posthumously by Bergman's student Hjelm in 1796. In the preface, Hjelm wrote that it had not been possible to take the antiphlogistic theory into account without rewriting the whole text and that there was still not enough proof against phlogiston to warrant such a rewriting [10].

It is clear that Bergman spent a lot of work editing Scheffer's lectures, as his own remarks constitute about half of the text. Many of Bergman's remarks are based on literature citations, Macquer and Marggraf being two authors that Bergman relied heavily on. Other comments are based on his own experience, but it is quite likely that some of this experience actually originated from Scheele. There are several examples were Bergman refers to unpublished experiments by Scheele, probably carried out on Bergman's request. In one instance, for example, Bergman wrote: "Mr Scheele has, on my request, repeated these experiments [...] and consequently the conclusion is completely different" [11]. When discussing the invisible text obtained by writing on paper with an aqueous solution of cobalt (II) chloride which becomes visible in blue on heating, Scheele had made an important experiment [12]. The text also became visible when the paper was placed in very dry air in a

[1] I have not examined this edition, and cannot asses its quality.

Medicinae Studiosum, Pereximium atq Praestantissimum Dominum Johannem Adolphum Dahlgren, Ostrogothum, in Chemia medica rite examinatum et approbatum testor.
Upsaliae d. 14 Martii anni 1771.

T. Bergman
Chemiae Prof. Reg. et Ord.

Fig. 11.1 A certificate that student Johan Adolf Dahlgren has passed the course in medical chemistry in 1771. *Photo* Uppsala University Library

closed jar over concentrated sulphuric acid and the blue colour was therefore not dependent on heat. We now know that the blue colour is due to the reversible dehydration of pink $CoCl_2{\cdot}6H_2O$ to blue $CoCl_2$. When discussing the solubility of metallic copper in ammonia, it was Scheele who had found that this was only possible after loss of phlogiston (i.e. by oxidation) [13]. In several cases, the book introduces concepts that Bergman was developing and that would be more formally announced in latter publications, such as his work on quantitative analysis (Sect. 23.2), chemical affinity (Chap. 20) and nomenclature (Chap. 26).

Bergman divided the original text into six main parts, each part further divided into chapters, thus using the same format he had used in his *Physical Description of the Earth*, the second edition of which appeared the year before *The Chemical Lectures of H.T. Scheffer*.The first part deals with salts. Salts were water-soluble substances with taste, and Bergman added that salts generally need less than 200 times their weight of water to dissolve (he later changed this definition to 500 times [14]). Scheffer divided salts into two classes, but Bergman's system of classification of salts was more elaborate. His scheme for classification is visually similar to the schemes in Linnaeus' *Systema Naturae*, which probably served as Bergman's inspiration. Bergman's definition of salts was much wider than the modern definition: it would not only include substances that we refer to as salts, but also included acids, bases and even sugar. The two main classes of salts were called "true salts" (egentlige salter) and "analogical salts" (analogiske salter). The true salts were divided into three subclasses: acids, alkalis and neutral salts. The neutral salts were composed of an acid and an alkali, and correspond to the sodium,

Framledne Direct: och Kongl. Vet. Acad. Ledamots
HERR H. T. SCHEFFERS
CHEMISKE
FÖRELÄSNINGAR,
Rörande SALTER, JORDARTER, VATTEN,
FETMOR, METALLER och FÄRGNING,
Samlade, i ordning ſtälde och
med Anmärkningar utgifne.

UPSALA,
Tryckte på Bokhandlaren M. Swederi
bekoſtnad hos Joh. Edman, Kongl. Acad.
Boktr. 1775.
)(ſe-

Fig. 11.2 Title page of the first edition of *The Chemical Lectures of H.T. Scheffer*. The laboratory scene was later reused when Swederus published Scheele's book on air and fire, *Chemische Abhandlung von der Luft und dem Feuer*, see Fig. 21.4. *Photo* Anders Lennartson

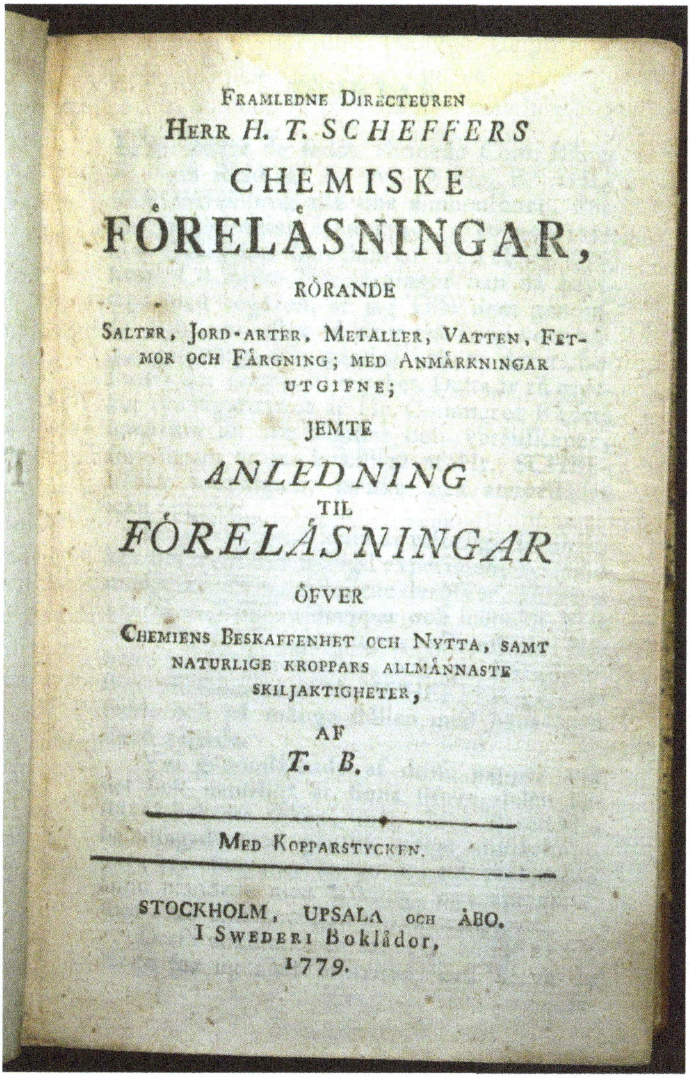

FRAMLEDNE DIRECTEUREN
HERR *H. T. SCHEFFERS*
CHEMISKE
FÖRELÄSNINGAR,
RÖRANDE
SALTER, JORD-ARTER, METALLER, VATTEN, FETMOR OCH FÄRGNING; MED ANMÄRKNINGAR UTGIFNE;
JEMTE
ANLEDNING
TIL
FÖRELÄSNINGAR
ÖFVER
CHEMIENS BESKAFFENHET OCH NYTTA, SAMT NATURLIGE KROPPARS ALLMÄNNASTE SKILJAKTIGHETER,
AF
T. B.

MED KOPPARSTYCKEN.

STOCKHOLM, UPSALA OCH ÅBO.
I SWEDERI Boklådor,
1779.

Fig. 11.3 The second edition of *The Chemical Lectures of H.T. Scheffer* appeared in 1779. The engraved title page was replaced with a simple title page set with types. *Photo* Anders Lennartson

potassium, and ammonium salts in modern terminology. The analogical salts were regarded as more complex by Bergman. They consisted of a "true salt" and a metal or earth. An earth was a solid substance that was insoluble in water and which could not be reduced to metal, i.e. the oxides of calcium, magnesium, aluminium, silicon and the recently discovered oxide of barium.

The second chapter treats the acids, which were recognised by a sour taste and the property to turn litmus or violet syrup red. The chapter describes the three mineral acids sulphuric acid, nitric acid and hydrochloric acid. The method of determining the concentration of nitric acid by addition of small portions of alkali carbonate solution until effervescence ceases [15] was a crude form of titration, and a rare example of the use of stoichiometry before Bergman. To the three mineral acids known to Scheffer, Bergman added hydrofluoric acid and arsenic acid discovered by Scheele (Chapter 9). Scheffer was only aware of one acid of vegetable origin, acetic acid (vinegar), but Bergman added tartaric acid and a less known acid (oxalic acid) present in *Sal acetocellæ* (potassium hydrogen oxalate, see Sect. 24.1). He also added two acids of animal origin: phosphoric acid which was obtained from urine and formic acid from ants. Bergman also described carbonic acid (aerial acid), which unlike the other acids was present in all three of nature's kingdoms. Given Bergman's interest in carbonic acid (Chapter 18), this section is remarkably short.

In Chapter 3, Bergman discussed the three alkalis: vegetable alkali, mineral alkali and volatile alkali. Vegetable alkali (*Alkali vegetabile*) was potassium hydroxide and potassium carbonate, which was obtained from wood ashes. Mineral alkali (*Alkali minerale*) was sodium hydroxide and sodium carbonate. The main source of sodium carbonate was deposits from soda lakes, hence the name, but it could actually also be obtained from vegetable sources, e.g. seaweed ashes, as well, volatile alkali (*Alkali volatile*), finally, was ammonia. The carbonates (fixed alkalis) differed from the hydroxides (corrosive alkalis) in that the former were saturated with carbonic acid (aerial acid).

The neutral salts are treated in Chapter 4. Of special interest is Bergman's inclusion of quantitative composition of salts (given as parts of alkali, acid and water per 100 parts of the salt). He even gives a table of the equivalents of different acids (sulphuric acid, nitric acid, hydrochloric acid and carbonic acid) corresponding to 100 parts of sodium- and potassium hydroxide, decades before the more famous and influential tabulation of chemical equivalents by Ernst Gottfried Fischer (1754–1831) in 1802 [16]. Bergman obtained his numbers by titrating the acid solution with alkali, with a reference to Homberg [6]. Compared to Fischer's values, Bergman's values were highly inaccurate (Sect. 23.2).

In Chapter 5, Bergman first discussed the salts of earths and metals with acids and alkali. A long table of metal salts introduces his system of systematic Latin nomenclature (Chapter 26), but without much discussion.

The second part concerns the earths. According to Scheffer, they were substances without taste or smell, and Bergman added the criterion of low solubility in water. Four earths were known: lime (CaO), magnesia (MgO), clay (Al_2O_3) and siliceous earth (SiO_2). The discovery of earth of heavy spar (BaO) (Sect. 15.2) by Scheele and Gahn is briefly mentioned in a comment. In the second chapter of part two, the melting of earths is discussed, including the production of glass and porcelain.

The topic of the third part is water. Neither Scheffer nor Bergman believed in the transformation of water to earth, which had been reported by previous authors. Bergman referred to the now famous experiment by Lavoisier, where he found that the earth formed on refluxing water in a glass apparatus was actually silicon species extracted from the glass. Scheele's experiments, which were known by Gahn in Spring 1770 (Sect. 13.2) are not mentioned, and seem to have been unknown to Bergman. A second chapter discusses the analysis of water samples (Chapter 17).

The fourth part discusses combustible materials: sulphur, alcohol, ether, oils and fats. Phosphorus is discussed under the section of sulphur, as the two substances are similar in the sense that they both were thought to consist of an acid united with phlogiston. The discovery of phosphorus in bones and horns was attributed to Scheele without mentioning the contribution of Gahn (Sect. 13.2); this was not corrected in the second edition. Oils were divided into two groups by Bergman. Essential oils were soluble in alcohol and volatile at 100 °C, unlike the fatty oils. The preparation of soap is also discussed.

The fifth part, on metals, is divided into three chapters: noble metals, full base metals and half base metals. The noble metals had the property to resist air, water and fire without any sign of oxidation (calcination). There were three of these metals: gold, platinum and silver. The criterion for dividing metals into full and half metals was based on malleability: the full metals could be shaped with a hammer while the half metals were brittle. The full metals were mercury (frozen mercury can indeed be shaped with a hammer), lead, copper, iron and tin. The half metals were bismuth, nickel, arsenic, cobalt, zinc and antimony. This division may seem strange to a modern chemist, but some of these metals were difficult to prepare in a pure state. For example, small amounts of impurities (especially arsenic) in zinc render the metal brittle and carbon-free nickel could not be obtained until the end of the nineteenth century. Arsenic and antimony are still usually referred to as semi-metals and are brittle in the pure state. The section on metals is largely devoted to the art of purifying the metals and testing their purity. This reflects Scheffer's background, but also suited Bergman's mining students. The first printed account of the isolation of metallic manganese by Gahn (Sect. 15.3), is also found here [17] as well as a discussion of Scheele's discovery that arsenic(III) oxide (white arsenic) can be oxidised to arsenic acid (Sect. 14.2), or as Bergman put it "Arsenic [As_2O_3] is nothing but a special acid, united with enough phlogiston to keep it in a dry state" [18].

The last part of the book concerns dyeing of textiles, a special interest of Scheffer. It is quite unlikely that Bergman would have given this topic a special section if he would have written the book by himself. A few years later, Bergman wrote an essay on indigo, [19] which may have been inspired by his editing of Scheffer's lectures.

Of some interest is also the topics *not* discussed in the book. Contrary to what one could have expected, there is no section or chapter devoted to air or combustion. Oxygen (fire air, pure air) is only introduced briefly by Hjelm in the 1796 edition, but without any discussion of its properties.

11.2 Instructions to the Lectures on the Nature and Benefit of Chemistry

Bergman's book *The Chemical Lectures of H.T. Scheffer* is not an introduction to chemistry, but requires some knowledge of the foundations of chemistry. The reader is, for example, assumed to be familiar with the concept of phlogiston, which is not explained anywhere in the text. To give the students an introduction to chemistry, Bergman published an additional 63-page text in 1779, *Instructions to the Lectures on the Nature and Benefit of Chemistry and the General Differences Between Natural Bodies* [20]. It was included as an appendix to the 1779 edition of *The Chemical Lectures of H.T. Scheffer*, but also published separately. It is not included in the 1796 edition of Scheffer's Lectures. Wallerius claimed in his autobiography that Bergman's text was based on his own *Letter, on the True nature, Benefit and Value of Chemistry*, [21] but the similarities of the two texts are only superficial. Wallerius, in his text, divided chemistry in *Chemia pura* and *Chemia applicata*, and although he stated that the objectives of *Chemia pura* is to determine the elements and principles of natural bodies, [22] he did not give any clue to what these elements would be. Instead, he devoted the whole text to applied chemistry. He was also less critical of his sources than Bergman. For instance, he attributed the discovery of wine fermentation to Noah, and also referred to the Egyptian deities Osiris and Isis, claiming they had invented some sort of alcoholic beverage [21, 23].

Of all Bergman's texts, his *Instructions* gives the best overview of his chemical views."Natural science", Bergman begins, "in the most general sentence, means all our knowledge about bodies, collected through observations and experiments" [24]. Bergman divided natural science into three fields. The first was *Natural history* (this would include, biology and geology in modern terminology) which focused on the outer shape of bodies. When a scientist from this field examined a plant, he would determine its genus and species and investigate its preferred environment. *Natural research* (physics), in Bergman's definition, investigated the nature of bodies more deeply and examined the laws controlling their properties. If a physicist examined a herb, he would measure the weight-loss due to evaporation, its vascular system, and the effects of sunlight:

> Chemistry constitutes the third degree and somehow the core, since it determines their [the bodies] composition and the reasons for their different nature. When a plant is examined chemically, we learn how much salt, oil, water, earth, etc., is found therein; one learns which parts causes its taste, smell and other properties. [25]

Bergman finally concluded: "Chemistry is thus a science that determines the elements of bodies to their nature, amount and combination".

In another text, Bergman gives another, more poetic introduction to chemistry:

> Nature may be compared not improperly to an immense book, written in an unknown language. In order to understand the text of which, it is necessary that the letters should be known, so that by attentively observing the resemblance and disparity of bodies, their distinguishing characters, and natural qualities, may be discovered. – This constitutes

> *Natural History*. Then the syllables are to be formed : – And this allegorical language points out general properties to be determined by proper experiments. – And, as in society, the genius and secret disposition of the mind and affections are rendered more conspicuous in situations of difficulty and distress; so, in the same manner, the secrets of nature are more unfolded by the molestations of art than when they are suffered to remain undisturbed. – From hence *Physics* arise. Next, the sense and connexion is to be gathered from the words, as to the proportion and various modifications : – Which constitutes *Chemistry*. [26]

After defining the science of chemistry, Bergman defined the concept of element, a topic which is discussed in detail in Chap. 22 of this book. Decomposition of a substance to its chemical elements was called *chemical analysis*, and the recombination of the elements was called *chemical synthesis*. To the best of my knowledge, this division between analytic and synthetic chemistry was not widely used by chemist at the time. Bergman wrote that one could only be sure of the composition of a body when the analysis could be confirmed by synthesis. Scheele always followed this rule in his published works. Bergman also stressed the importance of the quantitative composition of bodies: "The nature of a body depends not only on its elements, but also on their proportions and mode of combination. Emery has the same constituents as topaz, but none the less, there is a quite large difference in their properties" [27].

After discussing the elements, Bergman discussed the attractions between bodies: "All bodies in nature attract each other" [27]. The attractions of heavenly bodies follow only one simple law, while the attractions between small bodies were more complex. Bergman's theory of elective attractions is discussed in detail in Chap. 20 of this book. Bergman concluded by saying that the knowledge of the laws of attraction is in a way the key to the whole science of chemistry. Bergman realised that knowledge of the forces that held particles together could explain all chemical phenomena. This was true, but much more complicated than he could possibly imagine. It is a great misfortune that Bergman never could experience the rise of thermodynamics and quantum mechanics a century after his death.

Bergman then described the different branches of chemistry. He divided chemistry in philosophical or physical chemistry (*Chemia Philosophia* or *Chemia Physica*) and applied chemistry (*Chemia applicata*). Philosophical chemistry is what we would call basic research, aiming at an understanding nature and its elements from a chemical point of view. Alternatively, chemistry could be divided into general chemistry (*Chemia vulgaris*) and higher chemistry (*Chemia sublimior*). This division appears to be more arbitrary: the general chemistry searched for the cruder elements of bodies, while the higher chemistry searched for the finer elements.

As we use many chemical substances in our everyday life, it is easy to see the benefits of chemical knowledge, Bergman wrote, and therefore applied chemistry had to be discussed in some detail. For practical reasons, it had to be divided into medical chemistry (*Chemia medica*), agricultural chemistry (*Chemia oeconomica*) and technical Chemistry (*Chemia technica*).

Medical chemistry could be subdivided in *Chemia physiologica*, which described the constituents and chemical processes in living bodies. This science was still in its infancy, Bergman wrote, but advances in this area would lead to remarkable progress in medicine. For example, extensive works had been written about the generation of urinary calculi from lime, but chemists (actually himself and Scheele, see Sect. 24.2) had not detected any lime in urinary calculi. The other part of medical chemistry was *Chemia pharmaceutica*. This was the science of medications, including the science of mineral waters (Chap. 17) and of different gases for medical use. Here, the reader is for the first time presented with a brief description of oxygen, "the pure air", which constitutes ¼ of the atmosphere. This value is closer to the true value, 21%, than Scheele's value of 1/3. It seems like Bergman had determined the oxygen content of common air in his own laboratory.

One important object of *Chemia oeconomica* was to analyse the composition of soils in order to know which crops were optimal to grow at a certain field. In general, Bergman wrote, soils consist of clay, sand and sometimes lime, mixed with decomposing remains after animals and plants. The different constituents had different properties: clay, for example, was good for retaining water. Although there was much new knowledge, Bergman admitted that agriculture had made no progress since antiquity. Another field was the chemical processes involved in the preparation of wine, syrup, starch, cheese, butter, sugar, wax, etc.

Chemia technica was a very broad field, which was further subdivided. *Chemia halurica* was the chemistry of salts and their preparation. *Chemia geurica* was the chemistry of earths and included the preparation of lime, bricks, pottery, glass and porcelain. *Chemia thejurica* dealt with combustible materials such as sulphur, phosphorus, oils, perfumes, alcohol and gunpowder. Since Bergman believed that colour originated from phlogiston, this field also included pigments, dyes and inks. A related subject was the art of making materials fire-proof. Wood could be treated with clay while cloth and paper could be treated with potassium carbonate or alum. *Chemia metallurgica* was the art of preparation and purification of metals and the analysis of ores and metals by assaying. It also included the preparation of alloys, soldering and plating. *Chemica opificiaria*, finally, dealt with leather, parchment, wool, silk and linen and other materials and processes which were not entirely based on chemistry. Bergman discussed etching, removal of stains from cloth and protection of ship hulls from worms, and the disclosing of counterfeit wines and other products. Chemistry had gained knowledge not only from chemists but also from workshops and other businesses, and at the same time, chemistry could pay back with new useful knowledge, Bergman wrote.

The second half of the text is devoted to the nature and properties of natural bodies. On pondering over the vast diversity of natural bodies one cannot help but wondering over the wisdom and power of the creator, Bergman wrote in the introductory section. Bergman divided natural bodies into two main groups, organic and inorganic bodies, the difference being that the former have a vascular system for transporting fluids and supporting grows of the organism. Organic bodies could be divided into animals and plants while inorganic bodies were divided into salts, earths, combustible materials, metals, water and air.

Salts had a distinguishable taste and needed no more than 500 times their weight of water to dissolve (Bergman had stated 200 times in *The Chemical Lectures of H. T. Scheffer*). Water was "finer and more penetrating" at higher temperature, and thus the solubility of salts increased with temperature. While discussing the solubility, Bergman seems to have confused the concepts of solubility and rate of dissolution. Bergman then discussed the classification of salts after the same system that he presented in *The Chemical Lectures of H.T. Scheffer*.

Earths could be divided into primitive and compounded earths; five primitive earths were known corresponding to what we call the oxides of calcium, magnesium, barium, aluminium and silicon. The chemical similarity of calcium and barium was pointed out by Bergman. In addition, there were reasons to suspect a new earth, "earth of gems" to be present in diamonds (see Sect. 23.4).

In the introduction to the passage on combustible materials, Bergman introduced the concept of phlogiston: it was the cause of combustibility, but it was so fine and sublime that it eluded all our senses. Still it could be transferred between different bodies and was probably a component in all substances. The rest of the section is very similar to the corresponding section in *The Chemical Lectures of H.T. Scheffer*, discussing sulphur, phosphorus, oils, ether, alcohol and gunpowder.

Metals were shiny, opaque bodies heavier than all other bodies with densities between 6 and 20 times the density of water. They all consisted of phlogiston and particular earths, called metal calces. Metals in which the attraction between the calx and phlogiston was so strong that they could not be separated by fire, were called noble metals; they were gold, platinum and silver. Bergman discussed the melting points of alloys, and was surprised that certain mixtures of metals have lower melting points than the pure constituents (thus referring to what we call eutectic mixtures[2]). His claim that "all metals combine with sulphur except gold and zinc" is very difficult to understand for anyone who has explored the violent reaction between powdered zinc and flowers of sulphur. Also, zinc is commonly found in nature as zinc blende, ZnS. After discussing arsenic acid, he introduced his theory that all metal calces actually are metallic acids combined with phlogiston (Sect. 15.4). Bergman noted that metals do not dissolve in acids without the loss of phlogiston, meaning that they have to be oxidised.

When discussing water, he discussed his atomistic view of matter (Sect. 22.3): all liquids were composed of infinitesimally small particles. On cooling water to 0 °C, the particles lost their mobility and the water froze. Although Bergman supported the material view of heat, this statement brought him quite close to the truth (in the section on air, however, he clearly stated that "heat is not motion"). The claim that the volume increases 14,000 times when water turns to steam may simply be a printing error (the true value is 1,700). He also took the opportunity to discuss the highly debated question whether water can turn into earth and wrote that there were no proofs for such observations, but that it was not improbable.

[2]A eutectic mixture is a homogenous mixture which has a lower melting point than the pure components, such as soldering lead and the anti-freezing mixture of water and ethylene glycol used in car engines.

In the section on air, Bergman gave the account of oxygen that is missing in *The Chemical Lectures of H.T. Scheffer*. After discussing gases and vapours, he discussed oxygen (pure air), carbon dioxide (aerial acid) and nitrogen, which he did not give a name. He was of the opinion that the inner nature of nitrogen was unknown, but regarded it as "phlogisticated". Oxygen was consumed by respiration, putrefaction and combustion but Bergman believed in regeneration of oxygen in nature, and highlighted the role of green plants.

Bergman described hydrogen and the difference between different combustible gases (hydrogen and methane). Methane or marsh gas was evolved in swamps and Bergman believed that it collected in the atmosphere where it was ignited by electricity and gave rise to falling stars and other light phenomena. Hydrogen sulphide (hepatic air) was composed of sulphur, phlogiston and heat, while sulphur dioxide (acidic vitriolic air) was composed of phlogiston, heat and sulphuric acid deprived of water. This is in some sense correct, as sulphuric acid minus water is sulphur trioxide, which upon reduction gives sulphur dioxide. The other gases discussed are nitrogen oxide, hydrogen chloride, chlorine, hydrogen fluoride, ammonia and a gas that Bergman calls acetic air evolved on boiling concentrated acetic acid.[3]

11.3 Bergman's Lectures

In 1782, Spanish chemist and mineralogist Juan José d'Elhuyar (1754–1796) visited Bergman and attended his lectures. d'Elhuyar was later appointed inspector of mining in Spanish Colombia, and died in Bogotá. In the late 1950s, d'Elhuyar's notes of Bergman's lectures were owned by a descendent to d'Elhuyar in Buenos Aires and a copy on microfilm was presented to Stig Rydén, who travelled in South America on the behalf of the Swedish Museum of Ethnography in Stockholm. The notes were published (in its original French) by chemistry professor Fredga and Rydén [28], and in English translation by Schufle [29]. On investigating the notes, Rydén and Fredga found that the notes had been copied at a later date by someone who was not very familiar with chemistry. Also, neither Bergman nor d'Elhuyar were native French speakers, in fact, d'Elhuyar's French was quite poor [28]. This is important to remember, as it prevents a deeper analysis of the text.

The first part of the lecture deals with acids. It seems that d'Elhuyar only noted facts which were unfamiliar to him, since neither sulphuric nor carbonic acid are discussed. Carbonic acid (Chap. 18) was one of Bergman's favourite topics, and would definitely be included in the lecture. Perhaps Bergman had already discussed the topic with d'Elhuyar before the lecture. Most space is devoted to molybdic and tungstic acid, both recently discovered by Scheele (Chap. 15). It is thus not

[3]This seems to be an erroneous observation, and it is difficult to understand which gas it would correspond to other than acetic acid vapour.

surprising that d'Elhuyar went on to study tungstic acid and isolate metallic tungsten once back in Spain.

The section on alkali is very short, and includes the odd statement that alum does not crystallise without addition of sodium carbonate (mineral alkali). This is of course wrong and on the contrary to Bergman's findings (Sect. 15.2).

In the section of salts, Bergman's classification system is discussed followed by a discussion of "neutral salts" and "metallic salts". The green colour of hydrated iron(II) salts was attributed to phlogiston, since it was not observed after removal of phlogiston (oxidation to Fe^{3+}).

In the section of earths, Bergman speculated that the "heavy earth" (barium oxide) could possibly be reduced to a metal. Scheele was of the same opinion (Sect. 15.2), but the isolation of barium was not achieved until the nineteenth century.

After the earths, different combustible materials were discussed: sulphur, phosphorus, "pyrophore" (Sect. 28.5), oils, camphor, ether and charcoal. On discussing phosphorus, Bergman also discussed other phosphorescent materials such as barium sulphide and the theory of Beccaria[4] that phosphorescence is due to absorption of light. The discussion of ether is rather confusing and it appears that d'Elhuyar (or the person who copied the notes) did not understand the difference between diethyl ether and ethyl acetate. Bergman mentioned the discovery (by Scheele, Sect. 24.8) that mineral acid is needed for the formation of ethyl acetate, but the text suggests that ethyl acetate and diethyl ether were one and the same compound. Bergman would not have made such an error. On discussing charcoal, Bergman noted its similarity to coal and soot, and finally gave the recipe for an ink suitable for marking sheep. Finally, fulminant powder (a mixture of sulphur, alkali and potassium nitrate) and gun powder were discussed.

Bergman then turned to the metals and their characteristic properties. He discussed his investigation of the phlogiston content of metals (Sect. 22.5) and attempted to prove that the hydrogen gas evolved by dissolving metals in acids originates from the metal rather than from the acid.

Bergman then discussed water and heat, and stressed that water is composed of extremely small solid particles (Sect. 22.3). Bergman gave a long discussion on heat, and seems to adhere to Scheele's theory that it was composed of oxygen (vital air) and phlogiston and that light was composed of heat and phlogiston. Bergman explained the difference between temperature and specific heat, but the latter use of the term "specific heat" is rather confusing, and with Bergman's background as a physicist, it seems more likely that it is d'Elhuyar who was confused.

The last section discussed air and oxygen. He discussed the importance of oxygen for living organisms and the effect of oxygen on blood. He also discussed the evolution of oxygen by green plants, and assured that even toxic and bad-smelling plants were as effective in this regard.

[4]Giacomo Battista Beccaria (1716–1781). Italian physicist.

11.4 Bergman's Students

During his years as professor, 22 students defended their theses under Bergman's presidium; several of these men would enter the history books in one way or the other and are treated in the following sections. In addition to those who defended their theses under Bergman, a long list of other students is given by Hjelm [30], but the list is of course far from complete. It includes, e.g. Johan Gottlieb Gahn, Carl and Gustaf Rinman, sons of Bergman's friend Sven Rinman, and Anton Swab, nephew of Anton von Swab (Chap. 9).

With increasing international recognition, a number of international students travelled to Uppsala to learn more from Bergman; Hjelm mentioned nine students, [31] and Bergman mentioned six students in his autobiography. In 1768, Christian Ernst Heltzen (1745–1825) from Kongsberg, Norway, took a three-month course in chemistry and mineralogy; Rotheram from England took his M.D. in Uppsala; apothecary Friedrich Ehrhart (Sect. 14.1) from Switzerland stayed in Uppsala 1773–1776 and would correspond with both Bergman and Scheele after he left Sweden and published several of Scheele's letters in his journal. Isaak Grüno (1756–1783) from Hamburg visited Uppsala 1776–1780; Thordur Thoroddsen (1736–1796) from Iceland in 1775–1780; Nebelung from Altona; Ikonnikof from St Petersburg, de Virly from Dijon and d'Elhuyar from Biscaya. D'Elhuyar and de Virly, who visited Uppsala in 1782, were the last international students to visit Bergman [32].

11.4.1 Matthias Rydell

Matthias C Rydell was born in Yllestad in the Västergötland province, in 1738. He was Bergman's first student and defended his thesis in physics, *de attractione universali* in November 1758 (Chap. 6). The study of attractions between bodies would become one of Bergman's main fields as a chemist (Chap. 20). After his studies in Uppsala, Rydell appears to have made a military career and died in Karlskrona, the location of Sweden's major naval base, in 1798. He published no more scientific works, but published, e.g. a memorial lecture over a navy priest in 1771, a speech to celebrate the birth of Prince Gustav Adolf in 1779 and a memorial song over Bergman's patron Gustav III, who was assassinated in 1792.

11.4.2 Gustaf Swedelius

Swedelius was born in Västerås in 1748 and defended Bergman's thesis on alum refinement, *Disquisitio chemica de confectione aluminis*, in 1767 (see Sect. 9.2). This was Bergman's first work in chemistry. After leaving Bergman, Swedelius continued to study medicine in Berlin and back in Uppsala he obtained his M.D. degree in 1775 [33]. He practised as physician in Stockholm and Falun, where he died in 1804. Apart from the two dissertations, he published a paper on arsenic poisoning in 1800: [34] a 25-year-old maid in Falun had the habit of eating left-overs in the kitchen, and accidentally drank from a bottle of fly poison believing it to be mead. After treatment with potassium polysulphide (liver of sulphur) by Swedelius, the maid survived.

11.4.3 Carl Anders Plomgren

Carl Anders Plomgren [35] was one of Bergman's more extraordinary students. He was born in 1750 in Uppsala as son of merchant Anders Plomgren (1700–1766). He enrolled at Uppsala University in 1767 and in 1769 he was appointed auscultator at the Board of Mines. He had specialised in mining, and just like Bergman's students Gahn and Hjelm, he defended a thesis in law for Pehr Niclas Christiernin (1725–1799) [36]. The defence took place on December 13, 1769, and only 3 days later he defended his graduation thesis on fulminating gold, *de calce auri fulminante*, with Bergman as supervisor. This was the thesis that Wallerius attacked in *Lärda Tidningar* (Sect. 9.6). His career was initially impressive. He was ennobled in 1770 and took the name Plommenfelt. In 1772–1776 he travelled through Europe, where he made important contacts. He met, for example, Voltaire in Geneva. Plommenfelt developed a deep interest in occultism and became a free mason. This was an interest that he shared with Duke Charles (Hertig Karl), the younger brother of King Gustav III, and he became an important figure in the Royal Court. At his estate Huvudsta, he equipped a laboratory for "higher science", which impressed on the Duke, and he also held séances. In 1779, however, his career met some obstacles. The business house of Plomgren went bankrupt and Plommenfelt became involved in an infected legal case. The following year, it turned out that he did not have the connections to European free masons that he had claimed, and he found himself kicked out of the court. He started to live a bad life and when he, in 1782, wrote that the King was a deceiver on a pub window, he was sentenced to death and his nobility was withdrawn. He managed to flee to North America (possibly after receiving a hint from the King) and was last heard from in 1785. His crossed over coat of arms, the only of its kind, can still be seen in the House of Nobility in Stockholm.

11.4.4 Pehr Dubb

Pehr Dubb [37] (Fig. 11.4) was born on January 14, 1750 in Mariestad, Bergman's home town. He went to school in Skara, just like Bergman, before he came to Uppsala in 1768. Dubb studied under Linnaeus and Bergman, and defended his thesis on the water of Uppsala, *Chemical Investigation of the Spring Water in and Nearest Around Uppsala* (Sect. 17.3) in 1770. He practised as a physician at Sätra spa when merchant Claes Alströmer (1736–1794; son of Jonas Alströmer and brother of Patrik Alströmer) asked Bergman if he could suggest a private physician

Fig. 11.4 Physician Pehr Dubb. Engraving by Johan Fredrik Martin. Nationalmuseum, Stockholm

for him. Bergman recommended Dubb, who consequently moved to Gothenburg in 1776. In 1779–1780, Dubb followed count Sparre to Paris, where he studied the organisation of French hospitals. In 1782, Alströmer appointed Dubb manager of the newly founded Sahlgrenian hospital (present day Sahlgrenska University Hospital), opened with money donated by Niclas Sahlgren, one of the founders of the Swedish East India Company (Sect. 4.2) and cousin of Bergman's mother. Appropriately, the street outside Sahlgrenska University Hospital, Per Dubbsgatan, is named after Dubb.

Dubb became an influential figure in Gothenburg during late eighteenth century and early nineteenth century, with a special interest in health care, fire protection and fighting against poverty, begging, crime, drinking and prostitution. As a friend of wealthy merchant William Chalmers, he played a significant role in the founding of what is now Chalmers University of Technology in Gothenburg.

Fig. 11.5 Dubb's grave at the Stampen cemetery in Gothenburg is marked by a tall iron obelisk. *Photo* Anders Lennartson, July 2018

Pehr Dubb appreciated good food, and became seriously overweight. It has been said that his assistant had to tie a rope around him and pull him up the stairs in houses lacking handrail. He died on January 6, 1834 (Fig. 11.5).

11.4.5 Carl Henrik Wertmüller

Wertmüller [38] was born in Stockholm on July 3, 1752, the son of court apothecary and court physician Johan Ullric Wertmüller. He practised his pharmaceutical skills at his father's pharmacy, the Lion, in Stockholm and began his studies in Uppsala in 1770. In 1773 he defended his *pro exercitio* thesis *fonte acidulari Dannemarkensi* (Chap. 17) under the supervision of Bergman. After his chemistry studies, he travelled via Lübeck, Hamburg and Hannover to Göttingen, where he took his M.D. degree in 1779. He continued his travels in Dresden, Leipzig, Halle, Berlin, Hamburg and Copenhagen and finally returned to Stockholm in 1780. In 1782, he was appointed provincial physician in Stockholm County. In 1794 he took over his father's pharmacy which he sold in 1798. He died on April 9, 1829.

11.4.6 Johan Adolf Level

Level, from the Småland province, defended his thesis in pharmacy on antimony tartrate (Sect. 27.3) in 1773.

11.4.7 Peter Jacob Hjelm

Hjelm [39] was born on October 2, 1746 in Göteryd, in the Swedish province of Småland. He would be one of the most successful chemists among Bergman's students. After 5 years at Växsjö gymnasium he came to Uppsala in 1763. In December 1771 he defended a thesis in law under professor Christiernin [40] concerning the ownership of mineral deposits. The thesis contains nothing of chemical interest. Unlike Gahn, who also defended a thesis under Christiernin (Sect. 13.1), Hjelm continued his studies under Bergman and presented his thesis *Chemisk och mineralogisk afhandling om hvita järnmalmer* (*Chemical and Mineralogical Dissertation on White Iron Ores*) in 1774. He was not, however, awarded the master degree [41]. Hjelm left Uppsala for Stockholm in 1774 to take up a civil servant position at the Board of Mines and to study assaying under Sven Rinman. He was appointed assayer in 1782 and in 1784 he was elected a member of the Royal Swedish Academy of Sciences in Stockholm. In 1794, he succeeded von Engeström as manager of *Laboratorium chymicum*, which had been transferred to the Royal Swedish Mint.

In Uppsala, Hjelm had befriended Scheele, and they would continue to correspond by letters after they both left Uppsala. Hjelm's most significant chemical achievement, the isolation of metallic molybdenum (Sect. 15.4) was carried out on Scheele's request. His 24 papers in the *Transactions of the Royal Academy of Sciences* concern a wide range of subjects with a focus on metallurgy, chemistry and mineralogy, but his first paper dealt with the population of Uppsala diocese from 1749 to 1773 [42]. He published two important series of papers on manganese and molybdenum, translated Bergman's manual on the blow-pipe to Swedish (Sect. 23.4) and held the memorial lecture over his master in the Royal Swedish Academy of Sciences.

Hjelm remained loyal to the phlogiston theory, as can be seen, e.g. in the preface of his 1796 edition of *The Chemical Lectures of H.T. Scheffer*. Although conservative as a chemist, he had radical political views, and his diaries have revealed that he was against King Gustav III, and praised the French constitution of 1791. He died unmarried in Stockholm on October 7, 1813.

11.4.8 Johan Afzelius

Johan, or Jan, Afzelius [43] was born on June 13, 1753 in Larv in the Västergötland province. Like Bergman, he attended the Skara Gymnasium before enrolling atUppsala University in September 1769. He focused his studies on chemistry, mineralogy and physics and studied mainly under Bergman. He defended his thesis on nickel, *De nicolo*, in June 1773 and received a bachelor degree in December the same year. On June 13, 1776 he defended the thesis *de acido sacchari*, on oxalic acid and its salts. The results in this thesis were largely based on experiments by Scheele (Sect. 24.1), who is not mentioned in the thesis. Afzelius received his master degree the day after his defence. After a period at the Board of Mines, he was appointed *laborator* and became Bergman's assistant in February 1780. Finally, Bergman had an assistant that he could trust. Afzelius had a deep theoretical knowledge, and Bergman had very high thoughts of his former student. During Bergman's periods of illness from 1781, Afzelius was acting professor in chemistry and after Bergman's death Afzelius was appointed professor on December 13, 1784. Due to illness, however, he could not take up the position until October 1785. Unlike Bergman, who never travelled outside Sweden, Afzelius undertook mineralogical trips to Norway, Denmark and Russia in the 1790s. He was also elected Vice Chancellor of the University three times, in 1789, 1798 and 1806.

Afzelius, known in Uppsala as Sten-Jan (Stone Jan) due to his interest in minerals, was well known for his humour. He was an active teacher, but had unfortunately little interest in research (Sect. 28.4). He seems to have only gradually accepted the antiphlogistic theories, but preserved lecture notes show that he taught antiphlogistic chemistry in the nineteenth century. The statement that Afzelius never abandoned phlogiston [44] is in error. His main interests were maintaining the mineral collection and his garden. Afzelius retired in 1820 but managed the mineral collection until 1823. He died unmarried in 1837.

11.4.9 Carl Norell

Norell was born in Bergman's home province Västergötland and defended a thesis on *Magnesia alba* in 1775.

11.4.10 Andreas Pihl

Pihl [45] was born on February 26, 1759 in the Västmanland province. After defending his thesis on arsenic, *de arsenico*, in 1777 he worked as a mining expert in several Swedish mines, especially in Falun, where he came in contact with Gahn; he is occasionally mentioned in the correspondence between Hjelm and Gahn [46]. He died in *c*. 1835.

11.4.11 Anders Schedin

Schedin was born in the Uppland province and defended a thesis written in Swedish on the assaying of iron in solution in 1777.

11.4.12 Johan Peter Scharenberg

Scharenberg was born in 1759, the son of Johan Scharenberg, owner of the Raven pharmacy in Stockholm and Scheele's master during his years in Stockholm. Johan Peter Scharenberg was about 10 years old when Scheele worked for his father, so they were not likely to have discussed chemistry. He defended a thesis on water analysis (Sect. 23.3) under Bergman's supervision in 1778. Scharenberg went in his father's footsteps and became apothecary in Laholm in 1794, where his sister was married to a priest [47]. He died in Laholm on March 11, 1800 [48].

11.4.13 Bengt Reinhold Geijer

Geijer [49] (Fig. 11.6) was born on November 10, 1758 at Bofors ironworks, the son of ironmaster Salomon Gottschalk Geijer. He came to Uppsala as a student in February 1776 and defended his thesis *de mineris zinci* (on zinc ores) in March 1779. He was accepted as auscultator at the Board of Mines the same year. From 1782 he was an assayer at the Board of Mines, but was also teaching mineralogy at the land surveying office. During this time, he published four papers in the *Transactions of the Royal Swedish Academy of Sciences*, including papers on the use of oxygen in blow-pipe [50] and smelting experiments [51]. In 1796 he took a position at Jernkontoret, the Swedish Steel Association, but in 1797 he bought Rörstrands porcelain factory just outside Stockholm. He is regarded as one of the main figures in Swedish industry in the early nineteenth century. From 1785, in collaboration with count Erik Ruuth, he began searching for coal and heat-resistant clays in Southern Sweden by drilling. At Rörstrands, he used both his chemical knowledge to improve the process of porcelain production at the factory and

Fig. 11.6 Memorial coin of Geijer issued by the Royal Swedish Academy of Sciences. *Photo* Linköpings mynt och antikhandel

introduced new technology. For example, an English steam engine was installed in 1807. On January 10, 1788 he married Ulrika Castorin (1769–1818), the younger sister of another former Bergman student, Per Castorin, whom he met in Uppsala. Geijer died in Stockholm on November 12, 1815.

11.4.14 Jacob Paulin

Paulin [52] was born on June 25, 1752 in Mariestad, Bergman's home town. Paulin defended his thesis on the origins of chemistry, *de primordiis chemiae,* in June 1779 (Sect. 9.3). One and a half year later, on November 23, 1780 he took a bachelor degree in theology and was ordained on December 13, the same year. In 1787 he was appointed assistant vicar in Bro-Lossa in Uppland. He died on July 19, 1797.

11.4.15 König Alexander Grönlund

Grönlund defended his thesis *de terra silicia* (On siliceous earth; Sect.22.4) in 1779. Grönlund held the Hylting scholarship, established in 1752 and which is as of 2019 still available for students at Uppsala University.

11.4.16 Per Castorin

Castorin was born in Grythyttan on October 12, 1761. Castorin defended his thesis *de minerarum docimasia humida* (Sect. 23.4) in 1780. In 1794 he inherited his father's iron works in Älvestorp, complete with mining fields, blast furnace and factories. He died on January 20, 1808. His sister, Katarina Ulrika Castorin, married another of Bergman's students, Geijer, in 1788.

11.4.17 Andreas Niclas Tunborg

Tunborg, from the Dalarna province, defended his thesis on the determination of the phlogiston content in metals (Sect. 22.5) in 1780. He seems to have made a career as a civil servant of some kind, and is referred to as "notarie" by Hjelm [53].

11.4.18 Johan Gadolin

Gadolin [54] (Fig. 11.7) was born on June 5, 1760 in Åbo, Finland, which was then a Swedish province. His father, Jacob Gadolin, was bishop and professor in astronomy and physics at the University in Åbo, Åbo Academy. The name Gadolin was invented by his grandfather, and was derived from gadol (great), the Hebrew translation of Maunula, the farm where he lived [55]. Gadolin started his studies at the Åbo Academy in 1775, first focusing on mathematics, but his interest eventually turned towards chemistry. In chemistry, he was instructed by professor Gadd, but to gain a deeper knowledge he went first to Stockholm and finally, in June 1779, to Uppsala. In June 1781, he defended his thesis, *de analysi ferri* (on the analysis of iron) and obtained his master degree a year later.

Back in Åbo he held different positions at the University but without much success, and in 1786 he undertook a trip through Denmark, Germany, the Netherlands, England and Ireland. In 1789, he succeeded Gadd in Åbo, becoming full professor in 1797. As professor, Gadolin was very active in both teaching and research, publishing a large number of papers and doctoral theses. Many of his theses form series, where the text has been divided into a number of theses in an apparently arbitrary way. For instance, his book *Index Fossilium*, which is an alphabetic list of minerals and their composition, is made up of ten dissertations defended in 1823. In another long series of dissertations, *Historiam doctrinæ de affinitatibus chemicis*, defended in 1815–1819, he returned to Bergman's favourite

Fig. 11.7 Finnish stamp depicting Johan Gadolin. *Photo* Anders Lennartson

topic of chemical affinities. He devoted much time to the study of heat, but also conducted mineral analysis. In 1794, he found a new earth, yttria, in a mineral from the Ytterby mine in the Stockholm archipelago [56]. From this earth, yttrium was eventually isolated, the first in a series of elements named after the same small mine: yttrium, ytterbium, terbium and erbium. In 1800, Gadolin got the rare earth mineral gadolinite named after him, which in turn gave the name to the lanthanide element gadolinium.

Of special interest is Gadolin's opinions on phlogiston [57]. Influenced by Bergman, he was initially a phlogistonist. In 1788, however, he had started to move towards the theories of Lavoisier. In a paper published that year, he wrote that no one can question that metals upon calcination take up a substance which occurs in all metal claces and is present in air and water [58]. He was still, however, convinced of the existence of phlogiston as a component in light [59]. In 1798 he published a short textbook, *Inledning til chemien* (Introduction to Chemistry) based on Fourcroy's[5] book *Philosophie Chemique*, where he totally accepted the theories of Lavoisier. This is generally regarded as the first original textbook on antiphlogisic chemisty in Swedish, but a less well-known book which also treats antiphlogisic chemistry, *Inledning til kunskapen om de jordiske kropparne* (*Introduction to the Knowledge of the Earthly Bodies*) was published by Julius Segerstedt the same year. Fourcroy's *Philosophie Chemique*, had been translated to Swedish by Linnaean apostle Anders Sparrman in 1795. Of these book's, Gadolin's was by far the most influential.

Gadolin retired in 1822 and died on August 15, 1850. Born in the Swedish province Finland, he died in Russian Grand Duchy of Finland, as Sweden had lost the Finish war against the Russian Empire in 1809.

11.4.19 Pehr von Afzelius

Pehr Afzelius [60] (Fig. 11.8) was the younger brother of Bergman's student Johan Afzelius, born on December 14, 1760. Like Bergman and his older brother, he studied at Skara Gymnasium before enrolling at Uppsala University in October 1777. He defended his *pro exercitio* thesis, *In anervismata femoris observationes*, under the presidium of physician and professor of anatomy Adolph Murray (1751–1803). His master thesis, which constitutes the second part of Bergman's historical treatment of chemistry (Sect. 9.3) and covers the period between the eighth and seventeenth century, was defended in 1782. In 1784, finally, he presented his thesis *Historiæ morborum, observationibus practicis illustratæ* for his M.D. degree under the presidium of Jonas Sidrén (1723–1799), professor of medicine. He was awarded the M.D. degree in 1785 and after travelling through Europe he held several medical positions before becoming professor of medicine in Uppsala in 1799. Afzelius also held many important official positions and was ennobled in 1815.

While in Paris in 1785, Afzelius acquainted several of the prominent French chemists including Lavoisier and Fourcroy, leaders of the chemical revolution. Thus, he was likely more initiated in the new chemical theories than his older brother, Johan, the professor of chemistry. After returning to Sweden he would, however, mainly work in the medical field with one important exception. In 1795 he published an anonymous text, *An attempt to a Swedish Nomenclature in Chemistry*, written together with chemist Anders Gustaf Ekeberg (1767–1813). Ekeberg was Johan Afzelius' assistant, and unlike Johan Afzelius, Ekeberg carried out chemical research and is best remembered for his discovery of tantalum in 1802. The text that Afzelius and Ekeberg wrote introduced the Swedish words for oxygen (syre), nitrogen (kväve) and hydrogen (väte) still used today, and took a clear stand for the antiphlogistic chemistry. Thus, he made a more lasting impression on

[5]Antoine Françoise de Fourcroy (1755–1829); French chemist.

Fig. 11.8 Pehr von Afzelius. Oil painting by Johan Gustaf Sandberg. *Photo* Nationalmuseum, Stockholm

Swedish chemistry than his brother Johan. While Ekeberg was Jacob Berzelius main teacher in Chemistry, it was under Pehr Afzelius supervision that Berzelius obtained his M.D. degree in 1802 with a thesis on the medical effects of electricity. Pehr von Afzelius died on December 2, 1843.

11.4.20 Carl Didrik Hierta

Hierta [61] was born in 1758 in the Västergötland province and came from a noble family with strong military traditions. He defended his thesis *de analysi lithomargæ* (Sect. 25.6) in June 1782, and obtained his master degree the same year. By the time he presented his thesis, he held the Gyllenhjelm scholarship, established in 1629, and which as of 2019 is still awarded to students at Uppsala University. After Bergman's death, he was appointed assistant (laborator) to Bergman's successor, Afzelius in 1785. In 1798 he obtained an important administrative position at the

University as bookkeeper (räntmästare). His son, Lars Johan Hierta, born in 1801, became an important liberal politician, newspaper editor, book publisher and businessman. The newspaper he founded, Aftonbladet (Evening Post), is still one of the major Swedish tabloids.

11.4.21 Carl Gustaf Robsahm

Robsahm [62] was born in 1752. The Robsahm family had been involved in mining and iron production since the seventeenth century, [63] and a mining exam was a natural step in the career of Carl Gustaf. He defended a thesis on the analysis of asbestos, *de terra asbestina* (Sect. 25.6) in 1782. He went on to become bookkeeper at Letafors iron works near Torsby in the Värmland province. He died in 1821.

11.4.22 Fredrik Wilhelm Mannercrantz

Mannercrantz [64] was born in 1755 and studied pharmacy in Stockholm and Strängnäs before enrolling at Uppsala University. In 1782 he defended his thesis *de antimonialibus sulphuratis* and obtained a bachelor degree in medicine the same year. He practised a few years as physician in Strömstad before buying the pharmacy in Enköping and taking his apothecary exam in 1785. In 1792 he opened a pharmacy in Sigtuna [65]. He died in 1796. By 1782, Bergman's health started to decline, and Mannercrantz became the last student to present a thesis under Bergman's supervision.

References

1. Schück H (1916) Torbern Bergmans självbiografi, äldre svenska biografier 3–4, Uppsala, p 102
2. Wallerius JG (1751), Bref, om chemiens rätta beskaffenhet, nytta och wärde, til N.N. öfwersändt, och af honom til trycket befordradt, Gottfried Kiesewetter, Stockholm and Uppsala
3. Wallerius JG (1759) Chemia physica, första delen, föreställandes chemiens natur och beskaffenhet i gemen, dess historia, characterer, instrumenter, operationer och producter, Lars Salvius, Stockholm
4. Wallerius JG (1953) Självbiografi, Lychnos 235–259
5. Bergman T (1775) H T Scheffers Chemiske föreläsningar, rörande salter, jordarter, vatten, fetmor, metaller och färgning, samlade, i ordning stälde och med anmärkningar utgifne af T B, M Swederus, Uppsala, p II
6. Bergman T (1775) H T Scheffers Chemiske föreläsningar, rörande salter, jordarter, vatten, fetmor, metaller och färgning, samlade, i ordning stälde och med anmärkningar utgifne af T B, M Swederus, Uppsala, p 66

7. Trofast J (1994) Johan Gottlieb Gahn brev, vol 2, Lund, p 85
8. Bergman T (1775) H T Scheffers Chemiske föreläsningar, rörande salter, jordarter, vatten, fetmor, metaller och färgning, samlade, i ordning stälde och med anmärkningar utgifne af T B, M Swederus, Uppsala, p XIII
9. Bergman T (1779) Anledning til föreläsningar öfver chemiens beskaffenhet och nytta, samt naturlige kroppars almännaste skiljaktigheter, M Swederus, Uppsala and Åbo
10. Bergman T (1796) Framlidne directeuren herr HT Scheffers Chemiske föreläsningar, rörande salter, jordarter, metaller, vatten, fetmor och färgning; med anmärkningar utgifne af TB Andra tilökta uplagan, Johan Dahl, Stockholm, last (unnumbered) page of preface
11. Bergman T (1775) H T Scheffers Chemiske föreläsningar, rörande salter, jordarter, vatten, fetmor, metaller och färgning, samlade, i ordning stälde och med anmärkningar utgifne af T B, M Swederus, Uppsala, p 34
12. Bergman T (1775) H T Scheffers Chemiske föreläsningar, rörande salter, jordarter, vatten, fetmor, metaller och färgning, samlade, i ordning stälde och med anmärkningar utgifne af T B, M Swederus, Uppsala, p 138
13. Bergman T (1775) H T Scheffers Chemiske föreläsningar, rörande salter, jordarter, vatten, fetmor, metaller och färgning, samlade, i ordning stälde och med anmärkningar utgifne af T B, M Swederus, Uppsala, p 149
14. Bergman T (1779) Anledning til föreläsningar öfver chemiens beskaffenhet och nytta, samt naturlige kroppars almännaste skiljaktigheterM Swederus, Stockholm, Uppsala and Åbo, p 31
15. Bergman T (1775) H T Scheffers Chemiske föreläsningar, rörande salter, jordarter, vatten, fetmor, metaller och färgning, samlade, i ordning stälde och med anmärkningar utgifne af T B, M Swederus, Uppsala, p 18
16. Berthollet CL (1802) Über die Gesetze der Verwandtschaft in der Chemie, Berlin, p 232
17. Bergman T (1775) H T Scheffers Chemiske föreläsningar, rörande salter, jordarter, vatten, fetmor, metaller och färgning, samlade, i ordning stälde och med anmärkningar utgifne af T B, M Swederus, Uppsala, p 390
18. Bergman T (1775) H T Scheffers Chemiske föreläsningar, rörande salter, jordarter, vatten, fetmor, metaller och färgning, samlade, i ordning stälde och med anmärkningar utgifne af T B, M Swederus, Uppsala, p 366
19. Bergman T (1780) Analyse et examen chimique de l'indigo, tel qu'il est dans le commerce, pour l'usage de la teinture. Piece qui a concouru pour le prix sur la nature & l'usage de l'indigo, Mémoires de mathématique et de physique, présentés à l'Academie Royale des Sciences 9:121–164
20. Bergman T (1779) Anledning til föreläsningar öfver chemiens beskaffenhet och nytta, samt naturliga kroppars almännaste skiljaktigheter, M Swederus, Stockholm, Uppsala and Åbo
21. Wallerius JG (1953) Självbiografi, Lychnos, pp 235–259
22. Wallerius JG (1767) Bref, om chemiens rätta beskaffenhet, nytta och värde, till NN örfwersändt, Uppsala, p 4
23. Wallerius JG (1767) Bref, om chemiens rätta beskaffenhet, nytta och värde, till NN örfwersändt, Uppsala, p 36
24. Bergman T (1779) Anledning til föreläsningar öfver chemiens beskaffenhet och nytta, samt naturlige kroppars almännaste skiljaktigheter M. Swederus, Stockholm, Uppsala and Åbo, p 1
25. Bergman T (1779) Anledning til föreläsningar öfver chemiens beskaffenhet och nytta, samt naturlige kroppars almännaste skiljaktigheter M Swederus, Stockholm, Uppsala and Åbo, p 2
26. Bergman T (1791) Physical and Chemical Essays, vol III, G, Mudie, J & J Fairbairn and J Evans, Edinburgh, p 2
27. Bergman T (1779) Anledning til föreläsningar öfver chemiens beskaffenhet och nytta, samt naturlige kroppars almännaste skiljaktigheter M Swederus, Stockholm, Uppsala and Åbo, p 3
28. Fredga A, Ryden S (1959) Juan José de Elhuyars anteckningar efter Torbern Bergmans föreläsningar 1782. Lychnos 1959:161–208
29. Schufle JA (1985) Torbern Bergman a man before his Time. Cornado Press, Lawrence, Kansas, p 412

30. Hjelm PJ (1786) Åminnelse-tal öfver...Herr Torbern Bergman, Stockholm, p 87
31. Hjelm PJ (1786) Åminnelse-tal öfver...Herr Torbern Bergman, Stockholm, p 85
32. Aurillius PF (1785) Åminnelse-tal, öfver chemiæ professoren i Upsala...Herr magister Torbern Bergman...Uppsala, p 24
33. Sidrén J, Svedelius G (1775) Dissertatio inauguralis medica, de crisibus febrium perfectis, quam auspice deo, consensu experient. medic. ord. ups. præses Dom. Doct. Jonas Sidrén, ... et auctor Gustavus Svedelius, stip. reg. Arosia-Vestmannus. ... In audit. Gustav. die II Decemb. anni MDCCLXXV, Uppsala
34. Svedelius G (1800) Förgiftning af Arsenik, lyckligen botad. KVA Nya Handl 21:68–70
35. Burius A (1995–1997) Carl Anders Plommenfelt, Svenskt biografiskt lexikon, vol 29, p 382
36. Christiernin NP, Plomgren CA (1769) Några anmärkningar i svenska bergs-lagfarenheten om stångjärns verken, Uppsala
37. Josephson, CD, Hildebrand, B (1945) Pehr Dubb in Svenskt Biografiskt Lexikon, vol 11, Stockholm, p 478
38. Calissen ACP (1835) Medicinisches Schriftsteller-Lexicon, vol 21, Copenhagen, p 57
39. Franzén O (1971–1973) Peter Jacob Hjelm, in Svenskt biografiskt lexikon, vol 19, Stockholm, p 104
40. Christiernin PN, Hjelm PJ (1771) Academisk afhandling i svenska bergs-lagfarenheten om egande rätt till malmstrek och grufvor, Uppsala
41. Olsson H (1971) Kemiens historia i Sverige intill år 1800. Stockholm, Almqvist & Wiksel, p 153
42. Hjelm PJ (1776) Folkmängden i Upsala Stift, KVA Handl 37:49–59
43. Odén S (1918) Johan (Jan) Afzelius, Svenskt Biografiskt Lexikon, vol 1, Stockholm, p 222
44. Olsson H (1971) Kemiens historia i Sverige intill år 1800. Stockholm, Almqvist & Wiksel, p 156
45. Almquist JA (1909) Bergskollegium och bergsstaterna 1637–1857. Stockholm, Norstedts, p 255
46. Trofast J (1994) Johan Gottlieb Gahn brev, vol 2, Lund
47. Osbeck P (1796) Utkast til beskrifning öfver Laholms prosteri, in Sahlgren J (1922), Svenska bygder i äldre beskrivningar, CWK. Gleerups förlag, Lund, p 227
48. Osbeck P (1800) Inrikes Tidningar, onsdagen den Maji 1800, Stockholm, p 3
49. Rönnow S (1967–1969) Bengt Reinhold Geijer, in Svenskt Biografiskt Lexikon, vol 17, Stockholm, p 9
50. Geijer BR (1784) Sätt at nyttja eldsluft til blåsrörs-försök. KVA Nya Handl 5:193—197
51. Geijer BR (1784) Smältnings-försök med eldsluft, på några ädlare stenar samt andra jord- och stenarter. KVA Nya Handl 5:122—134
52. Fant JE, Låstbom AT (1842) Upsala ärkestifts herdaminne, part 1. Uppsala, Wahlström & Låstbom, p 181
53. Hjelm PJ (1786) Åminnelse-tal öfver...Herr Torbern Bergman, Stockholm, p 88
54. Mustelin O (1964–1966) Johan Gadolin, Svenskt Biografiskt Lexikon, vol 1, Stockholm, p 717
55. Pyykkö P (2015) Magically Magnetic Gadolinium. Nat Chem 7:680
56. Gadolin J (1794) Undersökning af en svart tung stenart ifrån Ytterby stenbrott i Roslagen. KVA Nya Handl 15:137–155
57. Lundgren A (1979) Berzelius och den kemiska atomteorin. Diss Uppsala University, Uppsala, p 20
58. Gadolin J (1788) Om kopparens förmåga at fälla tenn utur dess upplösning i vinstens-syra. KVA Nya Handl 9:186–194
59. Gadolin J (1788) Einige Bemerkungen über die Natur des Phlogiston. Chem Ann 1:1–17
60. Hult TO (1918) Pehr von Afzelius, in Svenskt Biografiskt Lexikon, vol 1, Stockholm, p 224
61. Anonymous (1971–1973) Hierta, släkter, Svenskt Biografiskt Lexikon, vol 19, Stockholm, p 15
62. Persson B (2016) Människor och miljöer i gamla Clara. Stockholm, BoD, p 237

63. Anonymous (1998–2000) Robsahm, släkt, Svenskt Biografiskt Lexikon, vol.30, Stockholm, p 251
64. Hjelt OEA (1893) Svenska och finska medicinalverkets historia 1663–1812 vol 3, Helsingfors, p 714
65. Anonymous (1847) Sigtuna det forna och närvarande, Wahlström & C, Uppsala, p 99

12 Bergman's Life as Professor

After being hired as professor, Bergman rapidly became a factor to count on in Swedish science. In May 1772, Gustav III founded the Order of Wasa, an order awarded to notable persons who could not be awarded the Order of Sword (Svärdsordern) or the Order of the Northern Star (Nordstjärneordern). Bergman was one of the first to be knighted in July, along with a number of his fellow members from the Royal Swedish Academy of Sciences, including Wallerius. The King, who regarded Bergman as his client, used Bergman for official assignments. In August 1777, for example, Bergman was appointed by the King to join a number of artillery officers to evaluate Strömberg's alleged improved method of gun powder production. As the trip did not take place until December, the boat trip to Skägga gunpowder works on Värmdö in the Stockholm archipelago was testing for Bergman's weak health [1]. Occasionally, Bergman was also consulted by civilians to solve problems. In July 1772, for example, he was visited by a painter from Stockholm who had run into problems [2]. A court painter, Meurling[1] had mastered a special painting technique but kept the method secret. Unfortunately, he had unexpectedly passed away and the formula was lost. Anyone applying for the well-paid vacant position was rejected by the King. The painter, who contacted Bergman had tried for years[2] to discover the secret before turning to Bergman for help. Fortunately, Bergman solved the problem. Bergman served two terms as Vice Chancellor of the University, the spring semester of 1771 and the autumn semester of 1779. Bergman's first term as Vice Chancellor has been described as very successful. Apparently Bergman was able to keep peace between the two contending parties the professors of the humanities, theology and law one side and the

[1]Alexander Meurling (1730–1771).

[2]Probably exaggerated, as Meurling had died the previous year.

A. Lennartson, *Carl Wilhelm Scheele and Torbern Bergman*, Perspectives on the History of Chemistry, https://doi.org/10.1007/978-3-030-49194-9_12

mathematicians' natural scientists on the other [3]. According to the records, the students, all of which were young men, are said to have behaved unusually well during the same period.

12.1 Contributions to Prize Questions

Bergman, who had successfully contributed in the prize question of finding a remedy against caterpillars (Sect. 6.3), also contributed to a prize question by the Royal Academy in Montpellier on the topic of classifying soils in order to determine suitable crops for different soils, and a way to improve infertile soil. Bergman's suggestion won, but unfortunately the ticket with Bergman's name that accompanied his contribution was lost, and at first the committee had no idea who the winner was [4]. Hjelm later wrote that Bergman's winning contribution was a pure pleasure to read, despite the fact that it contained no new information [4]. Bergman also contributed to a prize question with an attempt to analyse indigo, [5] where he attempted the task of determining its composition, a task impossible in Bergman's days, as well as a discussion on the best ways of dying silk and wool with indigo.

12.2 Bergman's Study of Bees

After his installation as professor of chemistry, Bergman had little or no time for his work in physics and astronomy. He would not, however, completely give up his interest in insects. He was still regarded as an authority on pest insects. In response to an anonymous paper on pine forests in 1769, [6] Bergman wrote a short paper on insects attacking pine trees [7]. The same year he submitted a contribution to the prize question of finding a means against caterpillar attacks on fruit trees (Sect. 12.1). His next paper on insects would appear ten years later, in 1779, and concerned beekeeping [8]. The nearly thirty pages long paper is a very good example of how Bergman attacked a problem. Bergman may have got some assistance from his wife in taking care of the bees [3].

Bergman's object was to study the profitability of beekeeping in Uppsala. He had studied bees for several years and had come to the conclusion that these hard-working creatures gave a high yield with a minimum of effort. The chemist Bergman, who was at the time working on the development of quantitative chemical analysis, constructed a balance that allowed him to safely weigh the hives. Bergman studied two colonies, denoted A and B. Colony A (about 25,000 bees) swarmed on June 30, 1778 from a healthy hive while colony B was about half that size and swarmed on July 6, from an unhealthy hive. Bergman regularly weighed the hives, and his tabulated results fill several pages of the paper. The weight of the hives increased until August, then it started to decrease. On August 16, 1779, both hives were broken up and the yield of honey, wax and waste was determined. He anticipated that some might criticise him for destroying the colonies rather than just harvesting a part of the honey: "I thus expect sharp criticism for having killed so many innocent hard-working creatures. With the following explanation, I anyway hope to receive some easement in the verdict". Bergman stressed the importance of avoiding overpopulation of bees in an area. The strategy that he had developed was to select the proper number of colonies among the healthiest young colonies in August, when the hives had achieved maximum weight. The remaining weaker colonies and the colonies from previous year were broken up. Bergman suggested planting trees and bushes to increase the yield of honey, and saw no reason why beekeeping would not be suitable also in the northern provinces of Sweden.

12.3 Bergman's Marriage

When Bergman moved in on the second floor of his laboratory building and started to receive salary from the University, his living conditions became more comfortable. It was not until now that Bergman could consider marriage.

On September 15, 1771, Bergman married Margareta Catharina Trast, born on July 4, 1735. Margareta was the daughter of Johan Trast, assistant vicar in Uppsala Cathedral. Her father had died aged just 39 when Margareta was only four years old, so Schufle's supposition that he attended the wedding cannot be true. Margareta's mother died when Margareta was 17 years old, not much more is known about her.

Little is known about Mrs Bergman, and Bergman's Swedish biographers only priced her for taking good care of her husband, an important qualification for a wife in the eighteenth century. It is also reported that she joined her husband on his trips [9]. Interestingly, Bergman does not mention his marriage or any other details from his private life in his autobiography. It is actually due to French sources, probably based on information from Bergman's French student Charles-André-Hector de Virely, that we have some scarce information about Bergman's wife [3]. Margareta is said to have shared her husband's interests for science, and when Bergman no longer had time for biological research, she kept his bees (Sect. 12.2), collected insects and grew plants. Their marriage was described as happy.

On March 5, 1773, Margareta gave birth to a son, named Torbern as his father. The christening ceremony seems to have been rather impressive. The ceremony was held in Uppsala Cathedral by Lars Hydrén,[3] vicar and professor in theology. The young Torbern had nine godparents, and many of them where professors at the University. Anders Berch (1711–1774) was professor in national economics, the first to hold such a position in Scandinavia. University librarian and historian Berge Frodin (1718–1783) was probably a personal friend of Bergman. Fredric Mallet was one of Bergman's friends from the Cosmographic Society and his time as an astronomer.

[3]Schufle thought that Hydrén was a physician attending the birth. This is not correct; Schufle may have been misled by Hydrén's doctoral title.

Mårten Strömer was professor in astronomy and Bergman's former teacher. Professor Daniel Melander (later ennobled Melanderhjelm; 1726–1810) was mathematician and astronomer. Miss Ullrika Klingenstierna was a daughter of Samuel Klingenstierna. Three godparents were absent: Nils Sahlgren from Gothenburg (apparently a relative to Bergman's mother), Bergman's former teacher Bengt Ferner, and Bergman's mother, Sara Bergman. Sadly, their son died on August 21 only five months old. According to Hjelm, the cause of death was dysentery [9]. He was buried three days later on the northern part of the cemetery reserved for university staff.

A second son, Carl Olof, was born on April 22, 1774 but died only two weeks later, on May 6. A more modest list of godparents suggests that the parents had little hope for their infant son's life. Bergman was now 39 and his wife 38; they would not get another child. Bergman seems to have worked hard during this traumatic period, possibly in order to distract his thoughts. He must have been working on the second edition of his *Physical Description of the Earth* and *The Chemical Lectures of H.T. Scheffer*. He also published a number of papers, including his paper on aerial acid, his student Level presented his thesis, and he also edited Scheele's paper on manganese. All of this happened during Scheele's period in Uppsala. Bergman also invited his mother, who lived her last years with him in Uppsala and died there in 1776, aged 80 [10].

References

1. Hjelm PJ (1786) Åminnelse-tal öfver...Herr Torbern Bergman, Stockholm, p 65
2. Trofast J (1994) Johan Gottlieb Gahn: brev, vol 2, Lund, p 54
3. Af Petersen H (1928) Om Torbern Bergmans och C. W. Scheeles franska förbindelser, Personhistorisk tidskrift, 191–201
4. Hjelm PJ (1786) Åminnelse-tal öfver...Herr Torbern Bergman, Stockholm, p 59
5. Bergman T (1780) Analyse et examen chimique de l'indigo, tel qu'il est dans le commerce, pour l'usage de la teinture. Piece qui a concouru pour le prix sur la nature & l'usage de l'indigo, Mémoires de mathématique et de physique, présentés à l'Academie Royale des Sciences, 9:121–164
6. Anonymous (1769) Om tall- eller furu-skogen KVA Handl 30:257–272
7. Bergman T (1769) Uplysning om skadelige tall-maskarne KVA Handl, 30:272–276
8. Bergman T (1779) Anmärkningar om bi, förnämligast i anledning af vägnings försök, KVA Handl, 40:300–329
9. Hjelm PJ (1786) Åminnelse-tal öfver...Herr Torbern Bergman, Stockholm, p 90
10. Hjelm PJ (1786) Åminnelse-tal öfver...Herr Torbern Bergman, Stockholm, p 7

13 Scheele in Uppsala

It was probably in order to get better access to a laboratory that Scheele moved to Uppsala in 1770. Others have suggested that it was in order to get into contact with Bergman and the University, but given the fact that Bergman had rejected the papers he had submitted to the Academy in Stockholm, and that it would take some effort to persuade Scheele to visit Bergman, it appears less likely. According to Wilcke/Sjöstén, Scheele moved to Uppsala in the autumn, [1] and this was generally accepted until Boklund investigated the matter in 1959 [2]. On August 6, 1770, Scheele signed a letter to Johan Gottlieb Gahn (more of him latter) in Uppsala. In the introduction of the letter, Scheele thanks Gahn for two previous letters, the first of which had been delayed at the post office since the staff did not know Scheele's address. Consequently, Scheele must have arrived in Uppsala at latest in the summer, more probably in the spring. Another clue is the chemical notes written by Gahn after his first meeting with Scheele, which are marked "Year 1770, in the spring" (although added in red ink, possibly at a latter point), but it is more likely that Scheele moved to Uppsala in spring than the autumn of 1770. A last piece of evidence is that Gahn took his degree on June 23, and probably left Uppsala soon after. On August 6, Gahn wrote to Bergman that he was *still* in Falun, indicating that he had been there for some time [3].

Scheele was hired by Christian Ludvig Lokk (1718–1800) at the Arms of Upland pharmacy (Uplands Wapen; Figs. 13.1 and 13.2). It was Lokk himself who established Arms of Upland in 1756. It was located in a wooden two-storey building at Stora torget (Great Square; see Fig. 4.1) and was the second pharmacy in Uppsala; the first pharmacy, Kronan (The Crown) belonged to the University. In 1868, the pharmacy changed its name to The Lion (Lejonet), after the name of the block where it was located, but it would remain in the same building until it was demolished in 1960 to give place for a department store. This building was, in turn, stripped down to its concrete frame in 2015 and rebuilt.

Lokk was German, as were so many other Swedish apothecaries. He was born in Hinter-Pommern and educated in Prussian Königsberg (present day Kaliningrad

A. Lennartson, *Carl Wilhelm Scheele and Torbern Bergman*, Perspectives on the History of Chemistry, https://doi.org/10.1007/978-3-030-49194-9_13

Fig. 13.1 The Arms of Upland pharmacy

Fig. 13.2 The historic building where Scheele made some of his most important discoveries was sadly enough demolished in 1960, to give place for a department store. *Photo* Anders Lennartson, June 2016

in Russia). Just like Kjellström, Scheele's principal from Malmö, Lokk had worked in Friedrich Ziervogel's pharmacy in Stockholm. This is not surprising, since the apothecaries in Sweden formed a well-connected network.

13.1 Scheele Befriends Johan Gottlieb Gahn

Soon after his arrival in Uppsala, Scheele met Johan Gottlieb Gahn (Fig. 13.3), and this would have a profound influence on his career. Gahn [4] was born on August 19, 1745, at Woxna works in the Hälsingland Province of Sweden. He graduated from the Gymnasium in Västerås in 1754 and enrolled at Uppsala University in

Fig. 13.3 Johan Gottlieb Gahn. Oil painting by Lorens Pasch the younger. *Photo* Nationalmuseum, Stockholm

1762, where he initially intended to study botany under Linnaeus. However, his interest soon shifted towards physics and chemistry, and he attended Bergman's lectures in chemistry in the autumn of 1768. Thought Gahn was one of Bergman's favourite students, he took his exam, not in chemistry, but in mining. His thesis [5] was on the subject "promotion of good economy at iron works" and was supervised by Pehr Niclas Christiernin. The thesis, 46 pages long, contains no discussions on chemistry or metallurgy, but contains many sound opinions on technical matters. While certainly a useful text in the mining districts, it is uncertain how many iron workers actually read Gahn's dissertation. In the introduction, Gahn,[1] if he was the true author, stresses that it is not primarily the supply of natural resources that makes a country rich, but the diligence and ability of the citizens to make proper use of the resources.

After graduating, Gahn moved to Falun, the location of one of the world's largest copper mines. Gahn planned either to make a foreign trip or to join Sven Rinman at the Garphyttan alum works in order to widen his knowledge, but was more or less forced into a collaboration with Samuel Gustaf Hermelin (1744–1820), a young mineralogist and cartographer. Hermelin, an acquaintance of Bergman, had been appointed by the Board of Mines to investigate the low quality of the Falun

[1]We cannot be sure Gahn is the actual author; in the eighteenth century it was common that the theses were actually written by the teacher rather than the student.

copper, and he attempted, with great persistence, to persuade Gahn to continue this work. Gahn was reluctant, as he wanted to continue his studies [6]. Bergman was aware of Hermelin's determination, and concluded that nothing could be done to change his mind. As matters unfolded, Gahn would remain in Falun during his entire career, but became a very skilled analytical chemist. In particular, he was a true master of blow pipe analysis (Sect. 23.5). Unlike Scheele and Bergman, however, he became more and more detached from academic chemistry and, surprisingly, he never published any of his experimental results. This is usually attributed to his very high expectations on his own work and the fear that his investigations were not thorough enough; all of Gahn's discoveries were made public by his friends. Bergman tried to persuade Gahn to publish, apparently in order to merit him for a membership in the Royal Swedish Academy of Sciences. Gahn answered, however, that he had little time for writing. Still, Gahn had a great reputation in Sweden, and was eventually elected a member of the Academy 1784.[2] At the end of his life, he befriended the young Berzelius, who was very impressed by Gahn's experimental skills, and persuaded Gahn to write a chapter on blow-pipe analysis for his textbook, and some essays for his *Afhandlingar i fysik, kemi och mineralogie* (*Memoires in Physics, Chemistry and Mineralogy*). Gahn married Anna Maria Bergström in 1784 and the couple had three children. As of 2019, dozens of descendent of the Gahn family lived in Sweden. Gahn himself passed away in Falun on December 8, 1818.

There are different versions of how Scheele and Gahn met. One story goes back to Carl Zetterström, who was hired as a private teacher for the children of Henric Gahn (1747–1816), physician and younger brother of Johan Gottlieb. In his diary, Zetterström describes a dinner at Johan Gottlieb Gahn's place in April 1794. The guests also included Anders Gustaf Ekeberg (Sect. 11.4.8) and Bergman's former student Bengt Reinhold Geijer. An excerpt from the diary reads:

> Gottlieb Gahn told how he discovered Scheele at Lokk's pharmacy, when he asked why it smelled of nitric acid[3] when Antimonium diaphoreticum [see Sect. 8.2] and Sal acetosellæ [potassium hydrogen oxalate] where mixed together, then one of the apprentices said: does not you Gentlemen understand that? Well, it works like this etc. And this pharmacy apprentice was Scheele. Gahn had difficulties to get him to Bergman, since Scheele had then already sent something to the Academy of Science in Stockholm [see Sect. 8.2], which Bergman, by not reading it correctly to the end, had given a negative review. [2]

As described in Sect. 8.2, Scheele had discovered how nitrous acid was liberated from nitrite in the mother liquor of *Antimonium diaphoreticum* by week acids. Gahn reportedly told the story to Bergman, who became curious and wanted to meet Scheele. A different version of the story is found in Thomson's *History of Chemistry* [7]. According to Thomson, who had met Gahn, Lokk had previously asked Gahn why potassium nitrate (saltpetre) that had been kept molten, gave off red

[2]According to preserved protocols, Gahn did not give any inaugural lecture in the Academy. These lectures had become more and more rare at the end of the century.

[3]Zetterström wrote"skedvatten", which has no direct English translation. The name came from German, Scheiden, to separate, and alludes to its property to separate silver from gold. "vatten" means water.

fumes when treated with vinegar. Gahn could not answer, and had promised to ask Bergman. When Gahn returned and said that not even Bergman new the answer, Lokk had already got the answer from Scheele. According to a third version of the story, Bergman, who bought his chemicals from Lokk, had sent Gahn to return a batch of saltpetre that Bergman claimed was impure as it gave red fumes with acetic acid after melting. Scheele then explained that nothing was wrong with the saltpetre. According to Sjöstén, Scheele befriended Johan Gottlieb's brother, physician Henric Gahn in Stockholm, and he claimed that it was Henric who introduced Scheele to his brother [8]. During this period Henric was both studying in Uppsala and practicing in Stockholm; he belonged to the same circles in Stockholm as Scheele, and was for example acquainted to Daniel Schulzenheim and at some point he befriended Scheele. Henric Gahn received one of the eight dedication copies of Scheele's book on air and fire (Sect. 21.3). Zekert claimed that Scheele and Gahn collaborated in Stockholm as early as 1768, [9] which seems unlikely. Cassebaum suggested that Scheele met Gahn in spring 1770, when he visited Uppsala to discuss his upcoming appointment [10]. It seems unlikely that Scheele would have made such a trip since he moved to Uppsala in spring 1770. Probably Cassebaum tried to explain why there are notes by Gahn dated in spring, when Scheele, as far as Cassebaum knew, moved in the autumn.

13.2 The Collaboration of Scheele and Gahn

As previously indicated, there is a preserved manuscript written by Gahn in the archive of the Royal Swedish Academy of Sciences containing notes made by Gahn after his meeting(s) with Scheele in spring 1770 (Fig. 13.4). The subjects include oxalic acid, tartaric acid, the transformation of water to earth, Prussian blue and animal earth. The note on animal earth is of special interest. Animal earth was the ashes left after calcining bones or horns. Scheele had found that the residue remaining after calcining dear horns consisted of calcium (lime) and a substance that he could not identify. Gahn has latter added a note reading:

> on account of the preceding, J.G.G. immediately undertook to investigate the animal earth and found soon that it consisted of lime [calcium] with acidum phosphori [phosphoric acid]. This, Mr Scheele initially refused to believe, even though he saw all the phenomena, and the identity of the reduced acid to correspond to phosphoric acid in every manner. He still had some doubts, until he saw phosphorus being prepared from it, which he first accomplished on a large scale in Uppsala in the summer of 1770, after I, in the spring,[4] with blow-pipe and in small glass tubes, had only showed him signs.

Previously, the only known source of phosphorus (Fig. 13.5) was urine, and the price of phosphorus was very high, not least since most workers had hard to stand the repulsive smell encountered in the extraction process [11]. The discovery by Gahn and Scheele would eventually greatly reduce the price of phosphorus, never

[4]This further proofs that Scheele arrived in Uppsala at the latest in spring.

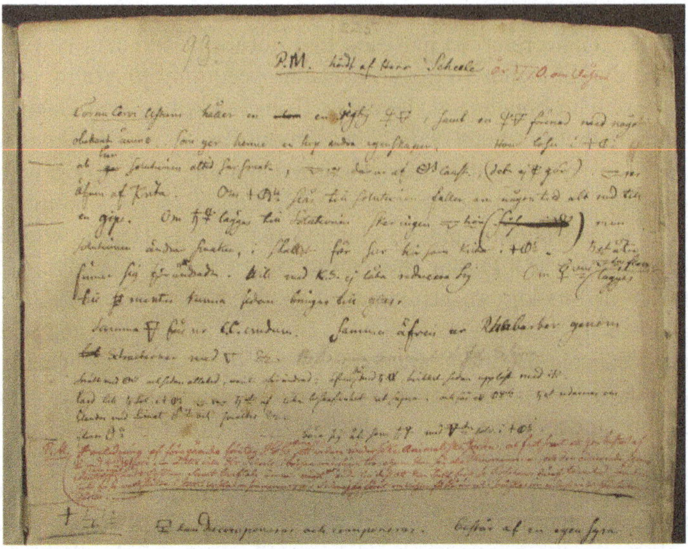

Fig. 13.4 Gahn's notes from his meetings with Scheele in Spring 1770. Carl Wilhelm Scheele's Archive, Centre for History of Science, Royal Swedish Academy of Sciences. *Photo* Anders Lennartson

Fig. 13.5 White phosphorus, typically yellowish due to small amounts of red phosphorus, is a highly flammable and toxic substance that must be kept under water to avoid autoignition. *Photo* Anders Lennartson

the less the two chemists did not initially publish their findings. Instead, the knowledge spread by more informal ways; Bergman mentioned it in a letter to Macquer in July the same year (before he met Scheele) and Scheele mentioned it in a footnote to his paper on fluorite the following year [12]. In 1774 the editor of *Nya Lärda Tidningar* published an excerpt from a letter written by Scheele giving a detailed account of the process [11]. Gahn's brother Henric published an account in a Scottish medical journal in 1775, which also appeared in German translation in 1776. During the following years, the refinement of the process received much international attention [13]. Usually, Scheele was credited for the discovery, while Gahn's contribution was unknown to most chemists. Berzelius was among the first to give Gahn proper credit [14].

13.3 Scheele Meets Bergman

Gahn tried to persuade Scheele to contact Bergman, but apparently Scheele was initially reluctant to do so. It should be kept in mind that the hierarchy of the eighteenth century made it difficult for a simple apothecary apprentice to seek contact with a professor. Also, Bergman's rejection of Scheele's papers is believed to have contributed to Scheele's reluctance. In a letter to Gahn, who had left Uppsala, dated August 6, Scheele wrote that Bergman had not yet returned from his summer vacation, but that he was expected to return within 10 days. In a letter from Gahn to Bergman dated August 16, Gahn had heard from his brother that Bergman had returned to Uppsala from Mariestad [15]. Scheele is believed to have visited Bergman soon after his return.

It was two rather young men that met; Scheele was 27 and Bergman 35. When it came to chemistry, there is no doubt that Scheele was the more experienced of the two, while Bergman had the broad academic training and was a talented writer. An indication of how fast their friendship developed is the fact that Scheele appears to have acted as Bergman's assistant when he was visited by Crown Prince Gustav and Prince Heinrich of Prussia in September (Sect. 13.6). Perhaps Bergman, who had taken all measures to make a good impression, did not trust his laboratory assistants, Tidström and Wibom, who were probably still loyal to Wallerius.

13.4 Scheele's First Papers

After Scheele had befriended Bergman, his situation changed dramatically, and he had no more problems publishing his results in the *Transaction of the Royal Swedish Academy of Sciences*. His first papers were submitted directly to Bergman, who read them through before sending them to the Academy. To what extent Bergman may have edited Scheele's early papers cannot be determined. The first paper published by Scheele [12] was an investigation on the mineral fluorite (fluorspar; CaF_2), which included the highly controversial discovery of hydrofluoric acid (Sect. 14.1) [16]. The investigation of fluorite may have been carried out on his own initiative, while the topic for his second paper, the mineral pyrolusite (MnO_2; Sect. 15.1), was suggested to him by Bergman. The long delay between the first and second paper, which did not appear until 1774, [17] had several reasons. One reason was that Bergman had high expectations and ambitions that stretched beyond those of Scheele [18]. He thus persuaded Scheele to continue the experiments although Scheele himself believed he had exhausted the subject. This was very fortunate for Scheele, as the resulting paper would be one of Scheele's most important works including the discovery of three chemical elements [19]. Scheele proved the elemental nature of manganese (manganese sulphate was known, but was confused with alum), he proved the elemental nature of barium and was the first to describe elemental chlorine. Other reasons for the delay were probably his concurrent experiments with air and fire and the 1772–1773 famine that kept Scheele busy in the pharmacy. The length of Scheele's manganese paper, Bergman referred to it as a small book in a letter to Gahn, was also a problem. According to a letter to Gahn, Bergman appears to have had problems getting the paper printed, and it had to be divided into two parts. It was this paper that made Scheele famous and was the main merit when he was elected a member of the Royal Swedish Academy of Sciences (Sect. 13.5) and offered the professorship in Berlin (Chap. 16). While in Uppsala, Scheele published only one more paper, on an improved method for preparing benzoic acid (Sect. 24.1), [20] but he had many unpublished results, including the discovery of oxygen, when he left the city after 5 years.

13.5 Scheele Elected a Member of the Royal Swedish Academy of Sciences

The ultimate proof of Scheele's growing reputation after the publication of the manganese paper is the fact that Bergius suggested Scheele as a member of the Royal Swedish Academy of Sciences on July 27, 1774 [21]. A second suggestion (also by Bergius) was submitted on October 26 and on February 4, 1775, Scheele was elected a member on a meeting in the presence of King Gustav III. His election is rather remarkable, given that Scheele, aged 33, had yet not even taken his apothecary exam. The news was brought to Scheele in a letter from Wargentin, dated February 6. Scheele did not, however, make the trip to Stockholm to take his seat in the Academy. Not until Bergman's term as president in 1777, was Scheele persuaded to deliver is inaugural lecture.

13.6 Prince Heinrich's Visit

On September 18, 1770, Crown Prince Gustav arrived in Uppsala with his guest Prince Heinrich of Prussia (1726–1802), brother of Frederick the Great (Friedrich II) and uncle of Gustav [22]. They arrived at 6 p.m., and their first stop was the University Library where they visited Professor of philology Johan Ihre. Next, the two princes visited the Professor of economy, Anders Berch, at the Department of Economy, *Theatrum oeconomico-mechanicum*. At about 8 p.m. they arrived at *Laboratorium chemicum*. Although Bergman had only known Scheele for a few weeks at most, it is generally agreed that Scheele appeared as Bergman's assistant. The source is a manuscript attributed to apothecary Helwig in Stralsund, a friend of both Scheele and his family [23]. According to Helwig, Heinrich was accompanied by the Duke of Södermanland, and consequently it has usually been supposed that he was accompanied by Prince Charles (Karl) who would become King Charles XIII (Karl XIII) in 1809. This is obviously wrong, as Charles did not become Duke until 1772, and spent the autumn of 1770 abroad [24]. Apparently, Wallerius was in the audience, probably hoping that Bergman would reveal his scarce knowledge in chemistry and embarrass himself. Bergman wrote to Gahn: "Both W [almost certainly Wallerius] and V arrived quite early, but I have been told by several [people], who had better opportunity than I to watch [them], that they finally left with very dissatisfied faces". An article in *Allmänna Tidningar* (Public Newspaper) from September 20 gives a clear account of the event. Bergman first showed a number of experiments. He demonstrated mixtures that "gave off quite elastic fumes" that exploded on ignition by the flame of a candle, and continued to burn for quite some time. This seems to have been hydrogen. He also showed mixtures that ignited spontaneously, and a mixture of which a pinch was heated on a metal sheet over a flame and exploded with such violence that the sheet was broken. This was probably a mixture of potassium nitrate, potassium carbonate and sulphur. After these spectacular experiments which would have caught the interest of the audience, Bergman turned to more practical matters and demonstrated how ores and metals could be examined with the aid of the blow-pipe (Sect. 23.5). This would have convinced the princes of the importance and usefulness of chemistry. Next, the two princes were taken to the Gallery, where the most spectacular specimens from the mineral collection were demonstrated. Bergman had planned the display carefully and placed two white wax candles on each cupboard for the visual effect [25]. Gustav and Heinrich left Bergman at 10 p.m. for a dinner with the Bishop. The following day, the two princes spent an hour in the Cathedral, made a trip to the Viking age tumuli at Old Uppsala 5 km from the city centre and on the way back they visited the botanical garden and the astronomical observatory. They left Uppsala at 2 p.m. Thus, a considerable part of the visit was devoted to Bergman's laboratory and Gustav seems to have attributed it great importance compared to the other items on the schedule. This is likely a clear evidence of Gustav's appreciation for (and expectations of) Bergman.

References

1. Sjöstén CG (1801) Åminnelse-tal, hållit för Kongl. Vetenskaps Academien öfver dess framledne ledamot herr Carl Vilhelm Scheele, den 14 oct. 1799. Stockholm, p 15
2. Boklund U (1959) När Gahn upptäckte Scheele på Lokks apotek. Lychnos, 217–222
3. Trofast J (1994) Johan Gottlieb Gahn brev, vol 2, Lund, p 16

4. Trofast J (1996) Johan Gottlieb Gahn—En bortglömd storhet, Lund
5. Christiernin PN, Gahn, JG (1770) Några anmärkningar i svenska bergslagfarenheten om författningar till befrämjande af god hushållning vid järnhyttor. Hvilka till en academisk läroöfning med den vidtlagfarne jurid. facultetens samtycke, vid kongl academien i Upsala, under ... Pehr Niclas Christiernins inseende d. 23 junii 1770. för middagen, uti den större Carol. lärosalen till offenteligt försvarande framgifne blifvit af Jan Gottlieb Gahn, ifrån Fahlun. Uppsala
6. Fors H (2003) Mutual Favours (diss.). Depatment of History of Science and Ideas, Uppsala University, Uppsala, p144
7. Thomson T (1831) The history of chemistry, vol 2. Henry Colburn and Richard Bentley, London, p 55
8. Sjöstén CG (1801) Åminnelse-tal, hållit för Kongl. Vetenskaps Academien öfver dess framledne ledamot herr Carl Vilhelm Scheele, den 14 oct. 1799. Stockholm, p 16
9. Zekert O (1963) Carl Wilhelm Scheele. Wissenschaftlige Verlagsgesellschaft, Stutgart, p 42
10. Cassebaum H (1982) Carl wilhelm scheele. Teubner Verlagsgesellschaft, Leipzig, p 14
11. Anonymous (1774) Bref, Nya lärda Tidningar, 1:108–110
12. Scheele CW (1771) Undersökning om Fluss-Spat och dess Syra, KVA Handl 32:120–138
13. Crell L (1785) Neuere Bereitungsart des Phosphors aus Knochen. Chem Ann 2: 503–509
14. Berzelius JJ (1808) Föreläsningar i djurkemien vol 2, Stockholm, p 122
15. Trofast J (1994) Johan Gottlieb Gahn brev, vol 2, Lund, p 17
16. Lennartson A (2017) The chemical works of carl wilhelm scheele. Springer, Cham, p 22
17. Scheele CW (1774) Om Brun-sten eller Magnesia och dess Egenskaper, KVA Handl, 35:89–116; 177–194
18. Oseen CW (1940) Torbern Bergman och Carl Wilhelm Scheele. KVA, Stockholm
19. Lennartson A (2017) The chemical works of carl wilhelm scheele. Springer, Cham, p 25
20. Scheele CW (1775) Anmärkningar om Bezoë-Saltet, KVA Handl 36:128–133
21. Hildebrand B (1936) Scheeleforskning och Scheelelitteratur Lychnos p 76–102
22. Annerstedt C (1913) Uppsala universitets historia, vol 3:1, Almqvist & Wiksell, Uppsala, p 453
23. Fredga A (1943) Carl Wilhelm Scheele. KVA, Stockholm, p 20
24. Fredga A (1946) Carl Wilhelm Scheele. Stockholm, KVA, p 20
25. Trofast J (1994) Johan Gottlieb Gahn: brev, vol 2, Lund, p 32

14 New Mineral Acids

Three mineral acids were known to the chemists in the seventeenth century: sulphuric acid, nitric acid and hydrochloric acid. In addition, phosphoric acid was known, but as the only known source of this acid at the time was urine, it was not recognised as a mineral acid. Scheele made important and controversial discoveries that forced chemists to give up the idea of three mineral acids.

14.1 Hydrofluoric Acid

Scheele's first paper [1] was an investigation of the mineral fluorite (fluorspar; CaF_2; Fig. 14.1) [2]. It may have been this mineral's property to glow in the dark that first attracted Scheele's attention, but Bergman has written in a letter to Macquer that the investigation was initiated on his suggestion [3]. The important discovery in this study occurred when he distilled a mixture of fluorite and sulphuric acid in a retort, with the receiver filled with water (Fig. 14.2). Scheele observed that the water turned into a fuming acid and that a white crust formed at the water surface. The acid showed similarities with hydrochloric acid, but had some distinct properties showing that this was a hitherto unknown acid. On heating fluorite and sulphuric acid, hydrogen fluoride (HF) is formed. Hydrogen fluoride is one of only a few substances that rapidly attack glass at room temperature, giving silicon tetrafluoride (SF_4) and hexafluorosilicic acid (H_2SiF_6). Upon contact with the water in the receiver, silicon tetrafluoride is hydrolysed to silisic acid, $Si(OH)_4$, which undergoes condensation to hydrated silicon dioxide species which formed the white crust that Scheele observed. Scheele's new acid was thus a solution of hydrofluoric acid (HF) and hexafluorosilicic acid. Unknown to Scheele, an artist from Nürnberg named Schwanhardt had found a mixture of fluorite and sulphuric acid useful for etching glass, and a similar recipe replacing sulphuric acid with nitric acid was introduced by Matthäus Pauli in 1725 [4]. Apparently unknown to Scheele, Marggraf had studied fluorite in the late 1760s, and also found that distillation of

A. Lennartson, *Carl Wilhelm Scheele and Torbern Bergman*, Perspectives on the History of Chemistry, https://doi.org/10.1007/978-3-030-49194-9_14

Fig. 14.1 Octahedral crystals of fluorite. The green colour originates from trace amounts of transition metal ions. *Photo* Petra Rönnholm

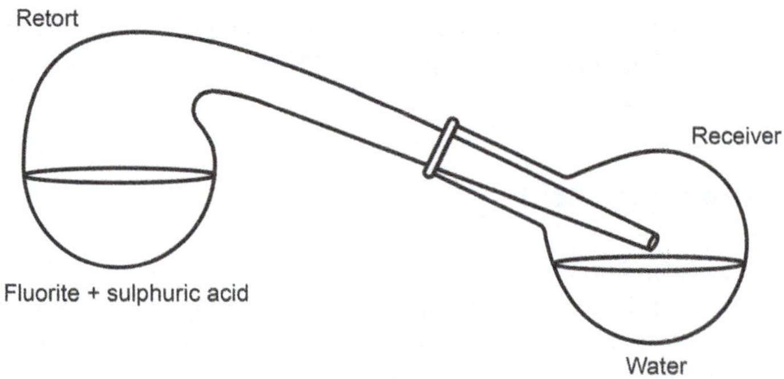

Fig. 14.2 The apparatus used by Scheele to isolate hydrofluoric acid. Anders Lennartson

fluorite with sulphuric acid in a glass retort etched the glass and that an earth collected in the receiver [5]. His conclusion was, however, that sulphuric acid extracted a volatile earth from fluorite.

Scheele's paper gained considerable attention and scepticism from the scientific community. A fourth mineral acid was unexpected and the fact that it always contained dissolved silica raised suspicions. Macquer found it so improbable that he asked Bergman for more information. Bergman could confirm Scheele's results and, by describing the experiments, he enabled Macquer to reproduce the experiments in Paris [3]. Priestley devoted 25 pages to the new acid in his book on gases and called it "Swedish acid": "This curious discovery was made by Mr. Scheele, a Swede; from which circumstances the acid is often distinguished by the name of the Swedish acid" [6]. The retired Wallerius, in a book about the formation and evolution of Earth published a few years later, took the formation of the white crust at the water surface as experimental proof that water can transform into earth; [7] this was important for Wallerius, who was searching the scientific means by which God created Earth. According to Genesis, there was first water, then earth, and Wallerius thought that Scheele's work could be a clue to the underlying mechanisms of the Creation. It is quite typical for Wallerius to pick out the facts that best fit his own system and reject the rest. Gustaf von Engeström in a lecture (Sect. 9.6) in the Royal Swedish Academy of Sciences, however, took Scheele's work as an example of what he saw as the unreliable nature of modern chemistry (Sect. 13.6) [8].

Scheele's claim to have discovered a new mineral acid was his most controversial discovery, and he had to write two more papers to defend his ideas. In 1780, he published the second paper on hydrofluoric acid disproving theories [9] put forward by Monnet and Boullanger [10]. Scheele showed that all calcium (lime) from the fluorite and all sulphuric acid added could be accounted for in the formed calcium sulphate (gypsum), and that the acid was free from hydrochloric acid [9]

It was Scheele's friend Johan Christian Friedrich Meyer (1739–1811)[1] in Stettin who first prepared silicon-free hydrofluoric acid by performing the operation in an apparatus made of lead in 1781, which he told Scheele in a letter. On March 9, Scheele wrote to Bergman admitting that he had been wrong on believing that the silicon was an integral part of the acid. He also admitted it in a letter to Ehrhart, who printed the letter in his journal [11]. In a note, Ehrhart added that "Finally my friend has surrendered and replaced his unintelligible idea with the truth". Scheele was probably not pleased by this comment since he did not send any more letters to Ehrhart (at least not for publication). Scheele had to return a final time to his hydrofluoric acid, in which he described the preparation of pure hydrofluoric acid [12] in a glass-free apparatus [13].

Silicon tetrafluoride was later isolated by John Davy, brother of Sir Humphry Davy, and when Berzelius passed silicon tetrafluoride over potassium, he became the first chemist to isolate elemental silicon [14]. The isolation of fluorine from hydrofluoric acid would be more troublesome, due to the extreme toxicity of hydrofluoric acid. Scheele was actually very fortunate to survive his studies. Many of the chemists who attempted to isolate fluorine had to pay a very high price, including Humphry Davy, Gay-Lussac, and Thénard [15]. The two Irish brothers George and Thomas Knox were seriously injured and both P Louyet and Jérôme Nicklès died. Elemental fluorine was finally isolated by French chemist Henri Moissan (1852–1907) in 1886 [16]. In part due this achievement, Moissan was awarded the 1906 Nobel Prize in chemistry.

[1]Not to be confused with chemist Johann Friedrich Meyer, who launched a theory of carbon dioxide and lime (Chap. 18).

14.2 Arsenic Acid

In part as a consequence of the discovery of chlorine (Sect. 15.2) during his studies of manganese, Scheele discovered another mineral acid, arsenic acid (H_3AsO_4). Salts of arsenic acid, arsenates, had been studied by Macquer, and arsenic acid had been isolated by Cavendish in 1767, but his unpublished results were unknown until 1840. It is also possible that Georg Brandt prepared the acid in 1728, but without any clear idea of its nature.

Scheele described two methods [17] to prepare the new acid. The first method involved dissolving arsenic (III) oxide (arsenic) in hot hydrochloric acid (*Spiritus salis*) and the addition of nitric acid (*Spiritus nitri*) to the hot solution. In the second method, chlorine was bubbled through a suspension of arsenic(III) oxide in water. Upon heating, hydrochloric acid and arsenic(III) chloride (*Butyrum arsenici*) distilled over, leaving arsenic acid in the retort. Before proceeding with his experiments, Scheele performed a crude toxicological study:

> Before I subjected this acid to my experiments, I was curious to know whether it is as fatal as arsenic itself. I therefore mixed a little of it with honey, and exposed it to the flies, when I found that it killed them in an hour. I gave 8 grs. [0.5 g] of the powdered acid, enclosed in a piece of meat, to a cat, which, two hours afterwards, seemed to be at the point of death; I then gave it some milk, upon which it vomited violently, and afterwards ran away. [18]

Scheele's results were very important for several reasons. He had isolated another new mineral acid, and as indicated earlier, there were only three mineral acids generally recognised. The discovery of arsenic acid was easier for the scientific community to accept than the discovery of hydrofluoric acid. Arsenic acid could be isolated in pure form, unlike hydrofluoric acid with its suspicious silica content, and the solid arsenic acid could hardly be confused with any other known acid. More importantly, Scheele now realised that, in modern terminology, arsenic (III) oxide (arsenic) could not only be reduced to elemental arsenic (arsenic regulus) but also *oxidised*. In Scheele's phlogistic terminology, he concluded that arsenic (i.e. arsenic(III) oxide) was not a true element, but it was actually composed of arsenic acid and phlogiston.

Scheele also found that he could obtain two different sodium salts of arsenic acid with different alkali content, i.e. NaH_2AsO_4 and Na_2HAsO_4, and had thus discovered that arsenic acid is a polybasic acid. The same year Bergman noted in his *Attractionibus electivis* [19] that tartaric and sulphuric acid behaved similarly. The existence of a normal and acidic sulphate of potassium had been indicated in 1754 by Rouelle, [20] but the idea of polybasic acids is usually attributed to Scottish chemist Thomas Graham (1805–1869) [21] and German chemist Justus Liebig (1803–1873) [22].

In a letter to Hjelm, dated March 6, 1775, while he still lived in Uppsala, Scheele wrote that he had given the arsenic manuscript to Bergman, who would read it and forward it to the Academy in Stockholm. The printed version appeared in the October to December issue of the *Transactions* for of 1775 [23].

Fig. 14.3 Scheele's green, a complex mixture of copper arsenite compounds. *Photo* Petra Rönnholm

While working with arsenic, Scheele discovered a green pigment (Fig. 14.3) obtained from potassium arsenite (itself obtained by dissolving arsenic (III) oxide in a solution of potassium carbonate) and copper sulphate [24]. Scheele first mentioned this pigment in a letter to Gahn, dated May 10, 1776, where he asked if Gahn knew of a pigment containing copper and arsenic. Scheele published the preparation of this pigment in 1778, when he had concluded that three-year old samples had retained their colour [25]. The pigment, now generally known as Scheele's green, consists of a mixture of compounds such as copper (II) metaarsenite ($Cu(AsO_2)_2$), copper (II) hydrogen arsenite ($CuHAsO_3$), copper (II) arsenite trihydrate ($Cu(AsO_3)_2 \cdot 3H_2O$), copper (II) ortoarsenite dihydrate ($Cu_3As_2O_6 \cdot 2H_2O$), and copper(II) diarsenite dehydrate ($Cu_2As_2O_5 \cdot 2H_2O$).

Scheele's green, and other related pigments, were used extensively in the nineteenth century for printing wallpaper. This use was unfortunate, as certain fungi can metabolise Scheele's green to highly toxic gaseous trimethyl arsane ($As(CH_3)_2$) which caused several cases of arsenic poisoning and death, possibly including that of Napoleon [26]. Although Scheele did not express any worries regarding the pigment itself, he actually expressed some environmental concerns regarding the waste: "The water with which the colour is lixiviated [i.e. the waste] contains a little arsenic, and must therefore be thrown out in a place to which cattle have no access" [27].

14.3 Prussian Blue and Hydrocyanic Acid

Prior to the eighteenth century, access to blue pigments was limited, and the blue pigments known (e.g. ultramarine prepared by powdering *Lapis lazuli*) were expensive. The knowledge of preparing Egyptian blue had, for instance, long been lost. The much cheaper pigment Prussian blue was discovered in 1704 by a German named Diesbach, who kept the preparation process secret. In 1724, John Woodward (1665–1728) could reveal the secret. Scheffer gives the following method for preparing Prussian blue: [28] sodium carbonate (soda), purified by dissolution in hot water, filtration and crystallisation, is mixed with the same weight of dried powdered ox blood. The mixture is heated, where upon the blood starts to burn and a viscose melt is obtained. The residue is cooled, powdered and extracted with hot water. This extract contained potassium cyanide and potassium hexacyanoferrates, and was known as *Lixivium sanguinis* or phlogisticated alkali. Prussian blue was finally precipitated by addition of a filtered solution of one part potassium aluminium sulphate (alum) and five parts iron(II) sulphate (iron vitriol). An early study of the chemical properties of Prussian blue was reported by Marggraf in 1745. Macquer isolated potassium hexacyanoferrate(II) from *Lixivium sanguinis* in 1752 (potassium hexacyanoferrate(III) was not isolated until 1822 [29]), otherwise little was known about the commercially very important pigment.

In a modern laboratory, potassium hexacyanoferrate(II) and hexacyanoferrate (III) can be prepared by addition of potassium cyanide to an iron(II) and iron(III) salt, respectively:

$$Fe^{2+}(aq) + 6CN^-(aq) \rightarrow [Fe(CN)_6]^{4-}(aq)$$
$$Fe^{3+}(aq) + 6CN^-(aq) \rightarrow [Fe(CN)_6]^{3-}(aq)$$

By adding an iron(II) salt to potassium hexacyanoferrate (III), or an iron(III) salt to potassium hexacyanoferrate(II), Prussian blue is obtained (Fig. 14.4), e.g.

$$4Fe^{3+}(aq) + 3[Fe(CN)_6]^{4-}(aq) \rightarrow Fe_4[Fe(CN)_6]_3(s).$$

The structure of Prussian blue is a framework of octahedral iron ions connected by cyanide ions. The intense blue colour arises from electron transfer between iron (II) and iron (III) sites in the lattice.

From Scheele's letters to Retzius, it is known that Scheele's interest in Prussian blue dates back at least to his time in Malmö, but it was not until his time in Köping (Chap. 19), that he managed to elucidate the composition of this pigment. By spring 1770, Scheele had (independently of Macquer and Sage) prepared potassium hexacyanoferrate(II), and he had also obtained ammonia by heating Prussian blue. In the summer of 1774, he discovered that *Lixivium sanguinis* lost its ability to precipitate Prussian blue after treatment with sulphuric acid. In 1780, he had concluded that the "colouring component" in Prussian blue was itself an acid; it was a very weak acid and even carbon dioxide could destroy *Lixivium sanguinis*. By the end of May 1782, he had evidently isolated this acid, known to modern chemists as hydrogen cyanide (HCN). The cyanide ions are very strongly bound to the iron ions in Prussian blue, and consequently hydrogen cyanide is not liberated by action of acids on Prussian blue, which is non-toxic. His method [30] of obtaining hydrogen cyanide was to boil Prussian blue with mercury(II) oxide (*Mercurius calcinatus*),

Fig. 14.4 Precipitation of Prussian blue by adding a solution of iron(III) nitrate to a solution of potassium hexacyanoferrate(II). *Photo* Petra Rönholm

which resulted in a solution with a "strong mercurial taste". This solution (containing mercury(II) cyanide) was treated with iron flings to give a solution of iron cyanide and metallic mercury. Since cyanide is less strongly bound to iron than to mercury, hydrogen cyanide could now be liberated by sulphuric acid and separated as an aqueous solution by distillation. The method that Scheele arrived at is rather complex and certainly impressive, and it is clear that Scheele must have put a lot of effort on the project before arriving at the final procedure.

Scheele accidently discovered that hydrogen cyanide is combustible:

> I met by accident with a very remarkable phenomenon: As I was one evening about to pour the liquor of the first distillation of the colouring matter out of the receiver into a bottle [...], and a burning candle happened to be standing near the orifice, the air contained in the receiver instantly took fire, without, however, any explosion" [31]. By combustion of hydrogen cyanide, Scheele obtained carbon dioxide (aerial acid) and by heating metal cyanides he obtained ammonia. From this he concluded that "the constituent parts of the colouring matter were volatile alkali [ammonia] and an oily substance [i.e. carbon and hydrogen]. [32]

When translated to modern terminology this is equivalent to saying that it was composed of carbon, nitrogen and hydrogen. He failed to combine oils or fats with ammonia, but when he added ammonium chloride to a red-hot mixture of potassium carbonate and charcoal or graphite, he obtained cyanide. Addition of iron(II) sulphate gave hexacyanoferrate from which Scheele obtained Prussian blue.[2]

Due to its length, the paper had to be split into two parts published in the *Transactions* in 1782 [33] and 1783 [34]. The manuscript for the second part was sent to Bergman on February 28, 1783, with the request that Bergman would forward it to Wargentin and the Academy in Stockholm.

Scheele was completely unaware of the toxicity of hydrogen cyanide and gave the following description: "This matter has a peculiar but not disagreeable smell, a taste somewhat approaching to sweet, and warm in the mouth, at the same time exciting cough" [35]. Scheele was quite lucky to survive his experiments!

References

1. Scheele CW (1771) Undersökning om Fluss-spat och dess syra KVA Handl 32:120–138
2. Lennartson A (2017) The chemical works of carl wilhelm scheele. Springer, Cham, p 22
3. Carelid G, Nordström J (1965) Torbern Bergman's Foreign Correspondence, Almqvist & Wiksell, Uppsala, p XXXV
4. Kopp H (1845) Geschichte der Chemie vol 3 Braunschweig, p 368
5. Marggraf, AS (1770) Observation concernant une volatilisation remarquable d'une partie de l'espece de pierre, à laquelle on donne les noms de flosse, flüsse, flus-spaht, et aussi celui d'hesperos; laquelle volatilisation a été effectuée au moyen des acides, Mémoires de l'Académie royale des sciences et belles-lettres, 24:3–11
6. Priestley J (1775) Experiments and observations on different kinds of air, vol 2, J Johnson, London p 187
7. Wallerius JG (1776) Tankar om verldenes, i synnerhet jordenes, danande och ändring, Stockholm, p 51
8. von Engeström G (1782) Tal, om vissa svårigheter och andra omständigheter, som möta vid utöfvandet af chymien, Stockholm p 3
9. Lennartson A (2017) The chemical works of carl wilhelm scheele. Springer, Cham, p 62
10. Scheele CW (1780) Anmärkningar om Fluss-Spat, KVA Nya Handl 1:18–26
11. Scheele CW (1782) Zweyter brief, Neues Magazin für Ärtze 4:292–295
12. Lennartson A (2017) The chemical works of carl wilhelm scheele. Springer, Cham, p 97
13. Scheele CW (1786) Neue Beweise der Eigentümlichkeit der Flussspatsäure Chem Ann 1:3–17
14. Berzelius J (1824) Undersökning af flusspatssyran och dess märkvärdigaste föreningar 46–98
15. Weeks ME (1956) Discovery of the elements, easton, p 762
16. Moissan H (1886) Action d'un courant électrique sur l'acide fluorhydrique anhydre. Compt Rend 102:1543–1544
17. Lennartson A (2017) The chemical works of carl wilhelm scheele. Springer, Cham, p 30
18. Scheele CW (1901) The chemical essays of Charles-William Scheele. Reissue of 1786 edition. Scott, Greenwood & Co, London, p 109

[2]This synthesis of potassium cyanide is certainly not the first approach that comes to the mind of a modern chemist, but it is easy to reproduce Scheele's results. A powdered mixture of charcoal and potassium carbonate is heated to near fusion. Ammonium chloride is added and the mixture stirred until the sublimation of ammonium chloride ceases. The mixture is allowed to cool, extracted with water and filtered. Addition of iron(II) sulphate solution to the colourless filtrate gives a precipitate of iron carbonate. Addition of an acidic solution of iron(III) chloride dissolves the precipitate, and if the green solution is left for a few minutes, Prussian blue precipitates.

19. Bergman T (1775) Disquisitio de Attractionibus Electivis. Nova Acta Regiae Societatis Scientiarum Upsaliensis 2:159–248. English translation by J. A Schufle: Bergman T (1968) Dissertation on elective attractions. Johnson Reprint Corporation, New York
20. Rouelle M (1754) Sur les sels neutres, Histoire de l'Académie royale des sciences, 572–588
21. Graham T (1833) Researches on the Arseniates, Phospates, and Modifications of Phosphoric acid. Phil Trans 123:253–284
22. Liebig J (1838) Ueber die Constitution der organischen Säuren. Ann 26:113–189
23. Scheele CW (1775) Om Arsenik och dess syra, KVA Handl 36:263–294
24. Lennartson A (2017) The chemical works of carl wilhelm scheele. Springer, Cham, p 54
25. Scheele CW (1778) Tilrednings-sättet af en ny grön Färg, KVA Handl 39:327–328
26. Emsley J (2005) Elements of Murder. Oxford University Press, Oxford, p 126
27. Scheele CW (1901) The chemical essays of Charles-William Scheele. Reissue of 1786 edition. Scott, Greenwood & Co, London, p 186
28. Bergman T (1775) H. T. Scheffers Chemiske föreläsningar, rörande salter, jordarter, vatten, fetmor, metaller och färgning, samlade, i ordning stälde och med anmärkningar utgifne af T. B., M. Swederus, Uppsala, p 158
29. Gmelin L (1822) Über ein besonderes cyaneisenkalium. Journal für Chemie und Physik 34:325–346
30. Lennartson A (2017) The chemical works of carl wilhelm scheele. Springer, Cham, p 75
31. Scheele CW (1901) The chemical essays of Charles-William Scheele. Reissue of 1786 edition. Scott, Greenwood & Co, London, p 285
32. Scheele CW (1901) The chemical essays of Charles-William Scheele. Reissue of 1786 edition. Scott, Greenwood & Co, London, p 287
33. Scheele CW (1782) Försök, beträffande det färgande ämnet uti Berlinerblå KVA Nya Handl, 3:264–265
34. Scheele CW (1783) Om det färgande Ämnet uti Berliner-blå, Fortsättning KVA Nya Handl, 4:33–43
35. Scheele CW (1901) The chemical essays of Charles-William Scheele. Reissue of 1786 edition. Scott, Greenwood & Co, London, p 244

15 New Metals

Seven metals had been known for millennia: gold, silver, copper, iron, lead, tin, and mercury. Zinc and bismuth were known, but not universally recognised as unique metals until the eighteenth century. Platinum had been known to the native South Americans and was brought to Europe in the 1730s, but many chemists believed it to be an alloy. The first new metals discovered in modern times were cobalt in *c.* 1735 and nickel in 1751. Scheele and Bergman contributed to the discovery of a number of new metals and contributed to the knowledge of several of the recently discovered metals.

15.1 The Investigation of Pyrolusite: The Discovery of Manganese

On Bergman's suggestion, Scheele undertook the study of a mineral known as manganese[1] or *Magnesia nigra* (Fig. 15.1) which consists mainly of manganese (IV) oxide (MnO_2). To avoid confusion with the metal now called manganese, the modern name pyrolusite will be used for the mineral throughout this text. The statement by Oseen[1] that Scheele had described experiments with pyrolusite to Gahn before he met Bergman is in error; Scheele is referring to *Magnesia alba* (basic magnesium carbonate) and not *Magnesia nigra* (pyrolusite).

Bergman may have suggested the study of pyrolusite already at his first meeting with Scheele, or soon after that meeting. It had been observed by German physician and chemist Johan Heinrich Pott (1692–1777) that pyrolusite absorbed phlogiston on heating and became soluble in acids, i.e. it is reduced to manganese (II).

[1]Scheele and Bergman used the Swedish name brunsten = brown stone.

A. Lennartson, *Carl Wilhelm Scheele and Torbern Bergman*, Perspectives on the History of Chemistry, https://doi.org/10.1007/978-3-030-49194-9_15

Fig. 15.1 A piece of pyrolusite. *Photo* Petra Rönnholm

Bergman, always interested in the larger contexts, wanted to have the chemical nature of pyrolusite established: although pyrolusite was used in pottery and glass making, its composition was unknown. Cronstedt suggested that it contained an unknown earth, while Pott and Westfeld claimed that it could be converted to alum (and thus would contain aluminium in modern terminology). Their conclusion is difficult to understand given the pink colour of hydrated manganese (II) sulphate.

Scheele rapidly disproved Pott's and Wesfeld's claims and communicated the results to Bergman. However, based on letters between Bergman and Gahn, it seems like Bergman was unsatisfied, and persuaded Scheele to continue the work until he had received the answers for which he was searching [2]. In December 1771, Scheele wrote to Gahn that he hoped that the study would be finished before the end of the year. In reality, it would take two more years until the experiments were finished and Bergman was satisfied. On November 7, 1773, Bergman wrote to Gahn: "Now I have finally received Scheele's work on Magnesia nigra. It is a small book, and it contains several curious things. The length causes some problems, and delays the printing". The paper was intended for the *Transactions*, which was restricted to *c*. 300 pages per year. Thus, Scheele's paper had to be divided into two parts, and it has been claimed that Scheele had to shorten his essay. The printed version appeared in the second [3] and third [4] quarter of 1774. The original manuscript (now lost) was written in German, and translated to Swedish by Hjelm [5]. Scheele's paper on manganese is unique in that it, by modern terminology, introduced three new chemical elements in a single paper.

Scheele showed that no alum could be obtained from pyrolusite, but that it indeed contained a new earth [6]. He prepared and studied a number of manganese compounds and was the first to systematically study the chemistry of manganese and should share the credit of discovering manganese with Gahn and Bergman. He found that manganese (IV) oxide (*Magnesia nigra*) was insoluble in acids (hydrochloric acid being the exception, see Sect. 15.2), unless phlogiston was present. This is true in the sense that Mn(IV) has to be reduced to Mn(II) in order to form soluble salts with acids; simple hydrated Mn^{4+} ions are unknown. Phlogiston could be supplied by sugar, but manganese (IV) oxide was also found to be soluble in sulphuric acid on intense heating. This is due to the oxidation of oxide ions to dioxygen by Mn(IV):

$$MnO_2 + H_2SO_4 \rightarrow MnSO_4 + H_2O + 1/2\,O_2.$$

Whether the formation of oxygen went unnoticed by Scheele is not known, at least he did not mention it. He took this experiment as evidence that the supplied phlogiston originated from the heat, the first published evidence that he had started to develop his views on combustion. Scheele found that manganese(II) hydroxide was spontaneously oxidised (in modern terminology) by air. Organic acids were decomposed into carbon dioxide on heating with pyrolusite.

Scheele's paper was directly followed by papers by Bergman [7] (see Sect. 15.3), von Engeström, [8] and Rinman [9]. von Engeström's paper reports some additional blow-pipe experiments, and in an undated letter to Gahn, Scheele acknowledged the results but strongly disagreed with von Engeström's conclusions. Rinman's paper reported on a new variant of pyrolusite found in Klapperud's iron mine in the Dalsland province of Sweden. Rinman was very impressed by Scheele's work and wrote in a letter to Gahn in March 1774: "That man is a razor in chemistry" [10].

15.2 The Discoveries of Chlorine and Barium

In spring 1771, Scheele found that a solution of pyrolusite in hydrochloric acid released a gas that smelled like *Aqua regia* and was "highly irritating for the lungs". This gas was chlorine, and Scheele was the first to describe elemental chlorine:

> On ½ *uns* [15 grams] of finely ground brownstone [Pyrolusite] was poured 1 *uns* [30 grams] pure Spiritus salis [hydrochloric acid]. After this mixture had been standing for 1 h in the cold, the acid had assumed a dark-brown colour. A portion of this solution was poured into a glass, which was left open in the heat. The solution gave off a smell similar to aqua regia, and after ¼ of an hour, it was clear and colourless as water, and the smell was gone. [3]

Scheele's explanation was that *Magnesia nigra* first forms a loose compound with hydrochloric acid (the brown solution). *Magnesia nigra* then attracts phlogiston from the acid to give a solution of *dephlogisticated Magnesia* bound to hydrochloric acid (manganese(II) chloride in modern terminology) and phlogisticated hydrochloric acid (chlorine). The idea that chlorine was an oxidised form of hydrochloric acid would still be defended by prominent chemists in the 1820s. The elemental nature of chlorine was first proposed by Humphry Davy in 1810 [11].

Scheele was actually not far from the truth, the reaction is indeed an oxidation of hydrochloric acid proceeding via a complex between manganese and hydrochloric acid. Manganese(IV) oxide first forms unstable hexachloromanganate(IV) with hydrochloric acid:

$$MnO_2(s) + 4H^+(aq) + 6Cl^-(aq) \rightarrow [MnCl_6]^{2-}(aq) + 2H_2O(l)$$

The hexachloromanganate (sometimes referred to as $MnCl_4$ in the literature) cannot be isolated but decomposes to manganese(II) chloride and elemental chlorine:

$$[MnCl_6]^{2-}(aq) \rightarrow Mn^{2+}(aq) + 4Cl^-(aq) + Cl_2(g).$$

Scheele noted that chlorine bleaches organic pigments, such as those in flowers and in litmus paper, giving hydrochloric acid as a by-product. Oils and fats were also attacked. This observation was important, and already by 1785, bleaching with chlorine had almost entirely replaced the old method of sun bleaching linen [12].

On February 28, 1774 (i.e. after the original manuscript had been sent to Bergman), Scheele wrote to Gahn and asked if he had observed small white crystals that occasionally occurred in pyrolusite specimens. These crystals were barium sulphate, and Scheele was about to discover yet another element.

The mineral baryte (barite, $BaSO_4$) was discovered by an Italian shoemaker and amateur alchemist, Vicentius Casciorolus, on Monte Paterno in the early seventeenth century [13]. He failed to extract gold from the heavy stone, but upon reduction he obtained phosphorescent barium sulphide, which inspired him to call the mineral *Lapis solaris* (solar stone). It underwent several studies by Scheffer, Cronstedt and Marggraf, but without significant result; it was still confused with gypsum ($CaSO_4$). Unknown to the Europeans, barium sulphate had been mined by the Chinese for two millennia and used for preparation of pigments, Han blue ($BaCuSi_4O_{10}$)[2] being the most well known. Although Scheele was first to recognise the distinct nature of barium compounds, Davy is still usually credited for the discovery of barium, on the grounds that he was the first to prepare metallic barium. This can be compared with the discovery of uranium, which is attributed to Martin Klaproth (1743–1829), although Klaproth never isolated the metal and his contribution to the discovery of uranium is very similar to Scheele's contribution to the

[2]Han blue is the barium analogue of Egyptian blue, $CaCuSi_4O_{10}$.

discovery of barium. The same holds for the discoveries of all of the lanthanides and many other metals, which are usually attributed to the chemist who first isolated the oxide.

It was Gahn who found that the earth isolated by Scheele was identical to the earth obtained from baryte (known in Swedish as *tungspat*, heavy spar). The discovery was made in spring 1774, and Scheele got the news in a now lost letter. Scheele replied on May 16, acknowledging the discovery. This was a major breakthrough, as baryte was a more convenient source for the new earth than the erratic appearance in pyrolusite. Gahn, as usual, never published anything, but the discovery was announced by Bergman: "The new earth, which Scheele mentions in his paper on [pyrolusite], is actually the basis in heavy spar, which Mr. J. G. Gahn discovered recently" [14]. Bergman introduced the name earth of heavy spar (Chap. 26), and was the first to use barium chloride as a reagent to detect sulphate (Sect. 23.3). In a paper on the analysis of sea water he wrote:

> Solution of heavy earth [BaO] in acid of salt [hydrochloric acid] soon precipitated a white powder [$BaSO_4$], which did not dissolve in boiling water, but had the appearance of heavy spar [$BaSO_4$], and showed that vitriolic [sulphuric] acid was present [in sea water]. This test is the most reliable of all known tests to reveal vitriolic acid, even when the amount is so small that it is unnoticed in any other manner. [15]

The discovery of a new elemental earth was remarkable, given that only a handful of earths had been recognised. Silica (SiO_2) and lime (CaO) had been known since centuries, but it was not until 1754 that Marggraf showed that alum contained a special earth (Al_2O_3) and the following year Joseph Black (1728–1799) distinguished magnesia (MgO) from lime [16]. Scheele realised the possibility that earth of heavy spar could be reduced to a metal, and several years later he wrote to Wilcke:

> Extraordinary, that my experiments with tungsten, molybdena and Magnesia nigra [MnO_2] have given reason to these new metals' discovery: let's see if not the earth in heavy spar [BaO] with some tweaking also will be added to the metals after reduction.

The low reduction potential and reactive nature of barium made such a reduction impossible in the eighteenth century. Early in the following century Berzelius and Pontin electrolysed moist barium hydroxide with a mercury cathode, which afforded barium amalgam. Metallic barium was, as already mentioned, first isolated by Davy in 1808 [17].

15.3 The Isolation of Metallic Manganese

It is possible that manganese had been prepared before Scheele's studies. It is, for example, possible that Kaim prepared manganese in 1770, [18] but his results are unverified and gained little recognition. Throughout his magnesia paper, Scheele referred to MnO_2 as an earth, but in a comment, following directly after Scheele's paper, Bergman suggested that it might instead be a metal oxide (metal calx): "on

Fig. 15.2 Metallic manganese prepared by electrolysis. Gahn's original samples would have had the appearance of small metallic grains. *Photo* Petra Rönnholm

my part I have to admit, that several circumstances appear to speak strongly for its metallic nature", Bergman wrote [7]. Among the evidence was the fact that manganese salts gave a precipitate with potassium hexacyanoferrate. Bergman thought that *Magnesia nigra* could be either an oxide (calx) of a previously unknown metal or, possibly, the oxide of platinum, as platinum oxide was unknown. He wrote that experiments were on-going, and by the time the paper was published in the summer of 1774, Gahn had already managed to reduce pyrolusite with charcoal to metallic manganese (Fig. 15.2). Bergman later wrote that Gahn worked unaware of his ideas [19]. Gahn told Scheele of the new metal in a letter written in May, and Scheele could verify the results by investigating samples sent to him by Gahn. He showed that manganese dissolve in sulphuric acid with evolution of hydrogen, and thus was a base metal, and summarised his experiments in a letter to Gahn dated June 27. Scheele lacked metallurgical training and working in a pharmacy laboratory he did not have access to a furnace hot enough to perform the reduction. Although MnO_2 is easily reduced to MnO, the last step, reduction to manganese metal, requires at least 1,400 °C [20]. Bergman had a better equipped laboratory and could confirm Gahn's results in his own laboratory, and called the new metal *magnesium* [19]. Gahn began writing up a paper for the *Transactions of the Royal Swedish Academy of Sciences*, but when Bergman asked him in August 1775, Gahn replied that he had not had time to finish the paper, and that additional experiments would have been required [21]. Thus, as usual, Gahn did not publish his results, and the news of the metal were spread by Bergman [19]. Hjelm also studied manganese, refined the reduction process and published a detailed account in 1785 [22]. He also studied manganese alloys (which are extremely important in modern metallurgy) and found traces of manganese in many iron ores [23].

15.4 The Discovery of Molybdenum

The Greek word $\mu\acute{o}\lambda\upsilon\beta\delta o\varsigma$ means lead, and the word *molybdæna* became synonymous with *galena*, a soft lead ore composed of lead(II) sulphide. Over time, however, the word *molybdæna* started to be used for other soft, dark minerals that were assumed to be lead ores. This included the mineral that is now called molybdenite (Fig. 15.3) which is composed of molybdenum(IV) sulphide, MoS_2. It is easy to understand the confusion this caused. Molybdenite had been studied by the Swedish mineralogists Cronstedt (Chap. 9) and Bengt Qvist (1729–1799). It was concluded that sulphur dioxide formed on roasting, and Cronstedt assumed that the mineral contained a new earth, while Qvist thought it contained a new metal. Qvist, who studied the mineral in the laboratory of the Board of Mines, published his results in 1754 [24]. By roasting, he obtained yellow samples of molybdenum (IV) oxide, but his conclusions are obscure, to say the least. His attempts to reduce *molybdæna* to metal failed. Although the nature of this mineral was more or less unknown in the eighteenth century, it found limited use as a lubricant in machines, a use that it still retains today.

Similarly, the word *plumbago* (from Latin *plumbum*, lead), was interchangeably used for different dark minerals assumed to be lead ores, mainly graphite, a mineral form of carbon. Graphite has been used to make pencils since the sixteenth century, and it is interesting to note that the graphite/clay mixture used for pencils is still commonly referred to as "lead", not only in English. Essentially nothing was known about the chemical composition of graphite in the eighteenth century. Scheele, who spoke German and Swedish, used the words *Blei* (lead in German) and *blyerts* in Swedish (bly = lead in Swedish) for any black, soft substance, be it *molybdæna* or *plumbago*. During his stay in Uppsala, Scheele had studied the

Fig. 15.3 Molydbenite or molybdæna. *Photo* Petra Rönnholm

blyerts he sold at the Arms of Upland pharmacy. This material was *plumbago*, i.e. graphite. In 1776, after moving to Köping, he met with Andreas Hofgaard (1748–1811; a Norwegian mineralogist, official at Kongsberg and an acquaintance of Bergman) [25]. Hofgaard presented Scheele with a small sample of *molybdæna*. This material was superficially similar to the *plumbago* he had studied, but chemically it was very different. Unfortunately, he had an insufficient amount to perform any thorough chemical investigation. In a letter to Gahn, dated February 9, 1777, Scheele asked his friend to send a sample of *molybdæna*, but as far as we know he never got any sample from Gahn. Scheele then turned to Bergman, Rinman and Hermelin, and by spring 1778 he finally had enough material for his study [26].

Scheele found that *molybdæna* was scarcely soluble in acids, hot nitric acid and hot arsenic acid being the exceptions [27]. Hot nitric acid gave a white powder, *Terra molybdæna* (Fig. 15.4). He found that this powder was slightly soluble in water, and that the solution had a peculiar sour taste; it also turned litmus red, and had all the properties of an acid. This powder was molybdic acid, hydrated molybdenum(VI) oxide, $MoO_3(H_2O)_x$ and Scheele could thus add another mineral acid (Chap. 14) to his discoveries. By heating *Terra molybdæna* with sulphur, Scheele could prepare synthetic *molybdæna*:

$$2\,MoO_3 + 5\,S \rightarrow 2\,MoS_2 + 3SO_2.$$

Confirming his analysis with a synthesis in this way was one of Scheele's standard procedures. Scheele gave a first summary of his results in a letter to Gahn, dated May 15, 1778; the official paper was printed in the July to September issue of the *Transactions of the Royal Swedish Academy of Sciences* [28].

Bergman proposed [29] that the new earth, just like MnO_2 investigated a few years earlier, actually corresponded to a new metal, but had no time to carry out any experiments. Scheele tried to reduce molybdic acid but failed. Unlike the reduction of MnO_2, the problem was probably not as much the generation of enough heat as the formation of molybdenum carbides. Reduction of MoO_3 with carbon occurs in

Fig. 15.4 Molybdenum (VI) oxide, Scheele's *Terra molybdæna* or molydenous earth. *Photo* Petra Rönnholm

two steps: at 420–630 °C, MoO_3 is reduced to MoO_2 and, at 820–870 °C, reduction to metallic molybdenum takes place [30]. At higher temperatures, however, the produced molybdenum reacts with carbon to generate molybdenum carbides and some metallurgic experience was required to optimise and control the reaction conditions. Thus, Scheele sent a sample of about 6 g (nearly ½ *lod*) of molybdic acid to Hjelm with the request that Hjelm attempted to reduce it to metal. [31]. Scheele corresponded with Hjelm in Stockholm about *molybdæna* and molybdic acid, and it is possible that Hjelm visited Scheele in Köping. At least Scheele wrote to Hjelm in June 1779 and thanked him for the honour of a forthcoming visit in Köping, and asked him to bring some different samples of *blyerts* (*molybdæna* or *plumbago*). It appears like they planned to carry out some experiments together. Hjelm, working in a metallurgical laboratory, was more successful, and his experiments resulted in the isolation of metallic molybdenum in 1781. The samples of molybdenum obtained by Hjelm were never very pure, but contained high levels of carbon since he used charcoal and linseed oil as a reductant; he never reported higher densities than 7.4, while pure molybdenum has a density of 10.2 gcm^{-3}. It would take another century until pure molybdenum could be prepared in any quantity. Hjelm brought the positive results to Scheele in September 1781, and Scheele answered in November: "It delights me that we now again have a new metal, molybdænum. I think I already can hear at least the French deny its existence, since they are not themselves the inventors of it". The first printed account of the new metal appeared in the preface of Hjelm's Swedish translation of Bergman's treatise on blow pipe analysis (Sect. 23.4) published the same year [32]. A more detailed account of the metal and its properties was published 1788–1791 in a series of seven papers by Hjelm in the *Transactions of the Royal Swedish Academy of Sciences*.

Scheele could show that *plumbago* was indeed very different from *molybdæna* [33]. Upon strong heating with potassium nitrate (saltpetre) or arsenic acid it gave carbon dioxide, and hence it was a form of carbon [34]. Similar conclusions were made about the same time by the French mineralogist Jean-Babtiste Romé de I'lse (1736–1790). Scheele was probably the first to realise that the difference between steel and cast iron is its carbon content, and that the powder remaining after dissolving cast iron in acid is graphite. This was an important discovery for the growing metal industry. Rinman had previously noted in a paper about etching of steel and iron that a "*blyerts*-like" residue remained after dissolving cast iron in acids, [35] but he made no further investigation of this material. From the correspondence between Bergman and Scheele, it can be concluded that Scheele's investigation of graphite was carried out in the summer of 1779; the manuscript for the paper was sent to Wargentin in September that year.

The difference between iron and steel was also studied in 1781 by Bergman and his student Gadolin, who examined 89 iron samples in an attempt to determine their phlogiston content [36]. In order to do so, they determined the amount of hydrogen evolved upon dissolution in dilute sulphuric acid, and found that pure malleable iron evolved more hydrogen than steel which in turn evolved more hydrogen than cast iron. The amount of insoluble residue (graphite) left after dissolution was higher in cast iron than steel. According to Thomson, "this study contributed very materially to advance the knowledge of the cause of the difference between iron and steel" [37].

15.5 Isolation of Tungstic Acid and the Discovery of Tungsten

After his successful investigation of molybdæna, Scheele turned his attention to a mineral now known as scheelite ($CaWO_4$; Fig. 15.5) [38]. To Scheele, however, the mineral was known by the Swedish name *tungsten*, literally meaning "heavy stone" due to its high density of *c*. 6 g cm^{-3}. The idea to study scheelite most likely came from Bergman, as he apparently had studied the same mineral a few years earlier. From the preserved correspondence between Scheele and Bergman, it appears likely that Scheele's investigations started during the summer of 1780; in August he wrote to Bergman and reported that the mineral was composed of calcium (lime) and a special acid. In November, he wrote to Hjelm: "I have investigated *tungsten*. It is composed of lime and an acid, which is similar to acidum molybdenæ [MoO_3], but in some circumstances different". Scheele had found that the mineral was partly soluble in nitric acid, giving a solution of calcium nitrate and a yellow residue, soluble in aqueous ammonia. The ammoniacal solution gave a precipitate with nitric acid, and this precipitate was Scheele's acid of tungsten (tungstic acid, $H_2WO_4 \cdot H_2O$; Fig. 15.6). It had a sour taste and coloured litmus tincture red. Scheele reported his results in a paper published in spring 1781; [39] this short 5 ½ page paper is important, since it is the first investigation of the chemistry of tungsten.

Directly following Scheele's paper on tungstic acid is a paper by Bergman, [29] who had studied the same mineral a few years earlier. Bergman had suspected the presence of barium (heavy earth) in the mineral, but had found it to contain calcium (lime), just like Scheele. The only additional experimental contribution by Bergman

Fig. 15.5 Scheelite or tungsten, as it was known to Scheele. *Photo* Petra Rönnholm

Fig. 15.6 Anhydrous tungstic acid is a bright yellow powder. *Photo* Petra Rönnholm

was a more thorough blow-pipe analysis. Although Bergman had little experimental information in addition to Scheele's findings, Bergman's tungsten paper is still of considerable interest. Here Bergman, for the first time, expresses the opinion that molybdic and tungstic acid could be reduced to metals. He had found that molybdenum and tungsten compounds gave precipitates with hexacyanoferrate (lye of blood) and gave colour to ammonium sodium phosphate (microcosmic salt) and sodium tetraborate (borax) during blow-pipe analyses, properties that he had hitherto only observed for metal compounds. He would soon be proven to be correct. Bergman also expressed the idea that all metal calces consisted of metallic acids and phlogiston (Sect. 22.4).

In June 1782, Scheele was visited by the Spanish chemist Juan José d'Elhuyar (1754–1796), who had been working in Bergman's laboratory for half a year. In Scheele's laboratory, they learned of Scheele's tungstic acid. Back in Spain, Juan José and his brother Fausto (1755–1833) investigated a mineral known as *wolfram* (now known as wolframite, $(Fe,Mn)WO_4$). It had been thought to be a tin ore, but all attempts to extract tin failed, and it was known to interfere with the smelting of genuine tin ores, and was referred to as *Spuma lupi*, [40] meaning "wolf foam". In German, this term was translated to Wolf Rahm or Wolfram and Wallerius called this material Volfram [41]. The eccentric British chemist and alchemist Peter Woulfe (1727–1803) had speculated about the possibility of a new metal in this mineral in 1779, but his ideas received no recognition. From this mineral, the d'Elhuyar brother isolated a substance which they found to be identical to Scheele's tungstic acid:

> This [—] made us suspect immediately that the matter, from which this colour proceeded, might be that particular product which Scheele, a Swedish chymist, has lately met with in a stone called Tungsten, or heavy stone [42]

The two Spanish chemists followed Bergman's suggestion, and reduced tungstic acid to obtain tungsten in the metallic state:

> Having put another hundred grains of this powder into a Zamora crucible, provided with charcoal, and well covered, and placed in a strong fire, where it remained an hour and a half, we found, on breaking the crucible after it waș cool (*q*), a button, which fell to powder between the fingers. Its colour was dark brown; and on examining it with a [magnifying] glass, there were seen a congeries of metallic globules, among which some were the bigness of a pin's head, and when broke had a metallic appearance at the fracture in colour like steel.

According to footnote q, the first attempt failed since they opened the crucible while still hot causing the contents to self-ignite. As in the case of molybdenum, the temperature must be controlled in order to obtain tungsten rather than tungsten carbides. The reduction of WO_3 with carbon starts at 783 °C, but at higher temperatures, metallic tungsten reacts with carbon [43].

In Germany, the name Scheel or Scheelium was suggested for the new metal but this name never stuck. In German and Swedish, the metal is known as Wolfram and volfram, respectively, explaining Berzelius' choice of the chemical symbol W. Several languages derive their name for the metal from the Swedish word tungsten tungsten (English), tungsténe (French), tungsteno (Italian) and tungstênio (Portuguese).

Scheele was enthusiastic over the isolation of metallic tungsten, and wrote to Bergman in April 1784: "I like that Mr. Luyarte has obtained tungsten regulus, I hope he has sent a sample to Mr Professor". A detailed account on tungsten and its properties was published by Crell, [44] and Bergman announced the discovery to his Swedish audience in his last original paper [45]. Given Scheele's and Bergman's contributions, it is not fair to give the d'Elhuyar brothers the full credit for the discovery of tungsten.

15.6 Establishing Platinum, Cobalt, Nickel and Manganese as Unique Metals

Platinum has been known to native South Americans for centuries, and it would be very difficult to name a discoverer for platinum. The discovery is usually attributed to Spanish general and explorer Antonio de Ulloa (1716–1795), who brought platinum to Europe, but as he was no scientist, he did not investigate the metal. Many believed it to be a gold alloy. A short account was given in a letter from British physician William Brownrigg (1712–1800) to William Watson in 1750 [46]. The first more extensive chemical investigation was reported in two papers by Scheffer published in 1752 [47, 48]. Scheffer's samples were unfortunately not pure, and platinum was still not universally accepted as a new metal. Bergman made important contributions in preparing pure platinum and establishing platinum as an individual metal.

Alströmer had presented Bergman with a large sample of about 1.7 kg ("nearly 4 *skålpund*") of platinum, and a few years later Bergman reported his experiments in order to establish the elemental nature of platinum [49]. It had, wrote Bergman, previously been reported by Marggraf and British chemist William Lewis (1708–1781) that solutions of platinum salts were not precipitated by sodium carbonate (alkali minerale), which Bergman found hard to believe. Most metals form insoluble carbonates. He found that platinum salts indeed were precipitated by sodium- and potassium carbonate, ammonia and, to his surprise, ammonium chloride (salmiac). This is due to the precipitation of sparingly soluble ammonium tetrachloroplatinate (II), $(NH_4)_2[PtCl_4]$.

Bergman devised a method for purification of platinum. Thin platinum flakes were boiled with hydrochloric acid to dissolve iron, the remaining metal dissolved in *Aqua regia*, ammonium chloride was added and the precipitate melted with ammonium sodium phosphate (microcosmic salt) to produce platinum. He investigated the effect of alkali on pure platinum chloride and showed that platinum indeed is a unique metal with a characteristic solution chemistry.

Cobalt compounds have been known and used for millennia for colouring of glass blue and painting porcelain. Metallic cobalt was, however, not reported until 1735 by Brandt [50]. It was Brandt's students, Crosnstedt, who reported the discovery of nickel in 1751 [51]. Decades later, some chemists still doubted the existence of cobalt and nickel. Experiments with impure materials had probably added to the confusion.

In a paper published in 1780, Bergman reported precipitation experiments with platinum, nickel, cobalt and manganese that finally showed that these metals where unique elements, but he also showed that samples were frequently contaminated [52]. Bergman mentioned that it had, for example, been suggested that platinum was an alloy of gold and iron, and that nickel and cobalt were merely modifications of iron. Among Bergman's results was that he found that platinum could be precipitated from its solution in *Aqua regia* with zinc, but when zinc was added to a solution of common nickel in nitric acid, arsenic was precipitated. He found that zinc did not precipitate nickel, cobalt or manganese from pure metal salt solutions. Bergman's observation was correct.

15.7 Cerium—A Missed Opportunity

In 1751, Cronstedt described a mineral which he called Bastnäs tungsten (now known as cerite), a heavy mineral discovered in the mines at Swedish Bastnäs [53]. In 1781, the 15-year old Wilhelm Hising (1766–1852), the son of a mine official at Skinskatteberg, not far from Bastnäs, sent samples to Scheele and Bergman. This mineral contains rare earth elements, mainly cerium and lanthanum, but Scheele erroneously concluded that it was composed of silicon (silica), aluminium (clay), carbonate (aerial acid) and traces of iron. d'Elhuyar was not more successful in Bergman's laboratory. He obtained iron, silica and lime [45]. Many years later, in 1803, Hisinger (as he was called after being adopted by his ennobled uncle) teamed up with his young friend Berzelius, and isolated cerium oxide from cerite [54]. Cerium oxide was independently isolated at about the same time by Klaproth in Berlin, [55] leading to an infected controversy over the priority of the discovery [56].

15.8 Bergman and the Discovery of Tellurium

Bergman also played a minor role in the discovery of tellurium. Bergman corresponded with a network of mineralogists throughout Europe exchanging not only letters but also minerals. The most important mineralogical correspondent of Bergman was Austrian mineralogists Ignaz von Born (1742–1793) and Franz Joseph Müller von Reichenstein (1742–1825 or 1826) [57]. In 1782, Reichenstein had found what he believed to be a new metal in a mineral from the Mariahilf gold mine near Zlatna (Zalatna) in Transylvania (in present day Romania) [58]. At first, he was told that the new material was antimony, but in April 1783 he contacted Bergman asking for his opinion [59]. Bergman's preliminary analysis of the sample he obtained via Born indicated that the material indeed was a new metal, but more material was needed for a conclusive decision. Unfortunately, when the new sample arrived in Uppsala, Bergman was already dead [60]. Reichenstein's discovery went unnoticed until Klaproth drew attention to it in 1798 naming the new element tellurium after the Latin word for Earth [61].

15.9 Hydrosiderum—The Metal that did not Exist

Cold-short iron is iron which is brittle due to a high phosphorus content. In the eighteenth century, it was recognised that certain ores gave cold-short iron, but the reason for the brittleness was not known.[3] Johan Christian Friedrich Meyer found that on dissolving such iron in acid, a white earth remained [62]. The earth could be reduced to what appeared to be a new metal, which Meyer called siderum or hydrosiderum, but which was actually an iron phosphide. In summer 1781, Meyer sent a sample of hydrosiderum to Scheele, who shared the material with Bergman, who was convinced of its nature as a new metal and reported his results in paper titled *On the cause of brittleness in cold-short iron*. A preprint appeared in 1782 [63] and a revised version in the third volume of his *Opuscula Physica et Chemica*. The official version was printed in 1784, [64] but by that time it was already out-dated.

[3]Compare Wallerius theory, Sect. 9.1.

In 1784, Meyer discovered that he could obtain a similar material from iron and phosphoric acid, but failed to extract phosphoric acid from hydrosiderum [65]. He correctly concluded that his new metal "was no more". Klaproth independently made the same discovery, but also failed to extract phosphorus from hydrosiderum [66]. It was Scheele who finally solved the last part of the mystery with the brittle, phosphorus containing iron.

Scheele dissolved cold-short iron in sulphuric acid, which gave a yellowish residue, which he treated with alkali [67]. This gave iron oxides (ochre) and a solution, from which he precipitated mercury phosphate with mercury (II) chloride. By heating the mercury phosphate with powdered charcoal in a retort, he observed phosphorous vapours, appearing like auroras in the receiver. Thus, Scheele had helped to prove that phosphorus-rich ores gave brittle iron, [68] an important discovery for the iron industry. In the same paper, Scheele included experiments on a sodium phosphate salt obtained from urine, *Sal perlatum*, which had been assumed to contain phosphoric acid, but which Joseph Proust (1754–1826) claimed he had disproved. Scheele showed that he could obtain phosphorus from *Sal perlatum* in the same fashion as from cold-short iron. The blue iron phosphate that Scheele described in his paper, and which Klaproth had called "natural Prussian blue" must be some sort of mixed valence compound. The mineral vivianite ($Fe_3(PO_4)_2 \cdot 8\ H_2O$) is green or blue, and is an example of such a naturally occurring mixed valence compound.

References

1. Oseen CW (1940) Torbern Bergman och Carl Wilhelm Scheele. KVA, Stockholm
2. Oseen CW (1940) Torbern Bergman och Carl Wilhelm Scheele. KVA, Stockholm, p 4
3. Scheele CW (1774) Om Brun-sten eller Magnesia, och dess Egenskaper KVA Handl 35:89–116
4. Scheele CW (1774) Om Brun-sten eller Magnesia nigra, och dess egenskaper KVA Handl 35:177–194
5. Hjelm PJ (1786) Åminnelse-tal öfver…Herr Torbern Bergman, Stockholm
6. Lennartson A (2017) The chemical works of carl Wilhelm Scheele. Springer, Cham, p 25
7. Bergman T (1774) Tilläggning om Brunsten KVA Handl 35:194–196
8. von Engeström (1774) Ytterligare Anmärkningar vid herr Scheeles Rön om Magnesia KVA handl 35:196–200
9. Rinman S (1774) Beskrifning på en ny art af spat-formig Magnesia eller Brunsten, ifrån Klapperuds Jern-Grufva i Fresko Socken på Dals Land KVA Handl 35:201–1205
10. Fors H (2008) Matematiker mot Linneaner. In: Widmalm S (ed) Vetenskapens sociala strukturer. Nordic Academic Press, Lund, p 184
11. Davy H (1811) The Bakerian Lecture, On some of the combinations of oxymuriatic acid and oxygene, and on the chemical relations of these principles, to inflammable bodies. Phil Trans 101:1–35
12. Cassebaum H. (1982) Carl Wilhelm Scheele. BSB B.G: Teubner Verlagsgesellschaft, Leipzig, p 84
13. (1932) Gemelin's Handbuch der Anorganischen chemie, 8th edition, Barium. Verlag Chemie, Weinheim, p 1

14. Bergman T (1775) Framledne Direct: och Kongl. Vet. Acad. Ledamots Herr H. T. Scheffers chemiske föreläsningar. M. Swederus, Uppsala, p. 165
15. Bergman T (1777) Hafs-vatten från ansenligt djup. KVA handl 38:26–29
16. Gowdie J, Black J (1754) Dissertatio medica inauguralis, de humore acido a cibis orto, et magnesia alba
17. Davy H (1808) Electro-chemical researches on the decomposition of the earths; with observations on the metals obtained from the alkaline earths, and on the amalgam procured from ammonia. Phil Trans 98:333–370
18. Kaim IG (1770) Disseratio de metallis dubiis, Wien
19. Bergman T (1775) Framledne Direct: och Kongl. Vet. Acad. Ledamots Herr H. T. Scheffers chemiske Föreläsningar, M. Swederus, Uppsala, p 390
20. (1973) Gmelins Handbuch der Anorganischen Chemie, Achte Auflage, Mangan, Teil B. Verlag Chemie, Weinheim, p 6
21. Trofast J (1994) Johan Gottlieb Gahn: brev, vol 2, Lund, p 87
22. Hjelm PJ (1785) Försök, at af Brunsten erhålla Magnesium och sammansmälta den med några andra Metaller KVA Nya Handl 6:141–156
23. Hjelm (1778) Försök om Brun-stens närvarelse i Järn-malmer KVA Handl 39:82–87
24. Qvist B (1754) Rön om Bly-Erts KVA Handl 15:189–209
25. Carelid G, Nordström J (ed), Torbern Bergman's foreign correspondence. Almqvist & Wiksell, Stockholm, p 160
26. Nordenskiöld AE (1892) Carl Wilhelm Scheele. Efterlämnade bref och anteckningar, Stockholm, p 202
27. Lennartson A (2017) The chemical works of carl wilhelm scheele. Springer, Cham, p 53
28. Scheele CW (1778) Försök med Blyerts, Molybdæna KVA Handl 39:247–255
29. Bergman T (1781) Tilläggning om Tungsten KVA Nya Handl 2:95–98
30. (1977) Gmelins Handbuch der Anorganischen Chemie, Achte Auflage, Molybdän, Ergränzungsband A1. Springer, Berlin, p 85
31. Hjelm PJ (1788) Försök med Molybdaena och med Reduction af dess Jord, KVA Nya Hand, 9:280–292
32. Bergman T (1781) Afhandling om Båsröret. Stockholm, p 4
33. Lennartson A (2017) The chemical works of carl wilhelm scheele. Springer, Cham, p 60
34. Scheele CW (1779) Försök med Blyerts, Plumbago KVA Handl 40:238–245
35. Rinman S (1774) Rön, Om Etsning på Järn och Stål KVA Handl 35:3–14
36. Bergman T, Gadolin J (1781) Dissertatio chemica de analysi ferri, Uppsala
37. Thomson T (1831) The history of chemistry, vol II. Henry Colburn & Richard Bentley, London, p 48
38. Lennartson A (2017) The chemical works of carl wilhelm scheele. Springer, Cham, p 67
39. Scheele CW (1781) Tungstens beståndsdelar KVA Nya Handl 2:89–95
40. Agricola G (1657) De re metallica, Basel p 609
41. Wallerius JG (1747) Mineralogia, eller Mineralriket, Indelt och beskrifvit, Lars Salvius, Stockholm, p 304
42. De Luyart, DJJ, De Luyart, DFA (1785) Chemical analysis of Wolfram and examination of a new metal, which enters its composition. Translated from the Spanish by Charles Cullen, Esq. To which is prefixed, A Translation of Mr. Scheele's Analysis of the Tungsten, or Heavy Stone; with Mr. Bergman's Supplemental Remarks. London, p 30
43. (1979) Gmelins Handbuch der Anorganischen Chemie, Achte Auflage, Wolfram, Ergränzungsband A1. Springer, Berlin, p 98
44. Crell L (1784) Ueber die Säure des Tungstein (Schwerstein); nebst ein Nachricht vom Hrn. Ritter Bergmann, über ein aus demselben erhaltenes neues Metall Chem Ann part 2:195–207
45. Bergman T (1784) Mineralogiske Anmärkningar KVA Nya Handl 5:109–122
46. Watson W, Brownrigg W (1750) Several Papers concerning a New Semi-Metal called Platina. Phil Trans 46:584–596

47. Scheffer HT (1752) Det hvita Gullet, eller sjunde Metallen, kallat i Spanien, Platina del Pinto, Pintos, små Silfver, beskrifvit til sin natur, KVA Handl 13:269–275
48. Scheffer HT (1752) Tilläggning om samma Metall, KVA Handl, 13:276–278
49. Bergman T (1777) Anmärkningar om platina, KVA Handl 38:317–328
50. Brandt G (1735) Disseratio de semimetallis, Acta Literaria et Scientiarum Sveciae 4:1–10
51. Cronstedt AF (1751) Rön och Försök Gjorde med en Malm-art, från Los Kobolt Grufvor i Färiöa Socken och Helsningland, KVA Handl 12:287–292
52. Bergman T (1780) Præcipitations försök med platina, nickel, cobolt och magnesium, KVA Nya Handl 1:282–293
53. Cronstedt AF (1758) Försök til Mineralogie, eller Mineral-Rikets upställning, Stockholm, p 183
54. Hisinger W, Berzelius, JJ (1804) Cerium, en ny metal, funnen i Bastnäs Tungsten från Riddarhyttan i Westmanland, Stockholm
55. Klaproth HM (1804) Chemische Untersuchung des Ochrits, Neues allgemeines Journal der Chemie 303–316
56. Trofast J (2014) Jac. Berzelius, Upptäckten av cerium, selen, kisel, zirconium och torium, Ligatum, Lund, p 25–72
57. Carelid G, Nordström J (1965) Torbern Bergman's Foreign Correspondence, Almqvist & Wiksell, Uppsala, p XXXI
58. Weeks ME (1956) The Discovery of the Elements, 6 edition, p 304
59. Carelid G, Nordström J (1965) Torbern Bergman's Foreign Correspondence, Almqvist & Wiksell, Uppsala, p XXXII
60. Carelid G, Nordström J (1965) Torbern Bergman's Foreign Correspondence, Almqvist & Wiksell, Uppsala, p XXXIII
61. Weeks ME (1956) The Discovery of the Elements, 6 edition, Easton, p 304
62. Meyer JCH (1781) Versusche mit der in dem Guβeisen entdeckten weiβen metallischen Erde Schriften der Berlinischen Gesellschaft naturforschender Freunde 2:334–348
63. Nergman T (1782) Commentationes e quarto novorum Reg. Scientarum Societatis Upsaliensis actorum tomo excerptæ pp. 51–136.Uppsala
64. Bergman T (1784) Commentatio chemica de causa fragilitatis ferri frigidi, Nova Acta Regiæ Societatis Scientiarum Upsaliensis, 4:51–62
65. Meyer JCF (1784) Das vermeyntliche neue Metall, das Wassereisen, vom Erfinder, Hr. Hofapotheker Meyer, selbst berichtigt. Chem Ann 1:195–197
66. Klaproth MH (1784) Von dem Wassereisen, als einem mit Phosphorsäure verbundenen Eisenkalke. Chem Ann 1:390–399
67. Lennartson A (2017) The chemical works of carl wilhelm scheele. Springer, Cham, p 87
68. Scheele CW (1785) Rön, om Ferrum phosphoratum och Sal perlatum, KVA Nya Handl, 6:134–141

The Invitation to Berlin 16

Königlich-Preußische Akademie der Wissenschaften (The Royal Prussian Academy of Sciences) in Berlin was one of the highest ranked scientific institutions in Europe in the eighteenth century. The Academy was founded in 1700 by Frederick I of Prussia and was a French language institution which flourished under the regime of his son Frederick the Great in 1740–1786. Since 1760, Andreas Sigismund Marggraf, one of the more prominent German chemists in the eighteenth century, held the position as director of the class of physics, which also included chemistry. This was one of the most prestigious positions a chemist could hold in the eighteenth century. In 1774, however, Marggraf's health started to decline; he had suffered a series of strokes that had left him partly paralysed. It was clear that a successor had to be found, but it was also decided that the search for a successor had to be conducted in secret. Marggraf was not to know of the process, and it was also feared that the Academy would be overwhelmed by applications if a vacancy was openly advertised [1]. The intrigues that followed have been studied in detail by Otto Zekert in his Scheele biography, [2] and by Eva Nyström in a M.Sc. thesis at Uppsala University [3].

Physicist Joseph-Louis de Lagrange (1736–1813), the director of the class of mathematics, took the first initiative. He first turned to Jean-Baptiste le Rond d'Alembert (1717–1783), French mathematician, physicist and closely associated with the Academy and King Frederick. d'Alembert had been co-editor of the famous Encyclopédie with Denis Diderot (1713–1784), and was well acquainted with the scientific circles in Paris. d'Alembert made some inquiries and somehow he learned of Scheele, whom he had not had any personal correspondence with. At this point, Scheele had only published two papers, the controversial paper on hydrofluoric acid and the investigation of pyrolusite. Still, Scheele was well known by chemists in Paris via Bergman's correspondence with Macquer. It can, for example, be noted that this was about the same time that Lavoisier sent a copy of his *Opuscules physiques et chymiques* to Scheele (Sect. 21.8). d'Alembert wrote to

A. Lennartson, *Carl Wilhelm Scheele and Torbern Bergman*, Perspectives on the History of Chemistry, https://doi.org/10.1007/978-3-030-49194-9_16

King Frederick and suggested him to recruit Scheele as Margraff's successor. It turned out, however, that the King had no intention to replace Marggraf.

In 1776, when Marggraf promoted his assistant, Franz Karl Achard (1753–1821), to collaborator and made it clear that he wanted him as his successor, the King reacted. He remembered d'Alembert's recommendation, but had forgotten Scheele's name, so when he contacted the Academy, he just said that there was a well-known chemist in Sweden that was to be recruited. The Academy turned to the Prussian embassy in Stockholm for advice. The embassy secretary, Jouffroy, wrote back to Merian[1] in Berlin with the answer that the two most important chemists in Sweden were Bergman and von Engeström, but that there was also a less well-known chemist named Scheele. Bergman was, according to Jouffroy, best known for his analysis of mineral waters and the artificial production of Spa and Pyrmont water, while von Engeström was described as more specialised in mineralogy and metallurgy. The Academy turned to Bergman in April 1776 and invited him to Berlin. Bergman politely declined due to his poor health. This was not the first time Bergman had been offered positions abroad; he had previously declined invitations to St Petersburg in 1768 and Spain in 1771 [4]. Lagrange then wrote to Merian suggesting him to wait for more information about von Engeström, and in particular Scheele. When d'Alembert learned that the Academy was about to offer the position to von Engeström, he wrote back and explained that it was Scheele who should be recruited. Nevertheless, the Academy once again contacted Bergman in February 1777. Bergman was now inclined to accept; the milder climate in Berlin would probably do him good and he thought the position in Berlin would give him more time for research, instead of teaching basic chemistry to students; apparently, he also saw the position in Berlin as a way to escape his enemies Wallerius and von Engeström [5]. He received little help from and distrusted his assistants, Tidström and Wibom, whom he had inherited from Wallerius, [6] so his working environment may have been rather strained, especially since Tidström had competed with him for the professorship. On the other hand, he was unsure how King Gustav III would react. He wrote to Count Nostiz, the new Prussian ambassador in Stockholm for more information, but he also turned to Wargentin for advice. Wargentin strongly recommended Bergman to stay, and so did Academy members Count Carl Fredrik Scheffer (1715–1786) and Johan Alströmer (1742–1788).[2] The information now leaked from the Academy in Stockholm to the Court, and on March 28 the University Chancellor, Carl Rudenschöld (1698–1783), wrote to Bergman and informed him that the King was deeply worried, and had asked whether something could be done with regard to Bergman's economic situation [5]. Bergman replied that he had struggled in Uppsala for 19 years, half of that time without salary and ruined his health on doing so. He had not even been able to put aside any money for his wife in the case of his death. Rudenschöld returned to Bergman on April 21 with the King's reply. The King regarded the Prussian offer as a personal insult, and

[1]Johann Bernhard Merian (1723–1807). Swiss philosopher.

[2]Scheffer was an influential diplomat and politician; Alströmer was son of industrialist Jonas Alströmer and brother of Patric Alströmer, whom Bergman was acquainted to.

reminded on his role in Bergman's appointment as professor. He was willing, however, not only to grant Bergman's wife a pension, but also to raise Bergman's salary. Thus, in April 1777, it was clear to Bergman that he would remain in Uppsala. Bergman declined the raised salary, and explained that he would be grateful if his wife was granted a pension of 100 riksdaler in the case of his death, and if he was given the right to retire with salary in case his health prevented him to continue his duties as professor. The King was relieved, and increased the pension for Bergman's wife to 150 riksdaler. The Royal Swedish Academy of Sciences in Stockholm was also relieved, and Bergman was granted annual funding of 150 riksdaler for his research (Sect. 19.4).

When Bergman had declined, Nostiz was asked by the Prussian Academy to persuade Scheele to come to Berlin. Strangely enough, he did not do so. Nyström suggested that the reason was that Bergius, whom Nostiz had consulted, had convinced him that Scheele was unlikely to accept the offer. Nostiz replied the Academy in July with a new summary of the three men's strong sides. Apparently, the position was offered to von Engeström. Unlike Bergman, von Engeström immediately accepted in August 1777. King Frederick was however not impressed. According to Nositz, von Engeström had made a bad general impression and his financial demands were very high. The King rejected Engeström as a candidate in September. Still at this point, no one had actually asked Scheele, whom d'Alembert had recommended in the first place. Count Nostiz turned to Bergius, who finally asked Scheele in October 1777. As expected, Scheele declined, although his letter to Bergius has not survived. In the end, Marggraf got his wish through, and Achard became his successor, but not until Marggrafs death in 1782. Achard had some interactions with Scheele.

In a study of fluorite, Achard launched a new, erroneous, theory of fluorite and hydrofluoric acid, a theory that Scheele disproved in one of his last papers [7, 8]. Scheele expressed himself very politely compared to his criticism of Weber (Sect. 18.2) and Desavive (Sect. 27.2).

The invitation to Berlin was regarded as a great honour for Bergman; he mentioned it in his short autobiography and it is included in early biographies as a proof of Bergman's great reputation. The truth, that it was actually Scheele who was intended for the position, was unknown to Bergman, Scheele and their contemporaries; as far as they knew, the offer was passed to Scheele after Bergman had declined.

References

1. Fors H (2003) Mutual Favours (diss.). Depatment of History of Science and Ideas, Uppsala University, Uppsala, p 192
2. Zekert O (193) Carl Wilhelm Scheele Sein Leben und seine Werke, 3–7. Theil, Gesellschaft für Geschichte der Pharmazie, Mittelwald, p 217–227
3. Nyström E (1971) Torbern Bergmans kallelse till Berlin. Uppsala University
4. Annerstedt C (1913) Uppsala universitets historia, vol 3:1, Almqvist & Wiksell, Uppsala, p 521
5. Annerstedt C (1913) Uppsala universitets historia, vol 3:1, Almqvist & Wiksell, Uppsala, p 522

6. Fors H (2003) Mutual Favours (diss.). Depatment of History of Science and Ideas, Uppsala University, Uppsala, p 48
7. Scheele CW (1786) Neue Beweise der Eigenthümlichkeit der Flußspathsäure, Chem Ann 1:3–17
8. Lennartson A (2017) The chemical works of carl wilhelm scheele. Springer, Cham, p 97

Bergman and the Chemistry of Mineral Waters

17

Mineral waters had been used in medicine since the Middle Ages, both for drinking and bathing. Rich people travelled long distances to the spas in Germany, Belgium and France, but these trips became dangerous or even impossible during the 30 years' war. As a result, bottled mineral water was exported and sold in pharmacies. The medical effects reported were often highly exaggerated, but in some cases mineral waters had a genuine positive effect: magnesium salts have a laxative effect, hydrogen carbonates neutralise stomach acid and dissolve mucus, and thus have relieving effect on ulcers, while iron rich mineral waters relieve iron deficiency. With an increasing use of mineral water in medicine, the interest in the properties of these waters grew as a natural effect. Early contributions [1] to the field of mineral water analysis were made by Boyle [2] and German physician and chemist Friedrich Hoffmann (1660–1742) [3].

British physician and pioneer of water analysis Peter Shaw (1694–1763), who classified the contents of mineral waters into salts, earths, sulphurs and spirits, stressed that there was nothing supernatural about mineral waters, but that they were simply solutions of different substances in water [4]. Shaw always begun his descriptions of mineral waters by describing the spring, followed by the physical characteristics of the water, followed by an analysis of the water by evaporation [1]. Bergman largely followed this scheme. In the 1740s, Brownrigg suggested that the volatile spirit in mineral waters (i.e. carbon dioxide) was identical to the "damps" in coal mines, [1] and Cavendish found that water from a well in London contained carbon dioxide (fixed air) [5]. He also observed the ability of carbon dioxide saturated water to dissolve calcium carbonate (lime). This is due to the formation of soluble calcium hydrogen carbonate:

$$CaCO_3(s) + CO_2(aq) + H_2O(l) \rightarrow Ca^{2+}(aq) + 2HCO_3^{2-}(aq).$$

Preparation of artificial mineral waters started as early as the 1680s, [1] but it would take until the 1770s until production of artificial mineral waters occurred in significant volumes. The production of artificial Pyrmont water by Priestley[6]

A. Lennartson, *Carl Wilhelm Scheele and Torbern Bergman*, Perspectives on the History of Chemistry, https://doi.org/10.1007/978-3-030-49194-9_17

gained international recognition, but his work was very qualitative, and his water had little resemblance with genuine water from Bad Pyrmont [7]. Priestley was convinced that mineral water could prevent scurvy, and for this hypothesis, he was awarded the Copley medal of 1772. Commercially most successful in the field was watchmaker and silver smith Jacob Schweppe (1740–1821) who started production of artificial mineral water in Geneva about 1780 and his company is still in the same business over two hundred years later.

17.1 Medevi—The First Swedish Spa

Baron Gustaf Soop (1624–1679), influential politician and one of Sweden's wealthiest men, was the owner of many estates, among them Medevi close to lake Vättern. The springs at Medevi had been used for sacrifices during the pre-Christian era, but had received little interest since then. In 1677 Soop noticed the water and sent a sample to Hiärne, who was the only authority on the subject in Sweden. Hiärne immediately noted that the water from Medevi had the same taste as the water from Aachen, and performed a crude chemical analysis. The following summer, Hiärne travelled to Medevi to perform new analyses, and could confirm that Medevi was a true *surbrunn*.[1] People in need for medical treatment started to arrive at Medevi (Fig. 17.1), but the rich people, (the costumers they wanted to attract) were harder to convince. As the interest in Medevi started to increase, new competing springs were reported all over Sweden. Hiärne published a number of texts, the most important being *Den lilla wattu-prowfaren* (The little water tester), to teach the difference between a true mineral water and ordinary water, and at the same time trying to protect and promote Medevi.

17.2 Bergman's Interest in Mineral Waters

It is not known with certainty when Bergman first became interested in mineral waters and when he first started to prepare artificial mineral water. In May 1769, Bergman asked Wargentin to lend him a book on mineral water written by Monnet [8].[2]

In his autobiography, Bergman claimed to have been drinking his own artificial mineral water as early as 1771, [9] a statement doubted by Boklund [10]. Boklund referred to a laboratory note made 11 p.m. on January 6, 1772 and which reports an obviously failed attempt to prepare Pyrmont water. Bergman would later learn that it was not simply possible to mix the required salts with water, but that more sophistication was needed to avoid precipitation. On the other hand, Bergman stated

[1]Literally "sour well", A mineral well with acidic carbonated water. I have found no English translation of the word.

[2]Presumably *Traité des Eaux Minerales*.

Fig. 17.1 In Medevi, a visitor is met by a genuine spa environment from the eighteenth and nineteenth centuries. *Photo* Anders Lennartson, June 2018

in 1775 that he had used the same method for carbonating water for 6 years [11]. One possible explanation of these contradictions would be that he initially used a simpler carbonated water free of iron.

17.3 Bergman's Early Work on Mineral Water

Bergman was not, as we have seen, the first in Sweden to pay interest to mineral waters, he was preceded by both Hiärne and, to some extent, by Wallerius, who had published a book in 1748 where he tried to classify different types of water, although in a completely arbitrary system [12].

Bergman had a more practical interest in mineral waters. He was of poor health, and after trying various cures, he turned to imported mineral water. During the winter 1769–1770, he consumed 80 *stop* (nearly 100 L). This improved his health, but was about to ruin his economy. As a solution, Bergman started to analyse mineral waters in order to prepare cheap substitutes. Therefore, Bergman always confirmed his analyses of mineral waters by preparing artificial water and comparing it to the genuine water. As Bergman most certainly suffered from anaemia (Sect. 28.1), iron rich mineral water could actually have had some positive effect on Bergman's health. From 1781 and to the end of his life, Bergman also travelled to Medevi every summer for recreation [13].

On Christmas Eve, 1770, Bergman's student Pehr Dubb defended a thesis on the waters in Uppsala (Fig. 17.2) [14]. It differs from the majority of Bergman's theses by being written in Swedish rather than Latin, and thus may have been written by Dubb rather than Bergman. In this thesis, the water from the public well Qvarn-källan (the Mill well) in Uppsala was described and analysed. Litmus was first used to determine the presence of acid. Addition of lead(II) acetate (sugar of lead) gave precipitates with chloride and sulphate, and the visual appearance of the precipitate would tell whether it was chloride or sulphate in the water. To detect

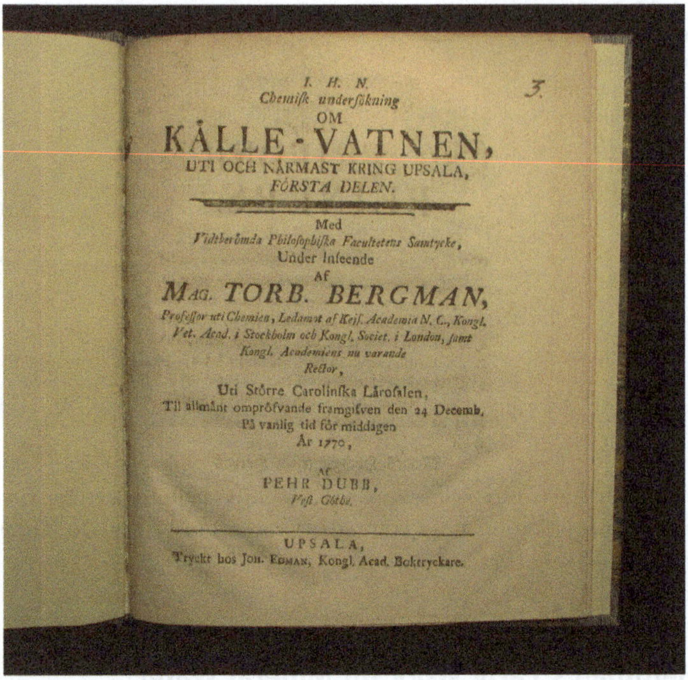
I. H. N.
Chemisk undersökning
OM
KÅLLE-VATNEN,
UTI OCH NÄRMAST KRING UPSALA,
FÖRSTA DELEN.

Med
Vidtberömda Philosophiska Facultetens Samtycke,
Under Inseende
Af
MAG. TORB. BERGMAN,
Professor uti Chemien, Ledamot af Keis. Academia N. C., Kongl. Vet. Acad. i Stockholm och Kongl. Societ. i London, samt Kongl. Academiens nu varande Rector,
Uti Större Carolinska Lårosalen,
Til allmänt ompröfvande framgifven den 24 Decemb.
På vanlig tid för middagen
År 1770,
Af
PEHR DUBB,
Vest Götha.

UPSALA,
Tryckt hos Joh. Edman, Kongl. Acad. Boktryckare.

Fig. 17.2 Pehr Dubb's dissertation on the spring water around and near around Uppsala. *Photo* Anders Lennartson

nitric acid, the water was evaporated and treated with potassium carbonate. As the dried residue could not be made to detonate after mixing with phlogiston (not specified in which form), no nitric acid was present. Carbon dioxide (aër fixus) was detected with calcium hydroxide solution (lime water). Next, the different earths in the water were analysed, which was done by evaporation. The residue effervesced with acids like calcium or magnesium carbonates (lime and white magnesia). A solution in hydrochloric acid gave a precipitate of calcium sulphate (selenit) on addition of sulphuric acid. No magnesium sulphate (English salt) was obtained from the mother liquor. A residue which was insoluble in acids was found to be silica. These analytic methods were rather crude compared to those Bergman would develop in the years to come.

Three years later, in 1773, Dubb's thesis was followed by another dissertation reporting chemical analysis of the water from Danmark,[3] a small village situated a few kilometres south east of Uppsala [15]. The sour water in Danmark was first discovered in 1733, but had been neglected until 1772. In the introduction, Bergman wrote that there are two ways of curing diseases: either using strong medicine with rapid effect or to give small doses of a mild medicine over longer time.

[3]Not to be confused with Denmark, which is also called Danmark in Swedish.

The medical powers of mineral waters did not lie in the water itself, but in the dissolved substances. On evaporation of the water, iron oxides precipitated and dissolved on addition of acid. Calcium sulphate and silicon dioxide remained. Upon slow evaporation of the water, crystals of sodium sulphate and sodium chloride were identified.

The same year, 1773, Bergman published his paper on aerial acid (Sect. 18.3) [16]. Here, Bergman noted that water saturated with carbon dioxide (*aër fixus*) got a pleasant, sour taste, similar to champagne or the water from Bad Pyrmont. Bergman wrote that he had prepared artificial mineral waters for himself and his friends for several years. He mentioned Priestley's method of preparing artificial Pyrmont water, but was not convinced of Priestley's formula, as genuine Pyrmont water contains other substances than carbon dioxide and iron. At this point Bergman did not, however, publish his own recipe. He later explained that he first wanted to carefully test the waters on himself, but when a foreigner had turned to Wargentin requesting a description of the method [17], Bergman finally published his method.

17.4 Bergman's Paper on Bitter-, Selzer-, Spa- and Pyrmont Water

In 1775, Bergman published a paper on the preparation of different types of artificial mineral waters, a subject he found particularly important: "Occasionally lies in this fluid substance [water] the best medicines for several diseases, especially the chronic, which without mineral waters rarely can be relieved or cured" [18]. Through experience, Bergman continued, it was known which water to use against a particular disease, but with the knowledge of the chemical composition of the water, one could even predict its properties. It was also possible to produce it artificially to a much lower cost, and make mineral waters accessible to people who otherwise could not afford them. There was also a national economic reason for artificial production of mineral waters. According to Bergman, 29,168 bottles to a value of 1,459 riksdaler and 19 Shilling was imported in 1772 and 23,405 bottles in 1773 to a value of 1,248 riksdaler and 32 Shilling [18]. Throughout the eighteenth century, the Swedish policy had been to try to reduce the import, and here Bergman saw an immediate benefit from his research.

Bergman focused on four types of water: bitter water (from Seyschütz in Bohemia), Seltzer water (from Selters in Germany), Spa water (from Spa in Belgium) and Pyrmont water (from Bad Pyrmont in Germany). The first part of the paper, published in the January-March issue of the *Transactions*, dealt primarily with analysis of the water samples [18]. For instance, he found 2.62 L (1 *kanna*) of bitter water to contain:

CO_2 (aerial acid) 12 ml (2 Swedish cubic inches)
$CaCO_3$ (lime saturated with aerial acid) 1.3 g (0.09½ *lod*)
$CaSO_4 \cdot 2H_2O$ (selenite or gypsum in crystals) 1.53 g (0.11½ *lod*)
$MgSO_4 \cdot 7H_2O$ (bitter salt in crystals) 51.87 g (3.90½ *lod*)
$MgCl_2 \cdot 6H_2O$ (hydrochloric acid united with magnesia) 0.73 g (0.05½ *lod*)
"$MgCO_3$" (Magnesia saturated with aerial acid) 0.4 g (0.03 *lod*)

The second part of the paper appeared in the subsequent issue, and describes the artificial production of mineral water [11]. A very important step in the preparation of artificial mineral water was, according to Bergman, to have pure ingredients.

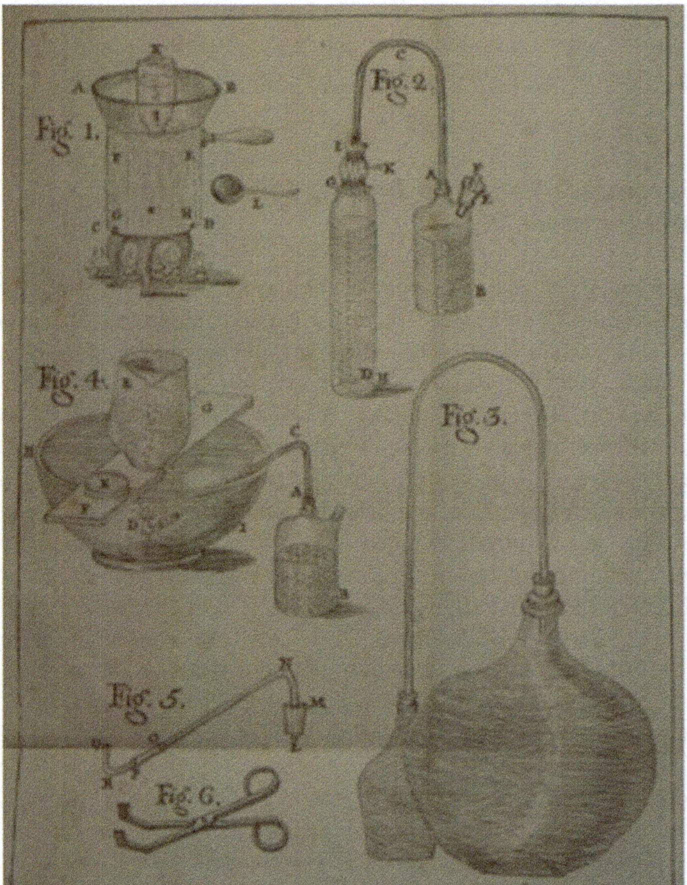

Fig. 17.3 Copper plate from Bergman's paper. Figure 3 shows the apparatus for carbonating water using a mixture of calcium carbonate and sulphuric acid in bottle B. Figure 3 depicts an apparatus used to carbonate water using a fermenting mixture. *Photo* Anders Lennartson

Pure water was obtained by distilling molten snow or spring water. Salts and earths were purified before use. To carbonate water, Bergman had used the apparatus in Fig. 17.3 for the last 6 years. Bergman also discussed apparatuses developed by Priestley and British apothecary Timothy Lane (1734–1807). The appropriate amounts of salts and earths were added to the carbonated water, the bottle was sealed and placed in a cellar. Usually, Bergman omitted calcium sulphate and calcium carbonate, to get a tastier water.

Carbon dioxide could be generated from calcium carbonate and sulphuric acid, but Bergman remarked that sulphuric acid is "highly corrosive, so by its use it should carefully be avoided to spill any thereof on hands, clothes or furniture". Bergman had also developed a method to carbonate water on a larger scale using the carbon dioxide from alcoholic fermentation.

The preparation of artificial mineral waters was not universally approved and Bergman's paper received some negative criticism [19]. An expanded trade edition of the paper [20] was published the following year. In the preface Bergman concluded that "Notwithstanding several unwilling remarks based on reasons, which I gladly keep secret, this art has been more practiced during one summer than I would have had expected during 10 [summers]". He had seen the method being used by women and very inexperienced men.

As mentioned in Sect. 9.6, Wallerius dismissed Bergman's work in a lecture for the Royal Swedish Academy of Sciences in 1783 by saying that no one had examined the water from Pyrmont more thoroughly and extensively than Hiärne had [21].

17.5 Hot Spa Water

The next paper on water was published in 1778, [22] and this time the topic was not water for internal use. Warm baths had been used for long times to cure illness, but Bergman remarked that it would not do with any warm water: "Here, like in the case of cold mineral waters, the presence of an elastic spirit is required, which increase the effect of the coarser dissolved particles, makes them finer and more penetrating". Bergman mentioned two types of water, one saturated with carbon dioxide, the other containing hydrogen sulphide. The water from Carlsbad (present-day Karlovy Vary in the Czech Republic) is an example of the first kind and the water from Aachen an example of the second kind. Since these waters lost their power on cooling, Bergman had not examined water from neither Carlsbad nor Aachen, and had to rely on published analyses. To carbonate hot water, Bergman used a pressure boiler of copper developed by Wilcke in Stockholm.

17.6 The Water of Medevi and Loka

Bergman had previously noted that the water from the Swedish spas was insufficiently analysed [11]. In two papers, Bergman discussed the water from Medevi [23] and Loka [24]. The water from Medevi had previously been examined by Hiärne a hundred years earlier. Bergman had received samples of Medevi water from Alströmer in 1778.

Bergman's analysis showed it to contain only small amounts of carbon dioxide, but in addition hydrogen sulphide, iron carbonate, calcium chloride and traces of sodium chloride. Compared to the Spa and Pyrmont waters, the carbon dioxide content was lower, but the hydrogen sulphide content was higher and it lacked the, in Bergman's opinion, harmful calcium carbonate. The water reportedly had a stronger taste 30 or 40 years earlier, but Bergman did not exclude that it was the perception of taste among the witnesses that had decreased with age. In any case, the water had been very good for Bergman's own health. Bergman's paper is much more than a chemical analysis and tells the history of Medevi and discusses the properties of the water.

The springs at Loka were exploited in 1725, but failed to reach the success of Medevi. By the mid-eighteenth century Medevi had become a meeting place for the Hat Party (Chap. 1), and supporters of the Cap Party and royalists thus had to meet elsewhere. When, in 1761, the King had to cure his health, his physician Nils Rosén (Sect. 9.1) suggested Loka. As the King's health improved, the trip was a success not only for the King but also for the owner of Loka and for Rosén, who was ennobled Rosén von Rosenstein.

Bergman's paper on the Loka springs is of little interest for a chemist, and is more concerned with the temperature of the water, the amount of water obtained per hour and the physical properties of the water. Perhaps the paper may have been-valuable in evaluating the commercial potential of new spas. Bergman reports no analysis: "Illness prevented me to evaporate the water from the other two wells". The paper also includes a long list of diseases treated with Loka water at the local hospital, and the outcome of the treatment. As of 2019, both Medevi and Loka are still open and, in addition, Loka is a major brand for bottled mineral water in Sweden.

17.7 Mineral Water Analysis in Sweden After Bergman

Bergman's successors in Uppsala did not continue Bergman's work on mineral waters. Berzelius, however, made some contributions. As a student, he spent a summer at Medevi working as physician and at the same time collecting material for his theses. His *pro exercitio* thesis [25] reports an analysis of the water and in his *pro gradu* thesis [26] he explored the effects of electricity on the patients at Medevi. Berzelius was later convinced by a physician, Lars Gabriel Werner, to invest in a mineral water factory in Stockholm, a venture that ended in bankruptcy and an economic catastrophe for Berzelius. For the promotion of this venture, Berzelius wrote an account of artificial mineral waters [27]. In 1834 Berzelius analysed the water from Porla [28] (a spa established in 1724), which suddenly made Porla fashionable.

References

1. Coley NG (1990) Physicians, chemists and the analysis of mineral waters: "the most difficult part of chemistry" In: Porter (ed), The medical history of waters and spas Medical history, Supplement No. 10, p 56–66
2. Boyle R (1684–1685) Short memoirs for the natural experimental history of mineral waters addressed by way of letter to a friend, London
3. Hoffmann F (1703) Methodus examinandi aquas salubres, Halle
4. Shaw P (1734) An enquiry into the contents, virtues, and uses, of the Scarborough spaw-waters: with the method of examining any other mineral-water. London
5. Cavendish H (1767) Experiments on Rathbone-Place Water, Phil Trans Roy Soc Lond 57:9–108
6. Priestley J (1772) Directions for Impregnating Water wit Fixed Air... J. Johnson, London
7. Boklund Uno (1956) Torbern Bergman som pionjär på mineralvattenområdet. In Torbern Bergman, om luftsyra..., Almqvist & Wiksell, Stockholm, p 121
8. Boklund Uno (1956) Torbern Bergman som pionjär på mineralvattenområdet. In Torbern Bergman, om luftsyra..., Almqvist & Wiksell, Stockholm, p 109
9. Schück H (1916) Torbern Bergmans självbiografi, äldre svenska biografier 3–4, Uppsala, p 103
10. Boklund Uno (1956) Torbern Bergman som pionjär på mineralvattenområdet. In Torbern Bergman, om luftsyra..., Almqvist & Wiksell, Stockholm, p 113
11. Bergman T (1775) Afhandling om Bitter- Selzer- Spa- och Pyrmonter-vatten, samt deras tilredande genom konst. Senare stycket. KVA Handl, 36:94–121
12. Wallerius JG (1748) Hydrologia eller Wattu-Riket... Lars Salvius, Stockholm
13. Aurillius PF (1785) Åminnelse-tal, öfver chemiæ professoren i Upsala...Herr magister Torbern Bergman...Uppsala, p 44
14. Bergman T, Dubb P (1770) Chemisk undersökning om källe-vatnen, uti och närmast kring Upsala. Upsala
15. Bergman T (1773) Dissertatio chemica de fonte acidulari Dannemarkensi, Uppsala
16. Bergman T (1773) Om Luftsyra, KVA Handl, 34:170–186
17. Bergman T (1776) Afhandling, om bitter- selzer- spa- och pyrmontervattens rätta hallt och tilredning genom konst; öfversedd och tilökt, Uppsala. Preface (unnumbered page)
18. Bergman T (1775) Afhandling om Bitter- Selzer- Spa- och Pyrmonter-vatten, samt deras tilredande genom konst. KVA Handl, 36:8–43
19. Aurillius PF (1785) Åminnelse-tal, öfver chemiæ professoren i Upsala...Herr magister Torbern Bergman...Uppsala, p 31
20. Bergman T (1776) Afhandling, om bitter- selzer- spa- och pyrmontervattens rätta hallt och tilredning genom konst; öfversedd och tilökt, Uppsala
21. Wallerius JG (1783) Tal, om nödig jämförelse emellan de chemiska undersökningar, och naturens verkningar, hållet i Kongl. Vetensk. Academien vid præsidii nedläggande, den 9 julii 1783. Stockholm, p 10
22. Bergman T (1778) Om Varma Hälso-vattens tilredning KVA Handl, 39:219–227
23. Bergman T (1782) Underrättelse, om Medevi Surbrunnar. KVA Nya Handl, 3:288–298
24. Bergman T (1783) Underättelse om Loka källor. KVA Nya Handl, 4:256–267

25. Berzelius J (1800) Nova analysis aquarum Medeviensium. Uppsala
26. Berzelius J (1802) De electricitatis galvanicæ apparatu cel. Volta excitæ in corpora organica effectu, Uppsala
27. Berzelius J (1803) Några underrättelser om artificiella mineral-vatten. Stockholm
28. Berzelius (1834) Undersökning af vattnet i Porla Källa. KVA Handl, 18–92

18 Research on Carbon Dioxide

Carbon dioxide, CO_2, is a compound of special interest in the history of chemistry. Long before the discussions on global warming, carbon dioxide was a highly controversial substance [1]. It was studied already by Johannes Baptista van Helmont (1580–1644), who coined the word gas and prepared carbon dioxide by several different methods. It was probably the first gas other than air to be described, although van Helmont did not realise that the carbon dioxide prepared by different methods actually was the same gas.

In aqueous solution, carbon dioxide is in equilibrium with carbonic acid, a weak, unstable acid that cannot be isolated:

$$CO_2(aq) + H_2O(l) \rightleftarrows H_2CO_3(aq).$$

Water saturated with carbon dioxide is only weakly acidic. At ambient temperature and pressure, water saturated with carbon dioxide contains only 0.17% carbon dioxide, of which less than 1% in the form of carbonic acid.

It was early recognised that gas was formed by fermentation and by addition of acids to carbonates, such as limestone ($CaCO_3$), potash (K_2CO_3) and soda (Na_2CO_3), e.g.

$$CaCO_3(s) + H_2SO_4(l) \rightarrow CaSO_4(s) + H_2O(l) + CO_2(aq).$$

Carbon dioxide was also expelled by heating limestone ($CaCO_3$) to produce quicklime (CaO):

$$CaCO_3(s) \rightarrow CaO(s) + CO_2(g).$$

Calcium oxide (quicklime) and calcium hydroxide ($Ca(OH)_2$; slaked lime), obtained from calcium oxide and water, readily absorb CO_2 from the atmosphere to reform calcium carbonate. This is the reaction occurring as mortar hardens, a process used for centuries. It was first believed that limestone absorbed fire to give

A. Lennartson, *Carl Wilhelm Scheele and Torbern Bergman*, Perspectives on the History of Chemistry, https://doi.org/10.1007/978-3-030-49194-9_18

the caustic quicklime on calcination, and the exothermic reaction with water to form slaked lime was believed to be due to water expelling the fire from quicklime. By adding slaked lime to a solution of sodium carbonate (soda) or potassium carbonate (potash), insoluble calcium carbonate precipitated, leaving strongly basic alkali hydroxides (caustic alkalis) in solution:

$$Na_2CO_3(aq) + Ca(OH)_2(aq) \rightarrow 2NaOH(aq) + CaCO_3(s).$$

The true nature of this reaction was first realised by Joseph Black in his work on magnesium carbonate (*Magnesia alba*) [2]. He regarded limestone as a compound of quicklime and *fixed air* (carbon dioxide). Fixed air had the property of making a strong caustic alkali (i.e. a hydroxide) mild (by conversion into carbonate). It was now found that the gas formed by calcining limestone, addition of acids to carbonates and fermentation was one and the same.

In 1764, however, Johann Friedrich Meyer (1705–1765),[1] an apothecary from Osnabrück, published a book of 418 pages introducing a competing, erroneous theory [3]. According to Meyer, limestone absorbed a fatty acid, *Acidum pingue*, from the fire upon calcination. It was this acid that gave the quicklime its caustic properties, and which was transferred to the soda and potash.

18.1 Scheele and the Acidic Properties of Carbon Dioxide

In a letter to Retzius, dated April 26, 1768, i.e. shortly before he left Malmö, Scheele discussed Meyer's theory of fixed air (carbon dioxide). Scheele returned a dissertation by Wallerius and Wibom, which he had borrowed from Retzius. Scheele was sceptic about Wallerius' dissertation, and put forward his ideas on carbon dioxide and alkali. Scheele supported Meyer's theory of an *Acidum pingue* which was indestructible by fire and which could be spontaneously released from potassium hydroxide (caustic lye) in the open air to give potassium carbonate (*Sal tartari*). At this point, Scheele was probably still unaware of Black's hypothesis [4]. He would, as will be seen, change his mind.

The acidic nature of carbon dioxide had been suggested by Hoffmann in 1708 [5, 6] and by Keir in his English translation of Macquers chemical dictionary in 1771 [7, 8]. This view of carbon dioxide as an acid did not receive any widespread recognition, however.

Soon after Scheele moved to Stockholm, he was joined by his friend Retzius (Sect. 8.1). In a letter from Retzius to Wilcke written in 1787, Retzius described how he had read Monnet's book on mineral waters, *Traité des Eaux Minerales* published in 1768,[2] and decided to translate it to Swedish.[3] This prompted him to carry out additional experiments, in which he engaged Scheele. Retzius carried out

[1]Not to be confused with Scheele's friend Johan Christian Friedrich Meyer in Stettin.
[2]Bergman asked Wargentin for this book in 1769.
[3]This translation was apparently never published.

experiments in order to find out whether pure water could dissolve iron, or if the solubility of iron was caused by another substance present in water. He came across Lane's paper claiming that carbon dioxide (fixed air) was required for water to dissolve iron [9]. "This", Retzius wrote, "I commissioned Scheele to repeat. He came to me completely happy one morning, and told me completely satisfied, that he evidently had found that the so called *Aer fixus* [carbon dioxide] had all properties of a weak acid". In Uppsala, Scheele told Gahn about his findings on carbon dioxide, and these experiments are found in Gahn's notes written in spring 1770: "If *Aer fixus* is caught in water, it reacts with litmus as an acid, despite all efforts to exclude any acidic vapour. After some time, this air evaporates away and the litmus turns blue again". Via Gahn, this knowledge may have reached Bergman, and may have put Bergman on the right track in his search for an acid in air (Sect. 18.3). On the other hand, Bergman was apparently unaware of Scheele's investigation of the transformation of water to earth in the same manuscript, so Bergman's work may be independent of Scheele's.

18.2 Scheele's Later Views on Carbon Dioxide

As Scheele learned about Black's hypothesis, he abandoned his support for Meyer, and joined the international discussion on the nature of carbon dioxide. In 1778, Jacob Andreas Weber published a book concluding that the theories of both Black and Meyer were wrong [10]. Weber's theory was that heat arises from electrical matter emanating from the sun, and it was this electric matter, in the form of phlogiston, which was removed from mild alkalis by quicklime [11]. Scheele was not impressed by this theory, and criticised it in a letter to J. C. F. Meyer[4], a letter which Crell published [12] with Scheele's permission. Weber's response in Crell's *Neuesten Entdeckungen in der Chemie* [13] caused Scheele to write a second letter where he (on good grounds) seriously questioned Weber's experimental skills: "I cannot understand, how he has performed his experiments" [14].

Kirwan had in 1782 concluded that carbon dioxide was composed of phlogiston and oxygen,[5] a subject that he discussed in letters with Bergman [15]. By 1783, he had convinced Bergman, but Scheele never accepted this theory. In a letter to Crell published in 1784, Kirwan wrote that Priestley had established the composition of carbon dioxide (fixed air) without all doubt: he had obtained carbon dioxide by burning iron in oxygen (dephlogisticated air) and by heating mercury(II) oxide (red precipitated mercury) with iron flings [16]. Later the same year, Crell published a letter from Scheele, including a comment on carbon dioxide (aerial acid), where Scheele maintained that it had not been proven that carbon dioxide was composed of oxygen (pure air) and phlogiston [17]. Scheele mentioned neither Priestley nor Kirwan. Another letter from Scheele to Crell followed, where Scheele correctly

[4]Not to be confused with J. F. Meyer, who launched *Acidum pingue* hypothesis.

[5]It would be correct for someone identifying phlogiston with carbon.

concluded that commercial iron contain carbon, which gives rise to carbon dioxide on combustion, thus explaining Priestley's observations [18]. In another letter, Scheele criticised Landriani's view that all acids owed their acidity to carbon dioxide [19]. As long as he could not prepare other acids from carbon dioxide, Scheele could not accept this theory. In the same letter, Scheele elaborated his view on Kirwan's experiments with carbon dioxide and iron. This caused Kirwan to respond: [20]

> The reason, why the red, by it self calcined, mercury [HgO] give fixed air [CO_2] by distillation with iron, is because the mercury calx [HgO] much easier deprive iron of its combustible [i.e. phlogiston], than the dephlogisticated air [O_2], which is combined with it in the fixed air.

Graphite (writing lead) did not give carbon dioxide with mercury(II) oxide according to Kirwan. This is true, but the reason is that mercury oxide decomposes to mercury and oxygen at lower temperature than would be required for the reaction with graphite to occur. Scheele found other reports on carbon dioxide in the literature that he did not believe; he could experimentally refute a report that carbon dioxide arise from distillation of lead amalgam or from nitrogen oxide (saltpetre air) and oxygen [21].

Bergman openly supported Kirwan's view in 1784, saying that he valued Kirwan's papers very high, and that he had reasons to favour Kirwan's views over Scheele's [22]. Thus, Bergman and Scheele did not always agree, and it is thus not fair to call Bergman the theoretician with Scheele merely performing his experiments. Scheele was truly an individual thinker and occasionally, such as in this case, Scheele who was closer to the truth.

18.3 Bergman and the Aerial Acid

Bergman had been investigating the presence of an acid in the atmosphere since at least 1769; in a paper published in spring 1773, he wrote that he had exposed a sample of pure alkali to the ambient atmosphere on an attic for 4 years. His original objective may have been to disprove Wallerius claim that air contains sulphuric acid (Sect. 9.6). By 1772, he had proved experimentally that carbon dioxide is an acid, and communicated these results in a letter to Priestley [23]. Priestley attributed the discovery of the acidic properties to Bergman [24]. In fact, both Black, [2] and especially Cavendish, [25] had found that carbon dioxide (fixed air) reacts with bases, but they did not explicitly state that carbon dioxide is an acid. Bergman did not refer to the works of Black or Cavendish in his paper, which irritated Thomson when he wrote his *History of Chemistry* [26]. As Bergman developed an interest in mineral waters (Chap. 17) about the same time, he had another strong reason to investigate the properties of carbon dioxide.

In 1773, Bergman published his results in an important paper entitled *On Aerial Acid* in the *Transactions* [27]. After keeping a sample of pure alkali (K_2CO_3 or Na_2CO_3) for 4 years on the attic, he had found no traces of sulphuric acid (vitriolic acid), and explained previous reports of sulphuric acid in the atmosphere by impurities in the alkali used or acid vapours present in laboratories, and he wrote that such experiments could not be carried out in a laboratory where acids were handled or stored. He then discussed *Aër fixus* (carbon dioxide), referring to Boyle and Hales, and then discussed the theories of Black and Meyer. Bergman summarised the general properties of acids, and set out to investigate if carbon dioxide fulfilled these criteria. A substance as sublime as carbon dioxide needed a *vehiculum*, i.e. it had to be dissolved in a solvent rather than being investigated in gaseous state. Bergman concluded that aqueous solutions were appropriate. Water saturated with CO_2 had a sour taste and was similar to Pyrmont water (Sect. 17.4) or Champagne. He found that CO_2 gave salts with bases, and that bases could be over-saturated, i.e. he had noticed the formation of HCO_3^-; he mentioned both $NaHCO_3$ and $Ca(HCO_3)_2$. Aqueous calcium hydroxide solution (lime water) first gives a precipitate ($CaCO_3$) with CO_2 that dissolves in the presence of excess CO_2 due to the formation of unstable water-soluble $Ca(HCO_3)_2$. When the solution is left in the sunshine, $CaCO_3$ is re-precipitated as CO_2 escapes. Bergman recognised the geological effect of carbon dioxide saturated water on limestone. He correctly defined lime ($CaCO_3$) as a salt, and found that CaO kept in a closed vessel is unaffected over time. Just as Lane had reported, Bergman found that aqueous solutions of CO_2 can dissolve iron. Bergman found that aqueous CO_2 solutions colour litmus solution red and that the solution turned blue with time, as the dissolved CO_2 escaped. Bergman described carbon dioxide as a glue that held the elements (*principa proxima*, see Sect. 22.2) together in substances. He attributed the alleged property of CO_2 to prevent scurvy and putrefaction to this fact. He concluded his paper with a number of precipitation experiments with metal salts.

In 1775, Bergman published an extended version of his paper in Latin [28]. A revised version of this paper appeared in the first volume of his *Opuscula*, and was translated to English, German, French, Italian and Spanish [29]. This paper starts with a discussion of fixed air, a word that may have two meanings according to Bergman: either any gas trapped in a solid, or more specifically the gas now known as carbon dioxide, or aerial acid (luftsyra) as Bergman called it. He then described the three main methods of preparing pure CO_2: heating of carbonates, addition of acid to carbonates and fermentation. Next, he described the general properties of acids and the properties of aqueous CO_2 solution: it has a sour taste and acts as a weak acid. The bulk of the paper discussed different carbonates. He determined their compositions and the amount of different mineral acids required to saturate them. In the case of Na_2CO_3, he was quite close to the truth. He found 100 parts of fresh crystals of aerated mineral alkali ($Na_2CO_3 \cdot 10\ H_2O$) to be composed of 16 parts CO_2 (fixed air), 64 parts water and 20 parts of pure alkali (NaOH). The true values are 15, 56 and 24, respectively.

The nature of aerial acid was further discussed by Bergman in 1779 [30]. Being a gas with acidic properties, Bergman compared carbon dioxide with sulphur dioxide and nitrogen oxides, and from this analogy he suspected carbon dioxide to contain phlogiston. Bergman continued the discussion by comparing aerial acid with hydrochloric acid, from which phlogiston can be removed (i.e. it can be oxidised to chlorine).[6] Aerial acid was something of a favourite topic of Bergman, and he actually wrote more about carbon dioxide than about oxygen in his preface to Scheele's *Chemische Abhandlung* (Sect. 21.3).

References

1. (1950) Gmelins Handbuch der Anorganischen Chemie, Achte Auflage, Calcium Teil A, Lieferung 1. Verlag Chemie, Weinheim
2. Black J (1756) Experiments upon Magnesia alba, Quicklime, and some other alkaline Substances, Essays and Observations, Physical and Literary read before a Society in Edinburgh, 2:157–225
3. Meyer JF (1764) Chymische Versuche, zur näheren Erkenntniß des ungelöschten Kalchs, der elastischen und electrischen Materie, des allerreinsten Feuerwesens, und der ursprünglichen allgemeinen Säure, Johann Wilhelm Schmidt, Hannover & Leipzig
4. Sjöstén CG (1801) Åminnelse-tal, hållit för Kongl. Vetenskaps Academien öfver dess framledne ledamot herr Carl Vilhelm Scheele, den 14 oct. 1799. Stockholm
5. Hoffmann F (1708) Dissertationes physic-medica. 2:183
6. Partington JR (1961) A history of chemistry, vol 2. Macmillan, London, p 694
7. Macquer J (1771) Dictionary of Chemistry vol 1 p 36; vol 2 p 838
8. Partington JR (1962) A history of chemistry, vol 3. Macmillan, London, p 140
9. Lane T (1769) A Letter from Mr. Lane, Apothecary, in Aldersgate-Street, to the Honourable Henry Cavendish, F. R. S. on the Solubility of Iron in Simple Water, by the Intervention of Fixed Air, Phil Trans, 59: 216–227
10. Weber JA (1778) Neuentdeckte Natur und Eigenschaften des Kalkes und der äzenden Körper, Berlin
11. Partington JR (1962) A history of chemistry, vol 3. Macmillan, London, p 152
12. Scheele CW (1781) Über das brennbare Wesen im rohen Kalk, Chem Ann, part 1, 30–41
13. Weber JA (1784) Bemerkungen über das brennbare Wesen im rohen Kalke, Neuesten Entdeckungen in der Chemie 12:94–111
14. Scheele CW (1785) Erläuterung über einige, den ungelöschten Kalk betreffende, Versuche, Chem Ann, part 2, 220–227
15. Carelid G, Nordström J (1965) Torbern Bergman's Foreign Correspondence, Almqvist & Wiksell, Uppsala, p L
16. Kirwan R (1784) Vom Hrn. R. Kirwan in London, Chem Ann, 1:36–38
17. Scheele CW (1784) Vom Hrn. Scheele in Köping, Chem Ann, 2:123–125
18. Scheele CW (1784) Vom Hrn. Scheele in Köping, Chem Ann, 2:328–329
19. Scheele CW (1785) Vom Hrn. Scheele in Köping, Chem Ann, 1:153–155
20. Kirwan R (1785) Vom Hrn. R. Kirwan in London, Chem Ann, 2:335–337
21. Scheele CW (1785) Vom Hrn. Scheele in Köping, Chem Ann, 1:455–457
22. Bergman T (1784) Vom Hrn. Profess. und Ritter Bergmann in Upsal, Chem Ann, 1:38–39
23. Priestley J (1772) Observations on Different Kinds of Air, Phil Trans 62:147–264

[6]It should be noted that carbon is in its highest oxidation state in carbon dioxide. It cannot be oxidised further.

24. Priestley (1774) Experiments and Observations on Different Kinds of Air, vol 1. J. Johnson, London, p 31
25. Cavendish H (1766) Three papers, containing experiments on factitious air Phil Trans, 56:141–184
26. Thomson T (1831) The history of chemistry, vol II. Henry Colburn & Richard Bentley, London, p 41
27. Bergman T (1773) Om Luftsyra KVA Handl 34:170–186
28. Bergman T (1775) Commentatio de acido aëreo, Nova Acta Regiæ Societatis Scientiarum Upsaliensis 2:108–160
29. Moström B (1957) Torbern bergman a bibliography of his works. Stockholm, Almqvist & Wiksel, p 32
30. Bergman T (1779) Anledning til föreläsningar öfver chemiens beskaffenhet och nytta, samt naturliga kroppars almännaste skiljaktigheter, M. Swederus, Stockholm, Uppsala and Åbo, p 62

Scheele in Köping

19

The pharmacy in Köping, a small trading town in the Västmanland province, was established in 1759 by Hindrich Pascher Pohl (1730–1775), an immigrant from Pomerania, just like Scheele [1]. In 1769, Pohl lost his wife of five years, Maria Molin and their two children. He remarried in 1771 to Sara Margaretha Sonneman, daughter of Emanuel Sonneman, a merchant and member of the city council. The Nordic museum in Stockholm still preserves a harrow and a lever for moving stones invented by Sonneman. Pohl died on April 12, 1775, aged 43. This left his widow Margaretha, aged only 24, in a difficult situation. She had inherited the ownership of the pharmacy and the valuable license to run the pharmacy in Köping, but in order to run a pharmacy, an apothecary exam was required. To get any income from the pharmacy, she had four alternatives: a common solution was to marry an apothecary, but she could also sell the pharmacy, rent it out, or hire a *provisor*, who would run the pharmacy for her. Margaretha Pohl chose the latter alternative and advertised for an apothecary. For Scheele, this was a good opportunity to become his own master. Working as an apprentice, he could never expect to save enough money to buy a pharmacy. He applied and was hired on a one-year contract. He moved to Köping sometime between July 15 and 20, 1775. Scheele signed his contract on July 21, and on July 28, he applied for respite from the requirement of an apothecary exam. It was not uncommon at the time for an apprentice to wait with the expensive exam until he was in charge of a pharmacy. Given Scheele's reputation and his prominent friends in *Collegium medicum*, the respite was probably granted him without hesitation. Gahn, writing to Bergman, was worried that all the new duties would prevent Scheele from performing research [2].

The choice to leave Uppsala was probably not easy for Scheele. Bergman wrote to Gahn that "Now Scheele has left. It did cost us both to separate: he is in charge of the pharmacy in Köping and may eventually fall for the widow". This seemed to be a probable long-term solution, provided that Margaretha was satisfied with

A. Lennartson, *Carl Wilhelm Scheele and Torbern Bergman*, Perspectives on the History of Chemistry, https://doi.org/10.1007/978-3-030-49194-9_19

Fig. 19.1 The street corner in Köping where the pharmacy was originally located. The current view shows no traces of the eighteenth century. *Photo* Anders Lennartson, June 2014

Scheele's work. As far as we know, Bergman and Scheele would only meet twice again, but exchanged letters until Bergman's death. They also sent samples to each other and Bergman sent Scheele books and dissertations.

The pharmacy was much simpler than his previous working places. It reportedly consisted of a single room where Scheele worked with an apprentice. The names of two of these apprentices are known, Carl Berg and Christian Hasselberg (1746–1823). The latter had worked at the Raven pharmacy in Stockholm 1769–1771, i.e. at the same time as Scheele, and the two men thus already knew each other. Hasselberg left Köping in 1783 and became apothecary in Mariefred in 1787. The Köping pharmacy had a small back yard with a shed. The shed was divided into two compartments by a simple wall made from boards. Scheele used one part as his laboratory; the other part was used for farming tools. Scheele rented a room at the inn, and Mrs Pohl cooked for him. The original pharmacy building is long since gone (Fig. 19.1).

19.1 The Ekelin Incident

After half a year in Köping, Scheele faced a serious problem. An apothecary named Peter Ekelin turned up in Köping and offered to rent the pharmacy. Mrs Pohl accepted the offer, and decided not to extend Scheele's contract. On December 22, Scheele wrote to Gahn:

> I am rather melancholic, as it seems I will not make my luck here. An apothecary from Landskrona, who ate with us at the table in the evening, has negotiated about the pharmacy with *rådman* Sonneman (the father of the widow). He has agreed with him and he will take over the pharmacy on July 22, more precisely as leaseholder. He refers to his money, and that has helped him. The whole contract was set up without mentioning a word to me. Oh, falseness! Should one necessarily have money to get his bread in this world?

Scheele rarely mentioned his private life in his letters to his colleagues, but he returned to the Ekelin incident in several letters, and it is clear that it made him very disappointed and unhappy.

Ekelin's life appears to have been surrounded by conflicts and trouble [3]. He was a pupil at the Split Eagle in Malmö from 1760, becoming apprentice in 1763. As Scheele apparently did not know him, he must have left before 1765. In 1769, he ran the Lion pharmacy in Landskrona in southern Sweden while the owner, Kollberg, was on leave. In 1771, Kollberg ran into financial problems and had to sell the pharmacy. Ekelin placed a bid, but since he had no exam and had only managed to pay 2,000 of the 6,500 silver dealers he was supposed to pay for the pharmacy, the bid was ultimately rejected. The following year, Ekelin returned with the exam and the remaining money. Soon, however, Ekelin went bankrupt and the pharmacy was sold on auction. Both Kollberg and Ekelin placed bids and after some complications, Ekelin could run it until 1775, when it was finally sold to merchant Eric Giörloff on the behalf of apothecary Peter Sjöqvist, who had been employed at the Split Eagle just like Ekelin and Scheele had been.

The response to the Ekelin incident clearly shows what an impression Scheele must have made on his friends, and what a great reputation he must have had after publishing only three papers. Bergman wrote to him in an undated letter:

> I have in these days, with considerable worry, heard that You have been betrayed in Köping. It hurts me deeply, and I hope I can be of any help on this occasion. Do You want to take on a position in Chemia oeconomica, that will probably be established soon in Alingsås? If You want to return to Uppsala while waiting for a suitable position, my home will be open for You. Let me soon know Your decision, and be certain that it will be my pleasure to do what is possible for your best.

Bergman and Scheele's former employer Lokk wanted Scheele to return to Uppsala, while Gahn wanted Scheele to move to Falun as his collaborator. Scheele wrote to Gahn and declined, but promised to visit Gahn, which he never did. In a letter from Bergius dated April, 17 1776, Scheele was offered the pharmacy in Alingsås, which had been put to his disposal by Bergman's merchant acquaintance Patrick Alströmer. The offer had been conceived at a dinner held by court apothecary Collin, where Bergius, Schultzenheim, Bäck, Wargentin and Linnæus attended. According to a letter to Gahn in July, Scheele appears to have declined, and from a letter to Bergman, it seems most likely that Scheele planned to return to Uppsala.

In Köping, the public opinion was clearly in Scheele's favour. The burghers of Köping had declared that they wanted no other apothecary in Köping than Scheele, and had written a letter to the county governor (landshövding), who in turn contacted *Collegium medicum*. The *Collegium* agreed that Scheele was more qualified than Ekelin, but had no authority to intervene, and asked the governor to make an arrangement with the widow. Scheele mentions this in his letter to Gahn of June 20.

Several biographies states that *Collegium medicum* granted Scheele the rights to open a competing pharmacy in Köping [4, 5]. This sound very unlikely, and Fredga found no evidence for such plans in the archives [6]. These rumours may originate in a misunderstanding based on letters from Scheele to Bergman and Gahn written in May 1776, where he wrote that the burghers had asked him to open a new pharmacy. At any rate, the contract between Pohl and Ekelin was cancelled that summer; on August 30, Scheele wrote to Gahn that Ekelin had withdrawn his offer after intervention by the county governor. Ekelin returned to Landskrona where he remained until 1777, where after he disappeared into the mists of time.

On October 18, 1776, Scheele bought the pharmacy for 1,333 riksdaler and 16 Shilling; this is noted in the city minutes of November 4, 1776. Scheele must have borrowed the money from his friends, but apparently, he must quite soon have been able to pay the money back. Scheele also agreed to support Mrs Pohl, her infant son, Emanuel (1775–1789), and to pay the pharmacy's remaining debts.

19.2 Scheele's Life in Köping

Köping was a small town with around 1,300 citizens, its main importance was as a harbour for shipping iron from the Bergslagen mining district to Stockholm. A fire in 1889 destroyed most of the city, and little of today's Köping resembles the city where Scheele lived. Preserved letters from his brothers and drafts of letters to his brothers, give us some information about Scheele's life in Köping.

Back at home in Stralsund, Scheele's father passed away in 1776 (very little is known about his mother, but according to a letter from his brother Friedrich Christoph, she was still in good health in 1783; she would survive her famous son by two years). Friedrich Christoph took over his father's business, but also had to support his two unmarried sisters and his mother. He ran into economic problems, and had to give up the business, but could resume it in 1778. To help his brother, Scheele agreed to take care of his youngest sister, Maria Juliana. She arrived in Köping in early 1779, and it was the first time Scheele met any of his relatives since a visit in Stralsund ten years earlier. Unfortunately, she appeared to have fallen ill during her trip, she suffered from cough and tiredness. She died at an age of 32 after 16 months in Sweden, on May 19, 1780, at 8 p.m., most probably from tuberculosis. Scheele assured his brother that he had provided the best care available.

Fig. 19.2 Scheele's pharmacy in Köping, as it appeared a century after Scheele's death. The building was demolished three years later. Wood cut by Amanda Maria Falander from Cleves biography [7]

During her short time in Sweden, she had learned Swedish and had become a Swede, as Scheele wrote. The funeral was held on May 25, and she was buried next to the grave of apothecary Pohl. Back in Stralsund, his brother got a daughter, named Catarina Eleonora Wilhelmina, the third name was an honour to his brother, the famous chemist. This increased the strain on his economy, and he asked Wilhelm to invite the remaining sister, Anna, aged 40, to Sweden. Unlike her younger sister, she was not so eager to go to Sweden, and Friedrich asked Scheele to persuade her, as neither his wife, nor his father-in-law, liked that he supported her economically. As will be seen later, other things came in between.

Scheele's economy improved and on February 7, 1781, after six years in Köping, he bought a house at the Great Square (Stora torget) where he moved his pharmacy (Fig. 19.2). The house was bought on auction, and Scheele's winning bid was 244 riksdaler specie and 20 skilling. It included a garden plot and pasture, and was renovated during the summer of 1782. By the time of his death, Scheele had two cows and a sow. In a letter to Bergman written in May 1782, Scheele wrote that the work with the house took up so much time, that he had not yet had time to study a sample of asbestos sent by Bergman.

Some insight into Scheele's life at the time is provided by a concept for a letter written by Scheele to his brother Friedrich Christoph. It is an answer to letter dated October 4, 1781, which gives an approximate dating of Scheele's letter. Scheele congratulates his brother, who just had a son, and thank him for apples that his brother had sent him from Stralsund. Mrs Pohl was supposed to have picked lingonberries in return, but due to the drought, there had been no lingonberries to pick. Then, he discussed his sister, Anna:

> What the other thing with our sister concerns, I have not been able to do anything with my newly bought house this summer; he who lives there was so persistent not to move before Michaelmas [September 29]. If the auction had been held two weeks earlier, he would have had to move out by last Easter. Now he is gone, but to build in the winter is no good, so I have to wait for the summer. It is true that I have a room free for Anna, but is it not possible that I would need the room before that? Could not I under these circumstances marry like you? If so, I plan to hold a wedding by autumn. Should not I then become a children manufacturer[1] just like you?

Whether Scheele actually had a particular woman in mind (Mrs Pohl for instance) is not known. It should be noted that Scheele clearly did not intend to devote his whole life to his science as is usually claimed, but that he intended to become a family father. This was unfortunately not to be. Scheele continued to argue for his sister to remain in Germany:

> It is also another issue that puts obstacles in the way. I have told Mrs Pohl about our sisters trip here, and she does not seem to be happy with it, since our deceased sister had told her too honestly about or sister's (Anna's) character, she is also convinced about the truth in this, based on the letter she [i.e. Anna] sent me last year, filled with dislike regarding the dividing of our deceased sisters belongings.

From a previous letter concept, it turns out that Scheele had spent 74 riksdaler, a considerable amount of money, on the funeral and wanted to pay with a part of his dead sister's belongings, while he would send clothes and most of what remained to his sister Anna. Apparently, Anna was not happy. An interesting detail is that when Scheele wrote "silver jewellery", he did not write "silver" but it's chemical symbol, a half-moon. Whether he changed this in the final letter is not known. Anna remained in Stralsund, but died not long after. It is also noteworthy how close Scheele and Mrs Pohl were, and that Scheele took her opinions into consideration, and told her about family matters.

Scheele used saved money to buy the new house, which gave him a much more convenient laboratory. The money from the Royal Swedish Academy of Sciences (Sect. 19.4) was probably more than welcome. The new pharmacy was a wooden building in two storeys and served as pharmacy for a century. In 1889, the out-dated building was demolished by apothecary Arbman (sad, but the 1889 fire would probably have destroyed it anyway). It was replaced by a stone building (Fig. 19.3), which was used as pharmacy until 1970 when it was converted to apartments. A set of wooden boxes (Fig. 19.4) used for drugs survived as they were sold to the pharmacies in Södertälje and Strömsund, from where they were later acquired by

[1]Scheele used the word "kindessfabrikatör" in the original manuscript. Note his odd spelling.

Fig. 19.3 The former pharmacy building in central Köping, located at the same place as Scheele's old pharmacy. *Photo* Anders Lennartson, June, 2014

the Nordic Museum in Stockholm [8]. They are now on display at the pharmacy museum at Skansen, Stockholm.[2] Contemporary visitors have described Scheele's laboratory as well equipped, and this is confirmed by the estate inventory. Although it is generally believed that Scheele had no interests besides chemistry, it has been reported that he—just for the pleasure—built a small shed and planted some trees around the Köping well [9].

19.3 Foreign Guests

In July 1782, Scheele was visited by the Spanish chemist Juan José d'Elhuyar (1754–1796) and the French chemist Charles-André-Hector de Virely, a friend of Guyton de Morveau [10]. The two had been visiting Bergman since late January (Chap. 11), and after leaving Uppsala they took the way via Köping just to visit Scheele. On arriving, they were welcomed by Scheele who was wearing a leather

[2]This house is sometimes described as Scheele's pharmacy, but the building has no relation to Scheele. The furniture and equipment is collected from different pharmacies. The building previously served as a café in central Stockholm before being moved to Skansen.

Fig. 19.4 Wooden boxes from Scheele's laboratory in Köping, now at Skansen, Stockholm. *Photo* Anders Lennartson, June 2015

apron. Scheele happily discussed his research with them, but did not interrupt his work in the laboratory. He spoke freely about his new discoveries, apparently without keeping unpublished work secret. When they invited him for dinner, he joined them, but immediately returned to his laboratory as soon as he had finished the food—without thanking his hosts. The two guests stayed for two days and left very impressed. From a letter to Bergman, it is clear that Scheele was very happy about the visit, but it seems that he never considered taking a break from his work.

In August 1784, Scheele was visited by Smithson Tennant (1761–1815), a British chemist who would be known as co-discoverer of osmium and iridium in 1803. In a letter to Wilcke, Scheele wrote that Tennant stayed for two days, and had written down all news Scheele told him. It was especially the discovery of hydrogen cyanide that Tennant found interesting. The trip to Sweden must have made a great impression on Tennant; when Berzelius met Tennant in England decades later, he was still carrying a map over Sweden in his pocket [11].

19.4 Scheele's Stockholm Visit

After moving to Köping, Scheele only left his hometown once, as far as we know, and he had two reasons for the trip. First of all, as an owner of a pharmacy, he finally had to take his apothecary exam. While in Stockholm, he had also agreed to take his seat in the Royal Swedish Academy of Sciences. During the third quarter of 1777, Bergman served as *preses* (president), a duty that included editing an issue of the *Transactions* and to give a lecture at the end of the term. As president, Bergman persuaded Scheele to finally take his seat in the Academy and deliver his inaugural lecture on October 29.[3] [12],

The lecture that Scheele delivered was rather unspectacular, and rather than giving a formal lecture on his views on contemporary chemistry or his role in the discovery of oxygen, as might have been expected, he simply read a paper describing a new and more convenient way to prepare mercury(I) chloride, a common medical ingredient at the time. The reason why Scheele gave this short lecture might be sought in his personal character. In 1782, Bäck attempted to persuade Scheele to come to Stockholm to give chemical lectures, but Scheele declined and told Bäck that he did not enjoy public performance. It could also be that Scheele did not want to waste his time writing a formal lecture. According to the diary of the Academy, Scheele held his lecture in German. The inaugural lectures were occasionally printed, but Scheele's lecture was published as a regular paper in the *Transactions* [13].

Scheele's lecture was followed by a reply by Bergman, the manuscript of which remains in the archive of the Academy, and is also quoted (in a slightly corrupted manner) by Sjöstén [14]. Bergman said:

> It is a true pleasure for the Academy to see such a hard working member taking his seat in its community, and for me, who has the honour of being president at this occasion, it is a doubled pleasure. I have for several years witnessed your incomparable diligence, your considerable skill to, by appropriate experiments, extort Nature its secrets and to your sharp conclusions from the performed experiments. What can then be more natural than that he, who deeply loves his science, with pleasure watches You take an honourable seat, to which only skill has paved your way. I may, on the behalf of the Royal Academy and myself, wish You[4] a lasting health and anything else that may be pleasant, since she [i.e. the Academy] is certain that pharmacy will win many unexpected pieces of information.

On Tuesday 11th of November, in the morning, Scheele turned up at *Collegium medicum* to take his apothecary exam (Fig. 19.5). The examination board consisted of 11 prominent physicians and botanists: Johan Gustaf Hallman (1726–1795), botanist Anton Rolandsson Martin (1729–1785), Johan Anders af Darelli[5] (1718–1780), physician Johan Lorentz Odhelius (1737–1816), Christian Ludvig Ramström (1740–1782), Henrik Sparschuch (1742–1786), physician Carl Ribben (1734–1803),

[3]The custom to deliver an inaugural lecture had been less common towards the end of the 18th century, and was more common in the early history of the Academy.

[4]Bergman said "Min Herre" which literally translates to "My Sir".

[5]It was af Darelli (Darelius before his ennoblement) who was the respondent of Wallerius infamous thesis against Linnaeus (Sect. 9.1).

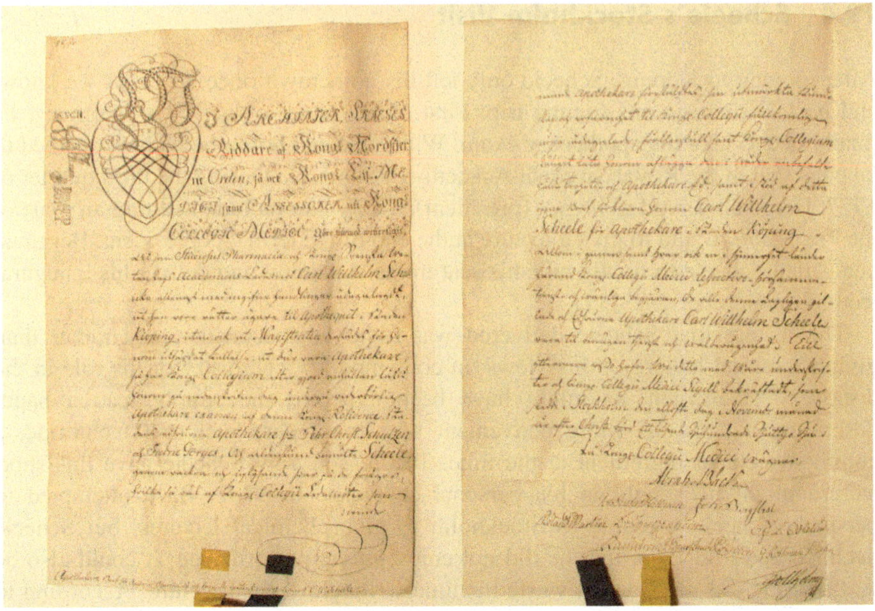

Fig. 19.5 Scheele's impressive apothecary diploma. Carl Wilhelm Scheele's Archive, Centre for History of Science, Royal Swedish Academy of Sciences. *Photo* Anders Lennartson

physician Göran Rothman (1739–1778) and Henric Gahn, all of which were apostles of Linnæus, Johan Bergstrahl (1715–1795) and Scheele's friend David von Schulzenheim. Chairman was Scheele's friend Abraham Bäck. According to the preserved minutes, Scheele answered all the questions in chemistry and pharmacy correctly. It has been claimed that the examination took the form of an informal celebration of the famous chemist; this has been questioned by Gentz [15], and no preserved document supports that view. However, six days after the exam, it was decided that Scheele would be freed from the costs of the exam, partly as he had promised to help with the new edition of the Swedish pharmacopeia, *Pharmacopoea Suecia* [16].

The day after the exam, Scheele once again attended a meeting of the Royal Swedish Academy of Sciences. This was Bergman's last meeting as president, and King Gustav III was attending. As a leaving president, Bergman's duty was to deliver a lecture. This lecture, *On Chemistry's Latest Progress* (Sect. 19.5), is a good insight into Bergman's views on chemistry.

In the election of new president, Scheele was one of those getting the largest number of votes, but lost when lots were drawn. In the end, Scheele never served as president in the Academy, and he was probably happy about that. On the initiative of Bergman, Scheele was granted an annual scholarship of 100 riksdaler specie for three years. This funding was later extended for the rest of his life, but was reduced to 60 riksdaler 1781–1784. On the same occasion, Bergman was also awarded a yearly scholarship of 150 riksdaler from a donation by Niklas Sahlgren to finance his experiments and in particular his geological work [17]. The initiative for this scholarship came from Wargentin [18], and was probably a way to compensate Bergman for the missed opportunity to go to Berlin.

19.5 Bergman's Lecture on the Progress in Chemistry

On November 12, 1777, Bergman held a lecture in the Royal Swedish Academy of Sciences, as his term as president ended. The lecture, entitled *On the Latest Progress in Chemistry*, was printed the same year including an answer from the secretary, Wargentin [19]. The printed version is 48 pages long, but the lecture that Bergman actually held was probably an abbreviated version. It was not uncommon that the lectures were extended by the author before printing [20]. Bergman's lecture was very different from the lectures delivered by Wallerius and von Engeström (Sect. 9.6), and as usual he did not fall for the temptation of criticising his competitors or their work in public. The first page of Bergman's printed lecture is a tribute to the King, Gustav III, were Bergman assured that the Academy would do everything, not only to maintain the Swedish scientific level, but to double it.

After a general introduction focused on the unique ability of human beings to examine nature and communicate their opinions to each other, Bergman gave a historical background to chemistry. He traced the history back to ancient Egypt and attributed the Egyptians a deep knowledge in chemistry, a knowledge that they kept secret and which was now largely lost. Next, chemistry saw a period of darkness, an insane quest for gold. This was a common view at the time, a view that is not uncommon to hear even today, but which finds little support among modern historians [21]. Bergman acknowledged that the alchemists' strive to make gold could not be proven to be impossible and that they accidentally made many important discoveries. During the reformation, a new light spread over science, Bergman continued. Chemistry, however, had only in recent years made significant progress.

Bergman's intention with the lecture was to focus on the recent development in chemistry, and for the audience to get an overview, it was important, Bergman said, to realise that bodies on Earth were mainly of two kinds. The largest class was composed of aggregated particles without any internal system of tubes. Bergman called these inorganic bodies, they included salts, earths, minerals, air, water and fire.[6] The second group—the organic bodies—had internal systems of tubes to transport fluids. This group included animals and plants. This was long before chemistry became divided into organic and inorganic chemistry.

First Bergman discussed the earths. In this context, he discussed minerals and crystalline substances. He stressed that the study of earths had been valuable for agriculture and in the manufacture of brick, glass and porcelain. Next, Bergman

[6] It is very remarkable that Bergman includes fire as an example of a substance. Later in the lecture, he clearly states that fire is not matter but a state when a substance loses phlogiston.

turned to the salts, a field where great progress had been made in recent years. The composition of old, well-known salts had been determined, and many new salts discovered, both in mineral, vegetable and animal materials. The "finest and most remarkable of all hitherto known acids", however, was common for all three kingdoms of nature. Bergman was of course talking about his favourite subject, carbonic acid (aerial acid), which he claimed could restore rotten meat, cure scurvy and other deceases. He also touched upon the subject of mineral waters. He talked about the phlogiston content of acids, and that the presence of phlogiston had only been verified in hydrochloric acid (i.e. it was the only acid that could be oxidised). Ammonia (volatile alkali) was composed of phlogiston (in this context hydrogen), and as soon as this was removed an "elastic vapour" remained, insoluble in water and which extinguished fire. This was nitrogen, and Bergman concluded that it had so far not been possible to unite nitrogen and hydrogen to ammonia. Large scale production of ammonia from hydrogen and nitrogen (the important Haber-Bosch-process) would not be realised until the twentieth century.

Next topic was combustible materials. Sulphurs were substances composed of an acid and phlogiston, e.g. common sulphur and phosphorus. Oils were probably composed of phlogiston, aerial acid and water (correct, using the phlogiston theory). The finest oil in the mineral kingdom, petroleum (naphta) occurred in such a plenty in the East that it often formed springs and streams; the source for that information is unfortunately not specified. Amber was found to consists of oil, earth and a special acid (now known as succinic acid).

The metals were the heaviest of all known bodies. The metals had been regarded as compounds of an earth or calx with phlogiston, but it had recently been found that arsenic consisted of an acid and phlogiston, and Bergman now suggested that all metals could be composed in a similar way (Sect. 15.5). Bergman warned the public of using copper ware for cooking; copper ware could be covered with tin, but the problem was that tin occasionally was adulterated with toxic lead. Bergman also mentioned a large lump of iron (weighing 680 kg) that had been found in Siberia. This was almost certainly the Krasnojarsk meteorite, discovered in 1749; Bergman believed that it had been formed by some kind of ancient volcanic activity.

After discussing water and its atomic nature (Sect. 22.3), Bergman turned his attention to the atmosphere, which was not homogenous but composed of three main components: foul air, good air and aerial acid. Bergman claimed that inhalation of aerial acid could cure lung deceases. Fresh air was always healthier to breath than indoor air, according to Bergman.

A number of substances from the vegetable kingdom were discussed by Bergman, e.g. sugar, which was used in "incredible amounts" and rubber. Bergman hoped that rubber trees could be planted in Europe, and that rubber-coated clothes and boots could protect people from rain in the future. He especially mentioned military applications, trying to point at the impact chemistry could have in a large number of important fields. Animal substances were different from substances extracted from plants; for example, acids were less common. Bergman mentioned formic acid, phosphorus, milk, cheese, fats and the study of bladder stones.

Finally, Bergman turned to the more theoretical parts of chemistry, but without much detail as this lecture was directed to a general audience. The number of true elements was unknown, and Bergman was most inclined to believe those speaking of two elements, one earthy, non-volatile, and on fiery, volatile. Bergman briefly mentioned elective attractions and the importance of performing thorough, well-planned experiments to avoid biased results. To round off, Bergman acknowledged a donation from Mrs Strömer, widow of his former teacher astronomer Mårten Strömer.

As customary, the Academy secretary, Wargentin, gave a short answer. He described chemistry as one of the most important branches of science for the society, and was pleased to see that Sweden had a number of chemists of such importance, that foreign countries attempted to recruit them (Chap. 16).

References

1. Levertin A, Schimmelpfennig CFV, Ahlberg KA (1910–1918) Sveriges apotekarhistoria. Vol 1 Stockholm, pp 1536–1538
2. Trofast J (1994) Johan Gottlieb Gahn: brev, vol 2, Lund, p 88
3. Levertin A, Schimmelpfennig CFV, Ahlberg KA (1918–1923) Sveriges apotekarhistoria. Vol 2 Stockholm, p 954
4. Sjöstén CG (1801) Åminnelse-tal, hållit för Kongl. Vetenskaps Academien öfver dess framledne ledamot herr Carl Vilhelm Scheele, den 14 oct. 1799. Stockholm, p 44
5. Cleve PT (1886) Carl Wilhelm Scheele. Ett minnesblad på årsdagen af hans död. M. Barkéns förlagsbokhandel, Köping, p 14
6. Fredga A (1946) Carl Wilhelm Scheele. KVA, Stockholm, p 21 note 20
7. Cleve PT (1886) Carl Wilhelm Scheele. Ett minnesblad på årsdagen af hans död. M. Barkéns förlagsbokhandel, Köping, p 20
8. Kockum A (1916) Nordiska museet, Farmaceutiska afdelningen, rum 113-114 : Vägledning. Medicinalia från Carl Wilhelm Scheeles apotek. Nordiska museet, Stockholm
9. Alfort P (1842) Handbok för brunnsgäster. H. 1, Beskrifning öfver Sveriges förnämsta helsobrunnar. Stockholm
10. af Petersens H (1928) Torbern Bergmans och Carl Wilhelm Scheeles franska förbindelser, Svensk Personhistorisk Tidskrift, 190–201
11. Berzelius J (1901) Självbiografiska anteckningar. KVA, Stokholm, p 191
12. Zekert O (193) Carl Wilhelm Scheele Sein Leben und seine Werke, 3–7. Theil, Gesellschaft für Geschichte der Pharmazie, Mittelwald, p 213
13. Scheele CW (1778) Sätt at tilreda mercurius dulcis, på våta vägen. KVA Handl 39:70–73
14. Sjöstén CG (1801) Åminnelse-tal, hållit för Kongl. Vetenskaps Academien öfver dess framledne ledamot herr Carl Vilhelm Scheele, den 14 oct. 1799. Stockholm, p 46
15. Gentz L (1758) Hur såg Scheele ut? Sv Farm Tidskr 373–394; 405–421
16. Sjöstén CG (1801) Åminnelse-tal, hållit för Kongl. Vetenskaps Academien öfver dess framledne ledamot herr Carl Vilhelm Scheele, den 14 oct. 1799. Stockholm, p 48
17. Hjelm PJ (1786) Åminnelse-tal öfver…Herr Torbern Bergman, Stockholm, p 66
18. Tiselius A (1958) Torbern Olof Bergman, Levnadsteckningar över K. Svenska Vetenskapsakademiens ledamöter, vol 9, p 134

19. Bergman T (1777) Tal, om chemiens nyaste framsteg, hållet, i Kongl. Maj:ts höga närvaro, för dess Vetenskaps-Academie, vid præsidii nedläggande, den 12 nov. 1777, KVA, Stockholm
20. Lindroth S (1967) Kungl. Svenska Vetenskapsakademiens Historia 1739–1818, vol. 1:1, Kungl. Vetenskapsakademien, Stockholm, p 40
21. Principe LM (2013) The secrets of alchemy. University of Chicago Press, Chicago

Bergman's Work on Elective Attractions

20

Early chemist noted that different substances appeared to have different affinities, or attractions, to each other. For example, silver nitrate was found to precipitate hydrochloric acid (formation of insoluble silver chloride) which was interpreted as a strong affinity of silver for hydrochloric acid. In 1718, French apothecary and chemist Étienne François Geoffroy (1672–1731) published a table where he attempted to summarise the affinities between chemical substances (Fig. 20.1; Table 20.1) [1]. Geoffroy's table consists of 16 columns, each column headed by a substance, and bellow is a list of other substances listed with decreasing order of affinity for the substance in the heading. To save space, Geoffroy used the alchemical symbols rather than writing out chemical names. In the first column, for instance, the heading is acids in general. Fixed alkali (K_2CO_3 and Na_2CO_3) is at the top with the greatest affinity for acids, followed by ammonia, absorbent earths (such as CaO) and finally metals. The table is rather arbitrary, both in the choice of substances as well as in the order. Some columns denote chemical reactions (e.g. the acid column), while for instance the last column (water) expresses the higher solubility of ethanol compared to sodium chloride in water, and the metal columns appear to describe formation of alloys. Despite its flaws, this table was an interesting step towards organising chemical knowledge. Geoffroy's table got considerable attention and was for instance reproduced (without modification) by Macquer in his *Elemens de Chymie Theorique* in 1749.

For Bergman, with his strong background in physics and his admiration for Newton, it was logical to attempt to apply the laws of physics on chemistry. It should also be recalled that "universal attractions" was the topic of the dissertation of his first student. In the 1770s, he would embark on an ambitious project trying to unify all chemical knowledge into a single system. He would attempt to bring all known substances together in a single table describing all possible reactions between what he considered simple substances. Bergman was familiar with Geoffroy's table, but was also aware of some of its major flaws and found it a useful way of representing the attractions between chemical substances.

A. Lennartson, *Carl Wilhelm Scheele and Torbern Bergman*, Perspectives on the History of Chemistry, https://doi.org/10.1007/978-3-030-49194-9_20

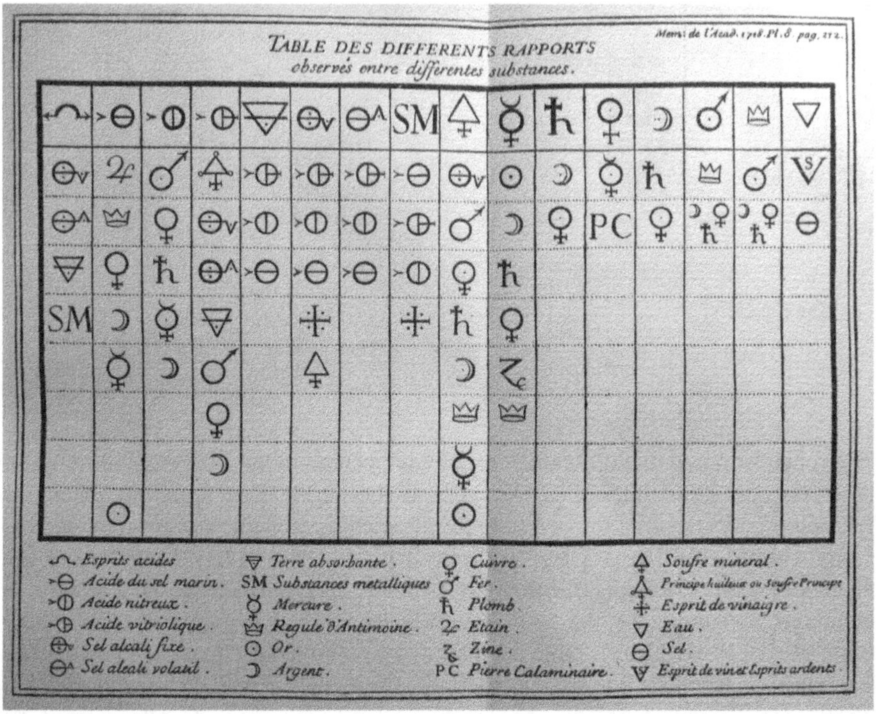

Fig. 20.1 Geoffroy's table of chemical affinities, published in 1718

Compared with Geoffroy's table, Bergman's table contains three major improvements: Bergman knew that the reactivity of a chemical substance was dependent on the reaction conditions. The most important factor was the temperature, which had previously been pointed out by Baumé in 1773 [2]. Thus, he divided his table into two parts, one listing reactions via *humida* (the wet way, i.e. reactions in solution) and one table listing reactions via *sicca* (the dry way, i.e. reactions between dry, heated solids, in the melt or dissolved in molten alkali sulphides). In this way, he eliminated an important factor of ambiguity, at least in the via *humida* table. The second improvement was that his table is a table of (what he considered) simple elective attractions (see next section), and thus (almost) all columns are based on the same basic principle. The third improvement was that he included a far larger number of substances, a few which were newly discovered, such as manganese and hydrofluoric acid. Consequently, Bergman had to modify the system of chemical symbols in order to include symbols for all of these new substances.

The first version of Bergman's table was published in his edition of *The Chemical Lectures of H.T. Scheffer* in 1775 and had 45 columns (Fig. 20.2, Table 20.2). The next version appeared in his seminal 90-page paper *De*

Table 20.1 Geoffroy's table translated to modern terms

Acid	HCl	HNO_3	H_2SO_4	Earth	Alkali carbonate	NH_3	Metal	S	Hg	Pb	Cu	Ag	Fe	Sb	H_2O
Alkali carbonate	Sn	Fe	Phlogiston	H_2SO_4	H_2SO_4	H_2SO_4	HCl	Alkali carbonate	Au	Ag	Hg	Pb	Sb	Fe	EtOH
NH_3	Sb	Cu	Alkali carbonate	HNO_3	HNO_3	HNO_3	H_2SO_4	Fe	Ag	Cu	?	Cu	Ag,Cu, Pb	Ag,Cu, Pb	NaCl
Earth	Cu	Pb	NH_3	HCl	HCl	HCl	HNO_3	Cu	Pb						
Metal	Ag	Hg	Earth		HOAc		HOAc	Pb	Cu						
	Hg	Ag	Fe		S			Ag	Zn						
			Cu					Sb	Sb						
			Ag					Hg							
	Au							Au							

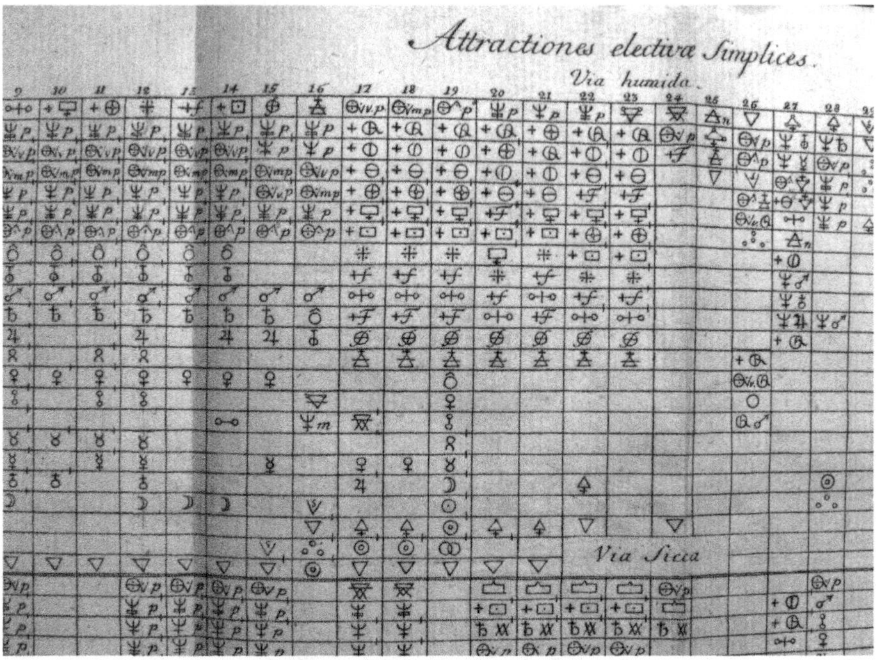

Fig. 20.2 A selection of Bergman's table of elective attractions from *The Chemical Lectures of H.T. Scheffer*, 1775. *Photo* Anders Lennartson

*attractonibus electivis [*3] printed in 1776 and had 50 columns. Later editions of *The Chemical Lectures of H.T. Scheffer* did not incorporate this updated table, but the original copper plate was reused. Bergman's paper on elective attractions gained much attention, and a second expanded and updated version (179 pages; 59 columns in the table) was included in the third volume of his collected works in 1783 [4]. Not only did Bergman include more substance in his updated tables, he also modified the order the substances appeared and in the last version he included another important modification. Rather than listing the metals in the acid columns, as he had done in his original table, he listed the metal oxides (metal calces). This was correct, since the metals appear as cations in oxidised states in salts. The idea for this change may have been inspired by a letter from Scheele, written in May 1777. It was the 1783 version that was translated to English [5], while an English translation of the original 1775 paper did not appear until 1968 [6]. French and Italian translations followed, but none in German or Swedish [7]. Bergman's work on elective attractions attracted much attention, mainly positive but gave also rise to some criticism [8]. Bergman considered his work on elective attractions as his most important contribution to science, and it is not a coincidence that he holds a scroll with an attraction table in his hand on the 1778 oil painting by Pasch (Fig. 3.4).

Table 20.2 Modern translation of 12 selected columns from Bergman's original affinity table. All columns are for reactions in solution (via *humida*)

1	3	5	16	17	19
H_2SO_4	HNO_3	HCl	H_2CO_3	KOH	NH_3
BaO	BaO	BaO	BaO	H_2SO_4	H_2SO_4
KOH	KOH	KOH	CaO	HNO_3	HNO_3
NaOH	NaOH	NaOH	KOH	HCl	HCl
CaO	CaO	CaO	NaOH	Oxalic acid	Oxalic acid
MgO	MgO	MgO	MgO	Tartaric acid	Tartaric acid
NH_3	NH_3	NH_3	NH_3	H_3PO_4	H_3PO_4
Zn	Zn	Zn		Acetic acid	Acetic acid
Mn	Mn	Mn		Formic acid	Formic acid
Fe	Fe	Fe	Fe	H_3AsO_4	H_3AsO_4
Pb	Pb	Pb	Zn	HF	HF
Sn	Co	Sn	Mn	Boric acid	Boric acid
Co	Cu	Co		H_2CO_3	H_2CO_3
Cu	Ni	Cu			Zn
Ni	As	Ni	Al_2O_3		Cu
As	Bi	As	Metallic calx	SiO_2	Ni
Bi	Hg	Bi			Co
Hg	Ag	Hg		Cu	Bi
Sb	Phlogiston	Sb		Sn	Ag
Ag	Sb_2O_3	Ag	ethanol		Au
Al_2O_3	SnO		H_2O	S	Unctuous oil
Fe_2O_3			ether	Unctuous oil	Burnt oil
H_2O	H_2O	H_2O	Unctuous oil	H_2O	H_2O
20	21	27	28	33	37
BaO	CaO	Phlogiston	S	Ag	Cu
H_2SO_4	Oxalic acid	MnO_2	PbO	HCl	Oxalic acid
Oxalic acid	H_2SO_4	HgO	KOH	H_2SO_4	Tartaric acid
HNO_3	HNO_3	N_2[a]	BaO	HNO_3	HCl
HCl	HCl	Cl_2	CaO	H_3AsO_4	H_2SO_4
HF	Tartaric acid	H_3AsO_4	MgO	H_3PO_4	HNO_3
H_3PO_4	H_3PO_4	Air		Acetic acid	H_3PO_4
Tartaric acid	Acetic acid	HNO_3		Formic acid	H_3AsO_4
Acetic acid	Formic acid	Fe_2O_3		HF	Acetic acid
Formic acid	H_3AsO_4	Sb_2O_3		Boric acid	Formic acid
H_3AsO_4	HF	SnO	CuO		HF
Boric acid	Boric acid	H_2SO_4			Boric acid
H_2CO_3	H_2CO_3				
			Unctuous oil		
			Ether	NH_3	
S	S				
H_2O	H_2O				

[a]Dephlogisticated, i.e. oxidised, ammonia

20.1 Bergman's Essay on Elective Attractions

Bergman's views on the interactions between bodies were largely inspired by Newton [9]. Bergman wrote that between all natural bodies, there is an attraction, the nature of which was unknown. Bergman's aim was not to seek to explain these attractions, only to determine their relative strengths. This was definitely a large enough challenge in the eighteenth century. For celestial bodies, Newton had found the attractions to be inversely proportional to the square of the distance, but the short-range attractions between molecules just out of contact seemed to obey different rules. Bergman realised that a main difference was that for celestial bodies, the diameter could be neglected when compared to the distance between the bodies, but this was not true for atoms in close proximity to each other. For short-range attractions, the surrounding environment was, according to Bergman, very important. He divided the short-range attractions into five types; the first three are described in a very ambiguous way by Bergman, and not further discussed in the paper:

Attractio aggregationis: homogenous materials are united to form a common homogenous mass. Here, Bergman possibly meant the attractive forces holding pure materials together.

Attractio compositionis: this may refer to the attractions between the elements in a compound.

Attractio solutionis and *Attractio fusionis*: here Bergman is certainly referring to the formation of solutions in solvents or melts.

The remaining two types are of more interest, however. Simple elective attractions (*Attractio simplex elective*) are the attractions between three chemical components, one compound and one simple substance. This is probably better explained by concrete examples rather than Bergman's more mathematical description. If, for example, zinc is exposed to copper(II) sulphate solution, metallic copper precipitates leaving a solution of zinc sulphate. Thus, zinc appeared to have a greater attraction to (or affinity for) sulphuric acid than copper, and thus appears higher up in the column than copper in Bergman's table. Another example would be if sulphuric acid was added to sodium chloride. This releases hydrogen chloride and leaves sodium sulphate (actually sodium hydrogen sulphate). Thus mineral alkali (sodium) appeared to have a stronger attraction to sulphuric acid than to hydrochloric acid. The column for *Alkali minerale* is therefore topped by sulphuric acid, with hydrochloric acid on the third place after nitric acid. Unlike Geoffroy's table, Bergman's table includes empty places. This is where he, by comparing with neighbouring columns, expected to find substances that he had not experimentally placed in the table.

In modern chemistry attraction tables are not used, as the elective attractions determined by Bergman actually is a mixture of several different factors such as solubility, volatility, acid/base strength and reduction potentials. The apparent affinity of silver for hydrochloric acid is, for example, due to the low solubility of silver chloride, while the apparently higher affinity of sodium for sulphuric acid than for hydrochloric acid is due to the volatility of hydrogen chloride and the increase in entropy accompanied by the evolution of gas. There is one important exception. Examining the order of the metals in the acid columns (i.e. the precipitation of one metal from its salts by a second metal) reveals a good correlation to the so-called electrochemical or electromotive series found in modern chemistry textbooks. Ordering the metals with *increasing*[1] standard reduction potential reveals a reasonable agreement with the order found by Bergman:

Bergman	Zn	Mn	Fe	Pb	Sn	Co	Cu	Ni	Bi	Hg	Ag
Modern	Mn	Zn	Fe	Co	Ni	Sn	Pb	Bi	Cu	Hg	Ag

Bergman's system works fairly well for acids, bases and metals but, unfortunately, he also included columns like water, ethanol, ether and oils. These columns do not describe chemical reactions and should not have been included.

Bergman had to account for several anomalies. For example, it had been reported by Baumé in 1760 that when nitric acid was added to an aqueous solution of potassium sulphate, potassium nitrate crystallised leaving sulphuric acid in solution. On the other hand, it was well known that heating potassium nitrate with concentrated sulphuric acid gave nitric acid and potassium sulphate (actually potassium hydrogen sulphate). While other authors had attributed this to what they called reciprocal attractions, Bergman clearly attributed this to a chemical equilibrium: "*Acidum vitrioli* [sulphuric acid] in sufficient amount decomposes *nitrum* [potassium nitrate] completely, even via *humida* [i.e. in solution] whence it appears to have more power of attraction than does *acidum nitri* [nitric acid]" [10]. This was an important statement, as it anticipated the discovery of the law of mass action by 25 years. The law of mass action was first clearly stated by Claude Louis Berthollet in 1801 [11] but already Boyle had pointed in the right direction in 1674: a large amount of a substance can compensate for a low affinity [12].

The fifth form of attraction discussed by Bergman was double elective attractions (*Attractio duplex elective*) which correspond to what a modern chemist would call metathesis reactions (or double decomposition in older literature). These are reactions between four components, e.g. two salts. One example would be the reaction between aqueous solutions of sodium sulphate and barium chloride, which yields insoluble barium sulphate and a solution of sodium chloride. Of course, Bergman did not attempt to represent these in a table.

An unexpected application of Bergman's theories of elective attractions is found in Goethe's novel *Elective Affinities* (*Die Wahlverwandtschaften*) originally published in 1809. This book tells the story of a couple, Eduard and his wife Charlotte. To bring some excitement to their slightly dull life, they invite Eduard's friend Otto and Charlotte's young niece Ottilie. Soon enough, attractions arise between Eduard and Ottile and between Charlotte and Otto. Written by Sturm und Drang author

[1]In modern textbooks the order would be reversed.

Goethe, everyone, of course, dies unhappy at the end. Goethe had most probably not read Bergman's original work, and Bergman's name is not mentioned in the book. It has been suggested that Goethe probably read about elective attractions in *Physikalisches Wörterbuch* written by Gehler in 1798 [13]. To make the story chemically more relevant (a case of double elective attractions), Goethe should have introduced Otto and Ottilie as a couple. Another similar reference to chemical affinities is found in Swedish author August Strindberg's autobiographical novel *Le plaidoyer d'un fou* (The Defence of a Fool) from 1895: "It reminded of chemical elective attractions, here active between living beings" [14].

20.2 Bergman's Representation of Chemical Reactions with Diagrams

Since the days of the alchemists, it was custom to represent chemical substances, operations and laboratory equipment with symbols. Bergman used a modified set of alchemical symbols in his notes and letters, and he used those symbols in his affinity tables. For new substances (especially those that Scheele supplied at an escalating rate), Bergman had to invent new symbols. Bergman's symbols are found, with Latin explanations, on fold out plates in *The Chemical Lectures of H.T. Scheffer* and *De attractionibus electivis*.

In the eighteenth century, there were no reaction formulae in the modern sense. Jean Beguin (1550–1620) had, however, used a diagram as early as 1615 to describe the reaction between antimony(III) sulphide and mercury(II) chloride, but he did not generalise the use of diagrams [15]. In the late 1750's, William Cullen used diagrams in his lectures to represent metatesis reactions [15], and according to Thomas Thomson [16], it was Cullen who introduced the idea of using reaction diagrams to Joseph Black. Neither Cullen, nor Black, who adopted the diagrams in his own lectures, used diagrams in any publication.

Bergman independently introduced a system for representing chemical reactions with diagrams. His system was introduced in *The Chemical Lectures of H.T. Scheffer* in 1775 and then used more extensively in *De attractionibus electivis*. These diagrams (the *Opuscula* version has a plate with 64 reactions) could express both simple and double elective attractions in uniform diagrams, but were a typographic nightmare, and are not found in any other of Bergman's publications. They received only modest use by other authors but were used by German chemist and pharmacist Johan Trommsdorff (1770–1837), who replaced Bergman's symbols with text [17, 18]. Similar diagrams were also used by British chemist William Brande (1788–1866), but without the pointing brackets [19].

Two examples of Bergman's reaction schemes are shown in Figs. 20.3 and 20.4. The first example (Fig. 20.3) is the reaction between calcium chloride and ammonium carbonate. The two reactants are shown to the left and right, respectively, and at the centre is the symbol for water, indicating a reaction in aqueous solution (via *humida*). It is a reaction between two salts, so the attractions involved

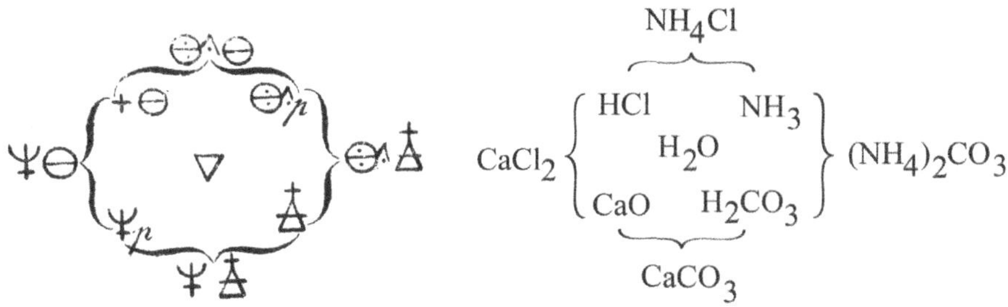

Fig. 20.3 Bergman's reaction scheme for the reaction of calcium chloride and ammonium carbonate in aqueous solution (left). To the right is a translation of the scheme with modern chemical symbols

Fig. 20.4 Bergman's reaction scheme for the reaction of lead(II) sulphide with metallic iron by heating (left). To the right is a translation of the scheme with modern chemical symbols

are examples of double elective attractions. Next to the brackets pointing at the reactants are the symbols for the simple substances that the reactants are built up from. Calcium chloride ($CaCl_2$) is composed of hydrochloric acid (HCl) and calcium oxide (CaO). The other reactant, ammonium carbonate ($(NH_4)_2CO_3$), is composed of by ammonia (NH_3) and carbonic acid (H_2CO_3). The lower bracket is pointing down, indicating that the product, calcium carbonate ($CaCO_3$), is precipitated. The upper bracket is pointing upwards showing that the other product, ammonium chloride (NH_4Cl), remains in solution.

The second example (Fig. 20.4) shows a reaction governed by simple elective attractions and heat (via *sicca*; indicated by the fire symbol at the centre). It is the reaction of lead sulphide (PbS), composed of lead (Pb) and sulphur (S) with iron (Fe). Here, both products, iron sulphide (FeS) and lead (Pb) are "precipitated" (obtained in solid form), so both brackets are pointing downwards.

References

1. Geoffroy EF (1718) Table des différents rapports observes en chimie entre différentes substances, Mémoirs l'Academie royale des sciences, 202–212
2. Schufle JA (1985) Torbern bergman a man before his time. Cornado Press, Lawrence, p 171

3. Bergman T (1775) Disquisitio de attractionibus electivis. Nova Acta Reg Soc Scient Upsaliensis 2:161–250
4. Bergman T (1783) Opuscula physica et chemica, pleraque antea seorsim edita, jam ab auctore collecta, revisa et aucta. Vol. 3, Uppsala, p 291–470
5. Bergman T (1785) A Dissertation on Elective Attractions, J. Murray, London
6. Bergman T (1968) A dissertation on elective attractions. Johnson Reprint Corporation, New York
7. Moström B (1957) Torbern bergman: a bibliography of his works. Almqvist & Wiksell, Stockholm
8. Carelid G, Nordström J (1965) Torbern Bergman's Foreign Correspondence, Almqvist & Wiksell, Uppsala, p XVII
9. Schufle JA (1985) Torbern bergman a man before his time. Cornado Press, Lawrence, p 165
10. Bergman T (1968) A dissertation on elective attractions. Johnson Reprint Corporation, New York, p 20
11. Partington JR (1964) The history of chemistry. MacMillan, London, p 576
12. Boyle R (1674) Tracts containing suspicions about some hidden qualities of the air, part 2, p 541
13. Broberg G (1997) in: Goethe JW (1997) Valfrändskap, Natur och Kultur, Stockholm, p 255
14. Strindberg A (1914) En dåres försvarstal, Samlade skrifter av August Strindberg, vol 26, Albert Bonniers Förlag, Stockholm, p 97
15. Crossland MP (1959) The use of diagrams as chemical 'equations' in the lecture notes of William Cullen and Joseph Black. Ann Sci 15:75–90
16. Thomson T (1814) On the discovery of the atomic theory. Ann Sci 3:330–338
17. Trommsdorff JB (1796) Lehbuch der pharmaceutischen Experimentalchemie. Altona
18. Trommsdorff JB (1809) Lehbuch der pharmaceutischen Experimentalchemie. Third edition. Vienna
19. Brande W (1819) A manual of chemistry. John Murray, London, p 17

21 The Discovery of Oxygen

It was well established that air was essential for fire and that a candle would burn only a finite time in a closed vessel. The works of Robert Boyle, Robert Hooke (1635–1703) and John Mayow (1641–1679) were of fundamental importance in this regard. Also, elemental oxygen had been isolated on a number of occasions throughout history, but its nature and relation with combustion had not been realised. Just to mention two examples, Stephen Hales (1677–1761) collected the gas (oxygen) obtained by heating potassium nitrate (saltpetre), but simply described it as "air". Pierre Bayen (1725–1798) obtained oxygen in 1774 by heating mercury oxide, but believed the gas to be carbon dioxide (fixed air). Scheele is reported to have been fascinated by fire since his childhood [1], and from his early letters to Retzius, it can be concluded that he conducted systematic experiments on redox-reactions already in Malmö.

21.1 Scheele's Discovery of Oxygen

The problem when trying to investigate Scheele's study of air and combustion is that he never discussed these experiments in his letters. From preserved letters, it is clear that Bergman knew nothing of Scheele's discoveries until early 1776, when the manuscript of his *Chemische Abhandlung von der Luft und dem Feuer* (Chemical Treatise on the Air and the Fire) was finished. In Bergman's edition of *The Chemical Lectures of H.T. Scheffer* published in 1775, the only reference to oxygen is a note on the last page, apparently added after the book was sent to the printer saying that "smoke of phlogisticated nitric acid" (nitrogen monoxide) gives a convenient way of determining the goodness of air, since it turns red and its volume is diminished (oxidation to nitrogen dioxide). Interestingly enough, Bergman's first insights into the chemical properties of air thus seem to originate from the works of Priestley, results that Bergman discussed in a letter to Gahn on August 21, 1775 [2]. It is not known why Scheele did not tell Bergman of his

A. Lennartson, *Carl Wilhelm Scheele and Torbern Bergman*, Perspectives on the History of Chemistry, https://doi.org/10.1007/978-3-030-49194-9_21

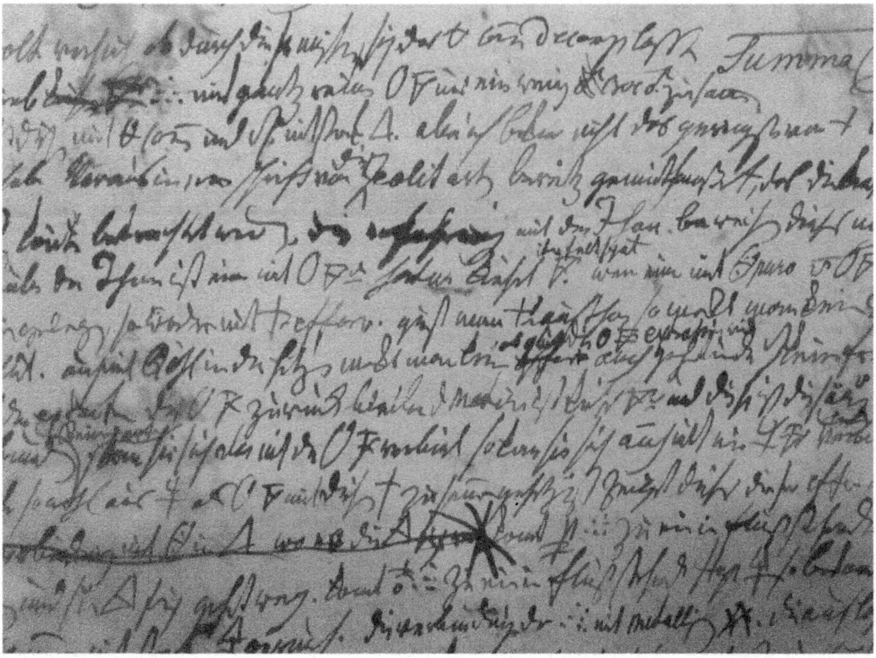

Fig. 21.1 Scheele's laboratory notes are undated and notoriously difficult to read. Carl Wilhelm Scheele's Archive, Centre for History of Science, Royal Swedish Academy of Sciences. *Photo* Anders Lennartson

discovery of oxygen. Oseen argues that oxygen did not fit into Bergman's chemical system, and that Scheele was unable to form his own opinion until after he left Uppsala and Bergman's influence on him weakened [3]. If so, this transition in Scheele's thoughts must have been very sudden, as Scheele left Uppsala in July, and had finished his manuscript by the end of December.

Another complicating factor is that Scheele never dated his laboratory notes (Fig. 21.1), meaning that the process leading to Scheele's discovery of oxygen has to be recreated from the various scattered documents available. The quite exact dating presented by Cassebaum [4] has to be regarded with suspicion.

In his short biography over Scheele, written on the occasion of the one hundredth anniversary of Scheele's death, Blomstrand noted that it must have taken Scheele a considerable time to perform all the experiments, work out the theories, and write the manuscript for his book on air and fire [5]. As Scheele moved to Köping in July 1775, with all the extra work and arrangements that required, it would have been close to impossible for Scheele to complete the manuscript by December 1775, if he had discovered oxygen later than Priestley, i.e. later than August 1, 1774 (Sect. 21.2).

Nordenskiöld brought further evidence for this view. He was the first to decipher parts of Scheele's laboratory notes, and by comparing the contents of these notes with the contents of dated letters, he concluded that Scheele had discovered oxygen before 1773 [6]. While the latest possible date for Scheele's discovery was set to November 16, 1772, Nordenskiöld did not attempt to establish the first possible date for Scheele's discovery, as his goal was to establish Scheele's priority over Priestley. Although Nordenskiöld's method of dating documents has been questioned, the conclusions were accepted by Boklund, the foremost expert on Scheele, and most others [7].

Oseen argued that Scheele's discovery of oxygen occurred before he came to Uppsala [8]. In 1784, Wilcke wrote to Scheele that he suspected that Bergman might have mentioned Scheele's discovery of oxygen in a letter to Priestley, and that Priestley thus had stolen Scheele's results.[1] Two letters from Priestley to Bergman are known, the last of which was dated October 21, 1772. That letter was, on Priestley's request, forwarded to Wilcke, so Wilcke was aware of the correspondence. There was no further correspondence between Priestley and Bergman. Oseen's interpretation of this is that Wilcke must have assumed that Bergman was aware of Scheele's discovery of oxygen in the summer of 1772, and that the reason for Wilcke to believe so was that Wilcke himself had learned of Scheele's discovery by that time and that he assumed that Scheele's other friends had the same knowledge [9]. Oseen found further support in a statement by Sjöstén that "apart from Wilcke several others had been fully informed about [the contents of Scheele's book] long before it saw the light of day" [10]. Since Retzius wrote to Wilcke after Scheele's death that he undertook the main part of his and Scheele's joint study of tartaric acid while Scheele focused on the experiments that formed the basis of his book on air and fire (Sect. 21.3), Oseen concluded that Scheele had discovered oxygen before he left Stockholm [11]. On one hand, the fact that Scheele experimented with air and fire in Stockholm does not necessarily mean that he had discovered oxygen. On the other hand, oxygen plays such an important role in Scheele's book that it is hard to imagine that the isolation of oxygen constituted the last stage of the study.

Boklund, finally, believed that Scheele may have isolated oxygen already in Malmö around 1767–1768 [7]. A. E. Nordenskiöld's father had met Gahn, who said that Scheele's first encounter with oxygen was when he observed how embers from the fire place burned with a brilliant flash when they drifted above crucibles with hot pyrolusite (MnO_2) [12]. This was due to oxygen evolved from pyrolusite on heating, and similar effects had been noted by Merret[2] in 1668, and by Pott in 1740, but had not been explained. Boklund argued that it is more likely that the observation was made using fused potassium nitrate (saltpetre) than MnO_2 [7]. When the flashing coal particles are mentioned by Scheele, it was in combination with molten potassium nitrate, not pyrolusite [13]. Scheele had investigated the formation of potassium nitrite (KNO_2) by heating potassium nitrate (KNO_3) in Malmö in 1767–1768,

[1]Wilcke's suspicion was of course unfounded.
[2]Christopher Merret (*c.* 1615–1695). British physician and scientist.

Fig. 21.2 Bubbles of oxygen formed during the decomposition of potassium nitrate. The yellow colour is due to nitrite ions in the melt. *Photo* Anders Lennartson

and anyone who has repeated this experiment will find it quite unlikely that the effervescence would have escaped Scheele's notice (Fig. 21.2). It seems unlikely that the constantly curious Scheele would not at least have attempted to identify this gas.

From his *Chemische Abhandlung*, one gets the impression that Scheele first obtained oxygen while distilling a mixture of potassium nitrate (saltpetre) and sulphuric acid, the standard procedure for preparing nitric acid in Scheele's days. At the end of the distillation, the hot residue gave of a gas, which Scheele collected in an ox bladder. He found that this gas sustained combustion and that a light burned in this gas with exceptional brightness. This does not, however, necessarily mean that this was Scheele's first encounter with oxygen. Scheele may have introduced oxygen in this way in his book, since the preparation of nitric acid was an experiment that every chemist was familiar with, unlike the experiment with the coal particles. Scheele soon found several methods to prepare oxygen by heating potassium-, magnesium—or mercury(II) nitrate, by heating pyrolusite (MnO_2) with sulphuric acid, or by heating gold- or mercury oxide. The most practical method, according to Scheele, was to heat potassium nitrate in a retort.

So, why did Scheele not mention oxygen in his letters to his friends? One can imagine several possible answers. (1) Scheele discovered oxygen in Malmö. First of all, we cannot be sure that the preserved report on nitrous acid (Sect. 8.2) is complete, and Nordenskiöld actually suspected that pages may be missing.

Theoretically, the observation of oxygen could have appeared in the original manuscript. If not, it may perhaps be that Scheele initially assumed that the formation of oxygen by heating saltpetre was well known. Of course, the evolution of gas from hot saltpetre had been noted by British priest and amateur botanist Stephen Hales in his *Vegetable Staticks* from 1727, but the nature of this gas had escaped notice. The fact that saltpetre supported combustion had also been known for centuries; it had been used in gunpowder since the Chinese Song dynasty. On the contrary, it could also be that Scheele found his discovery so remarkable, that he initially could not fit the discovery into his manuscript. This seems less likely, as Scheele always reported all his observations in an honest manner.

(2) Scheele discovered oxygen in Stockholm. As there are no letters preserved from his time in Stockholm, Scheele could simply have told Retzius and his friends of his discovery and there would have been little reason to discuss it in letters after he arrived in Uppsala. If Scheele discovered oxygen at such an early date, it is likely that it took him several years, just as Oseen assumed, to work out a theory of heat, combustion and phlogiston that included oxygen. The discovery is not, however, mentioned in Gahn's notes from spring 1770, and it still does not explain why it was never mentioned to Bergman.

(3) Scheele discovered oxygen in Uppsala. If Scheele discovered oxygen in Uppsala, it seems that he must deliberately kept the discovery a secret. There could be two reasons for doing so. Usually Scheele seems to have shared information freely with friends, so it is less likely that he was afraid of anyone stealing his idea. Perhaps the discovery confused him and he needed time to sort out his thoughts, as Oseen suggested. If he had told Bergman at an early stage, Bergman would have surely worked out a theory for Scheele, but one can assume that Scheele felt that would have deprived him from some of the excitement.

21.2 Priestley's Discovery of Oxygen

Joseph Priestley (Fig. 21.3) was born in Leeds, England, on March 24, 1733, and as a person he was, in many respects, the complete opposite of Scheele. As a child, he suffered from poor health and his education was not very organised: he focused on languages and religion. He initially worked as a teacher and preacher, and his interest in natural science did not arise until 1766; the following year he published his book *History of Electricity*. In 1772, he was hired as "literary companion" by Lord Shelburne (1737–1805), and it was at this point that Priestley became interested in chemistry and started investigating combustion phenomena. He isolated oxygen on August 1, 1774 by heating mercury oxide with a burning glass, an experiment carried out at Lord Shelburne's country residence Bowood House, near Calne, in Wiltshire. There is thus no doubt that Scheele isolated oxygen before Priestley. In 1780, Priestley moved to Birmingham, but his liberal religious and political views made him highly controversial and his support for the French

Fig. 21.3 Joseph Priestley. Engraving from William Ramsay's book. *The gases of the atmosphere*, 1896

revolution made him unpopular among both common people and the aristocracy. In 1791, his home and laboratory were raided by a mob, and after some time in London, he immigrated to America in 1794. He died in 1804, faithful to the phlogiston theory until the end. Priestley was a skilled experimentalist, but his theoretical knowledge was limited. As a chemist, he has to be regarded as an amateur and his theories gained little support. On the other hand, he freely admitted his mistakes.

21.3 Scheele's Book on Air and Fire

When leaving Uppsala in the summer of 1775, Scheele had largely finished his studies of oxygen and combustion, with only a few control experiments that may have been performed in Köping. After arriving in Köping, Scheele first finished the paper on arsenic acid, and it was most likely not until that autumn that he began writing on his *Chemische Abhandlung von der Luft und dem Feuer*. By then, Scheele was aware that Priestley was investigating the properties of air, and he probably realised that it was time to publish his own results. Priestley was, however, faster than Scheele. The second volume of Priestley's *Experiments and Observations on Different Kinds of Air* appeared in late 1775 (the dedication to John Pringle, President of the Royal Society, was signed in November), and in this book, Priestley gave the first printed account of oxygen:

> On the 1st of August 1774, I endeavoured to extract air from mercurius calcinatus per se [mercury(II) oxide prepared by heating mercury in air]; and I presently found that, by means of this lens [burning glass], air was expelled from it very readily [...]. But what surprized me more than I can well express, was, that a candle burned in this air with a remarkably vigorous flame [...] [14].

It took some time for Priestley to realise the significance of his results. In October, he travelled with Lord Shelburne to Paris where he met French chemist Lavoisier, and not until March 1775, he had developed his ideas of combustion [15]. The much more provocative style of Priestley compared to Scheele is apparent already in the preface of his book:

> I even think that I may flatter myself so much, if it be any flattery, as to say, that there is not, in the whole compass of philosophical writing, a history of experiments so truly ingenious as mine, and especially the Section on the discovery of dephlogisticated air [oxygen], which I will venture to exhibit as a model of the kind. I am not conscious to myself of having concealed the least hint that was suggested to me by any person whatever, any kind of assistance that has been given to me, or any views or hypotheses by which the experiments were directed, whether they were verified by result, or not. [16]

The rest of the preface is devoted to his religious ideas. Scheele most probably never saw Priestley's book. In February 1777, he wrote to Gahn: "Priestley's book, I have never seen. Is it in French, I would have liked to read it".

The length of Scheele's work on air and fire meant that it could not be published in the *Transactions of the Royal Swedish Academy of Sciences*, and thus Scheele made an agreement with book dealer Magnus Swederus (Sect. 9.7) in Uppsala. Swederus had previously published works by Bergman, but apparently Bergman was not entirely satisfied. In a letter to Gahn dated November 27, 1775, Scheele wrote that he was writing the last pages, and that he had been writing the whole nights. In the same letter, he was sceptic about Priestley's theories. In a letter to Gahn dated December 22, he wrote that he was about to send the manuscript to Swederus, and in a letter to Bergman dated January 12, 1776, the manuscript has already been sent. On January 19, Scheele wrote that the manuscript had been in the hands of Swederus for one month, but that he was unsure whether the printing had started. He also wrote that he had asked for eight dedication copies on high-quality paper. In March, Scheele wrote to Gahn that the printing had not started, but that Swederus had bought paper for the printing. Scheele had threatened Swederus to withdraw the manuscript and print it in Västerås instead. In an undated letter to Gahn from 1776, Scheele is quite annoyed, and had written "a quite sharp letter" to Swederus, who had been on a journey to England. As a result, Scheele had obtained three sheets full of printing errors. Scheele had met Swederus, who was on his way home from the fair in Örebro. Nordenskiöld found that there were fairs on January 19 and August 21, and assumed that it was in January that Scheele met Swederus. However, there is a letter to Bergman dated August 30, 1776 which does not mention the meeting with Swederus, but contains very similar wording as the undated letter to Gahn. There are several examples were Scheele wrote letters containing identical wording to Bergman and Gahn on the same day, so it is probably more likely that Scheele met Swederus in August.

In the undated letter to Gahn mentioned above, Scheele wrote that he had learned from Bergman about experiments in England: a gas had been obtained by heating mercury oxide with a burning glass; this gas supported combustion and respiration. Scheele was very annoyed, and blamed Swederus for the situation. In a letter to Bergman, Scheele was worried that, once his book was printed, the readers would suspect that he had stolen his results from Priestley. In addition, a new obstacle could cause further delays as a Mr Mohr, who had been responsible for handling the proofs for Swederus, was about to quit.

In a letter to Bergman dated January 25, 1777, Scheele complained that there had been little progress: he had received another sheet full of printing errors. Actually, a considerable number of printing errors remained in the final book. Scheele concluded that if it continued at the same speed, the book would not be published until 1780.

In May 1777, Bergman promised to write a preface for Scheele's book, and had borrowed the manuscript from Swederus. The preface which he wrote was signed on July 13. In June, the printing had made progress, and both Bergman and Gahn received printed sheets as the printing proceeded. On June 19, Gahn received seven sheets by mail from Samuel Troilius [17]. Bergman had a complete set of pages by June 27 [17]. On July 3, Gahn had read the whole book, but was rather sceptic to both the theories and the disposition of the text [18]. On August 21, Swederus passed Köping on his way to Leipzig, and handed over Scheele's eight dedication copies. These were presented to Bergman, Bäck, Bergius, J. G. Gahn, H. Gahn, Schulzenheim, Wargentin and Wilcke [19]. The illustration on the title page (Fig. 21.4) was reused from Bergman's edition of *The Chemical Lectures of H.T. Scheffer* (Fig. 11.2) published by Swederus in 1775, and was possibly engraved by Fredrik Akrel [20], who used to work for the Royal Swedish Academy of Sciences in Stockholm. The copper plate at the end of the book (Fig. 21.5) may also have been engraved by Akrel [21].

Based on the fact that the book was published soon after Bergman finished the preface, earlier biographies have blamed Bergman for the delay of Scheele's book, claiming that Bergman had delayed the preface for 2 years. This is wrong, and from

Carl Wilhelm Scheele's
d. Königl. Schwed. Acad. d. Wissenschaft. Mitgliedes,
Chemische Abhandlung
von der
Luft und dem Feuer.
Nebst einem Vorbericht
von
Torbern Bergman,
Chem. und Pharm. Prof. und Ritter; verschied.
Societ. Mitglied.

Upsala und Leipzig,
Verlegt von Magn. Swederus, Buchhändler;
zu finden bey S. L. Crusius.
1777.

Fig. 21.4 The title page of the rare first edition of Scheele's book on air and fire. This copy belongs to Gothenburg University Library. *Photo* Anders Lennartson

the preserved letters it is clear that Bergman did not offer to write a preface until spring 1777. Bergman did, however, borrow the manuscript, which caused Scheele to worry. On March 8, 1776, he wrote to Gahn:

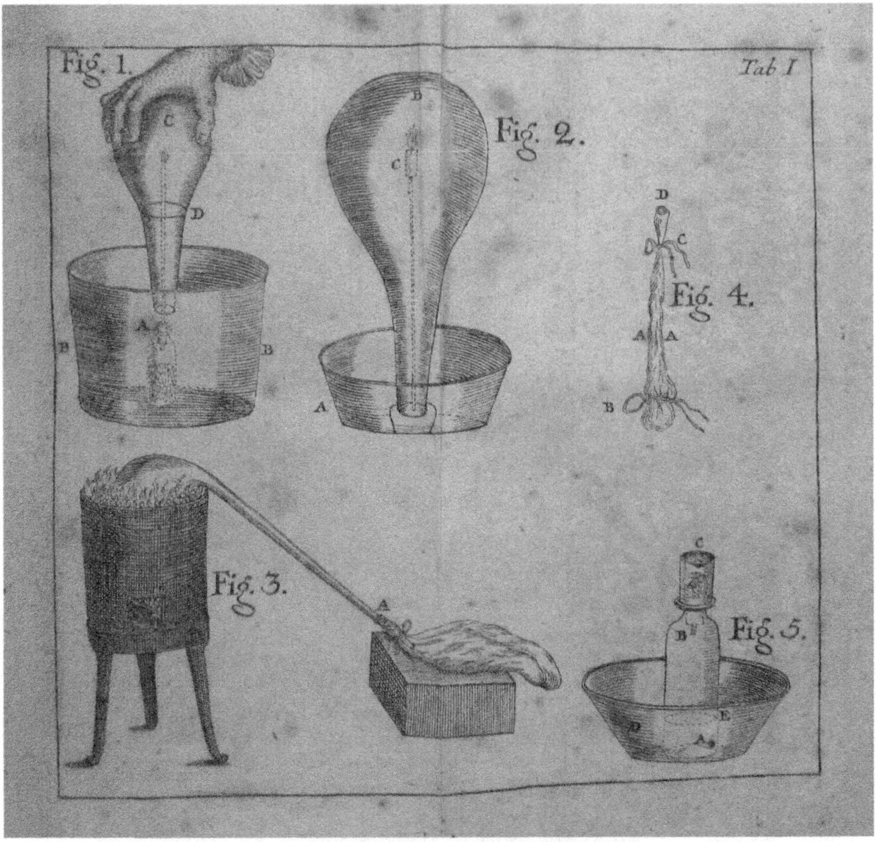

Fig. 21.5 Copper plate illustrating equipment used by Scheele to study air. Note how simple Scheele's equipment was, compared to the equipment used by Lavoisier. *Photo* Anders Lennartson

> I have learned that Prof. Bergman has my manuscript. I will hope that he is not the reason for the delays". On April 29, 1776, Bergman wrote to Scheele. "I have read the first sheets of your treatise on [fire] and hope to see it in print soon."

Instead, it may be that Swederus deliberately delayed the publication, at least Bergman suspected so in a letter to Gahn: "Svederus [sic!] has taken every measure so that Scheele's book would not be sent out before the autumn fair in Leipzig [...]

Although Swederus (and possibly the printer, Johan Edman) delayed the publication process by nearly 2 years, it was Scheele's own fault that his book was published later than Priestley's book. Priestley's *Experiments and Observations on Different Kinds of Air* was printed in late 1775, by the time that Scheele finished his manuscript. In retrospect, Scheele should have written a short communication of his results and submitted it to the *Transactions of the Royal Academy of Sciences* in order to secure his priority.

Actually, the first printed account of Scheele's results was in Bergman's paper *de attractionibus electivis* [22]. Bergman gave a clear summary of Scheele's work: nitrogen monoxide reacts with "pure air" (oxygen) to generate heat, while only one-fourth[3] of common air reacts with nitrogen oxide. Heat is a compound of "pure air" and phlogiston. Pure air can be obtained by heating mercury oxide and supports combustion and respiration. It has been suggested by Partington that this paper actually predates Priestley's book [23]. On the other hand, Cassebaum and Schufle claim that the printing of the second volume of *Nova Acta*, including Bergman's paper, was actually not finished until 1776, although it has the year 1775 printed on the title page [24]. This is supported by the preserved correspondence between Scheele and Bergman, which suggest that Bergman was unaware of oxygen until early 1776. In reality, this is of little importance, as few seem to have taken notice of Bergman's description of oxygen. In fact, the section on oxygen was removed in the second edition of his essay on elective attractions, published in Bergman's collected works in 1783 [25]. It was this edition that was eventually translated to English; [26] the original edition was not translated until 1968 [27]. The discovery of oxygen was most commonly attributed to Priestley, and not until Nordenskiöld's investigation over a century later, that Scheele's contribution was more widely acknowledged. No evidence has been found to back up Lavoisier's claim [28] of having discovered oxygen independently of Scheele and Priestley. In conclusion, it was through the works of Priestley that the world learned of oxygen, but it should be kept in mind that his ideas of air and combustion were very vague. Scheele made his discovery earlier and had a much clearer view on the subject than Priestley, but his theories still had little impact on the progress of chemistry, as Lavoisier already had made the first step towards our modern concepts of combustion by the time Scheele's book appeared. Priestley played the major role in spreading the knowledge of oxygen, but his vague theoretical speculations had no impact. Although Scheele and Priestley share the credit of having discovered oxygen, it was Scheele alone who realised that air is a mechanical mixture of oxygen, nitrogen, carbon dioxide, and various vapours such as water.

Scheele's book does not have the same logical disposition as the works by Bergman. It is composed of 97 numbered sections with experiments and theoretical discussions mixed. There is no concluding remarks or summary, so the reader is forced to read the whole book in order to get an overview. There is no strict division into chapters, but subtitles are inserted in the text in a slightly arbitrary manner. The order in which topics are discussed seem somewhat arbitrary. The book was probably written in haste, and would have gained much if Scheele would have planned his writing better or asked Bergman for advice.

Bergman's preface deals very little with Scheele's discoveries, but he assured that Scheele had worked independently, and that the manuscript had been finished 2 years earlier. In a letter to Gahn, Bergman wrote: "I have written a small preface to the work [i.e. Scheele's book] on the benefits of chemistry, especially the finer

[3]Scheele stated 1/3, so it seems that Bergman had repeated these experiments.

[chemistry]. The latter is despised by a W., a v.E. and other, who do not understand it" [29]. Bergman is of course referring to Wallerius and von Engeström.

The first section of Scheele's book is an introduction to the study of chemistry in general, and is followed by a single section on the general properties of air. Sections 8–29 describe Scheele's discovery of oxygen (fire air) and nitrogen (foul air), and the role of oxygen in combustion; §30 describes Scheele's methods for manipulating gases. Sections 31–58 treat the properties of heat and Scheele's attempts to prove that heat is a compound of oxygen and phlogiston. Sections 59–70 are devoted to the properties of light and are followed by a number of sections on phlogiston, fire, pyrophore, fulminating gold and the physiological actions of oxygen. The two last sections treat hydrogen and hydrogen sulphide, respectively. A detailed description of the contents of Scheele's book, section by section, is found elsewhere [30].

21.4 New Editions and Translations of Scheele's Book

In October 1780 [31], an English translation of *Chemische Abhandlung* appeared on the market [32]. The translation was made by John-Reinhold Forster[4] on initiative by Priestley [33]. The book is dedicated by Forster to Priestley. The English edition was expanded by a 55-page essay by Kirwan, possibly written on Priestley's initiative. Both Kirwan and Priestley had rather different views on phlogiston than Scheele, and the tone of Kirwan's essay is very negative. He more or less accused Scheele of copying Priestley's results. Two illustrative examples of Kirwan's style of writing are "Dephlogisticated Air [oxygen] and its properties were first discovered by Dr. Priestley [...]" [34] and "With regard to Air, (the knowledge of which indeed was not his principal object) he [i.e. Scheele] has done little, else than repeat, with some variation, in the manner, the Experiments published some years before his Treatise, by Dr. Priestley [...]" [35] When Scheele had arrived at different results compared to Priestley, Kirwan favoured Priestley's views. The essay was followed by a nine-page letter from Priestley to Kirwan, where he thanked Kirwan, but disagreed on some points. Priestley never accused Scheele for copying his results, but assumed that Scheele had worked independently of him.

A French translation followed in 1781. It was made by Baron Philippe Fréderich de Dietrich (1748–1793). On a meeting in the Académie des Sciences on May 19, 1781, Dietrich asked for assistance with the review of the translation and the Academy appointed Lavoisier and Berthollet as reviewers [36]. The French translation includes translations of Scheele's paper "*On the amount of pure air, that is daily present in our atmosphere*" [37] from 1779, Kirwan's essay and Priestley's answer to Kirwan.

[4]John-Reinhold Forster (1729–1798). Forster started his career as a Calvinist preacher before he became professor in natural history in Warrington and Halle. Forster took part in Cook's second expedition.

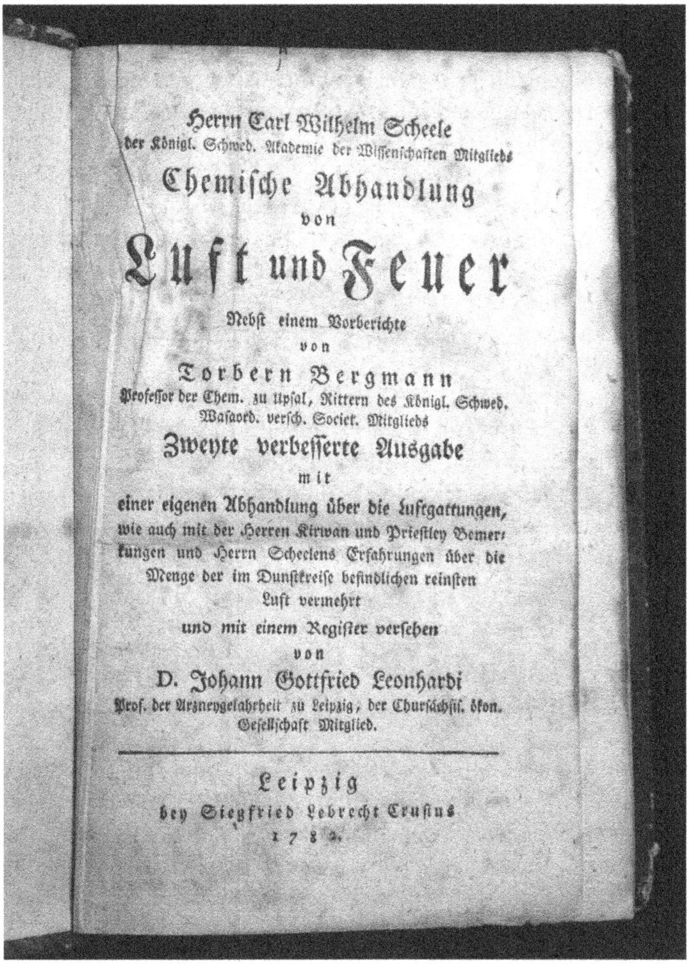
Herrn Carl Wilhelm Scheele
der Königl. Schwed. Akademie der Wissenschaften Mitglieds
Chemische Abhandlung
von
Luft und Feuer
Nebst einem Vorberichte
von
Torbern Bergmann
Professor der Chem. zu Upsal, Rittern des Königl. Schwed. Wasaord. versch. Societ. Mitglieds
Zweyte verbesserte Ausgabe
mit
einer eigenen Abhandlung über die Luftgattungen, wie auch mit der Herren Kirwan und Priestley Bemerkungen und Herrn Scheelens Erfahrungen über die Menge der im Dunstkreise befindlichen reinsten Luft vermehrt
und mit einem Register versehen
von
D. Johann Gottfried Leonhardi
Prof. der Arzneygelahrheit zu Leipzig, der Chursächsis. ökon. Gesellschaft Mitglied.
Leipzig
bey Siegfried Lebrecht Crusius
1782.

Fig. 21.6 The second, more widespread, edition of Scheele's book on Air and Fire. Note the slightly altered title. *Photo* Anders Lennartson

The original German edition published by Swederus (in collaboration with Siegfried Lebrecht Crusius in Leipzig) was small and rapidly sold out. Today, it is very rare and highly priced by collectors. A second edition (Fig. 21.6) was published by Crusius in 1782 on initiative by Gottfried Leonhardi.[5] Printing errors have been corrected and the book was expanded by inclusion of additional essays: "*Short account on the discovery of the airs*" by Leonhardi, German translations of Kirwan's essay, Priestley's letter and Scheele's paper "*On the amount of pure*

[5]Johann Gottfried Leonhardi (1746–1823). Professor in medicin in Leipzig and later in Wittenberg.

air...".. In addition, the second edition was supplied with an index and a foreword by Leonhardi. Altogether, the additions amount to 120 pages, nearly doubling the number of pages compared to the first edition. The copperplate is clearly printed from the same matrix as in the original edition, while a new copper plate had been engraved to illustrate the paper "*On the amount of pure air...*". It is quite remarkable that both the translations and the second edition contain Kirwan's essay, which cannot have pleased Scheele and which contains little of value.

In 1788, i.e. after Scheele's death, a Latin translation by Heinrich Schafer appeared in the Latin edition of Scheele's collected works, published on initiative by Ernst Benjamin Hebenstreit (1753–1803) [38]. The German text, including Kirwan's essay and Priestley's letter, was reprinted as the first volume in the German edition of Scheele's collected works in 1793 [39], and the second edition (excluding Bergman's preface and all additions) was reissued as volume 58 in *Ostwald's Klassiker der Exakten Wissenschaften* in 1894. The title page gives the impression that the first edition was used, which is not true. Probably the editor, physical chemist and later Nobel laureate Wilhelm Ostwald, did not have access to a copy of the rare first edition. The original 1777 edition was finally published in facsimile in 1970 with comments by Scheele scholar Uno Boklund.

21.5 Scheele's and Bergman's Views on Oxygen, Heat and Phlogiston

Scheele found that air was lost in a closed vessel containing air and potassium polysulphide (liver of sulphur), potassium sulphite, turpentine, iron(II) sulphate solution, nitrogen monoxide or copper(I) chloride solution. Of course, this is due to the absorption of oxygen by the readily oxidisable materials. Scheele first considered two possible explanations: either phlogiston caused the air to loose elasticity, and thus decrease in volume, or a part of the air was absorbed by the material (liver of sulphur, etc.). Scheele correctly excluded the first possibility, as he found that the air lost not only volume but also weight. Unfortunately, he also excluded the second, true explanation, since he failed to extract air from the oxidised materials. Scheele apparently had the erroneous impression that any gas absorbed by a solid material would behave as carbon dioxide (fixed air): it was well known that carbon dioxide could be absorbed—fixed—by strong bases such as potassium hydroxide or calcium oxide. When carbon dioxide reacts with potassium hydroxide it generates potassium carbonate which gives a precipitate with calcium hydroxide solution (lime water). When Scheele found that oxidised turpentine failed to give a precipitate with lime water, he concluded that it contained no "fixed air". This was the largest mistake in Scheele's entire career, and given Scheele's extensive knowledge, it is a quite remarkable error. At this point, Scheele could have been the one to realise the true role of oxygen in combustion and to overthrow the phlogiston theory; as Blomstrand put it: "A few seconds of thought in the right direction, would have been all needed" [40]. However, it is not likely that Scheele would have abandoned phlogiston altogether as it provided the only explanation for redox phenomena not involving oxygen. Thus, the role of revolutionising chemistry was left to Lavoisier.

Since Scheele believed he had proved that the consumed oxygen (or "fire air" as he called it) was neither present in the remaining air, nor in the oxidised product, the only remaining option for Scheele was that the oxygen escaped the vessel during the reaction. The logic conclusion was that oxygen took up phlogiston from the burning or oxidising material to form heat. As the concept of energy was not yet invented, the view of heat as a form of matter was well established among chemists, and heat was one of the elements listed by Lavoisier [41]. Not until about 1850 was the nature of heat reasonably well understood. Scheele proved that heat was evolved during every—what we would call—oxidation reaction, not only combustions. Scheele found further proof as he believed he could decompose heat by substances with a strong affinity for phlogiston. When mercury(II) oxide (mercury calx) decomposed to mercury and oxygen on heating, Scheele thought that the heat was decomposed by mercury(II) oxide into phlogiston and oxygen, the phlogiston reacting with the mercury calx to form mercury, while the oxygen (fire air) was liberated. Already in Malmö, Scheele had demonstrated that he could reduce silver nitrate and potassium nitrate by heat alone (Sect. 8.2). Scheele also reported a series of experiments that showed the difference between heat transfer by radiation and convection [42].

Light was also considered by Scheele as a compound of oxygen and phlogiston but richer in phlogiston than heat. He explained the decomposition of silver chloride and nitric acid by light by suggesting that silver chloride and nitric acid absorbed phlogiston from light. Thus, Scheele's theory of 1775 was that oxygen was a substance which readily absorbed phlogiston to form heat and light.

Bergman was convinced that heat was a substance with a small, almost immeasurable weight. In 1784, in one of his last contributions to science, Bergman proposed a method for determining the weight of heat [43]. The reason, according to Bergman, why previous attempts to determine the weight of heat by trying to measure the weight difference between a piece of metal when cold and when red-hot had failed was that the weight ratio of heat and metal is very small. Instead, Bergman suggested sealing ice in a metal container and determining the weight increase on melting. In this way, it would have been possible to measure the large amount of latent heat absorbed on melting. Although based on a false, material, view of heat, the suggestion was ingenious. Bergman introduced the term *eteric* to denote substances such as heat which could not be retained in any container, not even those made of dense gold [44].

In spring 1774, Lavoisier sent a copy of his *Opuscules physiques et chimiques* (Physical and chemical essays) to Scheele. In that book, he reported experiments conducted with a large burning glass. Scheele wrote a letter to Lavoisier, suggesting that he should heat silver carbonate (silver calx) by means of this burning glass. Lavoisier received the letter [45], but no letter from Lavoisier to Scheele is known, and nothing suggests that Lavoisier carried out the suggested experiment. Scheele's letter was published and discussed by Éduard Grimaux in 1890, but Grimaux's conclusions were disproved by Uno Boklund [46]. Grimaux concluded that if Lavoisier had carried out Scheele's experiment, Scheele would have been the discoverer of oxygen. This implies that Scheele could predict that heating silver calx would give rise to a previously unknown gas. Instead, Boklund concludes that Grimaux attempted to establish Lavoisier as an independent discoverer of oxygen. In reality, Scheele's letter (dated September 30) may, along with Priestley's visit in October, have been two factors putting Lavoisier on the right track.

According to Scheele, "the phlogiston is a true element and completely simple principle" [47]. It could be transferred between bodies, but "phlogiston can impossibly be obtained on its own". It should be recalled that Scheele had talked about hydrogen gas as "elastic phlogiston" in the late 1760s. Scheele also spoke vaguely about the effect of phlogiston on atoms (or particles as Scheele wrote): by the action of the heat that occupied the voids between atoms, it could cause a material to melt or to turn into a gas.[6] Phlogiston could also put the particles in such a situation that they absorbed all incoming light rays, or just a few. Thus, Scheele attributed colour to the interactions between light and atoms under the influence of phlogiston. In fact, oxidation/reduction have profound effect on light absorption and colour, both in organic compounds, where, i.e. reduction of conjugated π-systems blue-shifts the absorption, and in d-metal ions, which typically have very different colours in different oxidation states.

Bergman gave a definition of phlogiston in his *Instructions to Lectures in Chemistry* in 1779 (Sect. 11.2): [48] "The reason for such a nature [i.e. combustibility] lies actually in a very fine matter, which the ancients called Phlogiston. It is so subtle that, on its own, it escapes all our senses but can still be moved from one body to another according to the laws of affinity". Here, Bergman does not explicitly state that phlogiston cannot be isolated. Bergman wrote that phlogiston is probably a constituent of all matter, but in order to render a substance combustible, a large excess of loosely bound phlogiston is needed.

Bergman accepted Scheele's view on oxygen and combustion, and wrote in the second edition of his paper on elective attractions:

> The third system [of explaining the nature of heat] is that of my sagacious friend Mr. Scheele, who thinks that the matter of heat is not simple, but compounded of phlogiston and vital air [oxygen], closely combined, and that light consists of the matter of heat, with an excess of phlogiston. His Treatise on Air and Fire will best show how he arrived at these conclusions. This hypothesis is not without its difficulties, which I every where mention; it however seems to agree better with experiment than any other, and therefore I have often adapted my explanations to it. [49]

The anonymous translator (actually Thomas Beddoes, 1760–1808) was less impressed and added the following note: "To offer any arguments against Mr. Scheele's doctrine of composition of heat, would be now superfluous, since Mr. Kirwan (Notes to the treatise on Air and fire) and Fontana (Opusc. Litt. à S. Adolph. Murray) have abundantly confuted it" [50]. When Kirwan made a last attempt to defend the phlogiston theory with his book *An Essay on Phlogiston*, he

[6]In the English edition of Scheele's book, this discussion is abbreviated.

did not even bother discussing Scheele's theories: "I shall forbear entering on discussion of antiquated opinions long ago exploded, and also of that of Mr. Scheele, which has scarcely been embraced by any body, and has been sufficiently refuted by Mr. Lavoisier, and the experiments of Dr. Fordyce" [51]

While Scheele appears to have regarded nitrogen (ruined air) as an element, Bergman was more cautious. In his *Instructions to Lectures in Chemistry*, he wrote that its "true nature and inner properties" are still unresolved [52]. He wrote that because oxygen (pure air) can be completely destroyed by combustion or respiration, and become unsuitable to sustain fire and life, there are reasons to regard nitrogen as phlogisticated. Later on the same page, he wrote: "Since the pure air [oxygen] with something phlogistic appears to constitute such a harmful matter as ruined air [nitrogen]…" Here Bergman appears to regard nitrogen as phlogisticated air, and thus approach the views of Priestley (Sect. 21.6). Although Scheele's theories sometimes are somewhat naïve (e.g. his theories on the interactions of heat and atoms), this is an example where Scheele was closer to the truth than Bergman.

When Scheele learned of Lavoisier's work on the weight increase of sulphur and phosphorus on combustion, and of Cavendish's discovery in 1783 that combustion of hydrogen in pure oxygen gave rise to water, he was forced to modify his theory. In a paper in Crell's *Chemische Annalen*, published in 1785 [53], Scheele proposed oxygen to be a compound of *Principum salinum*, water and enough phlogiston to render the product gaseous. Upon combustion, the hypothetical *Principum salinum* gave heat and light with phlogiston, while the water was absorbed by the burning material, explaining the weight increase. Hydrogen was considered by Scheele as a compound of phlogiston and heat [54], and thus its combustion gave water and heat.

In Sweden, Scheele and Bergman were in good company holding on to phlogiston. von Engeström had also hard to believe in the new theory of Lavoisier [55]. He used arguments put forward by a de la Folie who argued against Lavoisier's initial theory that it was carbon dioxide (fixed air) that was absorbed by metals during calcination. Twelve *lod* (160 g) lead absorbed 700 cubic inches (4,300 L) of air on calcination, but von Engeström found it unlikely that such a large volume of air could be compressed into such a small volume of lead oxide (lead calx). von Engeström assumed that the absorbed air was fixed air, i.e. carbon dioxide. Another relevant point highlighted by von Engeström was the calcination of mercury. On heating mercury, it first forms mercury clax (it is oxidised by oxygen to mercury (II) oxide), but at a slightly higher temperature mercury is reformed (mercury (II) oxide decomposes to mercury and oxygen above 500 °C). This was difficult to explain by the prevalent theories of affinity.

Earlier authors have often been surprised or perhaps even embarrassed, that Scheele and Bergman did not convert to Lavoisier's theory, but it should be kept in mind that Lavoisier's theories did not receive widespread recognition until about 1785 [56], and by that time Bergman was dead and would soon be followed by Scheele. Also, Lavoisier's theory could not explain redox-reactions not involving oxygen, e.g. the reactions between metals and metal salt solutions investigated by

Bergman (Sect. 22.5). Such reactions would not get a satisfactory explanation until the discovery of valence electrons in the twentieth century.

Initially, Scheele assumed that the amount of oxygen in the atmosphere was constantly decreasing, as large amounts of oxygen were consumed by respiration and combustion. In 1778, he measured [57] the oxygen content in the atmosphere during a whole year in order to detect any variations, but found it to be constant [37]. He found this

> a very remarkable fact; and to assign the cause of it seems difficult, as a quantity of pure air [oxygen], in supporting fire, daily enters into a new union; and a considerable quantity of it is likewise corrupted, or changed into aerial acid, as well by plants as by respiration; another fresh proof of the great care of our Creator for all that lives. [58]

This view was of course rather naïve, as it should have been apparent to Scheele that a detectable decline in the atmosphere's oxygen content during a single year would have extinguished all life within a few years. It was Priestley who discovered that green plants could restore foul air [59]. This was before he discovered oxygen, but he concluded that "plants [...] reverse the effect of breathing and tend to keep the atmosphere sweet and wholesome". Scheele failed to reproduce Priestley's results; perhaps this was due to lack of sunlight in Scheele's laboratory.

21.6 Priestley's Theories of Oxygen and Combustion

Priestley's views on oxygen and air were fundamentally different from those of Scheele. While Scheele clearly stated that air is a mechanical mixture of oxygen, nitrogen, carbon dioxide and water vapour, Priestley seems to have regarded oxygen as air deprived of phlogiston, and nitrogen as phlogiston saturated air. He seems to have regarded oxygen and nitrogen as the two end points on a continuous scale, rather than two discrete substances. Throughout his life he called oxygen "dephlogisticated air" and nitrogen "phlogisticated air". In different sections of his book, *Experiments and Observations on Different Kinds of Air*, he gave different explanations of the nature of oxygen and, as Partington put it, "It is not easy to extract from his conflicting statements Priestley's considered views (if he had any) on the nature of oxygen and nitrogen" [60]. Since oxygen was easily obtained from nitrates, which in turn could be obtained from metal oxides, he arrived at the conclusion (unclear how) that air was composed of an earth (of undisclosed nature) combined with nitric acid and phlogiston:

> [...] there remained no doubt in my mind, but that *atmospherical air*, or the thing that we breathe, *consists of the nitrous acid and earth*, with so much phlogiston as is necessary to its elasticity; and likewise so much more as is required to bring it from its state of purity to the mean condition in which we find it. [61]

In some contexts, he seems to have expressed views closer to that of Scheele, i.e. that air would be a mixture of dephlogisticated (oxygen) and phlogisticated air (nitrogen).

Priestley studied the physiological effects of oxygen, and found that breathing oxygen was particularly easy: "Who can tell but that, in time, this pure air may become a fashionable article in luxury. Hitherto only two mice and myself have had the privilege of breathing it" [62]. Here, Priestley was obviously wrong, as Scheele, unknown to him, also was breathing pure oxygen in Sweden. Priestley added a caution:

> But, perhaps, we may also infer from these experiments, that though pure dephlogisticated air might be very useful as a *medicine*, it might not be so proper for us in the usual healthy state of the body: for, as a candle burns out much faster in dephlogisticated than in common air, so we might, as may be said, *live out too fast*, and the animal powers be too soon exhausted in this pure kind of air. A moralist, at least, may say, that the air which nature has provided for us is as good as we deserve [63]

21.7 The Discovery of Nitrogen

The discovery of nitrogen is usually attributed to a student of Joseph Black, Daniel Rutherford (1749–1819), who became a professor of botany in Edinburgh in 1786. In his doctoral thesis presented in 1772, he described an experiment where he left a mouse in a closed vessel until it died [64, 65]. A candle was then burnt in the remaining air until the flame went out. Rutherford then burnt phosphorus in the remaining air until the flame went out. He finally absorbed the carbon dioxide with alkali. Rutherford did not, however, give this gas a name, nor did he realise that it was a component in ordinary air. Scheele obtained pure nitrogen by heating ammonium chloride and manganese(IV) oxide before 1774, but he did not realise its identity with atmospheric nitrogen that he was studying about the same time [66].

21.8 Lavoisier and the Chemical Revolution

Our modern conception of chemical elements was introduced by the French chemist Antoine Lavoisier (1743–1794; Fig. 21.7). Formally, Lavoisier had a degree in law, but never practised as a lawyer. Among other official duties, he was responsible for inspecting the quality of gunpowder production in Paris, and his chemical research was conducted in his spare time. He got up at 5 a.m. every morning and spent three morning hours in his laboratory, and returned to the laboratory for three hours after concluding his official duties at 7 p.m.

In notes made in February 1772, Lavoisier expressed his intention to study air and carbon dioxide (fixed air) and on September 10, he described his first experiments on the weight increase of burning phosphorus. He found that when phosphorus or sulphur burned in closed vessels, the weight of the air in the vessel decreased. More importantly, he found that the phosphorus and sulphur increased in

Fig. 21.7 Antoine Lavoisier, the father of modern chemistry. Engraving from William Ramsay's book. *The gases of the atmosphere*, 1896

weight as they were converted to phosphorus and sulphur oxides (referred to as phosphoric and sulphuric acid, respectively, by Lavoisier). To conclude that the weight increase actually matched the weight decrease of the air took longer to establish. Phosphorus(V) oxide is highly hygroscopic and is deposited as a fine powder after the combustion. Sulphur dioxide is a gas and even more difficult to collect for weighing. Lavoisier realised the significance of this discovery, and realised that it could overthrow the phlogiston theory. Rather than giving off phlogiston, burning sulphur absorbed air. To secure his priority to this historical discovery, Lavoisier deposited a sealed note with the Académie des Sciences in Paris. This note read:

> It is about eight days since I discovered that sulphur while burning, far from losing weight, instead gained some, that is to say that from a pound of sulphur one could get much more than a pound of vitriolic acid, setting aside the humidity of the air. It is the same with

> phosphorus. This increase in weight comes from a prodigious quantity of air which is fixed during the combustion and which combines with the vapours. This discovery, which I have established by experiments which I regard as decisive, made me think that what is observed in the combustion of sulphur and phosphorus could well happen with regard to all bodies which gain weight by combustion and calcination, and I became convinced that the weight increase of metallic calces is related to the same cause. Experiment has completely confirmed my conjectures. I performed the reduction of litharge in closed vessels with Hales' apparatus, and I observed that at the moment of change from calx to metal a considerable quantity of air is released and that it makes up a volume at least one thousand times greater than the litharge used. This discovery seemed to me one of the most interesting that has been made since Stahl, and because it is difficult not to let one's friends in conversation catch a glimpse of something that might put them on the road to the truth, I deemed it necessary to make this deposition into the hands of the Academy's secretary, while waiting till I make public my experiments. [67].

It would, however, take Lavoisier several years to perfect his theories. After some preliminary experiments, Lavoisier's sealed note was opened on May 5, 1773, and in January 1774, he published his book *Opuscules physiques et chimiques*, where he reported some preliminary results. A copy of this book was sent to Scheele in Uppsala, and must therefore also have reached Bergman. In October 1774, Lavoisier got two important pieces of information. He both received Scheele's letter and met Priestley in the company of Lord Shelburne. Priestley told Lavoisier about his discovery of oxygen at a dinner, and the following month Lavoisier performed experiments on heating of mercury oxide. From his laboratory notes, it is clear that he still in March 1775 believed that the gas formed on heating mercury oxide was carbon dioxide.

On April 26, however, he read a paper at a meeting of the Académie des Sciences—the so-called Easter memoire. It is important to note that the original version was printed in Rozier's *Observatisons sur la Physique* [68], while the official version in *Mémoires de l'Académie Royale des Sciences [69]* was not actually printed until 1778. The final version was updated by Lavoisier and may give the impression that Lavoisier's theories were developed about 3 years earlier than was actually the case. In the original version, Lavoisier regarded air as a substance mixed with carbon dioxide and water vapour. Priestley criticised this theory [70]. In 1777, after many experiments, Lavoisier realised that air was not a substance, and in the final version of the Easter memoire, he stated that a burning substance reacts with the purest part of the air. The name oxygine (Greek ὀξύζ, acid, and γενής, former) was introduced in September 1777, when Lavoisier believed he could prove oxygen to be a constituent in all acids [71]. In his book *Traité élémentaire de chimie*, published in 1789 and regarded as the first modern textbook in chemistry, Lavoisier gave a clear description of oxygen as an element that reacts with burning substances. This book appeared in its first English edition, *Elements of Chemistry*, only a year later.

Lavoisier's theories were not immediately accepted, but gained more and more supporters during the years that followed. Some chemists, like Priestley, would stay true to the phlogiston theory well into the nineteenth century. Due to his involvement in a tax collecting firm, Ferme générale, Lavoisier was arrested during the French revolution and guillotined on May 8, 1794. Scheele and Bergman never accepted Lavoisier's interpretation of combustion phenomena, but had to modify their theories according to the new discoveries concerning the weight increase of a burning substance.

References

1. Sjöstén CG (1801) Åminnelse-tal, hållit för Kongl. Vetenskaps Academien öfver dess framledne ledamot herr Carl Vilhelm Scheele, den 14 Oct. 1799. Stockholm, p 7
2. Oseen CW (1940) Torbern Bergman och Carl Wilhelm Scheele. KVA, Stockholm, p p12
3. Oseen CW (1940) Torbern Bergman och Carl Wilhelm Scheele. KVA, Stockholm, p 22
4. Cassebaum H (1982) Carl Wilhelm Scheele. Teubner Verlagsgesellschaft, Leipzig, p 44
5. Blomstrand CW (1886) Minnesteckning öfver Carl Wilhelm Scheele, Stockholm, p 11
6. Nordenskiöld AE (1892) Carl Wilhelm Scheele. Efterlämnade bref och anteckningar. Stockholm, p XIX
7. Boklund U (1961) Carl Wilhelm Scheele, bruna boken, Stockholm, p 75
8. Oseen CW (1940) Torbern Bergman och Carl Wilhelm Scheele. KVA, Stockholm
9. Oseen CW (1940) Torbern Bergman och Carl Wilhelm Scheele. KVA, Stockholm, p 19
10. Sjöstén CG (1801) Åminnelse-tal, hållit för Kongl. Vetenskaps Academien öfver dess framledne ledamot herr Carl Vilhelm Scheele, den 14 oct. 1799. Stockholm, p 32
11. Oseen CW (1940) Torbern Bergman och Carl Wilhelm Scheele. KVA, Stockholm, p 21
12. Nordenskiöld AE (1892) Carl Wilhelm Scheele. Efterlämnade bref och anteckningar. Stockholm, p IV
13. Scheele CW (1777) Chemische Abhandlung von der Luft und dem Feuer, M Swederus & S. L Crusius, Uppsala & Leipzig, p 37
14. Priestley J (1775) Experiments and observations on different kinds of air, vol. 2. J. Johnson, London, p 34
15. Priestley J (1775) Experiments and observations on different kinds of air, vol. 2. J. Johnson, London, p 40
16. Priestley J (1775) Experiments and observations on different kinds of air, vol. 2. J. Johnson, London, p IX
17. Trofast J (1994) Johan Gottlieb Gahn: brev, vol 2, Lund, p 115
18. Trofast J (1994) Johan Gottlieb Gahn: brev, vol 2, Lund, p 116
19. Nordenskiöld AE (1892) Carl Wilhelm Scheele. Efterlämnade bref och anteckningar. Stockholm, p 200
20. Gentz L (1955) Carl Wilhelm Scheeles „Chemische Abhandlung von der Luft und dem Feuer"und seine Mitwelt. Internationale Gesellschaft für Geschichte der Pharmazie, Eutin, p 11
21. Gentz L (1955) Carl Wilhelm Scheeles „Chemische Abhandlung von der Luft und dem Feuer"und seine Mitwelt. Internationale Gesellschaft für Geschichte der Pharmazie, Eutin, p 19
22. Bergman T (1775) Disquisitio de attractionibus electivis. Nova Acta Reg Soc Scient Upsaliensis 2:161–250
23. Partington JR (1962) The discovery of oxygen. J Chem Educ 39:123–125
24. Cassebaum H, Schufle JA (1975) Scheele's priority for the discovery of oxygen. J Chem Educ 52:442–444
25. Bergman T (1783) Opuscula physica et chemica, pleraque antea seorsim edita, jam ab auctore collecta, revisa et aucta. Vol. 3, Uppsala, p 291–470
26. Bergman T (1785) A Dissertation on Elective Attractions, J. Murray, London
27. Bergman T (1968) A dissertation on elective attractions. Johnson Reprint Corporation, New York

28. Lavoisier A (1790) Elements of Chemistry, William Creech, Edinburgh, p 36
29. Trofast J (1994) Johan Gottlieb Gahn brev, vol 2, Lund, p 114
30. Lennartson A (2017) The Chemical Works of Carl Wilhelm Scheele. Springer, Cham
31. Letter from Schwediauer to Bergman daterd October 24, 1780
32. Scheele CW (1780) Chemical Observations and Experiments on Air and Fire. J. Johnson, London
33. Scheele CW (1780) Chemical Observations and Experiments on Air and Fire. J. Johnson, London. p iii
34. Scheele CW (1780) Chemical Observations and Experiments on Air and Fire. J. Johnson, London, p 222
35. Scheele CW (1780) Chemical Observations and Experiments on Air and Fire. J. Johnson, London, p 197
36. Gentz L (1958) Hur såg Scheele ut? Sv Farm Tidskr 373–394
37. Scheele CW (1779) Rön, om rena Luftens mängd, som dageligen uti vår Luft-krets är närvarande, KVA Handl 40:50–55
38. Gentz L (1955) Carl Wilhelm Scheeles "Chemische Abhandlung von der Luft und dem Feuer" und seine Mitwelt. Internationale Gesellschaft für Geschichte der Pharmazie, Eutin, p 43
39. Scheele CW (1793) Sämmtliche physische und chemische Werke, vol 1, Heinrich August Rottman, Berlin, p 1–264
40. Blomstrand CW (1886) Minnesteckning öfver Carl Wilhelm Scheele, Stockholm, p 28
41. Lavoisier A (1790) Elements of Chemistry, William Creech, Edinburgh, p 175
42. Scheele CW (1777) Chemische Abhandlung von der Luft und dem Feuer, M Swederus & S. L Crusius, Uppsala & Leipzig, p 53
43. Bergman T (1784) Ueber der Erforschung der Schwere des Feuers, Chem Ann 1:93–95
44. Bergman T (1784) Vom Hrn. Profess. und Ritter Bergmann in Upsal, Chem Ann, 1:149–151
45. Laitko H (1986) Carl Wilhelm Scheele und die Umwälzung des chemischen Denkens um die Wende vom 18. und 19. Jahrhundert. In: Carl-Wilhelm-Scheele-Ehrung, Berlin, p 40
46. Boklund, U (1957–1958) A lost letter from Scheele to Lavoisier. Lychnos, 39–62
47. Scheele CW (1777) Chemische Abhandlung von der Luft und dem Feuer, M Swederus & S. L Crusius, Uppsala & Leipzig, p 82
48. Bergman T (1779) Anledning til föreläsningar öfver chemiens beskaffenhet och nytta, samt naturliga kroppars almännaste skiljaktigheter, M. Swederus, Stockholm, Uppsala and Åbo, p 40
49. Bergman T (1785) A Dissertation on Elective Attractions, J. Murray, London, p 234
50. Bergman T (1785) A Dissertation on Elective Attractions, J. Murray, London, p 345
51. Kirwan R (1787) An essay on phlogiston, and the constitution of acids. P. Elmsly, London, p 95
52. Bergman T (1779) Anledning til föreläsningar öfver chemiens beskaffenhet och nytta, samt naturliga kroppars almännaste skiljaktigheter, M. Swederus, Stockholm, Uppsala and Åbo, p 53
53. Scheele CW (1785) Neuere Bemerkungen über Luft und Feuer, und die Wasser-Erzeugung, Chemische Annalen für die Freunde der Naturlehre, 1:229–238
54. Scheele CW (1777) Chemische Abhandlung von der Luft und dem Feuer, M Swederus & S. L Crusius, Uppsala & Leipzig, p 143
55. von Engeström G (1782) Tal, om vissa svårigheter och andra omständigheter, som möta vid utöfvandet af chymien; hållet för kongl. vetensk. academien, vid præsidii nedläggande, den 6 November 1782. Stockholm, p 8
56. Lundgren A (1979) Berzelius och den kemiska atomteorin. Uppsala University, Uppsala, p 11
57. Lennartson A (2017) The Chemical Works of Carl Wilhelm Scheele. Springer, Cham, p 57
58. Scheele CW (1901) The chemical essays of Charles-William Scheele. Reissue of 1786 edition. Scott, Greenwood & Co, London, p 194
59. Priestley J (1772) Observation on different kinds of air. Phil Trans 62:147–264

60. Partington JR (1962) A history of chemistry, vol 3. Macmillan, London, p 262
61. Priestley J (1775) Experiments and observations on different kinds of air, vol. 2. J. Johnson, London, p 55
62. Priestley J (1775) Experiments and observations on different kinds of air, vol. 2. J. Johnson, London, p 102
63. Priestley J (1775) Experiments and observations on different kinds of air, vol. 2. J. Johnson, London, p 101
64. Robertson G, Rutherford D (1772) Disseratio inauguralis de aer fixo dicto, aut mephitico. Edinburgh
65. Dobbin L 81935) Daniel Rutherford's inaugural dissertation. J Chem Educ 12:370–374
66. Scheele CW (1774) Om Brun-sten eller Magnesia nigra och dess egenskaper, KVA Handl 35:177–194
67. Finocchiaro MA (1977) Logic and Rhetoric in Lavoisier's Sealed Note: Toward a Rhetoric of ScienceAuthor. Philosophy & Rhetoric 10:111–122
68. Lavoisier A (1775) Mémoire sur la nature du principe qui se combine avec les Métaux pendant leur calcination, & qui en augmente le poids, Observations sur La Physique, 5:429–33
69. Lavoisier A (1778) Mémoire sur la nature du Principe qui se combine avec les Métaux pendant leur calcination et qui en augmente le poids, Histoire de l'Académie Royale des Sciences, 520–526
70. Priestley J (1775) Experiments and observations on different kinds of air, vol. 2. J. Johnson, London, p 322
71. Lavoisier AL (1777) Considérations générales sur la nature des acides et sur les principles dont ils sont composes, Histoire de l'Académie royale des sciences, 535–547

Bergman's and Scheele's Theories of Elements and Atoms

22

A chemistry student, who typically only comes into contact with the history of chemistry through short text boxes in textbooks, may get the impression that the concept of atoms was introduced in ancient Greece and then forgotten until Dalton rediscovered the atom in the early nineteenth century. This is far from the truth, and the idea of matter being composed of particles was widespread during the seventeenth and eighteenth centuries. This chapter discusses Bergman's and Scheele's views on the nature of matter.

22.1 Scheele's Views on Elements

The main, if not only, direct source for information on Scheele's ideas of elements is the first two sections of his *Chemische Abhandlung von der Luft und dem Feuer* (Sect. 21.3) written in 1775. In the first section, he wrote that the main goal of chemistry is to resolve bodies into their constituents, to study the properties of these constituents and to attempt to recombine them in different ways [1]. In the second section, he wrote that chemists had yet not agreed on the number of elements that build up matter. Scheele admitted that this was a difficult problem and some, Scheele wrote, claimed that there was no hope of investigating the elements. This opinion did not impress Scheele, who did not accept such limits on chemical research: "but far be it, that the noble science of Chemistry should be circumscribed in so narrow limits" [2]. Others, Scheele continued, claimed that earth and phlogiston were the constituents of all matter, while most scientists appeared to only accept the four Aristotelian elements (i.e. earth, water, air and fire). Here, Scheele did not state his opinion, but several other statements give a fairly good idea. It is highly unlikely that Scheele believed in the four Aristotelian elements, rather, he seems to have believed that substances were composed of metal oxides (metallic calces), alkalis, earths and acids, frequently in combination with phlogiston. In his *Chemische Abhandlung*, he wrote that combustible substances were usually

A. Lennartson, *Carl Wilhelm Scheele and Torbern Bergman*, Perspectives on the History of Chemistry, https://doi.org/10.1007/978-3-030-49194-9_22

composed of phlogiston combined with an acid, and if metal oxides (metallic calces) could be deprived of all their phlogiston, they would probably give rise to acids [3]. Bergman would elaborate on these ideas after Scheele's isolation of molybdic and tungstic acid [4]. Scheele seems to have regarded water as an element, while Bergman spoke of "the solid part of water" as a possible constituent in siliceous earth [5].

In his paper on pyrolusite (Sect. 15.1), Scheele noted that he obtained calcium sulphate by dissolving pyrolusite (MnO_2) in sulphuric acid in the presence of sugar. Surprisingly, Scheele did not propose an explanation but merely wrote that this could indicate a transmutation of—in modern terms—manganese to calcium. It is surprising that Scheele left such an important issue unresolved, but it may be that Scheele simply wanted to finish his work, which had already been extended by years on the request of Bergman. As Scheele was always very honest in his description of experiments, he could still not leave the observation out. It was not until after Scheele's death that his friend, Gadolin, found that the calcium compounds (lime) were contaminants in the sugar [6].

22.2 Bergman's Views on Elements

Bergman's discussion on the nature of the elements in his preface to Scheele's book on air and fire, Chemische Abhandlung von der Luft und dem Feuer, in 1777 has been studied previously; [7] there are, however, two better sources of information: his *Lecture on the Latest Progress in Chemistry* from 1777 and his *Instructions to Lectures on Chemistry's Nature and Benefits, and Natural Bodies General Differences* from 1779. The later was printed as an appendix to the second edition of *The Chemical Lectures of H.T. Scheffer*, but was also printed as a separate booklet (Sect. 11.2).

In the 1779 text, Bergman defined elements as "particles, which separates by splitting of bodies" [8]. Bergman distinguished between *mechanical elements* and *chemical elements*. The mechanical elements were simply small homogeneous particles obtained by mechanical division such as grinding. These could perhaps be compared to our molecules. The chemical elements were those "of different nature, which together constitute a body". For example, the chemical elements of calcium carbonate (chalk) were calcium oxide (calcareous earth), carbon dioxide (aerial acid) and water. The chemical elements formed different orders: the *proximate elements* were those which were obtained by the first decomposition, e.g. mercury and sulphur in mercury sulphide (cinnabar). If these could be further decomposed, *remote elements* were obtained. Since the proximate elements of mercury were mercury oxide (mercury calx) and phlogiston and the proximate elements of sulphur were sulphuric acid (vitriolic acid) and phlogiston, the remote elements of cinnabar were mercury calx, vitriolic acid and phlogiston. Elements that could not be further decomposed were referred to as "*primary elements*" by Bergman, but unfortunately, he did not explicitly state at this point whether he regarded mercury calx and vitriolic acid as primary elements. The decomposition of a body into its chemical elements was called *Analysis chemica*, and the reunion was called *Synthesis chemica* by Bergman. He stressed that it was not only the nature of the elements that determined the properties of a body, but also the proportions of the elements: marl and topaz had, according to Bergman, the same constituents but different properties. Actually, they are unrelated, but it is not absolutely clear what Bergman meant with "marl".

Some insight into Bergman's views of the primary elements is found in his 1777 lecture given in front of the Royal Swedish Academy of Sciences. Here, it is clear that he did not believe in the four Aristotelian elements, as he said that many fruitless attempts had been made to determine the elements of water [9] and that fire is nothing else than the state when combustible bodies are deprived of their phlogiston by oxygen (pure air) [9]. Later in the lecture, he entered into a rare speculation about the primary elements [10]. His own words clearly indicate his doubts and the hesitations he felt: "The hitherto mentioned results, concerning the composition of bodies, merely reveal the proximate elements: the remote elements are more difficult to find, and the primary or real elements are even more unknown. Several scientists have nevertheless thought themselves capable of, with certainty, stating their number. As this has been done more based on arguments than clear experiments, it is not surprising that the opinions are quite divided. Some call for only a single element, others two, three, four, yes even up to five. I need not now to examine them all, since several of them are most probably in error. The opinion of those supposing two elements is most probable: one earthy, passive and one fierily, active, of which, in different proportions mixed with each other and with other *Principia secundaria* of different orders, arises the endless number of different bodies that exist in nature. One should not, however, consider this guess as an established truth. What the first, earthy, element concerns, it cannot yet be proven by analysis that it is the same in all bodies, how probable it still may sound, but one should take things as they appear, and not determine their nature from our imagination. We know nothing a priori about bodies, everything must be determined through suitable experiments. What the other and active element concerns, it is not light, which is truly compounded, but phlogiston". In a paper on gems published in 1780 [11], he once again discussed the possibility that the different earths are modifications of only one earths:

> It cannot be denied that it is not altogether without probability some imagine that the number of these earths should be diminished, considering them all as modifications of one; but in the explanation of nature, we must not so far indulge conjecture, as to suffer the vain phantasms of imagination to prevail over phænomena confirmed by constant experience, and not impeached by a single experiment, when made with accuracy. [12]

As discussed in Sects. 24.7 and 22.4, both Bergman and Scheele came to the conclusion that metal oxides (metal calces) were not elements as many chemists supposed. After Scheele's isolation of arsenic acid, molybdic acid and tungstic acid, they came to the conclusion that metal calces were composed of metallic acids and phlogiston. As discussed in Sect. 22.4, what Bergman had actually discovered was that metals may appear in higher oxidation states than in the commonly encountered hydrated metal ions, and that such species frequently are acidic.

It is unlikely that Bergman would have believed in chemical transmutation, but on the other hand, when discussing the history of chemistry, he wrote that "...the alchemists work on a solution of a problem [i.e. transmutation of base metals to gold], that cannot be proved to be impossible [...]" [13].

22.3 Bergman's and Scheele's Views on Atoms

That matter was composed of particles seems to have been absolutely natural to both Scheele and Bergman, and they never discussed other possibilities. Bergman even used the word "atom" when describing water, and assumed that the particles were solid:

> The particles of water are, however, probably solid, since one can generally hardly imagine a fluid differently, than as a collection of such fine atoms, that they cannot be seen with the best armed eye, and so moveable, that their equilibrium requires a horizontal surface. [14]

The liquid nature of water was caused by heat, which was so firmly intermixed with the water particles that it could not be detected by a thermometer. The particles of matter attracted each other, giving rise to affinities (Chap. 20): "The elements union is caused by their mutual attractions [...]. All bodies in nature attract each other" [15]. With his background in astronomy, Bergman compared these attractions with the gravitational attractions between celestial bodies. In the gaseous state, Bergman found that "particles are not only easily separable, but seem in some degree to repel each other" [16]. It would take another century and the work of Maxwell, Boltzmann and others until the properties of gases could be understood at a molecular level. While Bergman described the water atoms as solid, he described air particles as compressible in 1764 when discussing the decrease in air pressure with altitude: "Now it is quite uncertain how many times an air particles volume has decreased by compression from its natural size at the Earth surface [...]" [17].

Scheele mentioned the particulate nature of matter on a few occasions, most notably in his discussion on heat and light in his book on Air and Fire. Heat, Scheele wrote, cannot penetrate matter but is absorbed in voids between the particles [18]. On heating a stone, it first absorbed heat into its pores, and by doing so, it expanded. On expanding, finer pores were made accessible and could accommodate light. This explained thermal expansion and incandescence (the emission of light by hot objects) [19].

It should be noted that neither Bergman's nor Scheele's views on atoms were clear enough to enable them to use atomic theories to explain chemical phenomena. Atoms or particles were used to describe physical properties of matter, but they rarely discussed the relation between elements and atoms. They did not state whether they believed that all atoms of an element were identical and different from atoms of other elements.

22.4 Scheele's and Bergman's Views on the Earths, Metals, Acids and Alkalis

Scheele strongly believed in the elemental nature of the primitive earths: siliceous earth, clay, magnesia, lime and earth of heavy spar. In his papers, Scheele never speculated over the possibility of these earths being composed of simpler bodies. Bergman, as will be seen, believed that the earths, or at least silica, could be decomposed into more primitive elements.

Bergman seems to have agreed with Scheele that the earths could not be interconverted. Frequently, however, there were reports from chemists claiming to have proven otherwise. Bergman and Scheele made great efforts in trying to prove or disprove these reports in order to reveal the true nature of the earths.

In 1776, Scheele published a paper proving the individual nature of aluminium and silicon oxides [20]. The reason was the opinion of Baumé that earth of alum (Al_2O_3) was actually siliceous earth (SiO_2), that clay was the same earth combined with sulphuric acid and that alum ($KAl(SO_4)_2$) was the same earth over-saturated with sulphuric acid. Scheele was sceptic: "With regard to chemical opinions, it is my custom not to credit any, till I have brought them to the test of experiment" [21]. Actually, Scheele discussed this issue with Gahn in a letter dated November 1772, the writing of this manuscript being the next task for Scheele after writing together the paper on arsenic acid and the book on air and fire.

At first, it seemed to Scheele that Baumé was correct: he melted powdered rock crystal (pure SiO_2) with potassium hydroxide in a crucible, extracted the residue (potassium silicates) with water and added sulphuric acid [22]. Siliceous earth precipitated and Scheele found small amounts of alum in the mother liquor. Still, this alum could have been a contaminant in the rock crystal or the alkali, and consequently Scheele repeated the procedure with the precipitated siliceous earth seven times, and each time he obtained alum. At this point Scheele noticed that the crucibles were corroded, and he found that the source of the alum was clay from the crucible: "But, behold! on examining the crucibles employed for these repeated fusions, I found them everywhere uneven in the inside, and full of little excavations, which they had not before the experiment" [23]. When the operation was performed in an iron crucible, no alum was obtained and Scheele could (probably to his relief) establish that "The siliceous earth, therefore, still remains a peculiar earth" [23]. In a similar fashion, Bergman criticised a paper claiming that cobalt could be obtained from iron and arsenic [24]. Bergman concluded that the cobalt must have been present as an impurity in the reagents.

Despite the work of Scheele, Bergman was not convinced of the elemental nature of siliceous earth (SiO_2), but in lack of evidence he had to consider it as elemental. What made Bergman suspicious were Scheele's experiments with hydrofluoric acid (Sect. 14.1), where siliceous earth appeared to arise upon contact of the acid with water, while it was in fact formed by the hydrolysis of volatile silicon tetrafluoride and condensation of the silicic acid formed:

$$SiF_4(g) + 4H_2O(l) \rightarrow Si(OH)_4(s) + 4HF(aq).$$

In his letter to Troil in June 1776 (Sect. 7.3) he wrote:

> Mr. Scheele has discovered the formation of the flint[1]; and I myself have found out, within these two year, a method of obtaining, with help of some fluor acid [hydrofluoric acid], thirteen precipitated crystals of sizes of small peas. Theses artificial pebble in all experiments, both in the wet and dry method, and even in focus of a burning-glass, in piece that I sent to Mr. Macquer, discovered exactly to him the same qualities as the natural one. All these circumstances, therefore, prove, that the pebble [*kisel* (= quartz) in the original Swedish text] is a saline earth, which is composed of fluor acid, and an original substance existing in the watry exhalations. It is not quite simple; but however, I have not been able to consider it as any other than an elementary earth: indeed my judgement is, that it cannot be compounded from any other principle. [25]

In a dissertation defended in November 1779 [26], Bergman argued that the formation of siliceous earth from hydrofluoric acid and water indicated that siliceous earth was composed of hydrofluoric acid and the principle of water. In his *Instructions to Lectures in Chemistry*, published the same year, he wrote "That acid of fluor-spar is present in silica is without doubt, but the basis in this earth, which appears to be the solid part of water, is still unknown" [5]. Still, the other primitive earths had not been decomposed, and until further experiments had proven the compounded nature of siliceous earth, he preferred to treat it as a primitive earth [27].

The nature of magnesium oxide (magnesia) was studied in a dissertation, *de Magnesia alba*, defended by one of Bergman's students in 1775, adding further evidence to the elemental nature of magnesium [28]. Two years later, a Johan Berger, apparently of military background, claimed in a paper on potassium nitrate (saltpetre) production that calcium (lime) could be converted to magnesium (magnesia) [29]. Berger's paper is followed by a short paper by Bergman, where he proved Berger wrong [30]. Bergman insisted that magnesia and lime were independent primitive earths, referring to the works of Black, Marggraf and his own dissertation on *Magnesia alba*. The problem was that when magnesium nitrate (*Magnesia nitri*) was removed from the potassium nitrate solution by precipitation with potassium carbonate, it yielded a precipitate which readily dissolved in sulphuric acid, but after calcination (i.e. decomposition to magnesium oxide), it became much less soluble in sulphuric acid, and was thus confused for calcium oxide (lime) by Berger. Bergman showed that there was no conversion of magnesium to calcium, but upon calcination the precipitate (magnesium carbonate hydroxide) lost carbon dioxide and water (turning to magnesium oxide) which dissolves more slowly in sulphuric acid.

In addition to the four generally accepted earths and the recently discovered barium oxide, Bergman also suspected the presence of sixth elemental earth, earth of gems, in diamonds [11], a view that he later had to abandon after realising that

[1] *kisel* in the original Swedish text (kisel = quartz).

the minute amounts he had obtained were merely a mixture of material [31] (presumably originating from his crucibles or reagents).

In a section of his book on Air and fire, *Chemische Abhandlung von der Luft und dem Feuer,* written in 1775, Scheele discussed the composition of combustible materials. He expressed the idea that they consist of an acidic element combined with phlogiston. Both sulphur and organic substances gave acids (SO_2 and CO_2, respectively) on combustion [32]. Strangely enough, he did not mention phosphorus but included zinc in the discussion.[2] He then turned to arsenic (actually arsenic(III) oxide) which he had found could not only be reduced to metallic arsenic but also deprived of phlogiston (oxidised) to arsenic acid. Scheele now suggested that all metallic calces and earths perhaps consisted of acids and phlogiston.

Scheele's 1781 paper on tungstic acid (Sect. 15.5) was followed by a paper by Bergman [4], where he suggested that molybdic and tungstic acid could be reduced to metals. He also saw a similarity with arsenic and arsenic acid, and put forward a similar idea on the composition of metals. In his phlogistic terminology, Bergman suggested that metal calces (metal oxides) were not true elements but were actually composed of metallic acids and phlogiston. In modern terminology, Bergman had realised that metals could occur in higher oxidation states than those frequently encountered in the common oxides. High-valent metal oxides are indeed often acidic, e.g. CrO_3, Mn_2O_7, WO_3 and MoO_3. Bergman developed a special interest in "acid of iron", and on doing so he was far ahead of his time. Still today, iron (IV) and iron(V) compounds are hot research topics, not least as such species constitute reactive intermediates during enzymatic oxidation reactions. Neither Bergman, nor Scheele, realised that Scheele had actually prepared salts of manganic and permanganic acids (H_2MnO_4 and $HMnO_4$, respectively) by oxidation of MnO_2 in his pyrolusite paper (Sect. 15.1). The following year, Bergman discussed the elemental acids in his *Meditationes de systemate fossilium naturali,* where he predicted that also vegetables and perhaps even animals contained a number of unknown acids.

Regarding the alkalis, there were clear indications that at least one of them, volatile alkali (ammonia), was not a true element. In 1779, Bergman wrote about alkaline air (gaseous ammonia): "when electric explosions go through this air, the volume is increased each time. In contrast to ammonia, this gas [a mixture of nitrogen and hydrogen] is not absorbed by water, and is found to be very flammable" [33].

22.5 Phlogiston Content of Metals—Determination of Equivalent Weights

As will be discussed in Chap. 23, Bergman was a firm believer in stoichiometry and the constant composition of chemical bodies. In a dissertation defended on December 13, 1780, by Andreas Tunborg, Bergman set out to determine the relative

[2]Zinc oxide is amphoteric, having both acidic and basic properties.

Fig. 22.1 Reduction of aqueous silver nitrate by metallic copper. Note the silver crystals growing on the copper spiral and the pale blue colour of the solution due to the formation of $[Cu(H_2O)_6]^{2+}$. *Photo* Anders Lennartson

amount of phlogiston in different metals [34]. The dissertation has been translated (in part) to English by Schufle [35]. Bergman's method was to determine the amount of different metals required to precipitate the silver from a solution of silver nitrate (Fig. 22.1), e.g.

$$2\,Ag^{+}(aq) + Cu(s) \rightarrow 2\,Ag(s) + Cu^{2+}(aq).$$

Bergman argued that the limiting factor was the amount of phlogiston required to reduce the silver ions to metallic silver (in modern terminology). This was in principle correct, as Bergman's phlogiston in reality was equivalent to our modern concept of valence electrons, and each silver ion requires one electron to be reduced to metallic silver. What Bergman actually determined, although he did not realise it, was equivalent weights. By doing so, he anticipated Dalton by several decades. Berzelius later wrote that Bergman was one of the first authors to imply that the proportions of the elements in a body followed a general rule [36]. The equivalent weight, a concept not used by modern chemists, is the atomic weight divided by the valency. For instance, while two silver ions are required to oxidise one copper atom, only one zinc atom is required to oxidise a copper atom. From Bergman's results, one can calculate atomic weights (Table 22.1).

In some cases, the results are fairly accurate, while other results deviate considerably from the true values. The case of mercury is complicated, as Bergman correctly noted, by the fact that it forms an amalgam with silver. The high values for manganese, nickel and cobalt are expected, as these metals could not be obtained in a pure state but would have contained carbon. When repeating Bergman's experiments, it becomes clear that the precipitated silver is far from pure. For example,

Table 22.1 Results of Bergman's determination of equivalent weights

Metal[a]	"Equivalents" of silver[b]	Corresponding atomic weight[c]	Modern atomic weight
Hg	74	292	200.59
Pb	43	502	207.2
Cu	323	66.8	63.55
Fe	342	63.1	55.85
Sn	114	119	118.71
Ni	156	138	58.69
Co	270	79.9	63.55
Zn	182	98,2	65.39
Mn	196	110	54.94

[a]Bi, Sb and As have been excluded, since these elements have a more complex behaviour in solution

[b]The amount of silver precipitated by 100 parts of the metal

[c]Atomic weight of the metal calculated assuming the metal is oxidised to M^{2+} and the atomic weight of silver is set to 107.87 u

silver precipitated with zinc has a very high zinc content and effervesces with hydrochloric acid.

The use of silver had the advantage that the precipitated silver flakes off the analysed metal fairly well and the mixture can be mechanically separated (unlike copper that forms a crust on the metal). Even so, the author of this book found it difficult, even in a modern laboratory, to obtain much more accurate values for the atomic weight of copper than Bergman did using his method.

References

1. Scheele CW (1777) Chemische Abhandlung von der Luft und dem Feuer, M Swederus & S. L Crusius, Uppsala & Leipzig, p 1
2. Scheele CW (1780) Chemical Observations and Experiments on Air and Fire. J. Johnson, London, p 2
3. Scheele CW (1777) Chemische Abhandlung von der Luft und dem Feuer. M. Swederus, Uppsala and Leipzig, p 86
4. Bergman T (1781) Tilläggning om Tungsten KVA Nya Handl 2:95–98
5. Bergman T (1779) Anledning til föreläsningar öfver chemiens beskaffenhet och nytta, samt naturliga kroppars almännaste skiljaktigheter, M. Swederus, Stockholm, Uppsala and Åbo, p 39
6. Gadolin J (1789) Undersökning, huruvida Brunsten kan förvandlas i Kalkjord. KVA Nya Handl 10:141–150
7. Cassebaum H, Kauffman GB (1776) The Analytical Concept of a Chemical Element in the Work of Bergman and Scheele, Ann Sci 33:447–45
8. Bergman T (1779) Anledning til föreläsningar öfver chemiens beskaffenhet och nytta, samt naturliga kroppars almännaste skiljaktigheter, M. Swederus, Stocholm, Uppsala & Åbo, p 2

9. Bergman T (1777) Tal, om chemiens nyaste framsteg, hållet, i kongl. maj:ts höga närvaro, för dess vetenskaps- academie, vid præsidii nedläggande, den 12 nov. 1777. KVA, Stockholm, p 25
10. Bergman T (1777) Tal, om chemiens nyaste framsteg, hållet, i kongl. maj:ts höga närvaro, för dess vetenskaps- academie, vid præsidii nedläggande, den 12 nov. 1777. KVA, Stockholm, p 36
11. Bergman T (1780) Disquisitio chemica de terra gemmarum, Nova Acta Regiæ Societatis Scientiarum Upsaliensis, 3:137–170
12. Bergman T (1788) Physical and Chemical Essays, vol 3, J Murray, William Creech, Edinburgh, p 91
13. Bergman T (1777) Tal, om chemiens nyaste framsteg, hållet, i kongl. maj:ts höga närvaro, för dess vetenskaps- academie, vid præsidii nedläggande, den 12 nov. 1777. KVA, p 10
14. Bergman T (1777) Tal, om chemiens nyaste framsteg, hållet, i kongl. maj:ts höga närvaro, för dess vetenskaps- academie, vid præsidii nedläggande, den 12 nov. 1777. KVA, Stockholm, p 26
15. Bergman T (1779) Anledning till föreläsningar öfver chemiens beskaffenhet och nytta, samt naturliga kroppars almännaste skiljaktigheter, M. Swederus, Stocholm, Uppsala & Åbo, p 3
16. Bergman T (1791) Physical and Chemical Essays, vol III, G, Mudie, J & J Fairbairn and J. Evans, Edinburgh, p 208
17. Bergman T (1764) Afhandling om nordskenens högd. Sednare stycket KVA Handl, 25:249–261
18. Scheele CW (1777) Chemische Abhandlung von der Luft und dem Feuer, M Swederus & S. L Crusius, Uppsala & Leipzig, p 79
19. Scheele CW (1777) Chemische Abhandlung von der Luft und dem Feuer, M Swederus & S. L Crusius, Uppsala & Leipzig, p 98
20. Scheele CW (1776) Rön och Anmärkningar om Kisel, Lera och Alun. KVA Handl 37:30–35
21. Scheele CW (1901) The chemical essays of Charles-William Scheele. Reissue of 1786 edition. Scott, Greenwood & Co, London, p 141
22. Lennartson A (2017) The chemical works of carl wilhelm scheele. Springer, Cham, p 32
23. Scheele CW (1901) The chemical essays of Charles-William Scheele. Reissue of 1786 edition. Scott, Greenwood & Co, London, p 142
24. Bergman T (1784) Vom Hrn. Profess. und Ritter Bergmann in Upsal, Chem Ann. 1: 38–39
25. Troil U (1780) Letters on Iceland J. Robson, London, p 346
26. Bergman T, Grönlund KA (1779) Dissertatio chemica de terra silicea, Uppsala
27. Bergman T (1788) Physical and Chemical Essays, vol II, J. Murray, London, 54
28. Bergman T, Norell C (1775) Dissertio [sic!] chemica de magnesia alba, Uppsala
29. Berger (1777) Tankar, om Salpeter, grundade på försök vid Salpeter-verket i Helsingfors, KVA Handl 38:193–213
30. Bergman T (1777) Anmärkningar om magnesia nitri, KVA Handl, 38:213–216
31. Hjelm PJ (1786) Åminnelse-tal öfver…Herr Torbern Bergman, Stockholm, p 57
32. Scheele CW (1777) Chemische Abhandlung von der Luft und dem Feuer. M. Swederus, Uppsala and Leipzig, p 84
33. Bergman T (1779) Anledning til föreläsningar öfver chemiens beskaffenhet och nytta, samt naturliga kroppars almännaste skiljaktigheter, M. Swederus, Stockholm, Uppsala and Åbo, 62
34. Bergman T (1780) Dissertatio chemica de diversa philogisti [sic!] quantitate in metallis, quam, venia ampl. facult. philosoph., praeside Mag. Torb. Bergman … publice ventilandam sistit Andreas Nicolaus Tunborg, Dalekarlus. In audit. gustav. maj. d. 13 dec. 1780, Uppsala
35. Schufle JA (1972) Torbern Bergman and Andreas N. Tunberg [sic!]: The different Quantities of Phlogiston in Metals. J Chem Educ 49:810–812
36. Berzelius J (1818), Lärbok i kemien, vol 3, Stockholm, p 2

Bergman as an Analytical Chemist 23

Before the eighteenth century, analytical methods were mainly applied to ores and mineral waters. Ores were analysed quantitatively by reducing them on small scale with charcoal and different fluxes to metal (assaying). This indicated the amount of metal ore would yield but did not accurately tell the actual metal content of the ore. Mineral waters were analysed mainly by fractionated crystallisation and identification of the deposited crystals by their visual appearance. Thus, analysis was almost exclusively performed in the solid state.

As we will see, the knowledge of solution chemistry grew during the first half of the eighteenth century and some of the discovered reactions found use for analytical purposes. However, this knowledge was scattered in the literature, and there was no general procedure for chemical analysis in solution. At the same time, the demands for analytical methods had increased dramatically, not least due to industrialisation. As Szabadváry put it: "The first person to attempt to overcome this problem was Bergman. [—] Chemical analysis had been practised for two thousand years before Bergman, but it was he who gave it the status of a separate branch of science—Analytical Chemistry" [1].

23.1 Chemistry in Solution

This section will give an indication of how the knowledge of chemical reactions in solution grew during the earlier phases of the history of chemistry. Basilius Valentinus[1] noted that, for instance, silver is precipitated from aqueous solution by

[1]The author and age of the works of Basilius Valentinus are disputed; he was alleged to be a Benedictine monk living in the second half of the fifteenth century. Some works have been attributed to Johann Thölde (c. 1565–c. 1624)

A. Lennartson, *Carl Wilhelm Scheele and Torbern Bergman*, Perspectives on the History of Chemistry, https://doi.org/10.1007/978-3-030-49194-9_23

copper or common salt and that wood ash (potassium carbonate) precipitates vitriol (iron, or possibly copper sulphate) [2, 3]. Libavius[2] used wet methods for analysis of mineral waters in the late sixteenth century. He evaporated the water samples and investigated the shape of the crystals formed to determine their nature (alum, saltpetre, vitriol), but he also used reagents to perform qualitative analysis in solution: oak apple (gall nut) extract gave a black colour in the presence of iron, and copper sulphate gave a blue colour with ammonia [4, 5]. Robert Boyle used oak apple extract to detect iron[6] and introduced the use of sulphide solutions to precipitate lead [5, 7]. One of Boyle's goals was to find a reliable reagent for the detection of arsenic.

Friedrich Hoffmann (Chap. 17) used iron to precipitate copper from water samples, and silver nitrate to detect sodium chloride. He used vegetable extracts as acid-base indicators to detect alkali [8, 9]. Vegetable extracts, most notably violet extracts, were also used as acid-base indicators by Boyle in his analysis of mineral waters [10]. In 1697, Eberhard Gockel (1636–1703) used sulphuric acid to detect lead in wine: in the presence of lead, insoluble lead sulphate precipitated [11].

Thus, by the end of the seventeenth century, the knowledge about chemical reactions in solution was very limited. Chemistry was mainly performed by heating, melting, and distillation. During the eighteenth century, however, much new knowledge was gained. One of the important figures in this field was Marggraf. He made extensive studies on the effect of alkali carbonates on metal salt solutions [12]. He also introduced the use of phlogisticated alkali (a solution containing potassium cyanide and potassium hexacyanoferrate (II)) to detect iron. Another important contribution was that he conclusively established the difference between sodium and potassium salts [13]. Bergman was well familiar with the works of Marggraf, which were his main source of inspiration in the development of his analytical methods:

> The illustrious Margraf [sic!] had no sooner discovered the true method of decomposition, the humid and menstrual, than he endeavoured, by his own exertions to render it easy and practicable. The new road into which he struck, was beset with thorns and briars; but it is certainly the only one that leads to a knowledge of principles, both as to quality and quantity; and therefore indispensably necessary in every enquiry into composition. [14]

Another source of inspiration may have been Scheffer, who devoted a chapter to water analysis in his lectures [15], which Bergman published in 1775. Scheffer made use of several reagents to qualitatively determine the constituents of water samples; for instance, a solution of silver nitrate could be used to detect common salt, lime could be detected with sulphuric acid, copper with ammonia and iron or zinc with oak apple tincture.

[2]Andreas Libau (1540–1616), better known as Libavius. German alchemist.

23.2 Quantitative Analysis

By the time Bergman was appointed professor of chemistry, the idea of conservation of mass in chemical processes was not yet universally accepted, and much less the idea of stoichiometry, i.e. that a certain amount of a particular acid, for example, needed a certain amount of a base for neutralisation. Chemists had to use trial and error to find the correct ratios for mixing of starting materials in chemical reactions. Sometimes, they came quite close, as, for example, when Basil Valentine (Basilius Valentinus) prepared metallic antimony by reducing antimony(III) sulphide with iron in his 1604 book *Triumph Wagen Antimonii* (*Triumphant Chariot of Antimony*). He used two parts of Sb_2S_3 and one part of Fe; the true stoichiometric ratio is 2:0.98. The term stoichiometry was coined by Jeremias Benjamin Richter (1762–1807), who explained the relationships between the amounts of substances in chemical reactions in mathematical terms in his book *Anfangsgründe der Stöchymometrie* (Introduction to Stoichiometry) 1792–1794. Still, the concept was not generally accepted until about 1805 and the controversy between French chemists Joseph Proust and Claude Louis Berthollet (1748–1822).

There are rare examples of the use of stoichiometry as early as in the seventeenth century, e.g. by van Helmont, Kunckel and Homberg [16]. Bergman referred to an experiment by van Helmont where he fused a weighed amount of silica (hydrated SiO_2) with alkali to form an alkali silicate. The same amount of silica was precipitated from the silicate solution by acids [17]. Homberg determined the concentrations of mineral acids in 1699 by determining the weight of alkali carbonate needed to neutralise the acid [18]. Henry Cavendish was another early pioneer of stoichiometry. In a paper dealing with water analysis, he introduced the word equivalent in 1767, writing "[...] as much fixed alkali [sodium or potassium carbonate], as was equivalent to 48 8/10 grains of calcareous earth [CaO]" [19]. For Bergman and Scheele, it seems that the concept of stoichiometry was so natural that they never discussed the matter, while Wallerius was a fierce opponent.

Bergman was familiar with at least the work of Homberg[20] and in his notes to *The Chemical Lectures of H.T. Scheffer*, published in 1775, he presented a table of the weights of sulphuric acid, nitric acid, hydrochloric acid and carbonic acid needed to neutralise 100 parts of sodium hydroxide and 100 parts of potassium hydroxide, respectively [21]. Alkali metal hydroxides are very hygroscopic and also absorb carbon dioxide from the air, which made the experiment complicated. Bergman removed moisture by heating the hydroxides, but they would still be contaminated with carbonate. Thus, Bergman dissolved the hydroxide in the minimum amount of water in a bottle (A) and closed it with a cork. The acid was put in a smaller bottle (B), and the two bottles were weighed; the weight of the hydroxide bottle being a and the weight of the acid bottle b.[3] Small portions of the acid were added to the base until the solution was neutral, and the bottles weighed again, the weights now being *a* and *b*, respectively. For a completely pure hydroxide, *a*-a

[3]With his mathematical and physical education, Bergman expressed himself in this algebraic manner whenever possible.

would equal b-*b*, but due to the loss of carbon dioxide, this was never the case. Although the experiment was well planned, his results are inaccurate, and there may be several reasons for that. First, the water content of the acids was not known, and this is a particular problem with hydrochloric acid and carbonic acids which are aqueous solutions.

Bergman reported that 100 parts of sodium hydroxide required 177 parts of sulphuric acid, 135½ parts of nitric acid or 125 parts of hydrochloric for neutralisation. The expected values are 123 parts of sulphuric acid, 158 parts of nitric acid or 260 parts of hydrochloric acid (assumed to 35%). He reported that 100 parts of potassium hydroxide required 78½ parts of sulphuric acid, 64 parts of nitric acid or 51½ parts of hydrochloric acid for neutralisation. The expected values are 87 parts of sulphuric acid, 112 parts of nitric acid or 186 parts of hydrochloric acid (assumed to be 35%). It is not easy to trace the origins of the low accuracy of Bergman's results. Partington suggested that Bergman's experiments may have been performed by his assistant, which could explain the lack of accuracy [22]. It should be noted that this was over two decades before the famous tabulation of equivalent weights by Ernst Gottfried Fischer (1754–1831) in his 1802 translation of Berthollet's *Statique Chemique* [23]. Fischer, who based his table on the work of Richter, tabulated the mass of different bases and acids that were equivalent to 1,000 parts sulphuric acid. He gave the values 859, 1605, 712, 1405 and 577, respectively, for sodium hydroxide (natron), potassium hydroxide (kali), hydrochloric acid, nitric acid and carbonic acid, respectively. Assuming that the alkalis are the pure hydroxides and the acids are water-free (carbonic acid thus being carbon dioxide), the expected values are 816, 1144, 744, 1285 and 449, respectively. Richard Kirwan also started to study equivalent weights in the 1780s, but initially believed that a certain amount of alkali needed the same weight of different acids for neutralisation.

Bergman's table of equivalent weights is also included in his paper on carbonic acid (Sect. 18.3) published the same year, 1775. In *The Chemical Lectures of H.T. Scheffer*, Bergman also gave the quantitative compositions of many salts given as parts alkali and parts acid per 100 parts of salt. It is unfortunately not easy to estimate the accuracy of these results, as the water content of the salts, acids and alkalis Bergman used are unknown.

Several other examples of Bergman's use of the concept of constant proportions and stoichiometry in quantitative chemical analysis will follow in this chapter; his accidental determination of equivalent weights while trying to determine the relative phlogiston content of metals was discussed in Sect. 22.5.

In his paper on the isolation of oxalic acid (Sect. 24.1), Scheele also used stoichiometry in an ingenious method to determine the proper amount of sulphuric acid required to decompose lead(II) oxalate to lead sulphate and oxalic acid without adding an excess of sulphuric acid: [24]: he precipitated lead oxalate from potassium oxalate solution using lead acetate, noting the amount of lead acetate required. He then measured the amount of sulphuric acid required to quantitatively precipitate lead sulphate from this amount of lead acetate. Finally, he added the same amount of sulphuric acid to his lead oxalate to liberate the oxalic acid quantitatively.

23.3 Water Analysis

From the early years of his chemical career, Bergman had been interested in water analysis (Chap. 17). This section will merely describe the analytical methods developed by Bergman, rather than his results. In 1775, he wrote an extensive paper on mineral waters [25], more precisely bitter water from Seidschutz (old spelling: Seydschutz) in Bohemia, Selzer water (from Selters in Hessen), Spa water (from Spa in Belgium) and Pyrmont water (from Bad Pyrmont in Niedersachsen). The challenge and the importance that Bergman attributed to the subject is indicated in the opening paragraph of the paper: "To test waters, and to discover the quantity and nature of the foreign substances, which hide therein, is one of the most important, and at the same time more difficult, problems in chemistry" [25].

To determine the amount of dissolved carbon dioxide, Bergman had designed a special apparatus (Fig. 23.1) consisting of a copper can (ABCD) standing on feet. A second can (EFGH) with a slightly smaller diameter was placed inverted inside the first can. This can have a small tube through the bulkhead. The apparatus was filled with the water to be examined, and a bottle (K) completely filled with hot water was inverted over the tube. The whole apparatus was heated over fire, and the water inside the inner can release its carbon dioxide through the tube into the bottle. When the bottle was filled to ¾, it was replaced with a new bottle. Taking the ambient pressure and the temperature of the water in the bottle into account, the volume of the expelled gas could be measured.

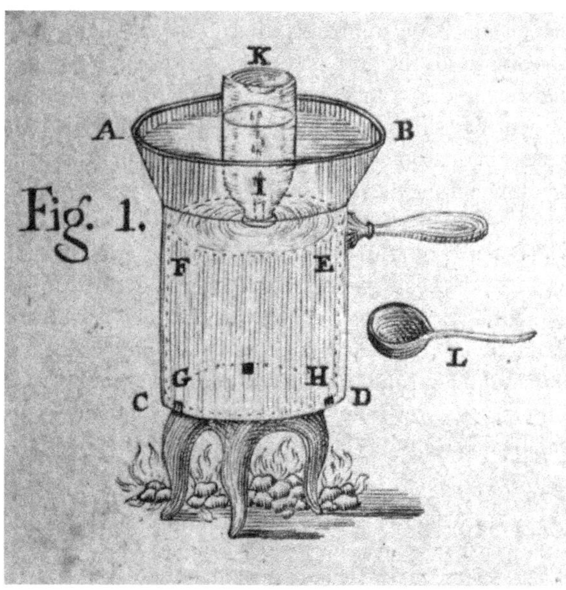

Fig. 23.1 Apparatus used by Bergman to measure the carbon dioxide content of mineral water

For analysis of the dissolved substances, the water samples were evaporated over red-hot charcoal in a soapstone kettle with a wooden lid. To illustrate the procedure, the analysis of bitter water is taken as an example here: as carbon dioxide was lost, any dissolved calcium carbonate precipitated first. This was filtered off and examined: it gave calcium sulphate (gypsum) with sulphuric acid. The evaporation of the water was continued, and the next salt to precipitate was calcium sulphate (gypsum) which was identified by its property to give calcium carbonate (lime) and potassium sulphate (*Tartarus vitriolatus*) with potassium carbonate (*Alkali vegetabile*). After filtration, the evaporation was continued, whereupon magnesium sulphate (bitter salt) crystallised. Reaction with calcium hydroxide solution (lime water) revealed its identity and excluded confusion with sodium sulphate (Glauber's salt): the presence of magnesium gave a white precipitate of magnesium hydroxide. Treatment of the crystallised salt with sulphuric acid revealed the presence of chloride (Bergman wrote hydrochloric acid, but he did not mean free acid) since liberated hydrogen chloride gives white fog in moist air.[4]

The next paper describing water analysis is a paper which appeared in the first quarter of 1777. Linnæus' student Anders Sparrman (1748–1820) had collected seawater from a depth of about 107 m (60 *famnar*). Where Sparrman collected the samples is not told, but Sparrman had participated in James Cook's second voyage 1772–1775, and after spending some time in southern Africa, he left Cape Town in April 1776 [26]. The samples were presented to Bergman in October 1776, but other duties prevented Bergman from analysing the samples until December the same year [27].

In this paper, Bergman used several chemical reagents to analyse the samples. The water did not change the colour of litmus tincture (an acid-base indicator; seawater is only mildly basic with a pH of about 8); it gave a precipitate with barium chloride solution ("solution of heavy earth in hydrochloric acid") indicating the presence of sulphate: "this is, of all known tests, the most reliable way of revealing vitriolic acid [sulphuric acid], even when its amount is so small, that it is undetectable by other means". This is the first mentioning of barium chloride as a reagent for sulphate. Oxalic acid (acid of sugar) also gave a precipitate with the water, indicating the presence of calcium (lime); alkali carbonate (*Alkali fixum*) precipitated magnesium (magnesia) and potassium hexacyanoferrate (lye of blood) revealed traces of iron.

To perform a quantitative analysis, Bergman evaporated 2.6 L (1 *kanna*) of water to dryness, yielding 113 g (8.46 *lod*) of residue. The residue was extracted with ethanol (*Spiritus vini rectificatissimus*), which dissolved magnesium chloride. Extraction with a small amount of boiling water revealed the presence of magnesium sulphate (bitter salt). The sodium chloride[5] was extracted with the minimum amount of water leaving calcium sulphate (gypsum) undissolved. Bergman obtained 86.2 g of sodium chloride, 23.5 g of magnesium chloride and 2.79 g of

[4]Bergman could have used a glass rod dipped in aqueous ammonia, which give a characteristic white smoke of ammonium chloride.

[5]Called koksalt in Swedish, meaning "cooking salt".

calcium sulphate. One would expect of 70 g NaCl, 29 g of $MgCl_2 \cdot 6H_2O$ ad 1,7 g of $CaSO_4 \cdot 2H_2O$, but it is difficult to judge Bergman's accuracy, since the conditions at which his salts were dried are not known. Also, Bergman's extraction procedure was not completely selective.

An important step in Bergman's development of analytical methods took the form of a dissertation defended by one of his students, Johan Peter Scharenberg, in June 1778 [28]. For some reason, the printing of the thesis was interrupted after the first seven sections. As it was custom to print the first word of the following page at the bottom of each page, it is clear that a §VIII was intended. The complete work appears in the first volume of *Opuscula* [29]. This publication is a complete manual for analysis of mineral waters, but the actual analytical procedures appear in the latter part that was omitted in the original thesis.

Bergman's procedure was, in short, first to separate and analyse any volatile components such as CO_2. The water was evaporated and consecutively extracted with ethanol, cold water and boiling water, finally leaving an insoluble residue. The ethanol extract mainly contained calcium and magnesium chloride and magnesium nitrate and occasionally barium chloride. This mixture was re-crystallised from water and the crystals examined with regard to shape, physical properties, taste and reactions with reagents. The cold water extract was treated the same way, while the hot water extract mainly contained calcium sulphate. Extraction of the insoluble fraction with acetic acid dissolved calcium and magnesium carbonates, which could be separated as sulphates (calcium sulphate is much less soluble than magnesium sulphate). The residue insoluble in acetic acid typically contained oxides of iron, silicon and aluminium, occasionally also barium and manganese. Iron and aluminium oxides are soluble in hydrochloric acid, leaving the silica undissolved. From the acid extract, iron could be precipitated with potassium hexacyanoferrate (II) and aluminia could be precipitated with alkali.

The second part of the publication makes extensive use of stoichiometry:

> The weight of the precipitate may often, indeed, be of considerable use even in that view, as shall presently be shewn, though it has not yet been employed for that purpose" [30].

For example, mineral waters frequently contain calcium and magnesium carbonates which are soluble in the presence of carbon dioxide. Bergman dissolved them in sulphuric acid and separated the soluble magnesium sulphate from the sparingly soluble calcium sulphate. To determine the amount of the corresponding carbonates that had originally been present in the water, the sulphates could be dissolved in water and precipitated with alkali carbonate, but "this tedious process may be avoided, if we recollect that 100 parts of gypsum [calcium sulphate dihydrate] contain about 34 [should be 33] of pure lime [calcium oxide], which are equivalent to nearly 62 [should be 58] of aerated lime [calcium carbonate]. [31]

This was the birth of gravimetric analysis and given the pioneering nature of this work, the accuracy is quite impressive.

To help the analyst, Bergman gave a description of all salts that frequently occur in natural waters, including their quantitative composition. It is noteworthy that all of Bergman's salts contain water, even salts such as sodium chloride and potassium nitrate which do not contain water of crystallisation. In parallel with Bergman, Richard Kirwan was working along similar lines determining the chemical composition of salts, and in the English translation of Bergman's manual, published in 1784 [32], the translator has added footnotes with analytical results reported by Kirwan for comparison. Bergman's way of expressing the composition of salts is ambiguous to a modern chemist. For instance, the composition of potassium sulphate is given as 52 parts of vegetable alkali, 40 parts of sulphuric acid and 8 parts of water per 100 parts of salt. In reality, potassium sulphate is not composed of potassium hydroxide and sulphuric acid, but water is eliminated when the acid and base react. Therefore, it is not absolutely clear what this actually means, or whether the values of Bergman and Kirwan are comparable. When it comes to non-hygroscopic metal salts, the values are easier to interpret, as we can compare the true metal content and the true water content with Bergman's and Kirwan's reported results. For example, copper(II) sulphate pentahydrate contains 25% copper and 36% water. Bergman reported 26% copper and 28% water, while Kirwan reported 27% copper and 43% water. Iron(II) sulphate heptahydrate contains 20% iron and 45% water. Bergman reported 23% iron and 38% water, while Kirwan reported 25% iron and 55% water. Zinc sulphate heptahydrate contains 23% zinc and 44% water. Bergman reported 20% zinc and 40% water, while Kirwan reported 20% zinc and 58% water. Thus, Bergman's values seem to be somewhat more reliable than those of Kirwan.

Bergman's methods of water analysis were employed by many chemists of the time [33] and were very influencing. However, Bergman's analyses suffered from one serious problem. He collected salts as hydrates, but since the water content in these salts often varied, so did Bergman's results. Unfortunately, Bergman's untimely death in 1784 prevented him from perfecting his methods. The work was, however, taken up by others. Bergman's methods were rationalised by Fourcroy and important contributions came from Klaproth, who heated the salts to constant weight, and isolated anhydrous salts rather than hydrates. Based on these contributions, the methods were perfected by Kirwan at the end of the century [34].

23.4 Mineral Analysis in Solution

Bergman's work on mineral water analysis led him to the idea of analysing minerals by wet methods, a very important step in the development of analytical chemistry. Bergman's first contribution to the field, a paper on the analysis of gems[35] was written in 1777, but the publication of the third volume of *Nova Acta*, including Bergman's paper, was delayed to 1780 [36, 37].

In this paper, Bergman analysed the composition of gems (emerald, sapphire, topaz, hyacinth, ruby and diamond) by a combination of dry methods (blow-pipe; Sect. 23.5) and wet methods. He found that acids (sulphuric, hydrochloric and nitric acid) extracted small amounts of calcium (lime) and traces of iron from all of the stones except diamond. As the main part of the samples was insoluble in acids, he proceeded by fusing them with sodium carbonate (fixed alkali). The method of fusing minerals with sodium carbonate is still used on large scale today in the extraction of chromium from its ores. The residue was treated with hydrochloric acid,[6] which left small amounts of silicon dioxide undissolved. Potassium hexacyanoferrate(II) (alkaline phlogisticated lixivium) was added to the solution to precipitate iron, and the earths were then precipitated with alkali carbonate (fixed alkali). The precipitate was treated with acetic acid (vinegar). In the cold, calcium, magnesium and barium carbonates were dissolved, leaving aluminium carbonate undissolved.[7] Magnesium, calcium and barium were re-precipitated with alkali carbonate and treated with dilute sulphuric acid. Barium sulphate (spatum ponderosum) is insoluble. Calcium sulphate (gypsum) is five hundred times less soluble in water than magnesium sulphate (Epsom salt) and the two are easily separated. As only minute amounts of samples could be used due to the high cost, the analysis of precious stones presented a great challenge.

Due to the delayed publication of the earth of gems paper, a paper on brown tourmalines[36] was actually published before the earth of gems paper, although written later. It describes brown tourmalines from Tyrol that Bergman recently had received and compared with tourmalines brought from Sri Lanka (Ceylon) by Linnaean apostle Carl Peter Thunberg (1743–1828). Thunberg visited Sri Lanka on the way home from his successful journey to Japan, where he compiled the first Japanese flora. After a crystallographic investigation, Bergman proceeded with a chemical analysis. He started with a traditional blow-pipe analysis and proceeded by noting that the stones were hardly soluble in nitric or hydrochloric acid. He then gave an account of the method described in the earth of gem paper, fusing the stones with sodium carbonate (alkali sodæ) and heating the residue with hydrochloric acid to bring the components into solution. He concluded that the tourmalines consisted of clay, silica, lime and iron, the reported quantitative analysis from the two kinds of tourmalines being quite similar. As tourmalines have very complex compositions, the accuracy of Bergman's analyses is hard to estimate. It should be recalled from Sect. 5.2 that Bergman's friend Rinman had examined tourmalines a few years earlier. Rinman's method was to observe the changes of the samples when fused using a blow-pipe, both alone and with different fluxes (e.g. borax, calcium fluoride and calcium carbonate). He observed that tourmalines did not dissolve in boiling sulphuric acid (*Oleo vitrioli*), nitric acid (*Aqvafort*) or hydrochloric acid (*Spiritu Sali*), but that they dissolved in nitric acid after fusion with borax leaving a gelatinous residue [SiO_2] which he did not

[6]The mixture must be heated to assure precipitation of silica. When I applied Bergman's method to a sample of feldspar, addition of hydrochloric acid at ambient temperature gave no precipitate, but the following day the whole solution had formed a gel.

[7]Indeed, it is found that the precipitate formed by mixing 0.1 M solutions of calcium nitrate and sodium carbonate is instantaneously dissolved upon addition of acetic acid at 0 °C, while the corresponding precipitate obtained from aluminium nitrate and sodium carbonate takes minutes to dissolve. The method is not perfect, as aluminium carbonate is not completely insoluble, but it is still an evidence of Bergman's ingenious ways of solving chemical problems. To find this method either required a deep chemical knowledge or a large number of trial-and-error experiments. If Bergman had heated the carbonate precipitate before treatment with acetic acid, virtually insoluble aluminium oxide would have formed.

examine. Rinman reported these tests without making any conclusions about the constituents of tourmalines. If we compare Bergman's methods with those of Rinman's, the magnitude of Bergman's contribution to mineral analysis becomes evident.

Bergman had now reached the point where a handbook in mineral analysis by wet methods could be written. This work, essentially equivalent to the thesis on analysis of waters defended by Scharenberg, was also presented as a thesis, defended publicly in June 1780 by Peter Castorin [38].

It was natural, Bergman concluded, that the first attempts to analyse ores were to simply carry out the reduction on smaller scale and determine how much metal a certain ore yielded. Traditional assaying by reduction with charcoal had its limitations, however, and Bergman listed three requirements for a successful outcome: the reduction must be quantitative, the metal must assemble in a single lump rather than small grains, and once reduced, the metal must be chemically inert to air. Bergman complained that the process was notoriously difficult to control. There was no way of measuring high temperatures, and as the reduction had to be performed in a closed crucible, the process could not be monitored. Estimation of losses was difficult. There were no methods to analyse minerals by using solution methods exclusively: "Chemistry has at length begun to examine the composition of ores by means of various menstrua; yet it must be confessed, that the fragments of the humid art of assaying, which have hitherto been published, are rather to be considered as instances of a mixed method, in which the mineral analysis is accomplished partly by the dry, partly by the humid method. The metallic part is indeed extracted by a menstruum, but is afterwards reduced by fire. In the following pages, however, I shall endeavour to point out means by which the end may be answered in the humid way alone, with out calcination or fusion" [39].

The following section of the thesis, which must be the result of a tremendous amount of laboratory work, is essentially a manual for the analysis of ores of gold, platinum, silver, mercury, lead, copper, iron, tin, bismuth, nickel, arsenic, cobalt, zinc, antimony and manganese. His methods will be illustrated by taking the silver ores as an example.

Native silver frequently contained gold and/or copper. To determine the composition of native silver, it was treated with nitric acid, which dissolved silver and copper, leaving solid gold. Copper was precipitated from solution with iron or alkali carbonate, but instructions for the isolation of silver are missing. One can assume that it was performed with metallic copper or sodium chloride. The next class of minerals were the silver sulphide ores ("silver mineralised by sulphur") which were boiled with dilute nitric acid, which left sulphur as a precipitate, mixed with any traces of gold. The loss of sulphur as H_2S was probably severe. Due to the difference in density, sulphur and gold were easily separated. Sodium chloride was added to the solution, which precipitated silver chloride. Bergman used stoichiometry to calculate the amount of silver, rather than reducing the silver chloride to metal and weighing it. The silver content was calculated from $100a/129$, where a is the weight of the silver chloride. This corresponds to a silver content of 77.5% in silver chloride, the true value being 75.3%. The sum of the silver and sulphur should sum up to give the original weight of the sample. The solution remaining after filtering off the silver chloride was treated with potassium hexacyanoferrate (phlogisticated alkali) to precipitate any traces of other metals. By finally adding alkali carbonate, any earths could be precipitated.

Silver ores may also be composed of silver, sulphur and arsenic. In this case, boiling with dilute nitric acid gave a white powder and a solution. As before, silver was precipitated from the solution with sodium chloride. The white powder was treated with *Aqua regia* to dissolve the arsenic (III) oxide, which could be re-precipitated with water (as As_2O_3). It is likely that the use of *Aqua regia* gave losses of arsenic, as a portion of arsenic was most probably oxidised to soluble arsenic acid. The sulphur remaining could contain traces of silver chloride, which was extracted from the residue with aqueous ammonia.

The next type of ores contained silver, sulphur, copper and arsenic. Boiling with dilute nitric acid gave a white powder and a solution containing copper and silver. Bergman noted that silver and copper could not be efficiently separated by sodium chloride, since the precipitate contained copper. Instead, silver was precipitated by addition of a known amount of copper. Copper could finally be precipitated with iron or alkali carbonate, taking care to subtract the amount of copper used to precipitate silver. The white powder obtained in the first step was treated with hydrochloric acid to dissolve the arsenic (this was probably safer than using *Aqua regia*, as stated above). Arsenic(III) oxide was re-precipitated with water. The sulphur was treated with aqueous ammonia to detect traces of silver chloride or copper salts (which give a deep blue colour with ammonia). If the silver ore contained antimony, the latter could be separated in the same way as arsenic.

The final type of ores was silver combined with hydrochloric and sulphuric acid. These ores were powdered and heated for a day with hydrochloric acid. Sulphate was precipitated from the solution with barium nitrate. The amount of sulphate could be calculated from the weight of the barium sulphate precipitate, but in this case Bergman's determination of the composition of barium sulphate was very far from the true value: Bergman reports that it contains 15% acid, while the true value is 41%. If Bergman failed to determine the composition of barium sulphate, he was more successful in other case, as can be seen from Table 23.1.

A mistake that Bergman made, and which was not so easily realised at the time, was that he precipitated several metals as carbonates (copper, tin, nickel and cobalt), but these metals may form hydroxycarbonates of varying compositions, and were therefore not suitable for accurate quantitative analysis. In these cases, it is not easy to judge Bergman's accuracy. Another problem faced when attempting to reproduce Bergman's experiments is the fact that any access to carbonate raises the pH, causing an increase in the solubility of copper due to complexation. The precipitation of iron as Prussian blue is also difficult to perform; due to the small particle size, the precipitate is often very difficult to collect quantitatively by filtration. Although latter authors managed to get more accurate results, Bergman's great authority caused his values to be trusted for a long time [40].

Table 23.1 Composition of metal content in simple salts reported by Bergman. Bergman did not state the content in per cent, but in the form 100*a*/*b*, where *a* being the weight of the precipitate and *b* a coefficient determined by Bergman. Compounds that have varying compositions, where the true value cannot easily be predicted, have been excluded

Salt	Metal content calculated from Bergman's results (%)	True value of metal content (%)
$AgCl$	77.5	75.3
Ag_2SO_4	68.75	69.1
$HgCl_2$	75.5	73.9
$MnCO_3$	55.6	47.8
$PbSO_4$	70.0	68.3
$ZnCO_3$	51.8	52.1

Bergman's analytical methods relied heavily on the accurate knowledge of the composition of the insoluble salts that he precipitated from the analyte. This subject was further explored in an essay called *De praecipitatis metallicis* (*On Metallic Precipitates*) which appeared in the second volume of Bergman's collected works published in 1780 [41]. This essay contains long discussions both on the dissolution of metals in acids and on precipitation phenomena. It ends with a section where Bergman discussed the advantages and disadvantages of analysis by wet methods. It contains a long table where Bergman has determined the composition of a large number of metallic precipitates, mainly those obtained by adding sodium carbonate, sodium hydroxide or potassium hexacyanoferrate (phlogisticated alkali) to solutions of metal salts. Bergman's method was to dissolve 100 parts of metal in acid and weigh the precipitate obtained by adding sodium carbonate to the solution. As indicated above, this is somewhat problematic. The precipitates with phlogisticated alkali are also problematic, since the phlogisticated alkali was probably not a pure substance, but a mixture containing potassium hexayanoferrate (II). The cases where the chemical nature of the precipitate can be predicted with certainty and thus compared to the true values are collected in Table 23.2.

Bergman also discussed the factors affecting the weights of the precipitates. As an example, he took lead. When 100 parts of lead were dissolved in nitric acid and precipitated with sodium carbonate, he obtained 132 parts of precipitate. Assuming the precipitate to be pure $PbCO_3$, one would expect to get 129 parts of $PbCO_3$ from 100 parts of lead.[8] Bergman added the precipitate to a weighed amount of nitric acid, where the precipitate dissolved with effervescence. He noted a weight decrease of 21 parts, which corresponds closely to the expected amount of carbon dioxide, 21.2. Bergman noted that the weight loss was less than would have been expected if the precipitate consisted of only metal and carbon dioxide (aerial acid), since 132–21 = 111. He also noted that calcination of 132 parts of precipitate left 110 parts; the theoretical yield of PbO being 108. Bergman's conclusion was that

[8]Precipitation of $Pb_3(OH)_2(CO_3)_2$ would have yielded 124 parts of precipitate from 100 parts of lead.

Table 23.2 Composition of metallic precipitates reported by Bergman. The majority of the results reported in Bergman's table have been excluded, as a comparison with true values cannot be made easily

Salt	Weight of precipitate obtained from 100 parts of metal reported by Bergman	Metal content calculated from Bergman's values (%)	True value of metal content (%)
AgCl	133	75.2	75.3
HgO	104	96.2	92.6
$Pb_2O(OH)_2$	116	86.2	89.3
$PbSO_4$	143	69.9	68.3
Approx. $Bi(OH)_3$	125	80.0	80.4
$Co(OH)_2$	140	71.4	63.4
$Zn(OH)_2$	161	62.1	65.8
$Mn(OH)_2$	163	61.3	61.8

the weight difference was due to the matter of heat. This was further supported by the fact that the precipitate obtained by adding sodium hydroxide to lead nitrate weighed more than the lead. As sodium hydroxide contained no aerial acid, only heat could be transferred to the lead. Another support for this theory came from another experiment: Bergman added alkali to acid and noted the temperature increase. If he saturated the same amount of acid with metal, and then added the alkali, he observed only a small or no temperature increase. Thus, a part of the heat must have been fixed by the metal. This is interesting, since Bergman had accepted Scheele's theory that heat was composed of oxygen and phlogiston. Although Bergman does not explicitly state it, it implied that a metal calx was composed of metal and heat, and thus of metal, phlogiston and oxygen. Lavoisier followed up on Bergman's research discussing the results based on his antiphlogistic views [42].

In two dissertations published in 1782, Bergman addressed the question of separating different kinds of earths. From his work on alum (Sect. 9.2), Bergman was aware that solutions of aluminium salts were acidic or, as Bergman put it, needed an excess of acid to be soluble. This is true; in neural solution, insoluble hydroxo salts precipitate. To analyse lithomerge [43], a kaolinite mineral, Bergman dissolved the powdered sample in hot sulphuric acid to obtain a solution of aluminium, iron, manganese and calcium sulphates, leaving silica undissolved. The solution was divided into two parts. The first part of the solution was carefully neutralised with calcium carbonate in order to selectively precipitate the more acidic trivalent Al^{3+} and Fe^{3+} ions, leaving Ca^{2+} and Mg^{2+} in solution. This was an elegant method requiring a deeper understanding of solution chemistry. As magnesium sulphate is much more soluble than calcium sulphate, the two salts were easily separated. To obtain the amount of calcium in the sample, the amount of calcium carbonate added for neutralisation had to be subtracted.

From the second part of the solution, iron was precipitated as Prussian blue with potassium hexacyanoferrate and aluminium precipitated with magnesium carbonate. A similar protocol was used for the analysis of asbestos, although asbestos is insoluble in acids and thus had to be fused with alkali to become soluble [44].

23.5 Bergman and the Blow-Pipe

Since centuries, metal workers had used a bent metal tube—a so-called blow-pipe—to achieve high temperatures for welding and brazing. The principle was simple: a stream of air was blown from the mouth through the tube into the flame of a lamp or a candle. The extra oxygen increased the temperature. The use of blow-pipes was adopted by glass-blowers and in a book on glass blowing, *Ars vitraria*, Kunckel introduced the idea of using the blow-pipe for chemical research in 1689: "There are even some other possibilities in this. Especially in the workshop of a chemist, it can be used for many things. I will just mention this among other things: It often happens that one has a small amount of metallic calx, or something similar, and wants to melt it and see what metal it contains. It can in no way be made easier than this, that one just takes a piece of charcoal, makes a cavity in it and takes the calx, or whatever one wants to melt, and puts it inside, and with the tube blows the flame of a strong candle against it" [45]. It appears to have been Stahl who first put this into practice and he was soon followed by Johann Cramer (1710–1770), Marggraf, Pott and von Swab [46]. According to Bergman, it was Swab who first used the blow-pipe for analysis of minerals in *c.* 1738 [47]. It is likely that Swab taught the method to Cronstedt, who frequently referred to it in his classical *Försök til Mineralogie* (An attempt to a mineralogy) in 1758, but without explaining the procedure. When von Engeström translated Cronstedt's book to English in 1765, he added an essay describing the use of the blow-pipe. Due to the influence from Swab and Cronstedt, the blow-pipe became very popular in Sweden and was used by Rinman and Qvist; although certainly not unknown, the blow-pipe found considerably less use on the continent. In the hands of chemists, the blow-pipe developed from a simple bent tube to include a spherical compartment to collect moister that otherwise condensed in the tube (Fig. 23.2). Bergman improved the design of the blow-pipe even further; he used a blow-pipe made of silver alloyed with a small amount of platinum. Rather than a spherical bulb to collect moister, Bergman's blow-pipe used a small box (Fig. 23.3).

One can suspect that it was Rinman or possibly Swab who taught Bergman how to use the blow-pipe. Bergman made extensive use of the blow-pipe and wrote a manual, *De tubo feruminatorio*, which he sent to von Born in Vienna in 1777, where it was published in 1779. According to Thomson, who had met Gahn, the experiments were performed by Gahn on Bergman's request [48]. This is not improbable, as Gahn was known to be a master of the blow-pipe but very reluctant

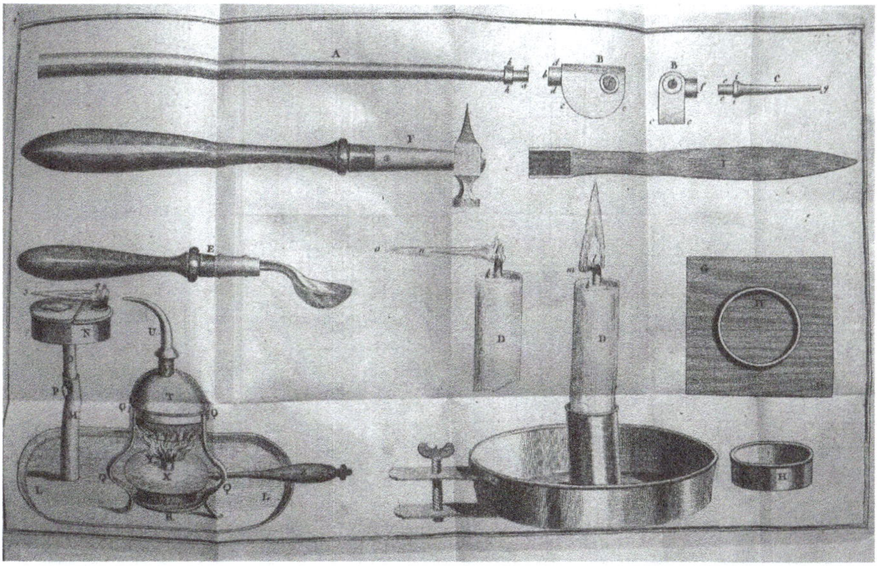

Fig. 23.2 Copperplate from Bergman's essay depicting equipment for blow-pipe analysis

Fig. 23.3 Bergman's original blow-pipe at Museum Gustavianum, Uppsala. *Photo* Anders Lennartson

Fig. 23.4 Original box for Bergman's equipment for blow-pipe analysis. Museum Gustavianum, Uppsala. *Photo* Anders Lennartson

to publish his results. A German translation by von Born appeared the same year, and a revised version appeared in the second volume of Bergman's *Opuscula* the following year, followed by translations to French, English and Italian. Bergman's student Hjelm translated the *Opuscula* version to Swedish and published it with additional comments by Bergman in 1781. According to Szabadváry, Bergman in his book on the blow-pipe became the first to clearly distinguish between quantitative and qualitative analyses [49].

To use the blow-pipe, Bergman assured, requires much practice, since a constant flow of air is required. Thus, the analyst needs to breathe through his nose while using the cheeks to provide the air pressure for the blowing.[9] The sample is placed on a piece of fir or birch charcoal (for reduction experiments) or in a silver (or even better golden) spoon for melting with fluxes. The fluxes Bergman used where ammonium sodium phosphate (microcosmic salt), sodium tetraborate (borax) and sodium carbonate (mineral alkali). The sample was powdered with a hammer (Figs. 23.3 and 23.4) on a steel plate and added to a small amount of flux. The sample was heated first with the outer part of the flame, then by the inner, blue, part. The analyst then observed any change in the sample. The remainder of the book describes in detail how minerals are analysed and how different metals are detected. As an example, the analysis of manganese is given here:

[9]This is, at least initially, very difficult.

> The regulus of manganese [manganese metal] scarcely yields to the flame; for a small particle is easily calcined, and a large one cannot be made sufficiently hot [to melt]. The black calx [MnO_2] imparts a bluish red colour to the fluxes; the tinge of borax, unless well saturated, is more yellow. The colour may be gradually altogether destroyed by the interior [reducing] flame, and again reproduced by a small particle of nitre, or the exterior [oxidising] flame alone; these changes may be alternated ad libitum.

As mentioned, Gahn made blow-pipe analysis his speciality. He developed such a skill that he could detect traces of metals far below the threshold of contemporary wet methods. Early in the nineteenth century, Berzelius befriended Gahn [50], who taught him to use the blow-pipe. Gahn also wrote the section on blow-pipe analysis for the second volume of Berzelius textbook [51, 52]. Later, Berzelius wrote a 300-page monograph on blow-pipe analysis [53], which was published in several languages, the most widespread being the German editions. Towards the end of the nineteenth century, the blow-pipe was largely replaced by emission spectroscopy, which was easier to use and relied to a lesser extent on the skill of the analyst.

23.6 Analytical Chemistry After Bergman

Unfortunately, Bergman died in the middle of his career, and the perfection of analytical chemistry was up to other scientists. The most famous of the next generation of analytical chemists were Kirwan, Klaproth and Vauquelin. Kirwan's main contribution was to simplify Bergman's methods for water analysis and at the same time increase the accuracy [54].

Louis Nicolas Vauquelin (1763–1829) was nearly 30 years younger than Bergman and as an assistant of Fourcroy, he belonged to the new antiphlogistic era. Vauquelin took up Bergman's work on mineral analysis by wet methods, and compared to Bergman, Vauquelin's analytical scheme was much more elaborate [55]. This increased the accuracy and enabled Vauquelin to discover new elements: beryllium and chromium.

The most prominent of the analytical chemists of the late eighteenth century was, however, Klaproth in Berlin. He was only 1 year younger than Scheele, but since his chemical career did not actually start until the 1780s, he had no background in the phlogiston theory. Klaproth's main interest was mineral analysis, while he was largely uninterested in the theoretical aspects of chemistry. He made important contributions to the procedure of analytical chemistry, but his methods are less easy to follow than those of Bergman; in fact he developed a specialised analytical scheme for each mineral [55]. Klaproth introduced the practice to dry samples to constant weight, and he reported the actual weights of his samples, and if his results did not sum up to 100%, he always took great care to find the reason. This led to the discovery that potassium can occur in minerals, and to the discovery of uranium, zirconium and (independently of Berzelius and Hisinger) of cerium. Thus, his results were much more accurate than those of Bergman. The analytical schemes developed by Bergman and his successors are still used, with modifications to the present day. It is also the basis for preparative separation of elements, for instance, the methods used by Marie Curie in her extraction of radium [56].

It is not unfair to call Bergman the father of analytical chemistry, as he summarised and organised the scattered information in the field and made several important innovations. He introduced pure solution methods for chemical analysis and, most importantly, he introduced gravimetry. Rather than reducing a silver ore to silver, the silver could be precipitated as silver chloride and the silver content calculated from the weight of the chloride. In many ways, Bergman's analytical works were imperfect, for instance, he never reported actual weights but adjusted all results to give a sum of 100%. Although unknown at the time, several of the substances used in Bergman's quantitative work had varying compositions or had varying water content, thus affecting the accuracy of Bergman's work.

Not only did the progress in analytical chemistry at the end of the eighteenth-century lead to knowledge about the composition of minerals and the discovery of new elements, it also paved the way for the development of stoichiometry, the atomic theory, and the determination of atomic weights in the next century. None of this work would have been possible if not someone had shown the way.

References

1. Szabadváry F (1966) History of analytical chemistry. Pergamon Press, Oxford, p 71
2. Valentin B (1712) Letztes Testament darinnen die Geheime Bücher vom grossen Stein der uralten Weisen, und anderen verborgenen Geheimnüssen der Natur… Strasbourg, p 244
3. Szabadváry F (1966) History of analytical chemistry. Pergamon Press, Oxford, p 28
4. Libavius A (1606) Commentariorum alchemiae, vol 2, p 139–192
5. Szabadváry F (1966) History of analytical chemistry. Pergamon Press, Oxford, p 30
6. Boyle R (1725) The philosophical works of the honourable Robert Boyle, J & W Innys, J. Osborne and T. Longman, London, p 508
7. Boyle R (1725) The philosophical works of the honourable Robert Boyle, J & W Innys, J. Osborne and T. Longman, London, p 506
8. Hoffmann F (1703) Methodus examinandi aquas salubres, Halle
9. Szabadváry F (1966) History of analytical chemistry. Pergamon Press, Oxford, p 31
10. Boyle R (1684) Short memoires for the natural experimental history of mineral waters, Samuel Smith, London, p 81
11. Szabadváry F (1966) History of analytical chemistry. Pergamon Press, Oxford, p 34
12. Marg(g)raf S (1762) Opuscules chymiques de M. Margraf, vol 1, Paris, p 72–85
13. Marg(g)raf S (1762) Opuscules chymiques de M. Margraf, vol 2, Paris, p 132–192
14. Bergman T (1791) Physical and chemical essays, vol III, G, Mudie, J & J Fairbairn and J. Evans, Edinburgh, p 225
15. Bergman T (1775) H. T. Scheffers Chemiske föreläsningar, rörande salter, jordarter, vatten, fetmor, metaller och färgning, samlade, i ordning stälde och med anmärkningar utgifne af T. B., M. Swederus, Uppsala, p 196
16. Partington JR (1957) A short history of chemistry. MacMillan & Co, London, p 153
17. Bergman T (1788) Physical and chemical essays, vol II, J. Murray, London, p 29
18. Homberg W (1699) Observation sur le quantité exacte des Sels Volatils Acides contenus dans les differens Esprits Acides, Histoire de l'Academie Royale des Sciences 1:44–51
19. Cavendish H (1767) Experiments on Rathone-place water, Phil Trans 57:92–108

20. Bergman T (1775) H. T. Scheffers Chemiske föreläsningar, rörande salter, jordarter, vatten, fetmor, metaller och färgning, samlade, i ordning stälde och med anmärkningar utgifne af T. B., M. Swederus, Uppsala, p 66
21. Bergman T (1775) H. T. Scheffers Chemiske föreläsningar, rörande salter, jordarter, vatten, fetmor, metaller och färgning, samlade, i ordning stälde och med anmärkningar utgifne af T. B., M. Swederus, Uppsala, p 67
22. Partington JR (1962) A history of chemistry, vol 3. Macmillan, London, p 187
23. Berthollet CL (1802) Über die Gesetze der Verwandtschaft in der Chemie, Berlin, p 232
24. Lennartson A (2017) The chemical works of carl wilhelm scheele. Springer, Cham, p 84
25. Bergman T (1775) Afhandling om Bitter- Selzer- Spa- och Pyrmonter-vatten, samt deras tilredande genom konst. KVA Handl, 36:8–43
26. Nyberg K (2007–2201) Svenskt biografiskt lexikon, vol 33, Riksarkivet, Stockholm, p 3
27. Bergman (1777) Hafs-vatten från ansenligt djup KVA Handl 38:25–29
28. Bergman T (1778) Dissertatio chemica de analysi aquarum frigidarum resp. Joh. Petr. Scharenberg, Uppsala
29. Bergman T (1779) Opuscula Pysica et Chemica, vol. 1, Leipzig, p 65–142
30. Bergman T (1784) Physical and chemical essays, vol. 1, J Murray, London, p 116
31. Bergman T (1784) Physical and chemical essays, vol. 1, J Murray, London, p 162
32. Bergman T (1784) Physical and chemical essays, vol. 1, J Murray, London, p 91–192
33. Coley NG (1990) Physicians, chemists and the analysis of mineral waters: "the most difficult part of chemistry" In: Porter (ed): The medical history of waters and spas Medical history, Supplement No. 10 p 56–66
34. Kirwan R (1799) An essay on the analysis of mineral waters. London
35. Bergman T (1780) Disquisitio chemica de terra gemmarum, Nova Acta Regiæ Societatis Scientiarum Upsaliensis, 3:137–170
36. Bergman T (1779) Bruna turmaliner til sina grundämnen undersökte KVA Handl. 40:224–238
37. Moström B (1957) Torbern bergman a bibliography of his works. Almqvist & Wiksell, Stockholm, p 57
38. Bergman T (1780) Dissertatio metallurgica de minerarum docimasia humida resp. Petrus Castorin, Uppsala
39. Bergman T (1784) Physical and chemical essays, vol. 2, J Murray, London, p 410
40. Kopp H (1843) Geschichte der Chemie, vol 1. Braunschweig, Friedrich Vieweg und Sohn, p 247
41. Bergman T (1780) Opuscula Pysica et Chemica, vol. 2, Uppsala, p 349–398
42. Lavoisier A (1782) Sur la precipitation des substances métalliques, les unes par les autres, Histoire de l'Académie royale des sciences, 512–529
43. Bergman T, Hjerta CD (1782) Dissertatio chemica de analysi lithomargæ, Uppsala
44. Bergman T, Robsahm CG (1782) Dissertatio chemica de terra asbestina, Uppsala
45. Kunckel J (1689) Ars Vitraria Experimentalis oder Vollkommene Glasmacher-Kunst, Christoph Riegel, Frankfurt and Leipzig, p 399–400
46. Szabadváry F (1966) History of analytical chemistry. Pergamon Press, Oxford, p 51
47. Bergman T (1781) Afhandling om blåsröret samt dess bruk och nytta wid kroppars undersökning, i synnerhet de minneraliske. Stockholm. p 5
48. Thomson T (1831) The history of chemistry, vol II. Henry Colburn & Richard Bentley, London, p 47
49. Szabadváry F (1966) History of analytical chemistry. Pergamon Press, Oxford, p 54
50. Trofast J (1996) Johan Gottlieb Gahn—en bortglömd storhet. Lund. p 122–133
51. Berzelius JJ (1812) Lärbok i kemien, vol 2, p 483–494
52. Trofast J (1996) Johan Gottlieb Gahn—en bortglömd storhet. Lund. p 105
53. Berzelius J (1820) Om blåsrörets användande i kemien och mineralogien. Henr, A Nordström, Stockholm

54. Szabadváry F (1966) History of analytical chemistry. Pergamon Press, Oxford, pp 115–117
55. Oldroyd DR (1973) Some eighteenth century methods for the chemical analysis of minerals. J Chem Educ 50:337–340
56. Sklodowska Curie M (1904) Radio-active substance, 2nd edn. Chemical News Office, London, pp 19–27

Scheele's Contribution to Organic Chemistry

24

By the time Scheele started his career, very few organic compounds were known in the pure state, and from this assembly of scattered compounds (*e.g.* sugar, ethanol, acetic acid and indigo) it was not possible to develop a field of organic chemistry. As described in Sect. 10.1, Scheele developed a method to isolate tartaric acid by precipitating the insoluble calcium salt, which was treated with sulphuric acid to give sparingly soluble calcium sulphate, which could be filtered off, leaving tartaric acid in the filtrate. Following the isolation of tartaric acid, Scheele continued to isolate a number of organic acids and other organic substances. This greatly enlarged the number of organic compounds known, and other chemists would soon follow in Scheele's steps. Related to Scheele's work on organic chemistry was a study on the preservation of vinegar (dilute acetic acid), where Scheele found that heating the acid and keeping it in airtight bottles prevented degradation [1], an early case of pasteurisation [2].

24.1 Oxalic Acid

After Scheele successfully isolated tartaric acid, probably in Malmö or possibly in Stockholm, he continued to study other sour vegetable substances, but his success was initially poor. First, he appears to have turned his attention to a salt known as *Sal acetosellæ* (potassium hydrogen oxalate), a sour-tasting salt extracted from wood sorrel (*Oxalis acetosella*). In the 1760s, Marggraf had showed that it contained alkali, and small amounts of acid (oxalic acid, Figs. 24.1 and 24.2) were obtained independently in 1733 by Savary, and by Johan Christian Wiegleb (1732–1800) in 1774, through pyrolysis of *Sal acetosellæ*. Scheele's attempts to isolate the acid by the same method as tartaric acid, i.e. via its calcium salt, failed, however. Scheele eventually found that the acid, *Acid acetosellæ,* had a stronger affinity than sulphuric acid for calcium (lime), i.e. sulphuric acid would not liberate *Acid acetosellæ* from its calcium salt. This is correct in some sense: calcium forms a strong network structure with

A. Lennartson, *Carl Wilhelm Scheele and Torbern Bergman*, Perspectives on the History of Chemistry, https://doi.org/10.1007/978-3-030-49194-9_24

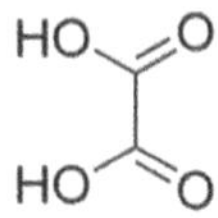

Fig. 24.1 Oxalic acid

Fig. 24.2 Crystals of oxalic acid. *Photo* Petra Rönnholm

oxalate ions, and calcium oxalate is much less soluble in water than calcium sulphate (0.00067 and 0.24 per 100 ml of water, respectively). The breakthrough came around 1784, when Scheele used the lead salt rather than the calcium salt, and as lead(II) sulphate is virtually insoluble in water, he was able to isolate *Acid acetosellæ* (oxalic acid; $C_2H_2O_4$) by the action of sulphuric acid on lead oxalate [3]:

$$KHC_2O_4(aq) + Pb(OOCCH_3)_2(aq) \rightarrow PbC_2O_4(s) + KOOCCH_3(aq) + CH_3COOH(aq)$$
$$PbC_2O_4(s) + H_2SO_4(aq) \rightarrow H_2C_2O_4(aq) + PbSO_4(s)$$

Using the same method, he managed to isolate the acid from earth of rhubarb (calcium oxalate), an insoluble earthy material found in the roots of medical rhubarb, which was used as a laxative and imported in large quantities from China. Around 1771, Scheele had noted that earth of rhubarb behaved similarly to calcium citrate on heating, and he therefore initially assumed that earth of rhubarb contained citric acid. Scheele now found, probably to his surprise, that the two acids from wood sorrel and rhubarb were identical.

Scheele also discovered that *Acid acetosellæ* was identical to acid of sugar, a substance he had obtained in 1772 by treating sugar with nitric acid (sugar is oxidised by nitric acid to oxalic acid). Scheele did not publish his discovery of acid of sugar, which was instead included in a thesis of one of Bergman's students (and eventually his successor as professor in chemistry in Uppsala) Johan Afzelius in 1776 [4]. It has been disputed whether Scheele agreed on this publication, and of the experiments in the thesis, it is not known which were performed by Scheele, Bergman, or Afzelius. Typically, when Scheele had discovered a new acid, he would try to prepare as many salts of the new acid as possible, and it is hard to imagine that acid of sugar would have been an exception. In the thesis, attempts to prepare salts with 22 metals and bases are described, and for instance, both the green water-soluble iron(II) oxalate and the yellow sparingly soluble iron(III) oxalate are described. In a letter to Wilcke written in December 1784, i.e. after Bergman's death, Scheele expressed his dissatisfaction with the omission of his name in Afzelius' thesis, and Thomas Thomson describes this omission as "one of the most remarkable facts in the history of chemistry" [5]. When Thomson met Gahn in 1812, Gahn assured him that Scheele was the true discoverer of acid of sugar, and that the omission of his name in the thesis was due to inadvertence on Bergman's side. This sounds probable, as deliberate omission of Scheele's name seems to contradict Bergman's nature. Since both sugar and nitric acid were expensive in the eighteenth century, acid of sugar was described in the thesis as the most expensive salt ever prepared. Unknown to Scheele and his contemporaries, oxalic acid had been described several decades earlier by Caspar Neumann, who had obtained sour crystals, *Spiritu* [sic!] *nitri dulcis* by the action of nitric acid on ethanol [6].

Scheele reported the isolation of oxalic acid from wood sorrel and earth of rhubarb, and the acid's identity as acid of sugar in 1784 [7]. By that time, October 1784, Bergman was dead, and the secretary of the Royal Swedish Academy of Sciences in Stockholm published the paper without reading it in the Academy, as no one would have any objections anyway. This means that Wilcke and the Academy now regarded Scheele as the undisputed authority on chemistry in Sweden, putting him before von Engeström and Bergman's successor Afzelius. The following year, Scheele published an investigation of the presence of calcium oxalate (earth of rhubarb) in different medical roots and barks [8]. As Scheele had examined 71 types of roots and 20 types of bark, there is a lot of work behind this short paper, which is not much more than a table.

24.2 Uric Acid

After Retzius' publication of the isolation of tartaric acid in 1770, it would take 6 years until Scheele published his next contribution to organic chemistry. Scheele's friend Bergius was studying bladder stones [9], hard lumps that form in the bladder, a painful condition that was often lethal in the eighteenth century. It was probably Bergius' hope that knowledge of the chemical composition of bladder stones could lead to a cure. Thus, he asked Scheele for help. Scheele found that bladder stones did not contain calcium (lime) but that they were composed mainly of a new acid, now known as uric acid (Fig. 24.3) [10].[1] This acid was only sparingly soluble in water and acids but dissolved in aqueous alkali. On investigating the acid, he also prepared a series of metal salts [11]. Treatment with nitric acid gave a substance

[1]Urinsyra in Swedish, literally meaning urine acid. It should be noted that the term urinsyra was used by Scheele for phosphoric acid, which traditionally was prepared from urine. Scheele called uric acid "bladder stone acid".

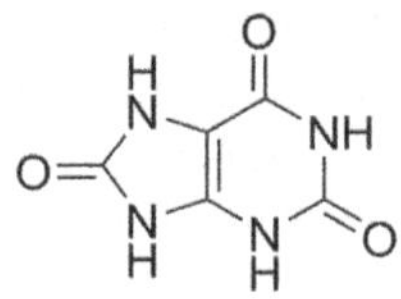

Fig. 24.3 Uric acid

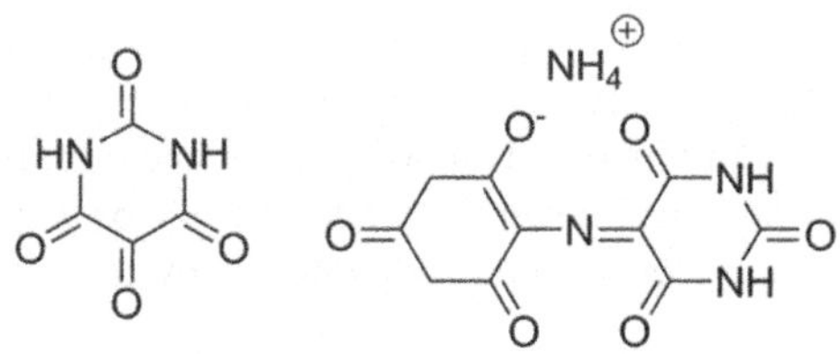

Fig. 24.4 Alloxan (left) and murexide (right)

(now known as alloxan) which gave bright purple colour to the skin. The purple colour is due to the formation of murexide (Fig. 24.4).

Bergman had also worked on bladder stones unaware that Scheele worked on the same topic. He published an addition to Scheele's paper where he presented his own results [12]. Bergman came to largely the same conclusions as Scheele but found traces of calcium (lime): while a solution of bladder stones gave no precipitate with oxalic acid (acid of sugar), he found that the minute amount of ashes remaining after burning the stones were white and composed of calcium carbonate. The lime did only constitute about half a per cent of the original stones, however. Bergman had also obtained alloxan and purple murexide. It may be noted that Scheele and Bergman believed that colours originated from phlogiston, and they should perhaps have been surprised that oxidation of (removal of phlogiston from) colourless uric acid gave rise to a purple colour.

24.3 The Investigation of Milk: Lactic and Mucic Acid

During the winter of 1779–1780, Scheele investigated milk and was able to extract lactic acid (Fig. 24.5) by evaporating whey from sour milk and extract the residue with ethanol [13, 14]. This acid was heat sensitive and rather difficult to handle. Lavoisier concluded that "The only accurate knowledge we have of this acid is from the works of Mr. Scheele" [15]. Other scientists confused the new acid with acetic acid, and Scheele's discovery was thus not immediately accepted. Many years later, Berzelius found that lactic acid also occurred in muscles [16]. The difference between the two acids was discovered later in the nineteenth century (the acid from

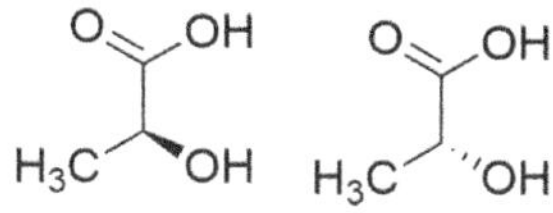

Fig. 24.5 (*S*)- and (*R*)-Lactic acid, respectively. The acid isolated by Scheele was a mixture of the two forms, while Berzelius' acid was the (*S*)-form

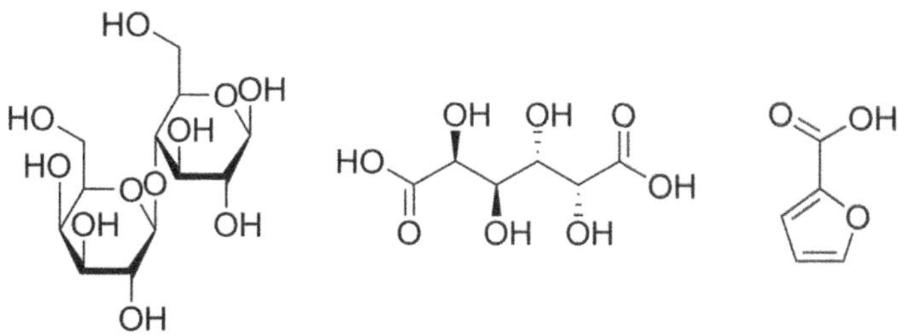

Fig. 24.6 Lactose, mucic acid and pyromucic acid, respectively

muscles is optically active, while the acid in milk is racemic), and in the hands of German chemist Johannes Wislicenus (1835–1902) lactic acid played an important role in the investigation of the three-dimensional structure of molecules [17].

While working with milk, Scheele studied lactose and investigated whether lactose, like common sugar, would yield oxalic acid (acid of sugar) on treatment with nitric acid [18]. Rather than obtaining oxalic acid, Scheele obtained yet another new acid, "acid of milk sugar", now known as mucic acid (Fig. 24.6) [19]. By pyrolysis of mucic acid, Scheele obtained pyromucic acid (2-furoic acid).

24.4 Citric Acid

Following his collaboration with Scheele on tartaric acid, Retzius also continued the work on vegetable acids and published his accumulated results in 1776 [20]. Retzius had studied lemon juice but had failed to isolate citric acid. Instead, it was Scheele who managed to isolate crystalline citric acid (Figs. 24.7, 24.8) by the same method he had used to isolate tartaric acid [21]. The manuscript was submitted to the Royal Academy in April 1784 and printed later that spring [22]. The paper contains little theory and mainly reports the method for isolation and crystallisation of citric acid.[2] This was done by neutralising boiling lemon juice with calcium carbonate (chalk) and decant the liquid from the precipitate. From the weight of chalk added, the correct amount of sulphuric acid could be determined. Scheele did

[2]Scheele's method is straightforward and easy to reproduce.

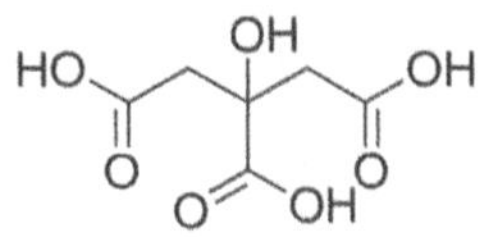

Fig. 24.7 Citric acid

Fig. 24.8 Crystalls of citric acid. *Photo* Petra Rönnholm

not write how he determined the correct amount of sulphuric acid but wrote that 146 g (11 *lod*) of sulphuric acid should be used for every 133 g (10 *lod*) of calcium carbonate. The correct value is 130 g of pure sulphuric acid, but Scheele's acid was not water-free and a slightly larger amount is expected. The precipitated calcium sulphate was filtered off, and the solution concentrated and crystallised.

24.5 Malic Acid

Scheele then studied a large number of unripe fruits and berries, and when he analysed the juice of unripe gooseberries, he found that the juice remained acidic after precipitating calcium citrate and concluded that gooseberries contained an unknown acid forming a water-soluble calcium salt[3] [23]. This acid could be precipitated as a lead salt and liberated with sulphuric acid in the same manner as oxalic acid [24]. Scheele soon found that unripe apples contained no citric acid or gum, and thus was a more convenient source of the new acid, which he consequently called acid of apples (now known as malic acid from Latin *malum*, apple; Fig. 24.9).

[3]The solubility of calcium maleate is 2.98 g/100 ml of water at 25 °C, compared to 0.85 g for calcium citrate (at 18 °C).

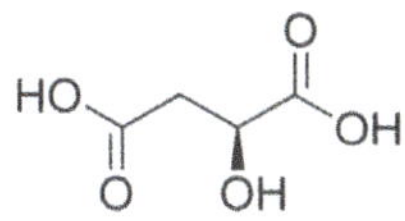

Fig. 24.9 Malic acid

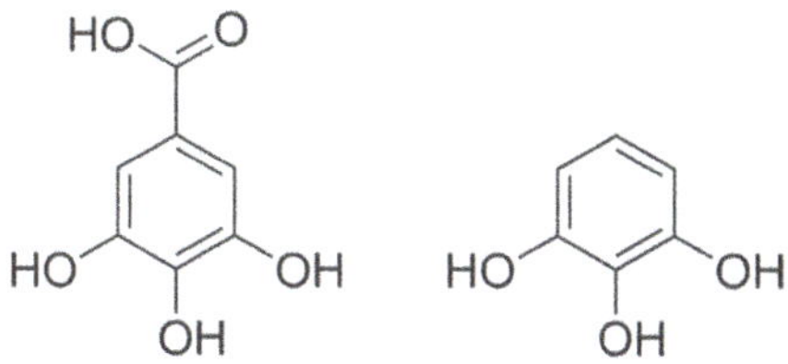

Fig. 24.10 Gallic acid and pyrogallol, respectively

24.6 Gallic Acid

In Gahn's notes from 1770, he wrote that Scheele had isolated an acid from fermented oak apples or oak galls, outgrows on oaks formed by the larvae of gall wasps. This acid, now known as gallic acid (Figs. 24.10, 24.11), would be the topic of Scheele's last full paper [25, 26], submitted on February 16, 1786. The black iron(III) salts of gallic acid had actually been used in black ink since the mid-fifth century. Through

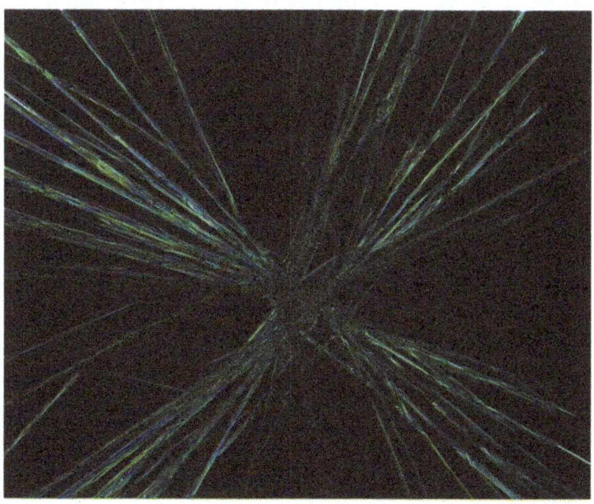

Fig. 24.11 Crystals of gallic acid vied through a polarising microscope. *Photo* Mattias Andersson

pyrolysis of gallic acid, Scheele obtained pyrogallol (Fig. 24.10), which a century later found use as a reducing agent for developing photographic plates.

24.7 The Discovery of Glycerol

In his shop, Scheele prepared common salve (*Emplastrum simplex*), a paste used to treat wounds and which was obtained by heating vegetable oil with lead(II) oxide (litharge) and water. The salve consisted of the water-insoluble antiseptic lead salts of fatty acids. Scheele realised the similarity of the salve with soap, which is a mixture of the corresponding sodium salts of fatty acids.

It was in the aqueous residue from the salve production that Scheele discovered glycerol (Fig. 24.12), an oily liquid with sweet taste [27]. Glycerol was the second pure alcohol, after ethanol, to be prepared. Scheele referred to it as the "sweetness" or the "sugar substance". The name "oil sweet", which is often attributed to Scheele, is not used in Scheele's papers. Scheele noted the sweet taste and attempted to crystallise and ferment glycerol, without success. As certain lead salts, like lead (II) acetate (sugar of lead), also have a sweet taste, Scheele investigated whether the glycerol contained lead, which it did not. He found that different types of fat such as olive oil or lard all yielded glycerol.

Scheele found that glycerol was more heat resistant than sugar and even survived distillation. Scheele attributed this difference to a higher phlogiston content in glycerol and proved this by measuring the amount of nitric acid needed to convert glycerol and cane sugar to oxalic acid. In Scheele's organic works, phlogiston is essentially equivalent to hydrogen, and glycerol ($C_3H_8O_3$) contains more hydrogen (8.76%) than cane sugar ($C_{12}H_{22}O_{11}$; 6.48% hydrogen). Therefore, more nitric acid is needed to oxidise glycerol to oxalic acid ($C_2H_2O_4$) than is needed for cane sugar. Heating glycerol at high temperature gave smoke with a pungent smell. This was due to decomposition into acroleine ($CH_2 = CHCHO$), a potent lachrymator. This was not an entirely new discovery, as Boyle had noted the acrid smoke formed on heating fats [28].

Scheele first wrote a paper on his discovery of glycerol in Swedish, published in 1783 [29], followed by a paper in German published by Crell [30]. The nature of fats was thoroughly studied by French chemist Michel Eugéne Chevruel (1786–1889) from 1813 and onward. This made fats the first class of biomolecules to be understood chemically.

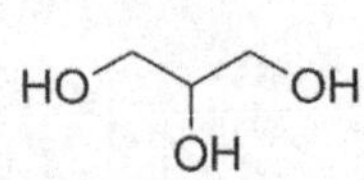

Fig. 24.12 Glycerol

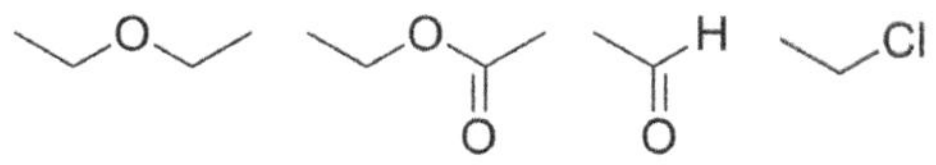

Fig. 24.13 The "ethers" examined by Scheele: diethyl ether, ethyl acetate, acetaldehyde and chloroethane, respectively

24.8 Esterification and Catalysis

To a modern organic chemist, an *ether* is a substance where two hydrocarbon groups are connected by a single oxygen atom. In the eighteenth century, however, the word ether referred to any flammable liquid insoluble in water. In September 1781, Scheele attempted to reproduce Count Lauraguais'[4] synthesis of "ether" (actually the ester ethyl acetate; Fig. 24.13) from acetic acid and ethanol, but failed. When he asked Bergman for advice, Bergman wrote back that he had failed as well. Scheele then repeated the experiment with the addition of mineral acid (hydrochloric or sulphuric acid), and by doing so, he obtained ethyl acetate [31]. Scheele had discovered acid-catalysed esterification: the reaction between ethanol and acetic acid is very slow, unless the acetic acid is activated by a strong mineral acid. This reaction is commonly known as Fischer-esterification, based on a paper by Nobel laureate Emil Fischer (1852–1919) and Arthur Speir from 1895 [32], but it would be more appropriate to call it Scheele esterification. Scheele realised the general nature of this reaction, and was also able to prepare ethyl benzoate from ethanol and benzoic acid.

In a letter to Bergman, dated October 3, 1781, Scheele clearly stated that the reaction between acetic acid and ethanol only takes place in the presence of mineral acid, and he appears to be the first to clearly describe the phenomenon of catalysis, a discovery of fundamental importance. The word catalysis was later coined by Berzelius [33], following the discovery of heterogeneous catalysis by Davy and Wollaston.

Scheele's paper on different kinds of ethers [34] is of great importance in the history of chemistry. He discussed not only the well-known diethyl ether but he also studied chloroethane, obtained from ethanol (*Spiritus vini*) with hydrochloric acid and zinc chloride ("Zinc calx dissolved in hydrochloric acid"), or ethanol and tin (IV)- or antimony(III) chloride (*Spiritus fumans Libavii* and *Butyrum antimonii*, respectively). This was the first compound in the important class of organic halides. Scheele found that chloroethane gave no precipitate with silver nitrate solution, but that a precipitate of silver chloride was obtained by igniting the mixture. Thus, the hydrochloric acid was a component in the "ether". From ethanol, sulphuric acid and manganese(IV) oxide Scheele isolated the first aldehyde, acetaldehyde (CH_3CHO). When distilling ethanol, manganese(IV) oxide, sodium chloride and sulphuric acid, or ethanol, manganese(IV) oxide and hydrochloric acid, he obtained both ether (chloroethane) and an oil. This oil was 2,2,2-trichloro-1-ethoxyethanol (chloral

[4]Louis Léon Felicité de Barclas, duc de Laugrais (1733 or 1735–1824); French general.

alcoholate; $CCl_3CH(OH)OCH_2CH_3$). Chlorine gives chloral (CCl_3CHO) with ethanol and in the presence of excess ethanol, a viscous solution of the hemiacetal chloral alcoholate is obtained. If Scheele would have studied this reaction more thoroughly, he would have been in the position to make an important discovery, since chloral reacts with water to chloral hydrate ($CCl_3CH(OH)_2$). Chloral hydrate was not isolated until 1869 and eventually [35] found to be a useful sedative and hypnotic but has now disappeared from the market due to its toxicity.[5]

References

1. Scheele CW (1782) Anmärkningar om sättet at conservera Ättika. KVA Nya Handl 3: 120–122
2. Lennartson A (2017) The chemical works of Carl Wilhelm Scheele. Springer, Cham, p 73
3. Lennartson A (2017) The chemical works of Carl Wilhelm Scheele. Springer, Cham, p 83
4. Bergman T, Afzelius J (1776) Dissertatio chemica de acido sacchari, Uppsala
5. Thomson T (1831) The history of chemistry, vol II. Henry Colburn & Richard Bentley, London, p 43
6. Neumann C (1755)Chymiae Medica Dogmatico-Experimentalis...der gründlischen und mit Experimenten erwiesenen Medicinischen Chemie, vol 4 part 2, p 359
7. Scheele CW (1784) Om Rhabarber-jordens bestånds-delar, samt sätt, at tilreda Acetosell-syran. KVA Nya Handl 5:180–187
8. Scheele CW (1785) Om Rhabarberjordens närvaro uti flera vegetabilier. KVA Nya Handl 6:171–172
9. Bergius P (1777) Anmärkningar om Blåsestenen. KVA Handl 38:304–309
10. Scheele CW (1776) Undersökning om Blåse-stenen. KVA Handl 37:327–332
11. Lennartson A (2017) The chemical works of Carl Wilhelm Scheele. Springer, Cham, p 33
12. Bergman T (1776) Tilläggning om Blåse-Stenen. KVA Handl 37:333–338
13. Scheele CW (1780) Om Mjölk, och dess syra. KVA Nya Handl 1:116–124
14. Lennartson A (2017) The chemical works of Carl Wilhelm Scheele. Springer, Cham, p 63
15. Lavoisier A (1790) Elements of chemistry. William Crech, Edinburgh, p 278
16. Berzelius JJ (1808) Föreläsningar i djurkemien vol 2. Stockholm, p 176
17. Ramberg PJ (2003) Chemical structure, spatial arrangement. Ashgate, Aldershot, p 42
18. Lennartson A (2017) The chemical works of Carl Wilhelm Scheele. Springer, Cham, p 65
19. Scheele CW (1780) Om Mjölk-Såcker-Syra. KVA Nya Handl 1:269–275
20. Retzius AJ (1776) Vidare försök med naturlig växt-syra. KVA Handl 37:130–140
21. Lennartson A (2017) The chemical works of Carl Wilhelm Scheele. Springer, Cham, p 80
22. Scheele CW (1784) Anmärkning om Citron-saft, samt sätt at crystallisera densamma. KVA Nya Handl 5:105–109
23. Scheele CW (1785) Om Frukt- och Bär-syran. KVA Nya Handl 6:17–27
24. Lennartson A (2017) The Chemical Works of Carl Wilhelm Scheele. Springer, Cham, p 85
25. Scheele CW (1786) Om Sal essentiale Gallarum, eller Galläple-salt, KVA Nya Handl 7:30–34
26. Lennartson A (2017) The Chemical Works of Carl Wilhelm Scheele. Springer, Cham, p 93
27. Lennartson A (2017) The chemical works of Carl Wilhelm Scheele. Springer, Cham, p 78
28. Boyle R (1661) The sceptical chymist. J Crook 257
29. Scheele CW (1783) Rön, beträffande ett särskildt Socker-ämne uti exprimerade Oljor och Fetmor. KVA Nya Handl 4:324–329

[5]In Sweden, the last commercial medicine containing chloral hydrate was withdrawn in 1969.

30. Scheele CW (1784) Entdeckung eines besondern süßen und flüchtigen Bestandtheils in den ausgepreßten Oelen und thierischen Fettigkeiten. Chem Ann 1:99–101
31. Lennartson A (2017) The chemical works of Carl Wilhelm Scheele. Springer, Cham, p 71
32. Fischer E, Speier A (1895) Darstellung der Ester. Ber 28:3252–3258
33. Berzelius J (1835) Årsberättelse om framstegen i Fysik och kemi afgifven den 31 mars 1835. Stockholm
34. Scheele CW (1782) Rön och Anmärkningar om Æther. KVA Nya Handl 3:35–46
35. Personne J (1869) Sur la préparation et properitétés de l'hydrate de chloral. Compt Rend 69:1363–1366

25 Bergman's Contributions to Mineralogy

As the majority of Swedish chemists up to the mid-nineteenth century, Bergman had a natural interest in mineralogy. In fact, due to the influence of the Board of Mines, he had to develop such an interest. This interest resulted in several papers on mineral analysis (Chap. 23) and two works on classification of minerals. To his help, Bergman had his extensive mineral collection (Sect. 9.4), but he also made several trips through Sweden during the summers to visit mines and other places of mineralogical interest. He also received mineral samples from colleagues around Europe. Since Bergman's analytical methods were discussed in Chap. 23, this chapter will focus on his analytical results and conclusions.

25.1 Semi-precious Stones and Gems

The October to December issue of the *Transactions* of 1777 featured a series of five papers in a row on a mineral called *Oculus mundi* (world's eye), now called hydrophane and recognised as a rare variety of opal. The last of these papers, *Tilläggning om Oculus mundi* [1] (addition on *Oculus mundi*), was written by Bergman. Bergman's addition to the discussion on this mineral was whether it was properly classified as an opal or carnelian. Bergman also used hydrochloric acid (*Acidum salis*) to extract aluminium (earth of alum) from *Oculus mundi*.

In his 1766 paper on the electrical properties of tourmalines (Sect. 5.2), Bergman did not discuss their chemical composition: that was Rinman's task. In 1779, however, he analysed brown tourmalines from Zillerthal in Tyrol sent to him by von Born, and compared them to tourmalines from Sri Lanka [2]. The problem was that the stones were not soluble in acids, and thus he used his method of fusing the powdered samples with alkali (Sect. 23.4). He concluded that the tourmaline of Tyrol was composed of 42 parts clay, 40 parts silica, 12 parts lime and 6 parts iron, while the stones from Sri Lanka were composed of 39 parts clay, 37 parts silica, 15 lime and 9 parts iron. As the composition of tourmalines varies, the accuracy of

A. Lennartson, *Carl Wilhelm Scheele and Torbern Bergman*, Perspectives on the History of Chemistry, https://doi.org/10.1007/978-3-030-49194-9_25

Bergman's results is difficult to estimate. Tourmalines contain boron, but as the nature of boric acid was much debated [3], there were no reasons for Bergman to search for boric acid in a mineral. Baumé even claimed that borax could be obtained from lard, which was disproved by Scheele.

The chemical nature of precious stones was examined by Bergman in a paper in 1780 [4]. These analyses were very challenging, as only very small samples could be examined for economic reasons. Thus, it is perhaps not surprising that the analytic results were less accurate and that he reported that both rubies and sapphires contain silica and lime, while they are in fact crystalline aluminium oxide. Emeralds, which actually contain silicon, were reported to contain less siliceous earth compared to rubies and sapphires, and the discovery of beryllium was left to Vauquelin in 1798. Bergman also analysed topaz and hyacinth. Here, Bergman could have discovered zirconium, an element discovered by Klaproth in 1789. When analysing diamonds, Bergman found, correctly, that intense heat gave soot, but he also erroneously reported a new earth, which he referred to as precious earth or earth of gems.

25.2 Tin Sulphide Minerals

Bergman had received a collection of minerals from Siberia[1], and among them there was a sample that had the appearance of antimony covered with a golden substance which was found to be tin(IV) sulphide (*Aurum mosaicum*) [5]. It was known that tin formed two compounds with sulphur, one black substance containing about 20% sulphur (SnS; 21% sulphur) and a gold-coloured substance containing about 40% sulphur (SnS_2; 35% sulphur) and although both sulphur and tin were frequently encountered in Earth's crust, no one had, according to Bergman, previously reported naturally occurring tin sulphides.

25.3 Mineralogial Remarks

Bergman's last paper is a collection of discussions and analyses of mineral samples sent to him by international friends and colleagues during the past couple of years. The paper has the subtitle "first part", but unfortunately, it would be Bergman's last published work. According to Crell [6], Bergman had finished the paper just 2 days before his death, but considering Bergman's condition the days before his death (Sect. 28.1), this seems unlikely.

Bergman had predicted [7] that barium carbonate would be discovered as a mineral, and English scientist William Withering (1741–1799), best known for discovering the medical effects of foxgloves (*Digitalis*), had found such a mineral in Scotland the previous summer. A sample sent to Bergman had not arrived, but

[1]There is no mention of who had sent him the samples.

Bergman had received another sample from Black via Schwediauer, which he had examined. At the end of the paper, Bergman mentioned, for the first time in Swedish, the isolation of metallic tungsten by the d'Elhuyar brothers (Sect. 15.5). He also mentioned the mineral (cerite) which Cronstedt had classified as a form of tungsten, which was chemically unrelated to tungsten. The discovery of cerium in this mineral would, however, have to wait until the analysis by Berzelius and Hisinger in 1803. Bergman also discussed, for the last time, the zeolites. These are porous minerals that have received widespread use today for large varieties of applications such as absorption and separation of molecules and catalysis.

25.4 Mineralogical Dissertations

A number of dissertations on mineral analysis and related subjects were defended by Bergman's students, and probably written by Bergman himself, although the authorship is difficult to establish. In June 1774, Peter Jacob Hjelm defended a thesis titled *Chemical and Mineralogical Dissertation on White Iron Ores* [8]. The dissertation is written in Swedish, and as Bergman's dissertations were typically written in Latin, the choice of language might suggest that the author is Hjelm. The thesis describes white iron minerals, *i.e.* minerals where iron occurs as iron(II) and lacks the dark colours of most iron(III) compounds. When exposed to air, these minerals darkened (due to oxidation to iron(III)) and finally disintegrated into black powders. The minerals effervesced with acids in the same way as lime stone, and the gas evolved was found to be carbon dioxide (aerial acid). All examined specimens were found to contain aerial acid, water, iron, lime and *Magnesia nigra* (manganese), and thus seem to be ankerite, $Ca(Fe,Mg,Mn)(CO_3)_2$. Another example of a white iron ore is siderite, $FeCO_3$.

A dissertation on zinc ores was defended by Geijer in 1779 [9]. It contains a historical introduction, as did many of Bergman's works (according to Partington, Bergman often relied on Pott as a source of historical information [10]), followed by analysis and discussion of different types of zinc ores. In a 1782 dissertation on lithomerge, a kaolinite clay mineral, Bergman found a method of selectively precipitating aluminium carbonate with magnesium carbonate from a solution of the clay in sulphuric acid [11]. This method of separating different kinds of earths also came to good use in a dissertation on asbestos defended the same year (Sect. 11.4). Bergman and his student Robsahm analysed 13 different varieties of asbestos [12], which were found to be composed of lime, magnesia, clay, silica and calx of iron, *i.e.* oxides of calcium, magnesium, aluminium, silicon and iron.[2] From a letter to Bergman from Scheele, dated in May 1782, it is clear that Bergman had asked Scheele to analyse asbestos, but that Scheele had found no time to do so.

[2]Asbestos is a heterogeneous group of fibrous minerals. They are all silicates of elements like calcium, magnesium and iron.

25.5 Outline of Mineralogy

Attempts had been made to classify minerals since the days of ancient Greece [13, 14]. The early systems of classification were based on physical properties such as colour, shape and texture. Since the days of Aristotle, nature was divided into three kingdoms: animals, plants and minerals. Thus, Linnaeus attempted to classify also the mineral kingdom in his *Systema naturae* in 1735. While Linnaeus system of classifying plants was very successful, he was less successful in classifying minerals, as he took the extreme position that only crystal form should be used for classification [14]. This was logical for Linnaeus, as he had classified the plants based on their anatomy rather than chemical composition. Wallerius published his book *Mineralogia, eller mineral-riket, indelt och beskrifvit* (Mineralogy, or the Mineral kingdom divided and described) in 1747, 3 years before becoming professor of chemistry in Uppsala. His book was soon translated into German and French and was his main qualification for the professorship. Wallerius put more emphasis on chemistry in his system. He divided minerals into four classes: earths, stones, ores and fossils. Each class was then divided into sections, which were further divided into subsections. For instance, the first section among the stones was limes, which was divided into limestone, marble, gypsum and spar. Under each subsection, several different minerals were listed.

The first true chemical classification of minerals was also provided by a Swedish mineralogist, Axel Fredrik Cronstedt. He published anonymously his *Försök til Mineralogie* (*Attempt to Mineralogy*) in 1758. Cronstedt's book was translated to German and English, and a second posthumous edition in Swedish with comments appeared in 1781. Cronstedt divided minerals into four classes: earths, salts, earth fats and metals. Each class was divided into sections and subsections. For example, the earths were divided into limes, silicas, garnets, clays, micas, fluxes, asbestoses, zeolites and "brownstones" (manganese(IV) oxide). The class of earth fats included combustible materials, *e.g.* amber, petroleum and sulphur. Despite Bergman's and other's attempts, it would take until the nineteenth century until the chemical classification system of minerals was universally accepted [15].

In 1782, Bergman published his *Sciagraphia Regni Mineralis* (*Outline of the Mineral Kingdom*), where he attempted to correct some mistakes by Cronstedt. The book was originally published by J. J. Ferber in Leipzig and Dessau but was re-issued three times the following year: in London, Florence; and again in Leipzig and Dessau. The third version is possibly a pirate edition [16]. The book was translated to German in 1782 and 1787, English in 1783, French in 1784 and 1792 and Portuguese in 1799–1800.

Bergman rejected the idea of classifying minerals based on physical properties: "Colour varies exceedingly, as does also the size of bodies. We cannot sufficiently wonder at the violence done to nature by the studied separation of earths from stones. The consequence is, that a stone of a certain size must constitute one genus, whilst the same thing, reduced to powder, must be placed under another genus, which shall not be found even in the same class" [17]. Bergman only accepted the use of physical properties to discriminate between different varieties of minerals: "Classes, Genera and Species are therefore to be formed upon the internal nature and composition, the varieties upon the external appearances. In such a system both methods conveniently agree" [18].

Bergman's system is similar to that of Cronstedt, and the minerals are classified in the same classes (although the salts constitute the first class in Bergman's system, followed by the earths, mineral resins and metals). Bergman's book is far less extensive than Cronstedt's work, and where ever possible, Bergman avoided descriptions of minerals by referring to Cronstedt's book. As mineral chemistry had made much advance since 1758, Bergman could in many cases correct Cronstedt by including more recent discoveries. As will be seen in Chap. 26, one of the most important points in Bergman's book was its use of rational chemical nomenclature.

25.6 Thoughts on a Natural System of Mineralogy

Also in 1782, Bergman wrote a long paper in Latin on classification of minerals, *Meditationes de systemate fossilium naturali* (*Thoughts on a Natural System of Minerals*). It was included in a preprint published in 1782 [19] and finally appearing in the fourth volume of the *Transactions of the Royal Society of Sciences in Uppsala* in 1784 [20]. A trade version was also published in Florence the same year [21].

This work is much more theoretical than *Sciagraphia*, dealing with the classification, analysis and naming of mineral substances rather than describing individual minerals. It is divided into two parts, the first dealing with classification and the second (which will be discussed in Sect. 26.8) dealing with nomenclature.

When justifying his works on mineralogical classification Bergman, perhaps surprisingly, used economic rather than scientific arguments: "As natural bodies may in various ways be rendered useful to man, a thorough knowledge of them becomes highly necessary; and it will, indeed, in general be found, that their utility encreases in proportion to the extent of that knowledge" [22]. Most probably this statement is an attempt to prove his research useful for Swedish industry. The true reason for Bergman's efforts was most likely found in his early contacts with the Linnaeus' systematic work on botany and his own main theme: the systematisation of chemistry and science.

After a discussion of the three kingdoms of nature, of organic and inorganic bodies and the three aggregation states, Bergman discussed different systems of classifying minerals. Bergman advocated a chemical classification, but such a system required a thorough knowledge of the chemical composition of minerals. In that context, Bergman went on to discuss different analytical methods, classical solid-state methods versus solution methods (Chap. 20).

Bergman devoted much effort to the identification of the classes and genera of minerals. As in *Sciagraphia*, Bergman adopted the classes devised by Cronstedt but changed their order. Salts, the first class in Bergman's system, were soluble in water, their water solutions had taste and their densities were generally less than twice that of water. Earths had neither taste nor solubility in water. They had higher densities than salts, but no known earth had a density exceeding 4.5 (Bergman is probably referring to barium carbonate which has a density of 4.43 gcm^{-3}). Metals were insoluble in water, had "a peculiar splendour" and high density, at least six times that of water. Phlogistic bodies, the fourth and last class, had low densities and were combustible. Bergman then discussed in detail different physical properties such as taste, colour and density and affinities (Chap. 20).

As most minerals are not pure elements, it was not trivial in which genus a specific mineral should be classified. Bergman suggested that the weights of the different components could be used; if a compound is composed of the proximate principles (Chap. 22) A and B, the mineral will belong to genus A if it contains more A than B. Thus, metal compounds, for instance, were classified under the corresponding metal. Some exceptions were possible; if B was more important or more valuable than A, the mineral was more conveniently classified in the genus of B. Bergman could, for example, have referred to ores of precious metals such as gold or silver. As in Cronsted's book, Bergman suggested putting mechanically mixed rocks in appendices.

When classifying minerals composed of different earths, Bergman devised a system of formulae for describing their composition. As in Berzelius' system from 1813, the system still used by chemists today, Bergman used Latin letters to denote the constituents: *p* for ponderous earth (barium), *c* for calcareous earth (calcium), *m* for magnesia (magnesium), *a* for argillaceous earth (aluminium) and *s* for siliceous earth (silicon). Berzelius had the advantage of being able to determine empirical formula and denoting the number of atoms with coefficients; Bergman did not have this luxury, let his formulae are semi-quantitative. The order of the letters in Bergman's formulae denoted the composition: the earth appearing in the largest proportion was given first and thus determined the genus, followed by the other constituents in decreasing order. In Bergman's system, *cas* would be a calcium aluminium silicate. He also devised a system for denoting rocks of mixed composition using upper case letters: *S* for salt, *T* for earth (Terra in Latin), *M* for metal and *I* for phlogistic bodies. Unfortunately, in the English translation (volume 3 of his *Chemical and Physical Essays*), lower case letters were used, making the meaning of symbols *s* and *m* ambiguous.

Shortly before his death, Bergman wrote to Crell that he was preparing a new edition of his mineralogy and was eager to get hold of any mineral that he had yet not studied [23]. This was a plea to the readers of Crell's journal to send him specimens.

25.7 Crystallography

Crystals have been studied since antiquity, but with no real progress until the re-emergence of atomic theory in the seventeenth century. Various schemes were proposed for an explanation of the crystal form based on atomic theory: British scientist Robert Hooke (1635–1703) regarded crystals as built up by spherical atoms, while Christiaan Huygens (1629–1695) preferred ellipsoidal atoms; Giovanni Battista Guglielmini (1763–1817) and Antoni van Leeuwenhoek (1632–1723) held that crystals were built up by polyhedral atoms.

Usually, the French mineralogist René Just Haüy (1743–1822) is credited as the father of crystallography, but Dutch historian of Science, Reijer Hooykaas, who studied the crystallographic works of Bergman and Haüy in great detail, has pointed out that Haüy based his early theories on Bergman's work [24].

Bergman's work on crystallography began with a discovery by Gahn during his time in Bergman's laboratory in the late 1760s. Gahn found that scalenohedral calcite ($CaCO_3$) crystals could be cleaved along cleavage planes to generate a rhombohedral kernel with the same shape as crystals of Iceland spar (another form of calcite; Fig. 25.1) [25]. From this finding, Bergman developed a theory where he attempted to classify a large group of crystals based on a common *primitive form*.

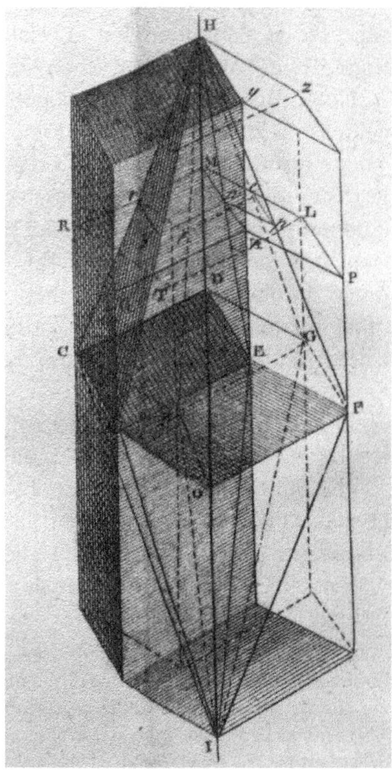

Fig. 25.1 A rather crowded figure illustrating Bergman's theory

His findings were first published in a paper in 1773 [26] and a considerably extended version appeared in the second volume of his *Opuscula Physica et Chemica* in 1780 [27]. The introduction to Bergman's paper gives a good indication of his views on crystallography:

> "Crystals are bodies which, though destitute of organic structure, yet externally resemble geometrical figures, more or less regular. If we attend to the numerous collections of these, we shall be ready to conclude, that nature has effectually eluded our research by the infinite variety; for frequently bodies widely differing in their nature and properties resemble one another in figure; and, on the contrary, those which are exactly alike in properties put on external appearances entirely different; yet, upon a careful examination and comparison of this variety of figures, we shall find, that a great number of them, though their surfaces differ with respect to their angles and sides, may be derived from and referred to a very small number of simple figures" [28].

Bergman's approach was to use the rhombohedral calcite kernel as a primitive form and stack thin lamella on its faces. First, Bergman used lamella of the same size and form as the kernel faces and, in that manner, he arrived at the crystal form of several crystals including garnet and hyacinth (yellow zircon; ZrO_2). Here, Bergman went too far, as he did not pay close attention to the angles between the crystal faces. Bergman's system was only valid for crystals of the same symmetry displaying different crystal habits,[3] such as the rhombohedral and scalenohedral forms of calcite [24]; both belong to the trigonal system while garnets belong to the cubic and zircon to the tetragonal systems, respectively. Inspired by his apparent success, he also attempted to relate the hyacinth crystal to a cruciform interpenetrating twin[4] of staurolite (monoclinic system).

Next, Bergman applied lamella with the same form but with continuously decreasing dimensions. By doing so, he correctly arrived at the scalenohedral calcite form. So far, Bergman's theory was based on the cleavage of crystals but the next step, introduced in order to describe an even larger number of crystals based on a common primitive form, was a purely geometrical construction: he now applied truncated lamellae. Although Bergman, through his qualitative approach, tried to relate unrelated crystals, his theory contained one important discovery: he could relate the morphologically very different rhombohedral and scalenohedral forms of calcite, and thus pointing in the direction of the basis for modern crystallography: the unit cell.

In the extended 1780 version of the paper, Bergman had come across a work on ice by French physicist and biologist Jean-Jacques d'Ortous de Mairan (1678–1771), published in 1716. By observing striation (lines that form on the faces of crystals; Fig. 25.2), de Mairan had concluded that crystals were built up by tiny fibres arranging themselves into hollow pyramids. As the direction of these fibres was not the same throughout the crystal, it would represent twinned crystals in

[3]Due to different growth rates of different crystal faces, crystals may have different forms although the atomic arrangements in the crystals are the same.

[4]A twinned crystal is a crystal where crystal domains are related by a symmetry element not described by the space group of the crystal.

Fig. 25.2 Striation on the surface of pyrite crystals. *Photo* Anders Lennartson

modern terms. Bergman adopted this fibre theory to explain the structure of his lamellae. By observing striation, Bergman concluded that all crystals could be derived from pyramids, neglecting that this contradicted his own lamellar approach to crystallography [24]. In fact, his second paper is essentially divided into two completely different parts giving two alternative explanations of crystal forms. Bergman built mineral crystals from lamellae (based on cleavage) but salt crystals from fibres (based on striation) and there is no unity in Bergman's theory [24]. Bergman incorporated more and more crystals into his system and, as Hooykaas put it, "It appears that once he has started partitioning crystals he cannot stop!"

In Bergman's system, there is no direct correlation between crystal form and chemical composition. For example, he found no correlation between the crystal form of salts and their composition, *e.g.* there were no similarities between different nitrates. The phenomenon of isomorphism[5] was not discovered until the nineteenth century. Bergman correctly attributed the crystal form to the attractions between atoms, and when he based crystal shape on pyramids, he did not mean that the atoms had a pyramidal shape. He devoted a section of his 1780 essay to crystallisation and stressed that the atoms must be freely moveable in order to form a crystal. Crystallisation from solution, melt and by sublimation was discussed.

In Bergman's days, it was commonly believed that crystallinity was due to the presence of salt in the substance, which Bergman denied:

> "As crystallization is the effect of attraction; and as all other matters, as well as salts, are subject to the laws of that attraction, we are not authorized to consider the regular and symmetrical figure as peculiar to saline bodies, although the assumption of such figure be more readily and frequently exercised by them, as being soluble in water" [29].

---

[5]Isomorphism occurs when the atoms or molecules in chemically similar compounds form analogous arrangements upon crystallisation, *e.g.* aluminium potassium sulphate and chromium potassium sulphate.

There were numerous examples of crystalline bodies that had no saline properties, *e.g.* metals. Both zinc and bismuth could be obtained as crystals, and gold, silver and iron could show beautiful crystals at the surface.

Hooykaas has pointed out that Haüy is credited as the founder of crystallography based on his influential publications of 1784 [30] and 1801 [31], while it in fact is important to pay attention to Haüy's first two essays on crystallography, which were read before the French Academy of Sciences in February and December 1781, respectively [24].

Haüy's original publication is based on lamellae in a similar fashion as in Bergman's work, and these lamellae did not necessarily have the same size. On the other hand, he corrected Bergman on several important points: he criticised Bergman for using identical primitive forms for calcite and garnet, he criticised Bergman's truncation of lamellae and his ignorance of the angles between crystal faces. Unlike Bergman, Haüy's theory relied more heavily on cleavage (but striation was used to indicate potential cleavage planes in hard crystals). He did not attempt to relate chemically unrelated crystals to each other and he paid closer attention to the angles between faces than Bergman did. For instance, the garnet was based on a regular rhombic dodecahedron is Haüy's system. This may partly be due to the fact that Haüy had a larger supply of crystals to study and base his theory on compared with Bergman who only studied naturally occurring minerals [32].

It was in 1784 that Haüy put forward a theory that would lay the foundation for modern crystallography. Haüy now proposed a system where a crystal was built up by stacking rhombohedral "crystal molecules" (molécules intégrantes), *i.e.* unit cells. These were stacked in rows, and as the crystal grew, rows could be omitted giving rise to different habits but always with the same inclination between faces.

In his 1784 essay, Haüy gave the impression that this theory was not new, but dated back to his original essays from 1781 and that it was solely his own invention [24]. However, as pointed out by Hooykaas, as he had not criticised Bergman's relation of hyacinth and garnet to Iceland spar in his original paper, this cannot be true. Actually, the very fact that he based his 1781 discussion on Iceland spar and garnet points directly to Bergman as his source of inspiration, since the only connection between these minerals is Bergman's theory. Haüy also constantly, and according to Hooykaas deliberately, referred to "Bergman's paper of 1779", rather than 1773, giving the impression that Bergman's publication appeared 6 years later than it actually did [24]. In his 1801 book, Haüy clearly stated that his original idea was to device crystal form from "crystal molecules". In Hooykaas words: "Haüy, as the creator of a new revolutionary theory, had a tendency to belittle his predecessors and forget his debts to them. He could not bear that anything should be detracted from his theory. This peculiarity does not necessarily imply dishonesty but it means that his sight was not very clear when it came to consideration of his own development".

Another French mineralogist, Romé de l'Isle (1736–1790) put forward another important theory of crystallography in 1783, the law of constant angles between crystal faces, a theory anticipated by Nicholas Steno 100 years earlier. Unlike Bergman, de l'Isle based his theory on exact measurements on angles, and was therefore not corrupted by superficial observations. Unlike Bergman, however, he did not relate crystal form with structure. In de l'Isle's view, there were as many crystal forms as there were substances. Unlike Haüy, de l'Isle cited Bergman correctly and correctly realised that Bergman had inspired Haüy's work [24].

References

1. Bergman (1777) Tilläggning Oculus mundi. KVA Handl 38:347–351
2. Bergman T (1779) Bruna turmaliner; til sina grundämnen undersökte. KVA Handl 40:224–238
3. Bergman T (1779) Framledne Directeuren Herr H.T. Scheffers chemiske föreläsningar...med anmärkningar utgifne... M Swederus. Stockholm, Uppsala and Åbo 1779 p 91
4. Bergman T (1780) Disquisitio chemica de terra gemmarum. Nova Acta Regiæ Societatis Scientiarum Upsaliensis 3:137–170
5. Bergman (1781) Försvafladt Tenn från Siberien KVA Nya Handl 1:328–332
6. Bergman T (1784) Mineralogische Anmerkungen. Chem Ann 1:387–400
7. Bergman T (1782) Sciagraphia regni mineralis, secundum principia proxima digesti. Leipzig and Dessau
8. Bergman T, Hjelm PJ (1774) Chemisk och mineralogisk afhandling om hvita järnmalmer. Uppsala
9. Bergman T, Geijer BH (1779) Dissertatio chemica de mineris zinci. Uppsala
10. Partington JR (1961) A history of chemistry, vol 2. Macmillan, London, p 718
11. Bergman T, Hjerta CD (1782) Dissertatio chemica de analysi lithomargæ. Uppsala
12. Bergman T, Robsahm CG (1782) Dissertatio chemica de terra asbestina. Uppsala
13. Povarennykh AS (1972) Crystal chemical classification of minerals. Springer, New York Chapter 1
14. Hazen RM (1984) Mineralogy: a historical review. J Geol Educ 32:288–298
15. Fors H (2003) Mutual favours (diss.). Department of History of Science and Ideas, Uppsala University, Uppsala, p 89
16. Moström B (1957) Torbern Bergman a bibliography of his works. Almqvist & Wiksell, Stockholm, p 70
17. Bergman T (1783) Outlines of mineralogy. Birmingham, p 7
18. Bergman T (1783) Outlines of Mineralogy. Birmingham, p 9
19. Bergman T (1782) Commentationes e quarto novorum Reg. Scientarum Societatis Upsaliensis actorum tomo excerptæ. Uppsala
20. Bergman T (1784) Meditationes de systemate fossilium naturali. Nova Acta Regiæ Societatis Scientiarum Upsaliensis 4:63–128
21. Bergman T (1784) Meditationes de systemate fossilium naturali. Florence
22. Bergman T (1791) Physical and chemical essays, vol III, Mudie G, Fairbairn J, Evans J. Edinburgh, p 209
23. Bergman T (1784) Vom Hrn. Profess. und Ritter Bergmann in Upsal. Chem Ann 1:149–151
24. Hooykaas R (1952) Torbern Bergman's crystal theory. Lychnos 21–54
25. Hjelm PJ (1786) Åminnelse-tal öfver...Herr Torbern. Bergman, Stockholm, p 46
26. Bergman T (1773) Variæ crystallorum formæ, e spatho ortæ. Nova Acta Regiæ Societatis Scientiarum Upsaliensis 1:150–155
27. Bergman T (1780) Opuscula physica et chemica, vol II. Uppsala, p 1
28. Bergman T (1788) Physical and chemical essays, vol II, Murray J. London p 1
29. Bergman T (1788) Physical and chemical essays, vol II, Murray J. London, p 24

30. Haüy RJ (1784) Essai d'une théorie sur la structure des crystaux. Paris
31. Haüy RJ (1801) Traité de minéralogie. Paris
32. Thomson T (1831) The history of chemistry, vol II. Henry Colburn & Richard Bentley, London, p 45

Bergman's Contribution to Chemical Nomenclature

26

Up to the eighteenth century, the number of known chemical substances was rather limited and their compositions were poorly understood. They were generally named after an important property, their physical appearance, discoverer or place of discovery. Chemists were talking about the fuming liquor of Libavius ($SnCl_4$), butter of antimony ($SbCl_3$), Glauber's salt (Na_2SO_4), liver of sulphur (K_2S_n) and Epsom salt ($MgSO_4$). Names were coined by chemists, alchemists, pharmacists, miners, dyers and physicians, leading to a very heterogeneous collection of names. Frequently, different professions had different terminologies resulting in vast numbers of synonyms. It was not uncommon that a substance had different names depending on how it was prepared. Different names could be used for the same substance in the same publication; Scheele and Bergman were no exceptions.

During the eighteenth century, the problems with this way of naming substances became evident. The number of known chemical substances, especially salts, started to increase at an unprecedented rate. With the discovery of new acids (Chaps. 14 and 24) and metals (Chap. 15), chemists faced a problem never experienced before: to devise names for predicted salts that had not yet been prepared. Thus, there was an immediate need for a systematic chemical nomenclature.

26.1 Macquer's Criticism

Critical voices were raised, but few concrete suggestions emerged. An important exception was Macquer who, aged 31, attacked names such as oil of vitriol (*huile de vitriol*; sulphuric acid) in his *Elémens de Chymie Théorique* in 1749, suggesting the name vitriolic acid (*acide vitriolique*) [1]. So far, he was far from alone. For instance, Wallerius wrote in his textbook "Concerning oil of vitriol, which with regard to the thick consistency, but not any other oily property, has received the name of oil […]" [2], Macquer, however, took his criticism one step further. In the second volume of his *Dictionnaire de Chymie* from 1766, he attempted to reform

A. Lennartson, *Carl Wilhelm Scheele and Torbern Bergman*, Perspectives on the History of Chemistry, https://doi.org/10.1007/978-3-030-49194-9_26

the nomenclature of salts by proposing related names for salts derived from the same acid. For example, he suggested that the name "vitriol" should be reserved for salts of sulphuric acid. Copper(II) sulphate and iron(II) sulphate were frequently referred to as vitriols, a word derived from Latin vitrum (= glass) referring to the glassy appearance of the crystals. There was, however, some inconstancy. For example, Lémery had used the name *Vitriol of the Moon* for silver nitrate [3]. Now Macquer proposed calling all sulphates "*vitriol de*..." followed by the name of the metal or earth. Similarly, he proposed *nitre* or *sel nitreux* for nitrates, *sel marin* or *sel* for chlorides, *sel tartreux* or *tartre soluble* for tartrates, *sel acéteux* for acetates, *sel phosphorique* for phosphates and *sel de borax* or *borax* for borates. He did not attempt to include the sodium and potassium salts in this system, and for the remainder of his book, he did not use the nomenclature he had devised. It was, however, picked up by Antoine Baumé.

26.2 The Reform of Nomenclature in Botany

Botanists and zoologists experienced a similar problem as the chemists, but some two centuries earlier, with a rocketing number of new species being described each year as new continents were explored. Still, there was no common system for deriving names for them.

A step in the right direction was taken in 1623, when Gaspard Bauhin (1560–1624) published his *Pinax Theatri Botanici*. Bauhin introduced the use of descriptive phrases which, although they were not strictly names, gave botanists a way to refer to plants in a standardised way. Unfortunately, the phrases were long and cumbersome to use. Similar phrases would latter appear in chemical works; Cronstedt used them in his *Försök til mineralogie* in 1758, where he, for instance, described silver chloride with the phrase *Acidum Salis Communis Argento Saturatum* (acid of common salt saturated with silver) [4, 5]. Macquer also used descriptive phrases, especially to describe alkali salts, *i.e. sel acéteux à bas d'alkali marin* (acetic salt with a base of marine alkali), for sodium acetate [6].

Order was finally established in the botanical chaos by Carl Linnaeus. The binomial nomenclature he introduced is still used today. Linnaeus' nomenclature was developed gradually over a period of several years. In 1735, he published his *Systema Naturae* where he introduced his sexual system of classification. In *Critica Botanica*, published in 1737, he set out the goal for a new nomenclature: all species of the same *genus* should have a common *generic name;* the name should be in Latin or Greek. Linnaeus initially referred to a species by giving the generic name followed by a descriptive phrase to distinguish the species from other species of that genus. Although shorter than those of Bauhin, the phrases could still in the worst cases consist of dozens of words. His solution to this problem was to replace the phrase with a reference to his previous work [7]. His next brilliant step was to introduce *trivial names* which were easier to remember than the literature reference. He now ended up with two-word names, the first defining the genus and the second the species: Oak was called *Quercus robur* and the brown bear *Ursus arctos*. This binomial nomenclature was used extensively in his *Species Plantarum*, published 1753, where he listed all known plants, and in the tenth edition of *Systema Naturae*, published in 1758–1759. Although Linnaeus was young and had no particular authority to suggest a new nomenclature when he published his *Critica Botanica*, he rapidly gained reputation and by the turn of the century, his binomial nomenclature had become a worldwide standard. This was simply because no one could provide a better system.

26.3 Bergman's Early Thoughts on Nomenclature

Bergman was well fitted to reform the language of chemistry. His biological background and early contacts with Linnaeus enabled him to take influences from botany, rather than alchemy, pharmacy, or mining, as other chemists probably would have done. In Linnaeus' system, he saw a possibility to name chemical substances based on their constituents rather than their properties. Bergman was of course also familiar with Macquer's *Dictionnaire*, and thus his proposals for a reformed nomenclature [8, 9].

Bergman first criticised the chemical nomenclature publicly in a paper on mercury chlorides published in 1772 [9]. Bergman objected to the name Calomel, an ancient name for mercury(I) chloride. Although the compound is white, the name suggests a black substance (καλός = beautiful and μέλας = black in Greek), possibly referring to the black colour arising on treatment of mercury(I) chloride with aqueous ammonia.

26.4 Bergman's Work on Nomenclature in 1775

It was in 1775 that Bergman started to work on a reformation in chemical nomenclature. That year, he published a number of publications where he criticised the chemical nomenclature and started to apply a Linnaean binomial nomenclature in chemistry.

In late 1775, Bergman published *The Chemical Lectures of H.T. Scheffer* of with his own additions, notes and comments. In this book, Bergman made several comments on nomenclature: "Absurd names are ought to be known to their meaning, but should not be retained, and after hand replaced by others, which have been selected on good grounds. Consequently, I call oil of vitriol [*i.e.* H_2SO_4] concentrated vitriolic acid, and its spiritus I call diluted vitriolic acid" [10]. Later in the book, in the chapter on neutral salts, Bergman wrote:

> Neutral and middle salts, that have been known since old times, have received absurd names, mostly due to false imaginations either due to their special use in medicine and alchemy, or their compositions etc., which will be obvious from many examples in the

> following. If we took some common denominations in consideration, could all other be named and designated on the same basis. One would then name first the base or the alkaline part with an addition designating with which acid it is combined. Consequently, could alkali vegetabile vitriolatum [K_2SO_4] be a useful designation for what otherwise has the absurd name tartarus vitriolatus [...]. [11]

Thus, Bergman devised a system of naming salts that was analogous to Linnaeus' system for naming plants and animals; a Latin name in two parts giving the name of the base and the acid. He included a table of names of metal salts of a number of different acids [12]. In the case of sulphuric acid, for some reason, he used Swedish names for the salts constructed from the name of the metal followed by the word vitriol. For the other acids, he used binomial names in Latin, *e.g. argentum nitratum*, *argentum salitum*, *argentum acetatum* and *argentum formicatum* for silver nitrate, silver chloride, silver acetate and silver formate, respectively. Still, he retained old names for a few salts, *e.g. saccharum saturni* for lead acetate.

In the second volume of *Nova Acta Regiæ Societatis Scientiarum Upsaliensis*, Bergman published an extensive Latin paper on carbonic acid (Chap. 18) [13]. Worth noting is that, although the title page says 1775, the printing was delayed to 1776 [14]. When Bergman described the salts of carbonic acid, he wrote:

> [...] they may not improperly be called aerated vegetable alkali [Alkali vegetabile aëratum in the original text, *i.e.* Na_2CO_3], as those salts which are saturated with vitriolic acid are distinguished by the term *vitriolated* ["vitriolata" in the original text], joined to the name of that base with which the acid is united. – All substances saturated with fixed air [CO_2] I shall call hereafter, for brevity sake, *aerated* ["aëratas" in the original text], thereby indicating that they contain that acid which is always present in common air. [15]

In other words, Bergman suggested taking the name of the acid, turn it into an adjective and add it to the name of the base. Thus, he arrived at the following names for the salts he studied: *Alkali vegetalbile aëratum* (potassium carbonate), *Alkali minerale aëratum* (sodium carbonate), *Alkali volatile aëratum* (ammonium carbonate), *Terra ponderosa aërata* (barium carbonate), *Calx aërata* (calcium carbonate), *Magnesia aërata* (magnesium carbonate), *Argilla aërata* (aluminium carbonate), *Ferrum aëratum* (iron carbonate), *Zincum aëratum* (zinc carbonate) and *Magnesium aëratum* (manganese carbonate). This type of names works well in Latin but is less suitable for translation to living languages (as is evident from the English translation above), but Bergman's intension was never to have the names translated but to use the Latin names as an international standard, just as in botany. Before Bergman, there was only one widespread example of such a binomial Latin name in chemistry: *Tartarus vitriolatus* (potassium sulphate).

Directly after the paper on carbon dioxide, the famous paper *On Elective Attractions* follows [16]. Here, Bergman extended his nomenclature to include other acids than carbonic acid, and one can find names such as *calx salita* (calcium chloride) and *calx vitriolata* (calcium sulphate), but there is still no systematic use of binomial nomenclature. A more systematic use of the new nomenclature is found in the dissertation *On Magnesia alba* which was defended by Carl Norell on December 23, 1775 [17]. The magnesium salts he studied were *Magnesia aërata* (carbonate), *Magnesia vitriolata* (sulphate), *Magnesia nitrata* (nitrate), *Magnesia salita* (chloride), *Magnesia fluorata* (fluoride), *Magnesia arsenicata* (arsenate), *Magnesia boraxata* (borate), *Magnesia saccharata* (oxalate), *Magnesia tartarisata* (tartrate), *Magnesia acetata* (acetate), *Magneisa formicata* (formate) and *Magnesia phosphorata* (phosphate).

26.5 Bergman's "Investigation of Truth"

Bergman's first more serious criticism on chemical nomenclature appears in an introductory essay called *Introitus de indagando vero* (*Investigation of Truth*) in the first volume of his collected works *Opuscula Physica et Chemica* published in 1779:

> I am not ignorant that words, like money, possess an ideal value, and that great danger of confusion may be apprehended from a change of names; in the mean time it can not be denied that chemistry, like the other sciences, was formerly filled with improper names. In different branches of knowledge, we see those matters long since reformed: why then should chemistry, which examines the real nature of things, still adopt vague names, which suggest false ideas, and savour strongly of ignorance and imposition? Besides, there is no doubt but that many corrections may be made without any inconvenience: – if, instead of oil of vitriol, and spirit of vitriol, we used the terms concentrated vitriolic acid, and dilute vitriolic acid, I think that no one would be thereby either confounded or misled. [18]

This first volume of his collected works contains revised versions of *Commentatio de acido aëreo* and *De Magnesia alba* and is filled with Bergman's binomial nomenclature. This was important, as we will see, since it more effectively spread Bergman's ideas on nomenclature internationally.

26.6 The Nomenclature in Bergman's Sciagraphia

In 1782, Bergman published a book called *Sciagraphia Regni Mineralis* (*Outline of Mineralogy*; Sect. 25.7) where he introduced a purely chemical classification of minerals. Here, Bergman made extensive use of his nomenclature but had the disadvantage of dealing with many substances of more complex composition compared to de Morveau (Sect. 25.7) who only considered simple salts. As the book dealt with mineralogy, the organic acids were not included in the discussions. While the French translator supported Bergman's nomenclature, the German translator replaced Bergman's systematic names with descriptive phrases. For instance, *Alkali vegetabile salitum* [19] was replaced with "Vegetabilischen Laugensalts mit Salzsaurem gesättigt" [20] (Vegetable alkali saturated with muriatic acid) and *Calx vitriolata* [21] was replaced with "Kalk mit Vitriolsaurem gesättigt" [22] (lime saturated with vitriolic acid) in the German edition of the book. Bergman himself used descriptive phrases when describing minerals of more complex composition, *i.e. Magnesia argillaceo, siliceo, et pyritaceo adunata* [23] (compound of magnesia, clay, silica and pyrite).

26.7 The Contributions of de Morveau 1780–1782

In 1777, de Morveau contributed with an article called *Hépar* in a supplemental volume of Diderot's and d'Alembert's influential *Encyclopédie*. Here, he criticised the term liver of sulphur, an old name for potassium polysulphide and suggested the Latin word *hepar* to avoid confusion. This criticism was by no means novel but indicates that de Morveau had an interest in nomenclature. In 1779, de Morveau came into contact with the first volume of Bergman's *Opuscula*. He was apparently very impressed and wrote to Bergman that summer and asked for permission to translate it to French [24]. The first letter is followed by another two letters the same year, five letters 1780, three letters 1781, one letter 1782 and two letters 1783 and 1784, respectively [25]. de Morveau even learned Swedish in order to be able to read Bergman's works [26], and his wife and fellow chemist Claudine Picardet (1735–1820), translated Scheele's works to French in 1785.

de Morveau's French translation of the first volume of Bergman's *Opuscula* appeared the following year including French translations of Bergman's chemical names. In a note to the translation of the dissertation on *Magnesia alba*, he recommended the widespread use of the new nomenclature [27]. Following his translation of Bergman's *Opuscula*, de Morveau took on an active role in the discussions of nomenclature. In 1781, he criticised the term *terre absorbante*, a view he had to defend against Romé de l'Isle [28].

In May 1782, de Morveau published an important paper on chemical nomenclature [29]. The subject was of course controversial, but de Morveau disarmed the critics by inviting to a scientific discussion rather than trying to change the chemical nomenclature single-handed [30]. In this paper, de Morveau identified five principles for a new nomenclature [31, 32]:

(1) "A phrase is not a name". A substance should have a unique name.
(2) "Denominations should, as far as possible, conform to the nature of things". He repeated the opinion held by Macquer and Bergman that all sulphates should have a name including vitriol, etc. Simple substances should have simple names; the names should only express the chemical constituents of the substance and not the name its discoverer.
(3) "When certain knowledge is lacking of the character which should principally determine the denomination, a name which expresses nothing should be preferred to a name which could express a false idea".
(4) Names should preferably be derived from Latin or Greek. de Morveau was unsatisfied with the term *terra pesante (heavy earth)* for barium oxide, as it had not been proved that it was heavier than other earths. He suggested the term *barote* from the Greek word *βάροσς* = heavy, hence our modern name barium.
(5) The names should fit into the language for which they are intended.

In addition to introducing the name barote, de Morveau also introduced one-word names for the alkalis: *potasse*, *soude* and *ammoniac*. These words have given rise to the modern English names potassium and sodium. Although not new, these names were criticised since the names potasse and soude had previously been used for impure commercial alkali carbonates.

The names of salts were derived from the name of the acid. An exception was chlorides which were defined by the name *muriate* although he preferred the name *acide marin* for hydrochloric acid. The other most common salts were *vitriol* (sulphate), *nitre* (nitrate), *méphites* (carbonate), *arseniate* (arsenate), *borax* (borate), *phosphate* (phosphate), *fluor* (fluoride), *acète* (acetate), *tartre* (tartrate) and *formiate* (formate). Interestingly, he included the name *régalte* for the salts of *acide régalte* (*Aqua regia*) although, of course, *Aqua regia* forms no unique salts. As he strived to form words compatible with French he was, compared with Bergman who used Latin, not very successful in forming a uniform system [33]. Compared with Bergman, de Morveau changed the order of the acid and the base in his names, and wrote *vitriol de fer* and *nitre alumineux*, a style that survived in English well into the nineteenth century and in French to the present day. The nomenclature of de Morveau gained important support from Macquer, Buffon[1] and from Mongez,[2] who translated Bergman's *Sciagraphia* to French and used de Morveau's terms in his translation [34].

26.8 Bergman's Revised Nomenclature

According to the traditional description of the development of Bergman's nomenclature[35], it would have taken Bergman 2 years to respond to Morveau's paper. In fact, he responded the same year. It is true that the paper *Meditationes de Systemate Fossilium Naturali* [36] was published in the fourth volume of *Nova Acta Reg Scient Upsaliensis* in 1784, but it appears that Bergman was prepared for a delayed publication (compare the publication of the third volume in Sect. 21.3) and a preprint of Bergman's paper appeared in 1782, 2 years earlier than commonly stated [37].

This paper, essentially an expanded version of his *Sciagraphia*, features a 21-page-long discussion on nomenclature, *De fossilibus denominandis* (Of Giving Names to Fossils).[3] Compared with *Sciagraphia*, this essay was intended to be more general, and thus also include organic acids not encountered in the mineral kingdom.

Bergman found the state of the chemical language remarkable: "Surely, it is highly improper that the noblest science, which constitutes, as it were, the very essence of natural philosophy, should deliver truths of the greatest importance in the most absurd of all languages" [38]. He welcomed the initiative of de Morveau:

[1]Georges-ouis Leclerc, Compte de Buffon (1707–1788). Influential naturalist and director of Jardin de Roi.
[2]Jean-André Mongez (1750–1788). French mineralogist.
[3]With *Fossil*, Bergman means *mineral*.

> Every real friend to chemistry, therefore, should wish for a happy issue to the plan of Mons. Morveau, to be attempted in the new Encyclopedia [Sect. 26.9]. In the mean time, it may be permitted me to offer a few cursory remarks, which I think are relating particularly to mineralogy, and submit them to the judgement of the public. [39]

Bergman repeated his opinion that absurd and false names should be removed, and that names should express "some essential property or composition" of the substance.

It is clear that Bergman tried to align his ideas with those of de Morveau in order to achieve a more universally accepted system. He accepted the term *Acidum mephiticum* suggested for carbonic acid by de Morveau in 1782, and scarified his *Acidum aeratum*. Bergman also replaced his *Terra ponderosa* with *barytes*, derived from de Morveau's suggestion *barote,* but for another reason than de Morveau: he wanted the names of the earths to be expressed by a single word. For the same reason, he suggested replacing the old *Alkali minerale*, *Alkali vegetabile* and *Alkali volatile* with *Natrum*, *Potassium* and *Ammoniacum*, respectively.

To distinguish the pure earths from naturally occurring minerals, he suggested altering the names *Calx*, *Magnesia*, *Argilla* and *Silex* to *Calcareum*, *Magnesium*, *Argillaceum* and *Siliceum*, respectively. To achieve a greater consistency, Bergman suggested that all names of metals should end with *-ium*, a principle that ever since has been applied in English (but ironically enough not in Swedish). The only change Bergman had to do was to switch gender from feminine to neuter for platina, thus establishing the name platinum.

As alkalis and earths were now expressed by a single word, he reduced the names of acids to a single word: *vitriolicum*, *nitrosum* and *muriaticum* rather than *Acidum* vitriolicum, *Acidum nitrosum* and *Acidum muriaticum*, respectively.

An important question for Bergman, in order to proceed to more complex compounds (neutral salts and analogical salts in Bergman's terminology), was whether the base or the acid determined the genus, *i.e.* if acid or base should appear first in the name. In the case of salts of metals and earths (analogical salts), Bergman believed that the genus was determined by the acid. In the case of the neutral salts, *i.e.* the sodium-, potassium- and ammonium salts, he was inclined to think that the genus was better described by the base, but in order to form a uniform system, he decided that the acid should determine the genus in this group as well. Thus, he switched the order of the acid in the base in his names, *i.e. Vitriolicum potassinatum* (K_2SO_4), *Muriaticum ammoniacum* (NH_4Cl), *Nitrosum argillatum* ($Al(NO_3)_3$), *Muriaticum barytatum* ($BaCl_2$) and *Nitrosum argentatum* ($AgNO_3$). By doing so, he had (accidently or on purpose) aligned his system with that of de Morveau. For some reason, he deviated from the consistency in the case of "Saline earths, with such an excess of earthy matter as nearly to obliterate their saline character" [40] and preferred to write *Barytes vitriolatus* and *Calcareum fluoratum* rather than *Vitriolicum barytatum* ($BaSO_4$) and *Fluoratum calcareum* (CaF_2).

Bergman's nomenclature could even account for salts of polybasic acids such as tartaric and oxalic acid: *tartareum potassini* was potassium hydrogen tartrate, while *tartareum potassinatum* was potassium tartrate and *oxalinum potassini* was potassium hydrogen oxalate while *oxalinum potassinatum* was potassium oxalate.

In the spirit of Linnaeus, he suggested the use of trivial names for substances of more complex composition such as *Tartarum Seignetti* for potassium sodium tartrate and *Phosphorum microcosmicum* for ammonium sodium hydrogen phosphate. These salts had previously been known as *Sal Seignetti* and *Sal microcosmicum*, respectively. While de Morveau had rejected any name derived from personal names, Bergman accepted such names for substances with more complex composition, referring to the use of personal names in botany and anatomy. By using trivial names, he obtained a larger agreement with botany and zoology, as he could express a chemical name with no more than two words. According to Bergman, the genus had to be described by one single word in Linnaean style.

By seeking to bring his nomenclature closer to that of de Morveau, Bergman probably hoped that the two systems eventually could converge into a single system of chemical nomenclature which would have a much greater chance of being accepted than a system developed by a single chemist. At one point, Bergman disagreed with de Morveau. In the conclusion, he wished that the chemical nomenclature should be written in Latin, and not be translated to living languages. In that way, the language of chemistry would be international and universally understandable by every chemist. To conclude, Bergman wrote:

> Thus, then, I have pointed out a method, as I apprehend, both easy and simple, by which all the known salts, about fifty in number, may be each denominated in one or at most in two words. According to the first division [*i.e.* acids, alkalis, metals and earths], we have the genus only. Of the second, the double salts completely saturated are indicated by the adjective of their base ending in *atus*. In the third, the imperfect salts are known by the genitive of their base. The fourth contains the triple salts and those of several principles, which are expressed by the trivial names; and as in them we neither find the adjective of the base *atus*, nor the genitive, it is not possible that any ambiguity can arise. [41]

26.9 The Development of Nomenclature After the Death of Bergman

As Bergman died in July 1784, he was unfortunately unable to continue his work in the reformation of chemical nomenclature. Thus, the task of winning acceptance for a systematic binomial chemical nomenclature was left to de Morveau, who by this time had established himself as an authority of the subject. Lavoisier wrote in the preface to his *Elements of Chemistry*: "In a letter which the learned Professor of Upsal, M. Bergman, wrote, a short time before he died, to M. de Morveau, he bids him *spare no improper names; those who are learned, will always be learned, and those who are ignorant will thus learn sooner*" [42]. Unfortunately, Bergman's important contributions are not universally recognised today. For example, in the official guidelines on nomenclature of inorganic compounds ("The Red Book") by IUPAC,[4] de Morveau receives the sole credit [43].

[4]International Union of Pure and Applied Chemistry.

de Morveau was commissioned to write the chemical articles for the new *Encyclopédie Méthodique*, the first part appearing in 1786. In *Encyclopédie Méthodique,* the subjects were separated, so all chemistry articles were published in a separate set of volumes. An encyclopaedia was a good forum for promoting the new nomenclature, just as it had been for Macquer two decades earlier. While preparing the first part of the first volume of *Encyclopédie Méthodique*, de Morveau was still a convinced phlogistonist, but during 1786 he started to doubt [44]. Around February 1787, he went to Paris to see Lavoisier and stayed there for several months [45]. In Paris, de Morveau entered collaboration with Lavoisier, Berthollet and Fourcroy, a collaboration resulting in the seminal book *Méthode de nomenclature chimique* (Method of Chemical Nomenclature), published in the summer of 1787. The book is composed of a number of essays, and it was de Morveau who was mainly responsible for developing the new nomenclature. In *Méthode de nomenclature chimique*, the name *acide méphitique* was changed to *acide carbonique* and its salts were called carbonates. In the same manner, we got acide sulfurique (H_2SO_4) and acide sulfureux (H_2SO_3). The salts were referred to as sulfate de potasse (K_2SO_4) and sulfite de potasse (K_2SO_3). This nomenclature has largely survived until the present day, although we write copper sulphate rather than sulphate of copper in English.

Thus, we have seen how criticism against the old chemical nomenclature started to grow stronger in the eighteenth century but initially without any constructive suggestions, an exception being the proposal by Macquer. It was Bergman who applied Linnaeus' successful binomial nomenclature on chemistry, and de Morveau who refined it. On doing so, however, one important aspect of Bergman's work was lost: Bergman worked hard for an international nomenclature, and instead of his *Vitriolicum cuprum* we now have copper sulphate (English), Kupfersulfat (German), sulfato de cobre (Spannish), sulfate de cuivre (French), kopparsulfat (Swedish), kuparisulfaatti (Finnish), solfato rameico (Italian), siarczan miedzi (Polish), θειικός χαλκός (Greek), сульфат меди (Russian) and 硫酸铜 (Chinese).

References

1. Macquer PJ (1749) Elemens Chymie Theorique. Jean-Thomas Herissant, Paris, p 38
2. Wallerius JG (1765) Chemiæ Pysicæ Andra del; första och andra afdelningen… Lars Salvius. Stockholm, p 15
3. Lemery N (1686) A course of chemistry, 2nd. Walter Kettilby, London, p 80
4. Cronstedt (1758) Försök til mineralogie. Stockholm, p 123
5. Crosland MP (1962) Historical studies in the language of chemistry. Heineman, London, p 125
6. Crosland MP (1962) Historical studies in the language of chemistry. Heineman, London, p 138
7. Headrick DR (2000) When information came of age. Oxford University Press, Oxford, p 24
8. Bergman T (1770) Historien om Qvicksilvers föreningar med Koksalts-syra. KVA Handl 31:79–110
9. Bergman T (1772) Slutet af Historien om Qvicksilvers föreningar med Koksalts-syra. KVA Handl 33:193–205

10. Bergman T (1775) Framledne direct: och Kongl: Vet. Acad. Ledamots Herr H.T. Scheffers Chemiske Föreläsningar..., M. Swederus. Uppsala, p 8
11. Bergman T (1775) Framledne direct: och Kongl: Vet. Acad. Ledamots Herr H.T. Scheffers Chemiske Föreläsningar..., M. Swederus. Uppsala, p 69
12. Bergman T (1775) Framledne direct: och Kongl: Vet. Acad. Ledamots Herr H.T. Scheffers Chemiske Föreläsningar..., M. Swederus. Uppsala, p 104
13. Bergman T (1775) Commentatio de acido aëreo. Nova Acta Reg Soc Scient Upsaliensis 2:108–160
14. Cassebaum H, Schufle JA (1975) Scheele's priority for the discovery of oxygen. J Chem Educ 52:442–444
15. Bergman T (1784) Physical and chemical essays, vol 1. Murray J. London, p 18
16. Bergman T (1775) Disquisitio de attractionibus electivis. Nova Acta Reg Soc Scient Upsaliensis 2:161–250
17. Bergman T, Norell C (1775) Dissertio Chemica de Magnesia alba. Uppsala
18. Bergman T (1784) Physical and chemical essays, vol 1. Murray J. London, p xxxvii
19. Bergman T (1782) Sciagraphia regni mineralis, secundum principia proxima digesti. Leipzig, p 38
20. Bergman T (1787) Grundriß der Mineralreichs. Wien, p 66
21. Bergman T (1782) Sciagraphia regni mineralis, secundum principia proxima digesti. Leipzig, p 45
22. Bergman T (1787) Grundriß der Mineralreichs. Wien, p 73
23. Bergman T (1782) Sciagraphia regni mineralis, secundum principia proxima digesti. Leipzig, p 78
24. Schufle JA (1985) Torbern Bergman a man before his time. Cornado Press, Lawrence, Kansas, p 287
25. Carelid G, Nordström J (1965) Torbern Bergman's foreign correspondence. Almqvist & Wiksel, Stockholm, pp 100–138
26. Carelid G, Nordström J (1965) Torbern Bergman's foreign correspondence. Almqvist & Wiksell, Uppsala, p XXXVIII
27. Bergman T (1780) Opuscules chimiques et physiques, vol 1. L N Frantin, Dijon, p 404
28. Crosland MP (1962) Historical studies in the language of chemistry. Heineman, London, p 155
29. Guyton de Mourveau LB (1782) Mémoir sur les dénominations chimiques, la necessité d'en perfectionner le système et le règles pour y pavenir. Obs sur la Physique 19:370–382
30. Crosland MP (1962) Historical studies in the language of chemistry. Heineman, London, p 156
31. Leicester HM, Klickstein HS (1965) A source book in chemistry 1400–1900. Harvard University Press, Cambridge, MA, p 182
32. Crosland MP (1962) Historical studies in the language of chemistry. Heineman, London, p 158
33. Crosland MP (1962) Historical studies in the language of chemistry. Heineman, London, p 161
34. Crosland MP (1962) Historical studies in the language of chemistry. Heineman, London, p 163
35. Crosland MP (1962) Historical studies in the language of chemistry. Heineman, London, p 148
36. Bergman T (1784) Meditationes de Systemate Fossilium Naturali. Nova Acta Reg Scient Upsaliensis 4:63–128
37. Bergman T (1782) Commentationes e quarto novorum Reg. Scientarum Societatis Upsaliensis actorum tomo excerpt. Uppsal
38. Bergman T (1791) Physical and chemical essays, vol III. Mudie G, Fairbairn J, Evans J. Edinburgh, p 298

39. Bergman T (1791) Physical and chemical essays, vol III. Mudie G, Fairbairn J, Evans J. Edinburgh, p 300
40. Crosland MP (1962) Historical studies in the language of chemistry. Heineman, London, p 150
41. Bergman T (1791) Physical and chemical essays, vol III, Mudie G, Fairbairn J, Evans J. Edinburgh, p 314
42. Lavoisier A (1790) Elements of chemistry. William Creech, Edinburgh, p xxxii
43. (2005) Nomenclature of inorganic chemistry. RSC publishing, Cambridge, p 2
44. Crosland MP (1962) Historical studies in the language of chemistry. Heineman, London, p 175
45. Crosland MP (1962) Historical studies in the language of chemistry. Heineman, London, p 174

Scheele's and Bergman's Contributions to Pharmaceutical Chemistry

27

The work in a European pharmacy shop was primarily regulated by the pharmacopoeia, the handbook which described which medicines should be sold and how they were prepared. As Scheele has begun his career, the old Stockholm pharmacopoeia, *Pharmacopea Holmiensis galeno-chymica* from 1686, was still in action. It contained many old traditional ingredients such as human skull, mummy, dear heart, wolf teeth, reindeer horns and dried snakes. These were of course completely inactive, and the need for a reformation based on scientific ideas was evident. Such a reformation was induced by Linnaeus, and the result was the 1775 *Pharmacopoea Suecica*. The work on the new pharmacopoeia was carried out by *Collegium medicum*, the work being initiated in the early 1770s. Among the members of the board was "apothecary apprentice S."—Carl Wilhelm Scheele. It is quite remarkable that a young apothecary apprentice was included in such an important and official work. In 1778, a second edition of the pharmacopoeia was published, and when Scheele underwent his examination at *Collegium medicum* in late 1777, he agreed to contribute also to this second edition. The second edition appeared in a German translation and Scheele, who reviewed this book, concluded that the translator neither understood the subject nor the Latin language [1, 2].

It is most probably not a coincidence that Scheele published a number of papers describing improved methods for the preparation of medicines around 1775 and 1778. These papers are almost certainly results of Scheele's work on the pharmacopoeia, although the archives of *Collegium medicum* do not reveal Scheele's contributions. It seems that Scheele's role was to work out new procedures on commission by *Collegium medicum*. This work was time-consuming, and in February 1778, for instance, Scheele wrote to Bergman that he had to delay the work on *molybdæna* (Sect. 15.4) due to the work on the pharmacopoeia.

A. Lennartson, *Carl Wilhelm Scheele and Torbern Bergman*, Perspectives on the History of Chemistry, https://doi.org/10.1007/978-3-030-49194-9_27

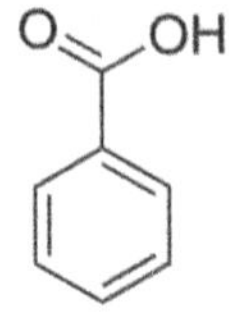

Fig. 27.1 Benzoic acid

27.1 Benzoic Acid

In 1775, Scheele published an improved method of preparing benzoic acid (Fig. 27.1) from benzoin [3], a resin from *Styrax* species. Traditionally, benzoic acid was obtained by pyrolysis of the resin, but Scheele showed that the product was impure. The hydrophobic resin was difficult to extract with water, but Scheele showed that the resin could be boiled with water and calcium oxide (unslaked lime) forming fairly soluble calcium benzoate.[1], Lennartson [4] Benzoic acid, being sparingly soluble in water,[2] could be precipitated from the solution with hydrochloric acid. Although benzoic acid (salt of benzoin) was known, it was Scheele who recognised it as an acid.

27.2 Mercurius Dulcis

The paper on mercury(I) chloride (*Mercurius dulcis*) [5] served as Scheele's inaugural lecture in the Royal Swedish Academy of Sciences (Sect. 19.4) and reports an improved preparation method without much theoretical discussion [6]. The traditional method was based on reduction of mercury(II) chloride (corrosive sublimate) with metallic mercury, but gave a product contaminated with mercury (II) chloride, which is more water soluble and thus more toxic than mercury(I) chloride. Instead, Scheele's method involved dissolution of mercury in nitric acid and precipitation of *Mercurius dulcis* with sodium chloride. The new method was, according to Scheele, simpler, cheaper and gave a more finely divided product. Also, the operator was not exposed to the toxic dust and vapours that inevitably formed during the trituration of $HgCl_2$ with Hg.

The method was criticised by a Desaive,[3] who claimed the product to be contaminated with mercury(II) sulphate. Scheele was apparently rather annoyed by this remark and published a semi-anonymous reply (signed "S…e") [7]. Scheele was far from impressed by Desaive's remark:

[1]Solubility 8.3 g/100 ml water at 80 °C.
[2]Solubility 0.27 g/100 ml water at 18 °C.
[3]The identity of Desaive is unknown to me.

> How well-founded does not this remark sound to someone who is not familiar with the basic foundations of chemistry. But what does a chemist say about this? Well, Mr Desaive is presumably not a chemist; he might have heard the bell ringing, but does not know in which church.

Scheele wrote that if Desavie only had taken a quick look in a table of affinities (Chap. 20), he would easily have found that mercury has a stronger affinity for hydrochloric acid than for sulphuric acid. Scheele disproved Desaive's "unnecessary remark" by showing that the product was free from mercury(II) sulphate [8]. As Scheele published his reply in a Swedish journal, it is perhaps not likely that his paper reached Desaive.

27.3 Emetic Tartar and Pulvis Algerothi

Bergman, being professor in chemistry and pharmacy, only presented a single dissertation related to pharmacy, a study of potassium antimonyl tartrate (emetic tartar) presented in 1773 [9]. Antimony compounds had been used to induce vomiting since the days of Basil Valentine (Basilius Valentinus) and his book *Triumph Wagen Antimonii* (*Triumphant Chariot of Antimony*), published in 1604.[4] There were several methods to prepare emetic tartar but unfortunately the strength of the emetic depended on its mode of preparation. The aim of this thesis was to find a reliable method to prepare an emetic of reproducible composition, or as Bergman put it: "In prescribing medicines, or in composing dispensatories, nothing is more necessary than that both the materials and method of preparation be so chosen as to be exactly alike in all cases and situations" [10]. The traditional source of antimony was antimony(III) sulphide, which was oxidised by ignition with potassium nitrate (saltpetre). This was the operation that led Scheele to the discovery of nitrous acid in Malmö a few years earlier (Sect. 8.2). Bergman concluded that the phlogiston content of the product was of crucial importance, that is, antimony had to be in its trivalent state. Not only did the ratio of Sb_2S_3 to KNO_3 matter, the method also affected the result. Either a mixture of the two substances could be ignited or the mixture could be poured into a red-hot crucible; the latter method gave the best result.

To find a suitable method, Bergman and his student Level treated several different antimony preparations with either tartaric acid, potassium tartrate (tartarised tartar) or potassium hydrogen tartrate (cream of tartrar). They found that antimony (III)oxychloride (SbOCl; *Pulvis Algerothi*) was the most uniform starting material, and that the best method of preparing emetic tartar was to boil antimony(III) oxychloride half an hour with either potassium tartrate or potassium hydrogen tartrate. A few years later, Scheele would perfect the method by introducing a new method of preparing *Pulvis Algerothi* [11].

---

[4]In Sweden, potassium antimonyl tartrate was used as an emetic until the 1960s.

Scheele's paper on *Pulvis Algerothi* is of considerable scientific interest [12]. *Pulvis Algerothi* was used as an emetic and laxative and as a precursor for the preparation of potassium antimonyl tartrate. The common procedure for the preparation of *Pulvis Algerothi* was to distill a mixture of antimony (*Regulus antimonii*) and mercury(II) chloride (*Mercurius corrosivus albus*) to afford antimony(III) chloride (*Butyrum antimonii*), which was hydrolysed with water to form the desired product:

$$2Sb + 3HgCl_2 \rightarrow 2SbCl_3 + 3Hg$$
$$SbCl_3 + H_2O \rightarrow SbOCl + 2HCl$$

Scheele noted that this procedure was awkward, expensive and dangerous. Scheele's analysis of the problem reveals a deep insight into the chemistry of antimony, he was, for instance, well aware that antimony occurs in three oxidation states.

Scheele explained that according to the prevailing theory, antimony had a higher affinity for hydrochloric acid than mercury, and therefore antimony would attract hydrochloric acid from mercury(II) chloride. Scheele did not accept this explanation as he knew that antimony in antimony(III) chloride is in the middle oxidation state or was "half-calcined" [13] to use Scheele's words and that hydrochloric acid could not oxidise (calcine) antimony. He continued: "it is well known that the mercury in the corrosive sublimate [$HgCl_2$] is not in a reguline form [i.e. in oxidation state 0], but in the state of a calx [i.e. in a higher oxidation state]". Scheele correctly concluded that antimony transferred phlogiston to (i.e. it reduced) mercury(II) chloride to mercury and the liberated hydrochloric acid united with the oxidised (calcined) antimony. Scheele concluded that the precursor should be "half-calcined", i.e. in oxidation state +III,[5] and finally derived at the following procedure: antimony(III) sulphide (*Antimonum crudum*) is mixed with potassium nitrate (saltpetre) and ignited in an iron mortar. The product is powdered and mixed with water, sulphuric acid and common salt. This was cheaper than using hydrochloric acid. The mixture is heated gently under stirring in a sand bath for 12 h. After cooling, the mixture (containing $SbCl_3$) is filtered and mixed with hot water, which precipitates *Pulvis Algerothi*.

27.4 A New Method for Preparing Magnesia alba

In 1785, Scheele published a short paper [14] describing an improved method for the preparation of *Magnesia alba*, $Mg_5(CO_3)_4(OH)_2 \cdot 5H_2O$, an important substance in eighteenth-century medicine [15]. The standard method was to precipitate *Magnesia alba* from a solution of magnesium sulphate (English salt) with

[5] "Fully calcined antimony" would have corresponded to antimony in oxidation state +V.

potassium carbonate (vegetable alkali). This gave potassium sulphate (*Alkali veg. vitr.*)[6] as a by-product which was not very useful. Scheele's method was to first treat magnesium sulphate solution with sodium chloride; from the solution sodium sulphate (*Sal Glauberi*) could be crystallised in the cold leaving magnesium chloride in solution. Sodium sulphate was far more useful than potassium sulphate. The stoichiometry in Scheele's procedure is quite accurate; for 1 mol of sodium chloride, he used 0.49 magnesium sulphate. The yield of sodium sulphate was around 34–48%.

Scheele noted that the operation had to be performed in the winter; in the summer it would invariably fail. At room temperature, a solution prepared according to Scheele's instructions is unsaturated. Furthermore, sodium sulphate is prone to form highly supersaturated solutions. The author of this book found that such a solution could be kept for hours at −18 °C without crystallisation. Upon seeding with sodium sulphate dekahydrate, it readily crystallises at 0 °C. Finally, *Magnesia alba* was precipitated in the usual manner from the magnesium chloride solution by the addition of an excess of potassium carbonate.

Bergman also studied *Magnesia alba* in a dissertation published in 1775 [16]. Bergman's objectives were, however, different than those of Scheele, and Bergman's work brought forward new evidence of the elemental nature of *Magnesia* and was of little interest for a pharmacist.

27.5 Other Works by Scheele Related to Medicine

In addition to the papers related to the pharmacopoeia, Scheele published two more papers where he was consulted in medicinal questions. Shortly prior to his death, Scheele wrote a letter to Bergius, which was subsequently published [17]. This letter reports the analysis of a medicine sent to him by the influential businessman Abraham Grill (1719−1799), who used the medicine to treat venereal diseases among his staff. The medicine was imported at great expense from Amsterdam, and Grill was keen to know if it could be prepared at a lower cost. Scheele found it to be a solution of mercury(II) chloride and iron(II) chloride and gave a recipe for its preparation [18]. Mercury, although toxic to humans, was a common cure for syphilis.

A second letter, also written to Bergius and subsequently published, Scheele [19] concerned a device for oxygen treatment of tuberculosis patients. Bergius and his colleague surgeon Herman Schützercrantz (1713−1802) was treating an apparently very wealthy and important but unnamed lady with oxygen delivered in small amounts from a pharmacy in Stockholm. The problem was that the patient wanted to be able to breathe oxygen for prolonged times, and a method for preparation of oxygen on a large scale was needed. Scheele suggested building an airtight cupboard connected to a retort where potassium nitrate (saltpetre) was heated [20].

[6]Note that Scheele is using Bergman's nomenclature.

Fig. 27.2 Scheele's apparatus for large-scale production of oxygen. Potassium nitrate was heated in an iron retort, the oxygen was bubbled through a potassium carbonate solution to remove nitric acid vapours and collected in balloons

Alternatively, the patient could breathe oxygen from large balloons (Fig. 27.2). The idea of building an airtight cupboard for the patient to sit in was rather unpractical, and would not have worked in reality if constructed according to Scheele's description. The turbulence encountered when the cupboard was opened would have dispersed the valuable oxygen in the room. Unfortunately, the patient died before the methods could be tested. The title of the paper (added by the editor), "*The effect of dephlogisticated air* [oxygen] *on diseases*", is misleading, as the paper is not concerned with the medical effects of oxygen, but only with the most practical method of preparing and administrating it on a larger scale.

References

1. Scheele CW (1778) Bref ifrån Hr. C. W. S. Til Hr. D. v. S. rörande Tyska Öfversättningen af Pharmacopoea Suecia, Stockholms Lärda Tidningar nr 22:287–291
2. Lennartson A (2017) The chemical works of Carl Wilhelm Scheele. Springer, Cham, p 56
3. Scheele CW (1775) Anmärkningar Om Benzoë-SaltetAnmärkningar Om Benzoë-Saltet. KVA handl 36:128–133
4. Lennartson A (2017) The chemical works of Carl Wilhelm Scheele. Springer, Cham, p 29
5. Scheele CW (1778) Sätt at tilreda Mercurius dulcis, på våta vägen. KVA Handl 39:70–73
6. Lennartson A (2017) The chemical works of Carl Wilhelm Scheele. Springer, Cham, p 51

7. Scheele CW (1782) Anmärkningar vid Herr Desaives Anmärkning, recenserad uti denna Veckoskriftens 2:dra Band, sid 158, 159, rörande præperation af Mercuris Dulcis efter Pharm. Svec. Ed. Alt. Vecko-skrift för läkare och naturforskare, 3:145–146
8. Lennartson A (2017) The chemical works of Carl Wilhelm Scheele. Springer, Cham, p 73
9. Bergman T, Level A (1773) Dissertatio pharmaceutica de stibio tartarisato, Uppsala
10. Bergman T (1784) Physical and Chemical Essays, vol 1. J. Murray, London, p 395
11. Scheele CW (1778) Ett beqvämare och mindre kostsamt sätt at tilreda Pulvis Algerothi. KVA Handl 39:141–145
12. Lennartson A (2017) The chemical works of Carl Wilhelm Scheele. Springer, Cham, p 52
13. Scheele CW (1901) The chemical essays of Charles-William Scheele. Reissue of 1786 edn. Scott, Greenwood & Co., London, p 162
14. Scheele CW (1785) Anmärkning vid tilredning af Magnesia alba. KVA Nya Handl 6:172–174
15. Lennartson A (2017) The chemical works of Carl Wilhelm Scheele. Springer, Cham, p 88
16. Bergman T, Norell C (1775) Dissertio [sic!] chemica de magnesia alba, Uppsala
17. Scheele CW (1786) Bref til Professor P. J Bergius från C. W. Scheele, dat. Köping den 10 Martii 1786 Vecko-skrift för Läkare och Naturforskare, 7:246–249
18. Lennartson A (2017) The chemical works of Carl Wilhelm Scheele. Springer, Cham, p 96
19. Scheele CW (1786) Dephlogisticerad Lufts värkan i SjukdomarDephlogisticerad Lufts värkan i Sjukdomar. Vecko-skrift för Läkare och Naturforskare 7:288–291
20. Lennartson A (2017) The chemical works of Carl Wilhelm Scheele. Springer, Cham, p 85

The End of the Story 28

Within 2 years of time, the chemical community lost two of its brightest minds, and not until Berzelius' days would Sweden excel in chemistry again. This is the end of the story of Scheele and Bergman.

28.1 Bergman's Illness and Death

As a child, Bergman had a relatively good health apart from suffering from problems with constipation throughout his life [1]. He survived the "Uppsala fever" in 1755 but the cold sessions in his laboratory in the winter of 1769 (Chap. 11) seriously degraded his health. He did not recover fully during the summer, and in the autumn he suffered from severe headache: he could feel the blood being drawn towards his head and feel the pulse in his temples [2]. This may very well be indicative of high blood pressure [3]. He was subjected twice to bloodletting, not surprisingly without any lasting relief. After another summer of poor health, he found that he suffered from haemorrhoids that started to bleed copiously for a few months. During this time, Bergman felt better, but as the bleeding stopped, he suffered from severe pain that kept him awake during the nights [4]. In December 1771, he wrote to Gahn that he had been "attacked by illness" the previous winter, and this winter his illness had set in even earlier [5]. A few months later, in March 1772, he wrote to Gahn that he had been forced to study medicine in order to find the cause of his illness [6]. By then, his health had improved with the milder weather. He advised Gahn to take care of his health since "Everything is to nothing, when the health is gone". Bergman's problems with constipation may have been the cause of his haemorrhoids [7]. In 1774, he suffered from another fever decease, but from 1775 his health was relatively good for some time. In 1781, his problems spread to his chest, as he put it, and he started coughing, especially in the mornings [8]. In cold weather, he suffered from fatigue and could not leave his house. In his last preserved letter to Gahn, dated February 24, 1782, he wrote that he had been in

A. Lennartson, *Carl Wilhelm Scheele and Torbern Bergman*, Perspectives on the History of Chemistry, https://doi.org/10.1007/978-3-030-49194-9_28

bed since August, spitting blood several times [9]. A triggering event seems to have been a leisure trip that summer, when he attempted to reach a small island in a small boat, which unfortunately capsized and Bergman fell into the water [10]. In the last minute, he got a grip on an adjacent, larger boat, but the chock and cold water had a very negative effect on his health. It is highly unlikely that Bergman could swim, so the chock is understandable. In 1782, he wrote down his autobiography ("in haste during my illness in January 1782") apparently fearing for his life. That summer he travelled to Medevi spa to recover his health and he also took the opportunity to travel via Loka spa [11]. Not only did he seek to cure his health, the trip also resulted in two publications on the water of Loka and Medevi, respectively (Sect. 17.6). At this point, his assistant and former student Johan Afzelius had to step in and take over Bergman's duties as teacher, and after 1782 no theses were defended by students of Bergman.

In the spring of 1784, Bergman was weak and thin and people in his surrounding had little hope of him recovering [12]. He decided to travel to the spa of Medevi earlier than usual in hope of some relief and he travelled via Köping to meet Scheele for, as it would turn out, the last time. Scheele described him as weak, anaemic and thin [13]. He arrived in Stockholm soon after Whitsun,[1] he was now so weak that he had to stay indoors for several days, where after he attended his former students Geijer's and Schwartz's smelting experiments with oxygen [14]. Before leaving Stockholm, the Finland Student Nation of Uppsala University, for which Bergman was inspector, presented him with a gold medallion with his portrait engraved by Ljungberger (Sect. 3.3) [15]. Accompanied by his wife, he finally undertook the last part of the trip [15].

The cause of Bergman's death has, to the best of my knowledge, not been discussed previously in print. There are, however, two sources that give enough information for an interpretation: Bergman's own autobiography and a report [16] written by spa curator and physician Anders Magnus Wåhlin (1731–1796), a former student of Linnaeus. This report describes Bergman's condition and treatment day by day during his stay at Medevi, as well as an account of the autopsy.

Bergman arrived at Medevi on June 6, and by then he was so weak that Wåhlin realised that a mineral water cure would make no difference. After 3 weeks in Medevi, Bergman's health started to turn to the worse, his appetite decreased and he had difficult to sleep. During his fourth week in Medevi, Bergman had to spend most time in bed. His cough got worse, and he was spitting blood. His haemorrhoids started to bleed again. On July 1, at about 1:30 pm, Bergman tried to rise from his bed to welcome a visitor, but most probably suffered a stroke [7]: he fell back to his bed suffering from a series of convulsions. He regained consciousness, but was very weak and needed full attention from his physicians. Crell's statement [17] that Bergman finished his last paper, *mineralogical remarks*, just 2 days before his death are obviously in error. His haemorrhoids were bleeding copiously, and he had neither appetite nor thirst and was delirious. In the afternoon of July 8, his weakness increased and he lost his speech; his breath became heavier and his sleep

[1]May 30.

deeper. At 11 pm, it was obvious that the end was near and half an hour later Bergman passed away at an age of 49: his body underwent a few weak twitches and finally relaxed.

At 9 am the following morning Wåhlin, Carl Ribben (1734 or 1738–1803)[2] and the spa feldsher conducted the autopsy, the results of which were also published [16], and strongly indicates that Bergman suffered from lung tuberculosis leading to severely weakened general condition [7]. The ultimate cause of death was probably a stroke, whether ischemic or haemorrhagic cannot be determined as the physicians performing the autopsy lacked a saw and thus could not access Bergman's brain [16].

28.2 Bergman's Funeral

Bergman's body was not brought back to Uppsala but buried under the floor in the Odencrantz mausoleum (Fig. 28.1) on the churchyard of Västra Ny Church, close to Medevi. This mausoleum was mainly intended for the Odencrantz family but was also used occasionally for more prominent guests who died during their stay at Medevi. The funeral service was held on July 13, in Västra Ny Church (Fig. 28.2) [16]. Among the guests were three members of the Diet (*riksråd*; two of which were Scheffer and Falkengren [18]) and the director of Medevi, Wåhlin, held a speech [19]. The speech, which was rather short and perhaps not a masterpiece, was published later that autumn [20]. Wåhlin spoke more generally over death, the deep hole that Bergman left after him and the great loss that Wåhlin assured would spread to the Royal Throne. Wåhlin did not speak about Bergman's accomplishments or his character. As director of the spa, Wåhlin took the opportunity to assure that Bergman's pains had been relieved on previous visits to Medevi, and that, although Bergman unfortunately had died, many people found their relief in Medevi. "This dust [Bergman's body]", Wåhlin concluded, "will be honoured by a posterity, and it should also be noted for the future, that it has been accompanied here by an honourable and tender company". Today, the mausoleum is used to store old furniture from the church, and there are no visible signs of any graves. An epitaph of Bergman is, however, found inside the church.

With only 4 days between Bergman's death and the funeral, it is unlikely that any guests from Uppsala attended the ceremony. Scheele was of course moved by the death of his friend. "Bergman's death has affected me very deeply", he wrote to Wilcke [13], who became Scheele's closest friend after Bergman's death.[3] Scheele sent another letter to Crell on July 19, informing him of Bergman's death. This letter is cited by Crell in a paper on tungstic acid and tungsten [21].

[2]Another student of Linnaeus.

[3]At least outside Köping; we now essentially nothing about his relationship to people that he did not have to write letters to.

Fig. 28.1 The Odencrantz mausoleum, where Bergman is buried. *Photo* Anders Lennartson June 2018

Fig. 28.2 Västra Ny Church, where Bergman's funeral was held. *Photo* Anders Lennartson June 2018

28.3 Bergman's Manuscripts and Collections

Bergman had paid much of his instruments with private money, and since they constituted the main part of his assets, he could not afford to donate them to the University, as he feared it would put his widow in a strained economic situation [22]. When Bergman died, his widow initially refused to hand over Bergman's private collections and papers to the University [23]. In 1777, she had been granted a pension of 150 riksdalers in the case of Bergman's death (Chap. 16), but she now requested the amount to be raised to 200 riksdalers and in exchange she would transfer Bergman's books and instruments to the University. The King approved this arrangement in February 1785. The University Chancellor now gave the order of an extensive inventory, but the committee found the task difficult to perform. For example, the catalogue of the mineral collection was missing, and many samples were unlabelled and some documents, e.g. those concerning gun powder manufacture, were categorised as classified by the widow. Clearly, Mrs. Bergman must have had some insight into her husband's work and documents, but one should not from this conclude that she acted as his assistant or that she was a scientist on her own rights. The hostility experienced by the committee was most probably in large due to the fact that Bergman's predecessor and sworn enemy Wallerius was one of the appointed experts, which Mrs. Bergman regarded as an insult. The inventory was finally finished in 1786. The book collection, comprising 1090 volumes, was first kept at the chemistry department at the request of Bergman's successor but was later transferred to the University Library [23]. Bergman's manuscripts were not included in the deal but were presented to the University by the widow in 1793, on the provision that his correspondence would remain unopened for 15 years [24]. Margareta Catharina Bergman died on April 22, 1800, in Uppsala, aged 64. She did not remarry after Bergman's death.

28.4 Bergman's Successor

Unlike Bergman's appointment as professor, the appointment of his successor was a very swift process. In fact, the Chancellor made the extraordinary decision on September 22 that the suggestion for the professorship had to be submitted without delay in order not to disrupt the chemistry teaching. Thus, the successor was appointed before the mandatory application time had run out. Bergman's student Johan Afzelius, who had been his assistant since 1780 and who had covered up for Bergman as a lecturer for two years while Bergman was ill, was unanimously suggested on October 8, and Afzelius was appointed professor by the King on December 13, 1784. This hasty appointment actually seems to have been completely pointless, as Afzelius could not assume his professorship until nearly a year later, October 4, 1785, due to poor health [25].

Although Bergman had very high thoughts of Afzelius, his appointment would turn out to be less fortunate for Swedish chemical research. Afzelius did not carry out any original research and during his 36 years as professor—he retired in 1820—only two theses were defended by his students [26]. He did not publish a single paper in the *Transactions of the Royal Swedish Academy of Sciences*. The claim [27] that Afzelius stayed true to the phlogiston theory until his death is, however, not true [28]. While Afzelius did not share Bergman's interest in chemical research or writing, he seems to have been an appreciated and knowledgeable teacher and it should be kept in mind that this was the main duty for a professor. Few professors were as ambitious researchers as Bergman at the time.

28.5 Scheele's Illness and Death

Unlike Bergman, Scheele had a good health, except for a short period of ache in a hip in 1782 [29]. In the fall of 1785, however, Scheele started to feel ill [30]. It started with swollen eyelids, which he treated with Spanish fly as he lacked leeches. This was followed by fever and aching joints. He was now treated with leeches and bloodletting, but with little effect: in January 1786, his condition got more serious. He had pain in arms and legs, he got weak and tired and his fingers got swollen. He started coughing and suffered from spasms in his respiratory passages. It became hard for him to run his shop and to carry out his experiments. He lost appetite and felt cold, a worrying sign as he had always hated warm rooms [30]. According to Margaretha Pohl, Scheele tried to continue with his work as normal, and when she asked him how he felt, he would ask sarcastically whether she was a doctor and could help him [31]. Despite his poor health, Scheele finished the manuscript of what would be his last paper, *On Sal essentiale Gallarum*, which was sent to Wilcke on February 16. In the letter following the manuscript, he wrote: "I have since last November been a gout patient. The foot, knee, arms and hands are in pain, and I feel no relief regardless of what medicine I use". This last manuscript was mainly based on old results; the isolation of gallic acid by Scheele is mentioned in Gahn's notes from spring 1770 (Sect. 13.2).

Scheele also had to defend a paragraph in his book *On Air and Fire* against German chemists Johan Friedrich Göttling (1755–1809) and Johan Christian Wiegleb. Scheele had performed experiments with "pyrophorus", also known as Homberg's pyrophorus, a self-igniting substance. It was initially obtained by Homberg during attempts to transmute mercury to silver, experiments that for some reason involved heating of excrements with alum [32]. It was later found that the excrements could be replaced by other combustible materials such as charcoal. Scheele had found that alum could be replaced by potassium sulphate and that the pyrophor consisted of potassium polysulphide (liver of sulphur) and charcoal. Both moisture and air were needed for spontaneous ignition. Scheele had to defend these findings and give clearer instructions for the new formulation [33]. Scheele was correct, in that no alum was required but that the reaction required potassium (fixed alkali).

Scheele's last documented chemical studies concerned the decomposition of nitric acid by sunlight [34]. When, in February 1786, Wilcke asked whether these experiments could be published, Scheele responded that he would repeat a few experiments the following summer, probably due to lack of sunshine in the winter [35]. Scheele would not live long enough, but a preliminary account was published by Crell [36].

On May 12, 1786, Scheele wrote to his old friend physician Abraham Bäck in Stockholm to ask for help, but it was too late. The letter, where Scheele describes his symptoms and the medication he had taken, is quite moving. A week later, Scheele had apparently realised that the end was close, since he married Margaretha Pohl in order to restore the ownership of the pharmacy to her. She was now aged 35 and had worked for Scheele during his years in Köping. Scheele passed away at 11:30 am on May 21, 1786, aged 43, with his friend Vicar Carl Johan Ahlström (1728–1790) by his side.

A few hours after Scheele's death, physician Birger Martin Hall (1741–1815), another former student of Linnaeus, arrived. He performed an autopsy [37], the results of which have not been preserved. Scheele himself described his condition as gout, while the official records state the cause of death as tuberculosis. In the 1930s, medical historian Birger Strandell studied Scheele's symptoms and concluded that he most likely suffered from rheumatic polyarthritis [38]. A common complication is different types of heart problems affecting the blood circulation to the lungs causing increased mucus production and coughing. The probable cause of death was heart failure. Scheele's widow, however, was convinced that it was exposure to chemicals that caused the death of both Scheele and Bergman: "Sure he experimented, through which he took his death, like Professor Bergman" [31]. Several biographies over Scheele have also suggested that Scheele's death was caused by his experiments but that seems not to be the case.

Scheele had no children, and there are thus no direct descendants of Scheele, although German chemist and spy Walter T. Scheele, who was arrested in the United States in 1916, claimed to be so [39].

28.6 Scheele's Funeral

The funeral of Scheele was held in the church of Köping on May 28, 1786. His widow assured Wilcke in a letter that it was the most honourable funeral held in the city, and that it had cost quite a bit of money [40]. The service was held by the vicar of Köping, Carl Johan Ahlström, who was a personal friend of Scheele. Ahlström held a speech, which was subsequently printed [41].[4] The speech is of much higher literary quality than Wåhlin's speech at Bergman's funeral.

> "How our blessed friend have been in the learned world, where he shined among the brighter stars, we will doubtless soon see described by the learned societies, where he was of the outermost value. It is above and outside our field of vision, to know and estimate his merits in this respect, neither I nor You, my sorrowful audience, could follow his tracks in the most hidden corners of Nature, where he resolved compounded bodies into their primary elements and from their actions searched his way to the causes. We have known him as a member of the parish, as a citizen and burgher. [—] He was like all other men, and consequently he carried the seeds for the same passions and desires for the worldly, but did he not show, during his entire life, [...] that he by religion and reason had found the invaluable method to kill these passions and harness these desires. Can anyone say that our blessed friend fell in love with anything but his science or his duties? He did not seek the honour, but fled from it, even when it searched for him [...]".

Scheele was put to rest at the Köping cemetery, but with time the exact location of his grave was forgotten. When the Swedish Pharmaceutical Society planned to erect a monument on his grave for the 40th anniversary in 1826, the location of the grave was already forgotten [42]. Some years later, however, when the church caretaker Lars Gillberg was digging a grave at the cemetery, he found a silver-plated copper plate with the inscription: "Carl Wilh: Scheele Apothecary in Köping Member of the Royal Academy of Sciences in Stockholm, Berlin, Turin etc. Born in Stralsund 9 Dec. 1742, married 19 May 1786, dead 21 May same year. [...]". The plate, which once was attached to Scheele's coffin, is now a part of the collections of the Swedish Pharmaceutical Society. In 1928 a memorial stone was erected at the place where Gillberg's son claimed that the plate had been found (Fig. 28.3) [43].

News did not always spread fast in the eighteenth century, and Scheele's death was initially an unconfirmed rumour until a chocked Lorenz Crell received more detailed news. "Carl Wilhelm Scheele is dead!! This says every experienced chemist everything" [44]. "It was a great misfortune that he should die away!!", Gahn wrote to Wilke in 1788 [45]. "I have again admired him this winter, as I read a newly published book by Kirwan on phlogiston and the constituents of acids".[5]

Scheele was survived by his mother, who died in 1788, and by two of his brothers who died in 1817 and 1825, respectively. After Scheele's death, his widow once again faced the same situation as in 1775 and had to find a new apothecary for the pharmacy. Mathias Georg Bölckou was recruited, and apparently the new apothecary was more in her taste, as she married him 2 years later, in 1788. Thus, three consecutive apothecaries in Köping were married to the same woman. Her son from her first marriage, Herman Emanuel Pohl, died in 1789, aged 14, and she passed away herself on November 9, 1793, aged 42.

[4]The speech was reprinted by the Swedish Pharmaceutical Society in 1936 on the 150 anniversary of Scheele's death.

[5]*An Essay on Phlogiston and the Constituents of Acids*, published in London 1787. As an opponent of Scheele, Kirwan ignored Scheele's work in this book, both his contribution to the discovery of oxygen and the new acids he had discovered.

Fig. 28.3 Scheele's supposed final resting place with its twentieth-century stone with the church in the background. *Photo* Anders Lennartson, June 2014

28.7 Scheele's Manuscripts

It was immediately realised that there could be notes on important unpublished discoveries among Scheele's papers. Physician Birger Martin Hall who arrived at Scheele's home a few hours after his death spent two hours collecting all papers, notes and letters he could find and in the presence of Mrs. Scheele and her father, he tied a string around the documents and was assured that they would not be thrown away or handed over to anyone without his permission. The widow was soon contacted by Wilcke in Stockholm, who offered economical compensation in exchange for the papers, which ever since have been stored in the archive of the Royal Swedish Academy of Sciences. Wilcke's intension was to use the papers for his memorial lecture over Scheele in the Academy, but upon his death 10 years later, the lecture was still not finished. It was finally completed in 1799 by physicist Carl Gustaf Sjöstén (1767–1817), acting Secretary of the Academy. The lecture was subsequently printed in 1801. In the introduction, Sjöstén, who unlike Wilcke did not know Scheele personally, wrote that he had only changed the beginning and end of the lecture, but Boklund found several alterations when he compared the printed lecture with Wilckes manuscript, where the text appears in several versions [46]. For example, Sjöstén claimed that Scheele's papers were close to useless, while Boklund's conclusion was that they were never thoroughly examined.

Scheele's notes are very difficult to read, even for an expert. They were mostly written in German, but after many years in Sweden, Scheele's language became less standard, and this in combination with many abbreviations and Scheele's peculiar handwriting means that few have tried to decipher them.

The first systematic study of Scheele's papers was made by Adolf Erik Nordenskiöld (1832–1901) for the 150th anniversary of Scheele's birth in 1892 [47]. Nordenskiöld is now best remembered as an arctic explorer and the first to cross the Northeast Passage in 1878–1879, but he was primarily a mineralogist and his father was an acquaintance of Berzelius and had met Gahn. Nordenskiöld was assisted in his work by Elin Bergsten from the National Swedish Archive who made the actual deciphering. Nordenskiöld's work, published in Swedish [47] and German [48], includes many of Scheele's letters, including those to Bergman (stored at Uppsala University Library), Gahn, Gadolin, Wilcke, Bergius and Wargentin and a selection of laboratory notes (stored at The Royal Swedish Academy of Sciences). Nordenskiöld made some alterations of the text to make it more accessible, he has, for example, replaced all chemical symbols with the old chemical names he found appropriate. During his work, Nordenskiöld was the first to prove Scheele's priority over Priestley for the discovery of oxygen (Chap. 21).

For the 200th anniversary of Scheele's birth in 1942, the Royal Swedish Academy of Sciences published a collection of Scheele's laboratory notes edited by physicist Carl Wilhelm Oseen [49]. This work consists of two volumes, one with facsimiles and one with transcriptions; it was not actually sold to the public but presented to selected institutions; still copies can occasionally be found in antiquarian book shops. An important reason for the publication was to ensure that the manuscripts would survive the war. This was a very real danger; the notebooks of John Dalton were, for example, destroyed during the bombing of Manchester in World War II. The title "Manuscripts 1756–1777" was criticised [50], as Scheele did not move to Sweden until 1757 and the oldest preserved documents originate from his time in Malmö. Also, the volume contains no manuscripts written after 1775. In addition to changing chemical symbols with names, Oseen corrected grammatical irregularities and spelled out abbreviated words.

A selection of papers known as "the Brown Book" (Fig. 28.4) since it was bound in brown leather boards in 1829 was left unexplored as it was regarded particularly difficult to read until Uno Boklund, after 10 years of struggle, published it in facsimile with transcriptions and comments in 1961 [51] with an English edition appearing in 1968 [52]. Boklund concluded that the Brown Book consisted of manuscripts (letter concepts, notes and concepts for papers) written after Scheele moved to Köping mixed with some older manuscripts that Scheele had picked out from a collection of papers he had brought with him from Uppsala. Boklund estimated that Scheele's laboratory journals contain around 15,000–20,000 experiments.

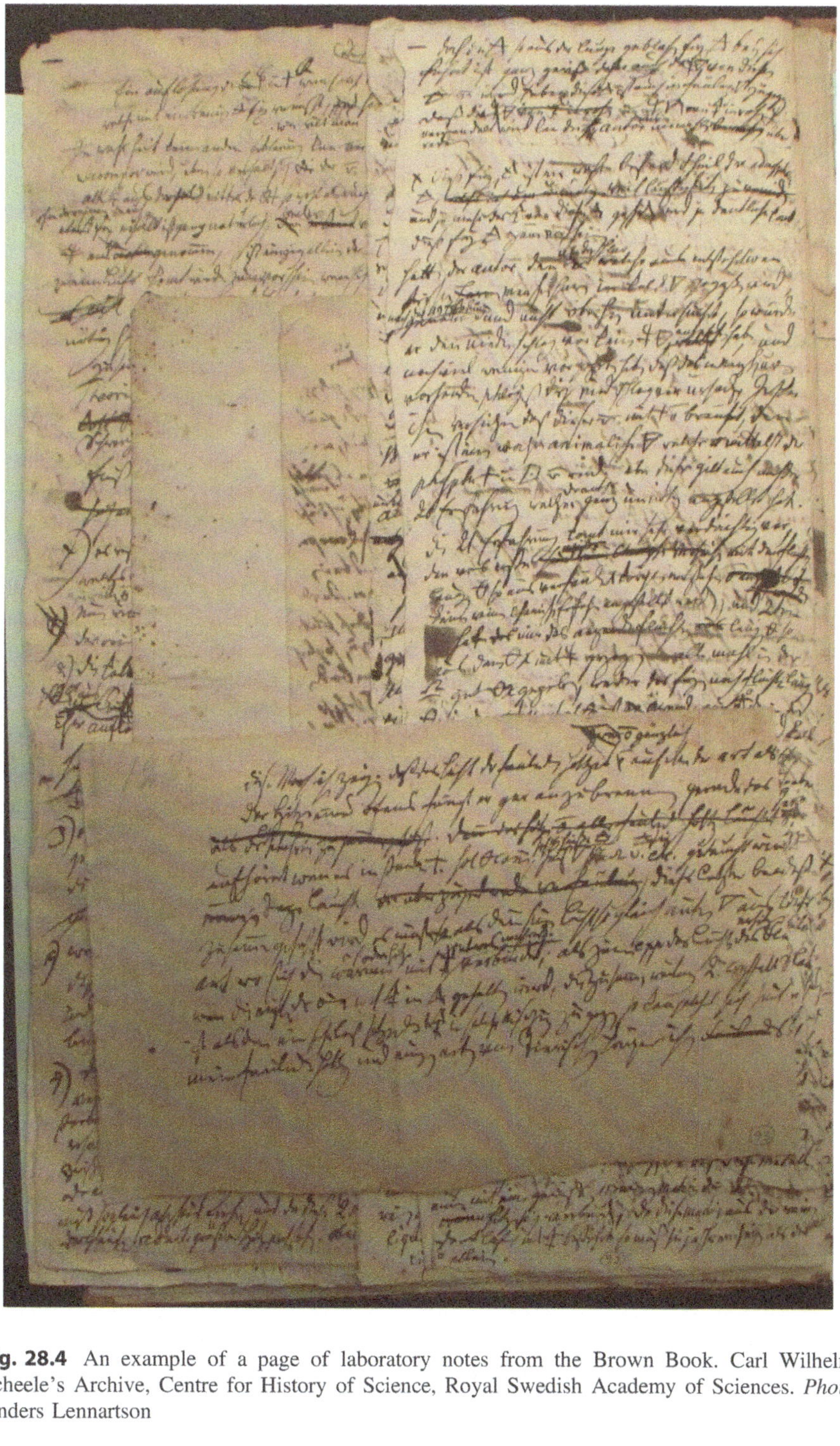

Fig. 28.4 An example of a page of laboratory notes from the Brown Book. Carl Wilhelm Scheele's Archive, Centre for History of Science, Royal Swedish Academy of Sciences. *Photo* Anders Lennartson

28.8 How We Remember Bergman and Scheele

Today, Scheele is by far the more well known of the two friends. In part, this is due to the nature of their work. Scheele discovered substances that are widely used today, while Bergman developed ideas and methods that were perfected by others after his premature death. Thus, Scheele's accomplishments have been easier to grasp and refer to compared with Bergman's more abstract work. Bergman's reputation also suffers from the fact that he worked in so many fields, while Scheele devoted his energy entirely to chemistry. At the best, Bergman is credited as the father of analytical chemistry and in Sweden he is sometimes credited as the father of the mineral water (or even soft drink) industry. In addition, Scheele's background as a simple apothecary apprentice has probably captured peoples' imagination and made him more interesting than Professor Bergman.

As a wave of patriotism swept over Sweden in the late nineteenth century, Scheele was turned into a national hero [53], entirely disregarding his German background. The 100th anniversary of Scheele's death in 1886 and the 150th anniversary of Scheele's birth in 1892 were thus celebrated with the erection of a statue, publication of a handful of short biographies and even a specially composed piece of music. It is a good guess that all of this would have made Scheele rather uncomfortable. Stralsund, where Scheele was born, became a city in East Germany in 1949, and for the 200th anniversary of Scheele's death, a symposium was held in his honour in East Berlin. The published lectures [54] contain several references to Friedrich Engels and even a quote from General Secretary Erich Honecker, who used Scheele as a good example of the good relationship between Sweden and East Germany. This is quite remarkable, as Scheele had no interest in politics whatsoever. Today, there are streets named after Scheele in about a dozen Swedish cities (Fig. 28.5), and there is a Scheele laboratory at the Karolinska Institute in Stockholm. He has a pharmacy in Stockholm and a School in Köping named after him. Every year since 1961, the Swedish Pharmaceutical Society awards the Scheele Award to prominent researchers in the field of drug discovery.

Compared to Scheele, Bergman has received far less recognition after his death. There are, for example, no streets or institutions named after Bergman, and there are no statues of Bergman, although the facade of the former Department of Chemistry at Uppsala (built in 1901–1904) is decorated with busts of Bergman, Scheele and Berzelius (Fig. 28.6); busts of Bergman and Scheele are also found on the facade of the physics building (which used to be a combined physics and chemistry building) of Chalmers Institute of Technology in Gothenburg, built in 1926 (Fig. 28.7). In addition, there is an urn with Bergman's portrait on the yard of the Västergötland student nation of Uppsala University (Fig. 28.8).[6] The Torbern Bergman medal (Fig. 3.7) is presented to researchers in the field of analytical chemistry by the Swedish Chemical Society every second or third year since 1967, and the winner is

[6]Bergman belonged to the Västergötland Student Nation, but the Nation was housed on a different address in Bergman's days.

Fig. 28.5 The Scheele street in Lund, close to the department of chemistry. *Photo* Anders Lennartson, May 2015

Fig. 28.6 The busts of Bergman, Berzelius and Scheele on the façade of the former Chemistry building of Uppsala University. *Photo* Anders Lennartson, June 2016

Fig. 28.7 The busts of Bergman and Scheele on the façade of the Physics building of Chalmers University of Technology, Gothenburg. The other busts (not in the photo) depict Wilcke, Ångström, Celsius, Berzelius and Nobel. *Photo* Anders Lennartson, September 2018

Fig. 28.8 Urn with Bergman's portrait on the yard of the Västergötland student nation of Uppsala University. *Photo* Anders Lennartson, April 2015

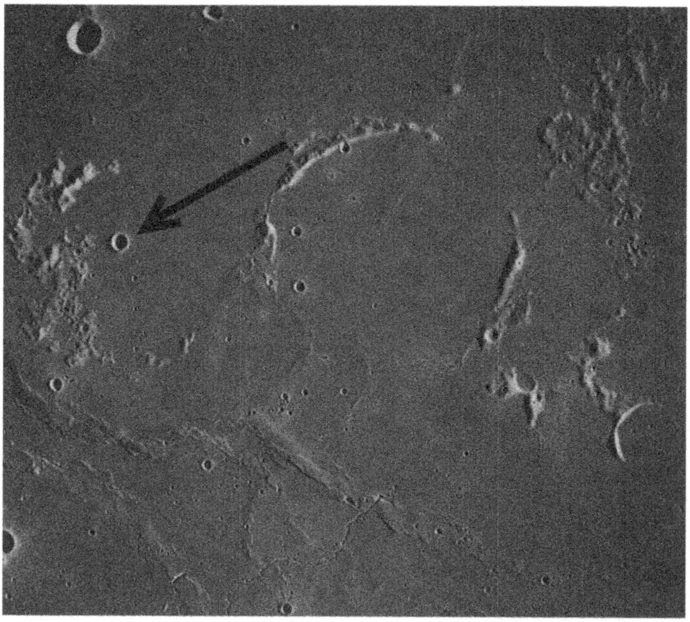

Fig. 28.9 Lunar impact crater Scheele (marked with an arrow) photographed from an altitude of 123 km in April 1972 by the metric camera in the Scientific Instrument Module on-board Apollo 16. *Photo* NASA. Image Reference AS16-M-2990. The image has been cropped

presented a Wedgewood porcelain plaque with Bergman's portrait in profile (Sect. 3.3).

Scheele is one of the characters in the play *Oxygen* by American chemist Carl Djerassi and Nobel laureate Roald Hoffmann which premiered in 1999. Scheele is also the main character in the play *Man Without a Face* by B. K. Cullen (pseudonym for Hungarian-American chemist Bela Kleiner) written in Hungarian in 2004 and translated to English in 2008, but which to the best of my knowledge have been never performed on stage. Bergman appears as a character in the 1979 novel *Flower King* (*Blomsterkungen*) about Carl Linnaeus by Swedish novelist Rune Pär Olofsson.

Both Bergman and Scheele have been honoured by having lunar impact craters named after them [55]. The Scheele crater (Fig. 28.9) is a 4 km crater located at Oceanus Procellarum on the lunar near side. The Bergman crater (Fig. 28.10) is located on far side of the moon and thus never visible from Earth. It is a 21 km impact crater close to the inner rim of the larger Mendeleev crater. Bergman and Scheele also have main-belt asteroids named after them. Asteroid number 12,356 Carlscheele (6 km in diameter) and asteroid number 29,307 Torbernbergman (unknown diameter) were both discovered by Belgian astronomer Eric Walter Elst at the European Southern Observatory in Chile in the autumn of 1993 [56]. Both

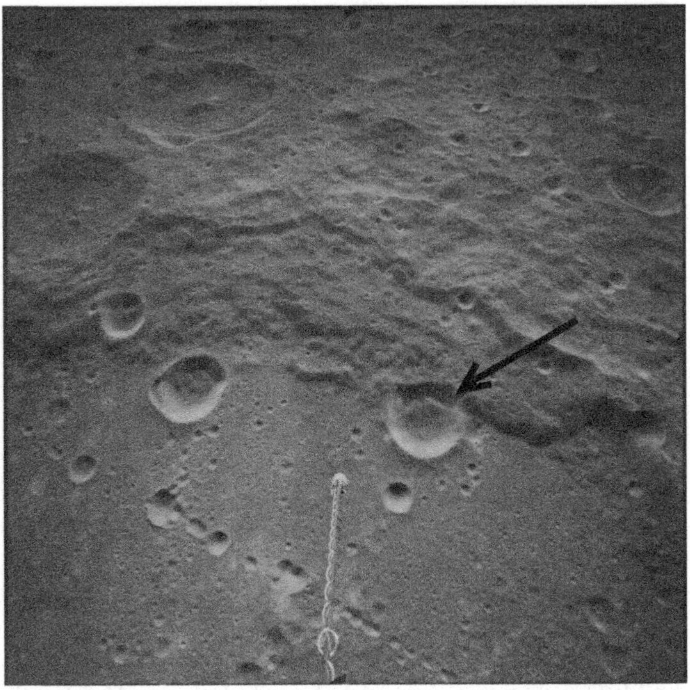

Fig. 28.10 Lunar impact crater Bergman (marked with an arrow) photographed from an altitude of 118 km in April 1972 by the metric camera in the Scientific Instrument Module on-board Apollo 16. The plane on the lower part of the image is the interior of the Mendeleev crater. The rod in the foreground is an antenna on the Apollo spacecraft. *Photo* NASA. Image Reference AS16-M-0477. The image has been cropped

Scheele and Bergman have minerals named after them. The mineral ($CaWO_4$) from which Scheele isolated tungstic acid and which was known in Sweden as tungsten, meaning "heavy stone" in the eighteenth century, is now called scheelite, and Bergman got two uranium minerals, torbernite ($Cu(UO_2)_2(PO_4)_2 \cdot 12\ H_2O$) and metatorbernite ($Cu(UO_2)_2(PO_4)_2 \cdot 8\ H_2O$) named after him.

Some may argue that the lives of individual scientists are of little or no interest: if one scientist fails to make a discovery, someone else will soon make the same discovery. Perhaps one should distinguish between scientists who are only linked to a single great discovery, and those who are still of interest if their greatest discovery is ignored. Take, for instance, Dmitri Mendeleev; if it were not for his contribution to the invention of the periodic table, he would only be a minor figure in the history of chemistry. Given the parallel work on the periodic law by Lothar Meyer and others, the periodic table would actually have developed along a similar line without Mendeleev's contribution. If we, on the other hand, suppose that Albert Einstein had not formulated the theory of special relativity, his other work would still have made him one of the most influential scientists of the twentieth century.

The same holds for Bergman and Scheele; both made a number of important discoveries that placed them among the most respected chemists of their time. For instance, Bergman is mentioned 15 times and Scheele 14 times in Lavoisier's classical textbook *Elements of Chemistry*, although they belonged to the phlogistic school.[7]

The research of Bergman and Scheele, although often tightly linked, are quite different. Most of Scheele's papers focused on the investigation of a single substance or material, such as a mineral or a drug. These investigations led him to the discovery of several new (what we call) chemical elements and a number of organic substances. The most important exceptions from this approach are his extensive investigations of air and combustion that lead to his discovery of oxygen. Bergman, on the other hand, rarely focused on individual chemical substances, but searched for larger contexts. The main exception was his thorough investigation of carbon dioxide, which led to the discovery of its acidic properties. In his published works, Scheele paid much closer attention to experimental detail, meaning that his experiments are easily reproduced. This is not always the case with Bergman's work.

Scheele's contributions to science are still widely recognised. He is best known as co-discoverer of oxygen, discoverer of chlorine and for his isolation of several organic acids. His contribution to the discovery of manganese, barium, molybdenum and tungsten, on the other hand, deserves more recognition. In these cases, Scheele was the first to distinguish their compounds from the compounds of other metals and to study these compounds in detail. Metallic molybdenum, for instance, was isolated by Hjelm, but on Scheele's request and using a sample of molybdic acid provided by Scheele. Although Scheele and his contemporaries had no means of isolating metallic barium, Scheele suspected that barium oxide could be reduced to a metal.

Bergman's contributions to science are not as widely recognised as one could wish. Following the steps of Marggraf, he laid the foundations of analytical chemistry, and his work on the quantitative composition of chemical substances was instrumental for the formulation of the laws of stoichiometry and the atomic theory that would transform chemistry in the early nineteenth century. It was also Bergman who created the first system of rational chemical nomenclature. Unfortunately, Bergman died young and others perfected his theories and methods and received the credit. Bergman's contributions to geology and crystallography are also usually overlooked.

[7]To put this in perspective, the name Mar(g)graf appears 5 times, Macquer 7 times, Priestley 6 times, Morveau 18 times, Kirwan 5 times, Berthollet 23 times and Fourcroy 8 times. When the name appears twice in close proximity, it has only been counted once.

References

1. Schück H (1916) Torbern bergman självbiografi, äldre svenska biografier 3–4, Uppsala, p 101
2. Schück H (1916) Torbern bergman självbiografi, äldre svenska biografier 3–4, Uppsala, p 102
3. Mikael Hoffmann. Private communication
4. Schück H (1916) Torbern bergman självbiografi, äldre svenska biografier 3–4, Uppsala, p 103
5. Trofast J (1994) Johan Gottlieb Gahn: brev, vol 2, Lund, p 42
6. Trofast J (1994) Johan Gottlieb Gahn: brev, vol 2, Lund, p 47
7. Hoffmann M. Private communication
8. Crell L (1784) Ueber die Säure des Tungstein (Schwerstein); nebst einer Nachricht vom Hrn. Ritter Bergmann, über ein aus demselben erhaltenes neues Metall, Chem Ann 2:195–207. This paper includes an account of Bergman's death, based on a letter from Scheele
9. Trofast J (1994) Johan Gottlieb Gahn brev, vol 2, Lund, p 131
10. Hjelm PJ (1786) Åminnelse-tal öfver...Herr Torbern Bergman, Stockholm, p 80
11. Hjelm PJ (1786) Åminnelse-tal öfver...Herr Torbern Bergman, Stockholm, p 81
12. Aurivillius PF (1785) Åminnelse-tal öfver chemiae professoren i Upsala...Herr Magister Torb. Bergman, hållet...vid Vestgöta nations sammankomst den 15 jun. 1785, Uppsala, p 3
13. Letter to Wilcke. Accessed 26 July 1784
14. Hjelm PJ (1786) Åminnelse-tal öfver...Herr Torbern Bergman, Stockholm, p 73
15. Hjelm PJ (1786) Åminnelse-tal öfver...Herr Torbern Bergman, Stockholm, p 74
16. Wåhlin AM, Ribben C (1784) Journal öfver professoren och riddaren m. m. Thorb. Bergmans sista sjukdom, död, m. m., insänd til K. Colegium Medicum, Vecko-skrift för läkare och naturforskare 5:386–396
17. Bergman T (1784) Mineralogische Anmerkungen von Torbern Bergmann. Chem Ann 2:387–420
18. von Beskow B (1860) Minne öfver kemie professoren Torbern Bergman. Svenska Akademiens Handlingar 32:23–70
19. Wåhlin AM (1784) Journal öfver Professoren och Riddaren m. m. Thorb. Bergmans sista Sjukdom, Död m. m., insänd til K. Collegium Medicum, Vecko-skrift för läkare och naturforskare 5:386–396
20. Wåhlin AM (1784) Få, men sanna ord, framförde wid Grafen då...Torbern Bergman Jordfästes i Wästra Ny kyrka..., Dagligt Allehanda, no 242, 20 October 1784
21. Crell L (1784) Ueber die Säure des Tungstein (Schwerstein); nebst einer Nachricht vom Hrn. Ritter Bergmann, über ein aus demselben erhaltenes neues Metall. Chem Ann 2:195–207
22. Hjelm PJ (1786) Åminnelse-tal öfver...Herr Torbern Bergman, Stockholm, p 91
23. Annerstedt C (1913) Uppsala universitets historia, vol 3, no 1. Almqvist & Wiksell, Uppsala, p 597
24. Annerstedt C (1913) Uppsala universitets historia, vol 3, no 1. Almqvist & Wiksell, Uppsala, p 598
25. Odén S (1918) Johan Jan Afzelius. Stockkholm, Svenskt biografiskt lexicon, p 222
26. Annerstedt C (1914) Uppsala universitets historia, vol 3, no 2. Almqvist & Wiksell, Uppsala, p 456
27. Olsson H (1971) Kemiens historia i Sverige intill år 1800. Stockholm, Almqvist & Wiksel, p 158
28. Lundgren A (1979) Berzelius och den kemiska atomteorin. Uppsala university, Diss, p 26
29. Letter to Abraham Bäck, May 12, 1786
30. Letter to Abraham Bäck, 12 May 1786
31. Letter from Margaretha Pohl to Wilcke, 15 September 1786
32. Jensen WB (1989) What happened to Homberg's pyrophorus? Bull Hist Chem 3:21–24
33. Scheele CW (1786) Berichtigende Bemerkungen über den Luftzünder. Chem Ann 1:483–486
34. Letter to Wilcke, 12 September 1785
35. Letter to Wilcke, 12 March 1786
36. Scheele CW 81786) Vom Hrn. Scheele in Köping. Chem Ann 1:439–440

37. Letter from Hall to Wilcke, 24 June 1786
38. Zekert O, Strandell B (1936) Scheele. Supplement II, Pharmazeutische Monatshefte 17:1–4
39. Emery T (2017) Hellfire Boys. Little brown and Company, New York, p 426
40. Boklund U (1961) Carl Wilhelm Scheele bruna boken del 2, Stockholm, p 403
41. Ahlström CJ (1786) Tal vid Herr Carl Wilhelm Scheeles graf, Stockholm
42. Ekström S (2006) Carl Wilhelm Scheele – liv och forskargärning. Ärat ditt namn, Köping
43. Segerström H (2006) Köping runt med Carl Wilhelm Scheele. Ärat ditt namn, Köping
44. Crell L (1786) No title. Chem Ann 2:287–288
45. Trofast J (1994) Johan Gottlieb Gahn brev, vol 2, Lund, p 275
46. Boklund U (1961) Carl Wilhelm Scheele bruna boken del 1, Stockholm, p 20
47. Nordenskiöld AE (1892) Carl Wilhelm Scheele. Efterlämnade bref och anteckningar, Stockholm
48. Nordenskiöld AE (1892) Carl Wilhelm Scheele. Nachgelassene Briefe und Aufzeichnungen, Stokholm
49. Oseen W (1942) Carl Wilhelm Scheele. Manuskript 1756–1777, KVA, Stockholm
50. Nordström J (1942) En edition av Scheeles efterlämnade manuskript Lychnos, p 254
51. Boklund U (1961) Carl Wilhelm Scheele bruna boken, Stockholm
52. Boklund U (1968) Carl Wilhelm Scheele: his work and life. The brown book, Stockholm
53. Fors H (2014) Hjälten utan ansikte. Om Carl Wilhelm Scheeles liv efter döden. In: Svensk snillrikhet? Nationella föreställningr om entrpenörer och teknisk begåvning 1800–2000. Nordic Academic Press, Lund, p 167–186
54. 1986) Carl-Wilhelm-Scheele-Ehrung 1986. Academie der Wissenschaften der DDR, Berlin
55. Andersson LE, Whitaker EA (1982) NASA Catalogue of Lunar Nomenclature. NASA RP-1097
56. JPL Small-Body Database Browser. https://ssd.jpl.nasa.gov/sbdb.cgi

Appendix A: A Bibliography of Torbern Bergman's Published Works

The following section is a list of Bergman's original publications in approximately chronological order. All translations have been omitted, but a complete bibliography, listing 307 titles, was compiled by librarian Birgita Moström in 1957 [1].

1. **Dissertatio de crepusculis, quam … præside … Martino Strömer … publico examini submittit stipendiarius regius Thorbern. Ol. Bergman, v-gothus, in audit. Carol. minori d. XIX. martii, anni MDCCLV**

16 pp. Uppsala 1755.

2. **Rön om Coccus aquaticus. Linn Faun. Suec. n. 725**

Kongl. Vetenskaps Academiens Handlingar, 17, 1756, 199–204.

3. **Utdrag af et bref från Upsala, dat. den 23 sistl. julii, til Herr observator Ferner i Stockholm**

Lärda Tidningar, 1756, 243–244.

4. **Afhandling om iglar**

Kongl. Vetenskaps Academiens Handlingar, 18, 1757, 304–314.

5. **Dissertatio gradualis, de interpolatione astronomica, quam, ex consensu ampliss. fac. philosoph. in academia Upsalensi, præside Benedicto Ferner … in audit. Gust. d. XV. Martii anni MDCCLVIII. H. A. M. S. publice ventilandam sistit alumnus regius Thorbernus Bergman, V. Gothus, adscriptus soc. scient. Upsalensi et academiæ Stockholmensi**

23 pp. Uppsala 1758.

A. Lennartson, *Carl Wilhelm Scheele and Torbern Bergman*, Perspectives on the History of Chemistry, https://doi.org/10.1007/978-3-030-49194-9

6. **Dissertatio physico-mathematica de attractione universali cujus partem priorem, cum consens. ampliss. facult. philos. in regia academia Upsaliensi, publico bonorum examini submittunt stipendiarius Thorbernus Bergman ... et Matthias Rydell, V. Gothi, In audit. Carol. maj. die XXIX. Nov. anni MDCCLVIII. H. A. M. S.**

27 pp. Uppsala, 1758.

7. **Vetenskaps historien om rägnbågens förklaring**

Kongl. Vetenskaps Academiens Handlingar, 20, 1759, 239–251.

8. **Anmärkningar om tysta eld-sken**

Kongl. Vetenskaps Academiens Handlingar, 21, 1760, 63–69.

9. **Vetenskaps historien om skymningarna**

Kongl. Vetenskaps Academiens Handlingar, 21, 1760, 239–251.

10. **A letter to Benjamin Wilson concerning electricity**

Philosophical Transactions, 51, 1760, 907–909.

11. **An account of the observations made on the same transit [of planet Venus] at Upsal in Sweden: In a letter to Benjamin Wilson**

Philosophical Transactions, 52, 1761, 227–230.

12. **Anmärkning om Islands krystalls electricitet**

Kongl. Vetenskaps Academiens Handlingar, 23, 1762, 62–65.

13. **Et sällsamt galläple, beskrifvit**

Kongl. Vetenskaps Academiens Handlingar, 23, 1762, 139–143.

14. **Anmärkningar om vild-skråpukar och såg-flugor**

Kongl. Vetenskaps Academiens Handlingar, 24, 1763, 154–175.

15. **Electriska försök med siden-band af åtskillig färg**

Kongl. Vetenskaps Academiens Handlingar, 24, 1763, 323–330.

16. **Observations on Auroræ Boreales in Sweden: In a letter to Benjamin Wilson**

Philosophical Transactions, 52, 1762, 479–486.

17. **Svar på den af Kongl. Vetenskaps Academien, år 1762, framstälda fråga: Huru kunna maskar, som göra skada på frukt-träd, medelst blommornas och löfvens affrätande, bäst förekommas och fördrifvas?**

Svar på frågan Huru kunna maskar, som göra skada på frukt-träd, medelst blommornas och löfvens affrätande, bäst förekommas och fördrifvas? Hvilken fråga, af Kongl. Vetensk. Academien blef upgifven, år 1762. Stockholm, 1763, 1–27.

18. **Afhandling om nordskenens högd. Förra stycket**

Kongl. Vetenskaps Academiens Handlingar, 25, 1764, 193–210.

19. **Afhandling om nordskenens högd. Sednare stycket**

Kongl. Vetenskaps Academiens Handlingar, 25, 1764, 249–261.

20. **Bref til Kongl. Vetenskaps Academiens secereterare, angående anmärkningarna, som utkommit öfver det svar på frågan om skadeliga frukt-träd-maskar, hvilket vunnit den af Kongl. Academien utlåfvda belöningen**

16 pp. Stockholm, 1764.

21. **Inträdes-tal, om möjeligheten at förekomma åskans skadeliga verkningar; hållit för Kongl. Vetenskaps Academien den 23 maji 1764**

103 pp. Stockholm, 1764.

22. **Observations in electricity and on a thunderstorm: In a letter to Benjamin Wilson**

Philosophical Transactions, 53, 1763, 97–100.

23. **A letter containing some experiments in electricity, to Benjamin Wilson**

Philosophical Transactions, 54, 1764, 84–88.

24. **Åminnelse-tal öfver framledne theol. professoren i Upsala, och Kongl. Vetenskaps Academiens medlem, herr doctor Nils Wallerius, på Kongl. Academiens befallning hållet uti stora riddarhus-salen den 2 februarii 1765**

28 pp. Stockholm, 1765.

25. **Electriska försök med sammangnidna glas-skifvor**

Kongl. Vetenskaps Academiens Handlingar, 26, 1765, 127–142.

26. **Landthushållningens nytta af noggranna meteorologiska observationer**

Oeconomiska tidningar, 1765, No: 23.

27. **Afhandling om tourmalinens electriska egenskaper**

Kongl. Vetenskaps Academiens Handlingar, 27, 1766, 57–68.

28. **Tilläggning om nordskenens högd**

Kongl. Vetenskaps Academiens Handlingar, 27, 1766, 224–227.

29. **Physisk beskrifning öfver jord-klotet, på Cosmographiska Sällskapets vägnar författad**

X, 431 pp. Uppsala, 1766.

30. **Disquisitio chemica de confectione aluminis, quam, consensu amplis. ordinis philosophici in regia academia Upsaliensi publico examini submittunt ... Torbernus Bergman ... et ... stipendiarius regius Gustavus Suedelius, Westmannus, in auditorio Carolino majori, die I Aprilis, anni MDCCLXII. Horis ante meridiem solitis**

16 pp. Uppsala, 1767.

31. **Förslag, at förbättra alun-luttringen**

Kongl. Vetenskaps Academiens Handlingar, 28, 1767, 73–80.

32. **J. F. Hartmann, Electrische Experimente im luftleeren Raume**

Kongl. Bibliotekets Tidningar om lärda saker, 1767, 77–78.

33. **Supplementum historiæ Reaumurianæ tenthredinum**

Nova Acta Physico-Medica Academiae Caesarae Leopoldino-Carolinae Naturae Curiosorum, 3, 1767, 166–179.

34. **Åminnelse-tal, öfver framledne bergs-rådet, öfver-directeuren vid Controll-verket, samt riddaren af Kongl. Nordstjerne-Orden, Herr Anton von Swab : på Kongl. Vetenskaps Academiens vägnar, Hållit i Stora Riddarhus-Salen den 29 Junii 1768**

54 pp. Stockholm, 1768.

35. **Anmärkningar om Vestgötha bergen**

Kongl. Vetenskaps Academiens Handlingar, 29, 1768, 324–336.

36. **Åminnelse-tal öfver framledne bergs-rådet och medicinae doctoren, samt K. Vetenskaps Sällskapets i Upsala, och Kongl. Academiens i Stockholm ledamot, Herr Georg Brandt : hållet för Kongl. Vetenskaps Acadamien, uti Stora Riddarehus-Salen, den 9 Aug**

32 pp. Stockholm, 1768.

37. **Disquisitio chemica, de calce auri fulminante, quam, cum consensu ampliss. ordinis philosophici, in reg. academia Upsaliensi, præside mag. Torberno Bergman ... publico examini submittit Carolus Andreas Plomgren ... in audit. Gustaviano die XVI Decemb. anni MDCCLXIX. H. A. M. S.**

26 pp. Uppsala, 1769.

38. **Svar, på Kongl. Vetenskaps Academiens förnyade fråga, om förekommande af maskar, som förtära blommor eller löf på fruktträd**

Svar på den af Kongl. Vetenskaps Academien andra gången framställda frågan, huru maskar, som göra skada på frukt-träd, medelst blommornas och bladens förtärande, bäst kunna förekommas och fördrifvas, Stockholm 1769, 1–8.

39. **Uplysning om skadelige tall-maskarne**

Kongl. Vetenskaps Academiens Handlingar, 30, 1769, 272–276.

40. **Chemisk undersökning om källe-vatnen uti och närmast kring Upsala, första delen. Med vidtberömda philosophiska facultetens samtycke, under inseende af ... Torb. Bergman ... uti större carolinska lärosalen, til almänt ompröfvande framgifven den 24 decemb. på vanlig tid för middagen år 1770, af Pehr Dubb, Vest-Göthe**

14 pp. Uppsala, 1770.

41. **Historien om qvicksilvers föreningar med koksalts-syra**

Kongl. Vetenskaps Academiens Handlingar, 31, 1770, 79–102.

42. **Tilläggning i föregående ämne** [comment on a paper by Wilcke on lightnings]

Kongl. Vetenskaps Academiens Handlingar, 31, 1770, 125–129.

43. **Swar, på de så kallade nödige och wälmente påminnelser, som finnas i N. 18 af Lärda Tidningarne, för innew. År**

Almänna tidningar, 1770, No. 39 (March 21). p. 156; No 2 (March 24), 159–160; No 41 (March 26), 162–163 and No. 4 (March 27), 166–168.

44. **Anledningar, at tilverka varaktigt tegel**

Kongl. Vetenskaps Academiens Handlingar, 32, 1771, 211–220.

45. **Fortsättning af historien om qvicksilvers föreningar med saltsyra**

Kongl. Vetenskaps Academiens Handlingar, 32, 1771, 294–301.

46. **Magnis litterarum patronis, patribus civibusque academicis et urbicis S.P.D.**

4 pp. Uppsala, 1771.

47. **Mémoires pour servir à l'historie des insectes par Charles de Geer**

Lärda Tidningar, 1771, 350–352.

48. **Classes larvarum, definitæ**

Nova Acta Regiæ Societatis Scientiarum Upsaliensis, 1, 1773, 58–65.

49. **Experimenta electrica, transitum commotionis per aquam illustrantia**

Nova Acta Regiæ Societatis Scientiarum Upsaliensis, 1, 1773, 111–118.

50. **Auroræ boreales, annis MDCCLIX. LX. LXI. et LXII. observatæ**

Nova Acta Regiæ Societatis Scientiarum Upsaliensis, 1, 1773, 118–150.

51. **Variæ crystallorum formæ, e spatho ortæ**

Nova Acta Regiæ Societatis Scientiarum Upsaliensis, 1, 1773, 150–155.

52. **Dissertatio chemica de fonte acidulari Dannemarkensi, quam consentiente amplissima facultate philosophica, publico examini submittunt, præses mag. Torb. Bergman, ... et respondens stipendiarius Stieglerianus Carolus Henr. Wertmüller, Stockholmiensis, in aud. Carolino majori die 15 Dec. 1773. Horis a. m. solitis**

14 pp. Uppsala, 1773.

53. **Dissertatio pharmaceutica de stibio tartarisato, quam, consentiente amplissima facultate philosophica, publico examini submittunt, præses mag. Torbernus Bergman ... et respondens stipendiarius regius Johannes Adolphus Level, Smolandus. In auditorio Gustaviano, die 22 dec. 1773. Horis a. m. solitis**

19 pp. Uppsala, 1773.

54. **Om luftsyra**

Kongl. Vetenskaps Academiens Handlingar, 34, 1773, 170–186.

55. **Slutet af historien, om qvicksilvers föreningar med saltsyra**

Kongl. Vetenskaps Academiens Handlingar, 33, 1772, 193–205.

56. **Physisk beskrifning öfver jord-klotet, på Cosmographiska sällskappets vägnar författad. Andra uplagan, tilökt och förbätrad**

Two volumes; XVIII, 470 pp.; 535 pp. Uppsala 1773–1774.

57. **Chemisk och mineralogisk afhandling om hvita järnmalmer, med den vidtberömda philosoph. facult. i Upsala samtycke, under inseende af mag. Torb. Bergman ... til almänt ompröfvande, f. m. d. 22 junii 1774 i gustavianska lärosalen, utgifven af Peter Jacob Hjelm, Smolänning**

42 pp. Uppsala, 1774.

58. **Dissertation qui a remporté, au judgement de la Société Royal des Sciences, le prix proposé par un des membres de cette compagnie, sur cette question: Quels sont les caractères principaux des terres en général? Assigner les défauts de celles qui sont peu propres à la productions des grains, & les moyes d'y remèdier**

Assemblée publique de la Société Royal des Sciences...le 8 décembre 1773. Montpellier, 1774.

59. **Tilläggning om brunsten**

Kongl. Vetenskaps Academiens Handlingar, 35, 1774, 194–196.

60. **Afhandling om bitter- selzer- spa- och pyrmonter-vatten, samt deras tilredande genom konst**

Kongl. Vetenskaps Academiens Handlingar, 36, 1775, 8–43.

61. **Afhandling om bitter- selzer- spa- och pyrmonter-vatten, samt deras tilredande genom konst. Senare stycket**

Kongl. Vetenskaps Academiens Handlingar, 36, 1775, 94–121.

62. **Commentatio de acido aëreo**

Nova Acta Regiæ Societatis Scientiarum Upsaliensis, 2, 1773, 108–160 [108–158 in a variant edition].

63. **Disquisitio de attractionibus electivis**

Nova Acta Regiæ Societatis Scientiarum Upsaliensis, 2, 1773, 161–250 [159–148 in a variant edition].

64. **Commentationes chemicæ, e secundo novorum Societatis Reg. Scient. Ups. actorum tomo excerptæ** [reprint of Commentatio de acido aëreo and Disquisitio de attractionibus electivis]

144 pp. Uppsala, 1775.

65. **Dissertio** [sic!] **chemica de magnesia alba, quam, consent. amplissima fac. philosophica, praeside mag. Torb. Bergman ... publice ventilandam sistet stipendiarius Kohreanus, Carolus Norell, V. Gothus. In aud. Gustav. d. 23 Dec. anni 1775; H. A. M. S.**

28 pp. Uppsala, 1775.

66. **Dissertatio chemica de niccolo, quam, consentiente amplissima facultate phil., præside Mag. Torb. Bergman ... publico examini submittit Johannes Afzelius Arvidsson, Vestrogothus, in auditorio Gustaviano die [12] Julii anni MDCCLXXV, horis ante meridem solitis**

22 pp. Uppsala, 1775.

67. **H. T. Scheffers Chemiske föreläsningar, rörande salter, jordarter, vatten, fetmor, metaller och färgning, samlade, i ordning stälde och med anmärkningar utgifne af T. B.**

XIV, 450 pp. Uppsla, 1775.

68. **Afhandling, om bitter- selzer- spa- och pyrmontervattens rätta hallt och tilredning genom konst; öfversedd och tilökt**

78 pp. Uppsala, 1776.

69. **Dissertatio chemica de acido sacchari, quam venia ampl. fac. philos. Ups. præside mag. Torb. Bergman ... pro gradu publico examini subiicit Iohannes Afzelius Arvidsson, Vestrogothus. In aud. Gustav. d. XIII iun. an. MDCCLXXVI**

22 pp. Uppsala, 1776.

70. **Ytterligare anmärkningar om aluntilverkningen**

Kongl. Vetenskaps Academiens Handlingar, 37, 1776, 177–189.

71. **Tilläggning om blåse-stenen**

Kongl. Vetenskaps Academiens Handlingar, 37, 1776, 333–338.

72. **Chemisk afhandling om järnmalms proberande på våta vägen, med den vidtberömde philosophiske facultetens samtycke, til almänt ompröfvande framgifven af praeses ... Torb. Bergman ... samt respondens Anders Schedin, uplänning ... uti gustavianska lärosalen d. 24 maji 1777**

14 pp. Uppsala, 1777.

73. **Dissertatio chemica de arsenico, quam, consentiente amplissima fac. philosophica, præside Mag. Torb. Bergman ... publice ventilandam sistit Andreas Pihl, Westmannus, in audit. gust. die 7 maji; anni 1777. H. A. M. S.**

24 pp. Uppsala, 1777.

74. **Extrait d'une lettre sur plusieurs points de physique & de minéralogie**

Observations sur la Physique, 9, 1777, 301–303.

75. **Hafs-vatten från ansenligt djup; undersökt af Torbern Bergman**

Kongl. Vetenskaps Academiens Handlingar, 38, 1777, 25–29.

76. **Anmärkningar om magnesia nitri**

Kongl. Vetenskaps Academiens Handlingar, 38, 1777, 213–216.

77. **Anmärkningar om platina**

Kongl. Vetenskaps Academiens Handlingar, 38, 1777, 317–328.

78. **Tilläggning om oculus mundi**

Kongl. Vetenskaps Academiens Handlingar, 38, 1777, 347–351.

79. **Ifrån Herr Professorn och Riddaren Bergman**

U. von Troil, *Bref, rörande en resa til Island MDCCLXXII.*
Uppsala, 1777, 327–376.

80. **Tal, om chemiens nyaste framsteg, hållet, i Kongl. Maj:ts höga närvaro, för dess Vetenskaps-Academie, vid præsidii nedläggande, den 12 nov. 1777**

48 pp. Stockholm, 1777.

81. **Vorbericht. (Aus dem Schwedischen)**

C. W. Scheele, *Carl Wilhelm Scheele's d. Köngl. Schwed. Acad. d. Wissenschaft. Mitgliedes, Chemische Abhandlung von der Luft und dem Feuer: Nebst einem Vorbericht von Torbern Bergman*
(4), 16, 155 pp. Uppsala and Leipzig, 1777, 1–16.

82. **Dissertatio chemica de analysi aquarum frigidarum, cujus partem priorem, consentiente amplissima facultate philos. praeside ... Torb. Bergman ... publico examini subjicit Joh. Petr. Scharenberg, Stockholmensis, in auditorio gustaviano, die 26 Junii, anni 1778, H. A. M. S.**

32 pp. Uppsala, 1778.

83. **Om varma hälso-vattens tilredning**

Kongl. Vetenskaps Academiens Handlingar, 39, 1778, 219–227.

84. **Åminnelse-tal, öfver Kongl. Maj:ts tro-man, hof-marschalken commendeuren af Kongl. Wasa-Orden med stora korset, riddaren af Kongl. Nordstjerne-Orden och ledamoten af Kongl. Vetenskaps Academien, högvälborne friherren, herr Carl de Geer, hållet för Kongl. Vetenskaps-Academien den 19 decemb. 1778**

40 pp. Stockholm, 1779.

85. **Anledning til föreläsningar öfver chemiens beskaffenhet och nytta, samt naturlige kroppars almännaste skiljaktigheter**

63 pp. Uppsala and Åbo 1779.

86. **Bruna turmaliner; til sina grundämnen undersökte**

Kongl. Vetenskaps Academiens Handlingar, 40, 1779, 224–238.

87. **Anmärkningar om bi, förnämligast i anledning af vägnings försök**

Kongl. Vetenskaps Academiens Handlingar, 40, 1779, 300–329.

88. **Commentatio de tubo ferruminatorio, eiusdemque usu, in explorandis corporibus praesertim mineralibus. Ex actorum Societatis literariae Bohemicae tomo IV**

64 pp. Vienna, 1779.

89. **Dissertatio chemica de mineris zinci, quam, consensu amplissimæ fac. philos. praeside mag. Torb. Bergman ... publice ventilandam sistit Benedictus Reinh. Geijer, Vermelandus. In auditorio Gustaviano, H. A. M. S. die 20 Martii, anni 1779**

30 pp. Uppsala, 1779.

90. **Dissertatio chemica de terra silicea, quam, venia amplissimae fac. phil. praeside Mag. Torb. Bergman ... publice ventilandam sistit stipendiarius Hyltingianus, König Alexander Grönlund, in auditorio Gustaviano die XXIV Nov. anno 1779. H. A. M. S.**

20 pp. Uppsala, 1779.

91. **Dissertatio gradualis de primordiis chemiæ, quam, venia amplissimæ fac. philos. Upsal. præside mag. Torb. Bergman ... publico examini submittit stipendiarius regius Jacobus Paulin, Vestrogothus, in auditorio Gustaviano die 4 jun. anno 1779**

58 pp. Uppsala, 1779.

92. **Framledne Directeuren Herr H.T. Scheffers chemiske föreläsningar, rörande salter, jord-arter, metaller, vatten, fetmor och färgning; med anmärkningar utgifne; jemte anledning til föreläsningar öfver chemiens beskaffenhet och nytta, samt naturlige kroppars almännaste skiljaktigheter**

XIV, 450, 63 pp. Stockholm, Uppsala and Åbo 1779.

93. **Magnis litterarum patronis, patribus civibusque academicis et urbicis, S.P.D.**

4 pp. Uppsala 1779.

94. **Opuscula physica et chemica, pleraque antea seorsim edita, jam ab auctore collecta, revisa et aucta. Vol I.**

XVI, 411 pp. Stockholm, Uppsala and Åbo, 1779.

95. **Opuscula physica et chemica, pleraque seorsim antea edita, jam ab auctore collecta, revisa et aucta. Vol II.**

510 pp. Uppsala, 1780.

96. **Analyse et examen chimique de l'indigo, tel qu'il est dans le commerce, pour l'usage de la teinture. Piece qui a concouru pour le prix sur la nature & l'usage de l'indigo**

Mémoires de mathématique et de physique, présentés à l'Academie Royale des Sciences, 9, 1780, 121–164.

97. **Dissertatio chemica de diversa philogisti quantitate in metallis, quam, venia ampl. facult. philosoph., praeside Mag. Torb. Bergman ... publice ventilandam sistit Andreas Nicolaus Tunborg, Dalekarlus. In audit. gustav. maj. d. 13 dec. 1780**

16 pp. Uppsala, 1780.

98. **Dissertatio metallurgica de minerarum docimasia humida, quam, venia amplissimæ fac. philos., praeside mag. Torb. Bergman ... publice ventilandam sistit Petrus Castorin, Vestmannus. In Auditorio Gustaviano majori, die 7 Jun. anno 1780**

40 pp. Uppsala, 1780.

99. **Præcipitations försök med platina, nickel, cobolt och magnesium**

Kongl. Vetenskaps Academiens Nya Handlingar, 1, 1780, 282–293.

100. **Producta ignis subterranei chemice considerata**

Nova Acta Regiæ Societatis Scientiarum Upsaliensis, 3, 1780, 59–136.

101. **Disquisitio chemica de terra gemmarum**

Nova Acta Regiæ Societatis Scientiarum Upsaliensis, 3, 1780, 137–170.

102. **Dissertatio chemica de analysi ferri, quam, venia ampliss. facult. philos., præside Torb. Bergman ... publice ventilandam sistit Johannes Gadolin, Aboa-Fenno. In auditorio Gustaviano majori d. 9 Jun. anno 1781**

74 pp. Uppsala, 1781.

103. **Tilläggning om tungsten**

Kongl. Vetenskaps Academiens Nya Handlingar, 2, 1781, 95–98.

104. **Försvafladt tenn från Siberien**

Kongl. Vetenskaps Academiens Nya Handlingar, 2, 1780, 328–332.

105. **Commentationes e quarto novorum Reg. Scientarum Societatis Upsaliensis actorum tomo excerptæ** [Preprint from the fourth volume of Nova Acta]

pp. 51–136. Uppsala, 1782.

106. **Dissertatio chemica de analysi lithomargæ, quam venia ampliss. ord. phil. præside mag. Torb. Bergman ... pro gradu philos. publice ventilandam sistit stipendiarius Gyllenhjelmianus Carolus Dieteric. Hjerta Westrogothus. In auditorio Gustaviano majori d. 15 Junii 1782**

14 pp. Uppsala, 1782.

107. **Dissertatio chemica de terra asbestina, quam venia ampl. fac. phil., præside mag. Torb. Bergman ... publice ventilandam sistit Carolus Gust. Robsahm, Vermelandus. In auditorio Gust. majori die 10 julii, anno 1782**

16 pp. Uppsala, 1782.

108. **Dissertatio gradualis sistens chemiæ progressus a medio sæc. VII ad medium sæc. XVII, cujus partem priorem venia ampliss. facultatis phil., præside mag. Torb. Bergman ... publice ventilandam exhibet Petrus Afzelius Arvidsson, stipend. Wictorin., Westrogothus. In audit. Gustav. maj. d. II junii 1782, H. A. M. S.**

40 pp. Uppsala, 1782.

109. **Extrait de trois mémoires sur le gas méphistique & sur les affinités chymiques de ce gas**

J. Fred. Corvinus, *Deux memoires sur les gas, et principalement sur le gas méphistique dit air fixe*. Lausanne, 1782, 211–244.

110. **Observationes chemicæ de antimonialibus sulphuratis, quas venia ampl. fac. phil., præside mag. Torb. Bergman ... publice ventilandam sistit Fred. Wilh. Mannercrantz. In auditorio Gust. majori die 10 julii, anno 1782**

14 pp. Uppsala, 1782.

111. **Sciagraphia regni mineralis, secundum principia proxima digesti**

166 pp. Leipzig and Dessau, 1782.

112. **Underrättelse om Medevi surbrunnar**

Kongl. Vetenskaps Academiens Nya Handlingar, 3, 1782, 288–298.

113. **Opuscula physica et chemica, pleraque antea seorsim edita, jam ab auctore collecta, revisa et aucta. Vol III.**

490 pp. Uppsala, 1783.

114. **Underrättelse om Loka källor**

Kongl. Vetenskaps Academiens Nya Handlingar, 4, 1783, 256–267.

115. **Underrättelse om Medevi surbrunnar**

13 pp. Stockholm, 1783.

116. **Commentatio chemica de causa fragilitatis ferri frigidi**

Nova Acta Regiæ Societatis Scientiarum Upsaliensis, 4, 1784, 51–62.

117. **Meditationes de systemate fossilium naturali**

Nova Acta Regiæ Societatis Scientiarum Upsaliensis, 4, 1784, 63–128.

118. **De ferro et stanno igne commixtis**

Nova Acta Regiæ Societatis Scientiarum Upsaliensis, 4, 1784, 221–228.

119. **Meditationes de systemate fossilium naturali**

125 pp. Florence, 1784.

120. **Mineralogiske anmärkningar**

Kongl. Vetenskaps Academiens Nya Handlingar, 4, 1784, 109–122.

121. **Vom Hrn. Profess. und Ritter Bergmann in Upsal**

Chemische Annalen für die Freunde der Naturlehre. 1784, part 1, 38–39.

122. **Ueber der Erforschung der Schwere des Feuers**

Chemische Annalen für die Freunde der Naturlehre. 1784, part 1, 93–95.

123. **Vom Hrn. Profess. und Ritter Bergmann in Upsal**

Chemische Annalen für die Freunde der Naturlehre. 1784, part 1, 149–151.

124. **Ueber die Entstehungsart der natürlichen hornartigen Metalle**

Chemische Annalen für die Freunde der Naturlehre. 1784, part 1, 377–378.

125. **von Hern. Profess. und Ritter Bergmann in Upsal**

Chemische Annalen für die Freunde der Naturlehre. 1784, part 2, 227–228.

126. **Opuscula physica et chemica, pleraque seorsim antea edita nunc collecta, revisa. Vo. IV**

392 pp. Leipzig, 1787.

127. **Opuscula physica et chemica, pleraque seorsim antea edita nunc collecta et revisa. Vo. V**

421 pp. Leipzig, 1788.

128. **Opuscula physica et chemica, pleraque seorsim antea edita nunc collecta et revisa. Vo. VI**

241 pp. Leipzig, 1790.

129. **Framlidne directeuren herr H.T. Scheffers Chemiske föreläsningar, rörande salter, jordarter, metaller, vatten, fetmor och färgning; med anmärkningar utgifne af T.B. Andra tilökta uplagan**

14 (unnumbered), 509 pp. Stockholm, 1796.

130. **Torbern Bergmans självbiografi**

Äldre svenska biografier, 3–4. Uppsala, 1916, 83–103.

Appendix B: A Bibliography of Scheele's Published Works

The following section is a list of Scheele's original publications in approximately chronological order. All translations have been omitted, but can be found in the list compiled by Nordenskiöld [2]. Nordenskiöld's bibliography is, however, not complete.

1. **Undersökning om Fluss-Spat och dess Syra**

Kongl. Vetenskaps Academiens Handlingar, 32, 1771, 120–138.

2a. **Om Brun-sten eller Magnesia och dess Egenskaper**

Kongl. Vetenskaps Academiens Handlingar, 35, 1774, 89–116.

2b. **Om Brun-sten eller Magnesia nigra och dess egenskaper**

Kongl. Vetenskaps Academiens Handlingar, 35, 1774, 177–194.

3. **Bref**

Nya Lärda Tidningar, 1, 1774, 108–110.

4. **Anmärkningar Om Benzoë-Saltet**

Kongl. Vetenskaps Academiens Handlingar, 36, 1775, 128–133.

5. **Om Arsenik och dess syra**

Kongl. Vetenskaps Academiens Handlingar, 36, 1775, 263–294.

6. **Rön och Anmärkningar om Kisel, Lera och Alun**

Kongl. Vetenskaps Academiens Handlingar, 37, 1776, 30–35.

7. **Undersökning om Blåse-stenen**

Kongl. Vetenskaps Academiens Handlingar, 37, 1776, 327–332.

8a. **Chemische Abhandlung von der Luft und dem Feuer. Nebst einem Vorbericht von Torbern Bergman**

16, 155 pp, Uppsala och Leipzig, 1777.

A. Lennartson, *Carl Wilhelm Scheele and Torbern Bergman*, Perspectives on the History of Chemistry, https://doi.org/10.1007/978-3-030-49194-9

8b. **Chemische Abhandlung von Luft und Feuer. Nebst einem Vorberichte von Torbern Bergmann. Zweite verbesserte Ausgabe mit einigen Abhandlung über die Luftgattungen, wie auch mit der Herren Kirwan und Priestley Bemerkungen und Herrn Scheelens Erfahrungen über die Menge der im Dunstkreise befindlischen reinsten Luft vermehrt und mit einem Register versehen von D. Johann Gottfried Leonhardi**

32, 286 pp, Leipzig, 1782.

8c. **Chemical observations and experiments on air and fire. With a prefatory introduction by Torbern Bergman. Translated from the German by J.R. Forster**

XL, 259 pp, London, 1780.

8d. **Traité chimique de l'air et du feu; avec une introduction de Torbern Bergma: ouvrage traduit de l'Allmand par le baron de Dietrich**

XLIV, 268 pp, Paris, 1781.

9. **Sätt at tilreda Mercurius dulcis, på våta vägen**

Kongl. Vetenskaps Academiens Handlingar, 39, 1778, 70–73.

10. **Ett beqvämare och mindre kostsamt sätt at tilreda Pulvis Algerothi**

Kongl. Vetenskaps Academiens Handlingar, 39, 1778, 141–145.

11. **Försök med Blyerts, Molybdæna**

Kongl. Vetenskaps Academiens Handlingar, 39, 1778, 247–255.

12. **Tilrednings-sättet af en ny grön Färg**

Kongl. Vetenskaps Academiens Handlingar, 39, 1778, 327–328.

13. ***Bref* ifrån Hr. *C. W. S.* Til Hr. *D. v. S.* rörande *Tyska Öfversättningen* af *Pharmacopoea Suecia***

Stockholms Lärda Tidningar, 1778, nr 22, 287–291.

14. **Erster Brief**

Hannnoverisches Magazin, October 30, 1778.

15. **Rön, om rena Luftens mängd, som dageligen uti vår Luftkrets är närvarande**

Kongl. Vetenskaps Academiens Handlingar, 40, 1779, 50–55.

16. **Dritter Brief**

Hannnoverisches Magazin, March 29, 1779.

17. **Försök, at decomponera Neutral-salter med osläckt kalk och järn**

Kongl. Vetenskaps Academiens Handlingar, 40, 1779, 158–160.

18. **Vierter Brief**

Hannnoverisches Magazin, August 6 1779.

19. **Försök med Blyerts, Plumbago**

Kongl. Vetenskaps Academiens Handlingar, 40, 1779, 238–245.

20. **Chemische Untersuchung der Schwer-Spatherde**

Beschäftigungen der Berlinischen Gesellschaft Naturforschender Freunde, 1779, part 4, 611–613.

21. **Anmärkningar om Fluss-Spat**

Kongl. Vetenskaps Academiens Nya Handlingar, 1, 1780, 18–26.

22. **Sechster Brief**

Hannnoverisches Magazin, March 24, 1780.

23. **Om Mjölk, och dess syra**

Kongl. Vetenskaps Academiens Nya Handlingar, 1, 1780, 116–124.

24. **Siebenter Brief**

Hannnoverisches Magazin, August 7, 1780.

25. **Achter Brief**

Hannnoverisches Magazin, August 28, 1780.

26. **Om Mjölk-Såcker-Syra**

Kongl. Vetenskaps Academiens Nya Handlingar, 1, 1780, 269–275.

27. **Einige beyläufige Bemerkungen über die Verwandschaft der Körper**

Chemisches Journal für die Freunde der Naturlehre, 1780, part 4, 78–86.

28. **Tungstens bestånds-delar**

Kongl. Vetenskaps Academiens Nya Handlingar, 2, 1781, 89–95.

29. **Über das brennbare Wesen im rohen Kalk**

Die neuesten Entdeckungen in der Chemie, gesammelt von L. Crell, 1781, part 1, 30–41.

30. **Rön och Anmärkningar om Æther**

Kongl. Vetenskaps Academiens Nya Handlingar, 3, 1782, 35–46.

31. **Anmärkningar om sättet at conservera Ättika**

Kongl. Vetenskaps Academiens Nya Handlingar, 3, 1782, 120–122.

32. **Anmärkningar vid Herr Desaives Anmärkning, recenserad uti denna Veckoskriftens 2:dra Band, sid 158, 159, rörande præperation af Mercuris Dulcis efter Pharm. Svec. Ed. Alt.**

Vecko-skrift för läkare och naturforskare, 3, 1782, 145–146.

33. **Erster brief**

Neues Magazin für Ärtze, 1782, part 4, 289–292

34. **Zweyter Brief**

Neues Magazin für Ärtze, 1782, part 4, 292–295.

35a. **Försök, beträffande det färgande ämnet uti Berlinerblå**

Kongl. Vetenskaps Academiens Nya Handlingar, 3, 1782, 264–265.

35b. **Om det färgande Ämnet uti Berliner-blå, Fortsättning**

Kongl. Vetenskaps Academiens Nya Handlingar, 4, 1783, 33–43.

36. **Rön beträffande ett särskildt Socker-Ämne uti exprimerade Oljor och Fettmor**

Kongl. Vetenskaps Academiens Nya handlingar, 4, 1783, 324–329.

37. **Entdeckung eines besondern süßen und flüchtigen Bestandtheils in den ausgepreßten Oelen und thierischen Fettigkeiten**

Chemische Annalen für die Freunde der Naturlehre, 1784, part 1, 99–101.

38. **Vom Hrn. Scheele, aus Köping, in Schweden**

Chemische Annalen für die Freunde der Naturlehre, 1784, part 1, 525–526.

39. **Anmärkning om Citron-saft, samt sätt at crystallisera densamma**

Kongl. Vetenskaps Academiens Nya handlingar, 5, 1784, 105–109.

40. **Ueber die krystallisirung der Citronensäure**

Chemische Annalen für die Freunde der Naturlehre, 1784, part 2, 3–4.

41. **Vom Hrn. Scheele in Köping**

Chemische Annalen für die Freunde der Naturlehre, 1784, part 2, 123–125.

42. **Vom Hrn. Scheele in Köping**

Chemische Annalen für die Freunde der Naturlehre, 1784, part 2, 328–329.

43. **Om Rhabarber-jordens bestånds-delar, samt sätt, at tilreda Acetosell-syran**

Kongl. Vetenskaps Academiens Nya Handlingar, 5, 1784, 180–187.

44. **Om Frukt- och Bär-syran**

Kongl. Vetenskaps Academiens Nya Handlingar, 6, 1785, 17–27.

45. **Rön, om Ferrum phosphoratum och Sal perlatum**

Kongl. Vetenskaps Academiens Nya Handlingar, 6, 1785, 134–141.

46. **Om Rhabarberjordens närvaro uti flera vegetabilier**

Kongl. Vetenskaps Academiens Nya Handlingar, 6, 1785, 171–172.

47. **Anmärkning vid tilredning af Magnesia alba**

Kongl. Vetenskaps Academiens Nya Handlingar, 6, 1785, 172–174.

48. **Vom Hrn. Scheele in Köping**

Chemische Annalen für die Freunde der Naturlehre, 1785, part 1, 59–62.

49. **Ueber die wahre Natur des Sauerklee-salzes, und seine künstliche Erzeugung**

Chemische Annalen für die Freunde der Naturlehre, 1785, part 1, 112–115.

50. **Vom Hrn. Scheele in Köping**

Chemische Annalen für die Freunde der Naturlehre, 1785, part 1, 153–155.

51a. **Neuere Bemerkungen über Luft und Feuer, und die Wasser-Erzeugung**

Chemische Annalen für die Freunde der Naturlehre, 1785, part 1, 229–238.

51b. **Neuere Bemerkungen über Luft und Feuer, und die Wasser-Erzeugung**

Chemische Annalen für die Freunde der Naturlehre, 1785, part 1, 291–299.

52. **Vom Hrn. Scheele in Köping**

Chemische Annalen für die Freunde der Naturlehre, 1785, part 1, 455–457.

53. **Vom Hrn. Scheele in Köping**

Chemische Annalen für die Freunde der Naturlehre. 1785, part 1, 549–550.

54. **Erläuterung über einige, den ungelöschten Kalk betreffende, Versuche**

Chemische Annalen für die Freunde der Naturlehre, 1785, part 2, 220–227.

55. **Vom Hrn. Scheele in Köping**

Chemische Annalen für die Freunde der Naturlehre, 1785, part 2, 437–439.

56. **Vom Hrn. Scheele in Köping**

Chemische Annalen für die Freunde der Naturlehre, 1785, part 2, 513–514.

57. **Om Sal essentiale Gallarum, eller Galläple-salt**

Kongl. Vetenskaps Academiens Nya Handlingar, 7, 1786, 30–34.

58. **Dephlogisticerad Lufts värkan i Sjukdomar**

Vecko-skrift för Läkare och Naturforskare, 7, 1786, 288–291.

59. **Bref til Professor P. J Bergius från C. W. Scheele, dat. Köping den 10 Martii 1786**

Vecko-skrift för Läkare och Naturforskare, 7, 1786, 246–249.

60. **Neue Beweise der Eigenthümlichkeit der Flußspathsäure**

Chemische Annalen für die Freunde der Naturlehre, 1786, part 1, 3–17.

61. **Vom Hrn. Scheele in Köping**

Chemische Annalen für die Freunde der Naturlehre, 1786, part 1, p. 332.

62. **Vom Hrn. Scheele in Köping**

Chemische Annalen für die Freunde der Naturlehre, 1786, part 1, 439–440.

63. **Berichtigende Bemerkungen über den Luftzünder**

Chemische Annalen für die Freunde der Naturlehre, 1786, part 1, 483–486.

64. **Vom Herr Professor Gadolin in London**

Taschen-Buch für Scheidekünstler und Apotheker aus das Jahr 1788, 9, 1788, 136–142.

B.1 Scheele's Collected Works

Publications 65−69 represent collections of Scheele's puplications, all but one published posthumously. There is no complete collection of Scheele's publications.

65. **Mémoirs de chymie. Tirés des Mémoirs de l'Academie Royale des Sciences de Stockholm, traduits du Suedois et de l'Allemand**

Two volumes. VI, 269 + VI, 246 pp, Dijon, 1785.

66. **The Chemical essays of Charles-William Scheele. Translated from the Transaction of the Acadamy of Sciences at Stockholm. With additions**

XVI, 406 pp, London, 1786.

67. **Opuscula chemica et physica Latie verit Godofr. Henr. Schæfer. Edit et præfatus est. Ern. Beniam. Gottli. Hebenstreit**

Two volumes. 284 + 316 pp, Leipzig, 1788–1789.

68. **Sämmtliche physische und chemische Werke, nach dem Tod des Verfassers gesammelt, und in Deutscher Sprache herausgeben von Sigism. Friedr. Hermbstädt**

Two volumes. XIV, 264 + 446 pp, Berlin, 1793.

69. **The collected papers of Carl Wilhelm Scheele. Translated from the Swedish and German originals by Leonard Dobbin.**

XV, 368 pp, London, 1931.

Appendix C: The Swedish Currency

At several places in this book, prices and salaries are discussed but unfortunately the history of the Swedish currency is rather complicated. The currency in Sweden 1624–1776 was the daler. To complicate matter, there were both copper dalers (daler kopparmynt), silver dalers (daler silvermynt) and riksdaler (Swedish for "national daler").

During the first period discussed in this book, 1 riksdaler (or riksdaler specie) = 3 silver dalers = 6 copper dalers = 96 silver öre (öre silvermynt). Prices were often stated in copper dalers, although the copper daler was only unit; copper daler coins were never minted. Silver dalers were minted, but not in silver as one could expect, but in copper; the world's largest "coin" was the Swedish 10 daler, a copper sheet weighing 19 kg. The riksdalers, however, were traditional silver coins.

In 1776, King Gustav III made a reform and introduced a new riksdaler, which was equivalent to 6 old silver dalers. 1 new riksdaler was divided into 48 shillings (skilling).

If one would attempt to translate the value of 1 silver daler in 1770 to a modern currency, one could obtain very different results depending on how the calculation was made. One could base the calculation on the prices of common goods, or one could base the calculations on salaries. A worker in 1770 typically had a very low salary, which was often not paid in money, but in food and accommodation. Thus, a Swedish worker in 1770 had to work a much longer time to afford 1 litre of milk than a Swedish worker of today. Below are approximate prices of some goods 1770 in silver dalers [3]:

1 *barrel* (150 l) of rye: 11 daler
1 ox: 36 daler, 13 öre
1 cow: 24 daler, 19 öre
1 lamb: 5 daler
1 hen: 14 öre
20 eggs: 20 öre
1 *lispund* (8.5 kg) dried fish: 2 daler, 11 öre
1 *lispund* dried meat: 2 daler 12 öre

A. Lennartson, *Carl Wilhelm Scheele and Torbern Bergman*, Perspectives on the History of Chemistry, https://doi.org/10.1007/978-3-030-49194-9

1 *lispund* butter: 7 daler, 12 öre
1 *famn* (3.14 m^3) firewood (birch): 3 daler, 20 öre
1 *barrel* (150 l) turnips: 1 daler
1 pair woolen socks: 1 daler
1 day of work for a worker (1 dagsverke): 22–48 öre.

Appendix D: Literature on Bergman and Scheele

Shortly after Bergman's death, two memorial lectures were given and subsequently printed. The first was held by Bergman's friend philologist Pehr Fabian Aurivillius (1756–1829) in the Västergötland Student Nation on 15 June 1785. The printed booklet consists of 51 p [4]. The second lecture was delivered in the Royal Swedish Academy of Sciences in Stockholm on 3 May 1786 by Bergman's student Peter Jacob Hjelm. The printed version consists of 104 p. and would remain the most extensive biographical work on Bergman for 200 years (Fig. D.1). Not until 1985 was a book-length biography over Bergman published. The book, *Torbern Bergman—a Man Before his Time* by American professor J. A. Schufle, contains biographical information on Bergman and discussion of his research mixed with Schufle's own experiences of Uppsala in the late 1970s. Unfortunately, the book is out of print and rather rare. In addition, its typesetting (it is reproduced directly from a typewritten manuscript) is not very inspiring to a modern reader. Although being a historian rather than chemist, Schufle's description of Bergman's work is quite accurate. A problem, however, is that as Schufle was not fluent in Swedish and there are many misspellings of Swedish names and words. In addition, Schufle's typewriter lacked the letters å, ä and ö, as correctly noted by George B. Kauffman in his review [5]. There are also some misconceptions due to mistranslations. The focus of the book is Bergman's research and some parts of his life, but not, for instance, his death. In some cases, Schufle has attempted to recreate the dialogues taking place during the defences of theses, etc. With its limitations in mind, this 547 p. book is a must-read for anyone interested in Bergman.

The memorial lecture over Scheele was delayed by Wilcke for unknown reasons and was finally delivered by his successor as secretary of the Academy, Carl Gustaf Sjöstén, on 14 October 1799 and printed in 1801 as an 84 p. booklet (Fig. D.1). The author is given as Sjöstén, while most of the text was written by Scheele's friend Wilcke, who had passed away in 1796. Another influential Scheele biography was the 54 p. long booklet [6] published by Uppsala professor Per Teodor Cleve (1840–1905; the discoverer of elements holmium and thulium) for the 100th anniversary of Scheele's death in 1886. The portrait of Scheele used as frontispiece (Fig. 3.17) is, although unhistorical, still the most reproduced image of Scheele. The professor of Chemistry in Lund, Christian Wilhelm Blomstrand and the professor of agricultural

A. Lennartson, *Carl Wilhelm Scheele and Torbern Bergman*, Perspectives on the History of Chemistry, https://doi.org/10.1007/978-3-030-49194-9

Fig. D.1 The memorial lectures over Bergman and Scheele, respectively, published by the Royal Swedish Academy of Sciences. For a long time, they were the primary biographies over the two scientists. *Photo* Anders Lennartson

chemistry at Ulltuna Agricultural Institute, Carl Erik Bergstrand (1830–1914) published short biographies for the same anniversary [7, 8].

The most extensive Scheele biography was for many decades the work published by Austrian Dr. Otto Zekert. His work *Carl Wilhelm Scheele, Sein Leben und Seine Werke* (Carl Wilhelm Scheele, His Life and His Works) was published in seven parts between 1931 and 1934 (with an index published in 1935). The first part (33 p.) is largely a genealogical study of the Scheele family containing some portraits of Scheele's ancestors, and the second part (39 p.) covers Scheele's life until he left Gothenburg. Part three to seven were published in a single volume (302 p) and is a mixture of biographical material and many long excerpts from letters and documents. For example, the 108 p. long third part is a transcription of Scheele's older brother's apothecary manual. Although an invaluable reference, as a biography it is close to unreadable. In 1963, however, Zekert published a 150 p. long biography simply called *Carl Wilhelm Scheele* (also in German) which is much more accessible for a reader. An East German biography by Cassebaum appeared in 1982 and is more focused on chemistry than Zekert's book. The only English biography over Scheele published as a separate book is the short 71 p. *A Pictorial Life of the Apothecary Chemist Carl Wilhelm Scheele* by German-born

American pharmacist Professor George Urdang (1882–1860) from 1942. A more extensive list of Scheele literature is found as an appendix to my Swedish biography of Scheele [9].

References

1. Moström B (1957) Torbern Bergman: a bibliography of his works. Almqvist & Wiksell, Stockholm
2. Nordenskiöld AE (1892) Carl Wilhelm Scheele. Efterlämnade bref och anteckningar, Stockholm
3. Lagerqvist LO (1976) Sveriges regenter, Bonniers, Stockholm, p 372
4. Aurillius PF (1785) Åminnelse-tal, öfver chemiæ professoren i Upsala...Herr magister Torbern Bergman...Uppsala
5. Kauffman GB (1987) Torbern Bergman: a man before his time (Schufle, J. A.) J Chem Educ 64:A58
6. Cleve PT (1886) Carl Wilhelm Scheele. Ett minnesblad på årsdagen af hans död. M. Barkéns Förlagsbokhandel, Köping
7. Blomstrand CW (1886) Minnesteckning öfver Carl Wilhelm Scheele, Stockholm
8. Bergstrand CE (1886) Tal vid minnesfesten öfver Carl Wilhelm Scheele, M. Barkéns Förlagsbokhandel, Köping
9. Lennartson A, Lindeke B, Ohlson B (2015) Ett kemiskt äventyr - Carl Wilhelm Scheele och hans värld, Apotekarsocieteten, Stockholm, pp 310–314

References

Index of Names

A. Lennartson, *Carl Wilhelm Scheele and Torbern Bergman*, Perspectives on the History of Chemistry, https://doi.org/10.1007/978-3-030-49194-9

Subject Index

A. Lennartson, *Carl Wilhelm Scheele and Torbern Bergman*, Perspectives on the History of Chemistry, https://doi.org/10.1007/978-3-030-49194-9

The manufacturer's authorised representative in the EU is Springer Nature Customer Service Centre GmbH, Europaplatz 3, 69115 Heidelberg, Germany. If you have any concerns regarding our products, please contact ProductSafety@springernature.com

Printed and bound by CPI Group (UK) Ltd, Croydon, CR0 4YY
22/09/2026
02220956-0012

Bioremediation and Biotechnology, Vol 2

Rouf Ahmad Bhat
Khalid Rehman Hakeem
Moonisa Aslam Dervash
Editors

Bioremediation and Biotechnology, Vol 2

Degradation of Pesticides and Heavy Metals

Springer

Editors
Rouf Ahmad Bhat
Department of Environmental Science
Cluster University, Sri Pratap College
Srinagar, Jammu and Kashmir, India

Khalid Rehman Hakeem
Department of Biological Sciences
King Abdulaziz University
Jeddah, Saudi Arabia

Moonisa Aslam Dervash
Department of Environmental Science
Cluster University, Sri Pratap College
Srinagar, India

ISBN 978-3-030-40332-4 ISBN 978-3-030-40333-1 (eBook)
https://doi.org/10.1007/978-3-030-40333-1

This Springer imprint is published by the registered company Springer Nature Switzerland AG
The registered company address is: Gewerbestrasse 11, 6330 Cham, Switzerland

Dedicated to our beloved Parents

Foreword

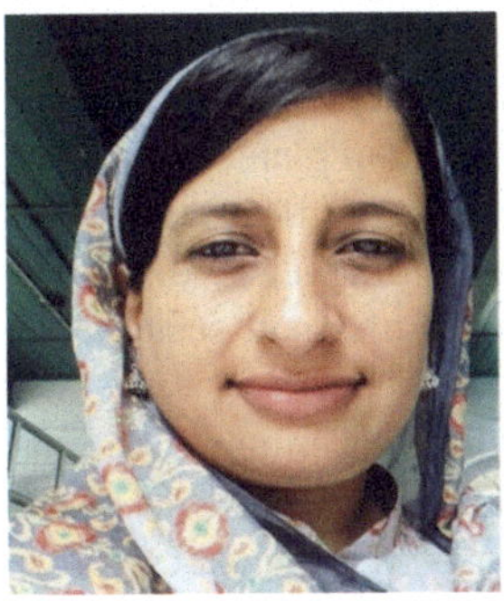

While the world is remarkably progressing each day towards an advanced age of science and technology, at the same time the pace at which the pollutants are being produced and released into the environment is alarming. The dilution in the values of sustainability has produced pollutants that have impacted the air, soil and water environments to such an extent that their restoration in the present scenario seems to be an impossible task. Diverse pollutants having varied and serious effects have created a hostile environment which is not fit for a healthy life to thrive in. The survival of humans on a polluted planet pumped with heavy metals, chemical fertilisers, pesticides, industrial chemicals and toxins is under threat. Continuous efforts have been made from time to time to combat these pollutants with the use of various techniques, however due to various reasons pertaining to ecological sustainability and economic viability; these technologies have failed to touch the mark of success. With the advent of biotechnology, a breakthrough has been achieved and many success stories have been recorded. The book titled *Bioremediation and Biotechnology: Degradation of pesticides and heavy metals in contaminated Environs* is a brilliant effort towards documenting the bio-techniques and technologies that are and can be used to address the problems of pollution in a sustainable manner. These technologies are not only eco-friendly but also cost-effective. The latest research pertaining to biotechnology and bioremediation has been presented in a lucid manner. The book has 15 chapters. Chapter 1 addresses the main causes and consequences of pollution through pesticides and heavy metals which are persistent pollutants capable of bio-magnification and accumulation in the food chain. Chapter 2 highlights the necessity for the development of mechanism or strategies to overcome metal exposure and to identify substances that devote significantly to overcome their toxic effects within living organism's body. In the present scenario of degraded environments particularly due to heavy metals and pesticides, there is a need to develop techniques which can keep a check on these pollutants. In this regard, Chap. 3 highlights the role of modern innovative techniques for monitoring the pollution due to heavy metals and pesticides using sensors and biosensors. A global scenario regarding the remediation techniques like biodegradation, bioventing, biosparging, phytoremediation, bioreactor, biopile and composting used to combat pesticide pollution is elaborated in Chap. 4 while Chap. 5 details the sustainable technique of mycoremediation for abatement of pesticide pollution and ecological reconstruction. Chapter 6 deals with the types of bio-pesticides and the mechanism of action, their advantages over the conventional and chemical pesticides as well as their role in maintaining the sustainability of the environment. With a mushrooming population in the world, there is a huge pressure on the agricultural sector for food production. However, the chemical fertilisers are reducing the soil fertility as well as crop productivity; therefore, there is an urgent need to use alternatives to meet the need for nutrient sources for crops. In this context, Chap. 7 elaborates the values of biofertilisers for sustainable management in agricultural industries. Macrophytes execute optimisation and self-purification tasks and are an important tool for the hyper-accumulation of micro- and macrotrace elements. Chapter 8 discusses in detail the role of macrophytes in phytofiltration thus helping in restoration of lacustrine environments. Certain plants are potent tools for the removal of pollutants from the environment. Chapter 9 discusses the role of *Salix* (Willows) for phytoremediation of heavy metals. Chapter 10 discusses biotechnological tools and applications as easy, time saving, safer and sustainable way to remediate pesticide pollution.

Industries produce chemical effluents that are toxic to the environment. A number of methods are used for the treatment of these chemicals, which however prove ineffective. Chapter 10 details the technique of photocatalysis as the most effective technique for the treatment of these industrial effluents which completely degrade the toxic organic pollutants and dyes. Chapter 11 focuses on another reliable technique, i.e. micellar-enhanced ultrafiltration, which sustainably removes dyes from wastewater. Another innovative modern technology for degradation and removal of heavy metals and recalcitrant pollutants from the polluted aquatic ecosystems is biofilms and elaborated in detail in Chap. 12.

Chapters 13 and 14 focus on the importance of using eco-friendly and economically viable techniques of biotechnology and bioremediation restore the degraded environments in a sustainable manner. The future prospects of these techniques are also delibrated upon in these chapters. The final chapter provides an overview of global environmental regulations framed for management of pesticides for global environmental safety, human welfare and sustainable development.

This book is a valuable resource material for researchers of biotechnology and allied sciences. The authors and editors have put in a lot of effort to make a noteworthy contribution to the area of biotechnology and bioremediation. I wish them all the best for their future endeavours.

Humaira Qadri
Head, Department of Environmental Science
Sri Pratap College, Cluster University
Srinagar, Jammu and Kashmir, India

Preface

The world is advancing towards an advanced age of science and technology, but at the same time the ever-increasing anthropogenic interferences are disrespecting the principles of sustainability causing serious pollution problems. These problems not only threaten the survival of humans on the earth but also make this planet a hostile place to live on. The seriousness of the pollution is directly related to the type of pollutant and its impacts. The pollutants of concern include heavy metals, pesticides, fertilisers and chemical dyes. Pollution from heavy metals and pesticides is a major problem as these pollutants can cause irreparable damage to health or end the life of living beings exposed to them. Chemical substances are introduced in the environment through fertilisers, herbicides, pesticides and phytohormones. Chemical pesticides lead to the depletion of the natural fertility of the soil and deterioration of the environment as a whole. The residual effects of these chemicals create an ecological imbalance as well as contribute to the promotion of modern diseases. Due to these and other associated anthropogenic stresses, the environment has changed drastically causing environmental problems including biodiversity loss and public health issues which has attracted the attention of environmentalists for the remediation and restoration of environment at an earliest mode. As compared to conventional and natural methods, the biotechnological tools and applications are an easy, safe and sustainable way to achieve the goal of remediating and restoring the contaminated environs. Environmental biotechnology is the integration of scientific and engineering knowledge that is employed to remediate and restore the degraded environment. This book focuses on the biotechnological approaches used to combat different environmental pollutions with new technologies to rejuvenate the degraded environments alongwith the future prospects and new developments for sustainable future. The book is an elaborate effort towards bringing together the knowledge related to the problems associated with the contamination of the environment and thereof the strategies to combat them.

The book highlights the necessity for the development of mechanism or strategies to overcome heavy metal exposure and to identify substances that devote significantly to overcome their toxic effects within living organism's body. During the present era, it is necessary to develop innovative techniques for assessing and monitoring the pollutants and develop new methodologies to battle their effects. The content of the book conveys how different types of pollution will be addressed and emphasises on the innovative ways of detecting these pollutants. The sustainable approach to reducing pesticide pollution by using mycoremediation has also been aimed at in an elaborate manner. Several techniques including photocatalysis as the most effective technique for the treatment of industrial effluents including toxic organic pollutants and dyes has also been discussed in detail.

The book elaborates bio-pesticides as an alternative to synthetic pesticides as well as a noteworthy part of numerous vermin control programmes. The future prospects for the improvement of new bio-pesticides are likewise discussed. The role and importance of biofertilisers as sustainable tool in the arena of modern agricultural sciences have also been delibrated on with emphasis on their role in agriculture by way of maintaining the soil sustainability and long-term fertility. A good portion of the book is dedicated to phytoremediation which during the present times is considered to be a a safe and eco-friendly mechanism to restore the degraded ecosystems. Some terrestrial plant species which are a potent tool to remove the contaminants from the environment are appropriately discussed. The macrophytes have been highlighted as potential metal and mineral removal tools acting as biofilters in the aquatic environments. Latest research studies have been presented in the book which reflect the importance of these biofilters in the eco-restoration of the degraded aquatic systems.

The use of bioremediation as an eco-friendly and economically viable technique has been discussed appropriately. Biofilm-based technologies have also been presented as tools with tremendous potential in cleaning pollutants with special focus on the potential role of innovative biofilm technologies to remediate aquatic pollution.

The book is a repository of advanced research in biotechnology and bioremediation. It shall act as a useful reference for the latest advances in research in the concerned area. The content of the book is diverse and shall attract the attention of students, researchers and scientists world over.

Srinagar, Jammu and Kashmir, India — Rouf Ahmad Bhat
Jeddah, Saudi Arabia — Khalid Rehman Hakeem
Srinagar, India — Moonisa Aslam Dervash

About the Book

This book addresses the grave concerns stemming out due to conventional treatment techniques. The main focus of this book revolves round the central kernel of novel technology (bioremediation and biotechnology) which has emerged as an independent warrior to clean up and restore the disturbed environs. Furthermore, this book is a coherent assortment of diverse chapters relevant to the role of biotechnology and bioremediation for restoration of the ecosystems degraded by pesticide and heavy metal pollution. The inaugural chapters deal with the quantification of problem and its magnitude due to pesticides and heavy metals, followed by innovative modern biotechnological and bioremediation treatment technologies and sustainable techniques to remediate the persistent pollutants. It is a detailed comprehensive account for the treatment technologies from unsustainable to sustainable. Academicians, researchers and students shall find it as a complete wrap up regarding biotechnological intervention for sustainable treatment of pollution and shall suffice for the diverse needs of teaching and research.

Contents

About the Editors

Rouf Ahmad Bhat, Ph.D. is working in Cluster University Srinagar (J&K), India, in the capacity of Assistant Professor and has his specialisation in Limnology, Toxicology, Phytochemistry and Phytoremediation. Dr. Bhat has been teaching graduate and post-graduate students of Environmental Sciences for the past 3 years. He is author of more than 50 research papers and 15 book chapters and has published more than 10 books with international publishers. He has presented and participated in numerous state, national and international conferences, seminars, workshops and symposium. Dr. Bhat has worked as an associate Environmental Expert in World Bank-funded Flood Recovery Project and also Environmental Support Staff in Asian Development Bank (ADB)-funded development projects. He has received many awards, appreciations and recognitions for his services to the science of water testing, air and noise analysis. He has served as editorial board member and reviewer of reputed international journals. Dr. Bhat is still writing and experimenting with diverse capacities of plants for use in aquatic pollution.

Khalid Rehman Hakeem, Ph.D. is Professor at King Abdulaziz University, Jeddah, Saudi Arabia. After completing his doctorate (Botany; specialisation in Plant Eco-physiology and Molecular Biology) from Jamia Hamdard, New Delhi, India, in 2011, he worked as a lecturer at the University of Kashmir, Srinagar, for a short period. Later, he joined Universiti Putra Malaysia, Selangor, Malaysia, and worked there as Postdoctoral Fellow in 2012 and Fellow Researcher (Associate Professor) from 2013 to 2016. Dr. Hakeem has more than 10 years of teaching and research experience in plant eco-physiology, biotechnology and molecular biology, medicinal plant research, plant–microbe–soil interactions as well as in environmental studies. He is the recipient of several fellowships at both national and international levels; also, he has served as the visiting scientist at Jinan University, Guangzhou, China. Currently, he is involved with a number of international research projects with different government organisations. So far, Dr. Hakeem has authored and edited more than 35 books with international publishers, including Springer Nature, Academic Press (Elsevier) and CRC Press. He also has to his credit more than 80 research publications in peer-reviewed international journals and 55 book chapters in edited volumes with international publishers. At present, Dr. Hakeem serves as an editorial board member and reviewer of several high-impact international scientific journals from Elsevier, Springer Nature, Taylor and Francis, Cambridge and John Wiley Publishers. He is included in the advisory board of Cambridge Scholars Publishing, UK. He is also a fellow of Plantae group of the American Society of Plant Biologists, member of the World Academy of Sciences, member of the International Society for Development and Sustainability, Japan and member of Asian Federation of Biotechnology, Korea. Dr. Hakeem has been listed in Marquis Who's Who in the World, since 2014–2019. Currently, Dr. Hakeem is engaged in studying the plant processes at eco-physiological as well as molecular levels.

Moonisa Aslam Dervash, Ph.D. is a young scholar who has earned her Ph.D. in Environmental Sciences from SKAUST-K, Shalimar (J & K), whereas, her B.Sc., M.Sc. and B. Ed. from Kashmir University. She has qualified many prestigious national and state level examinations like UGC-NET, ICAR-NET, JK-SET (twice) etc. Presently, she is working in All India Radio-Akashvani Srinagar (Vivedh Bharati) as a broadcaster on part time basis and also extends her services as a guest lecturer in J & K Higher Education Department. Her specialization is in Environmental monitoring, Ecology, Soil biology and Carbon sequestration. She has authored more than three books with national and International publishers. Moreover, she has published number of scientific research articles and book chapters in the journals of national & international repute. She has also participated and presented her papers in national and international scientific conferences.

She is recipient of many prestigious awards and felicitations like Bharat Excellence Award and Best Educationist Award for her dedicated accomplishments. Her focus has remained on many facets of society especially on environmental conservation and women empowerment. She has been felicitated by state government (J & K Department of Ecology, Environment and Remote Sensing) for her outstanding contribution in environmental awareness and conservation through electronic media. Moreover, she is blessed with the poetic tint which keeps her intact with literary world as well. She has participated in number of poetic symposia organized by Sahitya Akademi New Delhi, Akashvani Srinagar, Jammu & Kashmir Academy for Art, Culture & Literature and Adbi Markaz Kamraaz from time to time. She has been also awarded honorary D. Litt. (Vidya Sagar) by Vikramasheela Vidyapeeth (Bhagalpur, India) for her literary works. Recently, she represented J & K (Kashmiri language) in "All India Young Writers Meet (Young Harvest Programme in Annual Festival of Letters-2020)" which was held at Sahitya Akademi, National Academy of Letters Rabindra Bhawan, New Delhi.

Chapter 1
Causes and Effects of Pesticide and Metal Pollution on Different Ecosystems

Flávia Cristina Policarpo Tonelli and Fernanda Maria Policarpo Tonelli

1.1 Introduction

The interference of human beings on the environment is harmful if it does not respect the principles of sustainability (Qadri and Bhat 2019). It can end up inserting synthetic substances in the environment that become persistent pollutants or promoting the disposal of a large amount of naturally occurring elements in a quantity that represents risk to the living organisms. In the first described scenario, the use of pesticides, and in the second scenario the pollution promoted by heavy metals are examples (Mehmood et al. 2019; Rashid et al. 2019; Singh et al. 2019). So, respectively, organic and inorganic pollutants can threat ecosystems as consequence of human activities like the attempt to improve agricultural production, mine exploitation, industrial production of different products, etc. (Jafari et al. 2013; Morillo and Villaverde 2017; Barrios-Estrada et al. 2018; Mateos et al. 2018; Bilal et al. 2019; Mendes et al. 2019; Pesantes et al. 2019; Pu et al. 2019; Rosculete et al. 2019; Aruliah et al. 2019; Vázquez-Luna and Cuevas-Díaz 2019: Zhang et al. 2019).

1.2 Causes and Effects of Heavy Metal Pollution

Heavy metals are naturally present in trace amounts in the environment and they can reach living forms through the occurrence of natural processes like volcanic eruptions and forest fires. However, when it comes to pollution, anthropogenic processes, involving water, soil and/or air, as causes of environments contamination with these

F. C. P. Tonelli
Department of Biochemistry, Federal University of São João del Rei, Divinópolis, Brazil

F. M. P. Tonelli (✉)
Department of Cell Biology, Institute of Biological Science, Federal University of Minas Gerais, Belo Horizonte, Brazil

R. A. Bhat et al. (eds.), *Bioremediation and Biotechnology, Vol 2*,
https://doi.org/10.1007/978-3-030-40333-1_1

harmful elements are the most relevant source. Human interference can cause these elements to accumulate, in a way dependent on metal characteristics and climate factors (Zupancic 2017; Herath et al. 2018), reaching quantities that are considered unsafe.

These elements (for example: arsenic (As), cadmium (Cd), cobalt (Co), chromium (Cr), copper (Cu), mercury (Hg), nickel (Ni), lead (Pb), selenium (Se), and zinc (Zn)) with high atomic weight (Duffus 2002; Busairi and Syahir 2018) can, even at low concentration, threat health of living forms (Muszynska and Hanus-Fajerska 2015). The maximum allowable concentration for Hg, for example, is, according to the United States Environmental Protection Agency, 0.002 μg/mL. Zn is more efficiently tolerable possessing a limit of 5 μg/mL (Odobašić et al. 2019).

These inorganic contaminants represent a serious environmental problem due to the fact that their pollution is persistent, lasting for long time periods (long half-life) on contaminated sites. And besides that, they possess the ability to bioaccumulate through the food chain and cause biomagnification (Mehmood et al. 2019; Rashid et al. 2019; Singh et al. 2018).

The presence of organic matter can favor toxic metals pollution as it can improve the capacities of elements like Hg^{2+}, Ni^{2+}, and Pb^{2+} to binding and accumulate (Rikta et al. 2018).

Mining regions are a very relevant source of contamination where not only soil but also watercourses can be polluted. But different kind of industries (for example, dealing with dyes and pigments or petroleum) can also relevantly contribute by discharging their effluents containing heavy metals into water like rivers, for example (Salomons et al. 1995; Goyer 2001; Alkorta et al. 2004; Karthik et al. 2017; Zhu et al. 2017; Jacob et al. 2018; Ali et al. 2019; Wang et al. 2019).

Among the damages that heavy metals can cause to human health is cancer, once these contaminants are teratogenic and carcinogenic agents (Mushtaq et al. 2019). By poisoning different organs and tissues these elements can also affect the spleen, lungs (causing asthma, for example), skin, bones, liver, pancreas, cornea, gut/stomach, heart and vessels, reproductive organs, brain, and kidney (Farombi et al. 2012; Abarikwu et al. 2013; Tepanosyan et al. 2017; Liu et al. 2018; Boateng et al. 2019). However, they can also cause the organism death after poisoning once they can induce high rates of oxidative stress that, on its turn, can induce cells to die. They can accumulate in soft tissues after entering the body through mainly: skin absorption during industrial activities by adults or ingestion by kids (Jaishankar et al. 2014; Manzetti et al. 2014; Jacob et al. 2018; Masindi and Muedi 2018; Xiong et al. 2019; Yang et al. 2019).

Air can be highly polluted, especially in big cities with intense industrial activities and transportation based on the use of fossil fuel. Particulate matters can arise from natural events like volcanic eruptions, dust storms, and soil erosion but mainly from anthropogenic activities like waste incineration (Li et al. 2017; Weibel et al. 2017) causing the environment, especially near the incinerator, to contain, for example, As, Cd, Cu, Hg, Pb, and Zn (Ma et al. 2018). Eye irritation and respiratory problems are the most quickly harmful effects observed as a consequence of air pollution. Heavy metals present on air pollution can be delivered to the soil by acid rain, for example (Zheng et al. 2012; Ventura et al. 2017).

Soil can accumulate nondegradable heavy metals in high amounts being an important source of food contamination. The presence of heavy metals in soil also affect microorganism, impairing the degradation of biodegradable organic pollutants making the pollution scenario worse. Soil quality is also decreased as heavy metals can, for example, alter soil pH and natural chemistry (Musilova et al. 2016). Landfill is a relevant source of soil contamination with heavy metals risking the life of people who directly deal with the wastes and also people far from the pollution site who can be indirectly exposed through water collaterally contaminated or through air (Wittsiepe et al. 2017). Agriculture soils can also contain heavy metals like Mn, Zn, and Pb in high amounts exceeding the safe exposure values for humans (Hu et al. 2018; Shen et al. 2019; Yang et al. 2019).

A contaminated soil can pollute water nearby and/or groundwater through leachates, endangering the entire ecosystem (Jaishankar et al. 2014; Nagajyoti et al. 2010). As previously mentioned, the presence of organic matter can favor heavy metal pollution and it can also originate hidden pollution. Bottom sediments containing these elements adsorbed into matter can be generated (Liang et al. 2015; Ali et al. 2016; Petrosyan et al. 2019).

Therefore, water can also be polluted by heavy metals from diverse sources directly or indirectly and end up endangering all food chains and food webs (exposing human beings, that are commonly at the extreme of this chain, to a high level of pollutant, and causing serious health problems (Lee et al. 2002)). Water from mining areas, for example, experiences a reduction in biodiversity (Netto et al. 2013; Masindi 2016). Acid mine drainage containing heavy metals can induce lowering of water (and soil) pH making the environment unsuitable for the survival of organisms that are not acidophile (Tutu et al. 2008; Simate and Ndlovu 2014; Torres et al. 2014).

1.3 Causes and Effects of Pesticide Pollution

According to Food and Agriculture Organization (FAO 2019), pesticides are any substance or mixture of substances of chemical or biological ingredients used in agriculture to protect plants against weeds, pests, and diseases. They are capable of repelling, destroying, or controlling any pest, or regulating plant growth. In addition, it is also used against ectoparasites for animal protection. They have a variety of chemical structures.

These substances can be classified according to the target group they attack, their chemical structure, or origin: insecticides (chlorinated hydrocarbons, organophosphates, carbamate insecticides, pyrethroids, botanical and biological products, and others not elsewhere classified); mineral oils, fungicides, and bactericides (inorganic, dithiocarbamates, benzimidazoles, triazoles, diazoles, diazines, morpholines, others not elsewhere classified); herbicides (phenoxy hormone products, triazines, amides, carbamate herbicides, dinitroanilines, urea derivatives, sulfonylurea, bipiridils, uracil, others not elsewhere classified); disinfectants, plant growth regulators, rodenticides (anti-coagulants, cyanide generators, hypercalcemics, narcotics, others not elsewhere classified); seed treatment—fungicides (dithiocarbamates, benzimidazoles, triazoles, diazoles, diazines, morpholines, botanical products and biological, others not elsewhere classified), seed treatment—insecticides (organophosphates, carbamate insecticides, pyrethroids, others not elsewhere classified); and other pesticides not elsewhere specified. In addition, they can also be classified into synthetic (carbamate (nitrogen), chlorinated, phosphorus, and chlorophosphorus), botanical (based on nicotine, pyrethrin, sabadine, rotenone), and inorganic, as some elements (including heavy metals) can also possess pesticide's activity (arsenic, thallium, barium, nitrogen, phosphorus, cadmium, iron, selenium, lead, copper, mercury, and zinc) (FAO 2019).

In recent years their effects on the environment and living organisms have attracted great attention because they are highly toxic substances to human beings. According to the Pesticide Actions Network UK (PAN UK 2019), after a single episode of inhalation, skin contact or ingestion, the symptoms may be immediately evident or can arise within 48 h in the acute toxicity (allergic sensitization, nausea, vomiting, diarrhea, eye and skin irritation, respiratory tract irritation, sore throat and/or cough, headache, loss of consciousness, extreme weakness, seizures and/or death). The effects of long-term toxicity include depression and anxiety, Parkinson's disease, asthma, cancer, attention deficit and hyperactivity disorder (ADHD).

When pesticides are released or applied to the environment, their effects are affected by environmental conditions, such as soil interaction and properties, agricultural practice characteristics, and local climatology (Bhat et al. 2018a; Quaglia et al. 2019).

Agricultural chemicals, for example, consist of fertilizers and pesticides. The main cause of pollution by these substances is their improper use: due to their excessive use or lack of knowledge when they are used, problems were already reported in several plantations such as cotton, maize, wheat and rice (Jin et al. 2015). The amount of insecticides, fungicides, herbicides' use is also increasing to try to provide the growing population with food (Gavrilescu 2005).

Pesticide soil pollution affects soil quality in different ways. The microorganisms present in it, for example, may have their growth rate and activities modified by the chemical, which will lead to modification of the nutrient cycle in the soil. The accumulation of macro and micronutrients such as sulfur (S), manganese (Mn), magnesium (Mg), phosphorus (P), and potassium (K) present in the hybrid Bermuda grass leaf tissues can be reduced, for example, due to the use of isoxaben, prodiamine, and indaziflam (Brosnan 2015). In addition, the root mass of this plant was also reduced compared to plants not treated by these pesticides (FAO 2019).

Another effect to be highlighted is the influence of pesticides on nutrient availability in the soil. For example, sulfur is an essential nutrient especially for crop agricultural production. The monocrotophos application significantly changes the oxidation rate of this nutrient in black and red soils in India. Thus, this rate increases after 7–14 days of application of this substance alone or in combination with mancozeb in black soils (Srinivasulu et al. 2015).

In addition, the presence of pesticides in the soil may lead to changes in enzymatic activity. For example, the arylamidase and myrosinase enzymes may be stimulated by the acetamiprid and carbofuran insecticides. However, these same substances have a toxic effect on enzymatic activity in black and red clay soils (Mohiddin et al. 2015).

Thus, it is necessary to apply pesticide remediation techniques in order to try to eliminate these pollutants from the environment. However, it is also necessary to realize that remediation efficiency depends on factors such as chemical and physical properties of pesticides, climatic conditions, extent and concentration of pesticides, and the presence of other contaminants mixed in the soil/water. Remediation can be done by physical treatments such as soil washing or electro-remediation, or by bioremediation, through phytoremediation and microbial remediation in which plants and microorganisms, respectively, sequester, metabolize, or even release enzymes capable of destroying pesticides (Bhat et al. 2018a; Sun et al. 2018).

When it comes to water pollution, it can be divided into two types: diffuse sources and point sources (Vasilescu and Medvedovici 2005). Diffuse source pollution comes from unlicensed sources and dispersed land use activities. Examples of pollutants are: sediments, toxic substances, salts, acid sulfate soils in drained wetlands, raw pollutants, fertilizers, soil erosion by pathogens such as septic tank leaking bacteria. Pesticide spraying is one of the prime examples of diffuse source and, near the water surface, leads to its contamination. Pesticides enter the surface and soil through water runoff especially in agricultural areas, but also along roadsides, forested areas, golf courses, suburban, and urban landscape areas (Srivastava et al. 2019).

According to FAO, pesticide water contamination is related not only to agriculture but also to forestry in the forest industry, especially in Canada. In this industry, these substances are used to control forest pests, mainly insects, and are sprayed over large areas by aircraft. This use is mainly due to the generation of a habitat conducive to the creation of disease-bearing insects by modifying the water regime in irrigated agriculture in tropical and subtropical environments. In addition, to control disease vectors it is necessary in some cases to add insecticides such as DDT (FAO 2019).

Several factors are associated with the effects of pesticide on water quality. Highlights include impurities contaminating the active ingredient, additives mixed with the active ingredient such as buffers, preservatives, solvents, etc., active ingredients in the pesticide formulation, degraded by microbiological, photochemical, chemical degradation of active ingredients. Some factors are employed as criteria for measuring the effects of pesticides on the aquatic environment. They are: toxicity (expressed in Lethal Dose concentration), degraded, persistence (measured in half-life that may be influenced by metabolism and biodegradation in biotic factors, and hydrolysis, photolysis, and oxidation in abiotic factors), fate (this factor is influenced by the chemical's affinity with the environmental compartment: liquid, solid matter, gaseous form, and biota (FAO 2019)).

The effects of pesticides are varied and generally interrelated, and depend on the organism and the type of substance. Chronic, non-lethal effects are not noticeable, although they still have consequences in the food chain. The consequences are cancer, tumor, injury to fish and animals, suppression of the immune system, genetic and cellular damage, inhibition or failure of reproduction, death, disruption of the endocrine system, impoverishment of fish health marked by low red blood cells and white, intergenerational effects (effects are not apparent until subsequent generations of the organism) (Srivastava et al. 2019).

The main cause of air pollution by pesticides is the indiscriminate use of these substances especially in agriculture in an attempt to improve crop quality. Misuse of pesticides includes use of substances prohibited by local governments, lack of protective equipment, excessive spraying, reuse of washed pesticide containers such as food and drinking water containers, and improper storage. Pesticide use can be influenced by gender, age, chemical knowledge, cultural differences, and perceptions. In air, the main source of contamination is due to inhalation by crop sprays. Thus, the humans who are most affected by this pollution are those who live or work on farms (Özkara et al. 2015).

Organophosphates can enter animals through inhaled particles, ingestion, or skin contact. These pesticides are stored in their muscles and fat and can be found in the liver, lungs, and brain. In addition, these substances can be brought to humans through milk and other animal foods (Zhu et al. 2015).

Pesticide use is highly dangerous as poisoning can be lethal (Mushtaq et al. 2018). Organophosphate poisoning can lead to cholinergic syndrome with symptoms such as coma, blurred vision, delayed neuropathy, blockage of the respiratory center, headache, slurred speech. Although it is possible to reverse an acute bout of contamination, there are likely to be long-term nervous system effects. Other consequences of contamination with these substances are immunological endocrine disruption, function damage, female spontaneous abortions and preterm, and children neuro-developmental delays. In this way, it is essential to control the use of these substances and to study alternatives to their use and forms to revert their effect (Schierow and Esworthy 2012; Niti et al. 2013; Cheng et al. 2016; Liu et al. 2018; Reeves et al. 2019).

1.4 Conclusion

Human intervention on environment to economically explore it without considering sustainability induces an increasing concern and threat to life: pollution. As pollution increases in number of sites affected and its effects are increasingly noticed, researches have been dedicating attention to understand its causes and effects. Heavy metals and pesticides are two important persistent pollutants capable of biomagnifying and accumulating in the food chain. So, they are harmful to air, soil, and water, and consequently to the existence of life on the planet.

1.5 Future Perspectives

It is expected that humanity can understand the importance of interacting with the environment in a sustainable way to reduce the amount of pollutants accumulated. It is also necessary to find new ways to detect the presence of heavy metals and pesticides in water, air, and soil, and quantify their concentration to verify their compliance with the legislation parameters. Through multidisciplinary work, a strategy for monitoring these substances that is most economically viable, fast, reproducible, specific, and practical should be achieved. A good strategy is to adapt and refine existing techniques. Perhaps, this is the first step towards a sustainable environmental model, always remembering that each human being is part of it and by avoiding contamination he/she can protect his or her own health and the health from the others.

The great challenge of this process is the considerable diversity of pesticide's chemical structures, the interactions of heavy metals in each organism, and the interaction of contaminated living beings in the food chain. However, it is expected that in the future a viable alternative or, preferably, viable alternatives, will be reached.

References

Abarikwu S, Iserhienrhien B, Badejo T (2013) Rutin- and selenium attenuated cadmium-induced testicular pathophysiology in rats. Hum Exp Toxicol 32:395–406

Ali MM, Alia ML, Islamc MS, Rahmanda MZ (2016) Preliminary assessment of heavy metals in water and sediment of Karnaphuli River, Bangladesh. Environ Nanotechnol Monit Manag 5:27–35

Ali H, Khan E, Ilahi I (2019) Environmental chemistry and ecotoxicology of hazardous heavy metals: environmental persistence, toxicity, and bioaccumulation. J Chem 2019:1–14

Alkorta I, Herna'ndez-Allica J, Becerril JM, Amezaga I, Albizu I, Garbisu C (2004) Recent findings on the phytoremediation of soils contaminated with environmentally toxic heavy metals and metalloids such as zinc, cadmium, lead and arsenic. Rev Environ Sci Biotechnol 3:71–90

Aruliah R, Selvi A, Theertagiri J, Ananthaselvam A, Kumar KS, Madhavan J, Rahman P (2019) Integrated remediation processes toward heavy metal removal/recovery from various environments—a review. Front Environ Sci 7:1–66

Barrios-Estrada C, de Jesús Rostro-Alanis M, Muñoz-Gutiérrez BD, Iqbal HM, Kannan S, Parra-Saldivar R (2018) Emergent contaminants: endocrine disruptors and their laccase-assisted degradation–a review. Sci Total Environ 612:1516–1531

Bhat RA, Beigh BA, Mir SA, Dar SA, Dervash MA, Rashid A, Lone R (2018a) Biopesticide techniques to remediate pesticides in polluted ecosystems. In: Wani KA, Mamta (eds) Handbook of research on the adverse effects of pesticide pollution in aquatic ecosystems. IGI Global, Hershey, pp 387–407

Bhat RA, Dervash MA, Qadri H, Mushtaq N, Dar GH (2018b) Macrophytes, the natural cleaners of toxic heavy metal (THM) pollution from aquatic ecosystems. In: Environmental contamination and remediation. Cambridge Scholars Publishing, Cambridge, pp 189–209

Bilal M, Adeel M, Rasheed T, Zhao Y, Iqbal HM (2019) Emerging contaminants of high concern and their enzyme-assisted biodegradation–a review. Environ Int 124:336–353

Boateng TK, Opoku F, Akoto O (2019) Heavy metal contamination assessment of groundwater quality: a case study of Oti landfill site, Kumasi. Appl Water Sci 9:33–47

Brosnan JT (2015) Preemergence herbicides affect hybrid bermudagrass nutrient content. J Plant Nutr 38:177–188

Busairi N, Syahir A (2018) Recent advances in mercury detection: towards enabling a sensitive and rapid point-of-check measurement. J Toxicol Risk Assess 4:1–10

Cheng M, Zeng G, Huang D, Lai C, Xu P, Zhang C, Liu Y (2016) Hydroxyl radicals based advanced oxidation processes (AOPs) for remediation of soils contaminated with organic compounds: a review. Chem Eng J 284:582–598

Duffus JH (2002) "Heavy metals"-a meaningless term? (IUPAC Technical Report). Pure Appl Chem 74:793–807

FAO - Food and Agriculture Organization of United Nations (2019). http://www.fao.org/3/a-bt566e.pdf. Accessed 18 Aug 2019

Farombi E, Akintunde J, Nzute N, Adedara I, Arojojoye O (2012) Municipal landfill leachate induces hepatotoxicity and oxidative stress in rats. Toxicol Ind Health 28:532–541

Gavrilescu M (2005) Fate of pesticides in the environment. Eng Life Sci 5:497–526

Goyer RA (2001) Toxic effects of metals. In: Klaassen CD (ed) Casarett and Doull's toxicology: the basic science of poisons. McGraw-Hill, New York, pp 811–867

Herath D, Pitawala A, Gunatilake J, Iqbal MCM (2018) Using multiple methods to assess heavy metal pollution in an urban city. Environ Monit Assess 190:657–671

Hu W, Wang H, Dong L, Huang B, Borggaard OK, Bruun-Hansen HC, He Y, Holm PE (2018) Source identification of heavy metals in peri-urban agricultural soils of southeast China: an integrated approach. Environ Pollut 237:650–658

Jacob JM, Karthik C, Saratale RG, Kumar SS, Prabakar D, Kadirvelu K, Pugazhendhi A (2018) Biological approaches to tackle heavy metal pollution: a survey of literature. J Environ Manag 217:56–70

Jafari M, Danesh YR, Goltapeh EM, Varma A (2013) Bioremediation and genetically modified organisms. In: Goltapeh E, Danesh Y, Varma AM (eds) Fungi as bioremediators. Springer, Berlin, pp 433–450

Jaishankar M, Tseten T, Anbalagan N, Mathew BB, Beeregowda KN (2014) Toxicity, mechanism and health effects of some heavy metals. Interdiscip Toxicol 7:60–72

Jin S, Buemling B, Mol APJ (2015) Information, trust and pesticide overuse: interactions between retailers and cotton farmers in China. Wagening J Life Sci 72:23–32

Karthik C, Barathi S, Pugazhendhi A, Ramkumar VS, Thi NBD, Arulselvi PI (2017) Evaluation of Cr (VI) reduction mechanism and removal by Cellulosimicrobium funkei strain AR8, a novel haloalkaliphilic bacterium. J Hazard Mater 333:42–53

Lee G, Bigham JM, Faure G (2002) Removal of trace metals by coprecipitation with Fe, Al and Mn from natural waters contaminated with acid mine drainage in the Ducktown Mining District, Tennessee. Appl Geochem 17:569–581

Li T, Wan Y, Ben Y, Fan S, Hu J (2017) Relative importance of different exposure routes of heavy metals for humans living near a municipal solid waste incinerator. Environ Pollut 226:385–393

Liang A, Wang Y, Guo H, Bo L, Zhang S, Bai Y (2015) Assessment of pollution and identification of sources of heavy metals in the sediments of Changshou Lake in a branch of the Three Gorges Reservoir. Environ Sci Pollut Res 22(20):16067–16076

Liu X, Zhong L, Meng J, Wang F, Zhang J, Zhi Y, Zeng L, Tang X, Xu J (2018) A multi-medium chain modeling approach to estimate the cumulative effects of cadmium pollution on human health. Environ Pollut 239:308–317

Ma W, Tai L, Qiao Z, Zhong L, Wang Z, Fu K, Chen G (2018) Contamination source apportionment and health risk assessment of heavy metals in soil around municipal solid waste incinerator: a case study in North China. Sci Total Environ 631:348–357

Manzetti S, Spoel ER, Spoel D (2014) Chemical properties, environmental fate, and degradation of seven classes of pollutants. Chem Res Toxicol 27:713–737

Masindi V (2016) A novel technology for neutralizing acidity and attenuating toxic chemical species from acid mine drainage using cryptocrystalline magnesite tailings. J Water Process Eng 10:67–77

Masindi V, Muedi KL (2018) Environmental contamination by heavy metals. IntechOpen https://doi.org/10.5772/intechopen.76082

Mateos AC, Amarillo AC, Carreras HA, González CM (2018) Land use and air quality in urban environments: human health risk assessment due to inhalation of airborne particles. Environ Res 161:370–380

Mehmood MA, Qadri H, Bhat RA, Rashid A, Ganie SA, Dar GH, Shafiq-ur-Rehman (2019) Heavy metal contamination in two commercial fish species of a trans-Himalayan freshwater ecosystem. Environ Monit Assess 191:104. https://doi.org/10.1007/s10661-019-7245-2

Mendes KF, Régo APJ, Takeshita V, Tornisielo VL (2019) Water resource pollution by herbicide residues. IntechOpen https://doi.org/10.5772/intechopen.85159

Mohiddin GJ, Srinivasulu M, Maddela NR, Manjunatha B, Rangaswamy V, Koch ARK (2015) Influence of the insecticides acetamiprid and carbofuran on arylamidase and myrosinase activities in the tropical black and red clay soils. Environ Monit Assess 187:388–393

Morillo E, Villaverde J (2017) Advanced technologies for the remediation of pesticide-contaminated soils. Sci Total Environ 586:576–597

Mushtaq N, Bhat RA, Dervash MA, Qadri H, Dar GH (2018) Biopesticides: the key component to remediate Pesticide contamination in an ecosystem. In: Environmental contamination and remediation, Cambridge Scholars Publishing, Cambridge, pp 152–178

Mushtaq N, Singh DV, Bhat RA, Dervash MA, Hameed UB (2019) Freshwater contamination: sources and hazards to aquatic biota. In: Qadri H, Bhat RA, Dar GH, Mehmood MA (eds) Freshwater pollution dynamics and remediation. Springer, Singapore, pp 27–50

Musilova J, Arvay J, Vollmannova A, Toth T, Tomas J (2016) Environmental contamination by heavy metals in region with previous mining activity. Bull Environ Contam Toxicol 97:569–575

Muszynska E, Hanus-Fajerska E (2015) Why are heavy metal hyperaccumulating plants so amazing? Biotechnol J Biotechnol Comput Biol Bionanotechnol 96:265–271

Nagajyoti PC, Lee KD, Sreekanth T (2010) Heavy metals, occurrence and toxicity for plants: a review. Environ Chem Lett 8:199–216

Netto E, Madeira RA, Silveira FZ, Fiori MA, Angioleto E, Pich CT, Geremias R (2013) Evaluation of the toxic and genotoxic potential of acid mine drainage using physicochemical parameters and bioassays. Environ Toxicol Pharmacol 35:511–516

Niti C, Sunita S, Kamlesh K, Rakesh K (2013) Bioremediation: an emerging technology for remediation of pesticides. Res J Chem Environ 17:88–105

Odobašić A, Šestan I, Begić S (2019) Biosensors for determination of heavy metals in waters, vol 2019. IntechOpen, London, pp 1–15

Özkara A, Akyil D, Konuk M (2015) Pesticides, environmental pollution, and health, vol 2015. InTechOpen, London, pp 3–27

PAN UK - Pesticide Actions Network UK. https://www.pan-uk.org/health-effects-of-pesticides/. Accessed 18 Aug 2019

Pesantes AA, Carpio EP, Vitvar T, López MMM, Menéndez-Aguado JM (2019) A multi-index analysis approach to heavy metal pollution assessment in river sediments in the Ponce Enríquez Area, Ecuador. Water 11:590–601

Petrosyan V, Pirumyan G, Perikhanyan Y (2019) Determination of heavy metal background concentration in bottom sediment and risk assessment of sediment pollution by heavy metals in the Hrazdan River (Armenia). Appl Water Sci 9:102–110

Pu Q, Sun JQ, Zhang FF, Wen XY, Liu WH, Huang CM (2019) Effects of copper mining on heavy metal contamination in a rice agrosystem n the Xiaojiang River Basin, southwest China. Acta Geochim 38:753–773

Qadri H, Bhat RA (2019) The concerns for global sustainability of freshwater ecosystems. In: Qadri H, Bhat RA, Dar GH, Mehmood MA (eds) Freshwater pollution dynamics and remediation. Springer, Singapore, pp 1–15

Quaglia G, Jorisa I, Broekxa S, Desmeta N, Koopmansc K, Vandaeled K, Seuntjensa P (2019) A spatial approach to identify priority areas for pesticide pollution mitigation. J Environ Manag 246:583–593

Rashid A, Bhat RA, Qadri H, Mehmood MA (2019) Environmental and socioeconomic factors induced blood lead in children: an investigation from Kashmir, India. Environ Monit Assess 191(2):76. https://doi.org/10.1007/s10661-019-7220-y

Reeves WR, McGuire MK, Stokes M, Vicini JL (2019) Assessing the safety of pesticides in food: how current regulations protect human health. Adv Nutr 10:80–88

Rikta S, Tareq SM, Uddin MK (2018) Toxic metals (Ni^{2+}, Pb^{2+}, Hg^{2+}) binding affinity of dissolved organic matter (DOM) derived from different ages municipal landfill leachate. Appl Water Sci 8:1–8

Rosculete CA, Bonciu E, Rosculete E, Olaru LA (2019) Determination of the environmental pollution potential of some herbicides by the assessment of cytotoxic and genotoxic effects on *Allium cepa*. Int J Environ Res Public Health 16:75–85

Salomons W, Forstner U, Mader P (1995) Heavy metals: problems and solutions. Springer, Berlin, 401p

Schierow L, Esworthy R (2012) Pesticide law: a summary of the statutes. Report No.: RL31921. Congressional Research Service, Washington, DC

Shen F, Mao L, Sun R, Du J, Tan Z, Ding M (2019) Contamination evaluation and source identification of heavy metals in the sediments from the Lishui River Watershed, Southern China. Int J Environ Res Public Health 16:336–349

Simate GS, Ndlovu S (2014) Acid mine drainage: challenges and opportunities. J Environ Chem Eng 2:1785–1803

Singh DV, Bhat JIA, Bhat RA, Dervash MA, Ganei SA (2018) Vehicular stress a cause for heavy metal accumulation and change in physico-chemical characteristics of road side soils in Pahalgam. Environ Monit Assess 190:353. https://doi.org/10.1007/s10661-018-6731-2

Singh DD, Thind PS, Sharma M, Sashikanta S, John S (2019) Environmentally sensitive elements in groundwater of an industrial town in India: Spatial distribution and human health risk. Water 11: 2350

Srinivasulu M, Mohiddin GJ, Quimbiulco K, Ortiz DR, Tirado JO, Rangaswamy V (2015) Effect of pesticides alone and in combinations on sulphur oxidation in soils. Int J Curr Microbiol App Sci 4:412–418

Srivastava A, Vidyapith B, Jangid NK, Srivastava M, Rawat V (2019) Pesticides as water pollutants. In: Wani KA, Mamta (eds) Handbook of research on the adverse effects of pesticide pollution in aquatic ecosystems. IGI Global, Hershey, pp 1–19

Sun S, Sidhu V, Rong Y, Zheng Y (2018) Pesticide pollution in agricultural soils and sustainable remediation methods: a review. Curr Pollut Rep:1–11

Tepanosyan G, Maghakyan N, Sahakyan L, Saghatelyan A (2017) Heavy metals pollution levels and children health risk assessment of Yerevan kindergartens soils. Ecotoxicol Environ Saf 142:257–265

Torres E, Ayora C, Jiménez-Arias JL, García-Robledo E, Papaspyrou S, Corzo A (2014) Benthic metal fluxes and sediment diagenesis in a water reservoir affected by acid mine drainage: a laboratory experiment and reactive transport modeling. Geochim Cosmochim Acta 139:344–361

Tutu H, McCarthy TS, Cukrowska E (2008) The chemical characteristics of acid mine drainage with particular reference to sources, distribution and remediation: the Witwatersrand Basin, South Africa as a case study. Appl Geochem 23:3666–3684

Vasilescu MN, Medvedovici AV (2005) Herbicides. In: Worsfold P, Townshend A, Poole C (eds) Encyclopedia of analytical science. Elsevier, Oxford, pp 243–260

Vázquez-Luna D, Cuevas-Díaz MC (2019) In: Vázquez-Luna D, Cuevas-Díaz MC (eds) Soil contamination and alternatives for sustainable development. IntechOpen, London. https://doi.org/10.5772/intechopen.83720

Ventura LMB, Mateus VL, de Almeida ACSL, Wanderley KB, Taira FT, Saint'Pierre TD, Gioda A (2017) Chemical composition of fine particles ($PM_{2.5}$): water-soluble organic fraction and trace metals. Air Qual Atmos Health 10:845–852

Wang L, Yang D, Li ZT, Fu YH, Liu XM, Brookes PC, Xu JM (2019) A comprehensive mitigation strategy for heavy metal contamination of farmland around mining areas—screening of low accumulated cultivars, soil remediation and risk assessment. Environ Pollut 245:820–828

Weibel G, Eggenberger U, Schlumberger S, Mader UK (2017) Chemical associations and mobilization of heavy metals in fly ash from municipal solid waste incineration. Waste Manag 62:147–159

Wittsiepe J, Feldt T, Till H, Burchard G, Wilhelm M, Fobil JN (2017) Pilot study on the internal exposure to heavy metals of informal-level electronic waste workers in Agbogbloshie, Accra, Ghana. Environ Sci Pollut Res Int 24:3097–3107

Xiong J, Zhao T, Cheng H, Li S, Wang S, Chen G (2019) The assessment on the heavy metal pollution and health risks in the Liujiang River under the Xijiang River region. Desalin Water Treat 149:315–322

Yang S, Zhao J, Chang SX, Collins C, Xu J, Liu X (2019) Status assessment and probabilistic health risk modeling of metals accumulation in agriculture soils across China: a synthesis. Environ Int 128:165–174

Zhang Q, Yu R, Fu S, Wu Z, Chen HYH, Liu H (2019) Spatial heterogeneity of heavy metal contamination in soils and plants in Hefei, China. Sci Rep 1049:2045–2051

Zheng S, Zheng X, Chen C (2012) Leaching behavior of heavy metals and transformation of their speciation in polluted soil receiving simulated acid rain. PLoS One 7:e49664

Zhu Y, Huang B, Li QX, Wang J (2015) Organochlorine pesticides in follicular fluid of women undergoing assisted reproductive technologies from Central China. Environ Pollut 207:266–272

Zhu G, Guo Q, Xiao H, Chen T, Yang J (2017) Multivariate statistical and lead isotopic analyses approach to identify heavy metal sources in topsoil from the industrial zone of Beijing Capital Iron and Steel Factory. Environ Sci Pollut Res Int 24:14877–14888

Zupancic N (2017) Influence of climate factors on soil heavy metal content in Slovenia. J Soils Sediments 17:1073–1083

Chapter 2
Ecotoxicology of Heavy Metals: Sources, Effects and Toxicity

Maya Verma

2.1 Introduction

Environmental pollution is considered as a biggest challenge for whole world, mainly in developing countries. Though produced by anthropogenic activities, it has deleterious effects on environment and its resources. Aquatic habitats, mainly freshwater ecosystems are more prone to pollution as compared with others. Within limits, aquatic ecosystem can cope with pollution, but in severity its get reflected in community structure of flora and fauna. HMs play an important role in environmental pollution. Being toxic and persistent in nature, they are a major threat to the world and of course are of serious concern.

HMs' natural sources include weathering of rocks and volcanic eruptions. Mining, agricultural applications (pesticides, fertilizers especially phosphate fertilizers), industrial emissions come under anthropogenic sources. Being persistent in nature, they pose threats to food chains in ecosystem. Due to their toxic nature, they are also responsible for different health problems. HMs above threshold range heavily affect rhizosphere and reduce soil fertility. In an aquatic ecosystem, bioaccumulation of HMs creates adverse effect on aquatic animals.

HMs are those chemical elements which have density greater than 5 g/cm^3. They are also called trace elements. The toxic HMs commonly present in wastewaters are Hg, Cd, Cr, Pb, As, Cu, Ni, etc. The continuous release of HMs into water bodies creates many environmental problems. Their existence and amplification in environment is an outcome of natural and anthropogenic activities such as urbanization, industrial wastes, etc. Adverse effects of HMs on plants include reduction in seed germination rate and lipid content by cadmium, decreased plant growth and enzymatic activity by chromium, inhibition of photosynthesis by copper and mercury, reduction of chlorophyll production by lead (Gardea-Torresdey et al. 2005).

M. Verma (✉)
Kirori Mal College, University of Delhi, New Delhi, India

R. A. Bhat et al. (eds.), *Bioremediation and Biotechnology, Vol 2*,
https://doi.org/10.1007/978-3-030-40333-1_2

In animals, major effects include reduced growth and development, cancer, nervous system damage and multiple organ damage. To prevent toxicity in wastewater, there is an urgent need for enough treatment strategies of wastewater effluents before their discharge into aquatic water bodies.

2.2 Sources of HM Pollution

Natural sources include volcanic eruptions, soil erosion, urban wastes and aerosol particulates (Mushtaq et al. 2020; Qadri and Bhat 2020). Anthropogenic sources include electroporating processes, metal finishing, mining and smelting, textile industries and nuclear power activities. It is concluded that volcanic activities create hazardous impacts to environment, climate and health. During eruptions, many gases are released, i.e., CO_2, H_2S, CO and SO_2. Besides this, several organic compounds and HMs are also released. These HMs deteriorate the quality of water bodies. Volcanic activities have been regarded as responsible for release of toxic HMs (Hg, As, Pb, Cu, Ni) (Amaral et al. 2006).

Soil erosion is accomplished by wind and water. During rainfall, sediment bound HMs reach the soil. HMs during their run off are picked up and distributed in the environment. While in some cases, HMs reaches to poor drainage system and later on in nearby rivers (Taiwo et al. 2011). Aerosol particulates (fine colloidal particles) carry several kinds of contaminants (HMs). These particulates accumulate on leaf surfaces and enter interior of leaves through stomata. HMs enter groundwater and surface water through natural and anthropogenic sources. In present time, anthropogenic sources have exceeded the natural means of heavy metal pollution. The basic sources of metal pollution are the burning of fossil fuels, mining, smelting, sewage plant, fertilizers and pesticides (Fig. 2.1). Electroplating and metal finishing activities involve deposition of thin layers of metals using electrochemical method. During which, HMs are released into wastewater (Table 2.1). It can be either through rinsing of the product or dumping of process baths. It is also noted that cleaning of

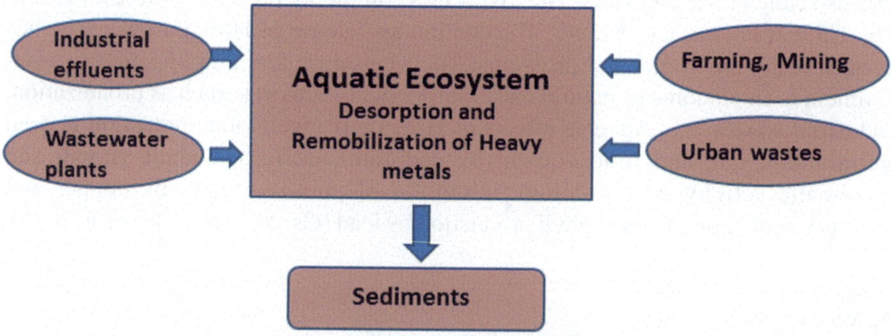

Fig. 2.1 Anthropogenic sources of HMs pollution in aquatic ecosystem

Table 2.1 Major sources of HMs in environment

HMs	Sources (natural and anthropogenic)
Cd	Alkaline batteries, fertilizers, plastic and electrical instruments
Fe	Refineries and textile industries
Cu	Fuel, pesticides, fertilizers, batteries and electrical instruments
Zn	Steel industries, coal burning, mining, pesticides, fertilizers and refineries
Pb	Electrical instruments, batteries, glass industries, fertilizers, refineries and fuel
Cr	Fertilizers and textile industries
Ni	Batteries, fuel and fertilizers
Co	Alloy, aircraft engines and magnets

tanks produces substantial quantity of wet sledge having high concentration of toxic metals (Cushnie 1985).

In the same way, mining activities release toxic metals. Smelting and mining have been considered as major sources of metals in environment. Similarly, textile industries effluents are also a major source of heavy metal pollution as they basically originate from dyeing process. The colouring compounds used in this process include Cu, Cr, Pb, Ni, etc. In nuclear power plant, effluents which are dumped into nearby aquatic bodies affect aquatic biota (Hagberg and Lofgren 2007; Begum et al. 2011).

2.2.1 *Mercury (Hg)*

Mercury exists in metallic form and inorganic salts in nature. It has been considered toxic HMs. Mercury enters air through mining and burning processes and in aquatic ecosystem through municipal wastewater discharge and industrial wastewater discharge (Jan et al. 2009; Chen et al. 2012). Mercury is used colossally in equipment, i.e., thermometer, mercury arc, fluorescent lamp. Its other applications include manufacturing of batteries and some dental applications. Mercury in nature prevails in both mercuric (Hg_2) and mercurous (Hg) forms. Being lipophilic in nature, it can permeate across cell membrane. Mercury in bound form such as methyl mercury (MeHg) enters food chain through fish. Another bound form ethylmercury enters human body via antiseptics and vaccine preparations (Farina et al. 2011; Geier et al. 2015). In soil, mercury (inorganic form) is acted upon by bacteria in soil and water which ultimately results its transformation into methylmercury, it accumulates in fishes and other animal tissue (Hintelmann 2010). Methylmercury induced toxicity includes lipid peroxidation, mitochondrial damages and microtubule destruction while ethylmercury is metabolized into inorganic salts rapidly (Patrick 2002; Carneiro et al. 2014). According to WHO, permissible limit for mercury in drinking water is 0.001 mg/L (WHO 2004). Mercury induced symptoms include lung damage, changes in mucous membrane, diarrhoea, skin rashes, hypertension, renal dysfunction and severe neurological abnormalities (Asano et al. 2000; Bates 2003).

2.2.2 Lead (Pb)

Lead has wide industrial applications such as smelting, refining, mining and battery manufacturing (Malekirad et al. 2010; Karrari et al. 2012; Mehmood et al. 2019; Rashid et al. 2019). It is also used in fertilizers and pesticides and for improving the octane rating of gasoline in vehicles. Due to rapid industrialization and urbanization, increase in effluent discharge from industries, sewage has resulted in contamination of aquatic bodies. Besides this, industrial and agricultural activities are a major cause for contamination of groundwater resources (Ebrahimi et al. 2011). In present times, prescribed limits for lead in drinking water is 0.01 mg/L (Karrari et al. 2012). Being a stable pollutant, its clinical manifestations include from subclinical to life threatening complications (Nelson et al. 2011). Mild exposure of lead results in loss of appetite, headache, abdominal pain, hypertension, renal dysfunction, arthritis, vertigo while severe exposure can cause autism, allergies, dyslexia, hyperactivity, paralysis and brain damage (Papanikolaou et al. 2005). Inhalation through airborne dust and ingestion of contaminated water and food have been considered the most probable routes of lead exposure. In comparison to adults, children are more prone to its toxicity. After absorption through intestine, lead is transferred to soft tissue (liver) and mineralizing tissue (bone) as part of constituent of blood. In evaluation of lead toxicity it was found that lead mobilization from bone to blood increases during lactation, pregnancy, chronic diseases and physiological stress (Gulson et al. 2003). Increased level of lead in pregnant women results in miscarriage, spontaneous abortions and children having low body weight.

2.2.3 Arsenic (As)

Arsenic holds the 20th rank among the most abundant elements in earth's crust (ASTDR 2013). Being inorganic and organic in nature, it is present throughout the world and is major cause for adverse effects on human beings and their environment (Hughes et al. 2011). It exists in more than 200 forms such as arsenate (As V) compounds, as sulphides/sulfosalt, arsenates, arsenides, oxides and silicates (Mandal and Suzuki 2002; Singh et al. 2007). Natural source of arsenic includes volcanic eruptions, rock weathering and geothermal water. Arsenic is primarily used in glass, ceramics, herbicides, pesticides, metals related applications. It is abundantly present in food originated from sea (Meharg 2004; Sohn 2014). Arsenic has been considered as group I carcinogen and in present times its contamination through drinking water has become a serious concern over worldwide (Kulshrestha et al. 2014). Arsenic occurs in two main oxidation states—arsenate (pentavalent) and arsenite (trivalent) anion in nature. After ingestion, it is absorbed by gastrointestinal tract. Their toxicity and level of cellular damage depend upon their valent states. As (III) is regarded as more toxic than As (IV) (Styblo et al. 2000). The permissible limit of this metal in drinking water is 10 μg/L (Smith et al. 2000). Arsenic induced toxicity includes wide range of health problems such as neoplasms, liver, kidney, skin and lymphatic cancer (Shi et al. 2004). Arsenic induced superoxides are also responsible for disrupting cell signalling pathway. Arsenic exposure at low intensity causes reduction in number of RBCs and leucocytes, nausea, vomiting and damage to blood vessels (Yoshida et al. 2004; Hopenhayn 2006).

2.2.4 *Cadmium (Cd)*

Cadmium is a toxic heavy metal which has its own considerable environmental concerns. Natural sources of cadmium are sedimentary rocks and marine phosphates (Gesamp 1987). It has many industrial applications such as manufacturing of alloys, batteries, mining and smelting (Wilson 1988). Basic route of cadmium exposure is either ingestion of contaminated food or inhalation. Exposure of cadmium in human beings is possible through many ways, e.g., smoking cigarettes, working in cadmium contaminated workplaces (IARC 1993; Paschal et al. 2000). Cadmium inhalation in chronic exposure results in emphysema, decrease in olfactory function. Certain research findings have documented the cadmium exposure in low intensity results in osteoporosis and reducing bone mineral density (Mascagni et al. 2003; Gallagher et al. 2008; Schutte et al. 2008). Cadmium has been regarded as severe pulmonary and gastrointestinal irritant. After acute cadmium ingestion, several health problems arise which include abdominal pain, salivation, muscle cramp, vertigo, shock, loss of consciousness within half an hour (Baselt et al. 1995). While chronic exposure of cadmium has a depressive effect (Singhal et al. 1976). Cadmium has been categorized as human carcinogen by the International Agency for Research on Cancer (IARC 1993). Lung is only the main established site of carcinogenesis in human beings. In animals, other targeted tissues are adrenal testes and haematopoietic system (Waalkes et al. 1995; Waalkes et al. 1996).

2.2.5 *Chromium (Cr)*

Chromium is naturally present in earth's crust. It occurs in many oxidation states such as Cr (II), Cr (III), Cr (IV), Cr (V) and Cr (VI). Out of these, Cr (VI) form is stable and is regarded as second most stable form (Patlolla et al. 2009). Chromium is produced from natural as well as anthropogenic sources, which includes metal processing industries, chemical industries and tanning industries. Hexavalent form of chromium is classified as human carcinogen by several regulatory authorities (ATSDR 1990; IARC 1990). WHO limit for chromium in drinking water is 50 μg/L (Velma et al. 2009). The main route of chromium exposure in human being is through inhalation and lung is considered as primary target organ (EPA 1992; Velma et al. 2009). More common incidence of dermatitis which has been noticed among workers, working in cement industries, is attributed to their exposure to chromium.

Cr (VI) may cause irritation to the lining of nose, ulcers in stomach and small intestine, anaemia. Incidental ingestion of extremely high doses of Cr (VI) results in cardiovascular, gastrointestinal, haematological, hepatic, renal and neurological effects. Toxicity of chromium related compounds depends upon their oxidation state and their solubility. Cr (VI) compounds are more powerful oxidizing agents. They tend to be corrosive, appear to be much more toxic systemically than Cr (III) compounds (Connett and Wetterhahn 1983; De Flora et al. 1990).

2.3 HMs in Aquatic Ecosystems

In aquatic water bodies, metals can exist either in dissolved, particulates or some other complex forms. Some common physiological processes, i.e., sedimentation, dispersion, absorption and desorption are governing factors for their distribution. Heavy metals exist either in suspended form or sediments in aquatic environment.

2.3.1 HMs Bioavailability to Aquatic Organisms

One of the unique features of HMs is that they are nondegradable. As soon as they enter in aquatic ecosystem, either they are accumulated in sediments or consumed by living organisms (Frémion et al. 2016). Further sediments act as a long term reservoir of contamination in trophic food chain. HMs bioaccumulate in aquatic plants and animals through which they reach up to human beings and cause many health problems (Varol and Şen 2012; Hasan et al. 2016). In sediments, metal accumulates through precipitation, binding and coprecipitation with some metal oxides, i.e., manganese oxide. The process of accumulation depends upon biochemical and physical status among sediment and associated water column (Vasile et al. 2005; Equeenuddin et al. 2013).

2.3.2 Trophic Transfer of HMs

Heavy metals bioaccumulation in trophic food chain depends upon their rate of accumulation rate of removal from body. Therefore all HMs possess different half lives in different species. HMs enter into organism body directly from the water, sediments, soil, etc. (Singh et al. 2018). The confinement of HMs in any organism depends upon speciation of the concerned metal and some physiological processes such as homeostasis, synchronization and detoxification of HMs. Methyl mercury accumulates in biotic environment readily and biomagnifies in food chain due to their lipophilic nature. Metallophytes have developed unique mechanism for copying with high concentration of HMs in soil and they have been categorized into three groups, i.e., (1) excluders, (2) indicators, (3) hyperaccumulators (Ali and Khan 2019). These HMs are taken up from contaminated soil and are transferred to successive trophic levels through food chain (Nica et al. 2012).

2.4 Assessment of HMs in Ecosystems

Specific terms are being used for quantification of HMs in environment, i.e., bioconcentration factor (BCF) and bioaccumulation coefficient (BCF). Bioconcentration factor is defined as the ratio of HMs in an organism tissue to its concentration in abiotic environment (specifically soil and water) (Ali and Khan 2017):

$$\text{Bioconcentration Factor} = \frac{C_{\text{organism tissue}}}{C_{\text{abiotic medium}}}$$

While bioaccumulation coefficient is the ratio of HMs concentration in plants to heavy metal concentration in soil (Ali et al. 2013):

$$\text{Bioaccumulation Coefficient} = \frac{C_{\text{plant}}}{C_{\text{soil}}}$$

2.5 Biomagnification of HMs

Ingestion and biomagnifications of HMs in abiotic environment depend upon variable factors. Ingestion of HMs in plants depends upon bioavailability of metals in soil, which is further dependent upon metal speciation, organic matter and pH of soil. Heavy metals accumulate in plants more simply which are absolutely bioavailable to soil and they would have increased bioaccumulation potential. Plants appear to be more responsive against environmental changes in comparison to soil (Shtangeeva 1995). Many plants and animal species have been regarded as bioindicators of heavy metal pollution, i.e., Lithophaga lithophaga has been reported as a valid bioindicator of heavy marine pollution (Miedico et al. 2016).

2.6 Human Exposure to HMs

HMs affect cell organelles and cellular component, i.e., mitochondria, lysosome, cell membrane, nucleus and endoplasmic reticulum. They are also involved in cellular metabolism and cell detoxification (Fernandes and Nayak 2012). They interact with DNA, nuclear proteins, causing DNA damage which results in cell cycle modulation, apoptosis and severe carcinogenicity (Zhao et al. 2014; Sanyal et al. 2015). Production of ROS and oxidative stress contribute an important role in carcinogenicity of toxic HMs, i.e., Hg, Pb, Cd, etc. (Semu and Singh 1996; Decena et al. 2018). The major HMs (Cd, Cr, Pb, As, Hg) are systemic toxicants which induce multiple organ damage even at lower concentrations (Mehmood et al. 2019). According to International Agency for Research on Cancer (IARC 1993), these metals have been regarded as human carcinogens on the basis of ecotoxicology and epidemiological studies on human beings and animals (Rashid et al. 2019). HMs induced toxicity involves many aspects which are still not clear. Some metals induce synthesis of RNA in nucleus which specifically codes for metallothioneins (MT). These are peptides having low molecular weight, high in cysteine residues (-SH group), found in cytoplasm, lysosome and nucleus. This thiol group induces MTs to bind HMs.

2.7 Conclusion and Suggestions

HMs are nondegradable and persistent in nature. Metals are environmental pollutant for aquatic and terrestrial ecosystem. They are considered toxic due to their unique properties, i.e., persistence in nature, bioaccumulation in successive trophic levels and hazardous in nature. Therefore it is important to access and monitor their concentration in abiotic and biotic environment. Study of environmental chemistry and ecotoxicology of HMs show that following steps should be taken to lessen the impact of toxic metals on human health and environment.

The following suggestions are made:

- There should be regular assessment and monitoring of potentially toxic HMs concentrations in soil, sediment and water.
- Considerable efforts should be made to reduce heavy metal concentration in aquatic and terrestrial ecosystem to protection of biota.
- Industrial effluents and wastewater from industries should be treated more effectively before their discharge.
- Public should be aware about toxic HMs effect on environment and health.
- Research on assessment and monitoring concerned with HMs should be encouraged for improving health of environment.

References

Agency for Toxic Substances and Disease Registry (ATSDR) (1990) Toxicological profile for chromium. U.S. Department of Health and Human Services, Public Health Service, Agency for Toxic Substances and Disease Registry, Atlanta

Ali H, Khan E (2017) Environmental chemistry in the twenty-first century. Environ Chem Lett 15:329–346

Ali H, Khan E (2019) Trophic transfer, bioaccumulation and biomagnification of non-essential hazardous heavy metals and metalloids in food chains/webs: concepts and implications for wild life and human health. Hum Ecol Risk Assess Int J 25:1353

Ali H, Khan E, Sajad MA (2013) Phytoremediation of heavy metals-concepts and applications. Chemosphere 91:869–881

Amaral A, Cruz JV, Cunha RT, Rodrigues A (2006) Baseline levels of metals in volcanic soils of the Azores (Portugal). Soil Sediment Contam Int J 15:123–130

Asano S, Eto K, Kurisaki E, Gunji H, Hiraiwa K, Sato M, Sato H, Hasuike M, Hagiwara N, Wakasa H (2000) Acute inorganic mercury vapour inhalation poisoning. Pathol Int 50:169–174

ASTDR (2013) Priority list of hazardous substances. http://www.atsdr.cdc.gov/spl/. Accessed 22 Sept 2015

Baselt RC, Cravey RH (1995) Disposition of toxic drugs and chemicals in man, 4th edn. Year Book Medical Publishers, Chicago, pp 105–107

Bates N (2003) Metallic and inorganic mercury poisoning. Emerg Nurse 11:25–31

Begum RA, Zaman MW, Mondol AT, Ismal MS, Hossain KMF (2011) Effects of textile industrial wastewater and uptake of nutrients on the yield of rice. Bangladesh J Agric Res 36:319–331

Carneiro MFH, Oliveira Souza JM, Grotto D, Batista BL, de Oliveira Souza VC, Barbosa F Jr (2014) A systemic study of the deposition and metabolism of mercury species in mice after exposure to low levels of Thimerosal (ethylmercury). Environ Res 134:218–227

Chen CW, Chen CF, Dong CD (2012) Distribution and accumulation of mercury in sediments of Kaohsiung River Mouth, Taiwan. APCBEE Proc 1:153–158

Connett PH, Wetterhahn KE (1983) Metabolism of carcinogenic chromate by cellular constituents. Struct Bond 54:93–24

Cushnie GC (1985) Electroplating wastewater pollution control technology. Noyes Publications, Park Ridge, pp 375–377

De Flora S, Bagnasco M, Serra D, Zanacchi P (1990) Genotoxicity of chromium compounds: a review. Mutat Res 238:99–172

Decena SC, Arguilles M, Robel L (2018) Assessing heavy metal contamination in surface sediments in an urban river in the Philippines. Pol J Environ Stud 27:1983–1995

Ebrahimi A, Amin MM, Hashemi H, Foladifard R, Vahiddastjerdi M (2011) A survey of groundwater chemical quality in Sajad Zarinshahr. Health Syst Res 6:918–926

Equeenuddin SM, Tripathy S, Sahoo P, Panigrahi M (2013) Metal behavior in sediment associated with acid mine drainage stream: role of pH. J Geochem Explor 124:230–237

Farina M, Aschner M, Rocha JBT (2011) Oxidative stress in methylmercury induced neurotoxicity. Toxicol Appl Pharmcol 256:405–417

Fernandes LL, Nayak GN (2012) Heavy metals contamination in mud flat and mangrove sediments (Mumbai, India). Chem Ecol 28:435–455

Frémion F, Bordas F, Mourier B, Lenain JF, Kestens T, Courtin-Nomade A (2016) Influence of dams on sediment continuity: a study case of a natural metallic contamination. Sci Total Environ 547:282–294

Gallagher CM, Kovach JS, Meliker JR (2008) Urinary cadmium and osteoporosis in U.S. women $\geq$ 50 years of age: NHANES 1988–1994 and 1999–2004. Environ Health Perspect 116:1338–1343

Gardea-Torresdey JI, Peralta-Videa JR, Rosa GD, Parsons JG (2005) Phytoremediation of heavy metals and study of the metal coordination by X-ray absorption spectroscopy. Coord Chem Rev 249:1797–1810

Geier DA, King PG, Hooker BS, Dorea JG, Kern JK, Sykes LK, Geier MR (2015) Thimerosal: clinical, epidemiologic and biochemical studies. Clin Chim Acta 444:212–220

Gesamp (1987) IMO/FAO/UNESCO/WMO/WHO/IAEA/UN/UNEP Joint Group of Experts on the Scientific Aspects of Marine Pollution: report of the seventeenth session. World Health Organization, Geneva

Gulson BL, Mizon KJ, Korsch MJ, Palmer JM, Donnelly JB (2003) Mobilization of lead from human bone tissue during pregnancy and lactation—a summary of long-term research. Sci Total Environ 303:79–104

Hagberg L, Lofgren E (2007) Soil and plant contamination by textile industries at ZFILM, Managua. Project work in Aquatic and Environmental Engineering, 10 ECTS, Uppsala University Project course, Swedish University of Agricultural Sciences
Hasan MR, Khan MZH, Khan M, Aktar S, Rahman M, Hossain F, Hasan ASMM (2016) Heavy metals distribution and contamination in surface water of the Bay of Bengal coast. Environ Sci 2:1–12
Hintelmann H (2010) Organomercurials: their formation and pathways in the environment. Met Ions Life Sci 7:365–401
Hopenhayn C (2006) Arsenic in drinking water: impact on human health. Elements 2:103–107
Hughes MF, Beck BD, Chen Y, Lewis AS, Thomas DJ (2011) Arsenic exposure and toxicology: a historical perspective. Toxicol Sci 123:305–332
IARC (1990) IARC monographs on the evaluation of carcinogenic risks to humans. IARC Scientific Publications, IARC, Lyon
International Agency for Research on Cancer (IARC) (1993) Monographs – cadmium. International Agency for Research on Cancer, Lyon
Jan AT, Murtaza I, Ali A, Haq QMR (2009) Mercury pollution: an emerging problem and potential bacterial remediation strategies. World J Microbiol Biotechnol 25:1529–1537
Karrari P, Mehrpour O, Abdollahi M (2012) A systemic review on status of lead pollution and toxicity in Iran; guidance for preventive measures. DARU J Pharm Sci 20:2
Kulshrestha A, Jarouliya U, Prasad GBKS, Flora SJS, Bisen PS (2014) Arsenic-induced abnormalities in glucose metabolism: biochemical basis and potential therapeutic and nutritional interventions. World J Translat Med 3:96–111
Malekirad A, Oryan S, Fani A, Babapor V, Hashemi M, Baeeri M, Bayrami Z, Abdollahi M (2010) Study on clinical and biochemical toxicity biomarkers in a zinc-lead mine workers. Toxicol Ind Health 26:331–337
Mandal BK, Suzuki KT (2002) Arsenic round the world: a review. Talanta 58:201–235
Mascagni P, Consonni D, Bregante G, Chiappino G, Toffoletto F (2003) Olfactory function in workers exposed to moderate airborne cadmium levels. Neurotoxicology 24:717–724
Meharg AA (2004) Arsenic in rice–understanding a new disaster for south-east Asia. Trends Plant Sci 9:415–417
Mehmood MA, Qadri H, Bhat RA, Rashid A, Ganie SA, Dar GH, Shafiq-ur-Rehman (2019) Heavy metal contamination in two commercial fish species of a trans-Himalayan freshwater ecosystem. Environ Monit Assess 191:104. https://doi.org/10.1007/s10661-019-7245-2
Miedico O, Ferrara A, Tarallo M, Pompa C, Bisceglia D, Chiaravalle AE (2016) Hazardous and essential trace elements profile in the different soft tissues of Lithophaga lithophaga (Linnaeus, 1758) from Southern Adriatic Sea (Italy). Toxicol Environ Chem 98:877–885
Mushtaq N, Singh DV, Bhat RA, Dervash MA, Hameed UB (2020) Freshwater contamination: sources and hazards to aquatic biota. In: Qadri H, Bhat RA, Dar GH, Mehmood MA (eds) Freshwater pollution dynamics and remediation. Springer, Singapore, pp 27–50
Nelson L, Lewin N, Howland MA, Hoffman R, Goldfrank L, Flomenbaum N (2011) Goldfrank's toxicologic emergencies, 9th edn. McGraw-Hill, New York, pp 1266–1280
Nica DV, Bura M, Gergen I, Harmanescu M, Bordean DM (2012) Bioaccumulative and chronological assessment of heavy metal transfer in a soil-plant-snail food chain. Chem Cent J 6:55
Papanikolaou NC, Hatzidaki EG, Belivanis S, Tzanakakis GN, Tsatsakis AM (2005) Lead toxicity update. A brief review. Med Sci Monit 11:329
Paschal DC, Burt V, Caudill SP, Gunter EW, Pirkle JL, Sampson EJ (2000) Exposure of the U.S. population aged 6 years and older to cadmium: 1988–1994. Arch Environ Contam Toxicol 38:377–383
Patlolla A, Barnes C, Yedjou C, Velma V, Tchounwou PB (2009) Oxidative stress, DNA damage and antioxidant enzyme activity induced by hexavalent chromium in Sprague Dawley rats. Environ Toxicol 24:66–73
Patrick L (2002) Mercury toxicity and antioxidants: part 1: role of glutathione and α-lipoic acid in the treatment of mercury toxicity. Altern Med Rev 7:456–471
Qadri H, Bhat RA (2020) The concerns for global sustainability of freshwater ecosystems. In: Qadri H, Bhat RA, Dar GH, Mehmood MA (eds) Freshwater pollution dynamics and remediation. Springer, Singapore, pp 1–15

Rashid A, Bhat RA, Qadri H, Mehmood MA (2019) Environmental and socioeconomic factors induced blood lead in children: an investigation from Kashmir, India. Environ Monit Assess 191(2):76. https://doi.org/10.1007/s10661-019-7220-y

Sanyal T, Kaviraj A, Saha S (2015) Deposition of chromium in aquatic ecosystem from effluents of handloom textile industries in Ranaghat-Fulia region of West Bengal, India. J Adv Res 6:995–1002

Schutte R, Nawrot TS, Richart T, Thijs L, Vanderschueren D, Kuznetsova T (2008) Bone resorption and environmental exposure to cadmium in women: a population study. Environ Health Perspect 116:777–783

Semu E, Singh BR (1996) Accumulation of heavy metals in soils and plants after long-term use of fertilizers and fungicides in Tanzania. Fertil Res 44:241–246

Shi H, Shi X, Liu KJ (2004) Oxidative mechanism of arsenic toxicity and carcinogenesis. Mol Cell Biochem 255:67–78

Shtangeeva IV (1995) Behaviour of chemical elements in plants and soils. Chem Ecol 11:85–95

Singh N, Kumar D, Sahu A (2007) Arsenic in the environment: effects on human health and possible prevention. J Environ Biol 28:359–365

Singh DV, Bhat JIA, Bhat RA, Dervash MA, Ganei SA (2018) Vehicular stress a cause for heavy metal accumulation and change in physico-chemical characteristics of road side soils in Pahalgam. Environ Monit Assess 190:353. https://doi.org/10.1007/s10661-018-6731-2

Singhal RL, Merali Z, Hrdina PD (1976) Aspects of the biochemical toxicology of cadmium. Fed Proc 35:75–80

Smith AH, Lingas EO, Rahman M (2000) Contamination of drinking-water by arsenic in Bangladesh: a public health emergency. Bull World Health Organ 78:1093–1103

Sohn E (2014) The toxic side of rice. Nature 514:62–63

Styblo M, del Razo LM, Vega L, Germolec DR, LeCluyse EL, Hamilton GA, Reed W, Wang C, Cullen WR, Thomas DJ (2000) Comparative toxicity of trivalent and pentavalent inorganic and methylated arsenicals in rat and human cells. Arch Toxicol 74:289–299

Taiwo AM, Adeogun AO, Olatunde KA, Adegbite K (2011) Analysis of groundwater quality of hand-dug wells in peri-urban areas of Obantoko, Abeokuta, Nigeria for selected physico-chemical parameters. Pac J Sci Technol 12:527–534

US EPA (1992) Environmental criteria and assessment office. United States Environmental Protection Agency. Integrated Risk Information System (IRIS), Cincinnati

Varol M, Şen B (2012) Assessment of nutrient and heavy metal contamination in surface water and sediments of the upper Tigris River, Turkey. Catena 92:1–10

Vasile G, Cruceru L, Petre J, Iancu V (2005) Complex analytical investigations regarding the bioavailability of heavy metals from sediments. Rev Chim 56:790–794

Velma V, Vutukuru SS, Tchounwou PB (2009) Ecotoxicology of hexavalent chromium in freshwater fish: a critical review. Rev Environ Health 24:129–145

Waalkes MP, Berthan G (eds) (1995) Handbook on metal-ligand interactions of biological fluids, vol 2. Marcel Dekker, New York, pp 471–482

Waalkes MP, Misra RR, Chang LW (1996) Toxicology of metals. CRC Press, Boca Raton, pp 231–244

WHO (2004) Guidelines for drinking-water quality. Sixty-first meeting, Rome, 10–19 June 2003. Joint FAO/WHO Expert Committee on Food Additives

Wilson DN (1988) Association cadmium. Cadmium-market trends and influences. In: Cadmium 87 proceedings of the 6th international cadmium conference, London, pp 9–16

Yoshida T, Yamauchi H, Fan Sun G (2004) Chronic health effects in people exposed to arsenic via the drinking water: dose-response relationships in review. Toxicol Appl Pharmacol 198: 243–252

Zhao S, Shi X, Li C, Zhang H, Wu Y (2014) Seasonal variation of heavy metals in sediment of Lake Ulansuhai, China. Chem Ecol 30:1–14

Chapter 3
Role of Modern Innovative Techniques for Assessing and Monitoring Heavy Metal and Pesticide Pollution in Different Environments

Fernanda Maria Policarpo Tonelli and Flávia Cristina Policarpo Tonelli

3.1 Heavy Metal and Pesticide Pollution

Heavy metals (Aruliah et al. 2019) and pesticides (Morillo and Villaverde 2017) are relevant environmental pollutants that require attention once they are prone to cause significant damage to various life forms on the planet, especially to humans (Mateos et al. 2018).

3.1.1 Heavy Metal Pollution

Heavy metals as arsenic (As), cadmium (Cd), chromium (Cr), copper (Cu), mercury (Hg), nickel (Ni), lead (Pb), zinc (Zn) consist in elements that are expected to be normally found on Earth's crust. However, when their concentrations exceed the naturally occurring safe range limit values they can accumulate (for example in water or soil) and offer high toxicity (even at low concentration), causing serious heath complications (Muszynska and Hanus-Fajerska 2015). Not all heavy metals possess the same toxicity. Actually, it varies from one heavy metal to another. According to United States Environmental Protection Agency, and as previously mentioned in Chap. 1, while the maximum allowable concentration (μg/ml) of Hg is 0.002, the one for Zn is 5 (Odobašić et al. 2019).

These metals possess relatively high atomic weight and specific gravity greater than 5.0 g/cm^3 (Duffus 2002; Busairi and Syahir 2018). Their deposition and the

F. M. P. Tonelli (✉)
Department of Cell Biology, Institute of Biological Science, Federal University of Minas Gerais, Belo Horizonte, Brazil

F. C. P. Tonelli
Department of Biochemistry, Federal University of São João del Rei, Divinópolis, Brazil

R. A. Bhat et al. (eds.), *Bioremediation and Biotechnology, Vol 2*,
https://doi.org/10.1007/978-3-030-40333-1_3

pollution distribution on environment depend upon metal characteristics but also environmental specificities like climate factors (Zupancic 2017; Herath et al. 2018). As relevant features, heavy metals possess long half-life persisting in contaminated sites. If not removed by anthropomorphic intervention, they can bioaccumulate and cause biomagnifications (affecting food chain and crop production, for example), present teratogenic and carcinogenic properties, and end up presenting lethal effect (Manzetti et al. 2014; Xiong et al. 2019; Yang et al. 2019).

Besides teratogenic and carcinogenic effects, by poising living organisms heavy metals can affect human beings causing asthma, cardiovascular, neurological, renal, liver diseases, gastrointestinal disorders, skeletal deformities, impairment to reproductive tissues among other severe damages (Farombi et al. 2012; Abarikwu et al. 2013; Tepanosyan et al. 2017; Liu et al. 2018; Mehmood et al. 2019; Rashid et al. 2019; Boateng et al. 2019).

The contamination of a small soil portion (Singh et al. 2018), for example, can end up polluting water. Leachates can drag pollutants to surface water nearby and to groundwater, causing severe undesirable effects not only in human health (Jaishankar et al. 2014), but also ecotoxicological damage (Nagajyoti et al. 2010). Leachates also contain dissolved organic matter which can favor toxic metals (like Ni^{2+}, Pb^{2+}, and Hg^{2+}) binding and accumulation, especially on young samples (Rikta et al. 2018). The kind of pollution that goes through adsorption into matter under water to become bottom sediments is called hidden pollution that is also studied, once it can become real pollution under external factors (Liang et al. 2015; Ali et al. 2016; Petrosyan et al. 2019).

However, what kind of phenomena can cause these elements accumulation in the environment? Human interference through activities involving these metals such as industrial activities related to the use of these elements (e.g., industries dealing with dyes and pigments, textiles, petroleum, fertilizer), mining, smelting, etc. (Goyer 2001; Zhu et al. 2017; Wang et al. 2019).

Waste incineration can also contribute to contamination especially from air, vegetation, and areas surrounding the incinerator (Li et al. 2017; Weibel et al. 2017). Ashes, vegetables, and soil samples collected in areas close to the incinerator can contain Cu, Pb, Zn, Hg, As, and Cd (Ma et al. 2018). Landfill is also a source of environmental contamination with heavy metals. It can contaminate soil, groundwater, water nearby, and plants growing on the area. The latter can accumulate high levels of these pollutants. *Tanacetum vulgare L.*, for example, accumulates Cd, Cr, Ni, and Zn mostly in leaves, and Co, Cu, Fe, Hg, Mn, and Pb mainly in the roots (Adamcová et al. 2017). This environmental pollution represents a wide risk to health, specially the health of the people who work directly with the wastes (Wittsiepe et al. 2017).

Agriculture soils can also contain high amounts of heavy metal (Hu et al. 2018). In China, for example, As, Cd, Cr, Ni, Pb, and Zn accumulate in a level that exceeds the exposure guideline values for children and adults, offering to these people an elevated cancer risk, especially regarding to As, Cd, and Cr (Yang et al. 2019). An interesting study that analyzed in China the surface sediments from an area with increased soil erosion determined, through statistical analysis, that Cr, Co, Ni, and

Cu present in the samples came from natural sources but Mn, Zn, and Pb originated from vehicular traffic, mining, and agricultural and industrial activities (Shen et al. 2019).

Human interference is also important to remove these pollutants from the environment and, consequently, it is necessary to develop innovative techniques for assessing and monitoring heavy metal (and also pesticide) pollution in order to determine pollution's hotspots.

When it comes to remediation of sites polluted by heavy metals physical, chemical, and biological methods (for which researchers continually try to develop improvements aiming cost reduction, more efficiency, and less generation of toxic sludge) can be combined and applied to decrease the amount or to promote total removal of pollutants. Using only one of remediation methods commonly offers worse results when compared to the results obtained by the combination of methodologies generating remediation processes.

In order to perform heavy metals remediation methodologies as generating genetically modified organisms to bioremediate or using species that already possess this ability, performing chemical extraction, bioleaching and/or electrokinetics or other method such as thermal treatment, chlorination, ion-exchange and membrane separation, can offer satisfactory results especially if combined. It is possible, for example, to isolate bacteria from contaminated environments (*Serratia marcescens* from tannery effluent contaminated sludge) and use them for metal removal (zinc) combining bioleaching and electrokinetic remediation in a bioleaching enhanced electrokinetic remediation (BEER) to achieve better results (93.08%) than the ones obtained using single methods as bioleaching (72.86%) or electrokinetic remediation (56.67%) (Selvi and Aruliah 2018; Aruliah et al. 2019).

3.1.2 Pesticide Pollution

Pesticide is the term by which is called a substance or a mixture of substances that can be mainly used to eliminate or control the presence of any pest (living organisms capable of causing damage to environments: commercial, agricultural, and/or domestic) (Schierow and Esworthy 2012; Bhat et al. 2018; Mushtaq et al. 2018).

After the Second World War the use of these substances (e.g., insecticides, fungicides, herbicides, and acaricides) increased in an attempt to improve agricultural production to provide food in a sufficient amount to the growing population (Gavrilescu 2005). However, not only benefit comes from pesticides' use. In fact, the damages that they can cause are various as they are an ecological risk and can threaten human health and life and the survival of other life forms. A large amount of pesticides sprayed may never reach their targets being dispersed instead, contaminating agricultural crop products, soil, water, and air (Niti et al. 2013; Liu et al. 2018; Reeves et al. 2019). For food and drinking water, there are legal limits considered safe (not harmful to human health) by government agencies as the Environmental

Protection Agency (in USA) and the European Environment Agency in European Union (Colmati et al. 2019).

The source of environmental contamination can be punctual (the application points themselves) or diffuse, and even if the soil is commonly the first spot of contamination, the pollutants transferring to waters can occur contaminating not only surface water but also underground water to a greater extent, the more pesticide properties favor their migration into the soil to deeper layers (Vasilescu and Medvedovici 2005).

As pesticides can take long time to be degraded and if metabolized their metabolites can, as same as heavy metals, accumulate in the environment, these substances represent, as previously mentioned, a risk. The less water-soluble the substance, the longer its half-life period and its partition coefficient, and the longer it tends to remain contaminating the environment in which it is present (Arias-Estévez et al. 2008).

Besides being toxic, they can also be biomagnified in the food chain (Ali et al. 2014). Thus, the remediation of sites contaminated by these pollutants is an important goal in order to preserve health and promote sustainable development (Cheng et al. 2016).

This kind of remediation is challenging because it is common that polluted sites contain a mixture of different pesticides and their metabolites. However, they are mainly organic compounds and can be reduced in quantity, eliminated, isolated, or stabilized by applying physical, chemical, and/ or biological processes.

3.2 Innovative Techniques for Assessing and Monitoring Heavy Metals and Pesticide Pollution

In order to detect the necessity of environment heavy metal and pesticide remediation and also to monitor the quality of the remediation process being carried out, and to determine if the present levels no longer represent risks allowing this process to be stopped, it is desirable the development of innovative techniques.

The main motivation to the development of these techniques to assess and monitor the level of environment contamination with pollutants are the disadvantages evidenced in methods conventionally used for this purpose such as the mass spectrometry (Hepel and Stobiecka 2008; Petrosyan et al. 2019; Shen et al. 2019; Xiong et al. 2019; Pesantes et al. 2019) and the chromatographic ones that are still being used by some research groups (Pan et al. 2019). In general, the development of simple methods (not requiring trained operators to apply) that can exhibit the desirable sensitivity within minimum time and also present low cost is the main goal.

When it comes to pesticide pollution, sensors and biosensors are interesting options (Muenchen et al. 2016; Mishra et al. 2018) (Table 3.1). Sensors can identify the presence of a target analyte and convert this information into a signal that can be measured (Hulanicki et al. 1991).

Table 3.1 Some examples of sensors developed in twenty-first century to detect pesticides in different environments

Sensor	Pesticide (s)	Reference
Fiber optic biosensor with acetylcholinesterase-immobilized	Organophosphorus	Choi et al. (2001)
Cantilever based biosensor functionalized with DDT hapten	Dichlorodiphenyltrichloroethane	Alvarez et al. (2003)
Optical microbial biosensor with *Flavobacterium* sp.	Methyl parathion	Kumar et al. (2006)
Carbon electrode modified with mercury film	Atrazine	Maleki et al. (2007)
Liposome nanobiosensors with acetylcholinesterase	Dichlorvos and paraoxon	Vamvakaki and Chaniotakis (2007)
Alkaline phosphatase electrochemical biosensor	2,4-Dichlorophenoxyacetic acid, 2,4,5-trichlorophenoxyacetic acid, carbofuran, α-endosulfan, Hg^{2+}, Cd^{2+}, Ag^{2+}, Zn^{2+}, and Cu^{2+}	Shyuan et al. (2008)
Electrochemical biosensor with multi-walled carbon nanotubes	Isoproturon, voltage and dicofol	Manisankar et al. (2008)
Fluorescent nanoimmunosensor	Chlorpyrifos	Zhou et al. (2010)
Biosensor containing photosystem II from spinach	Terbutryn	Ventrella et al. (2011)
Biosensor containing photosystem II from *Synechococcus elongatus f. thermalis*	Atrazine, isoproturon, diuron	Masojídek et al. (2011)
Amperometric biosensor containing tyrosinase	Chlorpyrifos, profenofos, and malathion	Liu et al. (2011)
Electrochemical (carbon electrode modified with copper nanowires)	Trifluralin	Mirabi-Semnakolaii et al. (2011)
Electrochemical biosensor with nanomaterials functionalized with acetylcholinesterase	Demalathion, chlorpyrifos, monocrotophos, and endosulfan	Chauhan and Pundir (2012)
AFM tip nanobiosensor with acetolactate synthase	Metsulfuron-methyl	Silva et al. (2013)
AFM tip nanobiosensor with acetolactate synthase	Imazaquin and metsulfuron-methyl	Deda et al. (2013)
Liquid crystal and paraoxonase enzymes	Paraoxon and other organophosphorus	Chen and Yang (2013)
Biosensor containing tyrosinase attached to a carbon electrode with carbon nanotubes	Organophosphorus pesticides	Vicentini et al. (2013)

(continued)

Table 3.1 (continued)

Sensor	Pesticide (s)	Reference
Biosensor containing alkaline phosphatase attached to a carbon nano-powder paste electrode	Carbofuran	Samphao et al. (2013)
AFM tip nanobiosensor with acetyl-CoA carboxylase	Diclofop	Amarante et al. (2014)
Biosensor containing thylakoids from spinach	Diuron	Rasmussen et al. (2014)
Immunosensor based on a nanocomposite	Chlorpyrifos	Ding et al. (2014)
Immunosensor based on printed carbon electrode modified	Atrazine	Deep et al. (2014)
Amperometric biosensor with carbon nanotubes and acetylcholinesterase	Organophosphorus	Kesik et al. (2014)
Tyrosinase/ZnONPs/SPCE biosensor	Chlortoluron	Haddaoui and Raouafi (2015)
Photoelectrochemical biosensor with nanomaterials and acetylcholinesterase	Paraoxon and dichlorvos	Li et al. (2015)
Acetylcholinesterase-functionalized gold nanoparticles biosensor	Dichlorvos	Wei and Wang (2015)
Quartz crystal microbalance sensor	Atrazine	Gupta et al. (2015)
Amperometric based on mushroom tyrosinase	Atrazine	Tortolini et al. (2016)
Amperometric biosensor containing the enzyme hydroxyphenylpyruvate dioxygenase	Sulcotrione	Rocaboy-Faquet et al. (2016)
Electrochemical biosensor with acetylcholinesterase functionalizing nanocomposites	Malathion and chlorpyrifos	Chauhan et al. (2016)
Electrochemical biosensor containing a nano-stabilizer and acetylcholinesterase	Organophosphorus pesticides	Xu et al. (2018)
Biosensor containing *Chlorella mirabilis*	Simazine and irgarol	Moro et al. (2018)
Modified screen-printed electrodes containing acetylcholinesterase	Chlorpyrifos and methamidophos	Hou et al. (2019)

It is possible to develop biosensors containing enzymes related to the photosynthetic process to detect the presence of herbicides although with a low selectivity. They are able to detect especially phenolic herbicides and they are useful as a first evaluation tool to determine if a sample is contaminated or not by the presence of these substances. Interesting examples are: a) a biosensor to detect Terbutryn through a colorimetric method using ultrathin film multilayers of Photosystem II (from spinach)-enriched photosynthetic membranes (Ventrella et al. 2011); b) a biosensor based in amperometry to detect the pesticides atrazine, isoproturon, diuron, thanks to the Photosystem II complex from *Synechococcus elongatus* and *F. thermalis*. The latter proved to offer more sensitivity to herbicides than the standard algal growth inhibition tests (Masojídek et al. 2011). However, entire organisms and thylakoids can also be successfully used as the bio part of biosensors to detect pesticides. Diuron presence could be detected by thylakoids from spinach-based system (Rasmussen et al. 2014) as same as simazine and irgarol presence by *Chlorella mirabilis* algae biosensor based on fluorescence (Moro et al. 2018).

The pesticide atrazine could be efficiently detected in wastewater using a quartz crystal microbalance sensor made on a gold chip surface (Gupta et al. 2015), and in soil and water samples by a sensor using a carbon electrode that, in the presence of thiocyanate, was modified to contain a mercury film. Atrazine protonates in acidic environment being able to establish an ionic interaction with complexes of mercury–thiocyanate on the surface of the electrode; atrazine adsorption could be determined by adsorptive stripping voltammetry (Maleki et al. 2007). Voltammetry also made possible the detection of the pesticides isoproturon, voltage and dicofol by using a glassy carbon electrode modified with multi-walled carbon nanotubes and with versions of this electrode functionalized with polyaniline or polypyrrole (Manisankar et al. 2008); the detection of trifluralin in soil samples could be performed using a composite of carbon with copper nanowires (that enhanced the conductivity potentializing the current) (Mirabi-Semnakolaii et al. 2011). Nanomaterials are commonly used in electrochemical sensors once they possess large surface-to-volume ratio improving desirable characteristics like being efficient transducing platforms, accepting well the immobilization of biomolecules, and offering the possibility of system miniaturization (Kurbanoglu et al. 2017).

Biosensors, in turn,differently from regular sensors possess a biological recognition element able to recognize the desirable analyte. After recognition, a signal is produced and can be detected and transformed into measurable information by a transducer. Then, this signal can finally be exhibited to the equipment operator as data (Korotkaya 2014). The biological recognition element is immobilized to other sensor components by different strategies as adsorption (noncovalent) or covalent binding (Pereira et al. 2018), and it is necessary to take into consideration during its development the concerns regarding stability, possibility of mass production, reproducibility, and if miniaturization is feasible (Uniyal and Sharma 2018).

As organophosphorus and carbamate are inhibitors of the enzyme acetylcholinesterase, it is common to use this enzyme to generate biosensors to detect these pesticides (biosensors using enzymes are normally based in inhibition). This enzyme catalyzes the cleavage of the neurotransmitter acetylthiocholine (the enzyme's substrate) into acetic acid and thiocholine; the latter can be oxidized by applying voltage allowing current detection by the sensor circuit. However, if a pesticide able to inhibit the enzyme is present, the higher the concentration of the inhibitor present in the sample, the lesser is the detectable current (Pundir and Chauhan 2012). An electrochemical biosensor can be constructed, for example, based on a glassy carbon electrode. It could be modified by iron oxide nanoparticles and poly (indole-5-carboxylic acid) to attach the enzyme, generating a sensing device to detect pesticides in water samples with satisfactory accuracy (Chauhan et al. 2016). An indium tin oxide-coated glass plate can also be used to generate a biosensor by attaching to it iron oxide nanoparticles decorated with carboxylated multi-walled carbon nanotubes and immobilized acetylcholinesterase. The obtained results regarding to the pesticides malathion, chlorpyrifos, monocrotophos, and endosulfan were in good agreement with the ones obtained through gas chromatography–mass spectrometry (Pundir and Chauhan 2012). An amperometric biosensor can also be constructed using a graphite electrode modified with the conducting polymer (poly(4-(2,5-di(thiophen-2-yl)-1H-pyrrol-1-yl) benzenamine)) and multi-walled carbon nanotubes attached to acetylcholinesterase. Electrochemical oxidation of acetylthiocholine was impaired by the presence of the pesticides paraoxon, parathion, and chlorfenvinphos, and it could be measured offering results comparable to the ones obtained using chromatographic methods (Kesik et al. 2014). The enzyme can also be attached to modified screen-printed electrodes. The modification mesoporous carbon-chitosan, for example, proved to be able to enhance the electrochemical reaction improving the biosensor sensitivity to detect chlorpyrifos and methamidophos in different samples (Hou et al. 2019). Acetylcholinesterase could also be attached to an electrochemical sensing platform in the presence of a nanostabilizer (*N*-carbamoylmaleimide-functionalized carbon dots that avoided signal loss of thiocholine) to detect the presence of multiple organophosphorus pesticides (parathion-methyl, paraoxon, dimethoate, and *O*, *O*-dimethyl-*O*-2,2-dichlorovinylphosphate), achieving ultrahigh sensitivity (Xu et al. 2018). Functionalized gold nanoparticles can also be used together with this enzyme for the development of a biosensor. ([BSmim]HSO_4)-AuNPs-porous carbon composite modified boron-doped diamond electrode attached to acetylcholinesterase could efficiently detect dichlorvos presence (Wei and Wang 2015). Biosensors that produce photocurrent could also be built using a glass electrode covered by graphene and CdSe-ZnS quantum dots attached to acetylcholinesterase. It could generate a sensitive photoelectrochemical biosensor capable to detect pollutant pesticides paraoxon and dichlorvos present in fruit samples (Li et al. 2015).

It is also possible to detect pesticides using acetylcholinesterase and optic methods. At a wavelength near 420 nm, *o*-nitrophenyl acetate is colorless. However, after this enzyme action over this substrate *o*-nitrophenyl is generated and this substance is yellow and there is absorbance at 420 nm. Therefore, contamination of organophosphorus in water could be detected by using a fiber-optic biosensor containing an acetylcholinesterase-immobilized Langmuir–Blodgett film and *o*-nitrophenyl acetate. The pesticides caused decrease on the yellow product generation due to the enzyme's inhibition which could be detected by this colorimetric method (Choi et al. 2001). The organophosphorus pesticides dichlorvos and paraoxon in water samples could be detected by liposome-based nanobiosensors. The liposomes contained the enzyme acetylcholinesterase, the substrate acetylcholine, and a pH-sensitive fluorescent indicator called pyranine. In the presence of pesticides able to inhibit the enzyme, it was observed a decrease in the hydrolysis of the acetylcholine and, consequently, a decrease in fluorescent signal (Vamvakaki and Chaniotakis 2007).

Besides acetylcholinesterase there are other enzymes that can be used to assess and monitor the presence of pesticides in samples, once these molecules are characterized as possessing high selectivity to substrates. To detect atrazine, for example, it is possible to use the enzyme tyrosinase from mushrooms immobilized in a carbon based electrode such as multi-walled carbon nanotubes. The pesticide can inhibit the catalysis of catechol's (enzyme's substrate) oxidation to *o*-quinone, allowing the detection of the pesticide pollution in water samples (Tortolini et al. 2016). An amperometric biosensor with a glassy carbon electrode modified with graphene containing tyrosinase immobilized on platinum nanoparticles could also detect organophosphorus pesticides (chlorpyrifos, profenofos, and malathion) based on the enzymatic inhibition of catechol oxidation; the nanomaterials improved the sensor's sensitivity (Liu et al. 2011).This enzyme could also be attached to a glassy carbon electrode modified with a nanocomposite of multi-walled carbon nanotubes-1-butyl-3-methylimidazolium chloride within a dihexadecylphosphate film, to detect the presence of catechol in water samples (Vicentini et al. 2013). The herbicide chlortoluron could also be detected using the enzyme tyrosinase (inhibited by the pesticide) attached to screen-printed carbon electrodes modified by ZnO nanoparticles (Haddaoui and Raouafi 2015). Paraoxonase 1, for example, can hydrolyze organophosphates (for example, the pesticide paraoxon) changing the medium pH as a consequence. This pH change could be detected as a change in color suffered by a biosensor containing also a pH-sensitive liquid crystal (Chen and Yang 2013). Alkaline phosphatase also allows not only pesticide but also heavy metal presence detection in samples and nowadays efforts are still being made to isolate these enzymes from different organisms like lactic acid bacteria (Chu et al. 2019). An electrochemical biosensor containing this enzyme immobilized in a screen-printed carbon electrode modified with hybrid sol-gel/chitosan film could detect alterations caused by the pollutants (2,4-dichlorophenoxyacetic acid, 2,4,5-trichlorophenoxyacetic acid, carbofuran, α-endosulfan, Hg^{2+}, Cd^{2+}, Ag^{2+}, Zn^{2+}, and Cu^{2+}) in the current generated by the oxidation of the cleavage product (ascorbic acid) from ascorbic acid 2-phosphate (Shyuan et al. 2008). This enzyme could also be attached to a carbon nano-powder paste electrode generating a biosensor capable of detecting carbofuran in spiked water and chilly samples (Samphao et al. 2013). Hydroxyphenylpyruvate dioxygenase is also an option of enzyme to be used to build a biosensor to detect the presence of the pesticide sulcotrione. An amperometric biosensor containing this enzyme could, in the absence of the pollutant, promote the oxidation of hydroxyphenylpyruvate in homogentisic acid, which could be oxidized generating a detectable current. In the presence of the pesticide the activity of the enzyme attached to poly(3,4-ethylenedioxythiophene) polystyrene sulfonate-modified screen-printed electrode and the current diminished, in a measurable way (Rocaboy-Faquet et al. 2016).

Antibodies can also be used as part of biosensors: the immunosensors. Anti-atrazine antibodies were attached to a carbon screen-printed electrode to perform a highly sensitive detection of atrazine in different concentrations (Deep et al. 2014). Anti-trichloropyridinol antibodies are also an interesting option to make an immunosensor once trichloropyridinol is a metabolite of chlorpyrifos: a highly used organophosphorus pesticides in agriculture. The sensor mechanism based on a competitive immunoreaction between the antibody and the pesticide (if present in a sample), or the antibody and 3,5,6-trichloropyridinol attached to quantum dot (for fluorescent readout) from the immunochromatographic test strip assay (Zhou et al. 2010). Antibodies anti-chlorpyrifos could also be used to detect this pesticide in agricultural products through a portable immunosensor based on gold nanoparticles and polyaniline/carboxylated multiwall carbon nanotubes-chitosan nanocomposite (Ding et al. 2014).

Not only molecules like antibodies and enzymes can be used in biosensors. Whole organisms are also viable options as pollution detectors. *Flavobacterium* sp., that contains the organophosphorus hydrolase enzyme, could convert the pesticide methyl parathion into a product (*p*-nitrophenol) detectable by the optic fiber system of the biosensor (Kumar et al. 2006).

It is also necessary to highlight the role of nanobiosensors of cantilevers (Lang et al. 2017). Cantilevers are movable beams, generally containing a tip at their free end, essential to atomic force microscope (AFM) operation. If these structures are functionalized (chemically modified), they can work as nanoscale sensors (with rapid response and high sensitivity) and not anymore as tools for topographic characterization (Peiner and Wasisto 2019). If the AFM tip is functionalized it also generates biosensor, but not cantilever based ones, AFM tip nanobiosensors, and both can be designed to detect the presence of pesticides in various samples (Table 3.1). Different types of functionalization generate sensors with different sensitivity and detection limit (Muenchen et al. 2016). A cantilever based biosensor was generated by immobilizing to the gold-coated side of the cantilever a synthetic hapten of the pesticide dichlorodiphenyltrichloroethane (DDT). The assay was performed in a competitive way comparing the differential surface stress causing nanometer-scale bending of a microcantilever exposed to a sample containing only monoclonal antibody anti-DDT or antibody and the pollutant pesticide DDT (Alvarez et al. 2003). AFM tip nanobiosensor functionalized with the acetolactate synthase enzyme could detect the herbicide metsulfuron-methyl; this detection was based on variation of relative adhesion force observed through the acquisition of force curves in the presence or absence of pesticide. In the presence of the pollutant the adhesion force values were higher due to the specific interaction enzyme-herbicide (Silva et al. 2013). It was also possible to develop this type of biosensor with specificity to detect diclofop by using the enzyme acetyl-CoA carboxylase and the measures obtained experimentally were in agreement with theoretical ones obtained through computer simulation (Amarante et al. 2014).The herbicides imazaquin and metsulfuron-methyl could also cause an increase of the adhesion force being detected using acetolactate synthase enzyme attached to a functionalized AFM tip (Deda et al. 2013).

When it comes to detecting pesticide's presence in water, surface-enhanced Raman spectroscopy (SERS) is a relevant tool. The technique is fast, possess a weak signal interference of water molecules, the sample pretreatment is simple and it can determine with high accuracy (fingerprinting characterization) the pollutant presence. Fonofos, phosmet, and sulfoxaflor (pesticides that if present in paddy water can damage rice production) could be detected efficiently (Weng et al. 2019). However, SERS is not limited to water analysis. Rape plants leaf could be analyzed for the presence of the pesticide thiabendazole, after mixed with acetonitrile and sodium chloride and the supernatant obtained (Lin et al. 2018). SERS also made possible the detection, for example, of residues of: isofenphos-methyl in corns (Liu et al. 2016), chlorpyrifos on spinach (without sample preparation, by just dropping silver colloids onto samples) (Zhai et al. 2016), carbaryl in Fuji apples (Fan et al. 2015), malathion in tomatoes' peels and damson plums (Albuquerque and Poppi 2015), and phorate and fenthion in apple skin (Li et al. 2014). Soil analyses through this technique were able to detect residues of deltamethrin and carbofuran by using gold nanoparticles to increase stability and sensitivity (He et al. 2019).

Another spectroscopy methodology has also proved to be useful to detect pesticide presence in samples. Laser induced breakdown spectroscopy (LIBS) when performed in the presence of metal nanoparticles acquired the capacity to detect not only trace pesticide (chlorpyrifos) but also heavy metal residues (cadmium) in apple and lettuce (Zhao et al. 2019b).

Regarding the heavy metal pollution LIBS could also be used to analyze industrial wastewater performing an online detection of Cd, Cr, Cu, Ni, Pb, and Zn (Zhao et al. 2019a) and the electrode graphite enrichment combined with plasma spatial confinement could improve the technique capacity to analyze Cd, Pb, and Ni faster in this kind of sample. However, if the electrode suffers aluminum enrichment the sensitivity increases and the detected concentration can be as low as ppb (Meng et al. 2017). Actually, LIBS could be used for analyzing different sample types like rock and sea salts detecting Cu, Cd, and Ni (Rehan et al. 2018).

It is possible to use Atomic fluorescence spectrophotometer to determine Hg and As soil samples, vegetables, and fly ashes (Ma et al. 2018). A fast and non-destructive technique called micro-X-ray fluorescence core-scanning was also useful to quantify Ti, Cr, Mn, Ni, Cu, Zn, Pb. Samples were first exposed to cation-exchange resin sachets that concentrated heavy metal cations; then the X-ray was performed and the results compared to the ones from scanned resin reference standards (Huang et al. 2019).

Functionalized cantilevers (modified, for example, with proteins or with chemicals to generate desirable functional groups on the surface) were also a viable option to detect heavy metals like cesium (Ji et al. 2000), mercury (Cherian et al. 2003; Kadam et al. 2006), zinc (Cherian et al. 2003), and cadmium (Velanki et al. 2007).

However, as same as for pesticide detection and measurement, biosensors (Table 3.2) represent a quick, reliable, and simple strategy to monitor heavy metal pollution in different samples (Asal et al. 2018). The antibodies able to detect EDTA–Cd^{2+}, DTPA–Co^{2+}, 2,9-dicarboxyl-1,10-phenanthroline–U^{6+}, and cyclohexyl–DTPA–Pb^{2+} could be used to build immunosensors to detect the presence of these heavy metals (Blake et al. 2001). Groundwater samples could be analyzed for the presence of U^{6+} thanks to an immunosensors with antibody to detect

Table 3.2 Some examples of biosensors developed in twenty-first century to detect heavy metals in different environments

Sensor	Heavy metal (s)	Reference
Immunosensor	Cd, Co, Pb, and U	Blake et al. (2001)
Biosensor containing synthetic phytochelatin fused to a maltose binding domain	Hg, Cd, Pb, Cu, and Zn	Bontidean et al. (2003)
Biosensor containing the enzyme urease	Cd	May and Russell (2003)
Biosensor containing mammalian metallothionein	Cd, Zn, and Ni	Wu and Lin (2004)
Biosensor containing the enzyme glucose oxidase	Hg	Malitesta and Guascito (2005)
Cardiomyocyte-based biosensor	Hg, Pb, Cd, Fe, Cu, and Zn	Liu et al. (2007)
Biosensor containing the enzyme 6His-Tag glutathione *S*-transferase	Zn, Cd, and Hg	Saatci et al. (2007)
Biosensor containing the enzyme glutathione *S*-transferase theta 2-2	Zn	Saatci et al. (2007)
Biosensor containing rodhamine-labeled metallothionein	Cd	Varriale et al. (2007)
Immunosensors: battery-operated or requiring grounded power source	U	Melton et al. (2009)
A colorimetric biosensor containing papain-functionalized gold nanoparticles	Hg, Pb, and Cu	Guo et al. (2011)
Biosensor based on structure-switching	Hg	Long et al. (2013)
Biosensor containing bioluminescent bacteria	Cr	Thacharodi et al. (2019)
Biosensor containing recombinant *Escherichia coli*	Cd	Lee et al. (2019)

U^{6+}-dicarboxyphenanthroline; one immunosensor battery-operated that took 5–10 min to analyze and the other that required a grounded power source and possessed the ability to, in a single experiment, autonomously analyze multiple samples (Melton et al. 2009). Not only immunosensors are able to detect heavy metals. Biosensor containing the enzyme urease (that can be inhibited by Cd) could detect cadmium through surface plasmon resonance determination of changes in the protein's conformation (May and Russell 2003). The mammalian metallothionein protein was also successfully used as part of a biosensor to detect, by surface plasmon resonance, Cd, Zn and Ni (Wu and Lin 2004). Metallothionein (MT) is a cysteine-rich protein capable of binding to Zn, Cu, and Cd, but the affinity for Cd was the highest one. Thus, other kind of biosensor could be developed after the binding equilibrium was reached among rodhamine-labeled MT exposed to Zn^{2+} –saturated chelex resin. In a reversed-displacement format the biosensor was exposed to contaminated water samples and could detect Cd when it was present (Varriale et al. 2007). A synthetic phytochelatin ($(Glu\text{-}Cys)_{20}Gly$) fused to a maltose binding domain was part of a biosensor capable of detecting Hg^{2+}, Cd^{2+}, Pb^{2+}, Cu^{2+}, and Zn^{2+} ions (Bontidean et al. 2003). A colorimetric biosensor containing papain-functionalized gold nanoparticles could detect Hg^{2+}, Pb^{2+}, and Cu^{2+} (Guo et al. 2011); nanomaterials are components with interesting characteristics to build sensors (Priyadarshini and Pradhan 2017). The enzyme glucose oxidase immobilized in poly-o-phenylenediamine could allow a biosensor to detect Hg^{2+} and to also respond to Cu^{2+} presence (Malitesta and Guascito 2005). 6His-Tag glutathione *S*-transferase in a biosensor made possible the detection of Zn^{2+}, Cd^{2+}, and Hg^{2+}; a different isozyme of glutathione *S*-transferase (the glutathione *S*-transferase Theta 2-2) was more selective to Zn^{2+} and also presented a wider dynamic range for it (Saatci et al. 2007). DNA can also be the bio portion of a biosensor to detect pollutants; an optical biosensor for Hg^{2+} detection based on structure-switching DNA was developed to perform rapid analyses. The biosensor surface contained DNA probes to bind to fluorescence-labeled DNA containing T-T mismatch. A decrease in fluoresce could be observed when, in the presence of the heavy metal in water samples, the DNA containing mismatch dehybridized from the sensor to generate T-Hg^{2+}-T complexes (Long et al. 2013). There are also biosensors that are cell-based: cardiomyocytes in a light-addressable potentiometric sensor could rapidly detect Hg^{2+}, Pb^{2+}, Cd^{2+}, Fe^{3+}, Cu^{2+}, Zn^{2+} (Liu et al. 2007); genetically modified *Escherichia coli* expressing a version of enhanced green fluorescent protein containing metal-binding loops could emit fluoresce indicating, in a biosensor, the presence of Cd in samples (Lee et al. 2019); the bioluminescent bacteria VIT-JAAK, isolated from the gut of sea urchin, could be used in a sensor (based on nutrient agar cubesand) to detect hexavalent Cr that inhibited the luminescence process when present in water samples (Thacharodi et al. 2019).

Sensors without the biological portion are also useful. Hg^{2+} was detected by layered titanate nanosheets containing sodium cations between them. Cation-exchange allowed voltammetric detection of mercury ions in real mushroom samples (Yuan et al. 2013). Silver nanoparticles and hydrogen peroxide in a colorimetric sensor were able to detect Hg^{2+}. Other metallic ions were not able to induce system's color change as mercury ions could (Wang et al. 2012).

Some conventional methods to detect heavy metals, other than the mentioned previously in this text, are not totally banned. Actually, they are still being used in a lot of different studies. Atomic absorption spectrophotometers are also useful to determine heavy metal's concentration in samples. Groundwater could be successfully analyzed for Cu, Cr, Cd, Pb, Zn, and Fe presence (Boateng et al. 2019) and soil samples, vegetables, and fly ashes for the presence of Cr, Pb, Cu, Ni, Zn, and Cd (Ma et al. 2018). This methodology also made possible to analyze tree barks of the species *Poincianella pluviosa*, after being digested in a nitroperchloric mixture, for the presence of Pb, Cd, Ni, Cu, Zn, Fe, and Mn (de Souza et al. 2019), and soil of steel limited dumpsite for the presence of Zn, Cu, Ni, and Cd (Bello et al. 2017). Water samples from industrial zone could also be analyzed for Pb, Cd, Cr, Fe, Cu, Ni presence, and for the presence of Hg through cold vapor atomic absorption (Pobi et al. 2019). Plasma optical emission spectroscopy also made possible the heavy metal analysis (Cd, Cu, Mn, Pb, Zn) on *Parmotrema arnoldii* and *Tillandsia usneoides* deposited by air pollution. Lichens and bromeliads are useful organisms to be used as biomonitors of air pollution (Benítez et al. 2019).

With the values detected through the techniques mentioned in this chapter are commonly performed pollution indices to assess the pollution status on the area and some statistical analysis, for example, multivariate analysis. Contamination factor, enrichment factor, and geo-accumulation index were recently performed to investigate the presence of Fe, Cd, Co, Cr, Cu, Ni, Zn, and Pb in soil surface sediments (Anbuselvan et al. 2018). These three parameters and also the pollution load index were performed to determine Cu, Pb, Zn, Ni, As, Cd, Sb, Sr, and Hg in sediments in Ecuador's artisanal and small scale gold mining area and compared to Canadian guidelines (Pesantes et al. 2019). To propose a new assessment method, the hazard load index, scientists compared the results with the ones obtained using pollution load index. The novel method could perform the assessment of Zn, Cu, Ni, and Cd pollution in top soils of Dana steel limited dumpsite concluding that the samples were in fact polluted. However, the pollution load index overestimated the risk. The authors attributed this fact to an improper control sampling that assessment through pollution load index requires (Bello et al. 2017).

3.3 Conclusions

Given the risk that heavy metals and pesticides pose to the existence of life on the planet, it is necessary to detect their presence in different environments. It is crucial that techniques make possible to determine their levels in order to discover if these pollutants are present and if so, if the concentration threatens health. Thus, new ways to assess and monitor the presence of heavy metals and pesticides in different samples continue to emerge. These new methods generally aim to save financial investment, reduce analysis time, facilitate the detection procedure, and make results more accurate and the analysis truer to reality.

3.4 Future Perspectives

New ways to access and monitor the presence of pollutants such as heavy metals and pesticides as many different samples as possible are expected to continue to be proposed. The work of multidisciplinary teams aims to gather knowledge from different areas so that these modern innovative techniques for assessing and monitoring these pollutants have advantages over the conventional ones: offering more reliable, reproducible, fast, economical, and practical analyzes. There is, for example, a tendency to adapt equipments (that allow this feature) to perform direct and continuous (called online) monitoring of soil, water, and air's quality and safety in polluted spots. It is also desirable that human beings develop their environmental awareness so that they can get as close as possible to a sustainable development model, and can convert the human action over nature into a less harmful and more harmonic one.

References

Abarikwu S, Iserhienrhien B, Badejo T (2013) Rutin-and selenium attenuated cadmium-induced testicular pathophysiology in rats. Hum Exp Toxicol 32:395–406

Adamcová D, Radziemska M, Ridošková A, Bartoň S, Pelcová P, Elbl J, Kynický J, Brtnický M, Vaverková MD (2017) Environmental assessment of the effects of a municipal landfill on the content and distribution of heavy metals in *Tanacetum vulgare L.* Chemosphere 185:1011–1018

Albuquerque C, Poppi R (2015) Detection of malathion in food peels by surface-enhanced Raman imaging spectroscopy and multivariate curve resolution. Anal Chim Acta 879:24–33

Ali U, Syed JH, Malik RN, Katsoyiannis A, Li J, Zhang G, Jones KC (2014) Organochlorine pesticides (OCPs) in South Asian region: a review. Sci Total Environ 476:705–717

Ali MM, Alia ML, Islamc MS, Rahmanda MZ (2016) Preliminary assessment of heavy metals in water and sediment of Karnaphuli River, Bangladesh. Environ Nanotechnol Monit Manag 5:27–35

Alvarez M, Calle A, Tamayo J, Lechuga LM, Abad A, Montoya A (2003) Development of nanomechanical biosensors for detection of the pesticide DDT. Biosens Bioelectron 18:649–653

Amarante AM, Oliveira GS, Bueno CC, Cunha RA, Ierich JCM, Freitas LCG, Franca EF, Oliveira ON Jr, Leite FL (2014) Modeling the coverage of an AFM tip by enzymes and its application in nanobiosensors. J Mol Graph Model 53:100–104

Anbuselvan N, Nathan DS, Sridharan M (2018) Heavy metal assessment in surface sediments off Coromandel Coast of India: implication on marine pollution. Mar Pollut Bull 131:712–726

Arias-Estévez M, López-Periago E, Martínez-Carballo E, Simal-Gándara J, Mejuto JC, García-Río L (2008) The mobility and degradation of pesticides in soils and the pollution of groundwater resources. Agric Ecosyst Environ 123:247–260

Aruliah R, Selvi A, Theertagiri J, Ananthaselvam A, Kumar KS, Madhavan J, Rahman P (2019) Integrated remediation processes toward heavy metal removal/recovery from various environments—a review. Front Environ Sci 7:1–66

Asal M, Özen O, Şahinler M, Polatoğlu I (2018) Recent developments in enzyme, DNA and immuno-based biosensors. Sensors 18:1924–1931

Bello S, Muhammad BG, Simon J, Bature B (2017) Development and validation of a new comprehensive method for heavy metal pollution assessment in soil media. Int J Swarm Intel Evol Comput 6:160–163

Benítez A, Medina J, Vásquez C, Loaiza T, Luzuriaga Y, Calva J (2019) Lichens and bromeliads as bioindicators of heavy metal deposition in Ecuador. Diversity 11:28–37

Bhat RA, Beigh BA, Mir SA, Dar SA, Dervash MA, Rashid A, Lone R (2018) Biopesticide techniques to remediate pesticides in polluted ecosystems. In: Wani KA, Mamta (eds) Handbook of research on the adverse effects of pesticide pollution in aquatic ecosystems. IGI Global, Hershey, pp 387–407

Blake DA, Jones RM, Blake RC, Pavlov AR, Darwish IA, Yu H (2001) Antibody-based sensors for heavy metal ions. Biosens Bioelectron 16:799–809

Boateng TK, Opoku F, Akoto O (2019) Heavy metal contamination assessment of groundwater quality: a case study of Oti landfill site, Kumasi. Appl Water Sci 9:33–47

Bontidean I, Ahlqvist J, Mulchandani A, Chen W, Bae W, Mehra RK, Mortari A, Csöregi E (2003) Novel synthetic phytochelatin-based capacitive biosensor for heavy metal ion detection. Biosens Bioelectron 18:547–553

Busairi N, Syahir A (2018) Recent advances in mercury detection: towards enabling a sensitive and rapid point-of-check measurement. J Toxicol Risk Assess 4:1–10

Chauhan N, Pundir CS (2012) An amperometric acetylcholinesterase sensor based on Fe_3O_4 nanoparticle/multi-walled carbon nanotube-modified ITO-coated glass plate for the detection of pesticides. Electrochim Acta 67:79–86

Chauhan N, Narang J, Jain U (2016) Amperometric acetylcholinesterase biosensor for pesticides monitoring utilising iron oxide nanoparticles and poly(indole-5-carboxylic acid). J Exp Nanosci 11:111–122

Chen CH, Yang KL (2013) A liquid crystal biosensor for detecting organophosphates through the localized pH changes induced by their hydrolytic products. Sens Actuators B Chem 181:368–374

Cheng M, Zeng G, Huang D, Lai C, Xu P, Zhang C, Liu Y (2016) Hydroxyl radicals based advanced oxidation processes (AOPs) for remediation of soils contaminated with organic compounds: a review. Chem Eng J 284:582–598

Cherian S, Gupta RK, Mullin BC, Thundat T (2003) Detection of heavy metal ions using protein-functionalized microcantilever sensors. Biosens Bioelectron 19:411–416

Choi JW, Kim YK, Lee IH, Min J, Lee WH (2001) Optical organophosphorus biosensor consisting of acetylcholinesterase/viologen hetero Langmuir-Blodgett film. Biosens Bioelectron 16:937–943

Chu YH, Yu XX, Jin X, Wang YT, Zhao DJ, Zhang P, Sunab GM, Zhang YH (2019) Purification and characterization of alkaline phosphatase from lactic acid bacteria. RSC Adv 9:354–360

Colmati F, Sgobbi LF, Teixeira GF, Vilela RS, Martins TD, Figueiredo GO (2019) Electrochemical biosensors containing pure enzymes or crude extracts as enzyme sources for pesticides and phenolic compounds with pharmacological property detection and quantification, vol 2019. IntechOpen, London, pp 1–5

de Souza MSPA, dos Santos FS, Magalhães LMS, de Freitas WK, de Gois G, Júnior JFO (2019) *Poincianella pluviosa* as biomonitor of heavy metals in the municipality of Volta Redonda, RJ, Brazil. Rev Bras Eng Agríc 23:71–76

Deep A, Saraf M, Bharadwaj SK, Sharma AL (2014) Styrene sulphonic acid doped polyaniline based immunosensor for highly sensitive impedimetric sensing of atrazine. Electrochim Acta 146:301–306

Deda DK, Pereira BBS, Bueno CC, da Silva AN, Ribeiro GA, Amarante AM, Franca EF, Leite FL (2013) The use of functionalized AFM tips as molecular sensors in the detection of pesticides. Mater Res 16:683–687

Ding J, Sun X, Guo Y, Jia H, Qiao L, Wang X (2014) A portable pesticide residues detection instrument based on impedance immunosensor. Sens Transd 172:27–33

Duffus JH (2002) "Heavy metals"—a meaningless term? (IUPAC Technical Report). Pure Appl Chem 74:793–807

Fan Y, Lai K, Rasco BA, Huang Y (2015) Determination of carbaryl pesticide in Fuji apples using surface-enhanced Raman spectroscopy coupled with multivariate analysis. LWT-Food Sci Technol 60:352–357

Farombi E, Akintunde J, Nzute N, Adedara I, Arojojoye O (2012) Municipal landfill leachate induces hepatotoxicity and oxidative stress in rats. Toxicol Ind Health 28:532–541

Gavrilescu M (2005) Fate of pesticides in the environment. Eng Life Sci 5:497–526

Goyer RA (2001) Toxic effects of metals. In: Klaassen CD (ed) Casarett and Doull's toxicology: the basic science of poisons. McGraw-Hill, New York, pp 811–867

Guo Y, Wang Z, Qu W, Shao H, Jiang X (2011) Colorimetric detection of mercury, lead and copper ions simultaneously using protein-functionalized gold nanoparticles. Biosens Bioelectron 26:4064–4069

Gupta VK, Yola ML, Eren T, Atar N (2015) Selective QCM sensor based on atrazine imprinted polymer: its application to wastewater sample. Sens Actuators B Chem 218:215–221

Haddaoui M, Raouafi N (2015) Chlortoluron-induced enzymatic activity inhibition in tyrosinase/ZnONPs/SPCE biosensor for the detection of ppb levels of herbicide. Sens Actuators B Chem 219:171–178

He MJ, Shen HR, Li ZT, Wang L, Wang F, Zhao KL, Liu XM, Wendroth O, Xu JM (2019) Ten-year regional monitoring of soil-rice grain contamination by heavy metals with implications for target remediation and food safety. Environ Pollut 244:431–439

Hepel M, Stobiecka M (2008) Interactions of herbicide atrazine with DNA. Nova Science Publishers, New York, p 66

Herath D, Pitawala A, Gunatilake J, Iqbal MCM (2018) Using multiple methods to assess heavy metal pollution in an urban city. Environ Monit Assess 190:657–671

Hou W, Zhang Q, Dong H, Li F, Zhang Y, Guo Y, Sun X (2019) Acetylcholinesterase biosensor modified with ATO/OMC for detecting organophosphorus pesticides. New J Chem 43:946–952

Hu W, Wang H, Dong L, Huang B, Borggaard OK, Bruun-Hansen HC, He Y, Holm PE (2018) Source identification of heavy metals in peri-urban agricultural soils of southeast China: an integrated approach. Environ Pollut 237:650–658

Huang J-JS, Lin S-C, Löwemark L, Liou SYH, Chang Q, Chang T-K, Wei K-Y, Croudace IW (2019) Rapid assessment of heavy metal pollution using ion-exchange resin sachets and micro-XRF core-scanning. Sci Rep 9:6601–6606

Hulanicki A, Glab S, Ingman F (1991) Chemical sensors definitions and classification. Pure Appl Chem 63:1247–1250

Jaishankar M, Tseten T, Anbalagan N, Mathew BB, Beeregowda KN (2014) Toxicity, mechanism and health effects of some heavy metals. Interdiscip Toxicol 7:60–72

Ji HF, Finot E, Dabestani R, Thundat T, Brown GM, Britt PF (2000) A novel self-assembled monolayer (SAM) coated microcantilever for low level caesium detection. Chem Commun 2000(6):457–458

Kadam AR, Nordin GP, George MA (2006) Use of thermally induced higher order modes of a microcantilever for mercury vapor detection. J Appl Phys 99:094905

Kesik M, Kanik FE, Turan J, Kolb M, Timur S, Bahadir M, Toppare L (2014) An acetylcholinesterase biosensor based on a conducting polymer using multiwalled carbon nanotubes for amperometric detection of organophosphorous pesticides. Sens Actuators B Chem 205:39–49

Korotkaya EV (2014) Biosensors: design, classification, and applications in the food industry. Foods Raw Mater 2:161–171

Kumar J, Jha SK, D'Souza SF (2006) Optical microbial biosensor for detection of methyl parathion pesticide using *Flavobacterium sp.* whole cells adsorbed on glass fiber filters as disposable biocomponent. Biosens Bioelectron 21:2100–2105

Kurbanoglu S, Ozkan SA, Merkoçi A (2017) Nanomaterials-based enzyme electrochemical biosensors operating through inhibition for biosensing applications. Biosens Bioelectron 89:886–898

Lang HP, Hegner M, Gerber C (2017) Nanomechanical cantilever array sensors. In: Bhushan B (ed) Springer handbook of nanotechnology. Springer, Berlin, pp 457–485

Lee W, Kim H, Kang Y, Lee Y, Yoon Y (2019) A biosensor platform for metal detection based on enhanced green fluorescent protein. Sensors 19:1846–1853

Li X, Zhang S, Yu Z, Yang T (2014) Surface-enhanced Raman spectroscopic analysis of phorate and fenthion pesticide in apple skin using silver nanoparticles. Appl Spectrosc 68:483–487

Li X, Zheng Z, Liu X, Zhao S, Liu S (2015) Nanostructured photoelectrochemical biosensor for highly sensitive detection of organophosphorous pesticides. Biosens Bioelectron 64:1–5

Li T, Wan Y, Ben Y, Fan S, Hu J (2017) Relative importance of different exposure routes of heavy metals for humans living near a municipal solid waste incinerator. Environ Pollut 226:385–393

Liang A, Wang Y, Guo H, Bo L, Zhang S, Bai Y (2015) Assessment of pollution and identification of sources of heavy metals in the sediments of Changshou Lake in a branch of the Three Gorges Reservoir. Environ Sci Pollut Res 22(20):16067–16076

Lin L, Dong T, Nie P (2018) Rapid determination of thiabendazole pesticides in rape by surface enhanced Raman spectroscopy. Sensors 18:1082–1089

Liu Q, Cai H, Xu Y, Xiao L, Yang M, Wang P (2007) Detection of heavy metal toxicity using cardiac cell-based biosensor. Biosens Bioelectron 22:3224–3229

Liu T, Xu M, Yin H, Ai S, Qu X, Zong S (2011) A glassy carbon electrode modified with graphene and tyrosinase immobilized on platinum nanoparticles for sensing organophosphorus pesticides. Microchim Acta 175:129–135

Liu D, Han Y, Zhu L (2016) Quantitative detection of isofenphos-methyl in corns using surface-enhanced Raman spectroscopy (SERS) with chemometric methods. Food Anal Method 10:1202–1208

Liu X, Zhong L, Meng J, Wang F, Zhang J, Zhi Y, Zeng L, Tang X, Xu J (2018) A multi-medium chain modeling approach to estimate the cumulative effects of cadmium pollution on human health. Environ Pollut 239:308–317

Long F, Zhu A, Shi H, Wang H, Liu J (2013) Rapid on-site/in-situ detection of heavy metal ions in environmental water using a structure-switching DNA optical biosensor. Sci Rep 3:1051–1059

Ma W, Tai L, Qiao Z, Zhong L, Wang Z, Fu K, Chen G (2018) Contamination source apportionment and health risk assessment of heavy metals in soil around municipal solid waste incinerator: a case study in North China. Sci Total Environ 631:348–357

Maleki N, Absalan G, Safavi A, Farjami E (2007) Ultra trace adsorptive stripping voltammetric determination of atrazine in soil and water using mercury film electrode. Anal Chim Acta 581:37–41

Malitesta C, Guascito MR (2005) Heavy metal determination by biosensors based on enzyme immobilised by electropolymerisation. Biosens Bioelectron 20:1643–1647

Manisankar P, Sundari PA, Sasikumar R, Palaniappan S (2008) Electroanalysis of some common pesticides using conducting polymer/multiwalled carbon nanotubes modified glassy carbon electrode. Talanta 76:1022–1028

Manzetti S, Spoel ER, Spoel D (2014) Chemical properties, environmental fate, and degradation of seven classes of pollutants. Chem Res Toxicol 27:713–737

Masojídek J, Souček P, Máchová J, Frolík J, Klem K, Malý J (2011) Detection of photosynthetic herbicides: algal growth inhibition test vs. electrochemical photosystem II biosensor. Ecotoxicol Environ Saf 74:117–122

Mateos AC, Amarillo AC, Carreras HA, González CM (2018) Land use and air quality in urban environments: human health risk assessment due to inhalation of airborne particles. Environ Res 161:370–380

May LM, Russell DA (2003) Novel determination of cadmium ions using an enzyme self-assembled monolayer with surface Plasmon resonance. Anal Chim Acta 500:119–125

Mehmood MA, Qadri H, Bhat RA, Rashid A, Ganie SA, Dar GH, Shafiq-ur-Rehman (2019) Heavy metal contamination in two commercial fish species of a trans-Himalayan freshwater ecosystem. Environ Monit Assess 191:104. https://doi.org/10.1007/s10661-019-7245-2

Melton SJ, Yu H, Williams KH, Morris SA, Long PE, Blake DA (2009) Field-based detection and monitoring of uranium in contaminated groundwater using two immunosensors. Environ Sci Technol 43:6703–6709

Meng D, Zhao N, Wang Y, Ma M, Fang L, Gu Y, Jia Y, Liu J (2017) On-line/on-site analysis of heavy metals in water and soils by laser induced breakdown spectroscopy. Spectrochim Acta B At Spectrosc 137:39–45

Mirabi-Semnakolaii A, Daneshgar P, Moosavi-Movahedi AA, Rezayat M, Norouzi P, Nemati A, Farhadi M (2011) Sensitive determination of herbicide trifluralin on the surface of copper nanowire electrochemical sensor. J Solid State Electrochem 15:1953–1961

Mishra GK, Barfidokht A, Tehrani F, Mishra RK (2018) Food safety analysis using electrochemical biosensors. Foods 7:141–152

Morillo E, Villaverde J (2017) Advanced technologies for the remediation of pesticide-contaminated soils. Sci Total Environ 586:576–597

Moro L, Pezzotti G, Turemis M, Sanchís J, Farré M, Denaro R, Giacobbe MG, Crisafi F, Giardi MT (2018) Fast pesticide pre-screening in marine environment using a green microalgae-based optical bioassay. Mar Pollut Bull 129:212–221

Muenchen DK, Martinazzo J, de Cezaro AM, Rigo AA, Brezolin AN, Manzoli A, Leite FL, Steffens C, Steffens J (2016) Pesticide detection in soil using biosensors and nanobiosensors. Biointerface Res Appl Chem 6:1659–1675

Mushtaq N, Bhat RA, Dervash MA, Qadri H, Dar GH (2018) Biopesticides: the key component to remediate pesticide contamination in an ecosystem. In: Environmental contamination and remediation. Cambridge Scholars Publishing, Cambridge, pp 152–178

Muszynska E, Hanus-Fajerska E (2015) Why are heavy metal hyperaccumulating plants so amazing? Biotechnol J Biotechnol Comput Biol Bionanotechnol 96:265–271

Nagajyoti PC, Lee KD, Sreekanth T (2010) Heavy metals, occurrence and toxicity for plants: a review. Environ Chem Lett 8:199–216

Niti C, Sunita S, Kamlesh K, Rakesh K (2013) Bioremediation: an emerging technology for remediation of pesticides. Res J Chem Environ 17:88–105
Odobašić A, Šestan I, Begić S (2019) Biosensors for determination of heavy metals in waters, vol 2019. IntechOpen, London, pp 1–15
Pan L, Feng X, Cao M, Zhang S, Huang Y, Xu T, Jing J, Zhang H (2019) Determination and distribution of pesticides and antibiotics in agricultural soils from northern China. RSC Adv 9:15686–15693
Peiner E, Wasisto HS (2019) Cantilever sensors. Sensors 19:2043–2045
Pereira AR, Sedenho GC, Souza JCP, Crespilho FN (2018) Advances in enzyme bioelectrochemistry. An Acad Brasil Ciênc 90:825–857
Pesantes AA, Carpio EP, Vitvar T, López MMM, Menéndez-Aguado JM (2019) A multi-index analysis approach to heavy metal pollution assessment in river sediments in the Ponce Enríquez Area, Ecuador. Water 11:590–601
Petrosyan V, Pirumyan G, Perikhanyan Y (2019) Determination of heavy metal background concentration in bottom sediment and risk assessment of sediment pollution by heavy metals in the Hrazdan River (Armenia). Appl Water Sci 9:102–110
Pobi KK, Satpati S, Dutta S, Nayek S, Saha RN, Gupta S (2019) Sources evaluation and ecological risk assessment of heavy metals accumulated within a natural stream of Durgapur industrial zone, India, by using multivariate analysis and pollution indices. Appl Water Sci 9:58–73
Priyadarshini E, Pradhan N (2017) Gold nanoparticles as efficient sensors in colorimetric detection of toxic metal ions: a review. Sens Actuators B Chem 238:888–902
Pundir CS, Chauhan N (2012) Acetylcholinesterase inhibition-based biosensors for pesticide determination: a review. Anal Biochem 429:19–31
Rashid A, Bhat RA, Qadri H, Mehmood MA (2019) Environmental and socioeconomic factors induced blood lead in children: an investigation from Kashmir, India. Environ Monit Assess 191(2):76. https://doi.org/10.1007/s10661-019-7220-y
Rasmussen M, Wingersky A, Minteer SD (2014) Comparative study of thylakoids from higher plants for solar energy conversion and herbicide detection. Electrochim Acta 140:304–308
Reeves WR, McGuire MK, Stokes M, Vicini JL (2019) Assessing the safety of pesticides in food: how current regulations protect human health. Adv Nutr 10:80–88
Rehan I, Khan MZ, Rehan K, Mateen A, Farooque MA, Sultana S, Farooq Z (2018) Determination of toxic and essential metals in rock and sea salts using pulsed nanosecond laser-induced breakdown spectroscopy. Appl Opt 57:295–301
Rikta S, Tareq SM, Uddin MK (2018) Toxic metals (Ni2+, Pb2+, Hg2+) binding affinity of dissolved organic matter (DOM) derived from different ages municipal landfill leachate. Appl Water Sci 8:1–8
Rocaboy-Faquet E, Barthelmebs L, Calas-Blanchard C, Noguer T (2016) A novel amperometric biosensor for ß-triketone herbicides based on hydroxyphenylpyruvate dioxygenase inhibition: a case study for sulcotrione. Talanta 146:510–516
Saatci E, Nistor M, Gaspar S, Csoregi E, Iscan M (2007) Comparison of two glutathione S-transferases used in capacitive biosensors for detection of heavy metals. Int J Environ Anal Chem 87:745–754
Samphao A, Suebsanoh P, Wongsa Y, Pekec B, Jitchareon J, Kalcher K (2013) Alkaline phosphatase inhibition-based amperometric biosensor for the detection of carbofuran. Int J Electrochem Sci 8:3254–3264
Schierow L, Esworthy R (2012) Pesticide law: a summary of the statutes; Report No.: RL31921. Congressional Research Service, Washington
Selvi A, Aruliah R (2018) A statistical approach of zinc remediation using acidophilic bacterium via an integrated approach of bioleaching enhanced electrokinetic remediation (BEER) technology. Chemosphere 207:753–763
Shen F, Mao L, Sun R, Du J, Tan Z, Ding M (2019) Contamination evaluation and source identification of heavy metals in the sediments from the Lishui River Watershed, Southern China. Int J Environ Res Public Health 16:336–349

Shyuan LK, Heng LY, Ahmad M, Aziz SA, Ishak Z (2008) Evaluation of pesticide and heavy metal toxicity using immobilized enzyme alkaline phosphatase with an electrochemical biosensor. Asian J Biochem 3:359–365

Silva ACN, Deda DK, Róz AL, Prado RA, Carvalho CC, Viviani V, Leite FL (2013) Nanobiosensors based on chemically modified AFM probes: a useful tool for metsulfuron-methyl detection. Sensors 13:1477–1489

Singh DV, Bhat JIA, Bhat RA, Dervash MA, Ganei SA (2018) Vehicular stress a cause for heavy metal accumulation and change in physico-chemical characteristics of road side soils in Pahalgam. Environ Monit Assess 190:353. https://doi.org/10.1007/s10661-018-6731-2

Tepanosyan G, Maghakyan N, Sahakyan L, Saghatelyan A (2017) Heavy metals pollution levels and children health risk assessment of Yerevan kindergartens soils. Ecotoxicol Environ Saf 142:257–265

Thacharodi A, Jeganathan C, Thacharodi D (2019) Biomonitoring of heavy metal pollution by bioluminescent bacterial biosensors. Indian J Sci Technol 15:12–20

Tortolini C, Bollella P, Antiochia R, Favero G, Mazzei F (2016) Inhibition-based biosensor for atrazine detection. Sens Actuators B Chem 224:552–558

Uniyal S, Sharma RK (2018) Technological advancement in electrochemical biosensor based detection of organophosphate pesticide chlorpyrifos in the environment: a review of status and prospects. Biosens Bioelectron 116:37–50

Vamvakaki V, Chaniotakis NA (2007) Pesticide detection with a liposome-based nano-biosensor. Biosens Bioelectron 22:2848–2853

Varriale A, Staiano M, Rossi M, D'Auria S (2007) High-affinity binding of cadmium ions by mouse metallothionein prompting the design of a reversed-displacement protein-based fluorescence biosensor for cadmium detection. Anal Chem 79:5760–5762

Vasilescu MN, Medvedovici AV (2005) Herbicides. In: Worsfold P, Townshend A, Poole C (eds) Encyclopedia of analytical science. Elsevier, Oxford, pp 243–260

Velanki S, Kelly S, Thundat T, Blake DA, Ji HF (2007) Detection of Cd(II) using antibody-modified microcantilever sensors. Ultramicroscopy 107:1123–1128

Ventrella A, Catucci L, Placido T, Longobardi F, Agostiano A (2011) Biomaterials based on photosynthetic membranes as potential sensors for herbicides. Biosens Bioelectron 26:4747–4752

Vicentini FC, Janegitz BC, Brett CMA, Fatibello-Filho O (2013) Tyrosinase biosensor based on a glassy carbon electrode modified with multi-walled carbon nanotubes and 1-butyl-3-methylimidazolium chloride within a dihexadecylphosphate film. Sens Actuators B Chem 188:1101–1108

Wang GL, Zhu XY, Jiao HJ, Dong YM, Li ZJ (2012) Ultrasensitive and dual functional colorimetric sensors for mercury (II) ions and hydrogen peroxide based on catalytic reduction property of silver nanoparticles. Biosens Bioelectron 31:337–342

Wang L, Yang D, Li ZT, Fu YH, Liu XM, Brookes PC, Xu JM (2019) A comprehensive mitigation strategy for heavy metal contamination of farmland around mining areas—screening of low accumulated cultivars, soil remediation and risk assessment. Environ Pollut 245:820–828

Wei M, Wang J (2015) A novel acetylcholinesterase biosensor based on ionic liquids-AuNPs-porous carbon composite matrix for detection of organophosphate pesticides. Sens Actuators B 211:290–296

Weibel G, Eggenberger U, Schlumberger S, Mader UK (2017) Chemical associations and mobilization of heavy metals in fly ash from municipal solid waste incineration. Waste Manag 62:147–159

Weng S, Zhu W, Dong R, Zheng L, Wang F (2019) Rapid detection of pesticide residues in paddy water using surface-enhanced Raman spectroscopy. Sensors 19:506–515

Wittsiepe J, Feldt T, Till H, Burchard G, Wilhelm M, Fobil JN (2017) Pilot study on the internal exposure to heavy metals of informal-level electronic waste workers in Agbogbloshie, Accra, Ghana. Environ Sci Pollut Res Int 24:3097–3107

Wu CM, Lin LY (2004) Immobilization of metallothionein as a sensitive biosensor chip for the detection of metal ions by surface plasmon resonance. Biosens Bioelectron 20:864–871

Xiong J, Zhao T, Cheng H, Li S, Wang S, Chen G (2019) The assessment on the heavy metal pollution and health risks in the Liujiang River under the Xijiang River region. Desalin Water Treat 149:315–322

Xu J, Yu C, Feng T, Liu M, Li F, Wang Y, Xu J (2018) N-Carbamoylmaleimide-treated carbon dots: stabilizing the electrochemical intermediate and extending it for the ultrasensitive detection of organophosphate pesticides. Nanoscale 10:19390–19398

Yang S, Zhao J, Chang SX, Collins C, Xu J, Liu X (2019) Status assessment and probabilistic health risk modeling of metals accumulation in agriculture soils across China: a synthesis. Environ Int 128:165–174

Yuan S, Peng D, Song D, Gong J (2013) Layered titanate nanosheets as an enhanced sensing platform for ultrasensitive stripping voltammetric detection of mercury (II). Sens Actuators B Chem 181:432–438

Zhai C, Peng Y-K, Xu T-F, Li Y-Y (2016) Detection of chlorpyrifos on spinach based on surface enhanced Raman spectroscopy with silver colloids. Spectrosc Spectr Anal 36:2835–2840

Zhao NJ, Meng DS, Jia Y, Ma MJ, Fang L, Liu JG, Liu WQ (2019a) On-line quantitative analysis of heavy metals in water based on laser-induced breakdown spectroscopy. Opt Express 27:1–12

Zhao X, Zhao C, Du X, Dong D (2019b) Detecting and mapping harmful chemicals in fruit and vegetables using nanoparticle-enhanced laser-induced breakdown spectroscopy. Sci Rep 9:906–910

Zhou Z, Du D, Wang J, Smith JN, Timchalk C, Li Y, Lin Y (2010) Quantum dot-based immunochromatographic fluorescent biosensor for biomonitoring trichloropyridinol, a biomarker of exposure to chlorpyrifos. Anal Chem 82:5125–5133

Zhu G, Guo Q, Xiao H, Chen T, Yang J (2017) Multivariate statistical and lead isotopic analyses approach to identify heavy metal sources in topsoil from the industrial zone of Beijing Capital Iron and Steel Factory. Environ Sci Pollut R 24:14877–14888

Zupancic N (2017) Influence of climate factors on soil heavy metal content in Slovenia. J Soils Sediments 17:1073–1083

Chapter 4
Global Scenario of Remediation Techniques to Combat Pesticide Pollution

Alina Butu, Ioana Grozea, Ioan Sarac, and Monica Butnariu

4.1 Background

Any agriculture is, in a way, an intervention against nature, which, when practiced unfairly, leads to a decrease in fertility, the modification of biosensors and ecosystems. It would be wrong to argue that chemicals used in agriculture would have no biological effect. Agriculture has, on the one hand, a positive role in the purification of the atmosphere; on the other hand, it sometimes has a negative role through its pollution. The positive effect of agriculture is due to green plants, which, by absorbing CO_2 from the atmosphere for its use in photosynthesis and O_2 removal, purify the air. A series of volatile substances that contaminate the atmosphere following industrial or other processes, $CaSO_2$ and NH_3 can be directly absorbed by green plants, which leads to its purification. Some of the industrial objectives also contribute to the purification of the atmosphere, affecting not only man and the biosphere, but also agriculture. Atmospheric pollution with SO_2, NH_3, NO_2, and cement powders can affect by inhalation or ingestion of animal organisms (Masbou et al. 2018).

Concentration above acceptable tolerance limits has harmful effects on the environment and hence on human life. Too many fertilizers, too many pesticides, as well as an excessively high concentration of livestock in a farm vitiate the quality of the environment. In the soil, excess nitrogen is transported in the form of nitrates in the groundwater canvas, which can then be used as drinking water or at the shelter of the animals. Increasing the water content in phosphorus stimulates the growth of algae and other aquatic vegetal organisms. It has been found that the intensive use of pesticides with nitrogen, phosphorus, and potassium leads to the increase of soil

A. Butu
National Institute of Research and Development for Biological Sciences, Bucharest, Romania

I. Grozea · I. Sarac · M. Butnariu (✉)
Banat's University of Agricultural Sciences and Veterinary Medicine "King Michael I of Romania" from Timisoara, Timis, Romania

R. A. Bhat et al. (eds.), *Bioremediation and Biotechnology, Vol 2*,
https://doi.org/10.1007/978-3-030-40333-1_4

content in some elements such as Zn, Pb, Ni, Cr, previously found in soluble forms only as traces (Song et al. 2019).

The term "pesticides" refers to any substance or mixture of substances, including mixtures with ingredients, for use in agriculture, forestry, storage, and other activities for the purpose of preventing, mitigating, removing, or destroying pests, phytopathogenic agents, weeds, and other forms of animal or plant life, including plant and domestic animal or plant viruses, insects, and rodents carrying diseases communicable to humans, and products used to regulate plant growth, defoliation, or desiccation. In fact, pesticides are toxic substances, meaning substances which, after entering the body at a relatively high dose once or more, in a very short period, or in small doses repeatedly for a long time, provoke temporary, transient, or permanent impairment of one or more functions, impairments that can go through to complete annihilation or death (James et al. 2019).

In the fight, pesticides cross different paths in ecosystems. A particular place is occupied by highly persistent, highly biodegradable pesticides, especially those organochlorinated. In the case of herbicides, the useful fraction is 5–40%. In fact, this extremely large, unused in terms of combating, creates all the known ecological problems, hence the need to ensure greater agro-availability of pesticides. Pesticides, once in the soil, besides their action on diseases, pests, and weeds, extend their action to microorganisms so that quantitative and qualitative changes occur both in the edaphic population structure and in physiological activities. The importance of knowledge of these changes is supported both by the requirements of scientific agriculture and by the protection of edaphic ecosystems.

In this context, knowing the influence pesticides have on the biochemical processes that take place in the soil is absolutely necessary. Also, negative influence on denitrifying microorganisms and soil denitrification process was highlighted (Alho 2008).

Referring to the possible damage of micro-organisms by soil active substance residues, some researchers show that this is an ecological feature of micro-organisms to be permanently exposed to natural disasters, losing them together. These occur, for example, in the case of rapid changes in temperature or soil saturation. Therefore, reductions in the population density of microorganisms by 50% and above are very common. When repairing favorable environmental factors, the population is doubled every 10 days, so that at a 12.5% drop in density from the initial level, within 30 days, a 100% restoration can be observed. Consequently, absolute population reduction and delay in regeneration of the initial population density were proposed as criteria, with the importance of delaying rather than the absolute value of population decline. In this case, it is indicated as the regeneration rule: a 30-day period can be neglected; duration of 60 days is tolerable; intervals longer than 60 days are critical, which in fact reflect forms of manifestation of soil pollution phenomena. The application of pesticides—with all the technological measures used in agriculture—represents human intervention in the biosphere and as it intensifies, the stronger and sometimes more aggressive it becomes. The task of agriculture has always been to ensure the need for food, and agricultural production has been energy-consuming since the beginning. In the past, when the population of our country did not exceed one million inhabitants, it averaged 3 ha of arable land per capita and another 6 ha were kept fallow land, food requirements through agricultural production were solved on account of solar energy and to some extent through human labor. Today, with the growth of the population, the need for food and, implicitly, the task of producing more have increased. Agriculture has intensified and its energy resources have changed with regard to the participation ratio: some energy resources are unrestricted, such as solar and wind power, which are considered inexhaustible; others are renewable: organic matter, human, and animal labor; some are limited: fossil energy (Rasheed et al. 2019).

The ratio of participation of different energy resources in the agricultural production process differs accordingly with the agricultural system. In intensive agriculture, this is in favor of limited resources given by fossil energy; the biconversion ratio ranging from 1 to 3–4 in mechanized, irrigated, and chemistry agriculture compared to the 1:129 ratio in extensive pastoral farming, when agricultural output was much lower.

At present, the applied pesticides represent a fossil energy consumption of 6.3 Gj/ha, meaning >50% of the fossil energy used to obtain the agricultural crop production. If we add the energy consumption incorporated in other substances used to fight diseases and pests and in fuels, consumption amounts to 2/3 of the total fossil energy used in the plants hoeing.

Therefore, in the context of the implications of the global energy crisis, in light of the tasks of the "Energy Research and Development Directive" program, it is necessary to rethink and improve the culture technologies, including the application of chemical pesticides, to be done only by certified humans and only as a result of a chemical control of the plant and soil to avoid any waste, which in essence means unnecessary fossil energy consumption.

There is also a need for a more rational choice of pesticide product because there are also differences in energy consumption used for their production. At present, for the synthesis of 1 kg of active substance, 73–91 Mj for nitrogen, 13–20 for Mj for phosphorus, and 8–14 Mj for potassium are consumed; consumption is different in relation to the basic energy source (Atma et al. 2017).

Another aspect related to the energy crisis is the ratio of biconversion in plant products of energy incorporated into pesticides. From the aspect of the biconvex ratio of fossil energy incorporated into pesticides, based on the synthesis of research data from our country for the main crop plants, they differ in relation to the species. The overall biconvex ratio of fossil energy incorporated into pesticides, in the energy contained in plant products, places phosphorus pesticides first, followed by potassium, the latter being nitrogenous pesticides, which are also the largest fossil energy consumers.

The biconversion must be judged also in terms of economic efficiency, which is due to the difference between the value of the yield harvest obtained and the cost of the fertilizer used, as well as the biological value of the vegetable products. Knowing this index is important because the same crop yield can be obtained with different doses and fertilizer ratios, or the fossil energy incorporated in it is not the same. Therefore, it is necessary to choose the right dosages according to the equally efficient curves and their application following a systematic control of the fertility status, the use of new types of fertilizers obtained with less energy consumption, as well as the recirculation of some vegetal residues, in order to increase the transformation coefficient of the energy incorporated in fertilizers into vegetable products, which on average is currently only 1: 3–4.

The tasks facing agriculture in the conditions of the increase of vegetal production imply the increase of the fertilizers used in the conditions of the energy economy, which means not a restriction of their use but an increase of the biconversion coefficient, by the rigorous pursuit of energy efficiency. For this, a new strategy is needed to attract natural organic fertilizers into the circuit of the nutrients in the household, as well as a series of plant and animal organic residues that, composted according to scientific rules, contribute to raising the rate of use of chemical fertilizers (Asano et al. 2014).

Integrated protection is a combination of chemical, physical, and biological methods and means in together with agro-technical (compulsorily compatible) methods applied harmoniously within agricultural ecosystems. Plant protection works are executed according to the emergency disposal technology (EDT) to minimize the number of phytosanitary interventions and to favor the natural control factors that regulate/limit the pest population.

It is necessary to radically reconsider the importance of various harmful and/or pathogenic species. Specifically, the economic importance of each species and the level of losses that could be produced must be known. Ecological and economic data has shown that most of the species recognized as harmful should not be treated. The following categories of species can be found in an agricultural crop: the main harmful species; secondary harmful species; harmful migratory species; indifferent species (Shi et al. 2017).

The evolution of modern agriculture is closely linked to changes in the traditional way of working. The resistance of the varieties to the main diseases and especially to pests is still weak. There are also situations where a variety known as resistance becomes sensitive after many years due to the emergence of new pathogen races. Combat control complexes are used against primary and secondary harmful organisms. At the same time, the measures applied must protect as many useful organisms as possible. Sometimes it is considered that combining methods means the same thing as integrated protection. In reality, combining means and methods is only part of the integrated struggle, but an essential part.

First of all, many prophylactic treatments, after the spraying calendar or healing treatments, have to be replaced with curative treatments. EDT is the most important means of regulating all types of phytosanitary interventions and is a concept of integrated combat that allows treatment to be applied only when the density and/or degree of attack can cause economically justifiable losses, 5% of the harvest, equivalent to the cost of phytosanitary treatment. EDT allows the treatment to be located only where the attack limits are exceeded and varies according to the variety, biological category, plant development stage, pedoclimatic conditions, technological conditions, etc. If all these are taken into account, a "critical situation" that requires phytosanitary intervention can be established in culture.

An assembly of methods and means of struggle structured with interdependent elements, an ensemble which also relies on natural factors to regulate the density of harmful organisms populations, must not exceed EDT (Shi et al. 2018).

From the point of view of the systemic approach, in the framework of the integrated struggle, the following structure can be made: the sub-site of the technological processes itself, which consists of all the plant protection works, together with the apparatuses, equipment, installations, plant protection machines, resistance varieties, methods of application of chemical, biological, physical, etc.; the control subsystem, of great importance for obtaining high quality agricultural products or mitigating the effects of pollution, and which is essentially based on the correct execution of the works; the handling site includes the technical processes of transport and storage of plant protection products, together with the machines specific to these processes, which can be considered as plant protection machines as well. To some extent, in this system, it is also possible to introduce biological products in commercial units; the management or organizational overhead ensures the correct application in time and space of different works, depending on the sequence of the technological links, depending on the EDT and the biological cycles of the plant and the harmful species. From a toxicological point of view, the highly toxic red and green group products (products with acute toxicity below 100 mg/kg) have to be phased out of the range of pesticides. Products with dangerous side effects (carcinogenic, embryotoxic), very persistent or high residue residues in the crop should also be excluded from the ICS (*integrated control systems*).

In their evolution, the models of intensive, superintensive, or agricultural type industrial technologies (using high productivity agricultural machines, new agrotechnical methods, high productivity varieties, and especially chemical means of combating, models which, through their content was approaching industrial production) have begun to be abandoned in favor of models of sustainable agriculture, organic farming, precision agriculture, etc.

The interdisciplinary approach is dictated by the complexity of the agro-system, especially the great variety of pests (phytopathogenic viruses, bacteria, phytopathogenic fungi, nematodes, mites, insects, rodents, weeds, etc.) and the complexity of the methods and the means of fighting. We must use results obtained in biology, physics, chemistry, ecology, systematics, agronomy, physiology, phytopathology, entomology, herbology research.

After a very brief review of the basic rules for the development of integrated protection systems we can see that the basic elements are the economic criteria and the ecological criteria (the ecological criteria being finally economic criteria, because it is now obvious that the environmental risk is an economic criterion, as defined above–danger–damage) (Munns et al. 2017).

4.2 Potential Risk Factors and Prevention of Contamination with Dangerous Chemical Substances and Mixtures

The European Chemicals Agency (ECHA) indicates the following categories of chemical substances and mixtures as high risk to human health and the environment (European Commission 2012). Biocides are products that contain or generate active substances and are used against harmful organisms in the domestic environment (disinfectants, rodenticides, repellents, and insecticides) or to protect natural or artificial products in industry and agricultural applications.

Considering that because of their intrinsic properties the use of biocidal products may pose health risks, the EU has set up a regulatory framework aimed at significantly increasing the safety of biocidal products used and sold in the EU, namely EU Regulation 528/1012. It stipulates the removal from the market of biocides that are carcinogenic, mutagenic, toxic for reproduction, endocrine disrupting, or environmentally hazardous.

Pesticides are products used to combat harmful organisms from agricultural crops such as weeds and insects. However, their use can endanger health and the environment (Bhat et al. 2018; Mushtaq et al. 2018).

Many studies have shown the link between exposure to pesticides and effects on the human body such as cancer, fertility and reproductive problems, respiratory diseases, disruption of the hormonal system (endocrine), damage to the immune system and the nervous system. EU pesticide legislation (EC Directive 128/2009) imposes prohibitions on the use of certain hazardous chemicals in these products. Endocrine disrupting substances are chemicals that interfere with hormones (the endocrine system) causing adverse health effects.

A wide range of substances, both natural and man-made, are thought to cause endocrine disturbances, including pharmaceuticals, pesticides, and organic industrial chemicals. The European Community Endocrine Disruptors Strategy, adopted in 1999, outlined the actions required to counteract the impact of endocrine disrupting substances on human health and the environment. Thus, endocrine disruptors have been introduced into the REACH Regulation lists and in January 2012, the European Commission launched an important study called "Endocrine Disrupting Substance Evaluation Status." The study, based on a thorough scientific documentation, was commissioned to inform the European Commission about the identification of endocrine disrupting substance regulatory criteria in line with EU legislation.

Of these, cadmium, lead, and mercury (from the content of some pesticides) are particularly toxic to the human body and therefore the EU has made considerable progress in addressing the contamination of these metals. Thus, in 2005, the EU Mercury Strategy was launched, providing restrictions on the sale of mercury-containing measuring devices, banning exports of metallic mercury from the EU, and new safety rules for its storage. This strategy was revised in 2010. The implementation of the EU's Industrial Emissions Directive, adopted in 2010, will further reduce mercury emissions from industrial processes.

The EU also firmly supports the international process of developing a global legal instrument on mercury under the aegis of UNEP. Cadmium, lead, and mercury levels are strictly controlled in accordance with the restrictions in the Dangerous Substances Directive (EC 95/2002). Strict limit values are required for these metals at certain specific types of electronic equipment. Also, the Vehicle Disposal Directive (EC 53/2000) further restricts the use of these metals. In addition, the use of cadmium in the manufacture of jewelry and plastic products has been banned under the

REACH Regulation, while in the Water Framework Directive (EC 60/2000) cadmium and mercury are indicated as priority hazardous substances, whereas lead as a priority substance. (Mech et al. 2019).

Persistent organic pollutants (POPs) are chemicals that remain intact in the environment for long periods of time, from where they are transported and penetrate into living organisms, including humans, where they accumulate in their fatty tissue. Due to their toxic characteristics, persistent organic pollutants have harmful effects on human health and the environment. The Stockholm Convention is the world's leading tool to protect against persistent organic pollutants. In doing so, the EU is committed to implementing the Stockholm Convention prescriptions and encouraging the additional listing of chemicals under this Convention. Furthermore, in the EU, the Persistent Organic Pollutants Regulation (EC 850/2004) supports and strengthens the provisions of the Stockholm Convention.

The Persistent Organic Pollutants Regulation contains provisions on the production, placing on the market and use of chemical substances of this type, management of stocks and wastes of persistent organic compounds, and measures to reduce unintended emissions of persistent organic compounds (Hennebert and Filella 2018). Statistics on the effects of pesticides include a physical, chemical, biological database collected through techniques and methods of comparative analysis of their impact on the ecosystem, as can be seen from Table 4.1.

Table 4.1 Effects of pesticides on specific subpopulations

The effect on human health	Pesticides causing the effect	Subpopulations at risk
Cancer	DDT, toxafen, 2,3,7,8-TCDD, mirex, HCH, PCB, HCB	First of all in adults (breast cancer, prostate cancer, and testicular cancer); cancer related to children
Effects on the device reproductively	PCB, dibenzodioxins, dibenzofurans	Fetuses, newborns, women of childbearing age
Growth retardation	PCB, dibenzodioxins, dibenzofurans	Fetuses, newborn babies, children (length, body weight, skull circumference)
Neurological disorders	PCB, dibenzodioxins, dibenzofurans	Fetuses and children (cognitive, memory, and attention problems); adults (Parkinson's and Alzheimer's)
Altering behavioral development	PCB, dibenzodioxine, PentaBDE	Children, continuing into adulthood (disorders with deficiency attention, learning difficulties)
System suppression immune	PCB, dibenzodioxins, dibenzofurans	Newborns, children (severe ear infections, impaired resistance to diseases); adults (suppressing the immune system)
Cardiovascular effects	PCB	Children and adults (variations in blood pressure and rhythm cardiac)
Effects on thyroid	PCB, PFOS, PBDES	Women (hypothyroidism)
Metabolic disorders	PCBs, POPs in general	Adult women and men (diabetes and obesity)
Bone disorders	PCB, dioxin, HCH	Adult women and men (osteomalacia, osteoporosis)

Nanomaterials are chemicals or materials that are manufactured and used in extremely small dimensions (nm, i.e., 10^{-9} m). At present, many products containing nanomaterials are already in use and this market is expected to flourish in the coming years. Nanomaterials have the potential to improve the quality of life and contribute to industrial competitiveness (Singh et al. 2020). However, new materials can also raise health and safety issues. Main technologies for bioremediation of pesticides-contaminated soils can be seen from Table 4.2.

These risks, and how these can be addressed through existing risk assessment measures in the EU, have been the subject of several scientific opinions issued by the European Commission's Scientific Committee on Emerging and Newly Identified Health Risks. Combinations of chemical substances and chemical mixtures, according to some studies, can have harmful effects on human health even when individual chemicals are below the "safety level." In the EU, the regulatory risk assessment for chemicals is based on individual products. However, there are concerns that this approach cannot provide sufficient protection against the risks generated by simultaneous exposure to several chemicals and that another approach is needed. Pharmaceutical residues have become an important source of environ-

Table 4.2 Bioremediation technologies for pesticides-contaminated soils

Method	Technologies	Advantages	Disadvantages
In situ	Biodegradation	Offers the possibility of simultaneous contamination of both soil and groundwater	The equipment requires a delicate operation of specialty
	Bioventing	The equipment is generally easy to place	During the operation it is difficult to appreciate the volume treated, its configuration, and the efficiency of the depollution process
	Biosparging	It can be applied to several types of pollutants	
	Phytoremediation	Can be used in combination with other remediation technologies	Removal efficiency may vary from site to site
	Natural attenuation controlled	It can be used for both long term and freshly contaminated sites	Long treatment time (6 months–3 years)
Ex situ	The bioreactor	Rapid and relatively complete removal of polluted components	High cost of transportation; the need for excavation and soil preparation
	Biopile	The possibility to continue the activity on site	The risk of partial dispersion of pollutants during evacuation work, loading, transporting, and unloading
	Composting	High depollution efficiency, offered by treatment in specialized plants	
	Land farming		Imposing limits of concentration in pollutants, before treatment

mental contamination. Many pharmaceutical chemicals are designed to be non-degradable and thus present a risk when they reach the environment where they persist. The European Commission recognizes that environmental pollution with pesticide residues is both an emerging issue and a public health problem and is committed to addressing this issue (Brinch et al. 2016).

4.3 Management of Plant Protection Products (Pesticides/Biocides)

The management of plant protection products is governed by the law, the rules of approval and placing on the market, storage, application, use of their specific technological means under the conditions of soil and water protection are presented in the Code of Good Practice. The document stipulates the specific rules for the approval and placing on the market, storage, and application of plant protection products.

4.3.1 The Risk of Soil Contamination with Persistent Organic Compounds and Its Prevention

As has been shown before, persistent organic compounds (POPs) are chemicals that remain intact in the environment for long periods of time, from where they are transported and penetrate into living organisms, including humans, where they accumulate in their fatty tissue. In our country, from the point of view of the pollution of soils with persistent organic compounds, significant was the pollution of agricultural soils with organochlorine insecticides (xx). Organochlorine insecticides, namely those based on HCH (hexachlorocyclohexane) and DDT (pp′-dichlorodiphenyl-trichloroethane), were introduced into use in the 1940s and have long been used to protect crops against vector insects. The organochlorinated insecticides HCH and DDT (isomers and metabolites) are very persistent, accumulating in the soil; half-life in soil is 2 years for HCH and decades for DDT. The widespread use of these pesticides and their high persistence have made their residues and metabolites highlighted in all the elements of the environment. Studies have been conducted that have shown that in areas where the soil is polluted with DDT, residues are present and its metabolites and in plants, in animal and human fat. Although it has been forbidden to treat these compounds throughout the country since 1985, HCH and DDT residues are now found in agricultural soils.

To remedy pesticide pollution, the methods of Fig. 4.1 are feasible:

Therefore, their monitoring is necessary; the following compounds are: α, β, γ, δ-hexachlorocyclohexane, p-dichlorodiphenyl trichloroethane (DDT), and the opium dichloride diphenyl trichloroethane (DDT), dichlorodiphenyl-dichloro-ethane, with

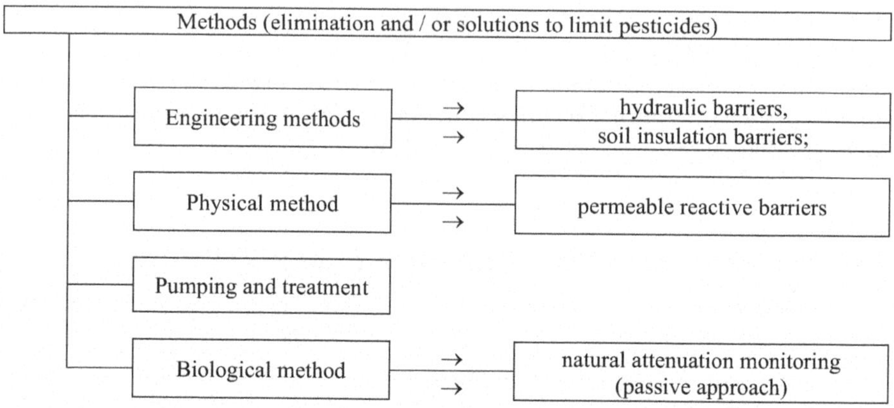

Fig. 4.1 Methods to reduce pollution with pesticide

the two isomers (op′-DDD and pp′-DDD), and dichlorodiphenyldichloro-ethene (DDE). Practically, by summing up these compounds the total concentration of HCH is obtained (De Rop et al. 2019).

4.3.2 European Policies and Instruments for the Protection of Human Health and the Environment Against Dangerous Chemicals and Mixtures

The European Union has developed policies and tools to protect human health and the environment against hazardous chemicals and mixtures such as: (European Commission 2012): REACH Regulation (EC 1907/2006) on the Registration, Evaluation, Authorization, and Restriction of Chemicals is a key element of the European Union's set of rules to make the use of chemicals as safe as possible. The REACH regulation addresses long-term exposure to chemicals and the implications for human health and the quality of the environment. The REACH Regulation creates a single regulatory system dealing with industrial chemicals and seeks to eliminate the information deficit related to them by placing accountability on chemical manufacturers who are required to provide safety information on their substances and to manage risks from chemicals.

The philosophy of the REACH regulation is that no chemical, in any form, should be placed on the market without proper documentation. Each manufacturer and importer of chemicals that distributes more than 1 ton per year is required to register them with the European Chemicals Agency (ECHA) and to provide information on their properties and uses and safe methods of handling them. The REACH Regulation also calls for the progressive replacement of the most dangerous chemicals with appropriate alternatives. The first registration deadline ended on November 30, 2010, with approximately 5000 chemicals registered with ECHA. This marks an important step towards safe chemicals management. CLP Regulation (EC 1272/2008) on Classification, Labelling, and Packaging of Chemicals that aligned the existing EU classification system to the United Nations international system, called the Globally Harmonized System of Classification and Labelling of Chemicals (UNGHS). By classifying, labeling, and packaging chemicals in accordance with the CLP Regulation, it is ensured that chemical users and manipulators are aware of the dangers of their use and handling (Jess et al. 2014).

All manufacturers and importers of hazardous chemicals have been compelled to label chemicals in accordance with the CLP Regulation and to notify the European Chemicals Agency (ECHA) as of 1 December 2010. Thus, more than 100,000 chemicals have been notified to ECHA, which allowed ECHA to make an inventory of chemical classification and labeling and to provide chemical users and manipulators with the information they need to limit the risks and to select safer chemicals.

The human biomonitoring HBM system involves the collection and analysis of human tissue or body fluids to identify the presence and levels of chemicals in the human body. Chemicals are taken directly or indirectly from the environment. The HBM system is an important tool for assessing human exposure to environmental substances. The EU Scientific Research Commission has funded a research program to develop a human biomonitoring framework at EU level called DEMOCOPHES.

It was tested the feasibility of a coherent approach to human biomonitoring in Europe by identifying biomarkers of mercury, cadmium, phthalates, and tobacco smoke in the human environment and in human urine. The project will collect biological samples and information from 120 parent–child pairs from each participating country within the EU (Lillebø et al. 2019).

4.4 Methodology for Managing Potentially Contaminated Sites

The historical contaminated area is: "A contiguous site (land and/or aquifer) where anthropogenic activities have determined the presence of pollutants in concentrations which are presenting and/or could be presenting an immediate risk to both the existing site and the surrounding areas or on long term for the health of the population and the environment." On a contaminated site there may be one or more sources of contamination. The potentially contaminated historical site is: "A contiguous site (land and/or aquifer) on which historical and/or present activities could have a significant impact on the health of the population or the environment for both the existing site and neighboring areas (Storck et al. 2017).

This definition focuses on contamination caused by past industrial practices that led to contamination as a result of inappropriate storage/handling/use of chemicals, soil, and/or aquifer. The management system for contaminated sites includes 4 phases: identification and registration; preliminary assessment; detailed assessment; and remediation. A potentially contaminated site may be classified as a "suspect"

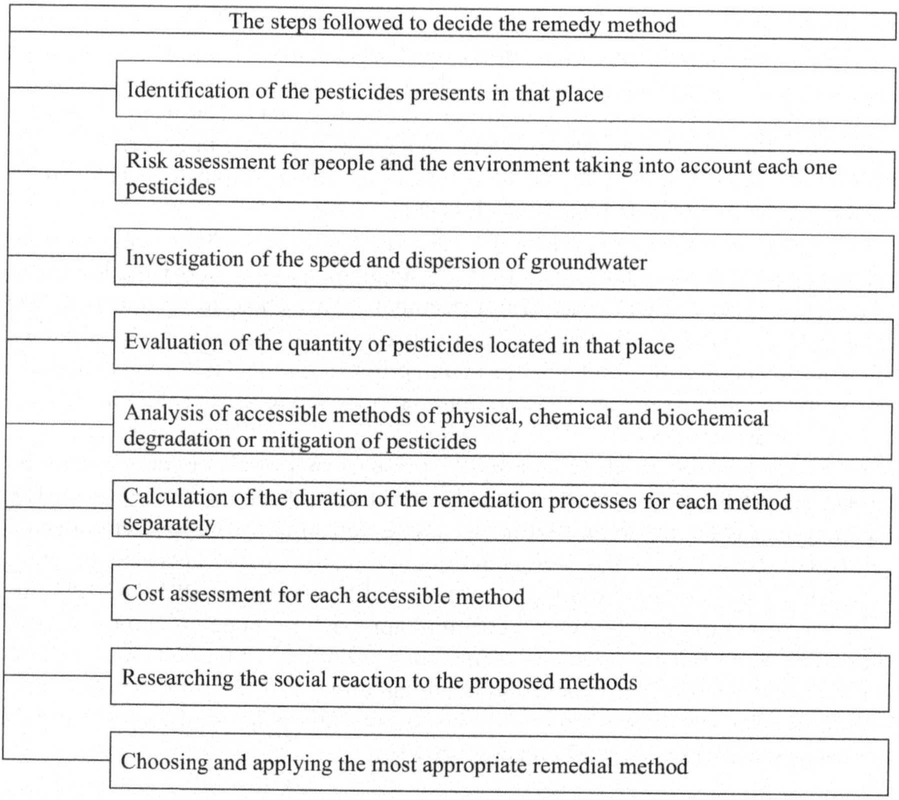

Fig. 4.2 The steps that need to be followed to decide which remediation method can be used

site following an analysis based on the following elements: assessment of the site based on owner's information; regular or irregular site inspections by local authorities and control bodies; analyzing the effects of natural disasters on the site; the owner's notification to the regulatory bodies. On all "potentially contaminated" sites, a preliminary assessment should be carried out. It is recommended that before deciding what remedial method can be used to go through the phases of Fig. 4.2.

The purpose of the preliminary assessment is to determine whether the degree of contamination is real or not and whether a detailed assessment is required on site. For this purpose, laboratory analyses will be performed for samples taken from sites. The corresponding indicator parameters for each industry must be listed. There are three possible outcomes of the preliminary site evaluation, namely: (a) No further investigation is required—the site will be kept in the inventory; (b) Additional investigations are required—detailed site assessment will be carried out; (c) Remedial—cleaning and remediation of the contaminated site will be initiated immediately.

The detailed assessment of the contaminated site requires the following steps: site characterization and contamination: on this occasion, information and location data will be collected; the Conceptual Placement Model (CPA) is being developed; all possible links are identified:

- the source–transfer–receiver path; samples of the surface and of the soil profile will be taken and laboratory contaminants will be determined by laboratory analysis;
- samples will be taken from the groundwater to know if it is contaminated or not; generic risk assessment; the measured contaminant concentration compares with the soil quality standard to define the transfer paths, and the generic scenario will include inhalation of fugitive particles, volatile inhalation, ingestion of groundwater, dermal absorption; if the values in the standard are exceeded, risk assessment will begin for all potential open transfer routes; Also, all the transfer paths will be identified;
- detailed risk assessment; information about site specificity is collected; the specific risk levels are calculated; if the level of risk is higher than the acceptable risk level, the site is listed as contaminated, requiring it to be rectified, the site cleaning level is established, all the alternatives to cleaning technologies will be identified and the technology selected appropriate to meet the remediation targets (Breysse et al. 2018).

During the remediation phase, the designing and execution of remediation works is carried out. The remedial effects will be monitored for both execution and post-execution.

Technologies for soil remediation of POPs that are more frequently applied are bioremediation technologies (in situ technologies presented above) as well as thermal desorption technologies which will be presented below:

- Thermal desorption is a thermal method that works on the principle of polluting soils so that pollutants can evaporate, so they can easily move through the soil to the area of a well system that collects and evacuates them. Thermal desorption can be applied in situ or ex situ.
- Thermal in situ desorption can be done by: steam injection—steam injection is injected into the soil by means of wells dug in the polluted area; the steam heats the area and mobilizes, evaporates and/or destroys the pollutants, the water and the gases produced being collected in special wells from where they are directed to purification plants; hot air injection—a procedure similar to steam injection except that hot air is injected through wells instead of steam, hot air warms the soil causing evaporation of pollutants; hot water injection—a procedure similar to the two above-mentioned processes, introducing hot water this time instead of steam or hot air; heating by electric resistance—a process by which an electric current is applied to the ground through a network of steel electrodes; the heat generated by the electrical resistance of the soil at the passage of the electric current vaporizes the underground water and the water from the soil, creating the conditions of the evaporation of the pollutants; radio waves heating—the process of placing an antenna that emits radio waves in a well; radio waves warm the soil causing the evaporation of pollutants. In some cases, thermal desorption is also

ex situ, which is a more expensive process but where the treatment process is much better controlled.

- Typical ex situ thermal desorption systems work in several successive phases, namely: pretreatment, desorption, and post-treatment of solid material and resulting gases; Pretreatment involves processes such as sorting, dehydration, neutralization, and blending; the desorption is done in thermal installations called desorbers, whose classification is usually made according to the heating system used;—post-treatment of waste gases—depends on equipment-specific factors and may include: combustion at high temperatures (over 1400 °C) followed by treatment and disposal, combustion at moderate temperatures (200–400 °C) using catalysts or gas purification and elimination burned (Butler 2018).

4.5 Monitoring of Soil Quality

The Soil Quality Monitoring System is characterized by four basic elements: spatial distribution; density of the observation network; set of indicators; and periodicity of the determinations. These basic elements of the monitoring system are tracked on three levels of detail:

- level I—consists of carrying out a minimum of investigations at all points of a fixed network, to identify the areas with degraded soils, in various stages and processes, periodically following their evolution through a set of mandatory indicators;
- level II—consists of the detailed investigation in some points of the level I network and in other representative additional points, in order to identify the causes of soil degradation processes (conducting intensive studies);
- level III—consists of sufficiently thorough investigations to verify the hypotheses, from detailed analyses of soil quality alteration processes, elaboration of forecasts of the evolution of soil quality alteration processes, and recommending possible remedial measures (Gorito et al. 2017).

The first level is characterized by the following elements:

- at the macroscale, the spatial distribution of the soil sampling points (sites) at the intersections of a fixed observation network, covering the entire country, with a density of 16 × 16 km (in accordance with the recommendations of the document "Convention on Long Transboundary Air Pollution"), resulting in a total of 942 sites, of which 670 sites in areas with agricultural soils and 272 sites in areas with forest soils;
- at the microscope, the choice and location of an observation point is made inside a square with a 400 m side, which has the center (the point of intersection of the diagonals) in the nodes of the observation network;
- from each observation point a set of samples will be collected to determine the basic indicators that contains: an average sample in the disturbed structure,

composed of 20–25 individual samples from the first 10 cm of soil depth, respectively, 3 individual samples for each soil horizon and 4 samples in the natural structure on the horizon;

- the periodicity of the determinations is of 4–10 years, for the observation points without special problems, and of 1–2 years for the observation points with problems (in which processes of degradation of the quality of the soil take place, as for example: pollution) (Panizzi et al. 2017).

4.6 Control of Residues in Plants and Plant Products, as a Result of Metabolism/Migration of Hazardous Chemical Substances and Mixtures in the Soil

Food security—refers to food availability and easy access to people. The World Health Organization defines the following criteria of food security: availability—sufficient quantity; access—to have sufficient resources (economic and physical) to have the foods corresponding to the individual diet; use—to have sufficient information about nutrition, water, and proper hygiene; stability—ensuring the first 3 conditions. Food safety—ensuring the handling (during the technological process) in hygienic conditions of food ingredients and food from the ground to the consumer so that they are not a source of disease of the body, moreover, to prevent the occurrence of diseases. The food has the capacity to accumulate polluting chemicals from the environment (accidental contamination) to which an additional contribution of chemical compounds from agricultural and/or food technologies (deliberate contamination) is accumulated. The toxicity of chemical compounds depends on a number of factors such as periodicity and amount of chemical contaminant carried; synergism or antagonism between chemical pollutant and food; attenuation or enhancement of the toxicity of the pollutant depending on the metabolic changes of the animal or plant tissue; alteration—modification of sensory and (sometimes) nutritional properties, possibly acquiring harmful properties; impurification—foreign compounds of the nature of the food (biological, chemical, physical contaminants); degradation—loss of nutritional (and/or organoleptic) properties that depend on: frequency of processing, non-compliance with the recipe, heat treatments, exposure to oxygen action, reheating, keeping of hot foods, presence of heavy metals, washing of fragmented foods, keeping in water, and removal of aqueous solution or brine; sanitation—transformation into a product harmful to the body; this does not imply other changes (alteration, degradation, impurification) (Handford et al. 2015).

Contamination of food with chemical substances or mixtures can come from the pollution of environmental factors: soil, water, atmosphere through the trophic chain soil–plant–animal–man, as well as from agricultural and food technologies (Singh et al. 2018).

According to the Codex Alimentarius Commission, the main types of contaminants and pollutants of food are classified into the following classes: pesticides: insecticides, herbicides, defoliants, rodenticides, fungicides, etc.; industrial chemicals: dioxins, polychlorinated biphenyls, chemical compounds from inappropriate packaging materials, etc.; heavy metals (Singh et al. 2018) and other elements; microbiological contaminants (Dar et al. 2020). In addition to these classes of contaminants and pollutants, some antinutritional toxic compounds can be added which can be formed under certain conditions during the processing or storage of foods. Also, parts of the contaminant's category are animal parasites. All these classes of contaminants and pollutants are under the influence of the legislation and the sanitary control. Pesticides are a special class of chemical contaminants, which, in modern agriculture, are widely used (Bhat et al. 2018). The control of pesticide residues and other hazardous chemicals, especially of organic nature from plants and plant products, derived from their migration/metabolization in soil, is based on modern analysis techniques that allow increasingly easy detection of the molecules found in infinitesimal quantities, namely chromatographic methods, such as gas chromatographic (GC) and liquid chromatographic (LC) methods, which will be described below (Adams et al. 2017).

Over 800 active substances are formulated in thousands of pesticides, these substances forming over 100 chemical classes of which the most important would be: benzoylureate, carbamate, organophosphorus compounds (OP), organochlorine (OC), pyrethroids (PYR), sulfonylureas, triazines, dithiocarbamates, azoles, phenoxy acids. Their physical and chemical properties differ greatly, which makes the determination of the residues of these chemicals quite difficult in any matrix and causes problems in the development of a universal method of analytical determination of residues, which would be the most desirable thing. In order to detect pesticide residues from plants and plant products, a very important first step is to prepare the samples for analysis. Most sample preparation methods for gas chromatographic (GC) and liquid chromatographic (LC) determination include the following steps:

- Homogenizing the sample to obtain a uniform matrix.
- Solvent based extraction of pesticide residues.
- Purification—eliminating interference between matrix components.
- Elution and/or fractionation of the extracted analytes.
- Concentration of the eluent and reconstitution in a solvent that is compatible with GC or LC conditions.
- Introduction of the solution containing pesticides in GC or LC. Methods of analysis to determine the content of pesticide residues in plants and plant products can be targeted or non-targeted (Núñez et al. 2012).

An example of targeted analysis is the inspection of MAL (maximum allowed limit) in vegetables, fruits, and foods; the relevant elements are fixed to be determined according to the definitions given by the MAL regulations. The most commonly used methods of analysis for the detection of pesticides and other chemical compounds, especially of organic nature, are gas chromatography and liquid chromatography. In gas chromatography the eluent (a chemically inert gas, e.g., nitrogen, helium, hydrogen) passes through the sample input device (where the sample is volatilized), takes it, and enters it into the chromatographic column, in which the separation process takes place. Due to the interactions between the sample components, with the stationary phase and/or the mobile phase, they migrate through the column at different speeds, due to the differences between the distribution coefficients of the 2 phases. The difference of the component migration speeds is significant so that they are separated at the exit of the column, from where they are taken by the eluent to the detector.

The gas chromatograph can be coupled with: flame ionization detector (FID), electron capture detector (GC–ECD), GC–MS mass spectrometer (quadrupole triple, TOF—"time of flight"—flight time or ion trap), or with high resolution mass spectrometer (HRGC–HRMS). Unlike gas chromatography, in which the mobile phase has the role only of transporting the components through the column, in the case of liquid chromatography the mobile phase, which is a liquid, participates directly in the separation process, through various interactions (adsorption, distribution, exchange, etc.). Also, the liquid mobile phase participates and directly influences the selectivity of the separation process. In the determination of pesticide residues, liquid chromatography (LC) can be coupled with UV–VIS detectors (HPLC UV–VIS) or coupled with a mass spectrometer (with triple quadrupole QQQ, ion trap, or TOF—flight time). In our country the most complete works in the field of pesticide residue analysis are by GC–MS and LC–MS techniques (Ferrer et al. 2007).

Another class of hazardous chemical residues from plants and plant products, derived from their migration/metabolism in soil, is the heavy metals that can contaminate food in the form of combinations of arsenic, cadmium, copper, lead, mercury, zinc, tin, and even of iron (Mehmood et al. 2019). Modern analytical techniques used especially at present that allow the detection of heavy metal residues in plants and plant products are flame atomic absorption spectrometry (AAS) and inductive coupled mass spectrometry (ICP–MS). Atomic absorption spectroscopy (AAS) can be performed in graphite furnace or flame. The principle of atomic spectroscopy is based on the property of the atom to emit or absorb electromagnetic radiation specific to a particular element under certain physical conditions. Thus the elements to be determined from a sample are released from their compounds by an energy absorption making them available as free particles. Since elementary mass spectrometry is based on the fact that electrically charged particles can be separated and detected in a mass spectrometer by energy absorption, the electromagnetic radiation that crosses the sample is provided by a special source called "hollow cathode lamp." The cathode of this lamp is made from the element to be determined (it follows that for each chemical element a specific lamp is therefore required).

Thus the elements to be analyzed are transformed into a free atomic state in an atomization device by thermal energy absorption. The free atoms have the ability to absorb the radiation specific to the analyzed element. Thus, depending on the concentration of the analyzed element, part of the radiation intensity of the hollow cathode lamp is absorbed by the free atoms formed from the sample. A detector consisting of two photomultipliers measures both the intensity of the unintended radiation and the radiation after leaving the atomizing device, the concentration of the element analyzed in the sample being obtained on the basis of the difference between the two intensities.

The preparation of the samples for analysis consists of calcining the samples at 450 °C with the gradual increase of the temperature, dissolving the ash in hydrochloric acid and evaporating to dryness the obtained solution, redissolving the final residue in nitric acid 1:6, and determining the metal content by atomic absorption spectrometry. Inductive coupled plasma mass spectrometry (ICP–MS) has now become one of the most powerful analytical techniques, highly sensitive, for analyzing a wide range of metals and some non-metals, at very low concentrations, at the level of 1–10 ppt. Inductive coupled plasma (ICP) was developed in 1960 by Reed and was used for the first time by Greenfield Group and collaborators for spectrochemical analysis. The first papers on the inductively coupled plasma mass spectrometer were published in the 1980s, and the first commercially available ICP–MS instrument being introduced in 1983. Since then, many improvements have been made for each generation of ICP–MS instruments. The ICP–MS method is based on the combination of inductively coupled plasma, as the ionization method, with mass spectrometry, as the ion separation and detection method. Combined with various chromatographic separation techniques, the ICP–MS method is a powerful and versatile method for analyzing elemental species, including isotopic species. Besides classical soil, water, plant products, food samples, currently a wide range of biological samples, both solids and liquids can be analyzed by ICP–MS (for example: blood, urine, plasma, serum, interstitial fluids, internal organs, tooth, hair, bones, and even cells) (Gui et al. 2016).

The presence of pesticides in food of plant origin is the result of their contamination from the atmosphere, water, and soil. In the atmosphere pesticides are present due to the fact that they are administered by air, as well as due to the volatilization of pesticides on the soil surface and plants. The binding of pesticides in soil is determined by Van der Waals forces, hydrogen bonds, hydrophobic bonds, chemical absorption, and ion exchange.

Some of the pesticides fall to the ground where they are solubilized by the rainwater and are entrained on the surface or in the depth of the soil where they are biodegraded, entrained in the groundwater, or absorbed by the lattice system of plants. The degree of soil migration in plants depends on the initial amount of pesticides, their properties, and the type of plants. In the body of animals, pesticides can enter through the respiratory tract, the skin, and the oral route.

When using pesticide combinations there are situations when the toxicity of the combination increases 5–10 times compared to the toxicity of the components. Among the most dangerous pesticides from the point of view of food contamination, organochlorine pesticides selectively accumulate in the animal body, the decreasing order being: mesenteric fatty tissue, depositional fat tissue, perirenal fat tissue, and then muscle tissue.

In ruminants, a greater accumulation is observed in pigs. Concentrated feed (cereals, meats, beets) is responsible for a significant accumulation of pesticides. It should be mentioned that the reduction of the amount of organochlorine in the animal body is done only after 3–4 weeks (Tran et al. 2019).

4.7 Regulation on Biocidal Products

Companies must request the approval of an active substance by submitting a file to ECHA. The active substances must be approved before the authorization of a biocidal product containing them. The active substances are first evaluated by the competent authority of the evaluating Member State, and the results of the evaluations are transmitted to the Committee for Biocidal Products of ECHA, which draws up an opinion within 270 days.

The opinion serves as the basis for the European Commission to adopt the decision regarding the approval of the product. The approval of an active substance shall be granted for a defined period, not exceeding 10 years, and may be renewed.

Biocidal Products Regulation introduces formal exclusion and replacement criteria, which is applied to the evaluation of active substances. After approval of an active substance, companies wishing to introduce biocidal products on the market of a Member State must apply for product authorization (Wynendaele et al. 2018).

4.7.1 Regulation Regarding the Prior Consent Procedure in the Knowledge of the Case

The Regulation on Prior Informed Consent (PIC Regulation) regulates the import and export of certain hazardous chemicals and imposes obligations on companies that intend to export such products to countries outside the EU. The PIC regulation contains three sets of provisions regarding the export of chemicals: The chemicals mentioned in the list in Annex I of the regulation are subject to the export notification procedure and the explicit consent requirement. This list is updated regularly as a result of regulatory measures under EU law, as well as evolutions under the Rotterdam Convention. The chemicals mentioned in the list in Annex V of the regulation are prohibited for export. All chemicals that are subject to export must comply with the rules on packaging and labeling in accordance with CLP (Classification, Labelling, and Packaging) Regulation (Pouokam et al. 2017).

4.7.2 The Effects of Pesticides/Dangerous Substances on Humans

In assessing the probability of survival during an accident occurring in a chemical installation, it is important to consider the following factors: information prior to the accident, information on the progress of the accident, staff reaction time, emergency procedures implemented, evacuation time and distance until at a safe place, the type of danger (toxic gas, thermal radiation, explosion, etc.), protection and mitigation of effects (shielding or reflection), the level of damage as a function of time, total exposure time. From the point of view of toxicity, the most widespread indicator is LC_{50}, respectively, the pollutant concentration which is lethal to 50% of the population after a certain period of time (usually 14 days). LC_{50} is specific for each species (mice, fish) and is determined by standard toxicity assays over a period of time (often under 24 h).

However, because ecosystems include species with different susceptibilities to various chemicals, it is necessary to test more than one species, and the tests to be repeated periodically.

In general, the LC_{50} indicator is specified in the hazardous substance sheet. As a rule, LC_{50} is expressed in mg substance/L air or ppm (1 ppm = 1 cm^3 of gas contained in 1 m^3 of air, or can be determined from the expression in mg/m^3 if gas density is known). In addition to toxicity, persistence and diffusion are considered, both for pure substances and for solutions or mixtures. This has resulted in a broad set of properties and risk indicators of substances, as well as multiple classification systems.

For example, MPC represents the maximum permissible concentration of a substance that appears in the form of gaseous, vapor, or suspended matter, which is not harmful to the health of the humans and does not cause any inconvenience, even in the case of prolonged and repeated exposure. The uncontrolled emission of pollutant represents the release into the environment of a toxic substance for humans and the environment as a result of an accident.

From the place from which it is released, a place called the source of pollution, the substance spreads in the environment to great distances, gradually being taken over by it. In order to estimate the effects of a pollutant, the following parameters must be followed: the concentration level at different source distances, respectively, the pollutant concentration at a specified distance relative to the source (concentration field)—the critical area in which the concentration exceeds an allowed threshold (action distance of the toxic).

In order to estimate the effects of a risk agent it is necessary to have the possibility to quantify the exposure, from the point of view of the intensity, the duration of exposure, and the consequences (Bravo et al. 2011).

This is usually done by estimating the received dose and comparing it with the statistical processing of the experimental data. Vulnerability criteria can be established to determine the exposure dose levels that have specific consequences. Criteria provide the threshold beyond which the protection is needed to prevent impairment of the functions necessary for an individual to be safe.

There are two approaches to determining the effects of the dose received: using the probit functions and determining the harmful dose (applied in case of toxic or thermal risks).

Probit functions take into account the variation of tolerance to harmful effects for an exposed population.

The death rate of humans exposed to harmful agents over a period of time can be calculated using the probit functions, which usually take the form:

$$Y = k_1 + k_2 \ln(V)$$

in which, Y is probit (with a range of values between 2.67 and 8.09 representing fatality between 1 and 99.9%, a measure of the percentage of vulnerable resources that can suffer damage. The probability of death can thus be determined by evaluating Y on a transformation diagram probit that of Finney (1971). k_1 and k_2 are constants, V—is the product of the intensity or concentration of the toxic compound at an exponent n with the exposure time (expressed in seconds or minutes).

The concentration can be expressed in parts per million (ppm) or milligrams per cubic meter (mg/m^3). Probiotics can be obtained for almost any dangerous agent. Probit analysis is an approximate methodology, but it allows quantifying the consequences of exposure.

It has to be taken into account. However, the estimated probability of death can vary significantly for the same toxic compound, depending on the choice of the probit equation. Also, care must be taken in choosing the units of measure suitable for the equation used (Oribhabor et al. 2016).

4.7.3 Determination of the Harmful Dose

The specified level of toxicity (SLOT) or significant likelihood of death (SLOD) is alternative approaches for estimating the level of fatality upon exposure to a toxic agent.

The SLOT approach involves the use of available toxicity data, which are then extrapolated for human use. The SLOD used in risk analysis estimates the dose needed to produce a significant probability of death. The consequences of exposure (inhalation of toxic substances, exposure to explosion pressure, or exposure to thermal radiation) are usually assessed from existing information, preferably data obtained from previous accidents to which humans have been exposed or from animal experiments.

Uncertainties in translating the data obtained from animals to data relevant to humans are large and, therefore, it is necessary to include "safety factors" in modeling. In general, SLOT is considered to be equivalent to LC1–5 derived from animal experiments. The SLOT and SLOD approaches extrapolate toxicity data to determine the hazardous toxic load (A), which causes a specific level of harm for a given dose. These values are calculated as the product of exposure and time and usually have the form:

$$C^{n \cdot t} = A$$

where t = the exposure time, in minutes, C = the concentration of the substance, in ppm, n = the exponent.

The exponent "n" is derived from the extrapolation of the toxicity data used. SLOT is usually defined as the dose that leads to the death of the vulnerable population and to severe suffering that requires medical attention for the rest of the exposed population and, as such, is equivalent to LD1 or LD1–5. SLOD is defined as the dose that determines the death of the exposed population in proportion of 50%, being therefore equivalent to DL50.

Knowing the SLOT or SLOD values and the exponent, one can calculate the concentrations needed to produce these effects, depending on different exposure time values.

Basically, the dynamic balance in an ecosystem/agrosystem is based on the complementarity of the genes of organisms placed in different ecological niches. Over time there is a re-adaptation of the genetic material of the harmful organisms but also of the plants attacked in regard to the virulence/avirulence gene system—genes of resistance or sensitivity, which is why the process of improvement in this direction is a continuous one.

In general, in plant protection, harmful organisms are not fully combated, an operation that would be exceptionally expensive and difficult to perform. However, for some harmful organisms whose development is rapid, in large areas producing adequate epiphytes or epizootics, the phenomenon of eradication can be practiced. To do this, we call on the methods and means of the operative sector of plant protection.

The close monitoring of the outbreaks in which such organisms appear and the prevention of the widespread of their propagates are elements of phytosanitary quarantine.

They are supported by an adequate legal framework and are within the competence of the government administration. Phytosanitary quarantine belongs to the operating sector of governmental structures but the cultural hygiene elements associated with preventing the spread of quarantine damage agents propagate are into the integrated protection structure (Eason and O'Halloran 2002).

The prediction of the onset of the attack of a certain harmful organism or groups of harmful organisms and the subsequent issuance of warnings for carrying out a plant protection intervention are of great importance for the correct conduct in time and space of the combat activities. It is good to know since harmful organisms exist in an area for agricultural crops in vegetation as well as potential levels of attack to know what spectrum of means of control should be used, what system of machines should be used for their application and optimal times of intervention so that the damages are as small as possible, the interventions as little as possible, the economic effects are maximum and the ecological ones are minimal.

4.8 Conclusions and Recommendations

Ideally, the pesticides used will run out once the purpose is achieved. But besides the undeniable advantages of pesticides, they also have a number of disadvantages. Being toxic to a life form, pesticides pose a risk of harm to humans, pets, hunting, birds, and microflora.

At the same time, the use of natural organic fertilizers and other organic residues contributes not only to raising soil fertility and to protecting the environment.

However, if we really want to reduce the risk posed by the growing diversity of synthetic chemicals released into the environment, we need to rethink some of the basics of industrial development.

This would allow the removal or at least the considerable diminution of one of the major causes that maintains and accentuates the processes of deterioration of the ecological system that produce a series of goods and services necessary for the development of the human society.

The use of pesticides to diminish their impact on the environment implies:

- the use of substances with lower toxicity and remanence, at which the dose of the active substance and the applied quantity per hectare will be optimal, to destroy pests but to affect less the useful species;
- decrease the frequency of pesticide application by using certain products or combinations of products, acting on several categories of pests;
- establishing the mode and surface of application, by editing on large surfaces and applying as much as possible to localized treatments;
- choosing when to use pesticides, which should take into account not only the most sensitive stage of the pest but also the period when the useful species are least exposed;
- increasing the weight of mechanical and physical methods in combating pests;
- the proper application of agrofitotechnical methodologies, emphasizing the components that disadvantage the pests and diminish their effects;
- maintaining or creating portions of land on the edge of crops or roads, on the springs of irrigation channels, in the woods, in parks, and nature reserves, in which the associations of spontaneous vegetation offer to the useful species places of wintering and of refuge during unfavorable time and also additional food;
- cultivation of honey plants, to attract the adults of the parasites, around the crops;
- spraying some nutrients on plants in order to provide additional food for parasites and predators;
- cultivation of nectariferous plants, pollinifers, to attract the adults of the parasites and predators, near the crops;
- recourse to granulated, encapsulated pesticides, with reduced remnant applied strictly localized only if there is no other solution, to minimize their negative effects.

Awareness of the danger posed by pesticides has led to encouraging consumers to develop organic farming. Organic farming does not use pesticide-like substances and the use of mineral fertilizers is much more rational.

This encourages crop diversification and extensive animal husbandry. By increasing the diversity of cultures, the possibility of spreading the attack and the impact of pests compared to the one recorded in monocultures is reduced.

Also, although the production of a particular crop may decrease, the total agricultural production obtained in organic, multifunctional farms is as high as in the intensive ones and the incomes can be even higher.

References

Adams S, Guest J, Dickinson M, Fussell RJ, Beck J, Schoutsen F (2017) Development and validation of ion chromatography-tandem mass spectrometry-based method for the multiresidue determination of polar ionic pesticides in food. J Agric Food Chem 65(34):7294–7304

Alho CJ (2008) Biodiversity of the Pantanal: response to seasonal flooding regime and to environmental degradation. Braz J Biol 68(4):957–966

Asano R, Kobayashi S, Sonobe K, Shime-Hattori A, Okazaki K, Ohtomo R (2014) Plant-available inorganic nutrient levels are increased in rice-derived distillery effluents inoculated with microbes. J Appl Microbiol 117(5):1412–1421

Atma W, Larouci M, Meddah B, Benabdeli K, Sonnet P (2017) Evaluation of the phytoremediation potential of Arundo donax L. for nickel-contaminated soil. Int J Phytoremediation 19(4):377–386

Bhat RA, Beigh BA, Mir SA, Dar SA, Dervash MA, Rashid A, Lone R (2018) Biopesticide techniques to remediate pesticides in polluted ecosystems. In: Wani KA, Mamta (eds) Handbook of research on the adverse effects of pesticide pollution in aquatic ecosystems. IGI Global, Hershey, pp 387–407

Bravo V, Rodríguez T, van Wendel de Joode B, Canto N, Calderón GR, Turcios M, Menéndez LA, Mejía W, Tatis A, Abrego FZ, de la Cruz E, Wesseling C (2011) Monitoring pesticide use and associated health hazards in Central America. Int J Occup Environ Health 17(3):258–269

Breysse N, Vial G, Pattingre L, Ossendorp BC, Mahieu K, Reich H, Rietveld A, Sieke C, van der Velde-Koerts T, Sarda X (2018) Impact of a proposed revision of the IESTI equation on the acute risk assessment conducted when setting maximum residue levels (MRLs) in the European Union (EU): a case study. J Environ Sci Health B 53(6):352–365

Brinch A, Hansen SF, Hartmann NB, Baun A (2016) EU regulation of nanobiocides: challenges in implementing the biocidal product regulation (BPR). Nanomaterials 6(2):E33. https://doi.org/10.3390/nano6020033

Butler D (2018) EU expected to vote on pesticide ban after major scientific review. Nature 555(7695):150–151

Dar GH, Bhat RA, Kamili AN, Chishti MZ, Qadri H, Dar R, Mehmood MA (2020) Correlation between pollution trends of fresh water bodies and bacterial disease of fish fauna. In: Qadri H, Bhat RA, Dar GH, Mehmood MA (eds) Freshwater pollution dynamics and remediation. Springer, Singapore, pp 51–68

De Rop J, Senaeve D, Jacxsens L, Houbraken M, van Klaveren J, Spanoghe P (2019) Cumulative probabilistic risk assessment of triazole pesticides in Belgium from 2011-2014. Food Addit Contam Part A Chem Anal Control Expo Risk Assess 36(6):911–921

Eason C, O'Halloran K (2002) Biomarkers in toxicology versus ecological risk assessment. Toxicology 181–182:517–521

Ferrer I, Thurman EM, Zweigenbaum JA (2007) Screening and confirmation of 100 pesticides in food samples by liquid chromatography/tandem mass spectrometry. Rapid Commun Mass Spectrom 21(23):3869–3882

Gorito AM, Ribeiro AR, Almeida CMR, Silva AMT (2017) A review on the application of constructed wetlands for the removal of priority substances and contaminants of emerging concern listed in recently launched EU legislation. Environ Pollut 227:428–443

Gui W, Tian C, Sun Q, Li S, Zhang W, Tang J, Zhu G (2016) Simultaneous determination of organotin pesticides by HPLC-ICP-MS and their sorption, desorption, and transformation in freshwater sediments. Water Res 95:185–194

Handford CE, Elliott CT, Campbell K (2015) A review of the global pesticide legislation and the scale of challenge in reaching the global harmonization of food safety standards. Integr Environ Assess Manag 11(4):525–536

Hennebert P, Filella M (2018) WEEE plastic sorting for bromine essential to enforce EU regulation. Waste Manag 71:390–399

James TK, Ghanizadeh H, Harrington KC, Bolan NS (2019) Effect on herbicide adsorption of organic forestry waste products used for soil remediation. J Environ Sci Health B 54(5):407–415

Jess S, Kildea S, Moody A, Rennick G, Murchie AK, Cooke LR (2014) European Union policy on pesticides: implications for agriculture in Ireland. Pest Manag Sci 70(11):1646–1654

Lillebø AI, Teixeira H, Morgado M, Martínez-López J, Marhubi A, Delacámara G, Strosser P, Nogueira AJA (2019) Ecosystem-based management planning across aquatic realms at the Ria de Aveiro Natura 2000 territory. Sci Total Environ 650(Pt 2):1898–1912

Masbou J, Drouin G, Payraudeau S, Imfeld G (2018) Carbon and nitrogen stable isotope fractionation during abiotic hydrolysis of pesticides. Chemosphere 213:368–376

Mech A, Rasmussen K, Jantunen P, Aicher L, Alessandrelli M, Bernauer U, Bleeker EAJ, Bouillard J, Di Prospero Fanghella P, Draisci R, Dusinska M, Encheva G, Flament G, Haase A, Handzhiyski Y, Herzberg F, Huwyler J, Jacobsen NR, Jeliazkov V, Jeliazkova N, Nymark P, Grafström R, Oomen AG, Polci ML, Riebeling C, Sandström J, Shivachev B, Stateva S, Tanasescu S, Tsekovska R, Wallin H, Wilks MF, Zellmer S, Apostolova MD (2019) Insights into possibilities for grouping and read-across for nanomaterials in EU chemicals legislation. Nanotoxicology 13(1):119–141

Mehmood MA, Qadri H, Bhat RA, Rashid A, Ganie SA, Dar GH, Shafiq-ur-Rehman (2019) Heavy metal contamination in two commercial fish species of a trans-Himalayan freshwater ecosystem. Environ Monit Assess 191:104. https://doi.org/10.1007/s10661-019-7245-2

Munns WR Jr, Poulsen V, Gala WR, Marshall SJ, Rea AW, Sorensen MT, von Stackelberg K (2017) Ecosystem services in risk assessment and management. Integr Environ Assess Manag 13(1):62–73

Mushtaq N, Bhat RA, Dervash MA, Qadri H, Dar GH (2018) Biopesticides: the key component to remediate pesticide contamination in an ecosystem. In: Environmental contamination and remediation. Cambridge Scholars Publishing, Cambridge, pp 152–178

Núñez O, Gallart-Ayala H, Ferrer I, Moyano E, Galceran MT (2012) Strategies for the multiresidue analysis of 100 pesticides by liquid chromatography-triple quadrupole mass spectrometry. J Chromatogr A 1249:164–180

Oribhabor BJ, Ikeogu GC (2016) Acute toxicity of the Pesticides, Dichlorvos and Lindane against the African air-breathing catfish, *Heterobranchus longifilis*, Valenciennes, 1840 (Siluriformes: Clariidae). Recent Pat Biotechnol 10(3):272–278

Panizzi S, Suciu NA, Trevisan M (2017) Combined ecotoxicological risk assessment in the frame of European authorization of pesticides. Sci Total Environ 580:136–146

Pouokam GB, Lemnyuy Album W, Ndikontar AS, Sidatt MEH (2017) A pilot study in Cameroon to understand safe uses of pesticides in agriculture, risk factors for farmers' exposure and management of accidental cases. Toxics 5(4):E30. https://doi.org/10.3390/toxics5040030

Rasheed T, Bilal M, Nabeel F, Adeel M, Iqbal HMN (2019) Environmentally-related contaminants of high concern: potential sources and analytical modalities for detection, quantification, and treatment. Environ Int 122:52–66

Rashid A, Bhat RA, Qadri H, Mehmood MA (2019) Environmental and socioeconomic factors induced blood lead in children: an investigation from Kashmir, India. Environ Monit Assess 191(2):76. https://doi.org/10.1007/s10661-019-7220-y

Shi B, Jiang J, Liu R, Khan AU, Wang P (2017) Engineering risk assessment for emergency disposal projects of sudden water pollution incidents. Environ Sci Pollut Res Int 24(17):14819–14833

Shi B, Jiang J, Sivakumar B, Zheng Y, Wang P (2018) Quantitative design of emergency monitoring network for river chemical spills based on discrete entropy theory. Water Res 134:140–152

Singh DV, Bhat JIA, Bhat RA, Dervash MA, Ganei SA (2018) Vehicular stress a cause for heavy metal accumulation and change in physico-chemical characteristics of road side soils in Pahalgam. Environ Monit Assess 190:353. https://doi.org/10.1007/s10661-018-6731-2

Singh DV, Bhat RA, Dervash MA, Qadri H, Mehmood MA, Dar GH, Hameed M, Rashid N (2020) Wonders of nanotechnology for remediation of polluted aquatic environs. In: Qadri H, Bhat RA, Dar GH, Mehmood MA (eds) Freshwater pollution dynamics and remediation. Springer, Singapore, pp 319–339

Song W, Li J, Wang Z, Zhang X (2019) A mini review of activated methods to persulfate-based advanced oxidation process. Water Sci Technol 79(3):573–579

Storck V, Karpouzas DG, Martin-Laurent F (2017) Towards a better pesticide policy for the European Union. Sci Total Environ 575:1027–1033

Tran TAM, Malarvannan G, Hoang TL, Nguyen VH, Covaci A, Elskens M (2019) Occurrence of organochlorine pesticides and polychlorinated biphenyls in sediment and fish in Cau Hai lagoon of Central Vietnam: human health risk assessment. Mar Pollut Bull 141:521–528

Wynendaele E, De Spiegeleer B, Gevaert B, Janssens Y, Suleman S, Cattoor S, Saunders JH, Veryser L (2018) Regulatory status of N-alkylamide containing health products. Regul Toxicol Pharmacol 98:215–223

Chapter 5
Mycoremediation: A Sustainable Approach for Pesticide Pollution Abatement

Marian Butu, Ramona Stef, Mihaela Corneanu, and Monica Butnariu

5.1 Background

Environmental problems related to pesticide pollution are complex (chemical, biological, technological, social) due to the high load of harmful substances but especially due to the imminent danger they pose to the environment and health, and their solution is only possible through cooperation of specialists from various fields that propose innovative technologies and services, the classical ones, that were applied until now, proving insufficient.

The sustainable approach contributes to the creation of an environmental product, materialized in the installation of ecosystems as close as possible to the natural ones on anthropic deposits polluted with pesticides and remedied by mycoremediation methods. The common denominator of all life forms in the soil is that every organism needs energy to survive.

While some bacteria, known as chemosynthesizers, get energy from sulfur, nitrogen, or even iron compounds, the rest must consume something that contains carbon to get the energy they need to sustain life. Carbon can come from organic material supplied by plants, residues produced by other organisms, or from the bodies of other organisms (Chanda et al. 2016).

Primarily, all life forms in the soil are getting carbon to fuel metabolism—it is an "eat-and-be-eaten" world, in and on the ground. Fungi are the main decomposition agent in the soil trophic network (Khanday et al. 2016; Bhat et al. 2017). The enzymes they release allow them to penetrate not only lignin and cellulose from plants (dead or alive) but also the harsh chitinous carapace of insects, animal bones,

M. Butu
National Institute of Research and Development for Biological Sciences, Bucharest, Romania

R. Stef · M. Corneanu · M. Butnariu (✉)
Banat's University of Agricultural Sciences and Veterinary Medicine "King Michael I of Romania" from Timisoara, Timis, Romania

R. A. Bhat et al. (eds.), *Bioremediation and Biotechnology, Vol 2*,
https://doi.org/10.1007/978-3-030-40333-1_5

and even hard nail proteins from the foot or hand. Bacteria can manage on their own, but require easily digestible foods, often derived from the decomposition made by fungi and only after such elements have been decomposed or broken by fungi or other organisms. Compared to fungi, bacteria are beginners in terms of decomposition abilities.

The fungi in the soil form two extremely important mutual relations with the plants. The first is the association of certain fungi with green algae, which leads to the formation of lichens. In this symbiotic relationship, the fungus receives food from the algae, which uses its power of photosynthesis, while the fungal filaments create the slope or body of the lichen in which the pair lives. The chemical substances secreted by the fungus break down the stone and the wood on which the lichen lives. It creates minerals and nutrients for soil, soil microbes, and plants (Chan-Cupul et al. 2016).

The second type of relationship is mycorrhiza (in Greek, meaning fungus root), symbiotic associations between fungi and roots. In exchange for the exudates from the roots of the plants, mycorrhizal fungi look for water and nutrients and then return them to the plant. The plant becomes dependent on the fungus, and this, in turn, cannot survive without the exudates of the plant. Through a too strong soil compaction, the fungal tubes are crushed and the fungi are destroyed. It is obvious that fungicides and pesticides, inorganic fertilizers, and soil physical modification (soil milling, deep digging) destroy fungal hyphae. Pesticides destroy fungal hyphae by extracting the cytoplasm from the fungal body (Hage-Ahmed et al. 2019).

5.2 Mycorrhizal Fungi Are of Two Types

Ectomycorrhizae grow close to the surface of the roots and can form networks around them. They are associated with conifers and hardwood trees. The second are endomycorrhizal. They actually penetrate and grow inside the roots and also extend outward into the soil. Endomycorrhizal fungi are favored by most vegetables, annuals, perennials, grasses, shrubs, softwood trees. Both types of mycorrhizal fungi can extend their range, as well as the surface of plant roots; the effective surface of the roots of a tree, for example, can be increased 700–1000 times by association (Song et al. 2016; Khanday et al. 2016; Bhat et al. 2017).

Mycorrhizae can be classified in at least 4 groups of mycorrhizae consisting of only three taxonomic classes, and the hosts are, at the moment, various gymnosperms and angiosperms (Table 5.1).

Ectomycorrhizae are usually found on the surface of the rhizoderms of the different forest trees between gymnosperms (*Pinus sylvestris, Picea excelsior, Larix decidua*), and among angiosperms, beech trees (*Fagus sylvatica*), evergreen (*Quercus petraea*), oak (*Quercus petraea*), and hornbeam (*Carpinus betulus*). The installation of an ectomitosis is done through a mycelial filament that approaches the rhizoderm and forms around the root an external hyphae mantle. Over time, the hyphae form a network in the space between cells that ultimately generates short

Table 5.1 Classification of mycorrhizae

The classification method	The type of mycorrhiza	Examples
After selectivity	Non-selective or generalist	The mushrooms *Amanita rubescens*, *Amanita pantherina*, *Boletus edulis*, or *Russula cyanoxantha* make mycorrhiza with several species of woody plants
	Preferentially selective	*Suillus luteus* preferentially pine mycorrhizal communities (trees), but also migrates to the oak tree; *Tricholoma sulphureum* establishes predominant symbiotic relationships with phage, but also passes on *Quercus species*
	Strictly selective	*Leccinum carpinus* makes mycorrhizae only with the hornbeam, *Boletus elegans* only mycorrhiza larch, *Boletus luteus* establishes symbiotic relations only with the pine forest
According to the relationships established between the fungal hyphae and the cells of the plant roots	Ectomycorrhizal or mycorrhizal ectotrophs	The hyphae penetrate intercellularly, so among the cortical cells of the plant root, in the interstitial space
	Endomycorrhizal or endotrophic mycorrhizae	The hyphae enter the cortical cells of the root, localizing intracellularly
	Ectoendomycorrhizal or mycorrhizal ectoendotrophs	Combination of the previous types
	Perimycorrhiza or mycorrhiza peritrofe	The mycelium of the fungus develops around the roots

intracellular hyphae. The entire construction serves to transport water and mineral salts from the root through the stem to the crown of the tree to feed the foliage. The species of fungi in this mycorrhiza group belong to the genera: Russula, Amanita, Lactarius, and other species with the hat formed of gills, consisting of the classes Basidiomycetes, Ascomycetes, and Zygomycetes.

Another group is the mycorrhizae that grow in the root rhizoderma of the order Ericales (fam. Ericaceae) with the following most well-known representatives: blueberry (*Vaccinium myrtillus L.*), cranberry (*Vaccinium macrocarpon A.*), *Erica tetralix*, bear grape (*Arctostaphylos uva ursi)*, *Andromeda polifolia*, *Calluna vulgaris, Rhododendron myrtifolium*, and others. In this group we can distinguish two categories of mycorrhizae: the type of arbutoid (living on the roots of shrubs in warm countries) and the type ericoid (living on semi-shrubs of low height) on the hedgehogs of boreal forests. The first type is composed of basidiomycetes, the second type of Ascomycetes class (Igiehon and Babalola 2017).

Mycorrhizae are specific to grassy plants in the Orchid family. These mycorrhizae are part of the ectomycorrhizal basidiomycetes class and are usually located within the root cortex (endomycorrhizal fungi). Most live in tropical forests. In the litters, gorunets, and oaks we find three species of orchids that possess inside the endomycorrhiza roots, these are: the saprophytic orchid *Neottia nidus avis* (bird's nest), on moist and humus-rich soil; *Epipogium aphyllum*, with green stalk and flowers, but without leaves and characteristic to our beech trees and *Corallorhiza trifida* living on the floor of the beech tree and our spruce, also devoid of leaves.

Mycorrhiza vesicular mycorrhiza that picks on the gymnosperms from us, the tis (*Taxus baccata*) or the exotic *Ginkgo biloba* (today only cultivated in the botanical gardens, brought to Europe from the Chinese Buddhist monasteries for the phytotherapeutic properties), then the *Sequoia sempervirens* and the shrubs, weed all the species of Fragaria from us and the bryophytes (leafy and ground muscles) in the forests. Fungi that produce such mycorrhizae belong to the class of Zygomycetes. Although these mycorrhizae live inside the root cortex, they form their sporangia and spores on the outside, as opposed to the other endomycorrhizae whose fruiting bodies are found within the roots (Mohamed et al. 2019).

Fungal arbuscular mycorrhizas form associations with the roots of a wide variety of vascular plants. The consequences of this association for the host plant vary continuously from positive (most often) to negative. In the classical literature, it has been suggested that the positive effects on plants are due to the increased nutritional supplement in one mycorrhizal plant compared to the other non-mycorrhizal one. Subsequently, it has been shown that plants can also benefit from mycorrhizal capacity in other ways. The presence of fungal associates can increase the resistance to stress, for example, under limited water conditions—hydric stress, in case of pathogenic attack, etc.

Tree mycorrhizal fungi are some of the most important symbionts for most plants. Under phosphate-limited conditions, mycorrhizal tree associations may influence the development of plant communities, nutrient, water intake, and aboveground productivity. They can also act as bioprotectors against pathogens or toxic stress. It has been suggested that, for any plant, there is a curvilinear relationship between the extension of the mycorrhizal fungal colony and the beneficial contribution to the plant exposed to infection. For some plants, a high density of fungal colonies may have a positive effect, while for others even at low levels, fungal colonies may result in decreased plant performance (Mallmann et al. 2018).

The reasons for the alleged negative effects of some mycorrhizal fungi species on some plant species are unclear, but they include loss of mycorrhizal photosynthesis products, immobilization of nutrients, alteration of root exudates to allelopathy but also negative effects of other components of the rhizosphere microflora. Fungal arbuscular mycorrhiza is recognized on the basis of specific characteristics such as biotrophy, asexual reproduction, large spores, and multinucleate with stratified walls, unexpected hyphae, and the formation of shrubs at the root of the plant. Research shows that the species of arbuscular mycorrhiza can grow up to the stage of spore production in vitro, in the absence of plant roots and in the presence of bacterial lines associated with the spores. Mycorrhizal tree fungi have asexual reproduction through spores. There is no evidence of sexual reproduction of arbuscular mycorrhizal fungi (AMF). A study reports the formation of zygospores sexed in Gigaspora but this has not been confirmed. Genetic recombinations, or only a low level, were not detected using molecular marker genes. Therefore, it is generally accepted that the spores of mycorrhizal fungi are asexually formed. The spores are relatively large (40–800 μm) with stratified walls having lipids in the cytoplasm. They are important in identifying mycorrhizal fungi. Traditionally, the taxonomy of arbuscular mycorrhizal fungi has been based on the morphology of the spores, in particular on the structure of their walls and on the methods of formation on the hyphae (Lekberg et al. 2017).

5.2.1 Life Cycle of Mycorrhizal Fungi

The life cycle of arbuscular fungi generally starts from the stage of spores lying in the soil or coming from the adjacent mycorrhizal roots. Hyphae (H) resulting from spores or mycorrhizal roots grow in the same direction as the root of the plant. At the root surface, the terminal portions of the hyphae swell and form a specific structure called apresor (Ap). Infective mycelial filaments start from the oppressors. The hyphae penetrate the cell walls of the root epidermis with the help of penetrating hooks. The point at which the hyphae form any kind of propagule and which enter the root for the first time is called the primary entry point (E). The number of primary points/moments made by the fungus at the root is equivalent to its inoculation potential. At the root, the hyphae grow intercellularly towards the internal cortical layers and in the internal region of the cortex, the hyphae begin to grow intracellularly. After that, the host cell membrane invaginates and envelops the fungus forming a new compartment called apoplastic space. This space allows the efficient transfer of nutrients between the two symbionts but prevents the direct contact between the fungal and plant cytoplasm. Inside the cortical cell, but outside the cytoplasm, the hyphae form shrubs and vesicles. Shrubs are intracellular hyphae structures, with dichotomous branching, which can be the place of exchange for phosphorus, carbon, or other nutrients. The vesicles are intracellular structures full of lipids, assuming they also store carbon, and also serve as reproductive propagules. It has been found that not all mycorrhizal fungi form vesicles, such as Gigasporaceae. The formation of the vesicles depends on both the fungal symbiont and the environmental conditions. With the beginning of the infection process, in the root colonization is accomplished both through the intraradicular mycelium inwards and through the extraradicular mycelium on the outside. Intraradicular mycelium colonizes the root in different ways. Depending on its structure, the mycorrhizae are separated into several types: Arum, Paris and Intermediate.

The Arum model implies a longitudinal growth of the intercellular hyphae as well as the penetration of the root cortex, followed by the formation of the shrubs. The shrubs form short branches on the intercellular hyphae, usually at right angles to the axis of the main root. Arum type is found in cultivated plants. In the Paris type, intercellular hyphae are absent and the existing hyphae are entirely intracellular and irregularly curled, some of them forming shrubs that are not terminal but are located in well-defined layers. Shrubs formed as intercalated spaces are called arboreal twigs. This type is usually found in plants in natural ecosystems. Sometimes both types of structures meet in the same root, as for example in pumpkin or tomatoes and this type has been called Intermediate (Spagnoletti et al. 2016).

Mycorrhizae are associations between fungi and the roots of grassy or woody plants. It is a symbiosis because it offers benefits to both parties. In this case, the plants increase their absorption range thanks to the fungi, while they receive nutrients synthesized by the plants. A single fungus can join several plants, thus forming an extended network. In fact, many forests are connected by mycorrhizae. In this regard, studies have been conducted that show that mycorrhizal trees are capable of living longer than those without mycorrhizal. There are different levels in which plants depend on mycorrhizae. Studies conducted in the area show that the plants may have high concentrations of mycorrhizae, intermediate concentrations, or their absence (Trocha et al. 2016).

Plants with high concentrations of mycorrhizae are dependent on fungi to survive. This is a mandatory relationship because the plant would die without mycorrhizal survival. Plants with intermediate concentrations of mycorrhizae do not depend on this type of association, but take advantage of this to improve their conditions (such as drought resistance and certain diseases). This relationship is optional.

Finally, there are plants whose roots resist the formation of associations with fungi. This happens when the soil conditions are sufficiently good (so the plant does not need the benefits of mycorrhiza) or when the plant is young and stable (so it does not require additional supplements). Mycorrhizae generate hundreds of advantages that can be highlighted in different fields. Thus, mycorrhiza benefits plants, soils, fungi, and human beings.

The mycorrhizae form a network of filaments that increase the degree of absorption of the plant roots. For example, the fungus attached to a plant can extend for hundreds of meters, which puts the plant in contact with more water and nutrients. Fungi are responsible for the breakdown of organic matter, from which they obtain essential minerals, such as nitrogen and phosphorus. These nutrients are sent to the plant through the roots.

By themselves, plants cannot degrade these substances. In addition, micro-germs make plants capable of withstanding certain conditions that they cannot withstand unless they have been connected to fungi. For example, thanks to the fungus, the plants survive drought and create resistance to certain soil pathogens.

Other benefits of micorrhiza for plants are: it helps plants grow faster and stronger, promotes flowering and fruit production in the plant, increases the tolerance of the plant to the salinity of the substrate, they reduce the incidence of diseases, protect the plant from predators, such as worms, increase the transfer of carbon between trees, even among different species, they extend the life of certain species of plants (Liu 2010).

Not only do mycorrhizae benefit from plants, but they also generate benefits for fungi.

Fungi are organisms that do not photosynthesize, so they cannot synthesize products independently. In this sense, they depend on other individuals to be able to feed. By forming mycorrhizae, the fungus receives nutrients produced by plants, such as

Table 5.2 The advantages of mycorrhizae

The major advantage	Intensifies the nutritional exchanges at the root level, can change the reaction of the soils, acting as living buffers. For example, Penicillium species in the rhizosphere change soil pH, making it possible to develop quails on alkaline soils (*Quercus species* do not support basic soils, in the absence of these fungi, the existence of trees on alkaline soils is compromised)
The main advantage is to increase the absorption of mineral nutrients from the soil by the plant	The extension of the external mycelium of the fungus over the rhizosphere is the main consequence of this effect, because it allows the absorption of nutrients outside the area of action of the roots
Other benefits	High resistance of plants to stress and water salinity
	Increased resistance and tolerance to soil pathogens
	Higher absorption of micro and trace elements
	Greater and more uniform production
	Improves plant growth
	More efficient use of fertilizers and water
	Better adaptation of plants to the environment
	Minimizing the influence of the state of fatigue of the land on the complaints

carbohydrates. In general, the presence of mycorrhizae makes the soil more fertile. It avoids erosion of the substrate when establishing connections in the basement. Increase nutrient circulation in the substrate, which creates more fertile soils. The benefits of mycorrhiza use are presented in Table 5.2.

Mycorrhizae are important in economics and medicine. Then, there are mentioned some aspects in which these associations intervene in human life. The fruits of ectomycorrhizae (fungi and truffles) offer nutritional values. For this reason, they are marketed, which makes them economically important. Many of these fruits also have medicinal properties, which is why they are used in the pharmaceutical industry. By increasing the nutrient absorption of plants, the mycorrhizae increase the nutritional value of products, such as fruits and nuts.

The salts absorbed by the mycorrhizal fungi are quickly transported to the upper symbiont, the fungus having a role in supplying the top plant with mineral salts. In turn, the upper symbiont supplies the mycorrhizal fungus with the organic substances needed.

Fungi hyphae are an important means available to the upper plants for the purchase of water and mineral salts. These associations are found in plants that live in poor resorts in mineral salts, in which mycorrhizal fungi—due to their chemotropism—can easily detect even very small amounts of mineral salts, so that only plants with a highly developed root system could enter into competition with them to obtain these salts. In addition, mycorrhizal fungi can absorb organic nitrogen from the soil, so that the upper symbiont is well supplied with nitrogen even in cases where inorganic combinations of this element are in small quantities in the substrate; this is the case of swamp plants, where the mycorrhizae are very widespread;

therefore, plants grow in conditions where the process of nitrification is very low in the soil, and the fertilizers with NH_3 have not proved effective (Koricheva et al. 2009).

5.3 Mode of Action of Pesticides

Pesticides are products that contain or generate active substances and are used against harmful organisms in the home environment (disinfectants, rodenticides, repellents, and insecticides) or to protect natural or artificial products in industry and agricultural applications. The mode of action of mycorrhizae on pesticides is shown schematically in Figs. 5.1 and 5.2. Given that their pesticide use intrinsic properties can pose health risks, the EU has set up a regulatory framework aimed at significantly increasing the safety of pesticides used and sold in the EU, namely EU Regulation 528/1012. It provides for the release of biocides that are carcinogenic, mutagenic, toxic to reproduction, which disrupt the endocrine system or which are dangerous to the environment.

Pesticides, in general, are organic compounds with low molecular weight and different water solubility. The chemical character, molecular shape and configuration, their acidity or basicity, water solubility, polarity of the molecule, size and polarizability of the molecule, all these properties, alone or together, influence the adsorption–desorption processes on soil colloids. Given these properties of pesticides and the character of the relationships involved in the adsorption–desorption process on soil colloids, they can be grouped into two major classes: polar and

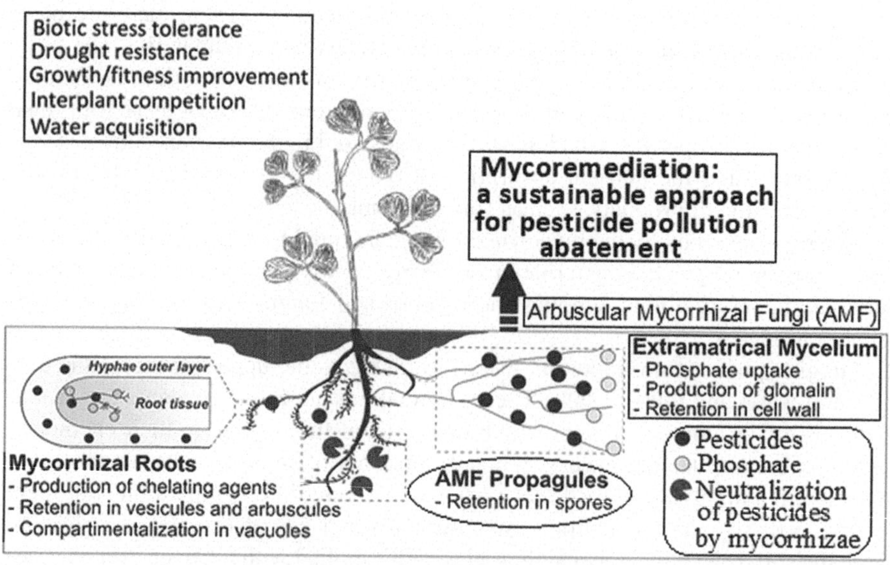

Fig. 5.1 Some processes that take place through mycoremediation: a sustainable approach for pesticide pollution abatement

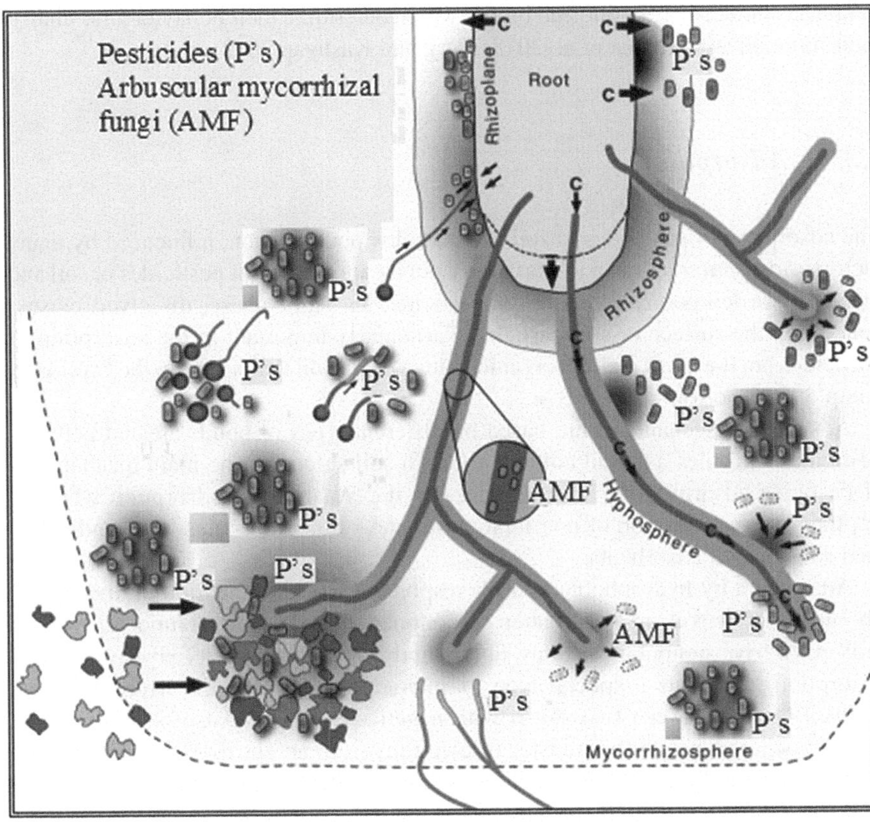

Fig. 5.2 Schematic view of possible interactions among different pesticides of the mycoremediation: a sustainable approach for pesticide pollution abatement

nonpolar. Pesticides containing compounds of an acidic, basic, or dissociative character act as cations, forming the group of ionic compounds. Products that have neither basic nor acidic character make up the group of non-ionic compounds.

Among the structural characteristics that determine the chemical character of the compounds and which influence their adsorption and desorption on soil colloids, the following are mentioned: the nature of the functional groups, the nature of the substitution groups, the position of the substitution groups against the functional groups, and the degree of unsaturation of the molecule. In high pH medium, acidic compounds, by dissociation, become anions. In contrast, in a low pH environment, weakly basic or neutral compounds behave similar to organic cations.

In general, the soil acts as a receptor and reservoir for pesticides, in which, either they degrade or are gradually dispersed into the environment or translocated into plants, some of which can still persist in the soil many years after application (Mushtaq et al. 2018). Pesticides, finely dispersed substances, in soil are subject to numerous influences, abiotic and biotic, which determine their behavior and, finally, their mineralization (Bhat et al. 2018; Garg and Kashyap 2019).

5.3.1 Adsorption

The adsorption of pesticides in soil is a complex phenomenon, influenced by many factors. Adsorption plays an important role in the movement of pesticides in soil and refers to the temporary retention of pesticides in vapor phase, dissolved or suspended on the surface of soil particles. Particularly important in the adsorption of pesticides are the clays and the organic matter in the soil that make up the "colloidal complex" of the soil.

Adsorption mechanisms are based on different types of bonds created between pesticide molecules and soil colloids through soil solution. The main mechanisms of pesticide adsorption by soil include: van der Waals force adsorption which is implicit in the adsorption of non-ionic pesticides of the molecules in an undissociated state on soil adsorbents.

Adsorption by hydrophobic bond is responsible for the retention on the hydrophobic areas of soil organic matter of nonpolar pesticides or compounds whose molecules have nonpolar portions significantly to the polar side. Hydrogen bond adsorption represents a special type of interaction in which the hydrogen atom forms a bridge between two atoms with negative electrons. Adsorption in charge transfer occurs through the transfer of electrons from an electron-rich donor to an electron deficient acceptor. Adsorption by ion exchange that takes place in the case of compounds present in the form of cations or those compounds which, by protonation, gain the positive charge. Adsorption by the formation of coordination bonds by exchange of ligands that takes place when the ions of the transition metals can form an adsorption center on the surface of the clays in the soil and is very important for the behavior of pesticides in the soil (Bruckner et al. 2019).

5.3.2 Circulation of Pesticides in Soil

The movement of pesticides into the soil occurs in solution, by volatilization or with the movement of the colloidal particles on which they are adsorbed. Soil migration, as a solution, is dependent on both the diffusion and mass transport processes. The summation of diffusion and mass transport processes determines the rate of migration of pesticides into the soil. Molecular diffusion motion—which occurs based on the concentration gradient. Movement by mass transport—which occurs with the migration of water or soil particles with which pesticide molecules are associated.

Degradation of pesticides in soil. Pesticides in the soil are subject to various factors that influence both the stage that ensures their effectiveness and the phase in which the product is no longer useful and becomes a residue. Once in the soil, pesticides are subject to degradation, a widespread phenomenon that plays an important role in the dissipation of many such substances. Pesticide degradation is determined by abiotic and biotic factors and processes.

5.3.3 Abiotic Degradation

In soil, pesticides undergo chemical transformations as a result of reactions with soil organo-mineral compounds. The physical and chemical properties of soil are the most important factors that influence the transformation of pesticides into soil.

5.3.4 Biotic Degradation

Numerous papers highlight the role of soil microorganisms in the decomposition of pesticides, as well as the fact that there are few active substances that are not biologically degraded. Some active substances in pesticides can serve as a source of nitrogen and carbon for soil microorganisms. In a synthesis on the biological degradation of herbicides, the role of microorganisms in soil is noted in a whole series of degradation processes, including beta oxidation, de-halogenation, and opening of cycles.

Persistence of pesticide residues in soil. A pesticide residue in the soil is considered to be any substance or mixture of substances that is found on or in the soil as a result of using a pesticide. Concern over the long-term effects of pesticide residues in soil has highlighted the idea that the persistence of a chemical is a quantifiable indicator that represents its resistance to degradation. For our purposes, the term "persistence" means the period of existence in the soil of a given pesticide, which may be expressed in units of time. In terms of persistence, the concept of half-life is used extensively, i.e., the time required for 50% of the pesticide to disappear. The persistence of a pesticide in the soil depends on a number of factors conditions such as content of organic matter, nature and content of clay minerals, pH, microflora and soil microfauna, temperature, ion exchange capacity, cultural practices, exposure to wind and light, precipitation. Above all, the most important role is played by the chemical nature of the pesticide (Zaller et al. 2018).

Due to the potential danger of soil and environmental pollution due to the high remnants of pesticides, we have used easily degradable pesticides such as carbamic insecticides, most organo-phosphoric insecticides, pyrethroid insecticides. However, some of the easily degradable pesticides are characterized by degradation products that can persist in the soil for up to several months and can be very toxic, which must be carefully considered in their use in crop production.

Translocation of soil residues into soil and plant fauna. In soils polluted with pesticide residues, many invertebrates from the soil absorb and accumulate residues in their bodies in concentrations several times higher than in the soil around them, thus having a bioconcentration process. Also, the presence of pesticides in the soil leads to the loading of plants with their residues. The absorption of insecticide residues in plants occurs only when they are in excessive quantities in the soil. Pesticides are absorbed by plants more easily from sandy soils and to a lesser extent from turbid soils that have a high content of organic matter.

Soil deposition of pesticide residues. Under normal conditions, decontamination takes place slowly due to soil cultivation or climatic conditions. For this purpose, different processes are used depending on the chemical composition of the pesticide, the degree of soil loading and its characteristics, the cultivated plants, and the cultivation technology. Thus, to reduce the activity of nitralin or fluometuron, activated charcoal is incorporated into the soil. Diffenamide, dicamba, and ambien products are degraded in soil if a certain microbial culture medium is administered. The fastest depollution processes are those based on the administration in soil of adjuvants, products that retain or degrade the pesticides. Other soil pollution measures are based on the ability of maize, sorghum, sugar cane, as well as some weeds, to pollute the soil of atrazine residues through absorption and metabolic degradation.

The process of atrazine degradation could be faster if the following factors were combined: early hybrid with high growth rate, high doses of organic and chemical fertilizers, suitable density to extract residual atrazine from soil. In this way, the depollution process could be intensified and reduced to a period of 60–70 days. Based on extensive research, mathematical methods have been developed to forecast the dynamics of pesticides in agro-ecosystems, taking into account the environmental factors that allow to know the concentration of a pesticide in the agro-ecosystem at a given time. For the same purpose, maps are developed on the application of pesticides and the conditions of depollution, which take into account a series of criteria, on the basis of which the following groups are established: the most toxic, totaling over 21 points; medium toxic, 21–24 points; relatively weak toxic, below 13 points (Buysens et al. 2015).

5.4 Mycoremediation Methods

Mycorrhizal fungi obtain the necessary carbohydrates from the exudates of the host plant and use them as energy to expand into the soil, pumping moisture and undermining nutrients in places that the plant is inaccessible to. These fungi are not lonely miners. They form complex networks and sometimes transport water and nutrients to the roots of different plants, not just the ones from which they started. The mechanisms by which the alternative agriculture system determines the manipulation of the plant metabolome to obtain the predominant expression of certain genes of agricultural interest.

The biotechnological management system of the agricultural culture was developed based on the new concepts of alternative agriculture by manipulating the plant metabolome for the preponderant expression of certain genes of agricultural interest and stimulating the photosynthesis and radicular translocation of the sap elaborated by using mycoremediation methods with near IR reflection. The use of mycoremediation methods for plant protection reduces the need for pesticide application. Pesticides also have an action to reduce the activity of mycorrhizal fungi (AM). The studies performed have shown to date a good compatibility between the bioproducts based on AM fungi and the antagonistic bacteria *Bacillus subtilis*.

The mechanisms by which the alternative agriculture system determines the manipulation of the plant metabolome for the predominant expression of certain genes of agricultural interest (Hashem et al. 2016).

By cultivating plants using mycoremediation methods as vegetative mulch/decaying plant residues of *V. villosa*, the following genes are predominantly activated: the genes responding to nitrogen—*NiR*, *GS1*, *rbcL*, *rbcS*, and *G6PD*; genes encoding chaperon proteins—*hsp70*, *BiP*; the genes of the plant defense system—chitinase and osmotine; cytokinin-dependent genes of *CKR*; GA 20 oxidase. The predominant activation of these genes in the crop plant leads mainly to the following effects: increasing the efficiency of the use of nitrogen; stimulation of photosynthetic carbon fixation; reducing the incidence of plant diseases; increasing the longevity of the culture. Apart from these effects (explained by the predominant activation of some genes) it is generally acknowledged that the use of mycoremediation methods by using the mulch–carpet of hairy moss has beneficial effects on the soil (reduces erosion, improves water infiltration, stimulates soil tillage) and on protection of plants (reduces the spread of pathogens specific to the main crop, promotes the development of predatory insects and thus reduces the attack of harmful insects, eliminates weeds) (Motaharpoor et al. 2019). All the effects resulted by applying this technological sequence of alternative agriculture are also useful for mycoremediation methods.

Of maximum interest for the final purpose of mycoremediation methods (respectively, increasing the amount of carbon dioxide fixed by sequestering photosynthesis in organic matter; from the soil) is the stimulation of photosynthesis, which, associated with the action of mycoremediation methods with IR reflection close to stimulating the radicular translocation of the sap elaborated, creates the premises to increase the secretion of glomalin in the soil (Wang et al. 2018).

Finding and returning phosphorus is so important for plants that it seems to be a major function of many mycorrhizal fungi; acids produced by mycorrhizal fungi can block, recover, and transport the chemically blocked phosphorus back to the host plant.

Mycorrhizal fungi also release copper, calcium, magnesium, zinc, and iron to be used by the plant. As always, any nutrient not delivered to the root of the plant is trapped in the fungus and is released when it dies and is broken down.

Mycoremediation is a form of bioremediation in which fungi technology is used to decontaminate the environment. Fungi have proven to be a cheap, efficient, and environmentally friendly way to help remove pollutants from damaged environments or wastewater.

Toxins include pesticides and herbicides. The by-products of remediation can themselves be valuable materials, such as enzymes (such as *Laccases* have copper-containing oxidase enzymes found in many fungi, microorganisms, and plants), edible or medicinal fungi, which makes the remediation process easier, even profitable. Fungi, due to nonspecific enzymes, are capable of decomposing many types of substances. Most plants can form symbiosis with fungi, a relationship called mycorrhiza.

Mycorrhizal fungi, in particular AMF, can improve the phytoremediation ability of some plants. This is mainly due to the fact that the stress caused by plants due to pollutants is very low in the presence of arbuscular mycorrhizal fungi, so that they can grow longer and produce more biomass. Fungi provide nutrition, especially phosphorus, and promote general plant health. Also, the expansion of the mycelium can greatly extend the area of influence of the rhizosphere (hyposphere), giving the plant access to more nutrients and contaminants. Increasing the general health of the rhizosphere means an increase in the bacterial population, which can also contribute to the bioremediation process.

This relationship has proven useful with many pollutants, such as *Rhizophagus intraradices* and *Robinia pseudoacacia* in lead-contaminated soil, *Rhizophagus intraradices* with *Glomus versiforme* incorporated in vetiver grass for lead removal, mycorrhizal fungi and *Calendula officinalis* and *Althaea officinalis*, in general, has been effective in increasing the bioremediation capacity of pesticide plants. In wetlands, shrub mycorrhizal fungi greatly benefit from biodegradation of organic pollutants such as benzene, methyl-tertiary butyl ether, and ammonia from groundwater when inoculated into *Phragmites australis* (Huang et al. 2017).

Perennials, shrubs, and trees prefer soils dominated by fungi. Annual plants, herbs, and vegetables prefer soils dominated by bacteria. The explanation for this is the way nitrogen is fixed in the soil. Depending on the soil environment, nitrogen may remain in the form of ammonium (NH_4) or be converted to nitrate (NO_3) by specific bacteria.

Soils dominated by bacteria tend to have an alkaline pH, which is a good living environment for bacteria that have this function of converting nitrogen into nitrates. In the fungi-dominated soils, most of the nitrogen remains in the form of ammonium. Fertilizers are salts that absorb water from bacteria, fungi, nematodes, and soil protozoa, destroying the entire feeding system of microscopic living creatures. Fungi are the main decomposers in the soil. Bacteria often consume matter that has already been processed by the fungus. Like bacteria, fungi should be regarded as containers with fertilizers (Sudová et al. 2018).

Phosphorus is almost always blocked in soils. Even when applied as a fertilizer, phosphorus becomes inaccessible to plants in seconds. The fungi not only seek it, but process it so that it can be absorbed by the roots of the plants. Also, similar processes take place with copper, calcium, magnesium, zinc, and iron. In soils dominated by fungi, nitrogen is in the form of ammonium. The fungi often come into symbiosis with the roots of plants.

In exchange for the exudates from the roots of the plants, the fungi look for water and nourishing substances that they bring to the plants. It was only in the 1990s that the term mycorrhiza (eng. Mychorriza) began to enter in the vocabulary of the agricultural industry, but not at all in the vocabulary of gardeners. Most conifers and hardwood species (birch, oak, beech) form mycorrhizal symbiotic relationships with ectomycorrhizal fungi (which live near the root surface and form networks around it). Most species of shrubs, softwoods, and perennials form mycorrhizal symbiotic relationships with endomycorrhizal fungi (which enter and grow in the root, but also outside it). However, there are exceptions—rhododendrons, azaleas, or blueberries need hedgehog fungi, which are not yet on the market. Some plants prefer the soil dominated by fungi, while others prefer the soil dominated by bacteria. Plants need nitrogen to produce amino acids; it is crucial in the development and survival of plants.

For this reason, soluble nitrogen-based inorganic fertilizers give such good results in plant growth, even if they are harmful to trophic networks. In aqueous solutions, these nitrates (NO_3^-) are immediately accessible to the roots of the plant, which absorb them like a sponge.

Like anions, they remain in aqueous solutions instead of fixing on humus or clay, as do cations loaded with positive charge. In a healthy soil trophic network we have two forms of nitrogen, nitrates and ammonium (NH_4) and, as most often in life when there are more options, some plants prefer nitrates nitrogen in the form of nitrates, others prefer ammonium. When nematodes and protozoa consume fungi and bacteria, they excrete nitrogen in the form of ammonium (Leberecht et al. 2016).

Ammonium oxidizes rapidly or, if found in sufficiently large amounts in the soil, the nitrifying bacteria turn it into nitrates. In the soil dominated by bacteria, things happen almost always after this scenario, because the viscous substance produced by them has a pH above 7, which represents the ideal life environment of the nitrifying bacterium bacteria.

In general, nitrifying bacteria thrive in soils dominated by bacteria. Fungi favor a lower pH because they use organic acids to break down organic matter into nutrients. If sufficient organic acids are available to annihilate the viscous matter produced by the bacteria, the soil pH drops below 7, making the acid environment to be acid and thus increasingly unfavorable for nitrifying bacteria. Therefore, more and more ammonium will remain unchanged ammonium. At first the soil is dominated by bacteria. But as more organic residues accumulate from debris discharged from these organisms and the plant life they support, fungi spores come in handy enough to provide enough nutrients to germinate. Having available a space to occupy and resources to maintain themselves, the fungi have increase emerged. The process is influenced by many other factors, but let us focus on what interests us: as plant life and soil trophic network diversify, the number of fungi populations increases, and plants with a short life span make room for plants with a longer existence, make room for existing plants. Longer, evergreen, shrubs (Castro-Rodríguez et al. 2017).

More organic matter is produced, ensuring the food of fungi populations is which always growing. The shrubs stabilize, then are followed by soft essence trees, evergreen shrubs that expand, hardwood trees that fully develop, and eventually coniferous forests as a whole. Throughout the process, the biomass of fungi grows in proportion to that of bacteria, but they cannot compete, being limited to digesting simple sugars and other carbohydrates—existing in limited quantities, given the increasing mass of complex plants, rich in lignin and cellulose. With the passage from the sandbank, so to speak, to the shrubs bushes, to old coniferous forests, the decisive influence of fungi on the soil is intensified step by step (Bargaz et al. 2018).

5.5 Mycorrhizal Relations/Intraspecific Relations

Organisms grow, reproduce, and die. They are affected by the conditions in which they live and the resources they obtain. But no organism lives in isolation. Each, at certain stages of life, is a member of a population. Individuals of a species have similar requirements for survival, growth, and reproduction, and their cumulative requirements may exceed at any given time the existing stock. As a result, competition for resources intervenes, which undoubtedly causes the deprivation of some of the organisms.

We can say that, the relationships between individuals (individual organisms) originate in the need for survival in the concrete environmental conditions. However, intraspecific relationships should not be viewed unilaterally (competition or relations of perfect harmony).

These relationships can be considered as individual–population relations or system relations (between the individual and the population system or between different subsystems of the population). Intraspecific relationships are appreciated for their role within the population system (Kranabetter et al. 2015).

The most important relationships are the trophic ones (which result from the necessities of getting the food), of protection (of defense against the variations of the ecological factors, of predators), of reproduction, of numerical adjustment of the population, of adjustment of the population structure. Apart from the inclusion of new biotechnological solutions within the agrotechnical sequences with a role in promoting the development of arbuscular mycorrhizal fungi and the accumulation of organic matter in the soil (including through the use of biofertilizers and biopesticides) the proposed technology has a pronounced "bio" and for that they are used, for monitoring the efficiency in the recovery of organic matter from soil, biochemistry methods (extraction and dosing of glomalin from soil; extraction and dosing of humic acids from soil) developed relatively recently based on new knowledge of cellular and molecular biology (Mei et al. 2019).

The proposed technology has the following stages: land management for stabilization and drainage in accordance with the recommendations of codes of good practice in land management; establishment on the degraded and landscaped land of a crop of *Vicia villosa* (the hairy vetch, fodder vetch, or winter vetch), by sowing a norm of 125 kg seeds per ha; the development until the spring of the vegetal carpet of *V. villosa*; mowing/rolling the vegetable carpet of *V. villosa* in strips 2 m wide; formation of a biocomposite mulch by depositing on the mulch of *V. villosa* of a biodegradable film formed by direct spraying of a peliculogen solution based on polyvinyl alcohol, protein isolate, glycerine/glycerol waters, and a red dye; seeding in the biocomposite mulch the seeds of the crop plant; maintaining the resulting crop according to the agricultural framework technology recommended for the crop area; harvesting of useful production and incorporation into soil of plant residues from both crops (*V. villosa* protective culture + basic culture).

This technology presents the following advantages: elimination of biological risks, the composition used containing exclusively the microorganisms useful for the growth of the crop plants and for the restoration of the organic matter from the soil, whose safety is generally recognized; restoration of organic matter from soil by accumulation of glomalin excreted by fungi from endomycorrhizal symbiosis on account of root exudates (whereby plants translocate up to 40% of photosynthetically fixed carbon dioxide); increasing the efficiency of the dark phase of photosynthesis due to the activation of Calvin cycle enzymes as a result of applying the alternative agriculture system of selective activation of agricultural genes by cultivation on *V. villosa* mulch; stimulation of the light phase of photosynthesis and radicular translocation of the sap produced by the use of mulches with red reflection and near IR, reflection produced by the red dye embedded in the biocomposite mulch (Pena et al. 2014); protection of plants against edaphic pathogens due to the presence of *Bacillus subtilis* bacteria, antagonists for phytopathogens; plant nutrition predominantly on the basis of biologically fixed nitrogen by nitrogen fixing bacteria, both naturally occurring in the soil that develops nitrogen fixing symbioses with *V. villosa* and bacterial endophytes of the genus *Azospirillum*, which fix nitrogen in association with the resins of plants, vegetables, or cereals.

Reducing the attack of diseases, pests, and weeds due to the intercalated culture, element of biodiversity that favors the development of populations of antagonists for phytopathogens and predators/parasitoids for harmful arthropods and reduces the development of weeds (Himmelstein et al. 2016).

Basically the technology described above is the contemporary (re)interpretation of some old agricultural practices. Protective crops transformed into vegetable mulch have been used since ancient times, by Romanians and ancient Greeks. Used as a protective crop, *V. villosa* determines nitrogen fixation, recycling of nutrients, reducing soil erosion and compacting, improving the quality of soil organic matter (by rhizodeposition).

When converted to mulch, *V. villosa* plant debris limits weed growth, increases soil organic matter content, reduces water loss, and acts as a controlled release fertilizer. The agricultural crop technology described above, which can also be used for the rehabilitation of degraded land, is part of the European priorities described above, and can be subsidized under the agri-environment measures (Smith et al. 2017).

5.5.1 Increasing Biodiversity

The system proposed obviously determined the increase of biodiversity in the field of vegetable production. In practice, the system involves the promotion and support of environmentally friendly agricultural practices that are directly beneficial to biodiversity (intercalated protective culture and main biocomposite mulch culture) or indirectly (*V. villosa* protective culture is a refuge for other organisms—species of feedable birds).

5.5.2 Reduction of Greenhouse Gas Emissions from Agriculture

The alternative system determines the growth of organic matter in the soil and will reduce the need to fertilize with nitrogen. In the original system a reduction of 50% of the demand of nitrogen fertilizers was demonstrated. Therefore, the proposed system will reduce the need for nitrogen fertilization and/or manure and reduce greenhouse gas emissions from agriculture: N_2O (nitrogen oxide) emissions from the soil, mainly due to nitrogen fertilization; CH_4 and N_2O emissions due to manure handling. In addition, the proposed system offers new solutions for sequestration of carbon in the soil, by re-depositing glomalin due to the symbiosis between the arbuscular mycorrhizal fungi and the root of the basic culture.

5.5.3 Conservation and Sustainable Use of Genetic Resources in Agriculture

The proposed system offers competitive advantages for the native varieties of tomatoes, much more resistant to the competition of the weeds and less productive in superintensive crops than the international varieties. In addition, the system offers a solution for using the genetic resources of microorganisms beneficial to crop plants (Zhang et al. 2016).

5.5.4 Protection of Soil from Erosion and Maintenance of Organic Matter and Soil Structure

The positive impact of the proposed system is evident in this case, resulting from the protective culture and from the rhizodeposition of organic matter by the basic culture in the symbioses formed with the beneficial microorganisms released controlled from the matrix of the biodegradable film. Rhizodeposition/elimination of photosynthetically fixed carbon from the root cells of the epidermis and root cortex of cultured plants leads to a proliferation of microorganisms inside (endorhizosphere), surface (rhizoplane), and outside root (ectorhizosphere).

The release of carbon fixed in the rhizosphere also causes important changes in the chemical, physical, and biological characteristics of the soil. The magnitude of these changes in soil properties is significantly determined by the amount and type of carbon released from the roots as well as by the intrinsic characteristics of the soil. Arboreal mycorrhizal fungi can secrete large amounts of an adhesive glycoprotein and high resistance to degradation (glomalin) in the surrounding soil, which represents for some soils (which increase their natural erosion resistance) a *long-lasting form of sequestration for carbon*. The R/FR reflexive mulch is included in the system not only due to the function of increasing the output by stimulating the light phase of photosynthesis, but also because of its effect on (phytochrome-mediated) stimulation of photosynthetically fixed carbon reallocation to the root system (Husna et al. 2016).

5.5.5 *Reduce the Use of Pesticides in Order to Reduce the Adverse Impact on the Environment*

This system aims to significantly reduce the use of pesticides. From the elimination from technology of the total herbicide in the initial phase of plant mulch formation to the reduction of the use of fungicides (due to the antagonistic activity of PGPR strains and due to the activation of the systemically acquired resistance mechanism, induced by *V. villosa* residues and mycorrhizal fungi and insecticides (due to the growth of predator/pest populations for insect pests).

5.5.6 *Reduce the Pollution of Waters with Nitrates from Agricultural Sources*

This is a consequence already presented of reducing the use of mineral nitrogen fertilizers for the basic crop (Lovett et al. 2009).

5.6 Reducing Pesticide Pollution

At EU level, the initiation of the process of integrating agriculture into the environment was determined by the complexity of the relationship between agriculture and the environment, even in terms of reducing pesticide pollution. In this complex relationship between agriculture and the environment are included aspects regarding greenhouse gases, and soil is the main factor of interconnection. In order to achieve this objective of agriculture in line with the environment, a set of good farming practices (GFP), considered necessary at European level, has been established, especially after the reform of the Community Agricultural Policies within the Agenda 2000.

European GFP priorities are practically related to soil protection: maintaining/increasing the organic matter content in the soil; maintaining/improving soil structure; reducing greenhouse gas emissions from agriculture; increasing biodiversity; conservation and sustainable use of genetic resources in agriculture; soil protection against erosion; decreasing the share of pesticide use to minimize their environmental impact; reducing pollution of waters with nitrates from agricultural sources (and above all from the soil).

All these priorities are supported by economic measures, as the main objective of the Community is sustainable development, with a balance between competitive agricultural production and the preservation/improvement of useful resources. The most spectacular economic measures were those included in the Community Agricultural Policy (CAP). According to the Agenda 2000 program document, the CAP has two pillars, the policy of ensuring income and a secure market for farmers ("the first pillar") and the sustainable development of rural areas ("the second pillar"). The 2003 CAP reform brought a new level of integration of agriculture into the environment, with new measures related to promoting the protection of the environment used in agriculture (and first of all soil) through both CAP pillars. These economic measures presented above have created a current stage at European level, with a significant demand for alternative agricultural practices, capable of meeting the requirements of a competitive agriculture in harmony with the environment (Torres et al. 2013; Asemaninejad et al. 2017).

Worldwide, commitments to reduce pesticide pollution began with the signing of the United Nations Framework Convention on Climate Change (Rio 1992), in agreement with Sosa-Hernández and collab. By this convention the signatory countries have assumed a global political commitment to reduce greenhouse gas emissions. In 1997, under the Convention, the Kyoto Protocol was signed, whereby the countries of Annex I (41 countries plus the European Union states) committed to reduce net greenhouse gas emissions (global reduction by 5% of CO_2 emissions) in the period 2008–2012 compared to 1990) in all economic-social activities on economically and socially efficient ways, by engaging the politics at national level, appropriate production technologies, and specific non-polluting practices. Added to these traditional ways were some new specifics, the so-called flexible instruments ("joint implementations," "clean development mechanisms," and "carbon market"). Through these instruments, the political commitment becomes effectively a financial one, materialized in concrete projects. At present, the Kyoto Protocol is only ratified by some of the signatory states, but the instruments created are functional and provide increasing financing in the field of energy efficiency and soil protection (by restoring the carbon recalcitrant stock to the use of land for agriculture and forestry (Sosa-Hernández et al. 2019).

The Kyoto Protocol implicitly refers to the field of carbon fixation in soil fixation in Art. 3.3 about net changes in emissions (sources) or absorption (deposits) of greenhouse gases, resulting from the direct human activity of land use in agriculture and forestry, starting in 1990 and in Art 3.4 about additional activities with net reduction effect on emissions (sources) or absorption (deposits) of greenhouse gases applied to the agricultural lands, changing land use, and forest land. Currently, projects that fall under Art 3.4 are accepted only for national emissions inventory reporting and are not financed through the Carbon Fund.

The reason is due to the fact that there are a number of major scientific difficulties in quantifying the carbon units actually stored through the different agricultural practices/crops (the net absorption from the atmosphere). The difficulty of quantification comes from the lack of scenarios associated with quantitative process models that allow a correct estimation of the addition of carbon units stored in all 720 soil types for over 100 potential crop plants. Under the incidence Art. 3.3 falls the afforestation activity, where the CO_2 storage is simulable by mathematical models based on the afforestation composition, the potential production class of the species in the composition and the allocation of biomass at the individual tree level in relation to the age, during the project implementation (which can be superimposed over the entire production cycle or only part of it) (Loeser and Treede 2008; Metcalfe et al. 2018).

An afforestation project funded for carbon fixing must meet a number of strict parameters defined by the Climate Change Convention: additionally fixed carbon (additional to current of business as usual practice), verifiability and transparency, replicability, durability, risk management and uncertainties and as a tool for regional and local sustainability by addressing multifunctionality.

The application of pesticides should be done only on warning. Treatment warnings are made when a pest has a tendency to develop above the economic damage threshold (EDT). The economic threshold of pest represents the level of the population of pests that produces a damage that exceeds the total costs (ecological and economic) of the treatment with means of protection of the plants (pesticides, biopreparados). Considering the central role of computer systems for forecasting and warning in good agricultural practices regarding pesticide use.

These expert systems were developed due to the development of microprocessors and reliable sensors, as well as due to advances in (radio) telecommunications. Thus, complex, flexible forecasting and warning systems have been developed, whose operation allows a rational use of methods and means of plant protection, with corresponding reduction of the impact produced by harmful organisms. The computer systems for forecasting and warning are actually expert systems, which work on the basis of (mathematical) models of specific biological processes.

The final objective of the forecast and warning programs, namely the efficient and sustainable use of agronomic resources, is subsumed by the economic and ecological objectives of sustainable agriculture and precision agriculture. The use of expert forecasting and warning systems leads not only to ecological effects (reducing environmental and food pollution) but also has direct economic consequences. The optimization of the treatments leads to significant savings of pesticides, fuel, labor force, which allow the rapid amortization of the investment in a forecasting and warning system. In EU countries, investments in forecasting and warning systems are considered environmental investments and benefit from fiscal facilities (Solarin et al. 2018).

5.7 Conclusions and Recommendations

In the long term, the evaluation of the expected effects on the environment is highlighted, with the follow-up of details regarding the efficiency of the micro-remedy, the use of biotests as tools for assessing the ecotoxicological risk, the real costs that support the implementation of these processes, and the modeling of the types of interactions between the species of fungi and the categories of pollutants.

No one has ever fertilized a forest.

The advantages of ecological reconstruction by means of mycoremediation of degraded soils are the following: it intervenes in the depollution of the soil by reducing hydrocarbons, herbicides, pesticides, heavy metals, and radionuclides; contributes to the aeration, decomposition, and transport of organic and inorganic substances, without leakage of alluvium; determine the reduction of soil erosion by protecting the useful fauna; favors the restoration of the indigenous mycobiota (mycorrhizae and saprophyte) with a role in the solving process; contributes to the filtration of parasitic protozoan species; are natural barriers for harmful insects that pierce the wood and improve the nutritional quality of the soil, favoring the development of the useful mycobiota; it can develop practical skills for educational purposes in the field of conservation and sustainable use of natural capital.

The ecological reconstruction by means of mycoremediation of soil degraded by pesticides is relevant in the following directions: remediation of soil quality through the use of mycelium in decomposing toxic residues and different categories of pollutants; forestry through the use of by-products resulting from the cultivation of fungi in the nutrition and protection of plant diversity; biological control of pests by the use of complex action substances produced by fungi; controlling the erosion of biodiversity through the microfiltration of biotic and abiotic components.

References

Asemaninejad A, Thorn RG, Lindo Z (2017) Experimental climate change modifies degradative succession in boreal peatland fungal communities. Microb Ecol 73(3):521–531

Bargaz A, Lyamlouli K, Chtouki M, Zeroual Y, Dhiba D (2018) Soil microbial resources for improving fertilizers efficiency in an integrated plant nutrient management system. Front Microbiol 9:1606. https://doi.org/10.3389/fmicb.2018.01606. eCollection 2018. Review

Bhat RA, Dervash MA, Mehmood MA, Bhat MS, Rashid A, Bhat JIA, Singh DV, Lone R (2017) Mycorrhizae: a sustainable industry for plant and soil environment. In: Varma A et al (eds) Mycorrhiza-nutrient uptake, biocontrol, ecorestoration. Springer, Cham, pp 473–502

Bhat RA, Beigh BA, Mir SA, Dar SA, Dervash MA, Rashid A, Lone R (2018) Biopesticide techniques to remediate pesticides in polluted ecosystems. In: Wani KA, Mamta (eds) Handbook of research on the adverse effects of pesticide pollution in aquatic ecosystems. IGI Global, Hershey, pp 387–407

Bruckner A, Schmerbauch A, Ruess L, Heigl F, Zaller J (2019) Foliar roundup application has minor effects on the compositional and functional diversity of soil microorganisms in a short-term greenhouse experiment. Ecotoxicol Environ Saf 174:506–513

Buysens C, Dupré de Boulois H, Declerck S (2015) Do fungicides used to control Rhizoctonia solani impact the non-target arbuscular mycorrhizal fungus *Rhizophagus irregularis*? Mycorrhiza 25(4):277–288

Castro-Rodríguez V, Cañas RA, de la Torre FN, Pascual MB, Avila C, Cánovas FM (2017) Molecular fundamentals of nitrogen uptake and transport in trees. J Exp Bot 68(10):2489–2500

Chan-Cupul W, Heredia-Abarca G, Rodríguez-Vázquez R (2016) Atrazine degradation by fungal co-culture enzyme extracts under different soil conditions. J Environ Sci Health B 51(5):298–308

Chanda A, Gummadidala PM, Gomaa OM (2016) Mycoremediation with mycotoxin producers: a critical perspective. Appl Microbiol Biotechnol 100(1):17–29. Review

Garg N, Kashyap L (2019) Joint effects of Si and mycorrhiza on the antioxidant metabolism of two pigeonpea genotypes under As (III) and (V) stress. Environ Sci Pollut Res Int 26(8):7821–7839

Hage-Ahmed K, Rosner K, Steinkellner S (2019) Arbuscular mycorrhizal fungi and their response to pesticides. Pest Manag Sci 75(3):583–590

Hashem A, Abd Allah EF, Alqarawi AA, Al-Huqail AA, Shah MA (2016) Induction of osmoregulation and modulation of salt stress in *Acacia gerrardii* Benth. By arbuscular mycorrhizal fungi and *Bacillus subtilis* (BERA 71). Biomed Res Int 2016:6294098

Himmelstein J, Maul JE, Balci Y, Everts KL (2016) Factors associated with leguminous green manure incorporation and Fusarium wilt suppression in watermelon. Plant Dis 100(9):1910–1920

Huang X, Wang L, Ma F (2017) Arbuscular mycorrhizal fungus modulates the phytotoxicity of Cd via combined responses of enzymes, thiolic compounds, and essential elements in the roots of *Phragmites australis*. Chemosphere 187:221–229

Husna BRSW, Mansur I, Kusmana C (2016) Growth and nutrient status of kayu kuku [Pericopsis mooniana (Thw.) Thw] with mycorrhiza in soil media of nickel post mining site. Pak J Biol Sci 19(4):158–170

Igiehon NO, Babalola OO (2017) Biofertilizers and sustainable agriculture: exploring arbuscular mycorrhizal fungi. Appl Microbiol Biotechnol 101(12):4871–4881

Khanday M, Bhat RA, Haq S, Dervash MA, Bhatti AA, Nissa M, Mir MR (2016) Arbuscular mycorrhizal fungi boon for plant nutrition and soil health. In: Hakeem KR, Akhtar J, Sabir M (eds) Soil science: agricultural and environmental prospectives. Springer, Cham, pp 317–332

Koricheva J, Gange AC, Jones T (2009) Effects of mycorrhizal fungi on insect herbivores: a meta-analysis. Ecology 90(8):2088–2097

Kranabetter JM, Hawkins BJ, Jones MD, Robbins S, Dyer T, Li T (2015) Species turnover (β-diversity) in ectomycorrhizal fungi linked to NH^{4+} uptake capacity. Mol Ecol 24(23):5992–6005

Leberecht M, Dannenmann M, Tejedor J, Simon J, Rennenberg H, Polle A (2016) Segregation of nitrogen use between ammonium and nitrate of ectomycorrhizas and beech trees. Plant Cell Environ 39(12):2691–2700

Lekberg Y, Wagner V, Rummel A, McLeod M, Ramsey PW (2017) Strong indirect herbicide effects on mycorrhizal associations through plant community shifts and secondary invasions. Ecol Appl 27(8):2359–2368

Liu W (2010) Do genetically modified plants impact arbuscular mycorrhizal fungi? Ecotoxicology 19(2):229–238

Loeser JD, Treede RD (2008) The Kyoto protocol of IASP basic pain terminology. Pain 137(3):473–477

Lovett GM, Tear TH, Evers DC, Findlay SE, Cosby BJ, Dunscomb JK, Driscoll CT, Weathers KC (2009) Effects of air pollution on ecosystems and biological diversity in the eastern United States. Ann N Y Acad Sci 1162:99–135

Mallmann GC, Sousa JP, Sundh I, Pieper S, Arena M, da Cruz SP, Klauberg-Filho O (2018) Placing arbuscular mycorrhizal fungi on the risk assessment test battery of plant protection products (PPPs). Ecotoxicology 27(7):809–818

Mei L, Yang X, Zhang S, Zhang T, Guo J (2019) Arbuscular mycorrhizal fungi alleviate phosphorus limitation by reducing plant N:P ratios under warming and nitrogen addition in a temperate meadow ecosystem. Sci Total Environ 686:1129–1139

Metcalfe DB, Rocha W, Balch JK, Brando PM, Doughty CE, Malhi Y (2018) Impacts of fire on sources of soil CO_2 efflux in a dry Amazon rain forest. Glob Chang Biol 24(8):3629–3641

Mohamed I, Eid KE, Abbas MHH, Salem AA, Ahmed N, Ali M, Shah GM, Fang C (2019) Use of plant growth promoting Rhizobacteria (PGPR) and mycorrhizae to improve the growth and nutrient utilization of common bean in a soil infected with white rot fungi. Ecotoxicol Environ Saf 171:539–548

Motaharpoor Z, Taheri H, Nadian H (2019) *Rhizophagus irregularis* modulates cadmium uptake, metal transporter, and chelator gene expression in *Medicago sativa*. Mycorrhiza 29(4):389–395

Mushtaq N, Bhat RA, Dervash MA, Qadri H, Dar GH (2018) Biopesticides: the key component to remediate pesticide contamination in an ecosystem. In: Environmental contamination and remediation. Cambridge Scholars Publishing, Cambridge, pp 152–178

Pena R, Lang C, Naumann A, Polle A (2014) Ectomycorrhizal identification in environmental samples of tree roots by Fourier-transform infrared (FTIR) spectroscopy. Front Plant Sci 5:229

Smith ME, Henkel TW, Williams GC, Aime MC, Fremier AK, Vilgalys R (2017) Investigating niche partitioning of ectomycorrhizal fungi in specialized rooting zones of the monodominant leguminous tree *Dicymbe corymbosa*. New Phytol 215(1):443–453

Solarin SA, Al-Mulali U, Gan GGG, Shahbaz M (2018) The impact of biomass energy consumption on pollution: evidence from 80 developed and developing countries. Environ Sci Pollut Res Int 25(23):22641–22657

Song F, Li J, Fan X, Zhang Q, Chang W, Yang F, Geng G (2016) Transcriptome analysis of Glomus mosseae/Medicago sativa mycorrhiza on atrazine stress. Sci Rep 6:20245. https://doi.org/10.1038/srep20245

Sosa-Hernández MA, Leifheit EF, Ingraffia R, Rillig MC (2019) Subsoil arbuscular mycorrhizal fungi for sustainability and climate-smart agriculture: a solution right under our feet? Front Microbiol 10:744

Spagnoletti FN, Balestrasse K, Lavado RS, Giacometti R (2016) Arbuscular mycorrhiza detoxifying response against arsenic and pathogenic fungus in soybean. Ecotoxicol Environ Saf 133:47–56

Sudová R, Kohout P, Kolaříková Z, Rydlová J, Voříšková J, Suda J, Španiel S, Müller-Schärer H, Mráz P (2018) Sympatric diploid and tetraploid cytotypes of *Centaurea stoebe* s.l. do not differ in arbuscular mycorrhizal communities and mycorrhizal growth response. Am J Bot 105(12):1995–2007

Torres JP, Fróes-Asmus CI, Weber R, Vijgen JM (2013) HCH contamination from former pesticide production in Brazil-a challenge for the Stockholm Convention implementation. Environ Sci Pollut Res Int 20(4):1951–1957

Trocha LK, Weiser E, Robakowski P (2016) Interactive effects of juvenile defoliation, light conditions, and interspecific competition on growth and ectomycorrhizal colonization of Fagus sylvatica and *Pinus sylvestris* seedlings. Mycorrhiza 26(1):47–56

Wang R, Wang M, Chen K, Wang S, Mur LAJ, Guo S (2018) Exploring the roles of aquaporins in plant microbe interactions. Cells 7(12):E267. https://doi.org/10.3390/cells7120267. Review

Zaller JG, Cantelmo C, Santos GD, Muther S, Gruber E, Pallua P, Mandl K, Friedrich B, Hofstetter I, Schmuckenschlager B, Faber F (2018) Herbicides in vineyards reduce grapevine root mycorrhization and alter soil microorganisms and the nutrient composition in grapevine roots, leaves, xylem sap and grape juice. Environ Sci Pollut Res Int 25(23):23215–23226

Zhang S, Wang L, Ma F, Zhang X, Fu D (2016) Reducing nitrogen runoff from paddy fields with arbuscular mycorrhizal fungi under different fertilizer regimes. J Environ Sci (China) 46:92–100

Chapter 6
Bio-Pesticides: Application and Possible Mechanism of Action

Javaid Ahmad Wani, Adil Farooq Wali, Sabhiya Majid, Saiema Rasool, Muneeb U. Rehman, Shahzada Mudasir Rashid, Shafat Ali, Sanah Farooq, Shabhat Rasool, Ajaz Ahmad, and Wajhul Qamar

6.1 Introduction

As per oxford dictionary, the word "pest" originated from the French word peste or Latin word pestis denoting the bubonic plague in fifteenth century. Pest is any destructive living organism that harms crops, food, livestock etc. Pesticides are synthetic chemicals that are applied usually to crops to destroy pests. The major group

J. A. Wani · S. Majid · S. Ali · S. Farooq · S. Rasool
Department of Biochemistry, Government Medical College (GMC), Srinagar, Jammu and Kashmir, India

A. F. Wali
RAK College of Pharmaceutical Sciences, RAK Medical and Health Sciences University, Ras Al Khaimah, United Arab Emirates

S. Rasool
Forest Biotech Lab, Department of Forest Management, Faculty of Forestry, University Putra Malaysia, Serdang, Selangor, Malaysia

M. U. Rehman (✉)
Department of Biochemistry, Government Medical College (GMC), Srinagar, Jammu and Kashmir, India

Department of Clinical Pharmacy, College of Pharmacy, King Saud University, Riyadh, Kingdom of Saudi Arabia

S. M. Rashid
Division of Veterinary Biochemistry, Faculty of Veterinary Science and Animal Husbandry, SKUAST-Kashmir, Shuhama, Jammu and Kashmir, India

A. Ahmad
Department of Clinical Pharmacy, College of Pharmacy, King Saud University, Riyadh, Kingdom of Saudi Arabia

W. Qamar
Department of Pharmacology and Toxicology, Central Lab, College of Pharmacy, King Saud University, Riyadh, Riyadh, Saudi Arabia

R. A. Bhat et al. (eds.), *Bioremediation and Biotechnology, Vol 2*,
https://doi.org/10.1007/978-3-030-40333-1_6

of pesticides comes under insecticides (e.g., phosphate esters, organochlorines, carbamate esters) and others are as rat poison (e.g., anticoagulants), herbicides (e.g., paraquat, diquat, 2,4-dichlorophenoxyacetic acid [2,4-D]), fungi killing substances (e.g., dithiocarbamates, ethyl mercaptan), and volatile poisonous gases (e.g., ethylene dibromide, methyl bromide) (Ellenhorn et al. 1997; Stamati et al. 2016). Meanwhile in 1960s, synthetic organic pesticides were the most common pesticides used for pest control, developed in the 1940 which begun with DDT usage then other pesticides like phosphate esters and carbamate esters pesticides (Nicholson 2007; Stamati et al. 2016). For successful enactment of Green Revolution technology through the application of pesticides to kill and control the pest population, pesticides get used in excess and their indiscriminate use becomes an important cause to several detrimental effects on plants, animals, and their environment. For instance, in human beings, pesticides induce free radical generation, lipid peroxidation, and inhibit the mechanism that generates total antioxidants for the body (Salazar-Flores et al. 2019; Abdollahi et al. 2004). Identifying the harmful effects of chemical pesticides such as development of pesticide resistant insect species, pest reappearance after application of pesticide, development of pests that were not considered in the category of pests before due to their natural death by application of pesticides, accumulation of pesticide residues in food such as fruits and vegetables consumed by humans, deteriorate their health and also causing ecological disturbances like accumulation in outer layer of lithosphere, atmosphere and hydrosphere (Salazar-Flores et al. 2019; Al-Zaidi et al. 2011). Thus without deteriorating environment and human health and also to generate high yield from diminished natural resources, it becomes need of the hour to develop environmental friendly alternatives for pest management.

Bio-pesticides are described as a broad spectrum of industrially generated products that are active agents of biological origin to control the population of insects, weeds, and disease causing microorganisms (Glare et al. 2012). As per FAO, bio-pesticides differ from biological control agents. Bio-pesticides are inactive agents of biological origin while biological control agents actively search for the pest and kill them such as parasites, nematodes pathogenic to insects. The US Environmental Protection Agency grouped bio-pesticides under microbial pesticides like bacteria, fungi, viruses, protozoa, and nematodes pathogenic to insects, second with genetically modified plants having herbicide, pesticide resistant traits and third are pheromones and other substances of natural origin (Fig. 6.1).

The first biocontrol agent, used as a rat poison was a *salmonella enteric* (Ratin) applied in Sweden and other European countries belonging to European continent in 1904s (Sundh and Goettel 2013) (Table 6.1).

Advantages of bio-pesticides over conventional pesticides/chemical pesticides (EPA 2006; Kumar 2012; Senthil-Nathan 2013; Bhat et al. 2018; Mushtaq et al. 2018):

- They are safe for consumers because of no residue problem specifically for edible fruits and vegetables.

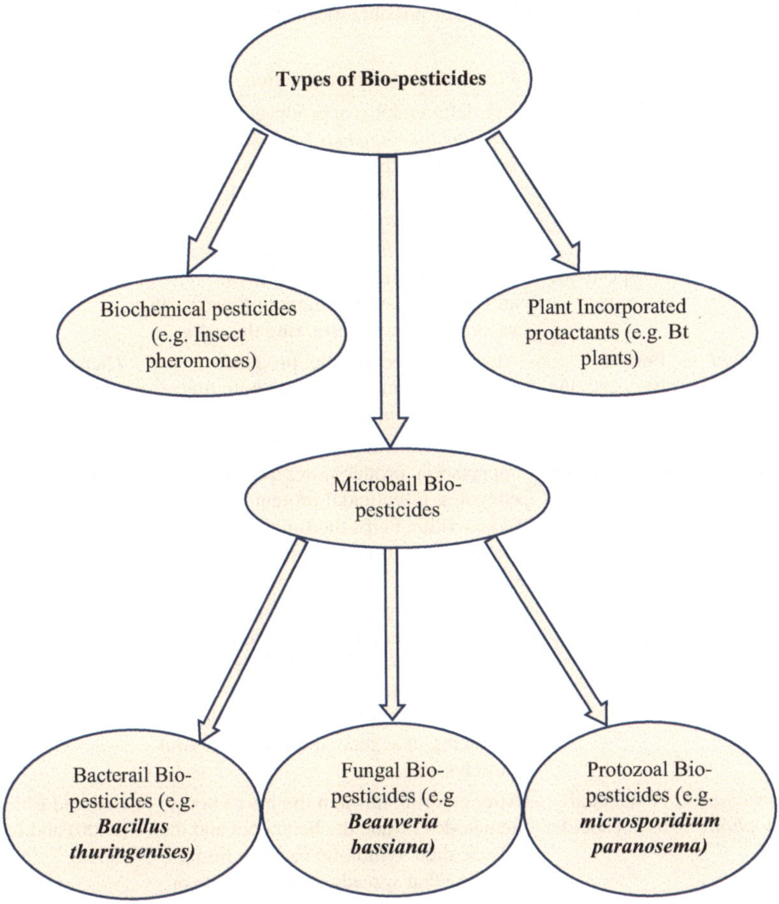

Fig. 6.1 Classification of bio-pesticides (US-EPA 2019)

- The efficacy of bio-pesticides is nearly same to chemical pesticides if we utilize them as one of the component of integrated pest management specifically for the cultivation of fruits, vegetables, and nuts.
- Since bio-pesticides have low pesticide resistance potential and are environment friendly thus do not face the same monitoring problems seen with chemical pesticides.
- Bio-pesticides do not cause environmental quality degradation problems, are gentle to beneficial insects, frequently target specific and also agricultural crops can be reentered soon after the treatment.
- When we want to cultivate organically, microbial bio-pesticides are a good alternative due to low detrimental effects to human health and low potential of resistance developed by pests.

Table 6.1 Types of bio-pesticides and their possible mechanism of action

Organism	Type of bio-pesticide	Possible mechanism of action	References
Bacillus thuringiensis	Bacterial bio-pesticide	Cry delta-endotoxin produced by *Bacillus thuringiensis* bind specifically insect gut epithelial cells, generate pores lead to cell content leakage and ultimately insect death	Jisha et al. (2013) and Shingote et al. (2013)
Bacillus subtilis	Bacterial bio-pesticide	*Bacillus subtilis* derived lipopeptides such as surfactin, Iturin A have antibacterial and antifungal properties by vacuolizing and necrotizing the cells	Ghribi et al. (2011, 2012)
Metarhizium anisopliae	Fungal bio-pesticide	Attach to insect cuticle, produce degradative enzymes which help the fungi to germinate, and grow on insect	Zheng et al. (2012)
Beauveria bassiana	Fungal bio-pesticide	After attachment to insect by appressoria, produces degradative enzymes, insecticidal proteins like 60 kD bassiacridin, helps the fungus to destroy the cuticle, germinate, and grow on the insect	Ortiz-Urquiza and Keyhani (2013) and Borgi and Gargouri (2016)
Baculovirus	Viral bio-pesticide	After ingesting polyhedral enveloped virus by the insect, outer protein matrix is dissolved in alkaline pH, releasing OBV that destroy peritrophic membrane, enter into mid gut cells, multiply and producing new generation of (OB) viral particles	Passarelli (2011)
Heterorhabditis bacteriophora	Nematodal bio-pesticides	After entering through the insect body, nematodes invade the hemocoel and then release their symbiotic bacteria into the intestine that spreads to whole body of insect causing septicemia, killing the host within 24–48 h	Han and Ehlers (2000) and Lu et al. (2016)
Bt plants	Plant-incorporated protectants (PIP)	Exotic DNA sequences that code for insecticidal toxins and others code for siRNA—20 nucleotide sequences that specifically inhibits mRNA translation leading to insect growth retardation or death are incorporated into plant genome	Ramaseshadri et al. (2013) and Fishilevich et al. (2016)
Azadirachta indica	Botanical pesticides	Azadirachtin acts on the prothoracic gland and insects, it is also a strong antifeedant of insects	Weinzierl (2000)

6.2 Microbial Pesticides

During the period of Green Revolution, indiscriminate use of synthetic chemical pesticides to produce maximum yield and this objective was fulfilled, but other problems arise especially the extinction of beneficial microbial and insect species. Hence the need for integrated pest management arose which persuaded us for the application of microbial biocontrol agent based pesticides as an effective and environmental sensitive approach (Chandler et al. 2011). As per US environmental agency (EPA), microbial pesticides are based on prokaryotic organisms like bacteria, eukaryotic organisms like protozoa, fungi, nematodes pathogenic to insects, microbial (entomopathogenic) agents, and acellular obligate parasites like virus.

6.2.1 Bacteria (Mechanism of Action and Applications)

6.2.1.1 *Bacillus thuringiensis*

Bacterial bio-pesticides contribute about 74% of the market (Thakore 2006). The first *Bacillus thuringiensis* subspecies named as *Bacillus thuringiensis* israelensis (Bti) was presented in the year of 1975 and found to be lethal to mosquito (Goldberg and Margalit 1977). *Bacillus thuringiensis* is a gram-positive spore generating bacteria, producing insect specific crystalline delta-endotoxins after lysis at the end of sporulation process. After ingestion by the insect, death occurs within 48 h (Jisha et al. 2013).

Several hours after ingestion of delta-endotoxin by the insect, a cascade of events starts in its mid gut leading to insect death by the increased pH. In mid gut of insect, delta-endotoxin is solubilized, and then fragmented by insect proteolytic enzyme into core toxic fragment (Jisha et al. 2013). The insect protein cleaving enzyme cleaves full protein into N and C terminus fragment which makes individual fragments active. These active fragments interact with the receptors present on the plasma membrane of epithelial cells lining the apical microvilli. This interaction causes conformational change in protein, leading its incorporation into the semipermeable cell membrane and the oligomeric complex forms on the cell membrane that develops an opening into the cell membrane causing leakage of cell contents, disruption of cell transport, endosmosis cause cell bursting and ultimately insect demise (Jisha et al. 2013; Schünemann et al. 2014). Another interesting binary toxin (vegetative insecticidal protein) produced from *Bacillus thuringiensis* strain is Vip1-Vip2 with coleopteran specificity such as store grain pest (*Sitophilus zeamais*) (*Curculionidae* family) causing 60% death (Yu et al. 2011; Shingote et al. 2013).

Lepidoptera is the major insect order containing species susceptible to *Bacillus thuringiensis* followed by Diptera and Coleoptera. Lepidoptera contains pest families that are of major concern in agriculture such as Cossidae, Thaumetopoeidae, Gelechiidae, Lymantriidae, Noctuidae, Pieridae, Pyralidae, Yponomeutidae, Tortricidae, and Pyralidae (Gathmann and priesnitz 2014). Diptera also contains agriculturally major pest families namely Tephritidae, Culicidae, Muscidae, Simuliidae, and Tipulidae whose some species are found susceptible to *Bacillus thuringiensis* toxicity (Azizoglu et al. 2017; Martins et al. 2018). In addition delta-endotoxin released by *Bacillus thuringiensis* which has antifungal activity, it produces acyl homoserine lactone lactonase enzyme that (blocks quorum sensing system in bacteria) and linear aminopolyol antibiotic called zwittermicin A (ZmA) having stronger lethal activity towards oomycota (water molds) and their relatives and also gram negative bacteria. As all fungi are covered by chitin layer, chitinase enzyme released by *Bacillus thuringiensis* has an essential role in biocontrol of phytopathogenic fungi such as 100% growth inhibition in *Sclerotium rolfsii*, 55–82% in *Aspergillus terreus*, *Aspergillus gillus flavus*, etc. (Reyes-Ramirez et al. 2004; Zhou et al. 2008; Lucon et al. 2010). Thus *Bacillus thuringiensis* may be used as a potential biocontrol in plant diseases.

6.2.1.2 *Bacillus subtilis*

Bacillus subtilis is a ubiquitous soil bacterium whose suitable habitat may be plant root system. It easily grows in plant root system by competing with root pathogens for food and shelter which attack the roots. *Bacillus subtilis* has been found as an effective biological agent for controlling population of plant virulent species of nematodes, bacteria, and fungi (Ongena et al. 2010). It has been found that *Bacillus subtilis* SPB1 strain causes death of an insect by bioactive lipopeptide. This bioactive lipopeptide actually enters into insects' tissue by vacuolizing and necrotizing of the epithelial cells by causing obvious histopathological changes like destroying the cells and their boundaries; its lipopeptide biosurfactant enters into insects' tissue. This has been found as an effective biological agent for the Olive moth (*Prays oleae*) population with a LC_{50} value of 142 ng/cm^2 (Ghribi et al. 2011, 2012).

It was found that the cells as well as spores of *Bacillus subtilis* are lethal to nematodes such as *Anguina tritici*, *Heterodera cajani M. incognita*, *Heterodera and cajani*, *Heterodera avenae*. As observed by Kavitha et al. (2012) plant dwelling (endophytic) *Bacillus subtilis* strains that produce a high amount of lipopeptide antibiotics effectively inhabit the egg laying and stop the metamorphosis process of *Meloidogyne incognita* at the second stage juvenile. This nematode is the most destructive pathogen, destroying a broad spectrum of crops under *in vitro* conditions.

Bacillus subtilis derived lipopeptides (e.g., surfactin, Iturin A) have been found to have biocontrol potential against several plant pests especially fungal phytopathogens (Ongena and Jacques 2007). A *Bacillus subtilis* derived lipopeptide (surfactin) has bioactive lethal activity against virulent fungi *Aspergillus flavus*, *Colletotrichum gloeosporioides*, and also exhibits growth inhibition of phytopathogenic fungi like Fusarium species, Aspergillus species, etc. (Mohammadipour et al. 2009; Velho et al. 2011). A novel linear bioactive lipopeptide called Gageotetrins A-C are obtained from a marine *Bacillus subtilis* that showed good time course motility inhibition and also possesses lytic activity towards late blight pathogen *Phytophthora capsici* that is a major pest causing damage to cucumber, chili, beans, and tomato at 0.02 mM concentration (Tareq et al. 2014). Some *Bacillus subtilis* strains synthesize mycolytic enzymes that cause lysis of pathogenic fungi cell wall, thus act as biological control agent of plant rot disease (Yan et al. 2011; Khanday et al. 2016; Bhat et al. 2017). *Bacillus subtilis* bacteria produces enzymes like chitinase, glucanase, and cellulase whose activities help in controlling anthracnose disease of chili caused by *Colletotrichum gloeosporioides* OGC1 (Ashwini and Srividya 2014).

6.2.2 *Fungi*

6.2.2.1 *Metarhizium anisopliae*

Metarhizium anisopliae is ubiquitous and found everywhere in the soil, and a large spectrum of insect pest species are susceptible to this fungi. Under natural conditions, *Metarhizium anisopliae* is induced by moist conditions of soil to produce infectious spores called conidia. This subspecies was first described in 1879 by Metschikoff as named *Entomophthora anisopliae* described as pathogen of wheat cockchafer and in 1883 name changed to *Metarhizium anisopliae* by Sorokin. A large variety of different strains come under this species (Tulloch 1976; Mnyone et al. 2010).

Entomopathogenic fungi attack the insect by infiltrating cuticle layer anywhere on the insect body, but other organisms like viruses, bacteria, and nematodes attack their host through some particular routes. Infection begins by attachment of entomopathogenic fungi with insect cuticle by the help of infectious conidia. After attachment to insect cuticle, it begins to release many degradative enzymes such as proteases that degrade proteins, chitinases that degrade chitin polysaccharide, lipases ultimately helps the pathogen to grow and germinate on the cuticle of insect (Zheng et al. 2012; Xiao et al. 2012).With the passage of time, novel structures appear from the growing fungi namely infiltration pegs or appressoria or both. These specific structures empower the rising hyphae to infiltrate deep into the insect integument. Brazil is one of the South American countries that uses this fungi on a large scale on their crop fields to protect them from spittle bugs (*Mahanarva fimbriolata*), its larval forms suck water and nutrients from the plant roots increasing stress while adults use leaves as food and insert toxins that cause leaf necrosis reducing production to 40% in Brazil (Dinardo-Miranda et al. 2002; Alves and Teixeira 2014). Above 100,000 ha of sugarcanes are applied with this fungi every year (Faria and Magalhaes 2001).

6.2.2.2 *Beauveria bassiana*

This fungus was discovered as bio-pesticide unintentionally by Agostino Bassi (1835) when he first investigated the disease of silk worms called white muscardine caused by this fungus. *Beauveria bassiana* is a diverse fungus targeting insects that are found in temperate and tropical regions. The suitable habitats of *Beauveria bassiana* range from desert soils to forests and cultivated soils.

The entomopathogenic infection begins with appearance of infection structures called appressoria and other virulent factors such as proteins, and enzymes, which facilitates *Beauveria bassiana* conidia to attach to the host arthropod, breach the cuticle, grow in number within the hemocoel as blastospores (hyphal bodies that escape from the surveillance of host immune system (Lewis et al. 2009) and finally the host is destroyed. Then *Beauveria* bassiana hyphae reappears, covering the whole body of insect corpse, and generates novel conidia, thus finishing the parasitic life cycle (Toledo et al. 2010; Ortiz-Urquiza et al. 2010; Ortiz-Urquiza and Keyhani 2013). Hydrophobins that are expressed on the surface of fungi make it hydrophobic

(Hyd1 and Hyd2).This hydrophobic surface easily adheres to the epicuticular layer of pest and thus contributing in its virulence (Cho et al. 2007; Zhang et al. 2011). After attachment, it produces degradative enzymes such as cytochrome $P_{450,}$ catalases, esterases, long-chain alcohols, and aldehyde dehydrogenases, which facilitates degradation of lipid, hydrocarbon, protein, and other compounds found in insect cuticle. *Beauveria bassiana* also generates insecticidal, 60 kD protein called bassiacridin that possess many enzymatic activities such as beta-glucosidase, beta-galactosidase, and N-acetylglucosaminidase activities (Quesada-Moraga and Vey 2004; Borgi and Gargouri 2016). In addition to this degradative and detoxifying enzyme cocktail, *Beauveria bassiana* also generates some of the organic acids mainly oxalic acid that has a role in *Beauveria* bassiana pathogenicity since oxalic acid facilitates to loosen the structural rigidity of insect cuticular surface (Bidochka and Khachatourians 1991).

In addition, smaller sized secondary metabolites synthesized in vitro and in vivo participate in parasitization process. They play important roles as (a) toxins that triggers insect's death, (b) immunomodulators that enable the fungus to evade from the surveillance of the host immune system, (c) generating some biochemicals that have an activity against rival microorganisms, and (d) defensive molecules against mycophagous organisms (Charnley 2003; Feng et al. 2015a, b; Lozano-Tovar et al. 2017). The known secondary metabolites produced by this entomopathogenic fungus include cyclic peptides, such as beauvericin, bassianolide, and beauverolides, and polyketide-derived pigments, such as oosporein, tenellin, and bassianin, but only those genes involved in the biosynthesis of beauvericin, bassianolide, tenellin, and oosporein have been functionally verified (Halo et al. 2008; Molnar et al. 2010; Feng et al. 2015a, b). Beauvericin is probably the most studied cyclic peptide compound produced by *Beauveria bassiana*. Beauvericin is an effective anti-pesticidal molecule (Hamill et al. 1969) but the exact mode of action is not fully understood (Wang and Xu 2012). In addition, this hexadepsipeptide has cytotoxic and proapoptotic activities in several human cell lines including leukemia cells (Wang and Xu 2012). Beauvericin seems to act as an ionophore, forming cation-selective channels and increasing intracellular Ca^{+2} (Kouti et al. 2003), for example, in case of erythrocytes in which increased intracellular calcium cause stimulation of KCL channels leading to water loss, cell shrinkage (Qadri et al. 2011) and also have been suggested to trigger calcium sensitive cell apoptotic pathways in some other cells (Wang and Xu 2012).

6.2.3 *Protozoa*

6.2.3.1 *Nosema*

There are almost 1000 entomopathogenic protozoans. They are lethal and act as biocontrol agent for many invertebrates. Grasshoppers and heliothine moth are common insect species susceptible to *Nosema,* for example, Henry et al. (1981) recently developed *microsporidium Paranosema* (*Nosema*) for grasshoppers control (Lange and Sokolova 2017). Nosema species and *Vairimorpha necatrix* are two well recognized entomopathogenic protozoan species. These two species come under Microsporidia (unicellular spore generating organisms) generally host specific and produce spores that are taken by the pest and germinate in its mid gut. The infection process is slow and chronic. Massive demolition of organs and tissues starts when the sporoplasm ruptures to release germinating spores. Sporulation process recycles from the infected tissues and upon ejection and assimilation by a susceptible host induces an epizootic (wide spread outbreak) within insect population (Brooks 1988; Milner et al. 2003). They are obligate parasites that thrive only in living hosts and some species need another host for the asexual development that is intermediate host. Microsporidia species have benefits over other entomopathogenic protozoa such as perseverance and repeating of their reproductive cycle in susceptible host populations and their debilitating effect on reproduction and overall fitness of susceptible host insects (solter and Becnel 2000).

Nosema locustae (entomopathogenic protozoan) is the only species that showed effective activity towards grasshoppers and later on was registered and recognized commercially (Henry et al. 1981). The first reported microsporidium is *Nosema bombycis* which is the causative agent of devastating silkworm disease called silkworm pébrine which continued in some parts of Asia, Europe, and North America, during the period of 1850 AD (Becnel and Andreadis 1999). Unfortunately, pébrine is still causing a wide spread outbreak among the silk-producing countries like China leading to large financial sufferings (Cai et al. 2012).

6.2.4 *Virus*

6.2.4.1 *Baculovirus*

Baculovirus is the most important class of arthropod virus (Herniou and Jehle 2007). The use of Baculovirus started in nineteenth century as a biocontrol agent for forestry and agriculture pests. They are the most deeply studied virus pathogenic to insect (Inceoglu et al. 2001). Baculovirus have two different observable phenotypes, one is the budded virus phenotype that is generated after initial infection by the budding through the plasma membrane of cells infected by the virus. Budded virus particles contribute to spread the viral particles to another region from mid gut of insect (Jehle et al. 2006). The other is occluded phenotype (OV) that is sealed in a crystalline protein matrix covering called occlusion body (OB) generated in the final phase of infection. Occlusion bodies enable the virus to remain alive even outside the host (Cory et al. 2000). Two groups of baculovirus are observed, one is called the nuclear polyhedrosis virus (NPVs) and the other one is granulovirus (GVs). In nuclear polyhedrosis virus, occlusion body contains many viral particles, but in granulovirus, occlusion body usually contains just single viral particle. Nuclear polyhedrosis virus constitutes three genera, Alphabaculovirus that is effective against lepidopteran, Gammabaculovirus which is effective towards hymenopteran, and last one is Deltabaculovirus effective for dipteran insects; granuloviruses constitute one genus Betabaculovirus that specifically controls the population of lepidopteran pests only (Jehle et al. 2006). Occlusion bodies in insect excrements spread to other insects when present on plants ingested by insect larva (Szewczyk et al. 2006).

In 1982 AgMNPV was developed which is a commercially well recognized viral bio-pesticide of Alphabaculovirus to be effective against *Anticarsia gemmatalis* that is a common and copious soybean pest in Brazil. It was used to treat around 2000 ha of soybean (Moscardi 1999) and its use starts growing. The application of AgMNPV to manage the insect population of *Anticarsia gemmatalis* in Brazil was a very effective program and is considered as the most important one through the world (Yang et al. 2012). Many other South American countries also followed the same program (Sosa-Gómez et al. 2008). Globally, *Cydia pomonella* is considered as a frequent pest of apples, pears, and walnuts (Arthurs et al. 2007). Since 1964 *Cydia pomonella* granulovirus (CpGV) was developed that is an effective viral bio-pesticide for insect control, safeguarding fruits from financial losses in integrated pome fruit production especially in European countries and some other countries (Moscardi 1999; Lacey et al. 2008; Kutinkova et al. 2012). CpGV viral bio-pesticide is possibly the well-established viral bio-pesticide on a commercial scale that is applied worldwide in terms of treated area (Vincent et al. 2007). Globally, the genera such as Heliothis and Helicoverpa signify important insect pests. They are accounted for million dollars economic losses every year (Moscardi et al. 2011). *Helicoverpa armigera* (cotton bollworm) is a highly polyphagous insect and one of the most serious pests recognized globally. This pest attacks numerous crops namely some kinds of beans, cotton, maize, tomatoes, tobacco, etc. belonging to wide range of plant families (Sun et al. 2002). HaSNPV viral bio-pesticide is manufactured, produced, and treated to the crops by many countries such as Australia, India, and in China, it is considered as one of the most important baculovirus bio-pesticides applying different NPVs on over 100,000 ha of crops per year (Moscardi et al. 2011; Yang et al. 2012).

Apart from wide application of baculovirus, the slow mode of killing mechanism becomes a hurdle in their wide spread practical use. So, many tactics are established to progress their killing action by altering the baculovirus genome using genetic engineering, to insert specific kinds of genes taken from insects such as genes coding for hormones or enzymes or toxins (Kamita et al. 2005; Li and Bonning 2007; Gramkow et al. 2010).

6.2.5 Entomopathogenic Nematodes

6.2.5.1 Steinernema and Heterorhabditis (Rhabditida)

There are 23 nematode families from which many nematodes are known to parasitize insects. Such nematodes are called entomopathogenic nematodes (EPN) (Koppenhofer 2007). Steinernema and Heterorhabditis were the first two active entomopathogenic nematode genera revealed and documented as a biocontrol agent towards insects during the beginning of 1990s (Copping and Menn 2000). The Heterorhabditidae and Steinernematidae entomopathogenic nematode families have received the greatest consideration because they keep many of the characteristics of an ideal biocontrol agent (Kaya and Gaugler 1993; Grewal et al. 2005; Koppenhofer 2007) and are applied as conventional, protective, and proliferative biocontrol agents. Steinernematidae have two genera represented by Steinernema and Neosteinernema while Heterorhabditidae has one genera indicated by the genus Heterorhabditis and used as biocontrol entities in agriculture for pest management (Kaya and Gaugler 1993; Kaya and Stock 1997; Lacey and Georgis 2012). The alimentary canal of all entomopathogenic nematodes is usually occupied with symbiotic bacteria like Xenorhabdus bacteria present in Steinernematidae, while Photorhabdus bacteria are found in Heterorhabditidae (Boemare 2002). Such nematode and bacterium complex works mutually as a biocontrol unit to destroy an insect pest. The above two nematode families are not closely related and belong to the order Rhabditida (Blaxter et al. 1998) but because they possess many biological similarities, they will be considered together. Entomopathogenic nematodes attack insects at soil-lodging stages and destroy them within 48 h by ejection of pathogenic bacteria (bacteria beneficial to nematode) that are present in alimentary canal of nematode. The infectious stages of the nematode become adult and an army of infective juveniles is produced from the insect cadaver. There have been various efforts to use nematode IJ to control insect pest population via spraying but little is known about the capability of native species of nematodes about their impact on the pest population (Peters 1996). However their capability can be enhanced by using in combination with other bio-pesticides such as entomopathogenic fungi, e.g., Wakil et al. showed that application of *Heterorhabditis bacteriophora* in combination with *Beauveria bassiana* exhibited greater effectiveness against the larval stages of *Rhynchophorus ferrugineus* (Olivier) commonly called red palm weevil (Wakil et al. 2017).

Once the non-feeding IJ have entered through the insects natural body pores like the egestion opening, ingestion opening, and respiratory openings (spiracles), nematodes invade the hemocoel and then release their symbiotic bacteria into the intestine that spreads to whole body of insect causing septicemia, killing the host within 24–48 h. The uptake of infective juveniles of nematode is quickly manipulated by the bacteria and decays the host tissues. Recently it has been also found that activated infective juveniles produce some lethal proteins called venom proteins that also contribute to its pathogenicity (Lu et al. 2017). Almost two to three progeny of the nematodes are produced within the host cadaver (Pomar Jr and Leutenegger 1968; Bird and Akhurst 1983; Grewal et al. 1997). Earlier it is found that the important insect pest families observed in stockpiled goods such as Pyralidae and Curculionidae can be efficiently controlled by entomopathogenic nematodes (Duncan and McCoy 1996; Shannag and Capinera 2000; Shapiro and McCoy 2000). Higher concentration of nematodes has been found to be susceptible to insect pests present in stockpiled goods like *Ephestia kuehniella Zeller* and *Tenebrio molitor L.* by Morris (1985). Georgis (1990) had also detected a field concentration of >2.5 billion nematodes per ha effective against some of the important insect pests of crops grown in rows (row crops).

6.3 Plant-Incorporated Protectants/Genetically Modified Plants (GMP)

Plant-incorporated protectants (PIPs) are bioactive ingredients having macromolecular nature, produced in transgenic plants whose genome is modified by introducing some exotic gene sequences. First-generation plant-incorporated protectants having insect killing activity are Cry genes (code for bt toxins) taken from bacterium *Bacillus thuringiensis* genome, introduced in specific plants like maize, cotton, etc. Now these plants are called genetically modified crops or transgenic plants. These *Bacillus thuringiensis* crops contain one or more than one gene sequences that code for insect killing Cry proteins chosen from the bacterium *Bacillus thuringiensis* (Bt) (Clark et al. 2005). Their size typically ranges from 63 to 68 kDa (Mendelsohn et al. 2003). A novel group of genetically modified crops are established that possess bioactivity by expression of double stranded RNA (dsRNA) PIPs that silences expression of specific mRNA which are essential for their survival. RNA interference mechanism protects the crops in which dsRNA is introduced which is a preserved eukaryotic gene expression regulation process revealed in 1998. The United States Environmental Protection Agency permitted the first RNA interference crop in June 2017 to biocontrol corn rootworm which is a major insect pest (Ramaseshadri et al. 2013; Fishilevich et al. 2016). This crop also produces a *Bacillus thuringiensis* delta-endotoxin protein and contains some specific genes that make it resistant to glyphosate herbicide (US, EPA 2015). Fifteen years after the dawn of genetic engineering, the first GM crops expressing PIPs were presented in 1987 (Barton et al. 1987; Fischhoff et al. 1987; Vaeck et al. 1987). Several Bt crops have been registered mainly cotton and maize from the year 1995 (GM crop database, EPA-PIPS registrations). Many Bt crops now show a "stacked" phenotype, expressing traits like two or more Cry proteins (Tabashnik et al. 2013, EPA-PIPS registrations) or Cry protein(s) expressed with other traits like immunity to the herbicide glyphosate (Que et al. 2010). In 2016, it is estimated that 98.5 million hectares of transgenic plants including 23.1 million ha of crops expressing cry protein and 75.4 million ha of crops expressing both anti-pesticidal and immunity to herbicide were planted globally, which is nearly equivalent to partial of the global area planted with cotton or maize (GM Crops 2016). Consent of genetically modified crops producing PIPs, developed on a commercial scale in Australia, Asia, and America, and, to a smaller level in parts of Africa. Nevertheless, the application of genetically modified crops producing plant-incorporated protectants is limited in Europe (GM Crops 2016).

Bt toxin protein and siRNA PIPs produce their desirable effects like blocking insect development or leading to insect death by their different mechanisms of action when an insect ingests the tissue of transgenic crop. Bt toxin protein interacts with definite receptors on the plasma membrane of epithelial cells lining the insect mid gut, incorporates into the plasma membrane, and finally generates transmembrane channels through which cell content is leaked leading to ultimately insect pest death (Bravo et al. 2007). There is structural diversity in cry proteins, governing their toxicities towards the insect pests that are specific to certain insect orders (Xu et al. 2014; Crickmore et al. 2016) like Cry1 proteins is lethal to Lepidoptera, e.g., the corn borer and Cry3 protein is lethal to Coleoptera, e.g., the corn rootworm (Clark et al. 2005).

After ingestion of tissue of transgenic plant, double stranded RNA is taken by the specific cells lining the insect mid gut. Within the cell, dsRNA is trimmed by RNase enzymes into siRNA that approximately 20 nucleotides in length which guide the cleavage of matching mRNA (complementary) into small pieces by its association with insect's endogenous RNAi machinery. The degradation of the targeted mRNA by siRNA-RISC complex blocks its translation pathway that is vital to the pest insect, causing sub-lethal consequences (e.g., reduced growth) or pest death. The first double stranded RNA plant-incorporated protectants that have received regulatory consent biocontrol the corn rootworm (*Diabrotica virgifera virgifera*) by preventing the translation of the Snf7 protein, an essential vacuolar sorting protein (Ramaseshadri et al. 2013)

6.4 Biochemical Pesticides

According to US Environmental Protection Agency biochemical pesticides are substances that control pest population by the non-lethal mode of action and also they should be naturally occurring substances. By contrast synthetic pesticides directly destroy or deactivate the pest like pheromones, plant extracts, natural insect growth regulators, etc.

6.4.1 Plant Extract/Botanical Pesticides

Botanical pesticides are usually secondary phytochemicals presented as a substitute to synthetic chemical formulations but they sometimes show little toxicity towards humans. The use of botanical pesticides are cited in Hieroglyph, Chinese, Greek, and Roman ancient times and also in India where the application of extracts obtained from the neem tree (*Azadirachta indica Juss* Meliaceae) was described in the Veda, a body of documents written in ancient Sanskrit dated at least 4000 years ago (Philogène et al. 2005; Regnault-Roger and Philogène 2008)

6.4.1.1 *Azadirachta Indica* (Neem)

Schmutterer (1990) first discovered neem tree as bio-pesticide when he found that crowded desert locusts defoliated almost all native flora but neem tree was resistant to its defoliating action in Sudan. It is found that neem seed oil is active against soft-bodied insects, mites, and management of plant pathogens due to the presence of disulfides contributing to its bioactivity (Dimetry et al. 2010; Dimetry 2012). The neem seed extract after separating oil from it contains a variety of bioactive secondary metabolites. The essential ingredient is a limonoid called azadirachtin, the remaining minor analogs which contribute little to overall efficacy of the extract. Seed extracts include substantial amounts of other triterpenoids such as salannin, nimbin, and their modified forms whose role is debatable but most observations suggest that azadirachtin as the most important active secondary metabolite (El-Sayed 1983a, b; Isman et al. 1996). Neem product activity is measured against 450–500 insect pest species in many countries worldwide, and from that, 413 insect pest species are found to be susceptible at various concentrations (Schmutterer and Singh 1995).

Azadirachtin acts on the prothoracic gland of insect, prevents the formation and exit of molting hormones from the gland cells (ecdysteroids) which lead to incomplete ecdysis in immature insects. A comparable mode of action cause infertility in adult female insects, in addition azadirachtin is an effective agent of antifeedant towards numerous pest insect species. Indeed, azadirachtin was first isolated based on its extraordinary antifeedant activity towards the desert locust, and this plant secondary metabolite becomes the most effective locust antifeedant discovered till now. Neem quickly became the modern standard for the progress of botanical pesticides in the USA (Weinzierl 2000).

6.4.1.2 *Annona muricata* (Custard Apple) and *Annona squamosa* (Graviola)

Botanical pesticides are conventionally developed from the seeds of tropical Annona species, namely (*Annona* squamosa) and soursop (*Annona* muricata) that are the species of the custard apple family (Annonaceae). They are major sources from which fruit juices are prepared in Southeast Asian countries. A detailed study in 1980 revealed that they contain a number of long-chain fatty acid derivatives termed acetogenins, later isolated, found accountable for anti-pesticidal activity. The important acetogenin isolated from *Annona* squamosa seeds is annonin I or squamocin. Later, an analogous compound, asimicin, was obtained which is observed in the bark of the American pawpaw tree (*Asimina triloba*) (McLaughlin et al. 1997; Johnson et al. 2000).

Earlier it was found that these secondary phytochemical compounds are sluggish stomach poisons and masticating pest insects namely Lepidopterans as well as Colorado potato beetle (*Leptinotarsa decemlineata*) are particularly susceptible. Additional study uncovered that the mechanism of action of acetogenins is undistinguishable from rotenone, i.e., it blocks ATP generation in mitochondria in mammals as well as in insects (Londershausen et al. 1991). Certain acetogenins in their pure form are lethal to mammalian species (LD_{50} is <20 mg kg^{-1}), an obstacle to regulatory consent; however, homogenous extracts obtained from the Annona seeds and Asimina plant tissues like seeds, bark show minimum toxicity. Annona seed extracts seem to prove more beneficial in countries where the seeds are left behind after fruits are consumed or used to produce the juice. For example, Leatemia and Isman (2004a, b) recently demonstrated that the diamondback moth is susceptible to crude ethanolic extracts or even aqueous extracts produced from seeds of Annona squamosa collected at several sites in eastern Indonesia (*Plutella xylostella*) (Isman 2006).

6.4.2 *Pheromones*

Insects release biochemicals in their environment called pheromones to evoke a specific behavioral response such as mating, aggregation, and alarm from other individuals. On the basis of their effects, they are categorized as sex hormones, aggregation hormones, alarm hormones, etc. Most of the pheromones stimulate the trail following behavior, egg laying behavior, and aggregation, whereas few other sex attractant pheromones attract partners of opposite sex for mating. Pheromones released by insects evoke the behavioral response only in the same species, for instance, maiden female insects are releasing sex pheromones in the surrounding environment that is recognized by the male of the same species. Aggregation pheromones are released by insects such as wood attacking beetles releases aggregation biochemicals when they found a rich food source to inform and aggregate the other members of the species (Copping and Menn 2000).

A large number of pheromones and sex attractants whose number exceeds 1600 have been found and many of them have a potential to become important tools to control and monitor the pest population that causes heavy losses to agricultural crops (Witzgall et al. 2004). Environmental protection agency (USA) registered 30 pheromone based products that have potential to disrupt mating in insect pest species belonging to order Lepidoptera that is responsible for heavy losses in agricultural crops (Copping and Menn 2000). Environmental conservation, food security, and pest resistance managing are some of the main issues guiding current pest management guidelines in commercial agriculture (Witzgall et al. 2010). While the needs of the rapidly growing world population and at the same time retaining sustainability development and ecological stewardship is an important issue faced by current agricultural system (Kogan and Jepson 2007). The usage of pheromones and/or plant defensive chemicals released against herbivores (allelochemicals) as pest behavioral disruption tools can substitute or counterpart existing pest management policies (Witzgall et al. 2008) resulting in a lesser dependence on pesticides that have wide range of targets including beneficial insects present in agricultural fields. At the present time, pheromones and other semiochemicals are applied to crops that monitor and control pests in millions of hectares. There are many benefits of utilizing pheromones as bio-pesticides over others, like less expensive, specific to target insect, easy to utilize, and highly sensitive compared to conventional pest management system (Wall 1990; Laurent and Frérot 2007; Witzgall et al. 2010). Plants and its derived attractants have been found to interact with the insects to a valuable food source and when combining with insect-obtained attractants can be developed an effective attraction to some insect pests (Copping and Menn 2000).

6.5 Conclusion

The use of natural products with business esteem is legitimately showed by the various compounds present in the market and that have stayed there much of the time after numerous years. These estimations of natural products are considered as a wellspring of new systems/ mechanisms and their ensuing fuse into high-yield screens is difficult to assess. As of late, a few works have been practiced and stretched out to be embraced to upgrade the timeframe of realistic usability, quick demise, the natural plan, proficient in the field and constancy, and the impact of average cost for basic items frameworks and there have been some striking achievements in circumstances where some disturbance to the yield is satisfactory.

Acknowledgement Special thanks are due to Research Centre, College of Pharmacy, King Saud University, Riyadh and Deanship of Scientific Research, King Saud University, Kingdom of Saudi Arabia.

References

Abdollahi M, Ranjbar A, Shadnia S, Nikfar S, Rezaie A (2004) Pesticides and oxidative stress: a review. Med Sci Monit 10(6):RA141–RA147

Almudena Ortiz-Urquiza, Nemat O. Keyhani, (2015) Stress response signaling and virulence: insights from entomopathogenic fungi. Current Genetics 61 (3):239–249

Alves R, Teixeira CGS (2014) Primeiro registro das espécies de cigarrinhas-da-raiz da cana-de-açúcar Mahanarva spectabilis (Distant) e *Mahanarva liturata* (Le Peletier & Serville) atacando canaviais na região de Goianésia (GO), Brasil. Agric Entomol 81:83–85

Al-Zaidi AA, Elhag EA, Al-Otaibi SH, Baig MB (2011) Negative effects of pesticides on the environment and the farmers awareness in Saudi Arabia: a case study. J Anim Plant Sci 21(3):605–611

Arthurs SP, Lacey LA, Miliczky ER (2007) Evaluation of the codling moth granulovirus and spinosad for codling moth control and impact on non-target species in pear orchards. Biol Control 41:99–109

Ashwini N, Srividya S (2014) Potentiality of Bacillus subtilis as biocontrol agent for management of anthracnose disease of chilli caused by Colletotrichum gloeosporioides OGC1. Biotech 4:127–136

Azizoglu U, Yilmaz S, Ayvaz A, Karabörklü S, Atciyurt ZB (2017) Mosquitocidal potential of native Bacillus thuringiensis strain SY49-1 against Disease Vector, Culex pipiens (Diptera: Culicidae). Trop Biomed 34(2):256–262

Barton KA, Whiteley HR, Yang NS (1987) Bacillus thuringiensis §-Endotoxin Expressed in transgenic Nicotiana tabacum provides resistance to lepidopteran insects. Plant Physiol 85(4):1103–1109

Bhat RA, Dervash MA, Mehmood MA, Bhat MS, Rashid A, Bhat JIA, Singh DV, Lone R (2017) Mycorrhizae: a sustainable industry for plant and soil environment. In: Varma A et al (eds) Mycorrhiza-nutrient uptake, biocontrol, ecorestoration. Springer, Cham, pp 473–502

Bhat RA, Beigh BA, Mir SA, Dar SA, Dervash MA, Rashid A, Lone R (2018) Biopesticide techniques to remediate pesticides in polluted ecosystems. In: Wani KA, Mamta (eds) Handbook of research on the adverse effects of pesticide pollution in aquatic ecosystems. IGI Global, Hershey, pp 387–407

Bidochka MJ, Khachatourians GG (1991) The implication of metabolic acids produced by Beauveria bassiana in pathogenesis of the migratory grasshopper, Melanoplus sanguinipes. J Invertebr Pathol 58:106–117

Bird AF, Akhurst RJ (1983) The nature of the intestinal vesicle in nematodes of the family Steinernematidae. Int J Parasitol 13:599–606

Blaxter ML, De Ley P, Garey JR et al (1998) A molecular evolutionary framework for the phylum Nematoda. Nature 392:71–75

Boemare N (2002) Biology, taxonomy and systematics of photorhabdus and xenorhabdus. In: Gaugler R (ed) Entomopathogenic nematology. CABI Publishing, Wallingford, pp 35–56

Borgi I, Gargouri A (2016) A novel high molecular weight thermo-acidoactive β-glucosidase from Beauveria bassiana. Appl Biochem Microbiol 52:602

Bravo A, Gill SS, Soberón M (2007) Mode of action of Bacillus thuringiensis Cry and Cyt toxins and their potential for insect control. Toxicon 49(4):423–435

Brooks FM (1988) Entomogenous protozoa. In: Ignoffo CM, Mandava MB (eds) Handbook of natural pesticides, vol V, microbial insecticides, part a, entomogenous protozoa and fungi. CRC Press, Boca Raton, pp 1–149

Cai SF, Lu XM, Qiu HH, Li MQ, Feng ZZ (2012) Phagocytic uptake of Nosema bombycis (Microsporidia) spores by insect cell lines. J Integr Agric 11:1321–1326

Chandler D, Bailey AS, Tatchell GM, Davidson G, Greaves J, Grant WP (2011) The development, regulation and use of Bio-pesticides for integrated pest management. Philos Trans R Soc Lond Ser B Biol Sci 366:1987–1998

Charnley AK (2003) Fungal pathogens of insects: cuticle-degrading enzymes and toxins. Adv Bot Res 40:241–321

Cho EM, Kirkland BH, Holder DJ, Keyhani NO (2007) Phage display cDNA cloning and expression analysis of hydrophobins from the entomopathogenic fungus Beauveria (Cordyceps) bassiana. Microbiology 153:3438–3447

Clark BW, Phillips TA, Coats JR (2005) Environmental fate and effects of bacillus thuringiensis (Bt) proteins from transgenic crops: a review. J Agric Food Chem 53(12):4643–4653

Copping LG, Menn JJ (2000) Bio-pesticides: a review of their action, applications and efficacy. Pest Manag Sci 56:651–676

Cory JS, Hirst ML, Sterling PH, Speight MR (2000) Native host range nucleopolyhedric virus for control of the browntail moth (Lepidoptera: Lymantriidae). Environ Entomol 29:661–667

Crickmore N, Bravo A, Narva, KLD; Sampson K, Schnepf E, Sun M, Ziegler DR Bacillus thuringiensis toxin nomenclature, 2016

Dimetry NZ (2012) Prospects of botanical pesticides for the future in integrated pest management programme (IPM) with special reference to neem uses in Egypt. Arch Phytopathol Plant Protect 45:1138–1161

Dimetry NZ, Abd El-Salam AME, El-Hawary FMA (2010) Importance of plant extract formulations in managing different pests attacking beans in new reclaimed area and under storage conditions. Arch Phytopathol Plant Protect 43:700–711

Dinardo-Miranda LL, Garcia V, Parazzi VJ (2002) Efeito de Inseticidas no Controle de *Mahanarva fimbriolata* (Stål) (Hemiptera: Cercopidae) e de Nematóides Fitoparasitos na Qualidade Tecnológica e na Produtividade da Cana-de-Açúcar. Neotrop Entomol 31:609–614

Duncan LW, McCoy CW (1996) Vertical distribution in soil, persistence, and efficacy against citrus root weevil (Coleoptera: Curculionidae) of two species of entomogenous nematodes (Rhabditida: Steinernematidae; Heterorhabditidae). Environ Entomol 25:174–178

Ellenhorn MJ, Schonwald S, Ordog G, Wasserberger J (1997) Ellenhorn's medical toxicology: diagnosis and treatment of human poisoning. Williams & Wilkins, Baltimore, pp 1614–1663

El-Sayed EI (1982–1983a) Evaluation of the insecticidal properties of the common Indian neem (Azadirachta Indica A Juss) seeds against the Egyptian cotton leaf worm (Spodoptera litoralis) (Boisd.). Bull Entomol Soc Egypt Econ Ser. 13:39–47S

El-Sayed EI (1982–1983b) Neem (Azadirachta indica A. Juss) seeds as antifeedant and ovipositional repellent for the Egyptian cotton leafworm Spodoptera littoralis (Boisd.). Bull Entomol Soc Egypt Econ Ser 13:49–58

EPA (Environmental Protection Agency) (2006) New biopesticide active ingredients. www.epa.gov/pesticides/Bio-pesticides/productlists/. Accessed 10 June 2019

Faria MR, Magalhães BP (2001) O uso de fungos ento-mopatogênicos no Brasil. Biotecnol Cienc Desenvolvimento 22:18–21

Feng P, Shang Y, Cen K, Wang C (2015a) Fungal biosynthesis of the bibenzoquinone oosporein to evade insect immunity. PNAS 112(36):11365–11370
Feng P, Shang Y, Cen K, Wang C (2015b) Fungal biosynthesis of the bibenzoquinone oosporein to evade insect immunity. Proc Natl Acad Sci U S A 112:11365–11370
Fischhoff DA, Bowdish KS, Perlak FJ, Marrone PG, McCormick SM, Niedermeyer JG et al (1987) Insect tolerant transgenic tomato plants. Nat Biotechnol 5(8):807–813
Fishilevich E, Vélez AM, Storer NP, Li H, Bowling AJ, Rangasamy M et al (2016) RNAi as a management tool for the western corn rootworm. Pest Manag Sci 72(9):1652–1663
Gathmann A, Priesnitz KU (2014) How susceptible are different lepidopteran/coleopteran maize pests to Bt-proteins: a systematic review protocol. Environ Evid 3:12
Georgis R (1990) Commercialization of steinernematid and heterorhabditid entomopathogenic nematodes. InBrighton Crop Protection Conference, Pests and Diseases-1990. 1;275-280, British Crop Protection Council
Ghribi D, Mnif I, Boukedi H, Kammoun R, Ellouze-Chaabouni S (2011) Statistical optimization of low-cost medium for economical production of Bacillus subtilis biosurfactant, a biocontrol agent for the olive moth Prays oleae. Afr J Microbiol Res 5:4927–4936
Ghribi D, Abdelkefi-Mesrati L, Boukedi H, Elleuch M, Ellouze-Chaabouni S, Tounsi S (2012) The impact of the Bacillus subtilis SPB1 biosurfactant on the midgut histology of Spodoptera littoralis (Lepidoptera: Noctuidae) and determination of its putative receptor. J Invertebr Pathol 109:183–186
Glare T, Caradus J, Gelernter W, Jackson T, Keyhani N, Kohl J et al (2012) Have bio-pesticides come of age? Trends Biotechnol 30:250–258
Global Status of Commercialized Biotech/GM Crops: 2016 (2016) ISAAA Brief No. 52. ISAAA, Ithaca
Goldberg LH, Margalit J (1977) A bacterial spore demonstrating rapid larvicidal activity against Anopheles sergentii, Uranotaenia unguiculata, Culex univittatus, Aedes aegypti and Culex pipiens. Mosq News 37:355–358
Gramkow AW, Perecmanis S, Sousa RLB, Noronha EF, Felix CR, Nagata T et al (2010) Insecticidal activity of two proteases against Spodoptera frugiperda larvae infected with recombinant baculoviruses. Virol J 29(7):143
Grewal PS, Lewis EE, Gaugler R (1997) Response of infective stage parasites (Nematoda: Steinernematidae) to volatile cues from infected hosts. J Chem Ecol 23:503–515
Grewal PS, Ehlers RU, Shapiro-Ilan DI (eds) (2005) Nematodes as biological control agents. CABI Publishing, Wallingford
Halo LM, Heneghan MN, Yakasai AA, Song Z, Williams K, Bailey AM et al (2008) Late stage oxidations during the biosynthesis of the 2-pyridone tenellin in the entomopathogenic fungus Beauveria bassiana. J Am Chem Soc 130:17988–17996
Hamill RL, Higgens GE, Boaz HE, Gorman M (1969) The structure of beauvericin, a new depsipeptide antibiotic toxic to Artemia salina. Tetrahedron Lett 10(49):4255–4258
Han R. Ehlers R.U. (2000) Pathogenicity, development, and reproduction of Heterorhabditis bacteriophora and Steinernema carpocapsae under axenic in vivo conditions. J. Invertebr. Pathol.75, 55–58.
Henry JE, Oma EA, Burges HD (1981) Pest control by *Nosema locustae*, a pathogen of grasshoppers and crickets. In: Microbial control of pests and plant diseases 1970–1980. Academic Press, New York
Herniou EA, Jehle JA (2007) Baculovirus phylogeny and evolution. Curr Drug Targets 8:1043–1050
Inceoglu AB, Kamita SG, Hinton AC, Huang Q, Severson TF, Kang KD et al (2001) Recombinant baculoviruses for insect control. Pest Manag Sci 57:981–987
Isman MB, Matsuura H, MacKinnon S, Durst T, Towers GHN, Arnason JT (1996) Phytochemistry of the Meliaceae. So many terpenoids, so few insecticides. In: Romeo JT, Saunders JA, Barbosa P (eds) Phytochemical diversity and redundancy. Plenum, New York, pp 155–178
Jehle JA, Blissard GW, Bonning BC, Cory JS, Herniou EA, Rohrmann GF et al (2006) On the classification and nomenclature of baculoviruses: a proposal for revision. Arch Virol 151:1257–1266

Jisha VN, Smitha RB, Benjain S (2013) An overview on the crystal toxins from *Bacillus thuringiensis*. Adv Microbiol 3(5):462
Johnson HA, Oberlies NH, Alali FQ, McLaughlin JE (2000) Thwarting resistance: annonaceous acetogenins as new pesticidal and antitumor agents. In: Cutler SJ, Cutler JG (eds) Biological active natural products: pharmaceuticals. CRC Press, Boca Raton, pp 173–183
Kamita SG, Kang KD, Hammock BD (2005) Genetically modified baculoviruses for pest insect control. In: Iatrou K, Gilbert LI, Gill SS (eds) Comprehensive molecular insect science. Elsevier, Oxford, pp 271–322
Kavitha PG, Jonathan EI, Nakkeeran S (2012) Effects of crude antibiotic of Bacillus subtilis on hatching of eggs and mortality of juveniles of Meloidogyne incognita. Nematol Mediterr 40:203–206
Kaya HK, Gaugler R (1993) Entomopathogenic nematodes. Annu Rev Entomol 38:181–206
Kaya HK, Stock SP (1997) Techniques in insect nematology. In: Lacey L (ed) Manual of techniques in insect pathology. Academic Press, San Diego, pp 281–324
Khanday M, Bhat RA, Haq S, Dervash MA, Bhatti AA, Nissa M, Mir MR (2016) Arbuscular mycorrhizal fungi boon for plant nutrition and soil health. In: Hakeem KR, Akhtar J, Sabir M (eds) Soil science: agricultural and environmental prospectives. Springer, Cham, pp 317–332
Kogan M, Jepson P (2007) Ecology, sustainable development and IPM: the human factor. In: Kogan M, Jepson P (eds) Perspectives in ecological theory and integrated pest management. Cambridge University Press, Cambridge, pp 1–44
Koppenhofer AM (2007) Nematodes. In: Lacey LA, Kaya HK (eds) Field manual of techniques in invertebrate pathology: Application and evaluation of pathogens for control of insects and other invertebrate pests, 2nd edn. Springer, Dordrecht, pp 249–264
Kouti K, Lemmens M, Lemmens-Gruber R (2003) Beauvericin induced channels in ventricular myocytes and liposomes. Biochim Biophys Acta 1609:203–210
Kumar S (2012) Bio-pesticides: a need for food and environmental safety. J Biofertil Biopestic 3:413
Kutinkova H, Samietz J, Dzhuvinov V, Zingg DP, Kessler P (2012) Successful application of the baculovirus product Madex® for control of Cydia pomonella (L.) in Bulgaria. J Plant Prot Res 52(2):205–213
Lacey LA, Georgis R (2012) Entomopathogenic nematodes for control of insect pests above and below ground with comments on commercial production. J Nematol 44:218–225
Lacey LA, Headrick HL, Arthurs SP (2008) Effect of temperature on long-term storage of codling moth granulovirus formulations. J Econ Entomol 101:288–294
Lange CE, Sokolova YY (2017) The development of the microsporidium Paranosema (Nosema) locustae for grasshopper control: John Henry's innovation with worldwide lasting impacts. Protistology 11(3):170–174
Laurent P, Frérot B (2007) Monitoring of European corn borer with pheromone-baited traps: review of trapping system basics and remaining problems. J Econ Entomol 100:1797–1807
Leatemia JA, Isman MB (2004a) Efficacy of crude seed extracts of Annona squamosa against Plutella xylostella L. in the greenhouse. Int J Pest Manag 50:129–133
Leatemia JA, Isman MB (2004b) Insecticidal activity of crude seed extracts of Annona spp., Lansium domesticum and Sandoricum koetjape against lepidopteran larvae. Phytoparasitica 32:30–37
Lewis MW, Robalino IV, Keyhani NO (2009) Uptake of the fluorescent probe FM4-64 by hyphae and haemolymph-derived in vivo hyphal bodies of the entomopathogenic fungus Beauveria bassiana. Microbiology 155:3110–3120
Li H, Bonning BC (2007) Evaluation of insecticidal efficacy of wild-type and recombinant baculoviruses. Methods Mol Biol 388:379–404
Londershausen M, Leight W, Lieb F, Moeschler H (1991) Molecular mode of action of annonins. Pestic Sci 33:427–438

Lozano-Tovar MD, Garrido-Jurado I, Quesada-Moraga E, Trapero-Casas MCR (2017) *Metarhizium brunneum* and *Beauveria bassiana* release secondary metabolites with antagonistic activity against *Verticillium dahliae* and *Phytophthora megasperma* olive pathogens. Crop Prot 100:186–195

Lu D, Macchietto M, Chang D, Barros MM, Baldwin J, Mortazavi A et al (2017) Activated entomopathogenic nematode infective juveniles release lethal venom proteins. PLoS Pathog 13(4):e1006302

Lu, D.; Baiocchi, T.; Dillman, A.R. Genomics of entomopathogenic nematodes and implications for pest control. Trends Parasitol. 2016, 32, 588–598

Lucon CMM, Guzzo SD, De Jesus CO, Pascholati SF, De Goes A (2010) Post harvest harpin or *Bacillus thuringiensis* treatments suppress citrus black spot in 'valencia' oranges. Crop Prot 29:766–772

Martins LN, de Lara APD, Ferreira MS, Nunes AM, Bernardi D, Leite FPL et al (2018) Biological Activity of *Bacillus thuringiensis* (Bacillales: Bacillaceae) in *Anastrepha fraterculus* (Diptera: Tephritidae). J Econ Entomol 111(3):1486–1489

McLaughlin JL, Zeng L, Oberlies NJ, Alfonso D, Johnson JA, Cummings BA (1997) Annonaceous acetogenins as new natural pesticides: recent progress. ACS Symposium Ser 117:133

Mendelsohn M, Kough J, Vaituzis Z, Matthews K (2003) Nat Biotechnol 21(9):1003–1009

Milner RJ, Lozano LB, Driver F, Hunter D (2003) A comparative study of two Mexican isolates with an Australian isolate of *Metarhizium anisopliae var. acridum*–strain characterisation, temperature profile and virulence for wingless grasshopper, *Phaulacridium vittatum*. Biocontrol 48:335–348

Mnyone LL, Koenraadt CJM, Lyimo IN, Mpingwa MW, Takken W, Russell TL (2010) Anopheline and culicine mosquitoes are not repelled by surfaces treated with the entomopathogenic fungi Metarhizium anisopliae and Beauveria bassiana. Parasit Vectors 3:80

Mohammadipour M, Mousivand M, Jouzani GS, Abbasalizadeh S (2009) Molecular and biochemical characterization of Iranian surfactin-producing Bacillus subtilis isolates and evaluation of their biocontrol potential against Aspergillus flavus and Colletotrichum gloeosporioides. Can J Microbiol 55:395–404

Molnar I, Gibson DM, Krasnoff SB (2010) Secondary metabolites from entomopathogenic Hypocrealean fungi. Nat Prod Rep 27:1241–1275

Morris ON (1985) Susceptibility of 31 species of agricultural pests to entomogenous nematodes Steinernema feltiae and Heterorhabditis bacteriophora. Can Entomol 122:309–320

Moscardi F (1999) Assessment of the application of baculoviruses for control of Lepidoptera. Annu Rev Entomol 44:257–289

Moscardi F, de Souza ML, de Castro MEB, Lara Moscardi M, Szewczyk B (2011) Baculovirus pesticides: present state and future perspectives. In: Ahmad I, Ahmad F, Pichtel J (eds) Microbes and microbial technology. Springer, New York

Mushtaq N, Bhat RA, Dervash MA, Qadri H, Dar GH (2018) Biopesticides: the key component to remediate pesticide contamination in an ecosystem. In: Environmental contamination and remediation. Cambridge Scholars Publishing, Cambridge, pp 152–178

Nicholson GM (2007) Fighting the global pest problem: preface to the special Toxicon issue on insecticidal toxins and their potential for insect pest control. Toxicon 49:413–422

Ongena M, Jacques P (2007) *Bacillus lipopeptides*: versatile weapons for plant disease biocontrol. Trends Microbiol 16:115–125

Ongena M, Henry G, Thonart P (2010) The roles of cyclic lipopeptides in the biocontrol activity of Bacillus subtilis. In: Recent developments in management of plant diseases. Springer, Dordrecht, pp 159–166

Ortiz-Urquiza A, Keyhani NO (2013) Action on the surface: entomopathogenic fungi versus the insect cuticle. Insects 4:357–374

Ortiz-Urquiza A, Riveiro-Miranda L, Santiago-Álvarez C, Quesada-Moraga E (2010) Insect-toxic secreted proteins and virulence of the entomopathogenic fungus *Beauveria bassiana*. J Invertebr Pathol 105:270–278

Passarelli LA (2011) Barriers to success: How baculoviruses establish efficient systemic infections. Virology 411 (2):383–392

Peters A (1996) The natural host range of *Steinernema* and *Heterorhabditis* spp. and their impact on insect populations. Biocontrol Sci Tech 6:389–402

Philogène BJR, Regnault-Roger C, Vincent C (2005) Botanicals: yesterday's and today's promises. In: Regnault-Roger C, Philogène BJR, Vincent C (eds) Bio-pesticides of plant origin. Lavoisier, Andover, pp 1–15

Pomar GO Jr, Leutenegger R (1968) Anatomy of the effective and normal third stage juveniles of *Steinernema carpocapsae* Weiser (Steinernematidae: Nematoda). J Parasitol 54:340–350

Qadri SM, Kucherenko Y, Lang F (2011) Beauvericin induced erythrocyte cell membrane scrambling. Toxicology 283:24–31

Que Q, Chilton MDM, de Fontes CM, He C, Nuccio M, Zhu T et al (2010) Trait stacking in transgenic crops: challenges and opportunities. GM Crops 1(4):220–229

Quesada-Moraga E, Vey A (2004) Bassiacridin, a protein toxic for locusts secreted by the entomopathogenic fungus Beauveria bassiana. Mycol Res 108:441–452

Ramaseshadri P, Segers G, Flannagan R, Wiggins E, Clinton W, Ilagan O et al (2013) Physiological and cellular responses caused by RNAi- mediated suppression of Snf7 orthologue in western corn rootworm (*Diabrotica virgifera virgifera*) larvae. PLoS One 8(1):e54270e

Regnault-Roger C, Philogène BJR (2008) Past and current prospects for the use of botanicals and plant allelochemicals in integrated pest management. Pharm Biol 46:41–52

Reyes-Ramirez A, Escudero-Abarca BI, Aguilar-Uscanga G, Hayward-Jones PM, Eleazar Barboza-Corona J (2004) Antifungal activity of *Bacillus thuringiensis* chitinase and its potential for the biocontrol of phytopathogenic fungi in soybean seeds. J Food Sci 69:131–134

Salazar-Flores J, Torress-Jasso JH, Rojas-Bravo D, Reyna-Villela ZM, TorresSánchez ED (2019) Effects of mercury, lead, arsenic and zinc to human renal oxidative stress and functions: a review. J Heavy Met Toxicity Dis 4(1):2

Schmutterer H (1990) Properties and potential of natural pesticides from the neem tree, *Azadirachta indica*. Annu Rev Entomol 35:271–297

Schmutterer H, Singh RP (1995) List of insect pests susceptible to neem products. In: Schmutterer H (ed) The neem tree: source of unique natural products for integrated pest management, medicine, industry and other purposes. Wiley-VCH, Weinheim, pp 325–326

Schünemann R, Knaak N, Fiuza LD (2014) Mode of action and specificity of Bacillus thuringiensis toxins in the control of caterpillars and stink bugs in soybean culture. ISRN Microbiol 2014:135675

Senthil-Nathan S (2013) Physiological and biochemical effect of neem and other Meliaceae plants secondary metabolites against Lepidopteran insects. Front Physiol 4:359

Shannag HK, Capinera JL (2000) Interference of *Steinernema carpocapsae* (Nematoda: Steinernematidae) with *Cardiochiles diaphaniae* (Hymenoptera: Braconidae), a parasitoid of melonworm and pickleworm (Lepidoptera: Pyralidae). Environ Entomol 29:612–617

Shapiro DI, McCoy CW (2000) Virulence of entomopathogenic nematodes to *Diaprepes abbreviatus* (Coleoptera: Curculionidae) in the laboratory. J Econ Entomol 93:1090–1095

Shingote PR, Moharil MP, Dhumale DR, Jadhav PV, Satpute NS, Dudhare MS (2013) Screening of vip1/vip2 binary toxin gene and its isolation and cloning from local Bacillus thuringiensis isolates. Sci Asia 39:620–624

Solter LF, Becnel JJ (2000) Entomopathogenic microsporida. In: Lacey LA, Kaya HK (eds) Field manual of techniques in invertebrate pathology: application and evaluation of pathogens for control of insects and other invertebrate pests. Kluwer Academic, Dordrecht, pp 231–254

Sosa-Gómez DR, Moscardi F, Santos B, Alves LFA, Alves SB (2008) Produçao e uso de vírus para o controle de pragas na América Latina. In: Alves SA, Lopes RB (eds) Controle microbiano de pragas na América Latina: abanicos e desafios. FEALQ, Piracicaba, pp 49–68

Stamati PN, Maipas S, Kotampasi C, Stamatis P, Hens L (2016) Chemical pesticides and human health: the urgent need for a new concept in agriculture. Front Public Health 4:148

Sun X, Chen X, Zhang Z, Wang H, Bianchi FJJA, Peng H, Vlak JM, Hu Z (2002) Bollworm responses to release of genetically modified *Helicoverpa armigera* nucleopolyhedroviruses in cotton. J Invertebr Pathol 81:63–69

Sundh I, Goettel MS (2013) Regulating biocontrol agents: a historical perspective and a critical examination comparing microbial and macrobial agents. Biocontrol 58:575–593

Szewczyk B, Hoyos-Carvajal L, Paluszek M, Skrzecz I, Lobo de Souza M (2006) Baculoviruses-re-emerging bio-pesticides. Biotechnol Adv 24:143–160

Tabashnik BE, Brévault T, Carrière Y (2013) Insect resistance to Bt crops: lessons from the first billion acres. Nat Biotechnol 31(6):510–521

Tareq FS, Lee MA, Lee HS, Lee YJ, Lee JS, Hasan CM et al (2014) Gageotetrins A-C, noncytotoxic antimicrobial linear lipopeptides from a marine bacterium *Bacillus subtilis*. Org Lett 16:928–931

Thakore Y (2006) The biopesticide market for global agricultural use. Ind Biotechnol 2(3):194–208. https://doi.org/10.1089/ind.2006.2.194

Toledo AV, de Remes Lenicov AMM, Lastra CCL (2010) Histopathology caused by the entomopathogenic fungi, *Beauveria bassiana* and *Metarhizium anisopliae*, in the adult plant hopper, Peregrinus maidis, a maize virus vector. J Insect Sci 10(35):1

Tulloch M (1976) The genus Metarhizium. Trans Br Mycol Soc 66:407–411

US Environmental Protection Agency (2019) EPA registers innovative tool to control corn rootworm, June 15, 2017. https://www.epa.gov/newsreleases/epa-registers-innovative-tool-control-corn-rootworm. Accessed 10 June 2019

Vaeck M, Reynaerts A, Höfte H, Jansens S, de Beuckeleer M, Dean C et al (1987) Transgenic plants protected from insect attack. Nature 328(6125):33–37

Velho RV, Medina LF, Segalin J, Brandelli A (2011) Production of lipopeptides among *Bacillus* strains showing growth inhibition of phytopathogenic fungi. Folia Microbiol (Praha) 56:297–303

Vincent CM, Andermatt M, Valéro J (2007) Madex® and VirosoftCP4®, viral Bio-pesticides for codling moth control. In: Vincent C, Goethel MS, Lazarovits G (eds) Biological control: a global perspective. CAB International, Cambridge, pp 336–343

Wakil W, Yasin M, Shapiro-Ilan D (2017) Effects of single and combined applications of entomopathogenic fungi and nematodes against *Rhynchophorus ferrugineus* (Olivier). Sci Rep 7:5971

Wall C (1990) Principle of monitoring. In: Ridgeway RL, Silverstein RM, Inscoe MN (eds) Behavior-modifying chemicals for insect management: applications of pheromones and other attractants. Marcel Dekker Inc., New York, pp 9–23

Wang Q, Xu L (2012) Beauvericin, a bioactive compound produced by fungi: a short review. Molecules 17:2367–2377

Weinzierl RA (2000) Botanical insecticides, soaps, and oils. In: Rechcigl JE, Rechcigl NA (eds) Biological and biotechnological control of insect pests. Lewis Publishers, Boca Raton, FL: CRC Press, pp 101–121

Witzgall P, Lindblom T, Bengtsson M et al (2004) The pherolist. online http://www. pherolist. slu. se.(accessed 2006-10-30).

Witzgall P, Stelinski L, Gut L, Thomson D (2008) Codling moth management and chemical ecology. Annu Rev Entomol 53:503–522

Witzgall P, Kirsch P, Cork A (2010) Sex pheromones and their impact on pest management. J Chem Ecol 36:80–100

Xiao G, Ying SH, Zheng P, Wang ZL, Zhang S, Xie XQ et al (2012) Genomic perspectives on the evolution of fungal entomopathogenicity in *Beauveria bassiana*. Sci Rep 2:483

Xu C, Wang BC, Yu Z, Sun M (2014) Structural insights into *Bacillus thuringiensis* Cry, Cyt and parasporin toxins. Toxins 6(9):2732–2770

Yan L, Jing T, Yujun Y, Bin L, Hui L, Chun L (2011) Biocontrol efficiency of *Bacillus subtilis* SL-13 and characterization of an antifungal chitinase. Chin J Chem Eng 19:128–134

Yang MM, Li ML, an ZY, Wang YZ, Qu LJ, Wang QH (2012) Baculoviruses and insect pests control in China. Afr J Microbiol Res 6(2):214–218

Zhang SZ, Xia YX, Kim B, Keyhani NO (2011) Two hydrophobins are involved in fungal spore coat rodlet layer assembly and each play distinct roles in surface interactions, development and pathogenesis in the entomopathogenic fungus, *Beauveria bassiana*. Mol Microbiol 80:811–826

Zheng P, Xia XL, Xiao GH, Xiong CH, Hu X, Zhang SW et al (2012) Genome sequence of the insect pathogenic fungus *Cordyceps militaris*, a valued traditional Chinese medicine. Genome Biol 12:R116. https://doi.org/10.1186/gb-2011-12-11-r116.

Zhou Y, Choi YL, Sun M, Yu Z (2008) Novel roles of *Bacillus thuringiensis* to control plant diseases. Appl Microbiol Biotechnol 80:563–572

Chapter 7
Values of Biofertilizers for Sustainable Management in Agricultural Industries

Sajad Ahmad Raina, Rouf Ahmad Bhat, Humaira Qadri, and Ashut Dutta

7.1 Introduction

Growing human population led to the dependence on pesticides and elemental fertilizers due to conventional agriculture that fulfills the food needs of population growth (Santos et al. 2012; Bhat et al. 2018; Mushtaq et al. 2018). Chemical fertilizers are elements composed of known amounts of nitrogen, phosphorus, and potassium, and their utilization causes adulteration in air and groundwater through the process known as cultural eutrophication (Youssef and Eissa 2014). In this sense, latest efforts must be more channelized to produce "high quality nutrient rich foods" in ecologically compatible manner to safeguard biosecurity. The inventive vision of agricultural production fascinates the increasing demand for biological fertilizers limited to agrochemicals (Raja 2013). In order to improve nutrient intake and preserve field management that helps to promote the alternative methods of soil fertilization is based on organic inputs (Araujo et al. 2008). Organic farming is one of the important strategies that increase soil biodiversity leading to the assurance of food safety (Megali et al. 2013). The additional benefits of biofertilizers include extended shelf life devoid of any postharvest adverse effects on the ecosystem (Sahoo et al. 2014).

The distinguished effects of the increased use of element enrichers in agriculture result in alterations in soil reaction, cause nutritional disproportions in plants, greater vulnerability to pests and diseases, decline of the root nodulation of legumes and mycorrhizal linkages existing among various crops which ultimately result in soil humus reduction and water pollution. The harmful effects of chemical fertilizers continuously used in large quantities without the use of organic components

S. A. Raina · A. Dutta
Department of Environmental Science, Bhagwant University, Ajmer, Rajasthan, India

R. A. Bhat (✉) · H. Qadri
Department of Environmental Science, Sri Pratap College, Cluster University, Srinagar, India

R. A. Bhat et al. (eds.), *Bioremediation and Biotechnology, Vol 2*,
https://doi.org/10.1007/978-3-030-40333-1_7

have aroused an interest to substitutes in which organic fertilizers play an important role to supply nutrients to plants in an integrated manner resulting in the Integrated Nutrient System for Plants (IPNS). Environmental problems, such as soil erosion, energy saving and freshwater pollution, might the farmers to acquaint with the new cultivation methods that have less effect on the environment. In this regard, the use of chemical fertilizers has reduced by the greater applicability of organic fertilizers that is considered a mandatory way to remove the pressure on the environment resulting from conventional agricultural practices. Increase in human population has led to a growing dependence on chemical fertilizers and pesticides in order to meet the food needs due to conventional agriculture practices (Santos et al. 2012). The exploitation of chemical fertilizers composed of known quantities of nutrients such as nitrogen, phosphorus, and potassium manufactured in industries causes air and groundwater pollution through the process known as cultural eutrophication (Youssef and Eissa 2014). In this regard, the new efforts have been channeled more in the direction of producing of "high quality nutrient rich foods" in sustainable behavior to safeguard biological security. The main environmental problems associated with the use of fertilizers lead to the pollution of water with nitrates and phosphates. These nitrates find their way into groundwater reservoirs through percolation and the consumption of such drinking water with high levels of nitrates is reflected as a pronounced disease (methemoglobinemia) that is lethal to human beings.

Phosphorus binds and moves with the soil particles and cannot be easily removed from the ground. Therefore, the soil that is eroded contains phosphorus can be washed in surface water together with the soil particles. Phosphorus stimulates the growth of algae in water bodies and is not considered dangerous. The algae causes the death of fishes due to removal of oxygen content in water bodies by death and decomposition, this process is called eutrophication.

There are different sources of these pollutants such as industrial waste, detergents, sewage disposal, and fertilizers. The decomposition of crop residues and organic residues produces nitrates and the main problems of high levels of nitrates in groundwater have been recorded since 1860, long before the use of fertilizers became common.

Synthetic fertilizers are an interesting alternative, but they have undesirable negative effects. Some are discussed as below.

7.2 Motility of Microorganisms and Plants

Some fertilizers possess highly acidic attributes which transform soil chemistry and thereby result in killing of earthworms and other microorganisms that are beneficial to plants. Earthworms perforate into the soil and result in increasing the passages in soil which help to ventilate the soil for boosting up the fertility. They also deposit their castings into the soil. The most important feature of earthworms and microorganisms is that organic matter decomposes. These microorganisms are difficult to find in soils treated with synthetic fertilizers. This soil is not rich, since it does not have the benefit of the nutrients released by these organisms (Shafi et al. 2018).

7.2.1 Increases Soil Toxicity

The chemicals contained in synthetic fertilizers are toxic to useful soil microorganisms. The distribution of earthworms is quite rare in a synthetically fertilized soil. And chemical toxicity of synthetic fertilizers makes the soil less beneficial for plants.

7.2.2 Increases Plant Diseases

Since plants grown in synthetically enriched soils do not have the nutrients that introduce useful microorganisms, they are more prone to plant diseases. They also have a weaker immune system.

7.2.3 Makes Waterways Impassable and Damage Marine Life

Nutrients from this soil penetrate into sewers and waterways. While doing so, they encourage algae growth. The proliferation of algae is sometimes large enough to prevent the flow of water through drainage and sewage systems.

7.2.4 These Fertilizers Have a Short Life

Most of these petroleum based fertilizers are water soluble. They release their nutrients quickly, but these nutrients disappear so quickly. A common complaint among gardeners using such fertilizers is that they often have to continue adding them, increasing their operating costs.

7.3 Synthetic Fertilizers Damage Humans and Pets

Synthetic fertilizers are not beneficial to humans and pets. Exposure can cause respiratory problems in humans. Studies also show that exposure to such fertilizers causes cancers in pets. Synthetic fertilizers produce vegetables with nutritional deficiencies that do not necessarily provide optimum vitamin concentration that our body requires for carrying day-to-day life.

7.4 Costs

"A professional lawn care contractor will charge on average between $ 50 and 80 to apply fertilizer on an 8000 square meter construction site." This number varies depending on whether one chooses an organic or synthetic fertilizer. Most contractors apply a fixed price based on the services provided and the size of the area. If the contractor charges hourly rates, then it is expected to pay an average of $ 65 per hour. An average of 8000 ft^2 of lawn will cost between $ 50 and 95 for the application of organic fertilizer. Futhermore, synthetic fertilizers are costilier and accounts around $ 25 and 80.

7.5 Environmental Concerns

Organic fertilizers are biodegradable, renewable and are ecologically sustainable. Organic fertilizers are made from unprocessed and natural ingredients as manure, cornmeal, compost and are considered one of the best solutions to produce a healthy environment for plant growth. These fertilizers are often acquired locally and mass scale production can be achieved through installing the greenhouses and farms, which further reduce the environmental impact.

Synthetic fertilizers are processed from a variety of different products, including mineral salts and petroleum products in order to ensure the stability of the nutrient content. The treatment and regular use of these fertilizers use more water and are potentially harmful to the environment due to additional chemical inputs that can lead to the accumulation of poisonous chemicals, such as arsenic and uranium, making adverse effects on people, pets, and plants.

7.6 Effects of Chemical Fertilizers on Soil

Chemical fertilizers are artificial fertilizers or industrially made fertilizers used to increase the nutrient level in the soil and are essential for plant growth as these nutrients are produced synthetically from inorganic materials and applied to the soil as similar to natural nutrients present in the soil such as nitrogen, phosphorus, and potassium. These fertilizers increase the yields in a short duration of time, but have some harmful effects on beneficial soil microbes and soil structure by the use of chemical fertilizers instead of organic fertilizers made from natural sources.

7.7 Groundwater Pollution

The continuous use of chemical fertilizers causes groundwater pollution or leaching. Chemical fertilizers with high solubility are absorbed by the soil more quickly than absorbed by the projected plants. Plants absorb a given level of nutrition and leaving the extra fertilizer to leach at a time. Leaching of chemical fertilizers is dangerous for groundwater, besides that it also affects the subsoil because these chemicals bind with the clay particles to form hard layers in soil called as hardpan.

7.8 Soil Friability Effect

The presence of various acids such as hydrochloric and sulfuric acids in the soil produces a harmful effect on the soil called soil friability. The various acids in the soil dissolve the crumbs in the soil that helps to keep the rock particles together. The crumbs of the soil are due to the combination of humus, or decomposed natural material, like dead leaves, with clay. These soil-rich mineral crumbs are essential for soil drainage and greatly improve air circulation in the soil. Since the chemicals contained in chemical fertilizers destroy the crumbs in the soil, the result is highly compacted soil with less drainage and air circulation.

7.9 Causes of Destruction of Microorganisms

The continuous use of synthetic chemicals as chemical fertilizers altering the pH of the soil thus has adverse effects on the health of beneficial soil microorganisms present naturally in the soil. Thus the altered level of acidity in the soil removes the beneficial microorganisms that are helpful to increase the defense mechanisms of plants against pests and diseases. These microorganisms are found in healthy soils to produce antibiotics and include bacteria, mycorrhizae, and other fungi. The continuous use of chemical fertilizers damages the bacterial community which helps in nitrogen fixation and balance the soil health. These bacteria arc accountable for converting atmospheric N_2 into nitrates that are easily available to plants.

7.10 Effect of Fertilizers on Human Health

The continuous use of fertilizers produces toxins into the soil that enter the humans through food chain by the consumption of contaminated vegetables and cereals as these fertilizers are made of a mixture of toxic chemicals absorbed in plants and also the runoff from agricultural fields finds their way into aquatic environ that eventually affects the human health (Table 7.1).

These fertilizers also cause dangerous diseases such as kidney diseases, lung and liver problems, and cancer because of the metal toxicity of fertilizers and it depends

Table 7.1 Toxic substances released by using synthetic fertilizers and their impacts on human health

S. no.	Pollutants associated with synthetic fertilizer	Impacts on human health
1	Contaminated water (runoff)	It may contain high levels of nitrates and nitrites, causing hemoglobin disorders
2	Heavy metals	Heavy metals such as mercury, lead, cadmium, and uranium have been found in fertilizers, which can cause kidney, lung, and liver problems and cause cancer. More than 29 popular fertilizers tested positive for 22 toxic heavy metals, including silver, nickel, selenium, thallium, and vanadium, all directly related to human health risks
3	Exposure to ammonium nitrate causes	Exposure to ammonium nitrate causes other health problems, such as eye and skin irritation, which produce a burning sensation. Exposure by inhalation may cause irritation to the nose, throat, and lungs. You can also experience nausea, vomiting, redness of the face and neck, headache, nervousness, uncontrolled muscle movements, fainting and collapse
4	Potassium chloride	It interferes with nerve impulses and virtually interrupts all bodily functions and mainly affects the functioning of the heart. It can cause all types of stomach and stomach pain, dizziness, bloody diarrhea, convulsions, headaches, mental disorders, redness or itching of the skin or eyes
5	Cadmium	It eventually enters human tissues and produces diseases such as tracheobronchitis, pneumonia, pulmonary edema, kidney failure, osteoporosis, and many others. Lettuce (*Latuca sativa* L.), which is found in large quantities in countries such as Brazil and others in South America, very easily absorbs Cd, with an average quantity of 870 mg Cd/kg. Through the gastrointestinal system, they absorb only about 5% of the Cd through food. However, those who do not have iron can absorb up to 6% more. Therefore, this can be harmful to them because cadmium poisoning, which results from their excessive consumption, can cause damage to the kidneys, bones, and lungs. The "Itai-Itai" (it hurts, hurts, in Japanese), discovered in Japan in 1912, is caused by cadmium poisoning, which produces acute pain in the joints and the spine
6	Major health risks (Cancer)	It increase the risk of cancer death by more than six times, including brain cancer, lymphoma (non-Hodgkin's lymphoma, NHL), prostate cancer, leukemia, and large bowel cancer
7	Methemoglobinemia	Methemoglobinemia is a blood disease in which an abnormal amount of methemoglobin, a form of hemoglobin, is produced. It is produced in place of essential hemoglobin, which is responsible for supplying oxygen to red blood cells. This causes difficulty in delivering sufficient oxygen to all body tissues and is related to birth and developmental disorders such as "blue baby syndrome"

upon the amount of consumption of fertilizers. Fertilizers remove essential nutrients from the soil, thus damages the soil and environment. The soil fertility decreases with the increasing use of fertilizers.

7.11 Why Biofertilizers?

Biological fertilizers more commonly known as biofertilizers that contain live soil microorganisms result in increasing the absorption and availability of mineral nutrients for plants (Vessey 2003). Their activities produce additional materials for plant growth and also support the soil ecosystem. They are available in the soil due to the interaction of microorganisms or its associations with plants and include organic fertilizers such as manure, etc. The seeds or soil inoculum multiply and participate in the nutrient cycles by the application of biofertilizers and increase the crop productivity. The process of nitrogen fixation, solubilization, or phosphate and potassium mineralization keeps the soil environment rich in all kinds of micro- and macronutrients by the application of biofertilizers, substances that are released by the plants regulate the growth, antibiotic production, and biodegradation of organic materials in soil which increases the tolerance to stress due to famine and humidity and also provides greater nutrient absorption.

The production of rice increases by the use of biofertilizers in comparison with chemical fertilizer. Thus, biofertilizers help to solve the problem of feeding growing world population when agriculture is facing several pressures and environmental changes by the use of chemical fertilizers.

7.12 Microbes as Biofertilizer

The microorganisms that have found application as a biofertilizer include *Rhizobium*, *Azospirillum*, *Cyanobacteria*, *Azolla*, liquid biofertilizers, *Azotobacter*, *Acetobacter*.

7.12.1 Rhizobium

Rhizobium, a symbiotic organism can easily inhabit the existing spaces in vegetable roots and its infection aids in fixation of atmospheric nitrogen. Their morphology and physiology will change from an open living condition to a bottom bacterium, they are the most productive biofertilizers based on the quantity of nitrogen deposited in question. They have seven genres and are very specific to shape the lump in vegetables, referred to as a cross-immunization group (Raja 2013).

7.12.2 *Azospirillum*

The essential pollutants of the soil are *Azospirillum lipoferum* and *A. brasilense*, the rhizosphere, and the intercellular spaces of the bark of the roots of herbaceous plants. They perform a harmonious cooperative connection with the grass. The microbes of the *Azospirillum* genus are creatures that alter nitrogen and are detached from the roots or other parts of the soil of a variability of plants and are gram negative, *Vibrio* or *Spirillum* which have an ample amount of poly-beta-hydroxybutyrate in the cytoplasm. To date, only five kinds of *Azospirillum* have been defined: *A. brasilense*, *A. lipoferum*, *A. amazonense*, *A. halopraeferens,* and *A. irakense*. The way of living is multiplied both in anaerobic conditions and in terms of oxygen consumption; however, it is above all the small scale aerofilia in the proximity or the non-occurrence of the nitrogen bond in the middle.

7.12.3 Cyanobacteria

Both cyanobacteria and free-living cooperatives have been affected by the development of rice in India. A composite of blue green algae with heterogeneous *Nostoc*, *Anabaena* and *Aulosira*, etc., is administered as an important inoculum in the plate and pots coated with polyethylene and then doubled in mass in the field used as soil chips for development of rice at the rate of 10 kg/ha.

7.12.4 *Azolla*

It is an aquatic fern that moves freely and floats in the surface of water and fixes the atmospheric nitrogen in relation to the blue green algae *Anabaena azollae*. The *Azolla* fronds are composed of sporophytes with a cabotage rhizome and bilobate leaves and roots that cover little. Rice development ranges in Southeast Asia and other underdeveloped nations have recently shown greater interest for the use of the valuable nitrogen *Azolla* sedimentation water plants as an another source of nitrogen or as a complement to commercial nitrogen fertilizers. *Azolla* is used as a biofertilizer for rice in wet areas and it is known to use 40 kg of N/ha at 60 kg N/ha for rice cultivation.

7.12.5 *Azotobacter*

It is imperative and certain that the high-impact nitrogen-free sedimentation bacteria are understood. In non-leguminous plants, it is used as a biofertilizer particularly in rice, cotton, vegetables, etc. *Azotobacter* cells are not present in the flat rhizosphere, but are abundant in the leguminous roots.

7.12.6 *Acetobacter*

This is a sacharophillic microbe and is associated with sugarcane, sweet potatoes, and sweet sorghum plants and can fix 30 kg/N/ha per year. Basically, the bacterium is marketed for the cultivation of sugarcane. It is known that it is used from 10 to 20 t/section of soil and the sugar content is about 10–15%.

7.13 Equipment Used in the Manufacture of Biofertilizer

In the field of biofertilizer creation, gear types (equipment and other manufacturing gadgets) represent the significant basis, which comprises 70% of capital assumption. Any negotiations on the usage of these kinds of equipment that are accompanied can finally reduce the nature of the biofertilizer. After observing the standards behind the use of all instruments, some part of the instruments can be changed by a company room equipped with a UV lamp. Hot air ovens, incubators, autoclaves, and fixing machines are manufactured locally with suitable specialized details. The correct use of the kinds of equipment will give a continuous demonstration with the quality inoculations.

7.14 Mechanism of Action of Numerous Biofertilizers

Mycorrhizal fungi are the association with the roots of the larger floors (Khanday et al. 2016; Bhat et al. 2017). While still a paradox, it assists as a model system for understanding the device behind stimulating radical cell growth as a result of the presence of mycorrhizae. Genome sequencing of two EMs (ectomycorrhizaes), bicolor L. *T. melanosporum* (Megali et al. 2013) and fungi (black truffle) (Molina-Favero et al. 2007), helps to identify factors that control the mycorrhizal development and its role in the plant cell (Bonfante and Genre 2010). Fifteen genes regulated during symbiosis were identified as putative hexose transporters in bicolor L. *T. melanosporum*. However, melanosporum has an invertase genetic factor and, unlike bicolor L., and can host sucrose directly (Bonfante and Genre 2010). Upregulation of carrier genes during symbiosis called transport action of beneficial complexes such as amino acids, oligopeptides, and polyamines through the symbiotic interface of one organism to another organism. The mycelium takes nitrates of the soil of life and ammonium for free. Subsequently, these compounds reach the soil and then transferred to the plants. A cysteine-rich protein (MISSP7) fungus plays an important role as effectors and mediators in the formation of symbiotic interfaces (Plett et al. 2011). Many genes that associated with the biosynthesis of auxins and root morphogenesis exhibited upregulation during mycorrhizal colonization (Ansari et al. 2013; Splivallo et al. 2009; Abdel-Raouf et al. 2012). Furthermore, *G. versi-*

forme transports of inorganic phosphate (Pi) in their hyphae that helps through absorption of soil phosphate and glutamine synthetase gene in *G. intraradices* found, which reinforces the opportunity of nitrogen metabolism in the fungal hifon that can be later available to the plant (Salvioli et al. 2012). It is suggested that bioactive compounds call for factors similar to *Rhizobium* Nod. Myc factors are secreted by mycorrhizae and *Rhizobium* and are perceived by the roots of host-activating signal transduction or by common symbiosis (SYM) (Kosuta 2003). The routes that prepare the plant for AM and rhizobium infection have some common characteristics. MGS commonly prepared by the host plant to produce changes at the molecular and anatomical level in the first contact of the fungal hyphae. Until now, it is assumed that calcium is the secondary messenger, increasing Ca^{2+} in the nuclear region of the absorbent hairs (Sieberer et al. 2009). *Leguminosarum biovarviciae* can induce *Rhizobium* various genes in plants (peas, alfalfa, and sugar beet) as shown by microarray studies (Ramachandran et al. 2011). The PGPR produces IAA, in turn, induces the production of nitric acid (NO), which serves as a second messenger to activate a complex signaling network that improves the processes of growth and development of the root (Molina-Favero et al. 2007). The expression of ENOD11 and many defense-related genes, as well as the root remodeling genes, is adjusted upward upon entry. Subsequently, this allows the formation of a pre-penetration device or PPA (Bucher et al. 2009). Once deposited, causes a decrease in arbuscular growth (Bapaume and Reinhardt 2012). It is known that many other genes, including the subtilisin 65 proteases, the transporter 66, or two phosphates of the transporter 67, are involved in the formation of arbusculi (Zhang et al. 2010). Today's scientists commonly use nitrogen fixer genes to create technical structures that can repair atmospheric nitrogen. The induction of genes in the case of nitrogen-fixing bacteria occurs at a low concentration of nitrogen and oxygen in the rhizosphere (Santos et al. 2012). Interestingly, sugarcane seedlings inoculated with a wild strain of *G. diazotrophicus* demonstrated radioactive N_2 fixation compared with *G. diazotrophicus* mutant having the mutant nif D gene demonstrated the importance of nif genes. The efficiency of nitrogen fixation depends on the use of carbon (Sevilla et al. 2001). A bacterium like *Bacillus subtilis* (UFLA285) can differentially induce 247 genes in cotton plants compared to the control in which PGPR was not supplied to the cotton plant (Baharlouei et al. 2011). Many disease resistance genes that work with jasmonate/ethylene signals, as well as osmotic regulation from proline synthesis genes, have been expressed differentially with the induction of UFLA285 (Baharlouei et al. 2011). Several differently uttered genes have been identified, including the type 1 metallothionein protein, an integral membrane protein similar to NOD26, ZmNIP2–1, a protein of the thionine family, a precursor of the gamma chain orycin, an associated protein to emphasize 1 (OsISAP1). They induce the PBZ1 and auxin protein and the probenazole genes that respond to ethylene (Brusamarello-Santos et al. 2012). It is also a well-established fact that the expression of the proteins related to the PBZ1 and thionine defense is suppressed in the seropedica association of rice-H, signifying the modulation of the defense responses of the plant in colonization (Brusamarello-Santos et al. 2012). Among PGPR species, it has been proposed that *Azospirillum* helps in regulation of gibberellins, auxins, and ethylene (Perrig et al. 2007). Some bacteria associated with plants can also induce the synthesis of phytohormones, as *Pinus contorta* when *Paenibacillus polymyxa* inoculated with IAA had elevated levels in roots (Bent et al. 2001). *Rhizobium* and *Bacillus* IAA can be synthesized under different growth conditions such as pH, temperature, and the presence of agricultural waste as a substrate (Sudha et al. 2012). Ethylene, unlike additional phytohormones, is accountable for inhibiting the growth of dicotyledonous plants (Ansari et al. 2013). It was initiated by Glick et al. (1998), PGPR could improve the growth of plants by suppressing ethylene expression. Interestingly, a model in which, the synthesis of ethylene from 1-aminociclopropane-1-carboxylate (ACC), an immediate precursor of ethylene, from the bacterial enzyme ACC deaminase hydrolyzes the need for nitrogen and carbon which ultimately acts as the source. It is also one of the mechanisms of induction of good conditions for growth. ACC deaminase activity has also been found in bacteria such as *Alcaligenes* sp., *Bacillus pumilus, Pseudomonas sp.,* and the *variovorax paradox* (Ansari et al. 2013). The participation of ACC deaminase in the indirect influence on plant growth has been demonstrated in canola, with mutations in the CAC deaminase gene that has resulted in the loss of the growth in *Pseudomonas putida*. Interestingly, the potential for PGPR has been further increased by introducing genes involved in the direct oxidation (DO) and phosphate solubilization (MPS) pathways in some useful PGPR strains. The gene encoding glucose dehydrogenase (gcd) involved in the route of the DO has been cloned and characterized by *Acinetobacter calcoaceticus* and *E. coli* and *Enterobacter asburiae* (Tripura et al. 2007). A soluble form of gcd was also cloned by *Acinetobacter calcoaceticus* and *G. oxydans* (Sashidhar and Podile 2010). In addition, it has been reported that glucose directed to dehydrogenase (GDH) and gluconate dehydrogenase (GADH) mutagenesis increases the activity of this enzyme. The simple S771 M substitution conferred thermal stability to *E. coli*, whereas the 742 mutation of the lysine glutamate improved EDTA tolerance of *E. coli* PQQGDH. The application of this technology has been carried out through the transfer of genes involved in the OD pathway, namely GDH, GADH, and pyrroloquinoline quinone (PQQ) for rhizobacteria and phosphoenolpyruvate carboxylase (PPC) of *P. fluorescens*, offering MPS functionality (Sashidhar and Podile 2010) (Fig. 7.1).

The host roots perceived bioactive ligands called Myco factors and NOD factors secreted by the mycorrhiza as well as *Rhizobium* trigger via signal transduction (Kosuta 2003), which initiates an additional indication of transduction pathway via unknown receptor (SYMRK and NORK) (Bonfante and Genre 2010), that causes the release of Ca^{2+} in the cytosol (Sieberer et al. 2009). All pathways involve receptor kinases like DMI and phosphorylating their SYM71 protein substrates (Maillet et al. 2011). The nuclear pore complex (NPC) and some of its proteins (NUP) play a role in the addition of calcium. The DM1 proteins play a role in maintaining the periodic oscillation of calcium ions inside and outside the nucleus. Many protein channels (Ca^{2+} protein channels) also facilitate this process with different carriers (Sieberer et al. 2009). CCaMK is calcium-dependent calmodulin, which phosphorylates the CYCLOPS protein product, thus initiating the activation of several genes that involve the formation of structures such as nodule and pre-penetration apparatus (PPA) protein kinase (Maillet et al. 2011).

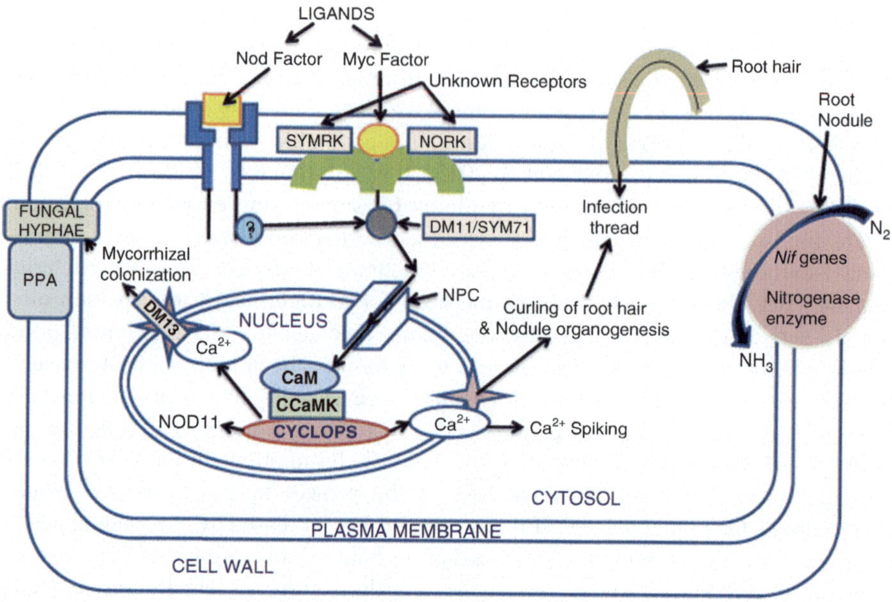

Fig. 7.1 Mechanism action of biofertilizers in the "root cell"

7.14.1 Mode of Action of Biofertilizer

1. They fix atmospheric nitrogen in the soil through the root nodules of the legume plants and make it accessible to the plant.
2. Solubilize the phosphate as tricalcium, iron, and aluminum phosphate that are available in insoluble form.
3. Produces hormones and antimetabolites that promote root growth.
4. Decompose organic matter.
5. By the application of biofertilizers on the soil and seeds, there is increase in the yield up to 10–20% and also an increase in the nutrients pertaining to plant growth without producing the adverse effects on the environment. Therefore, the height of the plant, the number of branches, the number of roots, the length of the root, the length of the shoot, the accumulation of dry matter in the plant organs significantly increase the growth parameters of the plant and the index of vigor, etc.

7.14.2 Application of Bio-Fertilizers

There are different four types of methods for the application of biofertilizers which are seed treatment, set treatment, seedling treatment, and soil treatment (Table 7.2).

Table 7.2 Methods and application of biofertilizers

S. no.	Method	Application
1	(a) The seeds needed to sow an acre in a pile are kept in a clean floor or cemetery (b) The culture suspension is prepared by mixing 1 pack (200 g) of biofertilizer in approx. 400 mL of water (1: 2) (c) The culture suspension is sprayed onto the seed stack and mixed by hand so that the thin coating is uniformly applied to the seeds (d) The seeds are spread in the shade for a while to dry and then sow (e) Instead of water, rice glue can also be used for better results	*Seed treatment* For the inculturation of cereals such as rice, wheat, sorghum, corn, etc. And oily seeds like peanuts, sunflower seeds, mustard, safflower, legumes such as cowpea, green gram, black gram, and soy. One pack (200 g) is sufficient to treat 10–12 kg of seed. On this basis, the dose of biofertilizer per acre can be calculated based on the seed rate
2	(a) The culture suspension is prepared by mixing 1 kg (five packs) of biofertilizer in 40–50 L of water (b) The cut pieces of sowing material that are required to sow one acre remain immersed in the suspension for 30 min (c) The cut pieces are removed and dried in the shade for some time before planting (d) After sowing, the field is watered in 24 h (e) For the setting treatment, the ratio of biofertilizer to water is about 1:50	*Set treatment* This method is generally recommended to treat sugarcane sets, chopped potato pieces, and the base of banana shoots
3	(a) The suspension is prepared by mixing 1 kg (5 packs) of biofertilizer culture in 10–15 L of water (b) The necessary seedlings are obtained for one hectare and small packages of seedlings are made (c) The root portion of these seedlings is immersed in the suspension for 15–30 min and immediately transplanted (d) In general, the proportion of inocula and water is 1:10 (approximately), i.e., 1 kg of biological fertilizer in 10 L of water (e) For vegetables like tomato, cabbage, cauliflower, one pack of biofertilizer is sufficient for 0.1 ha (10,000 m^2) of soil	Seedling treatment This method is recommended for crops such as rice, tobacco, tomato, cold, onion, cabbage, cauliflower, etc.
4	*Soil application* This method varies from one crop to another depending on its duration. In general, for a short-term harvest (<6 months), 10–15 packets (each of 200 g) are mixed with 40–60 kg of well-decomposed cattle manure or 40–60 kg of soil for an acre of earth. The mixture of biofertilizer and manure of livestock/soil sprayed with water is then transmitted to the soil at the time of sowing or at the time of irrigation in permanent crops. For long-term crops (perennial crops), 20–30 packs of biofertilizer (each with 200 g) are mixed with 80–120 kg of manure or soil per acre	For its application in permanent crops, perennials are pruned once a year. After pruning, the soil of the bed is dug up with a fork with due care to avoid damaging the roots. Then a mixture of biofertilizer and FYM/soil is applied, incorporating it into the soil, followed by irrigation

7.15 Advantages of Synthetic Fertilizers

7.15.1 Improve Soil Quality and Add Micronutrient Benefits to the Soil

Organic fertilizers improve the soil structure and soil quality. The creation of compost with a balanced combination of cooking residues, garden pruning, rodent dung, and straw encourages microorganisms to grow and enrich the soil. The vegetables produced are healthier and more beneficial for the body.

7.15.2 Organic Fertilizers Are Slow-Release

A lot of compost can take longer. It also releases its nutrients slowly. However, it does this constantly. Its nutrients are also more useful and stimulate the growth of healthier plants.

7.15.3 They Are Not Toxic

Chemicals in synthetic fibers are toxic, acidic and can burn the roots of seedlings. Organic fertilizers, without toxins, have a neutral pH level and do not produce this effect.

7.15.4 They Are Cheap

These fertilizers are also less expensive. Creating compost, with natural and organic ingredients from your kitchen, does not cost much. In contrast, some synthetic fertilizers are quite expensive.

With this persuasive drive to use organic fertilizers, it is prime time to start composting.

7.16 Advantages of Using Biofertilizers

Some of the advantages associated with the use of biofertilizers include:

- They are ecological and profitable
- Improves the soil enrichment and soil quality over time

- Although these fertilizers do not show immediate results, the results shown after some time are spectacular
- These fertilizers directly use atmospheric nitrogen and makes it available to plants
- Increasing the dissolution of phosphorus content of the soil and releasing the unavailable phosphorus
- Biofertilizers help to release the growth-promoting hormones and improve root proliferation
- The microorganism converts compound nutrients into simple nutrients available for plant growth
- Microorganisms that are present in biofertilizers promote the adequate supply of nutrients to host plants and to ensure its proper development of growth and regulation in its physiology
- Help increase the crop yields by 10–25%
- Biofertilizers can protect plants from ailments transmitted through the soil to some extent

7.17 Conclusion

Biofertilizers contain microorganisms that increase soil microbial activity. Often, organic matter is included to enhance the rate of production. And since "soil fertility is decreasing in most soils throughout the world due to soil erosion, loss of nutrition, accumulation of toxic elements, unbalanced accumulation of water and nutrients." Compost and organic fertilizers are alternatives to meet the need for nutrient sources for crops. Microbes which have the capacity in the production of biofertilizers include *Azotobacter*, *Azospirillum*, *Phosphobacter,* and *Rhizobacter*. Biofertilizers have great importance in agricultural production. They play a fundamental role in maintaining soil sustainability and long-term fertility by fixing atmospheric nitrogen, thus mobilizing macro- and micronutrients fixed in the soil. In this context, the excessive use and dependence on chemical fertilizers are not feasible on the long-term basis because of the costs and other negative implications on environment. In this context, the biofertilizers would be the feasible option for farmers to increase in productivity per unit area without harming the quality of the environment.

References

Abdel-Raouf N, Al-Homaidan AA, Ibraheem IBM (2012) Agricultural importance of algae. Afr J Biotechnol 11:1648–11658

Ansari MW, Trivedi DK, Sahoo RK, Gill SS, Tuteja N (2013) A critical review on fungi mediated plant responses with special emphasis to Piriformospora indica on improved production and protection of crops. Plant Physiol Biochem 70:403–410

Araujo ASF, Santos VB, Monteiro RTR (2008) Responses of soil microbial biomass and activity for practices of organic and conventional farming systems in Piaui state, Brazil. Eur J Soil Biol 44:225–230

Baharlouei K, Pazira E, Solhi M (2011) Evaluation of inoculation of plant growth-promoting rhizobacteria on cadmium. In: International conference on environmental science and technology IPCBEE, vol 6. IACSIT Press, Singapore

Bapaume L, Reinhardt D (2012) How membranes shape plant symbioses: signaling and transport in nodulation and arbuscular mycorrhiza. Front Plant Sci 3:223

Bent E, Tuzun S, Chanway CP, Enebak S (2001) Alterations in plant growth and in root hormone levels of lodgepole pines inoculated with rhizobacteria. Can J Microbiol 47:793–800

Bhat RA, Dervash MA, Mehmood MA, Bhat MS, Rashid A, Bhat JIA, Singh DV, Lone R (2017) Mycorrhizae: a sustainable industry for plant and soil environment. In: Varma A et al (eds) Mycorrhiza-nutrient uptake, biocontrol, ecorestoration. Springer, Basel, pp 473–502

Bhat RA, Beigh BA, Mir SA, Dar SA, Dervash MA, Rashid A, Lone R (2018) Biopesticide techniques to remediate pesticides in polluted ecosystems. In: Wani KA, Mamta (eds) Handbook of research on the adverse effects of pesticide pollution in aquatic ecosystems. IGI Global, Hershey, pp 387–407

Bonfante P, Genre A (2010) Mechanisms underlying beneficial plant-fungus interactions in mycorrhizal symbiosis. Nat Commun 27:1–48

Brusamarello-Santos L, Pacheco F, Aljanabi S, Monteiro R, Cruz L, Baura V, Pedrosa F, Souza E, Wassem R (2012) Differential gene expression of rice roots inoculated with the diazotroph Herbaspirillum seropedicae. Plant Soil 356:113–125

Bucher M, Wegmüller S, Drissner D (2009) Chasing the structures of small molecules in arbuscular mycorrhizal signaling. Curr Opin Plant Biol 12:500–507

Glick BR, Penrose DM, Li JA (1998) A model for the lowering of plant ethylene concentrations by plant growth-promoting bacteria. J Theor Biol 190:63–68

Khanday M, Bhat RA, Haq S, Dervash MA, Bhatti AA, Nissa M, Mir MR (2016) Arbuscular mycorrhizal fungi boon for plant nutrition and soil health. In: Hakeem KR, Akhtar J, Sabir M (eds) Soil science: agricultural and environmental prospectives. Springer, Basel, pp 317–332

Kosuta S (2003) Diffusible factor from arbuscular mycorrhizal fungi induces symbiosis-specific expression in roots of *Medicago truncatula*. Plant Physiol 131:952–962

Maillet F, Poinsot V, André O, Puech-Pagès V, Haouy A, Gueunier M, Cromer L, Giraudet D, Formey D, Niebel A, Martinez EA, Driguez H, Bécard G, Dénarié J (2011) Fungal lipochitooligosaccharide symbiotic signals in arbuscularmycorrhiza. Nature 469:58–63

Megali L, Glauser G, Rasmann S (2013) Fertilization with beneficial microorganisms decreases tomato defenses against insect pests. Agron Sustain Dev 34:649–656

Molina-Favero C, MónicaCreus C, Luciana Lanteri M, Correa-Aragunde N, Lombardo MC, Barassi AC, Lamattina L (2007) Nitric oxide and plant growth promoting rhizobacteria: common features influencing root growth and development. Adv Bot Res 46:1–33

Mushtaq N, Bhat RA, Dervash MA, Qadri H, Dar GH (2018) Biopesticides: the key component to remediate pesticide contamination in an ecosystem. In: Environmental contamination and remediation. Cambridge Scholars Publishing, Cambridge, pp 152–178

Perrig D, Boiero ML, Masciarelli OA, Penna C, Ruiz OA, Cassan FD, Luna MV (2007) Plant-growth promoting compounds produced by two agronomically important strains of *Azospirillum brasilense*, and implications for inoculant formulation. Appl Microbiol Biotechnol 75:1143–1150

Plett JM, Kemppainen M, Kale SD, Kohler A, Legue V, Brun A, Tyler BM, Pardo AG, Martin F (2011) A secreted effector protein of Laccaria bicolor is required for symbiosis development. Curr Biol 21:1197–1203

Raja N (2013) Biopesticides and biofertilizers: ecofriendly sources for sustainable agriculture. J Biofertil Biopestici 4:1000e112

Ramachandran VK, East AK, Karunakaran R, Downie JA, Poole SP (2011) Adaptation of *Rhizobium leguminosarum* to pea, alfalfa and sugar beet rhizosphere investigated by comparative transcriptomics. Genome Biol 12:106–109

Sahoo RK, Ansari MW, Pradhan M, Dangar TK, Mohanty S, Tuteja N (2014) Phenotypic and molecular characterization of efficient native *Azospirillum* strains from rice fields for crop improvement. Protoplasma 251:943. https://doi.org/10.1007/s00709-013-0607-7

Salvioli A, Zouari I, Chalot M, Bonfante P (2012) The arbuscular mycorrhizal status has an impact on the transcriptome profile and amino acid composition of tomato fruit. BMC Plant Biol 12:44

Santos VB, Araujo SF, Leite LF, Nunes LA, Melo JW (2012) Soil microbial biomass and organic matter fractions during transition from conventional to organic farming systems. Geoderma 170:227–231

Sashidhar B, Podile AR (2010) Mineral phosphate solubilisation by rhizosphere bacteria and scope for manipulation of the direct oxidation pathway involving glucose dehydrogenase. J Appl Microbiol 109:1–12

Sevilla M, Burris RH, Gunapala N, Kennedy C (2001) Comparison of benefit to sugarcane plant growth and 15n2 incorporation following inoculation of sterile plants with *Acetobacter diazotrophicus* wild-type and Nif–mutant strains. Mol Plant-Microbe Interact 14:358–366

Shafi S, Bhat RA, Bandh SA, Shameem N, Nisa H (2018) Microbes: key agents in the sustainable environment and cycling of nutrients. In: Environmental contamination and remediation. Cambridge Scholars Publishing, Cambridge

Sieberer BJ, Chabaud M, Timmers AC, Monin A, Fournier J, Barker DGA (2009) A nuclear-targeted cameleon demonstrates intranuclear Ca^{2+} spiking in *Medicago truncatula* root hairs in response to rhizobial nodulation factors. Plant Physiol 151:1197–1206

Splivallo R, Fischer U, Gobel C, Feussner I, Karlovsky P (2009) Truffles regulate plant root morphogenesis via the production of auxin and ethylene. Plant Physiol 150:2018–2029. https://doi.org/10.1104/pp.109.141325

Sudha M, Gowri RS, Prabhavati P, Astapriya P, Devi SY, Saranya A (2012) Production and optimization of indole-acetic-acid by indigenous micro flora using agro waste as substrate. Pak J Biol Sci 15:39–43

Tripura CB, Sudhakar Reddy P, Reddy MK, Sashidhar B, Podile AR (2007) Glucose dehydrogenase of a rhizobacterial strain of *Enterobacter asburiae* involved in mineral phosphate solubilization shares properties and sequence homology with other members of enterobacteriaceae. Indian J Microbiol 47:126–131

Vessey JK (2003) Plant growth promoting rhizobacteria as biofertilizers. Plant Soil 255(2):571

Youssef MMA, Eissa MFM (2014) Biofertilizers and their role in management of plant parasitic nematodes. A review. J Biotechnol Pharm Res 5:1–6

Zhang Q, Blaylock LA, Harrison MJ (2010) Two *Medicago truncatula* half-ABC transporters are essential for arbuscule development in arbuscular mycorrhizal symbiosis. Plant Cell 22:1483–1497

Chapter 8
Role of Macrophytes in Spontaneous Lacustrine Phytofiltration

Umar Nazir Bhat and Anisa B. Khan

8.1 Introduction

Aquatic plants mainly comprise hydrophytes like phytoplankton, periphyton and multicellular macrophytes occupying varied ecological niches in the water environment. Majorly fresh water vascular plants and macroalgae collectively form macrophytes (Wetzel 2001). Macrophytes consist principally of aquatic vascular flowering plants but also include the aquatic mosses, liverworts, ferns and the larger macroalgae (Chambers et al. 2008). Macrophytes being predominant primary production sources play an important role in energy input, nutrient budgeting and recycling (Algesten et al. 2003), biofiltration (Dhote 2007) and sedimentation processes in the water bodies. Macrophytes are ecologically dominant community of fresh water bodies especially lakes performing basic functions of energy and oxygen production, biogeochemical cycling, regulation of water quality, sediment stabilization and furnishing habitat to micro and macro water-organisms (Gurnell et al. 2012; Bhat et al. 2018). They influence hydrochemical processes and serve as substrate for periphyton and phytophilous invertebrates thus creating specific ecological conditions necessary for functioning of other aquatic biota (Raspopov et al. 2002; Vymazal 2002). Invertebrates and macrofauna including fish and waterfowl find them as suitable breeding, nesting and sheltering niches (Cvetkovic et al. 2010; Grabas et al. 2012; Jurca et al. 2012; Sierszen et al. 2012). The provision to act as fodder and green manure is an option too. Eutrophication is associated with transition from macrophyte dominated clear water state to turbid phytoplankton dominated state (Scheffer 1989; Bhat et al. 2017) and loss of submerged macrophytes (Körner 2002; Xiangcan 2003). Aquatic weeds scavenge inorganic and some organic compounds from water, incorporating the dissolved materials into their own structure and as such act as 'biological sinks' (Scholz 2003) for large

U. N. Bhat (✉) · A. B. Khan
Department of Ecology and Environmental Sciences, Pondicherry University, Puducherry, India

R. A. Bhat et al. (eds.), *Bioremediation and Biotechnology, Vol 2*,
https://doi.org/10.1007/978-3-030-40333-1_8

quantities of minerals which could otherwise cause 'algal blooms' (Mushtaq et al. 2020). The presence and dominant assemblages of macrophytes reflect differences in trophic status, therefore, are very good indicators of water quality. Macrophytes have been evaluated not only for stripping nutrients (Gottschall et al. 2007) but also in altering physicochemical environment of waters (Bhat et al. 2018). For instance, nutrient assimilation during photosynthetic productivity reflects alkaline pH (Hart et al. 2002). The relationship between water column nutrient concentrations and macrophyte growth is extensively documented (Moore and Wetzel, 2000). Harpole et al. (2011) studied colimitation of N and P for productivity, while Carrow et al. (2001) observed growth patterns at nutrient gradients exhibiting growth limited at low, improved at higher and toxic at the highest levels allied to osmotic stress and pH primarily. The quality and quantity of vegetation facilitate insight of environmental state (Lenzi et al. 2012). Vegetation reflects lower C:P and C:N in eutrophied waters. Either due to low physiological requirement or tolerance for nutrients, macrophytic group shifts occur in water bodies as the opportunistic species prevail (Lenzi et al. 2012).

'Green Clean'—a popular term for phytoremediation is an acceptable and eco-friendly in situ alternative to conventional remediation approaches (Rai 2008). It can be an ecologically responsible practice to detoxify, remove, sequester and stabilize persistent pollutants in aesthetic manner (Meagher, 2000) effectively in water bodies (Peng et al. 2009). This plant based removal, stabilization, transfer or decomposition technology utilizes natural or GM flora in varied ecosystems in eco-restoration programmes (Amaya-Chavez et al. 2006; Padmavathiamma and Li 2007; Tian et al. 2007). The widespread application of plants as toxic site (land, water or air) cleaners of all pollutant sorts of organic and inorganic nature are spearheading from lab to field and market place (Ibañez et al. 2016). Phytofiltration uses plants to remove contaminants from water (Arthur et al. 2005). Despite the nuisance character of aquatic plants, their capability to improve water quality has ecological significance for pollution abatement (Qadri and Bhat 2020). A growth (metabolic) property of aquatic macrophytes removes pollutants (Hasan et al. 2007; Mishra et al. 2008) as the knowledge range of cleanup duly applies on aquatic and semi-aquatic vascular species (Bhatia and Goyal 2014) like *Azolla*, *Carex*, *Ceratophyllum*, *Cyperus*, *Eichhornia*, *Elodea*, *Glyceria*, *Hydrocotyle*, *Juncus*, *Lemna*, *Phragmites*, *Pistia*, *Potamogeton*, *Sagittaria*, *Scirpus*, *Spartina*, *Typha* and the like. Hydrophytes entrap nutrients in the water column both by physical obstruction (biomass adsorption) and by uptake from soil solution (Lukács et al. 2009). Yang et al. (2007) list pollution tolerance, faster growth and high harvestable biomass as positive indices for a good phytoremediator. The frequent abundance of macrophytes in aquatic ecosystems prescribes them as preferred and potent bioagents for pollution abatement and biomonitoring. The phytoremediation technology addresses the global problem of pollution in air, water and soil. The macrophytic role in purification of aquatic ecosystems involves removal, detoxification or stabilization of contaminants in their growth matrix, viz. water and sediment (Becker 2000). Phytoextraction possesses several appealing attributes compared to the traditional established methods of remediation. It is natural, spontaneous, efficient both in situ and ex situ, eco-friendly, aesthetically pleasing, cost effective, solar driven, sustainable, least site invasive, applicable on large scale, demands minimal maintenance and amenable to amelioration of a broad range of contaminants (Lasat 2000; Morikawa and Erkin 2003; Tian et al. 2007). Further, phytoremediation offers maintenance of natural biochemical activity and geophysical structure along with sites' self-restoration. The bioaccumulation property of wetland plants has resulted in their widespread use as bioindicators and as bioremediative agents (Qian et al. 1999; Samecka-Cymerman and Kempers 2002). So the overall complex structural and functional integrity imparted to an aquatic ecosystem, indicative of its environmental quality conditions is regulated by macrophytes. Henceforth, an endeavour is made in the current study to evaluate the comparative spontaneous efficacy of unlike life-form macrophytes in the phytofiltration of nutrients and traces for their in situ role of self-remediative lacustrine function.

8.2 Methods

The city of Srinagar (33°59′14″–34°12′37" N latitudes and 74°41′06″–74°57′27″ E longitudes) sited on the banks of the Jhelum (*Vyeth*) spreads across the plains of Kashmir vale. It has a moderate physiography ≥1580 m a.s.l. representing a hill topography. The general climatic conditions resemble the sub-Mediterranean characterized by yearlong precipitation episodes except a few summer and autumn dry-periods, besides, both seasonal and diurnal extremes of temperature. The focuses of present limnological research remain two fluvial urban valley lakes, namely Anchar and Dal (average depth of 1.5 m) situated in the summer capital (Srinagar) of Jammu and Kashmir. A preliminary comparison indicates the Anchar as unmanaged, while the Dal as fairly managed lake with substantial differences in their physical appearance and open-water surface area (Fig. 8.1).

However, both the lakes are eutrophied and majorly engulfed in urban and agricultural transformations as evident from satellite imagery (Fig. 8.2). The inter-annual survey, sample collection and hydrobiochemical study of the two lakes are based on their five select stations. The five locations designated as A1, A2, D1, D2 and D3 were each further fragmented into ten sampling stations (indicative as red spots in Fig. 8.3) in order to obtain composite representative samples in triplicate for enhanced precision. The site selection criterion implemented here represents the distinct spatial features within the lakes. A1 adjoins the urban locale highly infested with vegetation and diffuse sewage inputs; A2 is situated closer to the lakes' exit and recipient of run-off from paddy fields; D1 vicinity is contributed with key inlet and STP discharges; D2 typifies the floating garden area; and D3 nearby the outlet has hotel and house-boat zone.

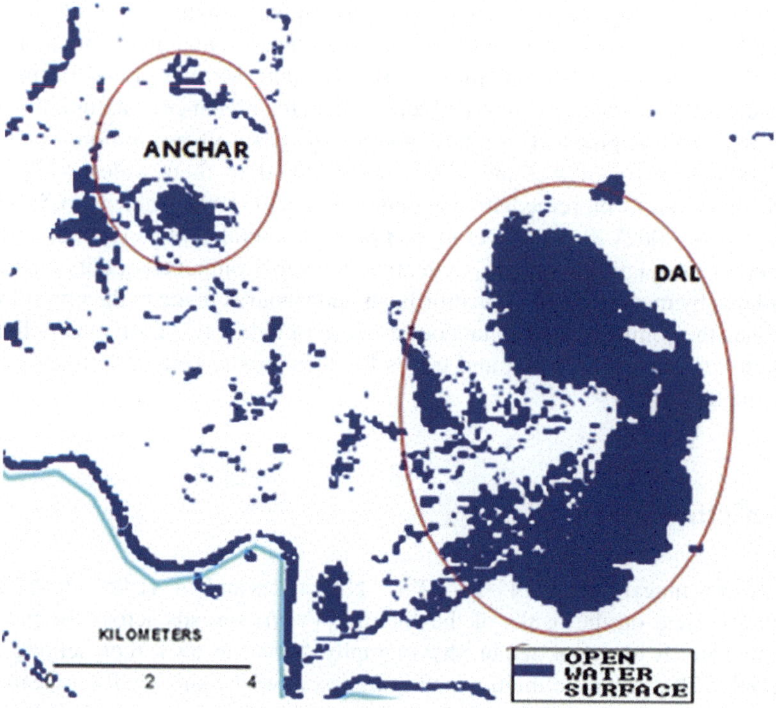

Fig. 8.1 Open water expanse of the study sites (Source: NRSC/ISRO—Bhuvan)

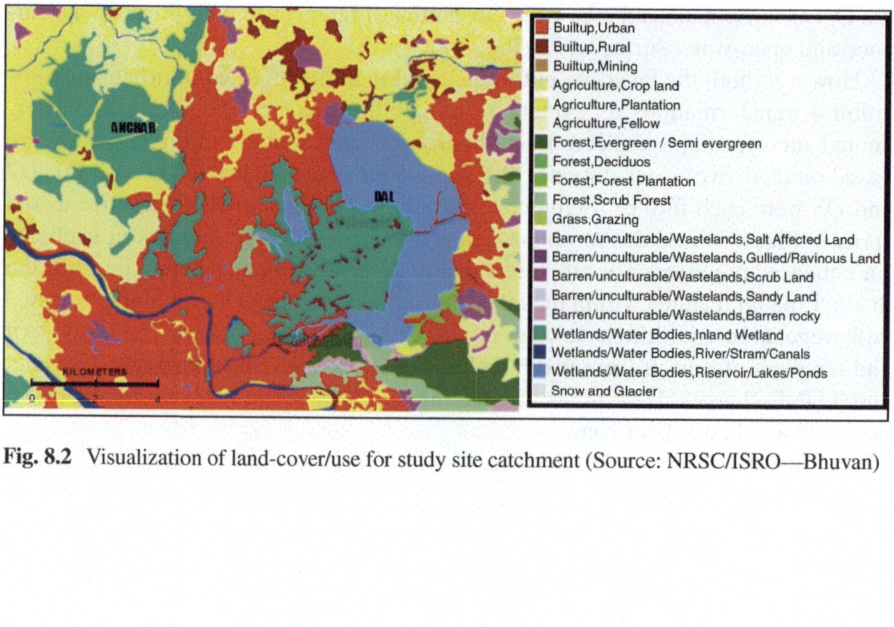

Fig. 8.2 Visualization of land-cover/use for study site catchment (Source: NRSC/ISRO—Bhuvan)

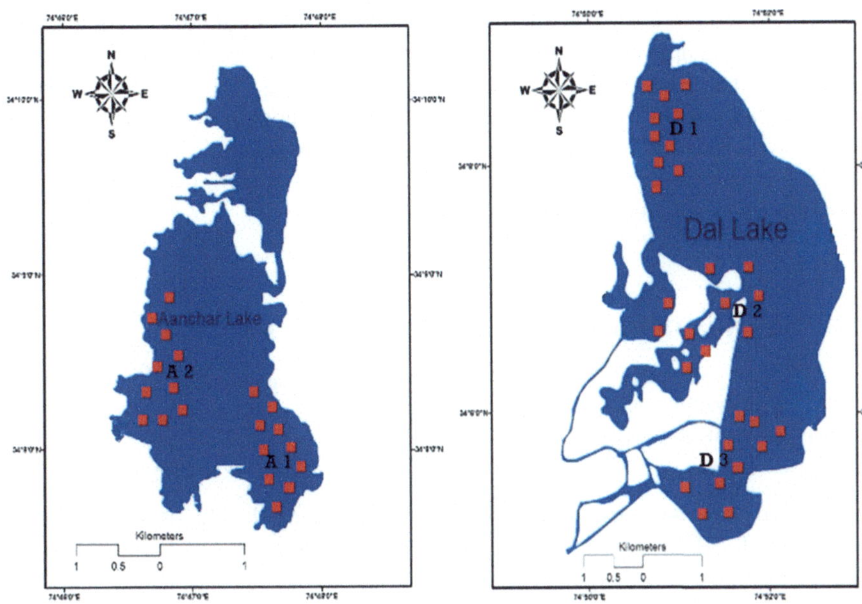

Fig. 8.3 Sample collection stations

8.3 Macrophytes

The biological component of macrophytes employed in the study includes select species chosen from emergent, free-floating, rooted-floating and submerged categories, namely *Myriophyllum aquaticum*, *Salvinia natans*, *Nelumbo nucifera* and *Ceratophyllum demersum,* respectively. Their selection represents omnipresent abundance at each sampling location besides being high biomass yielding, eutrophication tolerant and hyperaccumulators. The preferred producer species are recognized while referring to standard taxonomic scheme of Cook (1996), Ghosh (2005) and Arshid et al. (2011). A brief account is provided herewith species wise indicating taxonomic and morphological aspects.

8.3.1 Myriophyllum aquaticum *(Parrot-Feather Watermilfoil)*

- Family = Haloragaceae
- Genus = Myriophyllum
- Species = aquaticum

Commonly referred to as 'Parrot-feather Watermilfoil', *Myriophyllum aquaticum* is an amphibious, invasive, herbaceous and perennial rooted shallow water plant with more emergent than submerged shoots (Arshid et al. 2011). Its feather-like pinnate leaves appear whorled on an elongated stem (40–90 cm) growing into a population of dense stands. It is a tolerant heterophyllous colonizer of diverse habitat ranging from fresh to saline waters exhibiting phenoplasticity in response to extreme environmental gradients (Trémolières 2004; Wersal et al. 2011). Its growth is chiefly regulated with temperature and light conditions. It shows vegetative reproduction via fragmentation of shoots.

8.3.2 Salvinia natans *(Floating-Moss or -Fern)*

- Family = Salviniaceae
- Genus = Salvinia
- Species = natans

Floating-moss is a noxious annual perennial weed. It is a deciduous, free-floating, rootless and clonal aquatic fern emerging on water surface. It has horizontal stem with opposite hair covered water resistant leaves (1–3 cm). Amongst its leaf whorl of three, two elliptic leaves perform the floating function, while the profoundly dissected third one submerges to perform as a root alike. Its strong ability to duplicate mats of foliage makes it capable to engulf all the water. Its sexual reproduction occurs after a winter diapause by fertilization of male microspore and female macrospore on the water surface at temperatures $>12.4 \pm 0.2$ °C (Gałka and Szmeja 2013). The spores release gradually from the reproductive organs called sporocarps lying at the bottom as their walls disintegrate. The resultant zygote develops into a structurally clonal sporophyte. Again the vegetative means entail the mature clone fragmentation.

8.3.3 Nelumbo nucifera *(Asian Lotus)*

- Family = Nelumbonaceae
- Genus = Nelumbo
- Species = nucifera

Lotus is an attractive ornamental and sacred aquatic plant associated with Indian cultural heritage besides being accorded the status of National Flower. Its phenotypic diversity and thermoplasticity grants it wide naturalization globally (Yang et al. 2013). It is a rooted-floating type perennial macrophyte possessing ≥ 3 cm long petioles. The leaves are orbicular, palmate but slightly concave, peltate and tough leathery structured with waxy cuticle (glaucous) on their upper side. The circular leaf blades range from 10 to 100 cm in diameter. Its flowering period is June to September displaying large (10–25 cm across) solitary bisexual pinkish-red blossoms. Its fruit called torus is an enlarged receptacle enclosing embedded seeds.

It multiplies by seed and vegetative rhizome propagation. It has commercial, medicinal and pollution management utility in addition to palatable flowers, leaves, petioles, rhizomes and seeds (Sinha et al. 2000). It can tolerate acidic conditions but grows luxuriantly in alkaline nutrient rich waters.

8.3.4 Ceratophyllum demersum *(Hornwort or Coontail)*

- Family = Ceratophyllaceae
- Genus = Ceratophyllum
- Species = demersum

Ceratophyllum demersum is commonly referred to either as 'Coontail' due to its resemblance to a racoon's tail or 'Hornwort' attributed to its horny plant appearance. It has a brittle branched body bearing long, thin and flexible stem (1–3 m) with nodes and internodes. Its leaves (8–40 mm) are flat, forked, vertical, denticulate at the apex, whorled (6–12 leaves) and more crowded near tip of stem. It is a monoecious plant owning both male and female flowers (2 mm long). The spiny fruit (a small nut) about 5 mm long contains a single seed. It is a rootless submerged free-floating herb but can develop rhizoids (modified leaves) to anchor. Calm, deep and permanent waters serve as its habitat. It is a fast grower nutrient sponge (Khang et al. 2012) in semi-hard or hard waters. It is almost undemanding for light levels and CO_2 supplements. It is known to combat phytoplankton and blue-green algae (cyanobacteria) by producing allelochemicals. This invasive plant shows proliferated growth and dense mat formation (Keskinkan et al. 2004). It develops turions—the buds forming new plants in spring after residing at the lakebed during winters.

8.4 Bioassays

The survey, sample collection and limnological analysis of macrophytes, sediment and water from selected sites of the two lakes were carried out from March, 2011 to February, 2013. For the purpose of present research, macrophytes, sediment and water are collected simultaneously at the same sampling locations. Aquatic plant samples are obtained manually on growth-phase basis during the stretch of the annual macrophytic progress from emergence to decay distributed into three durations: (1) March to May as sprouting (sp.), (2) June to October as peak growth (pg.) and (3) November to February as senescence (sn.) phases. Quadrat method (Havens et al. 2004; Gunn et al. 2010) is followed for macrophytic sample collection. Hooks and Ekman dredge are used in case of the submerged plants falling within a floatable quadrat. The specific bioassays of macrophytes include:

8.4.1 Photosynthetic Pigments

Photosynthetic pigments, viz. total chlorophyll, are estimated in 1 g fresh weight of macrophytic sample applying Sadasivam and Manikam (2005) procedure. The isolated pigment extracts' colour intensity is measured in visible region on Elico SL-27 spectrophotometer and the chlorophyll content measured according to the equation:

$$\text{Total Chlorophyll}\left(\text{mg g}^{-1}\right)=\left[6.45\times A_{665}+17.72\times A_{649}\right]/1000\times V/W$$

where V = volume of the total extract (50 mL); W = fresh weight of plant material (1 g) and A_{649}/A_{665} are optical densities of the extract at 649 nm and 665 nm.

8.4.2 Biomass, Productivity, NPP, Turnover and Growth Rate

Biomass of the selected species is estimated by harvest method (Johnson and Newman 2011) as dry weight per unit area (1 m^2 quadrat) after drying adherent free collection in an oven at 80 °C for 24 h to attain constant weight. At this temperature besides averting thermal decomposition, complete moisture removal ensures insistent reference value (Mishra et al. 2008; Peng et al. 2009). The values expressed as g m^{-2} denote mean biomass for five random quadrats for each species. The rate of production is measured as a net positive increment in dry biomass at seasonal intervals by deducting the initial biomass value (B_1) at time, t_1 from final biomass value (B_2) at time, t_2.

Therefore, productivity (g m^{-2} day^{-1}) is change in biomass (ΔB) between the time interval (Δt), i.e., productivity (g m^{-2} day^{-1}) = $\Delta B/\Delta t = B_2 - B_1/t_2 - t_1$. Net primary production (NPP) for each species is calculated by employing Caçador et al. (2007) formula: NPP (g m^{-2}) = [Maximum Biomass − Minimum Biomass]. Turnover period of every species is the ratio of its biomass parameters of NPP and maximum dry weight biomass (Duarte et al. 2010), viz. Turnover Rate = NPP (g m^{-2})/ Maximum Biomass (g m^{-2}). And growth rate describes weight gain in a definite phase interim relative to mean weight figures (APHA 2005), i.e., growth rate (day^{-1}) = Weight Gain, g/Time Interval, d × 1/Mean Weight, g where mean weight = [initial weight, g ÷ final weight, g ÷ 2].

8.4.3 BCF, Removal Potential and Element Turnover

Bioconcentration factor (BCF) designates apportionment fraction of an element concentration in biomass at harvest, E_B and initial element concentration in its medium, E_M (Wang et al. 2008; Nguyen et al. 2011) given by:

BCF = E_B/E_M. BCF exceeding the reference value of 1 implies hyperaccumulation. Nutrient storage [macro- and micro-elements in mg while traces' in μg] is defined as the product of harvest biomass '*B*' and plant nutrient quantum 'C_N' per unit area '*A*' (Wu et al. 2011), i.e., nutrient storage = $B \times C_N/A$. So, a plant's element composition, *C* (mg g^{-1} or μg g^{-1}) multiplied by the yield, *Y* (g m^{-2}) gives removal potential (mg m^{-2} or μg m^{-2}) of a specific species. Therefore, removal potential = $Y \times C$. Element turnover is given as the quotient of two element accumulation parameters (Duarte et al. 2010), i.e., g of net primary accumulation (NPA) and the g of element pool (EP). EP can provide effective instantaneous amount of metal retention as the same stands as sample biomass = $^{[\mathrm{BIOMASS}]}{}_t$ times the metal content per gram = $^{[\mathrm{METAL}]}{}_t$. So, element turnover = NPA/EP where NPA = maximum metal pool minus minimum metal pool and EP = $^{[\mathrm{BIOMASS}]}{}_t \times {}^{[\mathrm{METAL}]}{}_t$.

Besides, the water samples were analysed for temperature, pH, conductivity, bicarbonates, chloride, calcium, magnesium, sodium, potassium, nitrate nitrogen and total phosphorus on monthly basis, while trace elements of Al, As, Cd, Co, Cr, Cu, Fe, Hg, Mn, Ni, Pb, Se, Sn and Zn were measured for spring, summer and winter seasons. The collection, preservation, preparation, storage and estimation of water samples follow standard methods of APHA (2005), Radojevic and Bashkin (2006) and Estefan et al. (2013). The sediment analysis includes H-ion concentration, electrical conductivity, bicarbonates, organic-C, organic matter, total-N, C/N ratio, elemental composition of Ca, Cl, K, Mg, Na, P, S, Si and the given trace elements for each season. Again, the macrophytic biomass were analysed for mineral composition (Ca, Cl, K, Mg, N, Na, P, S, Si) and the trace elements. For plant and sediment samples, the common parametric evaluations are performed following Gupta (2004), Ryan et al. (2007), Radojevic and Bashkin (2006) and Estefan et al. (2013). The multi-elemental quantification of Ca, Cl, K, Mg, Na, P, S and Si in sediments and plant biomass is performed using wavelength dispersive X-ray fluorescence spectrometer (WD-XRF) at central instrumentation facility (CIF), Pondicherry University. However, the trace elements in all samples were determined at sophisticated analytical instrument facility (SAIF) IITM, Chennai using ICP-OES (Perkin Elmer Optima 5300 DV) instrument. Further, the descriptive and illustrative statistical analysis is performed using MS-Excel 2010, PAST 3 and SPSS 19.

8.5 Findings

The biochemical assays of select macrophytes for suitable parameters carried out in tandem with water and sediment associates divulge their cohesive and codependent networking. A marked difference in mean total chlorophyll content (mg g^{-1}) of fresh weight across the hydrophyte species and their growth phases of sprouting (sp.), peak growth (pg.) and senescence (sn.) is shown in Fig. 8.4. The maximum total chlorophyll concentration is noted in *Ceratophyllum demersum* (0.92 mg g^{-1}) during its peak growth at site D3, while a minimum 0.02 mg g^{-1} is recorded both in case of *Myriophyllum aquaticum* at site A2 and *Salvinia natans* at site D3 throughout senescence. The individual ranges of chlorophyll content (mg g^{-1}) for *Myriophyllum*

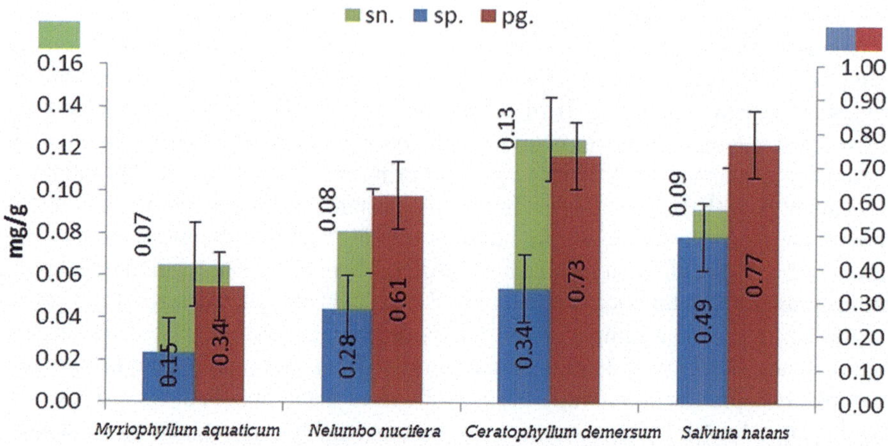

Fig. 8.4 Average total chlorophyll concentration of the macrophytes. *sn.* senescence phase, *sp.* sprouting phase, *pg.* peak growth phase

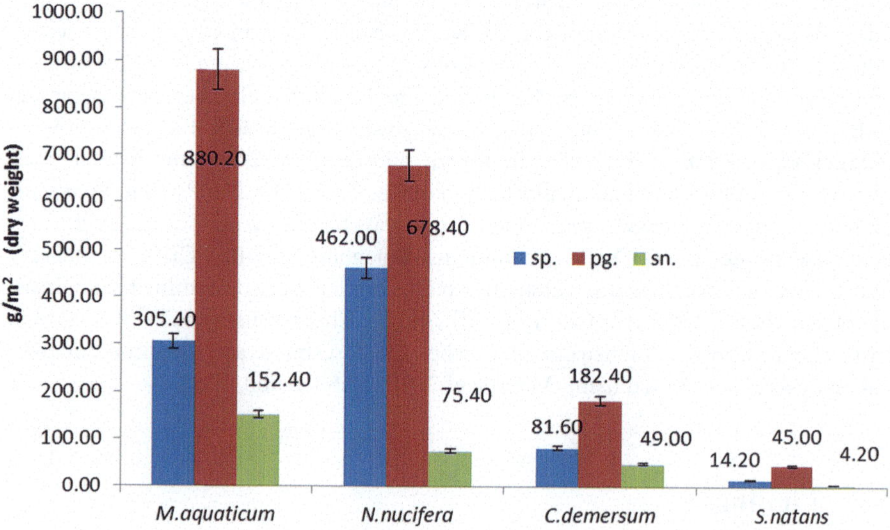

Fig. 8.5 Average biomass values (g m^{-2} dry weight). *sp.* sprouting phase, *pg.* peak growth phase, *sn.* senescence phase

aquaticum, *Nelumbo nucifera*, *Ceratophyllum demersum* and *Salvinia natans* include 0.02–0.37, 0.05–0.72, 0.10–0.92 and 0.02–0.87, respectively.

The productive capacity of the investigated macrophytes calculated on the basis of dry weight mean biomass (g m^{-2}) for each experimental species in the three major growth phases is presented in Fig. 8.5. The biomass calculations reflect subsequent ranges for each species as 107–960 g m^{-2} for *Myriophyllum aquaticum*, 66–750 g m^{-2} for *Nelumbo nucifera*, 39–231 g m^{-2} for *Ceratophyllum demersum* and 1–55 g m^{-2} for *Salvinia natans*. *Myriophyllum aquaticum* (880.2 g m^{-2}) dominated *Nelumbo nucifera* (678.4 g m^{-2}), *Ceratophyllum demersum* (182.4 g m^{-2}) and *Salvinia natans* (45 g m^{-2}) in terms of mean peak dry weight biomass values. A similar fashion low occurred in dry weight (g m^{-2}) during their senescence period ranking by way of 152.4 > 75.4 > 49 > 4.2, respectively.

The percentage contribution of various life-form classes among the studied macrophytes towards production (dry weight biomass, g m^{-2}) as well as specific percent relative difference during sprouting, peak growth and senescence phases is described in pie-chart diagrams accessible in Fig. 8.6.

Other biomass parameters like productivity, net primary productivity and specific growth rate establish uniform variations in the experimental species of the order: emergent *M. aquaticum* > rooted-floating *N. nucifera* > submerged *C. demersum* > free-floating *S. natans* but species turnover is highest in case of *S. natans* and lowest for *C. demersum* as illustrated in (Table 8.1).

The average tissue concentration of analysed elements fluctuated not only between the select species (Figs. 8.7 and 8.8) but also depicted growth-phase variations within the same species. However, no significant site contrasts were noticed.

The ANOVA test conducted between the species ('df' = 11) for the mineral composition of macrophytes conveys significance for Cu ($p < 0.05$), K ($p < 0.001$), Na ($p < 0.05$), Se ($p < 0.01$) and Si ($p < 0.01$) only, whereas ANOVA for their growth phases ('df' = 11) reflect significance in case of Al ($p < 0.05$), As ($p < 0.001$), Ca ($p < 0.05$), Co ($p < 0.05$), Cr ($p < 0.05$), Fe ($p < 0.05$), Mg ($p < 0.05$), Mn ($p < 0.001$), Ni ($p < 0.01$), Pb ($p < 0.01$) and Zn ($p < 0.001$).

The bioconcentration factor (BCF) criterion adopted indicates hyperaccumulation phenomenon for most of the micro- and trace-elements in *C. demersum* and *S. natans* except for Pb, Sn and Zn, and *M. aquaticum* hyperaccumulates Cd and Mn while *N. nucifera* Mn only. The removal potential of the macrophytes clearly stipulates species-specific variation (Figs. 8.9 and 8.10) ensuring a generalized arrangement of emergent *M. aquaticum* > rooted-floating *N. nucifera* > submerged

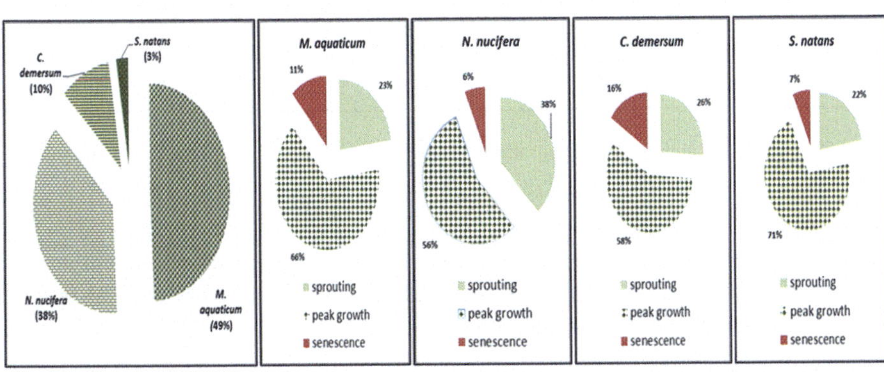

Fig. 8.6 Comparative of percent dry weight biomass (g m^{-2}) across growth phases and species

Table 8.1 Trends in productivity, NPP, specific growth rate and species turnover

Biomass parameter	*M. aquaticum*	*N. nucifera*	*C. demersum*	*S. natans*
Productivity (g m^{-2} day^{-1})	4.78 ± 0.32	1.81 ± 0.73	0.84 ± 0.24	0.26 ± 0.07
NPP (g m^{-2})	853	684	192	54
Specific growth rate (day^{-1})	29.68 ± 8.13	7.39 ± 2.18	3.96 ± 1.64	1.83 ± 0.87
Species turnover	0.89	0.91	0.83	0.98

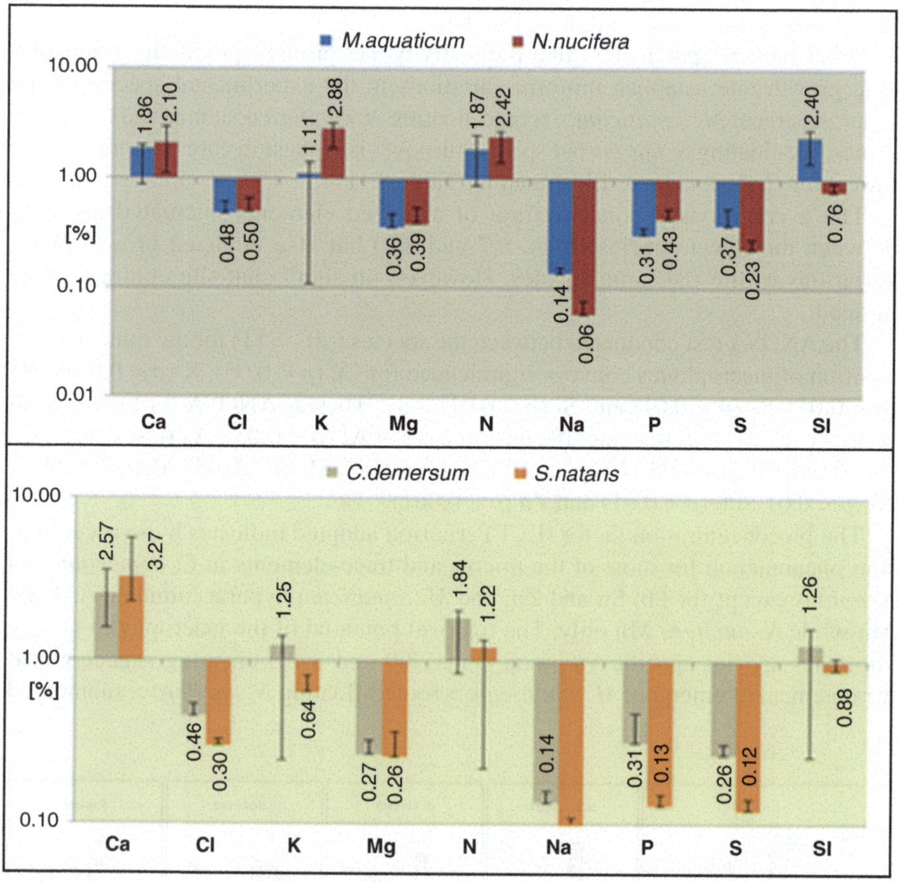

Fig. 8.7 Average elemental bioconcentration (%) in *M. aquaticum*, *N. nucifera*, *C. demersum* and *S. natans*

C. demersum > free-floating *S. natans*. Nevertheless, element-wise removal potential of each studied macrophyte for accumulation of (a) nutrients and (b) traces, respectively, follow the following pattern:

1. *M. aquaticum*: (a) Si > N > Ca > K > S > Cl > Mg > P > Na and (b) Fe > Al > M n > Zn > Cu > Cr > Ni > As > Co > Sn > Pb > Se > Cd.

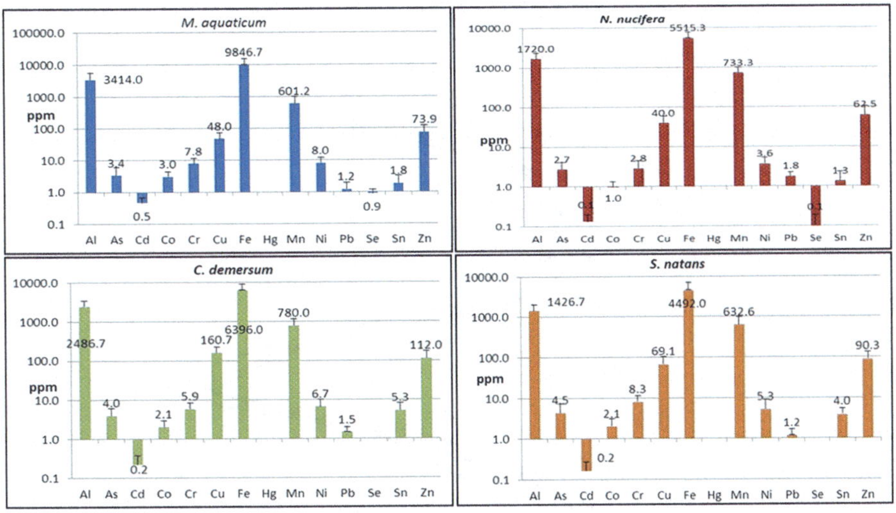

Fig. 8.8 Mean trace element bioaccumulation (ppm)

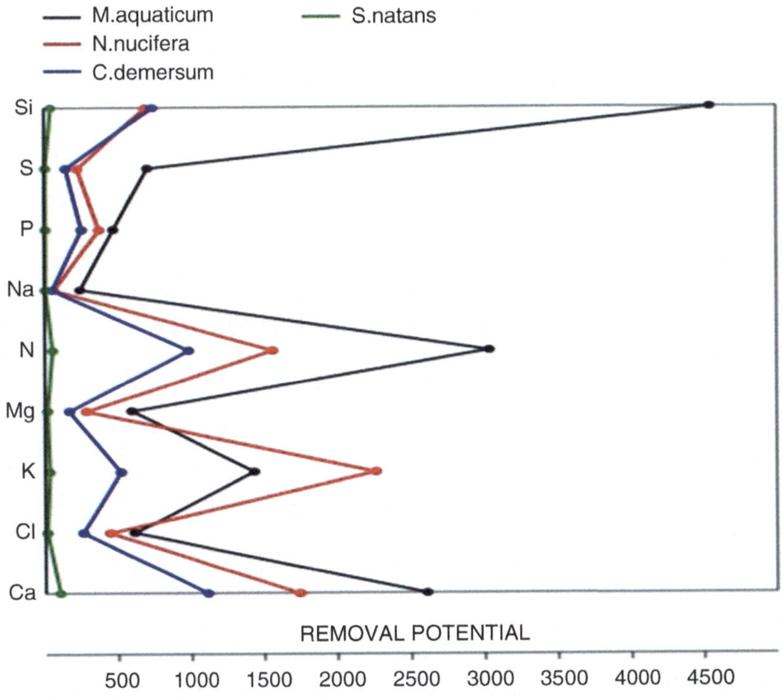

Fig. 8.9 Removal potential for major or minor elements (mg m^{-2})

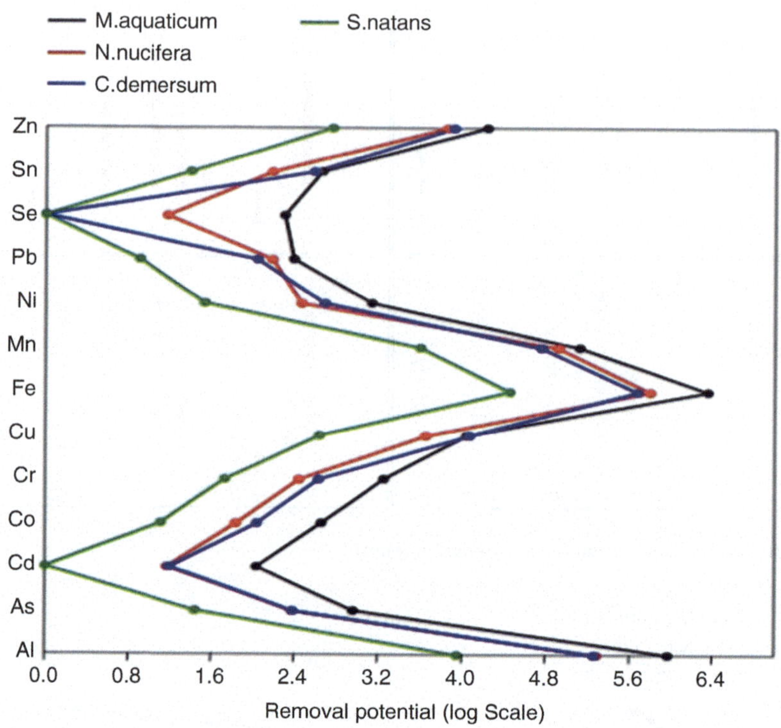

Fig. 8.10 Removal potential for traces (μg m^{-2})

2. *N. nucifera*: (a) K > Ca > N > Si > Cl > P > Mg > S > Na and (b) Fe > Al > Mn > Zn > Cu > Ni > Cr > As > Pb > Sn > Co > Cd > Se.
3. *C. demersum*: (a) Ca > N > Si > K > Cl > P > Mg > S > Na and (b) Fe > Al > Mn > Cu > Zn > Ni > Cr > Sn > As > Co > Pb > Cd > Se.
4. *S. natans*: (a) Ca > N > Si > K > Mg > Cl > P > S > Na and (b) Fe > Al > Mn > Zn > Cu > Cr > Ni > As > Sn > Co > Pb > Cd > Se.

Lastly, significant element turnover rate is observed for almost all minerals except Mg and Na in case of *M. aquaticum*. However, *N. nucifera* has significant turnover for Ca, Cl, Mg, As, Cd, Co, Cr, Cu, Ni, Pb, Se, Sn and Zn. Similarly, less significance is reflected for K, Mg, Na, Al and Se turnover by *C. demersum*. And *S. natans* has lesser turnover rate for Mg, N and Se only.

8.6 Discussion

The phyto-assessment of aquatic environments involving vascular plants determine phytostimulation as well as phytotoxicity due to nutrient inputs from sewage, sediments and catchment surface run-off and toxins from commercial chemicals, industrial effluents, agrochemical and municipal wastes, hazardous wastes, etc., respectively. The in situ biomonitoring technique utilizing vascular aquatic vegetation gives information about environmental quality and plant responses to toxins and nutrients. The various parameters of study during phyto-assessment include measurement of biomass and pigment content, determination of bioaccumulation pattern, productivity, physiological effects and community composition. All of these diagnostic attributes reflect the trophic status and effect of all sources and types of contaminants present in the study area. The selected biomonitors among the macrophyte community manifest both contribution (command) and subservience to its abiotic allies. Macrophytes dominate mineral regulation between sediment and water (Kissoon et al. 2013). They are capable to remove nutrients even at low loading rates (Gottschall et al. 2007). Their adaptive advantage to lock up minerals from nutrient pool is helpful to check degradation primarily and restore mineral loading later on (Ismail et al. 2014). N and P are intimately related to biological productivity of aquatic ecosystems (Kalff 2002). The productivity also lies in consonance with developmental stage, ambient nutrient medium and physiognomy of a species (El-Otify 2015). Higher temperature favours bio-production and elevates carbonate driven pH significantly (Hasler et al. 2016). Among the evaluated species *M. aquaticum* accumulated 49% of the dry weight biomass annually, whereas *S. natans* contributed only 3% but it outdid in species turnover comparison. Biomass parameters like dry weight, productivity, NPP and specific growth rate establish similar variations in the experimental species but species turnover is highest in case of *S. natans* and lowest for *C. demersum*. Substantial OM productions by macrophytes contribute towards nutrient immobilization and provide OC requirement for denitrification while decaying (Lin et al. 2002; McElarney et al. 2010). The upgraded biomass configuration property establishes an additional nutrient and trace element interception feature of hyperaccumulation from the growth medium (Shaltout et al. 2014).

Macrophytes are tested ecological indicators when compared against watershed (Søndergaard et al. 2010, 2013), catchment development (Sass et al. 2010) and even climate (Beck et al. 2014) in order to validate ecosystem changes. Pertinently submergeds are least pollution and stress resistant species (Pulido et al. 2011; Sierszen et al. 2012). Environmental variations are more influential compared to interspecific competition in governing the community structure (Grabas et al. 2012). A combination of limnochemical and physicomorphic factors of water bodies describes the macrophyte community characterization (Schneider et al. 2015) but nitrate commonly influences their growth, height and biomass profile as against ammonium supplements (Wersal and Madsen 2011). The gaining momentum for comprehending phyto-nutrient dynamics and their implications for water quality management or eco-restoration is a subject of site specific (habitat) application of a suitable phytoremediator (Udeigwe et al. 2015). A number of determinants like light, temperature, redox, pH, solubility, water flow, chelation of metal compound, metal concentration, metal species, exposure duration and cation exchange capacity altogether affect extent and rate of bioaccumulation for decontamination. The mere presence of nutrients is not the condition for rising or reducing uptake of toxic metals (Göthberg et al.

2004). Though the solubility of target element is not a problem in aqueous media but its bioavailability matters which consequently controls plant accumulation and translocation (Zaier et al. 2010). The peak nutrient uptake and bioconcentration coincided with peak biomass in summer and autumn (Garbey et al. 2004). The diminishing conductivity at improved temperature conditions suggest higher bioaccumulation rate of nutrients (Liu et al. 2014). The structural (physiognomy) difference between the species and the micro-habitat occupied within aquatic ecosystem are major factors related to bioconcentration of various elements (Lukács et al. 2015). The removal potential pattern of the emergent accrued more Si, the rooted floating retained greater K, while the rootless submersed and free floating ones amassed Ca. Although the quantum of removal potential for different elements in the analysed species is divergent but the pattern is related which suggests unselective absorption (Sarwar et al. 2010). The elemental turnover rates in select macrophytes closer to the reference value of 1 are significant. The vascular aquatic plants possess higher productivity (C:N and C:P ratios) with organic polymer body capacitating their slowed decomposition that improves nutrient and C capture potential and as a result net autotrophy (Cotner et al. 2009). The concurrence of peak bioaccumulation and biomass production empowers the concept of high yield harvesting (Quilliam et al. 2015). The harvesting of macrophyte biomass suitably during peak growth phase is perfect to avoid nutrient and metal remobilization from belowground vegetative parts or during senescence leaching on ultimate OM mineralization (Wang et al. 2015). Similarly, a mono annual harvesting practice improves height, shoot density and biomass of macrophytes, although leading to nominal increment for mass TN and TP removal rates (Zheng et al. 2015). The ANOVA comparing the species shows uptake preference mode for Cu, K, Na, Se and Si indicating specific role in the unlike macrophytes (Sistla et al. 2015). Contrarily, the growth-phase variance for Al, As, Ca, Co, Cr, Fe, Mg, Mn, Ni, Pb and Zn possibly implies their short-term raised presence or bioavailability during the annual developmental cycle (Wiik et al. 2015). BCF criterion indicates hyperaccumulation for most of the metals in case of *C. demersum* and *S. natans*. The BCF apportionment clearly summarize *M. aquaticum* and *N. nucifera* to dissipate the PTE in their rhizosphere zone but the rootless *C. demersum* and *S. natans* concentrate them in their foliage parts (Harguinteguy et al. 2014).

Anthropogenic disturbances arising from land-use change, altered global biogeochemistry and biodiversity interferences are clearly reflected by community structure and function with macrophytic communities being no exception. The ecological disturbances experienced by freshwater biota vary in magnitude, frequency, predictability, duration and on spatial-scale as well (Lake et al. 2000). Macrophytes react to the ambient physicochemical environment changes not only by altering species composition but also by displaying plant biomass variations. Their presence, density and morphology are determined by factors like sediment type, water turbidity, water current, nutrient concentration, water depth, shoreline disturbance, herbivore grazing and human activities (Bornette and Puijalon 2011; Wood et al. 2012). They exhibit a stress response of depletion in diversity (genetic erosion) associated with predominantly excessive proliferation of a selected few (Michelan et al. 2010; Chappuis et al. 2011). Although the tissue concentration of nutrients and trace elements differs species wise but well correlates with ambient water and sediment media (Amari et al. 2014). The biological sinks scavenge the deleterious proportions of nutrient supply from lakes preventing them from being overtaken by blooms. Even if slight phyto-extraction of inorganics would have occurred physically compared to large-scale quanta of the in situ media, still the biogenic calcification evidenced in the summers is capable of lowering them mediated via temperature, pH, HCO_3^- and Ca changes (Boyd et al. 2016). Hence, a selective and scaled deweeding practice can thereby work wonders in lake management and restoration programmes (Novak and Chambers 2014).

References

Algesten G, Sobek S, Bergström AK, Ågre A, Tranvik LJ, Jansson M (2003) Role of lakes for organic carbon cycling in the boreal zone. Glob Chang Biol 10(1):141–147

Amari T, Ghnaya T, Debez A, Taamali M, Youssef NB, Lucchini G, Abdelly C (2014) Comparative Ni tolerance and accumulation potentials between *Mesembryanthemum crystallinum* (halophyte) and *Brassica juncea*: metal accumulation, nutrient status and photosynthetic activity. J Plant Physiol 171(17):1634–1644

Amaya-Chavez A, Martinez-Tabche L, Lopez-Lopez E, Galar-Martinez M (2006) Methyl parathion toxicity to and removal efficiency by *Typha latifolia* in water and artificial sediments. Chemosphere 63:1124–1129

APHA (2005) Standard methods for the examination of water and wastewater, 21st edn. APHA, AWWA, WEF, Washington

Arshid S, Wani AA, Ganie AH, Khuroo AA (2011) On correct identification, range expansion and management implications of *Myriophyllum aquaticum* in Kashmir Himalaya, India. Check List 7(3):299–302

Arthur EL, Rice PJ, Rice PJ, Anderson TA, Baladi SM, Henderson KL, Coats JR (2005) Phytoremediation—an overview. Crit Rev Plant Sci 24(2):109–122

Beck MW, Tomcko CM, Valley RD, Staples DF (2014) Analysis of macrophyte indicator variation as a function of sampling, temporal, and stressor effects. Ecol Indic 46:323–335

Becker H (2000) Phytoremediation: using plants to clean up soils. Agric Res 48(6):4

Bhat RA, Shafiq-ur-Rehman, Mehmood MA, Dervash MA, Mushtaq N, Bhat JIA, Dar GH (2017) Current status of nutrient load in Dal Lake of Kashmir Himalaya. J Pharmacogn Phytochem 6(6):165–169

Bhat RA, Dervash MA, Qadri H, Mushtaq N, Dar GH (2018) Macrophytes, the natural cleaners of toxic heavy metal (THM) pollution from aquatic ecosystems. In: Environmental contamination and remediation. Cambridge Scholars Publishing, Cambridge, pp 189–209

Bhatia M, Goyal D (2014) Analyzing remediation potential of wastewater through wetland plants: a review. Environ Prog Sustain Energy 33(1):9–27

Bornette G, Puijalon S (2011) Response of aquatic plants to abiotic factors: a review. Aquat Sci 73(1):1–14

Boyd CE, Tucker CS, Somridhivej B (2016) Alkalinity and hardness: critical but elusive concepts in aquaculture. J World Aquacult Soc 47(1):6–41

Caçador I, Costa AL, Vale C (2007) Nitrogen sequestration capacity of two salt marshes from the Tagus estuary. Hydrobiologia 587(1):137–145

Carrow RN, Waddington DV, Rieke PE (2001) Turf grass soil fertility and chemical problems: assessment and management. Ann Arbor Press, Chelsea

Chambers PA, Lacoul P, Murphy KJ, Thomaz SM (2008) Global diversity of aquatic macrophytes in freshwater. Hydrobiologia 595(1):9–26

Chappuis E, Gacia E, Ballesteros E (2011) Changes in aquatic macrophyte flora over the last century in Catalan water bodies (NE Spain). Aquat Bot 95(4):268–277

Cook CD (1996) Aquatic and Wetland Plants of India: a reference book and identification manual for the vascular plants found in permanent or seasonal fresh water in the subcontinent of India south of the Himalayas, vol 198548214. Oxford University Press, Oxford, pp 1–385

Cotner JB, Kenning J, Scott JT (2009) The microbial role in littoral zone biogeochemical processes: why Wetzel was right. Verh Int Ver Limnol 30:981–984

Cvetkovic M, Wei A, Chow-Fraser P (2010) Relative importance of macrophyte community versus water quality variables for predicting fish assemblages in coastal wetlands of the Laurentian Great Lakes. J Great Lakes Res 36(1):64–73

Dhote S (2007) Role of Macrophytes in improving water quality of an aquatic eco-system. J Appl Sci Environ Manag 11(4):133–135

Duarte B, Caetano M, Almeida PR, Vale C, Caçador I (2010) Accumulation and biological cycling of heavy metal in four salt marsh species, from Tagus estuary (Portugal). Environ Pollut 158(5):1661–1668

El-Otify AM (2015) Evaluation of the physicochemical and chlorophyll-a conditions of a subtropical aquaculture in Lake Nasser area, Egypt. Beni-Suef Univ J Basic Appl Sci 4(4):327–337

Estefan G, Sommer R, Ryan J (2013) Methods of soil, plant and water analysis. International Center for Agricultural Research in the Dry Areas (ICARDA), Beirut

Gałka A, Szmeja J (2013) Phenology of the aquatic fern *Salvinia natans* (L.) all in the Vistula Delta in the context of climate warming. Limnologica 43(2):100–105

Garbey C, Murphy KJ, Thiebaut G, Muller S (2004) Variation in phosphorus content in aquatic plant tissues offers an efficient tool for determining plant growth strategies along a resource gradient. Fresh Water Biol 49:1–11

Ghosh SK (2005) Illustrated aquatic and wetland plants in harmony with mankind. Standard Literature, Kolkata, pp 1–225

Göthberg A, Greger M, Holm K, Bengtsson BE (2004) Influence of nutrient levels on uptake and effects of mercury, cadmium, and lead in water spinach. J Environ Qual 33(4):1247–1255

Gottschall N, Boutin C, Crolla A, Kinsley C, Champagne P (2007) The role of plants in the removal of nutrients at a constructed wetland treating agricultural (dairy) wastewater, Ontario, Canada. Ecol Eng 29(2):154–163

Grabas GP, Blukacz-Richards EA, Pernanen S (2012) Development of a submerged aquatic vegetation community index of biotic integrity for use in Lake Ontario coastal wetlands. J Great Lakes Res 38(2):243–250

Gunn ID, O'Hare M, Carvalho L, Roy DB, Rothery P, Darwell AM (2010) Assessing the condition of lake habitats: a test of methods for surveying aquatic macrophyte communities. Hydrobiologia 656(1):87–97

Gupta PK (2004) Soil, plant, water and fertilizer analysis. Agro Botanica, Bikaner

Gurnell AM, Bertoldi W, Corenblit D (2012) Changing river channels: the roles of hydrological processes, plants and pioneer fluvial landforms in humid temperate, mixed load, gravel bed rivers. Earth Sci Rev 111(1):129–141

Harguinteguy CA, Cirelli AF, Pignata ML (2014) Heavy metal accumulation in leaves of aquatic plant *Stuckenia filiformis* and its relationship with sediment and water in the Suquía river (Argentina). Microchem J 114:111–118

Harpole WS, Ngai JT, Cleland EE, Seabloom EW, Borer ET, Bracken ME, Smith JE (2011) Nutrient co-limitation of primary producer communities. Ecol Lett 14(9):852–862

Hart BT, Van Dok W, Djuangsih N (2002) Nutrient budget for Saguling reservoir, West java, Indonesia. Water Res 36(8):2152–2160

Hasan SH, Talat M, Rai S (2007) Sorption of cadmium and zinc from aqueous solutions by water hyacinth (*Eichhornia crassipes*). Bioresour Technol 98(4):918–928

Hasler CT, Butman D, Jeffrey JD, Suski CD (2016) Freshwater biota and rising pCO2? Ecol Lett 19(1):98–108

Havens KE, Sharfstein B, Brady MA, East TL, Harwell MC, Maki RP, Rodusky AJ (2004) Recovery of submerged plants from high water stress in a large subtropical lake in Florida, USA. Aquat Bot 78(1):67–82

Ibañez S, Talano M, Ontañon O, Suman J, Medina MI, Macek T, Agostini E (2016) Transgenic plants and hairy roots: exploiting the potential of plant species to remediate contaminants. New Biotechnol 33(5):625–635

Ismail Z, Othman SZ, Law KH, Sulaiman AH, Hashim R (2014) Comparative performance of water hyacinth (*Eichhornia crassipes*) and water lettuce (*Pistia stratiotes*) in preventing nutrients build-up in municipal wastewater. Clean (Weinh) 43(4):521–531

Johnson JA, Newman RM (2011) A comparison of two methods for sampling biomass of aquatic plants. J Aquat Plant Manag 49(1):1–8

Jurca T, Donohue L, Laketić D, Radulović S, Irvine K (2012) Importance of the shoreline diversity features for littoral macroinvertebrate assemblages. Fundam Appl Limnol 180(2):175–184

Kalff J (2002) Limnology: inland water ecosystems, vol 592. Prentice Hall, Upper Saddle River

Keskinkan O, Goksu MZL, Basibuyuk M, Forster CF (2004) Heavy metal adsorption properties of a submerged aquatic plant (*Ceratophyllum demersum*). Bioresour Technol 92(2):197–200

Khang HV, Hatayama M, Inoue C (2012) Arsenic accumulation by aquatic macrophyte coontail (*Ceratophyllum demersum* L.) exposed to arsenite, and the effect of iron on the uptake of arsenite and arsenate. Environ Exp Bot 83:47–52

Kissoon LT, Jacob DL, Hanson MA, Herwig BR, Bowe SE, Otte ML (2013) Macrophytes in shallow lakes: relationships with water, sediment and watershed characteristics. Aquat Bot 109:39–48

Körner S (2002) Loss of submerged macrophytes in shallow lakes in North-Eastern Germany. Int Rev Hydrobiol 87(4):375–384

Lake PS, Palmer MA, Biro P, Cole J, Covich AP, Dahm C, Verhoeven JOS (2000) Global change and the biodiversity of freshwater ecosystems: impacts on linkages between above-sediment and sediment biota all forms of anthropogenic disturbance—changes in land use, biogeochemical processes, or biotic addition or loss—not only damage the biota of freshwater sediments but also disrupt the linkages between above-sediment and sediment-dwelling biota. BioScience 50(12):1099–1107

Lasat M (2000) Phytoextraction of metals from contaminated soil: a review of plant/soil/metal interaction and assessment of pertinent agronomic issues. J Hazard Subst Res 2(5):1–25

Lenzi M, Gennaro P, Renzi M, Persia E, Porrello S (2012) Spread of *Alsidium corallinum* C. Ag. in a Tyrrhenian eutrophic lagoon dominated by opportunistic macroalgae. Mar Pollut Bull 64(12):2699–2707

Lin YF, Jing SR, Wang TW, Lee DY (2002) Effects of macrophytes and external carbon sources on nitrate removal from groundwater in constructed wetlands. Environ Pollut 119(3):413–420

Liu JL, Liu JK, Anderson JT, Zhang R, Zhang ZM (2014) Potential of aquatic macrophytes and artificial floating island for removing contaminants. Plant Biosyst 150(4):702–709

Lukács BA, Gy D, Tóthmérész B (2009) The function of macrophytes in relation to environmental variables in eutrophic backwaters and mortlakes. Phytocoenologia 39:287–293

Lukács BA, Tóthmérész B, Borics G, Várbíró G, Juhász P, Kiss B, Erős T (2015) Macrophyte diversity of lakes in the Pannon ecoregion (Hungary). Limnologica 53:74–83

McElarney Y, Rasmussen P, Foy R, Anderson N (2010) Response of aquatic macrophytes in Northern Irish soft water lakes to forestry management; eutrophication and dissolved organic carbon. Aquat Bot 93(4):227–236

Meagher RB (2000) Phytoremediation of toxic elemental and organic pollutants. Curr Opin Plant Biol 3(2):153–162

Michelan TS, Thomaz S, Mormul RP, Carvalho P (2010) Effects of an exotic invasive macrophyte (tropical signalgrass) on native plant community composition, species richness and functional diversity. Freshw Biol 55(6):1315–1326

Mishra VK, Upadhyaya AR, Pandey SK, Tripathi BD (2008) Concentrations of heavy metals and nutrients in water, sediments and aquatic macrophytes of GBP Sagar an anthropogenic lake affected by coal mining effluent. Environ Monit Assess 141:49–58

Moore KA, Wetzel RL (2000) Seasonal variations in eelgrass (*Zostera marina* L.) responses to nutrient enrichment and reduced light availability in experimental ecosystems. J Exp Mar Biol Ecol 244(1):1–28

Morikawa H, Erkin ÖC (2003) Basic processes in phytoremediation and some applications to air pollution control. Chemosphere 52(9):1553–1558
Mushtaq N, Singh DV, Bhat RA, Dervash MA, Hameed UB (2020) Freshwater contamination: sources and hazards to aquatic biota. In: Qadri H, Bhat RA, Dar GH, Mehmood MA (eds) Freshwater pollution dynamics and remediation. Springer Nature, Singapore, pp 27–50
Nguyen THH, Sakakibara M, Sano S, Nhuan MT (2011) Uptake of metals and metalloids by plants growing in a lead–zinc mine area, Northern Vietnam. J Hazard Mater 186(2):1384–1391
Novak PA, Chambers JM (2014) Investigation of nutrient thresholds to guide restoration and management of two impounded rivers in south-western Australia. Ecol Eng 68:116–123
Padmavathiamma PK, Li LY (2007) Phytoremediation technology: hyper-accumulation metals in plants. Water Air Soil Pollut 184:105–126
Peng JF, Song YH, Yuan P, Cui XY, Qiu GL (2009) The remediation of heavy metals contaminated sediment. J Hazard Mater 161(2):633–640
Pulido C, Lucassen EC, Pedersen OLE, Roelofs JG (2011) Influence of quantity and lability of sediment organic matter on the biomass of two isoetids, *Littorel la uniflora* and *Echinodorus repens*. Freshw Biol 56(5):939–951
Qadri H Bhat RA (2020). The concerns for global sustainability of freshwater ecosystems. Freshwater pollution dynamics and remediation. Qadri H, Bhat RA & Dar GH, Mehmood MA. Springer Nature Singapore: 1–15
Qian JH, Zayed A, Zhu YL, Yu M, Terry N (1999) Phytoaccumulation of trace elements by wetland plants: III. Uptake and accumulation of ten trace elements by twelve plant species. J Environ Qual 28(5):1448–1455
Quilliam RS, Van Niekerk MA, Chadwick DR, Cross P, Hanley N, Jones DL, Oliver DM (2015) Can macrophyte harvesting from eutrophic water close the loop on nutrient loss from agricultural land? J Environ Manag 152:210–217
Radojevic M, Bashkin VN (2006) Practical environmental analysis, 2nd edn. The Royal Society of Chemistry, Cambridge
Rai PK (2008) Heavy metal pollution in aquatic ecosystems and its phytoremediation using wetland plants: an ecosustainable approach. Int J Phytoremediation 10(2):133–160
Raspopov IM, Adamec L, Husák S (2002) Influence of aquatic macrophytes on the littoral zone habitats of the Lake Ladoga, NW Russia. Preslia Praha 74(4):315–322
Ryan J, Estefan G, Rashid A (2007) Soil and plant analysis laboratory manual. ICARDA, Aleppo
Sadasivam S, Manikam A (2005) Biochemical methods, 2nd edn. New Age International, New Delhi
Samecka-Cymerman A, Kempers AJ (2002) Aquatic macrophytes as biomonitors of pollution by textile industry. Bull Environ Contam Toxicol 69(1):82–96
Sarwar N, Saifullah Malhi SS, Zia MH, Naeem A, Bibi S, Farid G (2010) Role of mineral nutrition in minimizing cadmium accumulation by plants. J Sci Food Agric 90:925–937
Sass LL, Bozek MA, Hauxwell JA, Wagner K, Knight S (2010) Response of aquatic macrophytes to human land use perturbations in the watersheds of Wisconsin lakes, USA. Aquat Bot 93(1):1–8
Scheffer M (1989) Alternative stable states in eutrophic, shallow freshwater systems: a minimal model. Hydrobiol Bull 23(1):73–83
Schneider B, Cunha ER, Marchese M, Thomaz SM (2015) Explanatory variables associated with diversity and composition of aquatic macrophytes in a large subtropical river floodplain. Aquat Bot 121:67–75
Scholz M (2003) Performance predictions of mature experimental constructed wetlands which treat urban water receiving high loads of lead and copper. Water Res 37:1270–1277
Shaltout KH, Galal TM, El-Komi TM (2014) Biomass, nutrients and nutritive value of *Persicaria salicifolia* Wild in the water courses of Nile Delta, Egypt. Rendiconti Lincei 25(2):167–179
Sierszen ME, Morrice JA, Trebitz AS, Hoffman JC (2012) A review of selected ecosystem services provided by coastal wetlands of the Laurentian Great Lakes. Aquat Ecosyst Health Manag 15(1):92–106
Sinha S, Mukherjee PK, Mukherjee K, Pal M, Mandal SC, Saha BP (2000) Evaluation of antipyretic potential of *Nelumbo nucifera* stalk extract. Phytother Res 14(4):272–274

Sistla SA, Appling AP, Lewandowska AM, Taylor BN, Wolf AA (2015) Stoichiometric flexibility in response to fertilization along gradients of environmental and organismal nutrient richness. Oikos 124(7):949–959

Søndergaard M, Johansson LS, Lauridsen TL, Jørgensen TB, Liboriussen L, Jeppesen E (2010) Submerged macrophytes as indicators of the ecological quality of lakes. Freshw Biol 55(4):893–908

Søndergaard M, Phillips G, Hellsten S, Kolada A, Ecke F, Mäemets H, Oggioni A (2013) Maximum growing depth of submerged macrophytes in European lakes. Hydrobiologia 704(1):165–177

Tian DL, Xiang WH, Yan WD, Kang WX, Deng XW, Fan Z (2007) Biological cycles of mineral elements in a young mixed stand in abandoned mining soils. J Integr Plant Biol 49(9):1284–1293

Trémolières M (2004) Plant response strategies to stress and disturbance: the case of aquatic plants. J Biosci 29(4):461–470

Udeigwe TK, Teboh JM, Eze PN, HashemStietiya M, Kumar V, Hendrix J, Kandakji T (2015) Implications of leading crop production practices on environmental quality and human health. J Environ Manag 15:267–279

Vymazal J (2002) The use of sub-surface constructed wetlands for wastewater treatment in the Czech Republic: 10 years experience. Ecol Eng 18(5):633–646

Wang KS, Huang LC, Lee HS, Chen PY, Chang SH (2008) Phytoextraction of cadmium by *Ipomoea aquatica* (water spinach) in hydroponic solution: effects of cadmium speciation. Chemosphere 72(4):666–672

Wang CY, Sample DJ, Day SD, Grizzard TJ (2015) Floating treatment wetland nutrient removal through vegetation harvest and observations from a field study. Ecol Eng 78:15–26

Wersal RM, Madsen JD (2011) Influences of water column nutrient loading on growth characteristics of the invasive aquatic macrophyte *Myriophyllum aquaticum* (Vell.)Verdc. Hydrobiologia 665(1):93–105

Wersal RM, Cheshier JC, Madsen JD, Gerard PD (2011) Phenology, starch allocation, and environmental effects on *Myriophyllum aquaticum*. Aquat Bot 95(3):194–199

Wetzel RG (2001) Land–water interfaces: larger plants. In: Limnology, 3rd edn. Academic Press, San Diego

Wiik E, Bennion H, Sayer CD, Davidson TA, McGowan S, Patmore IR, Clarke SJ (2015) Ecological sensitivity of marl lakes to nutrient enrichment: evidence from Hawes Water, UK. Freshw Biol 60(11):2226–2247

Wood KA, Stillman RA, Clarke RT, Daunt F, O'Hare MT (2012) Understanding plant community responses to combinations of biotic and abiotic factors in different phases of the plant growth cycle. PLoS One 7(11):e49824

Wu H, Zhang J, Li P, Zhang J, Xie H, Zhang B (2011) Nutrient removal in constructed microcosm wetlands for treating polluted river water in northern China. Ecol Eng 37(4):560–568

Xiangcan J (2003) Analysis of eutrophication state and trend for lakes in China. J Limnol 62(1s):60–66

Yang Q, Chen ZH, Zhao JG, Gu BH (2007) Contaminant removal of domestic wastewater by constructed wetlands: effects of plant species. J Integr Plant Biol 49(4):437–446

Yang M, Liu F, Han Y, Xu L, Juntawong N, Liu Y (2013) Genetic diversity and structure in populations of Nelumbo from America, Thailand and China: implications for conservation and breeding. Aquat Bot 107:1–7

Zaier H, Ghnaya T, Rejeb KB, Lakhdar A, Rejeb S, Jemal F (2010) Effects of EDTA on phytoextraction of heavy metals (Zn, Mn and Pb) from sludge-amended soil with *Brassica napus*. Bioresour Technol 101(11):3978–3983

Zheng Y, Wang XC, Ge Y, Dzakpasu M, Zhao Y, Xiong J (2015) Effects of annual harvesting on plants growth and nutrients removal in surface-flow constructed wetlands in northwestern China. Ecol Eng 83:268–275

Chapter 9
Phytoremediation of Heavy Metals Using Salix (Willows)

Khursheed Ahmad Wani, Zuber M. Sofi, Junaid Ahmad Malik, and Jahanger Ahmad Wani

9.1 Introduction

Due to demotechnic growth, naturally occurring potential toxic elements in soils recorded an upward trend, subsequent to which these areas have been rendered contaminated (Singh et al. 2018). To quote, increased pyrite ore roasting in Torviscova area leads to contamination which creates various effects on living entities and their health aspects (Moosavi and Seghatoleslami 2013). In fact such areas remain devoid of vegetation which cause soil degradation and pollution off-site (Ali et al. 2013). High bioavailability is the other problem of industrial pollutants, contrary to metalloids are naturally present in soils which may not be available for the use of plant (Park et al. 2011).

Phytoremediation is the implementation of plants for the degradation and decontamination of soils (Bhat et al. 2018) and render them harmless. In fact phytoremediation is a low cost alternative remediation technique (Vishnoi and Srivastava, 2008) which is environmentally acceptable. Since this technique is not toxicant specific, hence phytoremediation could be used for many pollutants (Moosavi and Seghatoleslami 2013).

The present chapter aims at the extent of phytoremediation of *Salix* spp. which grows on the polluted soils under natural factors.

K. A. Wani (✉) · J. A. Wani
Department of Environmental Science, Government Degree College, Bijbehara, J&K, India

Z. M. Sofi · J. A. Malik
Department of Zoology, Government Degree College, Bijbehara, J&K, India

R. A. Bhat et al. (eds.), *Bioremediation and Biotechnology, Vol 2*,
https://doi.org/10.1007/978-3-030-40333-1_9

9.2 Ecology of *Salix alba*

Salix spp. has about 450 species distributed all over the globe but mostly dominated in the Northern Hemisphere (Argus 1997). They are mostly found in temperate and arctic zones. However, willows have also been recorded from tropical and subtropical areas which comprises of trees, shrubs and groundcovers. Geographically they are found in every continent of the world with the exception of Antarctica and Australia. Of the total flora available, trees offer a costless, eco-friendly and sustainable sound solution for phytoremediation of trace metal contaminated land. Moreover, the use of plants such as willow and poplar species with trace metal contamination have also been found to be promising in phytoremediation as they accumulate Cd and Zn in stem. So, poplar and willows could be used for harvesting not only the nutrients but also the heavy metals as they:

1. Are easy to propagate with an extensive root system (Aasamaa et al. 2010) with high evapotranspiration (Kowalik and Randerson 1994) that can stabilize pollutants
2. Can thrive on soils with polluted conditions (Justin et al. 2010)
3. Have high biomass production (Kowalik and Randerson 1994)
4. Can resist the reasonable extent of PTE in the roots (Borišev et al. 2009).

Realizing their importance in heavy metal harvesting, Belarus planted willow trees for the phytostabilization and phytoremediation of land with polluted soils, in mining landscapes and on sanitary landfills. But in India, studies on the potential of phytoremediation have remained restricted under the field conditions. Vervaeke et al. (2004) premeditated that the degraded sediments with raised hydraulical aspects are viable with *Salix* species for phytoextraction or stabilization of sediment disposal sites. Needless to mention poplar and willow offer better mechanisms for the effective dexotification of overall metal compositions to a less effective level. A paramount concern with planting trees for metal defiled soils will be the potential impacts with respect to biological community to remove metals.

Lebrun et al. (2017) observed that *Salix alba* and *Salix viminalis* that have been raised on arsenic and lead concentrated mine technosol concentrate higher concentrations of arsenic and lead in the underground system compared to aboveground parts. Since such soils have extreme pH, poor nutrient content and high PTE concentrations, it has been established that application of organic amendment like Biochar improves the nutrient value of soil through an increase in pH, dissolved organic carbon (Beesley et al. 2010), nutrient availability (Agegnehu et al. 2016), water retention capacity (Omondi et al. 2016), electrical conductivity (Kloss et al. 2014), plant productivity (Hossain et al. 2010) and decrease in greenhouse gas emission (Janus et al. 2015) while decreasing bulk density.

Moreover, branches of Salix and related species revealed comparatively elevated contents of elements at early stages of development, as they uptake higher nutrients, while the effect of dilution resulted in very less levels of metals till the flowering stage. Needless to mention the natural process of senescence often resulted in an augment in the absorption of metals in the vegetation due to loss of fluid.

9.3 *Salix* and Ecological Niches

The willows and poplar trees are referred to colonize new environments, for example, marshes, sand rises and rock bars, roadside trench, deserted farming fields, railways, old mine tailings and rock pits just as glaciated or overwhelmed territories (Argus 1986; Skvortsov 1999) because of their capacity to create harmonious relationship with mycorrhizal parasites that use phosphorus and natural nitrogen (Schramm 1966; Lodge 1989; van der Heijden and Kuyper 2003), yearly generation of various seeds and a successful arrangement of seed dispersal for long separation travel (light weight and hair connection of willow seed), expanding the odds of finding an opening for germination and development; foundation into vegetation holes supporting those plants that can endure full daylight (willows are not adjusted to shade) and high relative development rate with moderately short future (Raven 1992). Needless to say that Salix needs a situation with consistent dampness supply for quick seed germination (Dorn 1976) as seeds of Salix are feasible for just half a month (McLeod and McPherson 1973; Maroder et al. 2000). So moisture or water accessibility is a constraining element in districts with dry conditions in pre-summer. The invasion of disturbed sites by Salix trees serve to recuperate the harmed environments by expanding surface shade, action of roots and humus formation for the improvement of soil structure and soil productivity (Stott 1992).

One of the most significant ecological niche for Salix are the wetlands that are currently disappearing at a very fast rate due to urbanization, encroachment, soil erosion, ill planned development, industrial growth, disposal of wastes, etc. Hence there is a need for the conservation and mitigation of altered wetlands due to anthropogenic intervention and subsequently *Salix* sps., which have a high significance for wildlife, providing rich habitat and food for diverse organisms (Kennedy and Southwood 1984; Hightshoe 1998; Sommerville 1992). The invertebrate herbivores from aphids to caterpillars nourish on willows, and maintain a large food web of complex trophic level organisms. Animals, viz. livestock, caribou, muskrats deer, elk, moose and, porcupines, etc., depend on willows for food and shelter like beavers and birds (Bates 1951; Smith et al. 1978). Salix species thrive well near aquatic ecosystems especially at wetland sites (Fig. 9.1a, b).

9.4 Land Reclamation

Since land surfaces have been degraded by human interferences hence there is an urgent need to address the same. Beyond all doubts, the stabilization of ecosystems degraded by chemicals could be achieved by using Salix species, which have a tendency to survive in environmentally disrupted areas owing to their drought resistance (Logan 1992), less nutrient requirements and capability to bud, root, and sprout in extreme conditions like degraded soils with scarce top soil, extremely eroded soils, waste areas and roadsides (Hartwright 1960; FAO 1979; White 1992; Bungart and Huttl 2001), heavy metal tolerance (Eltrop et al. 1991; Sennerby-

Fig. 9.1 (**a**) *Salix alba* in a wetland near Degree College Bijbehara. (**b**) *Salix alba* growing in wetland near Railway station Srinagar

Forsse et al. 1993; Punshon and Dickinson 1997), etc. Since willows inhabit open areas (Schramm 1966; Clewell 1999) in a spatially random fashion that cannot be predicted, the strategic plantation of willows into degraded landscapes has been found to restore such lands more quickly. For the introduction of plant cover, willows in the forms of mats, stakes, rooted cuttings and stumps with root masses could be used for reclamation and management of wastelands.

9.5 Phytoremediation

Owing to immense potential of Salix in nutrient harvesting and phytoremediation, extensive researches across globe are underway for establishing its degraded lands through synthetic chemical and metabolic processes like phytoextraction (the expulsion of overwhelming metals from land because of the plants' take-up and translocation of metals into aboveground organs), phytodegradation (the aggregation and biochemical change of natural contaminations by plants and related microorganisms), rhizostimulation and rhizofiltration (the expulsion of poisons from fluid arrangements through direct take-up by plants roots) and phytostabilization (substrate drying out and counteractive action of toxin transport). As willows are quickly developing with broad root framework other than having capacity to amass target poisons, *Salix* is the most favoured plant for phytoremediation.

9.6 Phytoextraction

Voluminous literature has emanated on the confrontation of willows to some metals (Cd, Cu, Fe Ni, Pb, Zn) and its ability to decontaminate the polluted soils (Dickinson et al. 1994; Landberg and Greger 1994; Punshon and Dickinson 1997; Ali et al. 1999; Greger and Landberg 1999; Robinson et al. 2000; Klang-Westin and Perttu 2002; Watson et al. 2003; Ali et al. 2003; Keller et al. 2003; Klang-Westin and Eriksson 2003; Lunackova et al. 2003; Kuzovkina et al. 2004). Although heavy metal resistance in *Salix* has been elaborated by many workers but the biochemical mechanisms for tolerance of metals is yet to be examined. Ali et al. (2003) revealed that tolerance of metal in Salix may be due to phytochelatin synthesis. Riddell-Black (1994) and Pulford et al. (2002) observed that different clones and species of willow have variable metal translocation patterns and confrontation to heavy metals. Landberg and Greger (2002) projected that metal tolerance of some clones may be due to variation in lipid peroxidation and enzyme concentration against the levels of metal levels. But Eriksson and Ledin (1999) suggested that there is a need of more exploration to be carried out to optimize the uptake of metals through replicas and management alternatives. Although new willow clones are introduced by plant breeders but the potential impact of metal contamination on vegetation needs to be monitored as it can have devastating effects on ecosystem in general and food web in particular (Mertens et al. 2001; Granel et al. 2002; Watson et al. 2003). Nevertheless, role of *Salix* in phytoremediation has been well-documented technique that can be used as a remedy for highly polluted sites and exhausted farmlands (Perttu and Kowalik 1997; Pulford et al. 2002).

9.7 Phytodegradation

While the capacity of poplars to expel a huge range of natural pollutants by assimilation into the roots of plant is very much recorded, new research demonstrates an ideal effect of *Salix* spp. on the fate of natural contaminations. Prairie Cascade willow (Salix x' Prairie Cascade') has demonstrated overwhelming development on darkened soil created by an oil slick and capacity of soil decontamination through activation of oil-degrading microorganisms related with their root system (Thompson 1998). *Salix* spp. are required to be cultivated in oil-mining territories in the Siberian taiga. *Salix* spp. in degraded areas without soil amendments immediately shaped green cover increased the removal of synthetic chemicals contrasted with exposed land soils. Quicker sequestration of metals than in past endeavours, the plants, when planted on cost effective imported soil layer, had been noticed. It is evident that critical abatement of 57% in mineral oil focus in the land having willow species instead of 15% on neglected plots, occurred on disposal sites for dredged sediment (Vervaeke et al. 2004). The capability of willows' to transport oxygen by root zone through aerenchyma is known to provide better conditions to bacterial development. An investigation of willow plantation on the soil near landfill (Maurice et al. 1999) has demonstrated that those destinations are having more oxidation rates for methane contrasted with soils having no vegetation. Trees with broad root systems will give a superior domain to methane oxidizing microscopic organisms accordingly lowering its outflow into the air. Promising outcomes in remediation of ethanol-mixed pollutants utilizing willow and its resistance to expanded degrees of ethanol, has been accounted for (Corseuil and Moreno 2001). In a research centre examination, cuttings of sobbing willow (*Salix babylonica*) had the option to diminish ethanol and benzene focus in watery arrangement by over 99% in fewer than 7 days through root take-up and plant biomass sorption. It was proposed that the take off by plants was essentially identified with the plant's transpiration. Willows show conceivable utility for the remediation of sites polluted with cyanide mixes. Transport and digestion of cyanide and ferrocyanide by willow has added to the debasement of these contaminants in the wastewater from gold mining (Ebbs et al. 2003). Cyanides (glycosides) are produced by most of the plants for chemical defence and plant cells by and large have a high ability to wipe out free cyanide, as it is a well-established fact that willow roots and leaves have the quickest rates of cyanide removal capacity (Larsen et al. 2004). High biomass generation and proficient control on water in soils, because of high rates of evapotranspiration makes willow a reasonable potential "bioreactor" for cyanide expulsion.

9.8 Rhizofiltration and Rhizostimulation

Recently, in America and Eurasia the utilization of constructed wetlands for the removal of pollutants is gaining much importance. A need is felt to incorporate vegetation filters to enhance nitrogen removal due to hydraulic mass flux by root zone technology. *Salix* sps have been utilized in these types of ecosystems that have reduced the pollution load to greater extent and hence these plants are considered a good source for improving the quality of effluent before being drained into any aquatic system (Martin and Reddy 1997). *Salix* sps is a potential tool for improving the quality of domestic, industrial, municipal wastewater and agricultural runoff (Elowson and Christersson 1994; Kirt 1994; Perttu and Kowalik 1997; Rosenqvist et al. 1997). The ability of Willow spp. having high rates of evapotranspiration, effective nutrient uptake, tolerance of flooded conditions and better biomass productivity are some of the physiological characteristics of *Salix* sps. that make it potential candidate as vegetation filters.

Salix alba has a potential to transport oxygen deep down the root zone via aerenchyma formation that will contribute to suitable conditions for bacterial growth. They have a composite biomass that provides lignin for composting operations and hence used for wetland construction.

9.9 Phytostabilization

The drying of swampy soils by Salix is quite old and well known among farmers. Most of the farmers in malaria affected areas prefer to plant Salix trees as it helps them to dry the accumulated water very quickly that reduces the chances of malaria in these areas. The introduction of willow vegetation buffer zones for coping waste dumping sites has been successfully applied in phytostabilization projects. A general argument is that strategic planting and positioning of willow around the contaminated area will help to decrease the water entering landfill. The type of Salix vegetation may also solve the problems of landfill as most of these species are tolerant to landfill sites.

The planting of trees will be a best remedy to keep away the linked pollutants from soil, because of diminishment of surface water percolating into the groundwater. Planting of *Salix* on municipal sewage deposits enhances the nitrification processes and contaminants in the sludge apparently do not affect the growth of willows (Scheirlink et al. 1996; Wielgosz 2000). The process of nitrification will be enhanced by planting more *Salix* trees on municipal sewage deposits that will decrease contaminants to great level. Extensive root systems and copious litter fall stabilize soil structure, augment cycling of nutrient and endorse biotic communities and creation of forest microclimate.

9.10 *Salix alba* and High Metal Biomagnifications

Pulford et al. (2002) argued that due to luxuriant biomass productivity *Salix* sps have performed goof in the acidic medium of soil in terms of heavy metal accumulation in their shoots, although no benefit was observed by addition of elemental sulphur. However, Cd and Zn levels were highly accumulated in leaves than in stems.

Hinchman et al. (1995) demonstrated that approximately 9.5% of the lead and 0.1% of the arsenic were removed by willows, respectively, from heavy metal contaminated sites. However, newly planted willow trees removed only 1% and 0.1% of the available lead and arsenic, respectively, from the same soil. Similarly, when the experiments were conducted in sand culture amended with 40% of the lead and Arsenic, willows were able to remove 30–40% of the lead and Arsenic from it.

Pulford et al. (2002) argued that one group of *Salix* sps has very less amount of Ni and Cu in the bark and very high Cd and Zn in the wood, with better survival rates and high productivity rates. On the other hand, other group of willow had relatively good content of Ni and Cu in the bark and very little Cd and Zn in the wood and these *Salix* sps were very poor in terms of endurance and biomass production. This is an indication that different factors play an important role in accumulating heavy metals.

Vandecasteele et al. (2005) advocated that willow foliar Cd concentrations in *Salix* sps were highly associated with soil and soil water Cd levels. Both clones exhibited very high accumulation levels of Cd and Zn in shoot system. Roots accumulated Cu, Cr, Pb, Fe, Mn and Ni. Bioaccumulation factors of Cd and Zn in the leaves were the maximum for the treatments with the least soil Cd and Zn levels.

Vervaeke et al. (2004) showed that Cu and Pb, but not Cd, were more available in the root zone by amending it with water and ammonium acetate extraction when compared with the bulk sediment. Sediment of the root zone was better structured and aggregated and thus more porous for downward flow of water that caused leaching of a fraction of the metals and significantly lower total contents of Cd, Cu and Pb.

Vyslouzilova (2003) found that As and Cd uptake enhanced in Suchdol-Zn soil compared to Suchdol-Pb soil very slowly and the element removal from soil was significantly high in Suchdol-Pb soil due to a significant decrease in biomass yield in Suchdol-Zn soil. The yield reduction had decreased the uptake of plant-available elements by biomass which indicates that higher plant-available portions of As and Cd were present in Suchdol-Zn soil.

Mleczek et al. (2010) revealed significant differences among the clones of Salix. The clones most effective in concentrating metals were *S. viminalis* '1154' and '1054'. The *Salix* population under study was considerably varied as regards to accumulation efficiency. The removal efficiency in terms of different heavy metals was Cd 84%, Cu 90%, Hg 167%, Pb 190% and Zn 36%.

Marius et al. (2017) presented the impact of simultaneous variations of lead and magnesium levels on growth of radicular and leaf system of *Salix* alba that were cultured in hydroponic cultures. Hydroponic solutions were amended with lead and magnesium in equal concentrations. Willow offshoots were placed in hydroponic solutions having different types of water with varied levels of pollutants for a period of 35 days. The evolution of leaves and roots number in time showed that phytotoxicity limit for simultaneous adding of lead and magnesium ranges from 0.1 % to 0.5 % (w / w).

Popoviciu and Ticuţa (2018) reported highest average concentrations of heavy metals in *Salix alba* L. species: 8.89 mg/kg Cu, 43.94 mg/kg Mn and 94.10 mg/kg Zn, values that are not consistent with hyperaccumulation. *Salix alba* L. was the only species with BAC > 1 for Cu (constant in all samples) and Zn (only as average value).

Bajraktari et al. 2019 determined that the content different heavy metals in *Salix alba*. The different pollutants in water and soil samples are above maximum permissible level and it is recommended that *Salix alba* plantations will reduce the pollutant load (Table 9.1).

Table 9.1 Phytoremediation characteristics of *Salix* sp.

Salix sp.	Metal(s)	Results	Author(s)
Salix alba	Cu, Zn and Mn	Highest accumulation of heavy metals	Popoviciu and Ticuţa (2018)
Salix alba	Pb and Mn	Reduced concentration in hydroponic system	Marius et al. (2017)
S. viminalis x miyabeana, S eriocephala	As	Total As accumulation was greater	Puckett et al. (2012)
S. geyeriana, S. drummondiana, S. planifolia	Cd, Cu, Pb and Zn	Metal concentration accumulated in roots	Meiman et al. (2012)
S. Purpurea, S. alba, S. fragilis, S. japonica	Cd, Co, Cr, Cu, Ni, Pb and Zn	Salix sps accumulated more and other species were effective	Mleczek et al. (2010)
S. Purpurea, S. alba, S. fragilis, S. japonica, S. nigra	Cd, Co, Cr, Cu, Ni, Pb and Zn	In general, highest concentration of these metals	Mleczek et al. (2009)
S. Purpurea, S. eriocephala, S. Sachaliensis	Zn, Mn, Fe, Cu and Al	More Zn, Fe and Al were accumulated than populous	Zalesny and Bauer (2007)
S. Purpurea, S. viminalis x miyabeana, S. Sachaliensis	AS	By amendment in soil with phosphate As accumulation increased	Purdy and Smart (2008)
S. geyeriana, S. monticola	Cd, Mn and Pb	*S. monticola* is better for Cd accumulation as compared to *S. geyeriana*	Boyter et al. (2009)
S. geyeriana, S. monticola	Mn and Zn	Good for Mn resistance and better for phytostabilization	Shanahan et al. (2007)
S. nigra	Mn, Ni and Fe	Metals were found to be present in almost all parts	Punshon et al. (2005)
Salix sps	Cd and Zn	These metals were found more in leaves than in stems	Pulford et al. (2002)

9.11 Conclusions

Heavy metals are a cause of concern both for the terrestrial and aquatic ecosystems. The removal of heavy metals from these ecosystems is a priority of experts as heavy metals may render these ecosystems unsuitable for the functioning for which they are meant. Different methods have been tried to recover heavy metals from these ecosystems and phytoremediation is considered to be a safe and an eco-friendly approach. Willows are excellent phytoremediation agents as they accumulate and tolerate heavy metals and follow dominant vegetation even at higher elevations. They have been found to be quite useful in abandoned mine sites across the year world. This is considered a super phytoextraction plant that requires high translocation rate from roots to shoots. They are useful due to their fast growth, high biomass and high metal translocation ability. They are also highly propagated due to highly developed root system that quite often reaches deep down the soil and accumulate heavy metals over there.

References

Aasamaa K, Heinsoo K, Holm B (2010) Biomass production, water use and photosynthesis of Salix clones grown in a wastewater purification system. Biomass Bioenergy 34(6):897–905

Agegnehu G, Bass A, Nelson P, Bird M (2016) Benefits of biochar, compost and biochar–compost for soil quality, maize yield and greenhouse gas emissions in a tropical agricultural soil. Sci Total Environ 543:295–306

Ali MB, Tripathi RD, Rai UN, Pal A, Singh SP (1999) Physico-chemical characteristics and pollution level of Lake Nainital (U.P., India): role of macrophytes and phytoplankton in biomonitoring and phytoremediation of toxic metal ions. Chemosphere 39(12):2171–2182

Ali MB, Vajpayee P, Tripathi RD, Rai UN, Singh SN, Singh SP (2003) Phytoremediation of lead, nickel, and copper by *Salix acmophylla* Boiss.: role of antioxidant enzymes and antioxidant substances. Bull Environ Contam Toxicol 70:462–469

Ali H, Khan E, Sajad M (2013) Phytoremediation of heavy metals—concepts and applications. Chemosphere 91(7):869–881

Argus GW (1986) The genus *Salix* (Salicaceae) in the Southeastern United States. Syst Bot Monogr 9:170

Argus GW (1997) Infrageneric classification of *Salix* (Salicaceae) in the New World. Syst Bot Monogr 52:121

Bajraktari D, Biljana B, Zoran K, Lulzim Z (2019) Environmental pollution and heavy metals accumulation in *Salix alba* L. (Fam. Salicaceae), along the river stream of Sitnica. Agric Conspec Sci 84:95–101

Bates J (1951) Trailside botany. Pfeifer-Hamilton, Duluth, p 227

Beesley L, Moreno-Jiménez E, Gomez-Eyles J (2010) Effects of biochar and green waste compost amendments on mobility, bioavailability and toxicity of inorganic and organic contaminants in a multi-element polluted soil. Environ Pollut 158(6):2282–2287

Bhat RA, Dervash MA, Qadri H, Mushtaq N, Dar GH (2018) Macrophytes, the natural cleaners of toxic heavy metal (THM) pollution from aquatic ecosystems. In: Environmental contamination and remediation. Cambridge Scholars Publishing, Cambridge, pp 189–209

Borišev M, Pajević S, Nikolić N, Pilipović A, Krstić B, Orlović S (2009) Phytoextraction of Cd, Ni, and Pb using four willow clones (*Salix* spp.). Pol J Environ Stud 18(4):553–561

Boyter MJ, Brummer JE, Leininger WC (2009) Growth and Metal Accumulation of Geyer and Mountain Willow Grown in Topsoil versus Amended Mine Tailings. Water Air Soil Pollut 198 (1–4):17–29

Bungart R, Huttl RF (2001) Production of biomass for energy in post-mining landscapes and nutrient dynamics. Biomass Bioenergy 20:181–187

Clewell A (1999) Restoration of riverine forest at hall branch on phosphate-mined land, Florida. Restor Ecol 7(1):1–14

Corseuil HX, Moreno FN (2001) Phytoremediation potential of willow trees for aquifers contaminated with ethanol-blended gasoline. Water Res 35(12):3013–3017

Dickinson NM, Punshon T, Hodkinson RB, Lepp NW (1994) Metal tolerance and accumulation in willows. In: Aronsson P, Perttu K (eds) Willow vegetation filters for municipal wastewater and sludges. Swedish University of Agricultural Sciences, Uppsala, pp 121–127

Dorn RD (1976) A synopsis of American Salix. Canadian J Bot 54:2769–2789

Ebbs S, Bushey J, Poston S, Kosma D, Samiotakis M, Dzombak D (2003) Transport and metabolism of free cyanide and iron cyanide complexes by willow. Plant Cell Environ 26:1467–1478

Elowson S, Christersson L (1994) Purification of groundwater using biological filters. In: Aronsson P, Perttu K (eds) Willow vegetation filters for municipal wastewater and sludges. Swedish University of Agricultural Sciences, Uppsala, pp 219–223

Eltrop L, Brown G, Joachim O, Brinkmann K (1991) Lead tolerance of Betula and Salix in the mining area of Mechernich/Germany. Plant Soil 131:275–285

Eriksson J, Ledin S (1999) Changes in phytoavailability and concentration of cadmium in soil following long term Salix cropping. Water Air Soil Pollut 114:171–184

FAO (1979) Poplar and willows in wood production and land use, Forestry series no. 10

Granel T, Robinson B, Mills T, Clothier B, Green S, Fung L (2002) Cadmium accumulation by willow clones used for soil conservation, stock fodder, and phytoremediation. Aust J Soil Res 40(8):1331–1337

Greger M, Landberg T (1999) Use of willow in phytoextraction. Int J Phytoremediation 1(2):115–123

Hartwright TU (1960) Planting trees and shrubs in gravel working. Sand and Gravel Association of Great Britain, London, p 72

Hightshoe G (1998) Native trees, shrubs and vines for urban and rural America. Wiley, New York, p 819

Hinchman RR, Negri MC, Gatliff EG (1995) Phytoremediation using green plants to clean up contaminated soil, groundwater, and wastewater. Argonne National Laboratory Hinchman, Applied Natural Sciences, Inc., Hamilton. http://www.treemediation.com/Technical/Phytoremediation_1998.pdf

Hossain M, Strezov V, Yin Chan K, Nelson P (2010) Agronomic properties of wastewater sludge biochar and bioavailability of metals in production of cherry tomato (*Lycopersicon esculentum*). Chemosphere 78(9):1167–1171

Janus A, Pelfrêne A, Heymans S, Deboffe C, Douay F, Waterlot C (2015) Elaboration, characteristics and advantages of biochars for the management of contaminated soils with a specific overview on *Miscanthus biochars*. J Environ Manag 162:275–289

Justin M, Pajk N, Zupanc V, Zupančič M (2010) Phytoremediation of landfill leachate and compost wastewater by irrigation of *Populus* and *Salix*: biomass and growth response. Waste Manag 30(6):1032–1042

Keller C, Hammer D, Kayser A, Richner W, Brodbeck M, Sennhauser M (2003) Root development and heavy metal phytoextraction efficiency: comparison of different plant species in the field. Plant Soil 249:67–81

Kennedy CEJ, Southwood TRE (1984) The number of species of insects associated with British trees; a re-analysis. J Animal Ecol 53:455–478

Kirt E (1994) Vegetation filter experiment in Estonia. In: Aronsson P, Perttu K (eds) Willow vegetation filters for municipal wastewater and sludges. Swedish University of Agricultural Sciences, Uppsala, pp 79–82

Klang-Westin E, Eriksson J (2003) Potential of Salix as phytoextractor for Cd on moderately contaminated soils. Plant Soil 249:127–137
Klang-Westin E, Perttu K (2002) Effect of nutrient supply and soil cadmium concentration on cadmium removal by willow. Biomass Bioenergy 23:415–426
Kloss S, Zehetner F, Wimmer B, Buecker J, Rempt F, Soja G (2014) Biochar application to temperate soils: effects on soil fertility and crop growth under greenhouse conditions. J Plant Nutr Soil Sci 177(1):3–15
Kowalik PJ, Randerson PF (1994) Nitrogen and phosphorus removal by willow stands irrigated with municipal waste water—a review of the Polish experience. Biomass and Bioenergy 6(1–2):133–139
Kuzovkina YA, Knee M, Quigley MF (2004) Soil compaction and flooding effects on the growth of twelve Salix L. species. J Environ Hort 22:155–160
Landberg T, Greger M (1994) Cadmium tolerance in *Salix*. Biol Plant 361:280
Landberg T, Greger M (2002) Differences in oxidative stress in heavy metal resistant and sensitive clones of *Salix viminalis*. J Plant Physiol 159:69–75
Larsen M, Trapp S, Pirandello A (2004) Removal of cyanide by woody plants. Chemosphere 54:325–333
Lebrun M, Macri C, Miard F, Hattab-Hambli N, MotelicaHeino M, Morabito D, Bourgerie S (2017) Effect of biochar amendments on As and Pb mobility and phytoavailability in contaminated mine technosols phytoremediated by *Salix*. J Geochem Explor 182:149–156
Lodge DJ (1989) The influence of soil moisture and flooding on formation of VA-endo-and ectomycorrhizae in *Populus* and *Salix*. Plant Soil 117:243–253
Logan TJ (1992) Reclamation of chemically degraded soils. Adv Soil Sci 17:13–35
Lunackova L, Masarovicova E, Kral'ova K, Stresko V (2003) Response of fast growing woody plants from family Salicaceae to cadmium treatment. Bull Environ Contam Toxicol 70:576–585
Marius B, Gabriela B, Cristina M, Nicolescu A, Irina G, Ivona D, Costel B (2017) Influence of lead toxicity on growth and antioxidant enzyme activity for *Salix Alba* offshoots in hydroponic cultures with different levels of pollutants. J Sci Arts 4(41):803–814
Maroder HL, Prego IA, Facciuto GR, Maldonado SB (2000) Storage behaviour of Salix alba and Salix matsudana seeds. Ann Bot 86:1017–1021
Martin JF, Reddy KR (1997) Interaction and spatial distribution of wetland nitrogen processes. Ecol Model 105:1–21
Maurice C, Ettala M, Lagerkvist A (1999) Effects of leachate irrigation on landfill vegetation and subsequent methane emissions. Water Air Soil Pollut 113:203–216
McLeod KW, McPherson JK (1973) Factors limiting the distribution of Salix nigra. Bull Torrey Bot Club 100:102–110
Mertens J, Luyssaert S, Verbeeren S, Vervaeke P, Lust N (2001) Cd and Zn concentrations in small mammals and willow leaves on disposal facilities for dredged material. Environ Pollut 115:17–22
Meiman PJ, Davis NR, Brummer JE, Ippolito JA (2012) Riparian shrub metal concentrations and growth in amended fluvial mine tailings. Water Air Soil Pollut 223(4):1815–1828
Mleczek M, Magdziak Z, Rissmann I, Golinski P (2009) Effect of different soil conditions on selected heavy metal accumulation by Salix viminalis tissues. J Environ Sci Health A 44(14): 1609–1616
Mleczek M, Rutkowski P, Rissmann I, Kaczmarek Z, Golinski P, Szentner K, Stachowiak A (2010) Biomass productivity and phytoremediation potential of Salix alba and Salix viminalis. Biomass and Bioenergy, 34(9):1410–1418
Moosavi SG, Seghatoleslami MJ (2013) Phytoremediation: a review. Adv Agric Biotechnol 1(1):5–11
Omondi M, Xia X, Nahayo A, Liu X, Korai P, Pan G (2016) Quantification of biochar effects on soil hydrological properties using meta-analysis of literature data. Geoderma 274:28–34
Park J, Lamb D, Paneerselvam P, Choppala G, Bolan N, Chung J (2011) Role of organic amendments on enhanced bioremediation of heavy metal(loid) contaminated soils. J Hazard Mater 185(2–3):549–574

Perttu KL, Kowalik PJ (1997) Salix vegetation filters for purification of water and soils. Biomass Bioenergy 12(1):9–19
Popoviciu DR, Ticuţa N (2018) Copper, manganese and zinc bioaccumulation in three common woody species from black sea coastal area. UPB Sci Bull Ser B 80(2):49–56
Pulford D, Riddell-Black D, Stewart C (2002) Heavy metal uptake by willow clones from sewage sludge-treated soil: the potential for phytoremediation. Int J Phytoremediation 4(1):59–72
Puckett EE, Serapiglia MJ, DeLeon AM, Long S, Minocha R, Smart LB (2012) Differential expression of genes encoding phosphate transporters contributes to arsenic tolerance and accumulation in shrub willow (*Salix* spp.). Environ Exp Bot 75:248–257
Punshon T, Dickinson N (1997) Acclimation of Salix to metal stress. New Phytol 137:303–314
Punshon T, Jackson BP, Lanzirotti A, Hopkins WA, Bertsch PM, Burger J (2005) Application of synchrotron X-ray microbeam spectroscopy to the determination of metal distribution and speciation in biological tissues. Spectrosc Lett 38(3):343–363
Purdy JJ, Smart LB (2008) Hydroponic screening of shrub willow (*Salix* spp.) for arsenic tolerance and uptake. Int J Phytoremediation 10(6):515–528
Raven JA (1992) The physiology of Salix. In: Watling R, Raven JA (eds) 1992 willow symposium. Proceedings of the Royal Society of Edinburgh, vol 98. The Royal Society of Edinburgh, Edinburgh, pp 49–62
Riddell-Black D (1994) Sewage sludge as a fertilizer for short rotation energy coppice. In: Aronsson P, Perttu K (eds) Willow vegetation filters for municipal wastewater and sludges. Swedish University of Agricultural Sciences, Uppsala, pp 91–100
Robinson BH, Mills TM, Petit D, Fung LE, Green SR, Clothier BE (2000) Natural and induced cadmium-accumulation in poplar and willow: implications for phytoremediation. Plant Soil 227:301–306
Rosenqvist H, Aronsson P, Hasselgren K, Perttu K (1997) Economics of using municipal wastewater irrigation of willow coppice crops. Biomass Bioenergy 12(1):1–8
Scheirlink H, Lust N, Nachtergale L (1996) Transpiration of two willow species (*Salix viminalis* and *Salix triandra*) growing on a landfill of dredged sludge. Silva Gandav 61:33–45
Schramm JR (1966) Plant colonization studies on black wastes from anthracite mining in Pennsylvania. Trans Am Phil Soc N S 56:6–194
Sennerby-Forsse L, Melin J, Rosen K, Siren G (1993) Uptake and distribution of radiocesium in fast-growing Salix viminalis L. J Sustain For 1(3):93–103
Shanahan JO, Brummer JE, Leininger WC, Paschke MW (2007) Manganese and zinc toxicity thresholds for mountain and Geyer willow. Int J Phytoremediation 9(5):437–452
Singh DV, Bhat JIA, Bhat RA, Dervash MA, Ganei SA (2018) Vehicular stress a cause for heavy metal accumulation and change in physico-chemical characteristics of road side soils in Pahalgam. Environ Monit Assess 190:353. https://doi.org/10.1007/s10661-018-6731-2
Skvortsov AK (1999) Willows of Russia and adjacent countries. Taxonomical and geographical review. Univ Joensuu Fac Math Nat Sci. Rept Ser 39:307
Smith FF, Smith DK, Argus GW (1978) 'Willows for pleasure and benefit', Amer. Horticulture 57(2):22–25
Sommerville AHC (1992) Willows in the environment. In: Watling R, Raven JA (eds) 1992 willow symposium. Proceedings of the Royal Society of Edinburgh, vol 98. The Royal Society of Edinburgh, Edinburgh, pp 215–225
Stott KG (1992) Willows in the service of man. In: Watling R, Raven JA (eds) 1992 willow symposium. Proceedings of the Royal Society of Edinburgh, vol 98. The Royal Society of Edinburgh, Edinburgh, pp 169–182
Thompson W (1998) Botanical remedies. Landsc Archit 8:38–43
van der Heijden EW, Kuyper THW (2003) Ecological strategies of ectomycorrhizal fungi of Salix repens: root manipulation versus root replacement. Oikos 103:668–680
Vandecasteele B, Meers E, Vervaeke P, Vos BD, Quataert P, Tack FMG (2005) Growth and trace metal accumulation of two Salix clones on sediment-derived soils with increasing contamination levels. Chemosphere 58(8):995–1002

Vervaeke P, Tack FMG, Lust N, Verloo M (2004) Short- and longer-term effects of the willow root system on metal extractability in contaminated dredged sediment. J Environ Qual 33(3):976–983
Vishnoi SR, Srivastava PN (2008) Phytoremediation – green for environmental clean. In: The 12th world lake conference. pp. 1016–1021
Vyslouzilova M, Tlustos P, Szakova J, Pavlikova D (2003) As, Cd, Pb and Zn uptake by Salix spp. clones grown in soil enrich by high load of this elements. Plant Soil Environment 49(5):191–196
Watson C, Pulford ID, Riddell-Black D (2003) Development of a hydroponic screening technique to assess heavy metal resistance in willow (*Salix*). Int J Phytoremediation 5(4):333–349
White JEJ (1992) Ornamental uses of willow in Britain. In: Watling R, Raven JA (eds) 1992 willow symposium. Proceedings of the Royal Society of Edinburgh, vol 98. The Royal Society of Edinburgh, Edinburgh, pp 183–192
Wielgosz E (2000) Aktywnosc biochemiczna w osadach posciekowych poddanych czteroletniej transformacji roslinnej. Annales Universitatis Mariae Curiie-Sklodowska LV 20:185–193
Zalesny Jr, RS, Bauer EO (2007) Selecting and utilizing Populus and Salix for landfill covers: implications for leachate irrigation. Int J Phytoremediation 9(6):497–511

Chapter 10
Photo Catalysis: An Effective Tool for Treatment of Dyes Contaminated Wastewater

Muhammad Saeed, Shahid Adeel, Majid Muneer, and Atta ul Haq

10.1 Introduction

Today the world is facing a number of problems; however, the environmental pollution is the most significant problem which increases continuously and gives rise to irreversible and serious damage to the life of living organism. Environmental pollution can be classified into three major types:

1. Water pollution
2. Soil pollution
3. Air pollution

Among these pollutions, research on water pollution got much attention. There are number of factors including domestic wastes, pharmaceutical residue, fertilizers in agriculture sector, excess use of pesticides and industrial effluent that are responsible for water pollution (Legrini et al. 1993; Lianos 2011; McLaren and Williams 2015). Wastewater contains a variety of pollutants like dyes, chlorophenols, polychlorinated biphenyls (PCBs), pesticides, dibenzofurans, dioxins, and solvents. Among these pollutants, the dyes are considered as major contributors to water pollution because a number of industries use the dyes in various processes. Hence the discharge of these industries contains a large concentration of these toxic and carcinogenic dyes. Most of these dyes have complex aromatic structures due to which they are highly stable and resistant to biological decomposition (Jo and Tayade 2014; Shinde et al. 2014; Nguyen and Juang 2015; Reddy and Kim 2015).

The natural dyes, dyes obtained from natural resources like plants and animals, have been used for dying of fabrics since 5000 years back. However, the synthetic dyes discovered in the nineteenth century have replaced the use of natural dyes. These synthetic dyes are industrially produced in huge amount which are used in

M. Saeed (✉) · S. Adeel · M. Muneer · A. ul Haq
Department of Chemistry, Government College University Faisalabad, Faisalabad, Pakistan
e-mail: msaeed@gcuf.edu.pk

R. A. Bhat et al. (eds.), *Bioremediation and Biotechnology, Vol 2*,
https://doi.org/10.1007/978-3-030-40333-1_10

different industries like electroplating, pharmaceuticals, agricultural research, cosmetics, food, paper, leather, and fabrics. Synthetic dyes can be classified into different groups like direct dyes, disperse dyes, basic dyes, acid dyes, pigments, mordant, metallic, sculpture, vat dyes, solvent dyes, and reactive dyes. These dyes reflect their behavior and functionalities. Presently, vast varieties consisting of more than 100,000 of these synthetic dyes are commercially available worldwide. It is estimated that about one million tons of these synthetic dyes are produced on yearly basis. It has been reported that 10–15% of dyes employed in different industries are lost in industrial effluent to environment during manufacturing and application processes (Saeed et al. 2015a, b; Natarajan et al. 2013a, b; Ajmal et al. 2014).

A huge amount of synthetic dyes released in industrial effluent generate a threat to living organisms and environment. As most of these dyes are brightly colored, therefore a small amount of these dyes can be easily recognized in wastewater. The wastewater contaminated with these dyes is badly affected in terms of transparency and aesthetic merit. These dyes present in water bodies adsorb and reflect the solar radiation resulting in inhibition of penetration of solar rays to interior of the water body. This kind of inhibition of solar radiations into the water body has effects on aquatic life, growth of various aquatic species, and photosynthesis. Furthermore, these dyes have chronic and acute effects on aquatic organisms. The severity of these effects depends on concentration of dyes in wastewater and exposure time of living organisms to contaminated water. Hence, the removal of these dyes is a major concern toward environmental pollution abatement.

In order to minimize the environmental impact of dyes present in wastewater, a number of methods and techniques have been developed for treatment of dyes contaminated wastewater. The developed methods are divided into four groups as given below.

1. Physical methods
2. Chemical methods
3. Biological methods
4. Acoustical methods

Physical methods include filtration (reverse osmosis, nanofiltration, ultrafiltration, and microfiltration), adsorption, and sedimentation. Chemical methods include oxidation, coagulation, electrochemical oxidations, and advanced oxidation processes. Aerobic and anaerobic treatments are included in biological methods. Acoustical methods include radiation and electric processes (Delee et al. 1998; Rai et al. 2005; Barragan et al. 2007; Maezawa et al. 2007; Hong et al. 1999; Aguedach et al. 2005; Bandala et al. 2008; Zhao et al. 2008; Shi et al. 2007; Al-Bastaki 2004; Marmagne and Coste 1996; Linsebigler et al. 1995; Avlonitis et al. 2008; Gupta and Suhas 2009; Hage and Lienke 2006; Namboodri et al. 1994; Dogan and Turkdemir 2005; Wang 2008).

Physical methods (like electro-dialysis, reverse osmosis, nanofiltration, and adsorption) and chemical methods (like electro flotation, flocculation, and precipitation) are suitable techniques for elimination of dyes in aqueous medium; however, these methods are effective in decreasing the concentration of dyes only without being degrading the dyes. These methods transform the pollutants from one form to another form which result in secondary pollution. Similarly, the biological methods which use the enzymes for removal of dyes from wastewater are also not suitable due to longer time. Furthermore, some dyes are very resistant to biological degradation. Additionally, the biological degradation of dyes produces aromatic compounds which are highly carcinogenic (Freeman 1989).

10.2 Photo Catalysis

Recently, the advanced oxidation processes, AOPs, which are highly effective in complete degradation of dyes and economically favorable, have got extra attention for dyes contaminated wastewater treatment. These advanced oxidation processes are performed under irradiation of ultraviolet and visible light. These processes are also assisted with hydrogen peroxide and ultrasonication (Aleboyeh et al. 2012; Srivastava et al. 2013; Rao et al. 2009). In photo catalysis the dyes and other organic pollutants are degraded over semiconductor metal oxides catalysts like cadmium sulfide (CdS), zinc oxide (ZnO), zinc sulfide (ZnS), titanium dioxide (TiO_2), tin oxide (SnO_2), strontium titanate ($SrTiO_3$), bismuth oxide (Bi_2O_3), vanadium oxide (V_2O_5), iron oxide (Fe_2O_3), etc. (Natarajan et al. 2013a, b; Lee and Mills 2004; Mills and Lee 2002; Tayade et al. 2007; Huang et al. 2013; Sobana and Swaminathan 2007).

Photo catalysis is the increase in rate of reaction or initiation of reaction in the presence of a semiconductor metal oxide under irradiation of light which cause chemical transformation that come into contact with it and regenerates its chemical composition after each cycle of such interactions by absorption of light quanta. The semiconductor metal oxide is called as photo catalyst (Fox and Dulay 1993).

Two approaches are used to promote the efficiency of metal oxide semiconductors photo catalysts (Tayade et al. 2008, 2011; Saeed et al. 2018a, b; Mogal et al. 2012).

1. The metal oxide semiconductor photo catalysts are doped with metal atoms like Ag, Zn, etc. These metal islands deposited on semiconductors photo catalyst prevent the recombination of positive hole and photo excited electron. These positive holes and photo excited electrons are generated by irradiation of catalysts under visible or ultraviolet light. These positive holes and photo excited electrons take part in degradation of pollutants.
2. The semiconductor metal oxide photo catalysts are anchored on the surface of a non-catalytic material having high surface area, the support, such as clay, zeolite, silica, activated carbon, etc. These high surface area supports provide maximum dispersion to the active catalytic material. Additionally, these non-catalytic supports have better sorption characteristics due to charge density and high surface area. These semiconductor metal oxide photo catalysts are easily separated from reaction mixture.

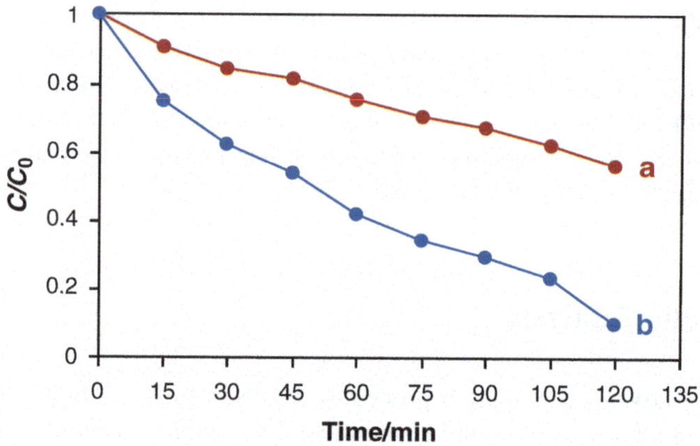

Fig. 10.1 Comparison of performance of (a) Co_3O_4 and (b) Ag-Co_3O_4 on photo degradation of rhodamine B dye. This study was conducted by suspending a 0.05 g of catalyst in 200 ppm solution of rhodamine B. The suspension was agitated at 40 °C under irradiation of visible light

The improvement in catalytic activity of photo catalyst by deposition of metal is the commonly used method. In the first method, dispersion of active catalyst on solid support, some of the dye molecules may be adsorbed on the solid support which will remain un-degraded. In the second method, the dye molecules are completely degraded. Therefore, the latter technique is the commonly used method for enhancement of photo catalytic activity. Saeed et al. (2018a) have enhanced the performance of Co_3O_4 by deposition of Ag on Co_3O_4 for photo degradation of rhodamine B dye. They reported that activity of Co_3O_4 increased from 24% to 57% by deposition of 3%Ag on Co_3O_4. Figure 10.1 shows the increase in catalytic activity of Co_3O_4 by deposition of Ag.

Similarly, in another report the performance of manganese oxide has been enhanced by deposition of silver on manganese oxide (Saeed et al. 2018b). They found that catalytic activity of manganese oxide increased from 45 to 91% for degradation of rhodamine B dye by deposition of manganese oxide with silver. Figure 10.2 shows the enhancement of photo catalytic activity of manganese oxide with deposition of silver.

10.3 Mechanism of Photo Catalytic Degradation Reactions

A photocatalytic process can be described as follows.

1. Generation of pair of hole and electron by irradiation of photo catalyst with ultraviolet or visible radiations.
2. Separation of generated pair of hole and electron by traps.

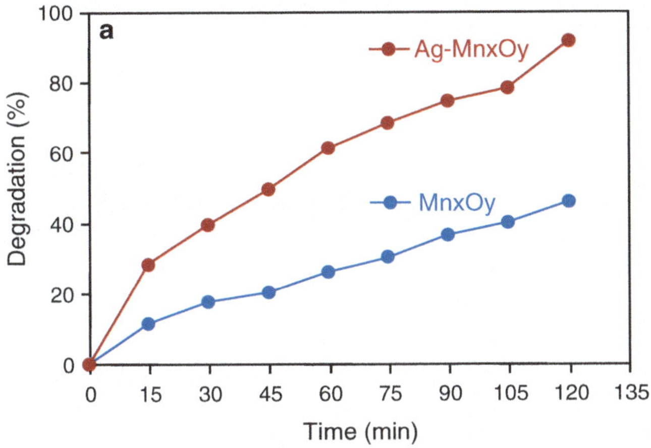

Fig. 10.2 Comparison of performance of Mn_xO_y and Ag-Mn_xO_y towards rhodamine B dye degradation. This study was carried out by suspending a 0.05 g of catalyst in 40 mL of 200 mg/L of rhodamine B dye solution. The reaction mixture was stirred under irradiation of visible light at 40 °C

3. Series of secondary reactions of hole and electron with water and oxygen to produce OH radicals.
4. Degradation of pollutants by reaction with OH radical and regeneration of the catalyst.

These steps are explained in Fig. 10.3 (Natarajan et al. 2018).

The reaction mechanism of photo catalytic reaction can be described as redox mechanism and the role of irradiation of semiconductor photo catalyst is its excitation. The band gap of semiconductor metal oxide photo catalysts has a prominent effect on degradation of pollutants and dyes. The catalytic efficiency of semiconductor metal oxide photo catalyst towards degradation of pollutants varies with catalyst band gap position. The band gap also called as energy gap is the energy difference between valence band and conduction band of the semiconductor metal oxide photo catalyst. Band gap of semiconductor metal oxide is the energy range where electron cannot exist. The position of band gap determines the power of oxidation or reduction of photo catalysts. Most of the semiconductor metal oxides have band gaps from 0.7 to 5 eV. The semiconductor metal oxides are used as photo catalysts on the basis of band gap of these materials. Those semiconductor metal oxides which have band gap from 2.0 to 3.3 eV are generally selected as photo catalysts for treatment of dyes contaminated wastewater.

The mechanism of photo catalytic degradation reactions can be classified as direct photo catalytic mechanism and indirect photo catalytic mechanism.

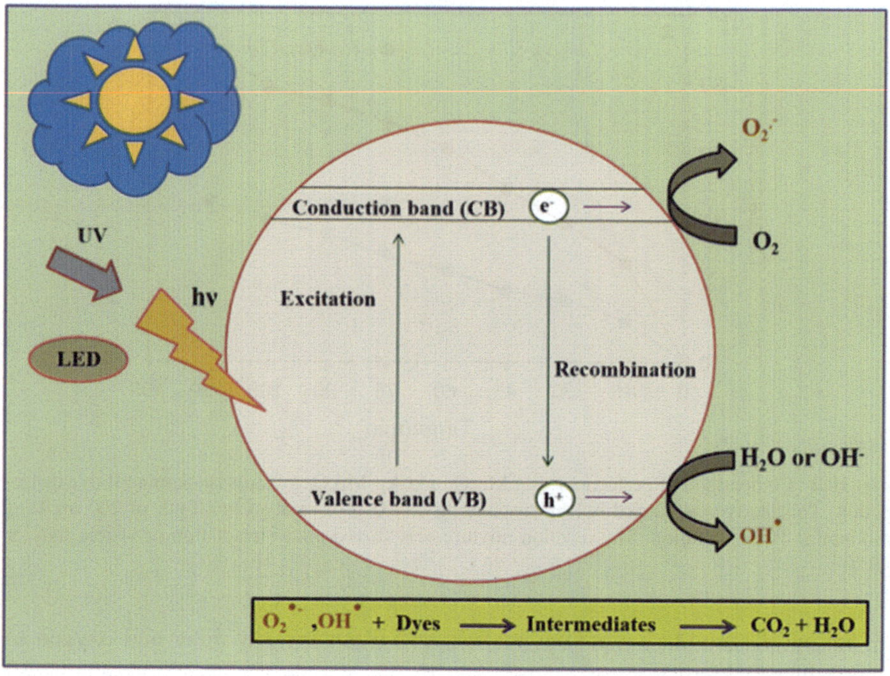

Fig. 10.3 Mechanism of ultraviolet or visible light assisted photo catalytic degradation of dyes (Natarajan et al. 2018)

10.3.1 Direct Photo Catalytic Mechanism

The direct photo catalytic mechanism is also called as Langmuir–Hinshelwood mechanism. According to direct pathway, the Langmuir–Hinshelwood mechanism, the dye molecules gets adsorbed at the surface of catalyst in the first step. In the second step the degradation of dye molecules takes place. According to this mechanism, when catalyst is irradiated with ultraviolet of visible light then a pair of positive hole and photo excited electron takes place. The positive hole is generated in valence band of catalyst. The photo excited electron is generated in conduction band of catalyst. The positive hole generated in valence band of catalyst is trapped by surface defect of the catalyst. This trapping of positive hole by surface defect gives rise to surface-active center (S). Then, the surface-active center reacts with molecules of dye and produces an adduct species $(\text{S-Dye})^{+}$. This adduct species decomposes to produce degradation products or it can re-combine with photo excited electron. This process can be outlined as:

$$\text{Catalyst} + \text{Dye} \rightarrow \text{Dye}_{\text{ads}} \tag{10.1}$$

$$\text{Catalyst} + h\nu \rightarrow h^{+} + e^{-} \tag{10.2}$$

$$S + h^{+} \rightarrow S^{+} \tag{10.3}$$

$$S^{+} + e^{-} \rightarrow S \tag{10.4}$$

$$S^{+} + \mathrm{Dye}_{\mathrm{ads}} \rightarrow (S - \mathrm{Dye})^{+} \tag{10.5}$$

$$(S - \mathrm{Dye})^{+} + e^{-} \rightarrow S + \mathrm{Products} \tag{10.6}$$

The rate expression for Langmuir–Hinshelwood mechanism can be written as (Saeed et al. 2015a, b, 2016, 2018c):

$$r = k\frac{KC}{1 + KC} \tag{10.7}$$

$$\frac{1}{r} = \frac{1}{kKC} + \frac{1}{k} \tag{10.8}$$

where r, k, K, and C is the rate of reaction, rate constant, adsorption equilibrium constant, and concentration of dye, respectively.

Saeed et al. (2018c) have reported Langmuir–Hinshelwood mechanism for degradation of dyes catalyzed by $CoFe_2O_4$. Langmuir–Hinshelwood model (Eq. 10.7) was applied with Curve Expert software to degradation data of dyes. The concentrations of dyes were calculated using Curve Expert software. Both the experimental and calculated concentrations of dyes were close to each other as given in Fig. 10.4. The values of k, the rate constant and K, the equilibrium constant are listed in Table 10.1.

10.3.2 Indirect Photo Catalytic Mechanism

The indirect mechanism is also called as Eley–Rideal mechanism which states that the dye reacts in fluid phase and forms the degradation products. This mechanism also consists of generation of positive hole and photo excited electron under irradiation of catalyst. The positive hole which is produced in valence band of catalyst is trapped by water molecule and produces OH radical and H^{+}. Similarly, the photo excited electron in conduction band is trapped by molecular oxygen to form superoxide radical which can enter into a chain reaction to produce HO radicals. Finally, the OH radicals, which are highly reactive species, attack the dye molecules and produce intermediates and end products (Yue et al. 2002).

This process can be outlined as:

$$\mathrm{Catalyst} + h\nu \rightarrow h^{+} + e^{-} \tag{10.9}$$

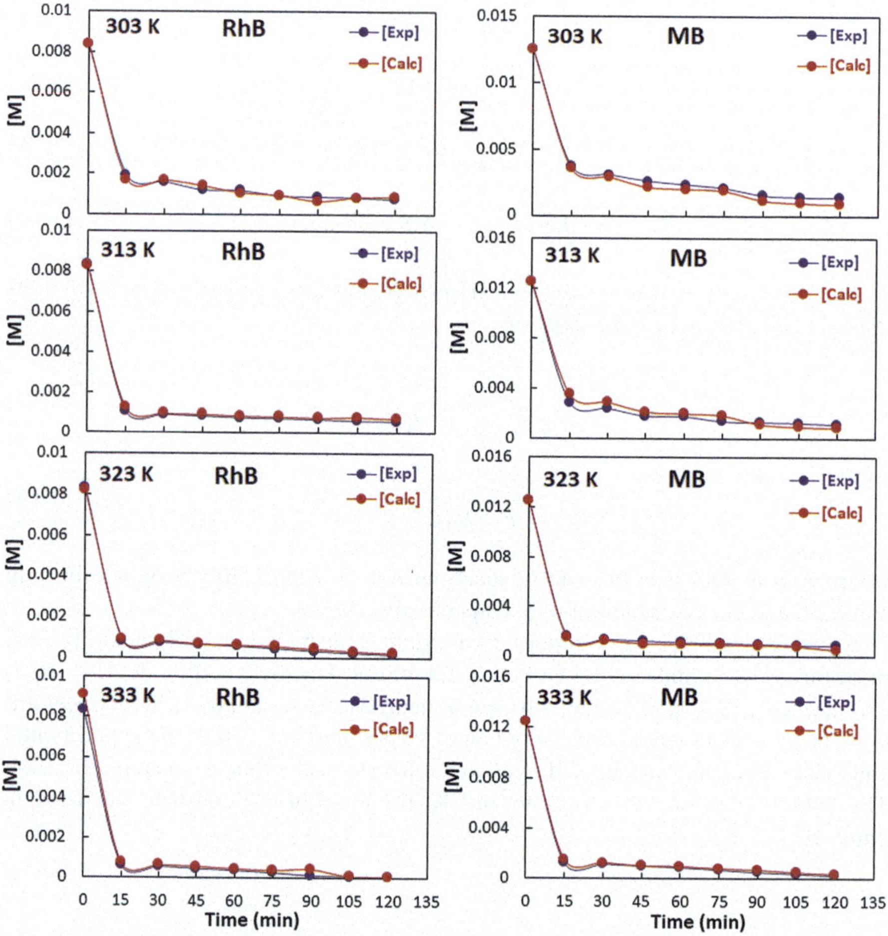

Fig. 10.4 Langmuir–Hinshelwood mechanism of degradation of rhodamine B and methylene blue dyes catalyzed by $CoFe_2O_4$

Table 10.1 Langmuir–Hinshelwood kinetics parameters of $CoFe_2O_4$ catalyzed degradation of dyes

	RhB		MB	
T (K)	*k* (/min)	*K* (L/mol)	*k* (/min)	*K* (L/mol)
303	0.028	1.61	0.031	1.74
313	0.066	1.60	0.069	1.71
323	0.098	1.58	0.098	1.69
333	0.128	1.55	0.131	1.67

$$H_2O + h^+ \rightarrow OH + H^+ \tag{10.10}$$

$$O_2 + e^- \rightarrow O_2^- \tag{10.11}$$

$$H^+ + O_2^- \rightarrow HO_2 \tag{10.12}$$

$$HO_2 \rightarrow O_2 + H_2O_2 \tag{10.13}$$

$$H_2O_2 \rightarrow OH \tag{10.14}$$

$$\text{Dye} + \text{OH} \rightarrow \text{Degradation products} \tag{10.15}$$

The reaction of methylene blue dye in fluid phase can be expressed by kinetic expression:

$$r = -\frac{dC}{dt} = kC \tag{10.16}$$

$$\ln\frac{C_o}{C_t} = kt \tag{10.17}$$

Saeed et al. (2018a) have explained the Ag-Co_3O_4 catalyzed photo degradation of rhodamine dye in terms of indirect photo catalytic mechanism. The cited mechanism is given in Fig. 10.5. The rate constants determined according to given mechanism were 0.0198, 0.0140, and 0.0121 at 323, 313, and 303 K, respectively.

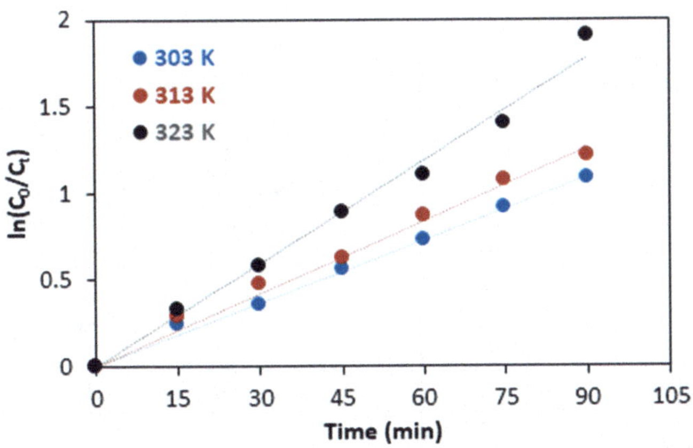

Fig. 10.5 Eley–Rideal mechanism for degradation of rhodamine B dye catalyzed by Ag-Co_3O_4

10.4 Simultaneous Adsorption and Photo Catalysis

Originally the dyes and other pollutants in aqueous medium were removed by adsorption process. In last few decades the semiconductor catalyzed photo degradation of dyes and other pollutants have gained much attention. Up till now, a number of researchers have reported a vast range of catalysts for degradation of dyes. On the basis of vast data on photo catalytic degradation of dyes, it can be concluded that penetration of radiation through catalyst, electron and hole recombination rate and phase composition of the catalyst, crystallinity of the catalyst, band gap and surface area of the catalyst, and ability of pollutants to adsorb on the surface of catalyst are the parameters which play significant role in the photo catalytic degradation performances (Corma and Garcia 2004). Therefore, the researchers have tried to produce photo catalysts with significant surface area and able to adsorb dyes to a significant extent (Anandan and Yoon 2003). To get catalysts with higher surface area, the catalysts like CdS, ZnO, TiO_2, etc., care synthesized needle like, rod like and tubular with nanocrystalline structures. It is reported that nano tube TiO_2 has much higher surface area compared to bulk TiO_2 (Tayade and Key 2010). Different researchers have reported the synergistic effect of carbon and TiO_2 (Matos et al. 2007; Areerachakul et al. 2007). Similarly, Sun et al. (2016) studied the carbon-TiO_2 as photo catalyst to investigate the synergistic effect. The incorporation of adsorbent with photo catalyst provides a new approach to photo catalytic system. Similarly, Li et al. (2014) have explained the synergistic effect by using g-C_3N_4-ZnO composite. They used this composite for removal of rhodamine B dye in aqueous system. Figure 10.6 explains the whole process.

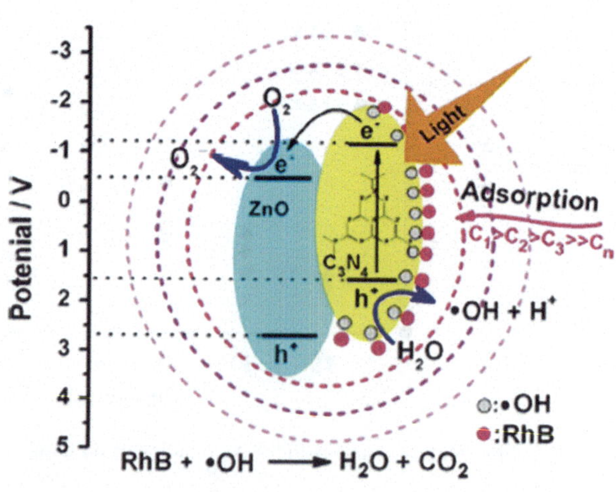

Fig. 10.6 Reversible separation of electron and positive hole and adsorption at composite of g-C_3N_4/ZnO (Li et al. 2014)

References

Aguedach A, Brosillon S, Morvan J, Lhadi EK (2005) Photocatalytic degradation of azo-dyes reactive black 5 and reactive yellow 145 in water over a newly deposited titanium dioxide. Appl Catal B 57:55–62

Ajmal A, Majeed I, Malik RN, Idriss H, Nadeem MA (2014) Principles and mechanisms of photocatalytic dye degradation on TiO_2 based photocatalysts: a comparative overview. RSC Adv 4:37003–37026

Al-Bastaki N (2004) Removal of methyl orange dye and Na_2SO_4 salt from synthetic waste water using reverse osmosis. Chem Eng Process 43:1561–1567

Aleboyeh A, Kasiri MB, Aleboyeh H (2012) Influence of dyeing auxiliaries on AB74 dye degradation by UV/H_2O_2 process. J Environ Manag 113:426–431

Anandan S, Yoon M (2003) Photocatalytic activities of the nano-sized TiO_2 supported Y-zeolites. J Photochem Photobiol C 4:5–18

Areerachakul N, Vigneswaran S, Ngo H, Kandasamy J (2007) Granular activated carbon (GAC) adsorption-photocatalysis hybrid system in the removal of herbicide from water. Sep Purif Technol 55:206–211

Avlonitis SA, Poulios I, Sotiriou D, Pappas M, Moutesidis K (2008) Simulated cotton dye effluents treatment and reuse by nanofiltration. Desalination 221:259–267

Bandala ER, Pelaez MA, Garcia-Lopez JS, Maria DJ, Moeller G (2008) Photocatalytic decolourisation of synthetic and real textile waste water containing benzidine-based azo dyes. Chem Eng Process 47:169–176

Barragan BE, Costa C, Marquez MC (2007) Biodegradation of azo dyes by bacteria inoculated on solid media. Dyes Pigments 75:73–81

Corma A, Garcia H (2004) Zeolite-based photocatalysts. Chem Commun 13:1443–1459

Delee W, O'Neill C, Hawkes FR, Pinheiro HM (1998) Anaerobic treatment of textile effluents: a review. J Chem Technol Biotechnol 73:323–325

Dogan D, Turkdemir H (2005) Electrochemical oxidation of textile dye indigo. J Chem Technol Biotechnol 80:916–923

Fox MA, Dulay MT (1993) Heterogeneous photocatalysis. Chem Rev 93:341–357

Freeman HM (1989) Standard handbook of hazardous waste treatment and disposal, 2nd edn. McGraw-Hill Book Company, New York

Gupta VK, Suhas (2009) Applications of low-cost adsorbents for dye removal — a review. J Environ Manag 90:2313–2342

Hage R, Lienke A (2006) Applications of transition-metal catalysts to textile and wood-pulp bleaching. Angew Chem Int Ed 45:206–222

Hong Q, Hardcastle JL, Mckeown RAJ, Marken F, Compton RG (1999) The 20 kHz sonochemical degradation of trace cyanide and dye stuffs in aqueous media. New J Chem 23:845–849

Huang H, Leung DYC, Kwong PCW, Xiong J, Zhang L (2013) Enhanced photocatalytic degradation of methylene blue under vacuum ultraviolet irradiation. Catal Today 201:189–194

Jo W, Tayade RJ (2014) New generation energy-efficient light source for photocatalysis: LEDs for environmental applications. Ind Eng Chem Res 53:2073–2084

Lee S, Mills A (2004) Detoxification of water by semiconductor photocatalysis. J Ind Eng Chem 10:173–187

Legrini O, Oliveros E, Braun AM (1993) Photochemical processes for water treatment. Chem Rev 93:671–698

Li X, Li M, Yang J, Li X, Hu T, Wang J, Sui Y, Wu X, Kong L (2014) Synergistic effect of efficient adsorption g-C_3N_4/ZnO composite for photocatalytic property. J Phys Chem Solids 75:441–446

Lianos P (2011) Production of electricity and hydrogen by photocatalytic degradation of organic wastes in a photoelectrochemical cell: the concept of the photo fuel cell: a review of a re-emerging research field. J Hazard Mater 185:575–590

Linsebigler AL, Lu G, Yates JT (1995) Photocatalysis on TiO_2surfaces: principles, mechanisms and selected results. Chem Rev 95:735–758

Maezawa A, Nakadoi H, Suzuki K, Furusawa T, Suzuki Y, Uchida S (2007) Treatment of dye wastewater by using photo-catalytic oxidation with sonication. Ultrason Sonochem 14:615–620

Marmagne O, Coste C (1996) Color removal from textile plant effluents. Am Dyestuff Rep 85:15–20

Matos J, Laine J, Herrmann JM, Uzcategui D, Brito J (2007) Influence of activated carbon upon titania on aqueous photocatalytic consecutive runs of phenol photodegradation. Appl Catal B Environ 70:461–469

McLaren J, Williams ID (2015) The impact of communicating information about air pollution events on public health. Sci Total Environ 538:478–491

Mills A, Lee S (2002) A web-based overview of semiconductor photochemistry-based current commercial applications. J Photochem Photobiol A 152:233–247

Mogal SI, Mishra M, Gandhi VG, Tayade RJ (2012) Metal doped titanium dioxide: synthesis and effect of metal ions on physico-chemical and photocatalytic properties. Mater Sci Forum 734:364–378

Namboodri CG, Perkins WS, Walsh WK (1994) Decolorizing dyes with chlorine and ozone: part II. Am Dyestuff Rep 83:17–26

Natarajan TS, Natarajan K, Bajaj HC, Tayade RJ (2013a) Enhanced photocatalytic activity of bismuth-doped TiO_2nanotubes under direct sunlight irradiation for degradation of rhodamine B dye. J Nanopart Res 15:1669

Natarajan TS, Natarajan K, Bajaj HC, Tayade RJ (2013b) Study on identification of leather industry waste water constituents and its photocatalytic treatment. Int J Environ Sci Technol 10:855–864

Natarajan S, Bajaj HC, Tayade RJ (2018) Recent advances based on the synergetic effect of adsorption for removal of dyes from waste water using photocatalytic process. J Environ Sci 65:201–222

Nguyen AT, Juang R (2015) Photocatalytic degradation of p-chlorophenol by hybrid H_2O_2 and TiO_2 in aqueous suspensions under UV irradiation. J Environ Manag 147:271–277

Rai HS, Bhattacharya MS, Singh J, Bansal TK, Vats P, Banerjee UC (2005) Removal of dyes from the effluent of textile and dyestuff manufacturing industry: a review of emerging techniques with reference to biological treatment. Crit Rev Environ Sci Technol 35:219–238

Rao AN, Sivasankar B, Sadasivam V (2009) Kinetic studies on the photocatalytic degradation of direct yellow 12 in the presence of ZnO catalyst. J Mol Catal A Chem 306:77–81

Reddy VLP, Kim KA (2015) Review of photochemical approaches for the treatment of a wide range of pesticides. J Hazard Mater 285:325–335

Saeed M, Adeel S, Shahzad MA, Haq A, Muneer M, Younas M (2015a) Pt/Al_2O_3 catalyzed decolorization of Rhodamine B dye in aqueous medium. Chiang Mai J Sci 43(3):730–744

Saeed M, Ilyas M, Siddique M (2015b) Kinetics of lab prepared manganese oxide catalyzed oxidation of benzyl alcohol in the liquid phase. Int J Chem Kinet 47(7):447–460

Saeed M, Haq A, Muneer M, Adeel S, Hamayun M, Ismail M (2016) Degradation of direct black 38 dye catalyzed by lab prepared Nickel hydroxide in aqueous medium. Global Nest J 18:309–320

Saeed M, Muneer M, Mumtaz N, Siddique M, Akram N, Humayun M (2018a) Ag-Co_3O_4: synthesis, characterization and evaluation of its photo-catalytic activity towards degradation of rhodamine B dye in aqueous medium. Chin J Chem Eng 26(6):1264–1269

Saeed M, Ahmad A, Boddula R, Din I, Haq A, Azhar A (2018b) Ag@Mn_xO_y: an effective catalyst for photo-degradation of rhodamine B dye. Environ Chem Lett 16(1):287–294

Saeed M, Mansha A, Hamayun M, Ahmad A, Haq A, Ashfaq M (2018c) Green synthesis of $CoFe_2O_4$ and investigation of its catalytic efficiency for degradation of dyes in aqueous medium. Z Phys Chem 232:359–371

Shi B, Li G, Wang D, Feng C, Tang H (2007) Removal of direct dyes by coagulation: the performance of preformed polymeric aluminium species. J Hazard Mater 143:567–574

Shinde SS, Bhosale CH, Rajpure KY (2014) Photodegradation of organic pollutants using N-titanium oxide catalyst. J Photochem Photobiol B 141:186–191

Sobana N, Swaminathan M (2007) Combination effect of ZnO and activated carbon for solar assisted photocatalytic degradation of direct blue 53. Sol Energy Mater Sol Cells 91:727–734

Srivastava P, Goyal S, Tayade R (2013) Ultrasound-assisted adsorption of reactive blue 21 dye on TiO_2 in the presence of some rare earths (La, Ce, Pr and Gd). Can J Chem Eng 9999:1–11

Sun Z, He X, Du J, Gong W (2016) Synergistic effect of photocatalysis and adsorption of nano-TiO_2 self-assembled onto sulfanyl/activated carbon composite. Environ Sci Pollut Res Int 23:21733–21740

Tayade RJ, Key DL (2010) Synthesis and characterization of titanium dioxide nanotubes for photocatalytic degradation of aqueous nitrobenzene in the presence of sunlight. Mater Sci Forum 657:62–74

Tayade RJ, Surolia PK, Kulkarni RG, Jasra RV (2007) Photocatalytic degradation of dyes and organic contaminants in water using nanocrystalline anatase and rutile TiO_2. Sci Technol Adv Mater 8:455–462

Tayade RJ, Surolia PK, Lazar MA, Jasra RV (2008) Enhanced photocatalytic activity by silver metal ion exchanged NaY zeolite photocatalysts for the degradation of organic contaminants and dyes in aqueous medium. Ind Eng Chem Res 47:7545–7551

Tayade RJ, Bajaj HC, Jasra RV (2011) Photocatalytic removal of organic contaminants from water exploiting tuned band gap photocatalysts. Desalination 275:160–165

Wang SA (2008) Comparative study of Fenton and Fenton-like reaction kinetics in decolourisation of waste water. Dyes Pigments 76:714–720

Yue B, Zhou Y, Xu J, Wu Z, Zhang X, Zou Y, Jin S (2002) Photocatalytic degradation of aqueous 4-chlorophenol by silica-immobilized polyoxometalates. Environ Sci Technol 36:1325–1329

Zhao D, Chen C, Wang Y, Ma W, Zhao J, Rajh T, Zang L (2008) Enhanced photocatalytic degradation of dye pollutants under visible irradiation on Al(III)-modified TiO_2: structure, interaction, and interfacial electron transfer. Environ Sci Technol 42:308–314

Chapter 11
Removal of Dyes from Waste Water by Micellar Enhanced Ultrafiltration

Amnah Yusaf, Shahid Adeel, Muhammad Usman, Nadia Akram, and Tanveer Hussain Bokhari

11.1 Introduction

11.1.1 Dyes

The dyes are major component of textile industries effluent. The low biodegradability of dyes and other synthetic compounds do not let the biological treatment to get rid of it in effective way and, thus, their presence opposes aerobic biological treatment (Hassani et al. 2008). Lot of work has been done recently on dyes like the microwave radiation effect on the dyeing e.g. with reactive blue 21 dye (Ghaffar et al. 2019), Disperse Yellow 211 dye (Adeel et al. 2018a, b, c) and Reactive Violet H3R (Kiran et al. 2019), Microwave assisted modulation of vat dyeing of cellulosic fiber (Adeel et al. 2018a, b, c), Ultrasonic assisted improved dyeing of cellulosic fabric using Vat Blue 4 (Saeed et al. 2018), and the eco-friendly disperse dyeing of Ultraviolet treated polyester fabric using Disperse Yellow 211 dye (Adeel et al. 2018a, b, c) have been studied. A large quantity of other organics like aromatic hydrocarbons, alkylphenols, aliphatic alcohols, chloroaromatic compounds, and aromatic amines are also present in waste water, in addition to dyes (Dehghani et al. 2011; Pramauro and Prevot 2007). Production of dyes on the earth has been estimated to be more than 10,000 tons each year being consumed in leather industries, additives, cosmetics, paints and ink industry (Abedi and Nekouei 2011). The dyes, being colored compounds, are responsible for far-reaching environmental damage and pose serious threat to aquatic life because suspended dye particles block sunlight and cause decrease in dissolved oxygen in water (DO) and increase in the biological oxygen demand (BOD) of the waste water (Rawal et al. 2018).

Approximately 15% of quantity of used dye is lost during the mechanical procedures for example dyeing, dye assembling, textile finishing, pulp and paper

A. Yusaf · S. Adeel · M. Usman (✉) · N. Akram · T. H. Bokhari
Department of Chemistry, Government College University, Faisalabad, Pakistan

R. A. Bhat et al. (eds.), *Bioremediation and Biotechnology, Vol 2*,
https://doi.org/10.1007/978-3-030-40333-1_11

production (Chen et al. 2011). The strict environment protection rules have invited research to find new ways for treatment of waste water (Das et al. 2008).

11.2 Removal Methods

Several techniques have been utilized till now to remove dyes contaminants from waste water, for example, utilizing adsorption by activated carbon, ozonations, changing pH of solution, nanofiltration, (Purkait et al. 2004c) cloud point separations, polyglycol systems (Korzystka et al. 2003), physical and chemical techniques such as oxidation, coagulation, chlorination, and adsorption onto different supports, reverse osmosis, solvent extraction, and ultrafiltration. Direct eradication of organic compounds by ultrafiltration of aqueous solutions is not achievable. Such a separation can be accomplished utilizing surfactant solutions. Micellar enhanced ultrafiltration (MEUF) is one of the conceivable membrane techniques to remove traces of organic contaminations from aqueous streams (Bielska and Prochaska 2007).

11.3 Ultrafiltration Process

The uses of ultrafiltration membranes are picking up a ton of interest nowadays because of their higher selectivity, thermal stability, and permeation rate. In recent years, different UF membranes were prepared with generally excellent structural integrity, high selectivity, and fouling resistance. The composite UF membranes establish various sorts of thin polymeric layers, for example, polysulfone, cellulose acetate, polyvinyl acetate, polyvinyl pyrrolidone, polydimethylsiloxane, etc. (Jana et al. 2011). The sulfonated and aminated derivatives of polyethersulfone can likewise be applied for the separation of dyes through MEUF process (Pozniak et al. 2009).

11.3.1 Micellar Enhanced Ultrafiltration

This technique has been recommended as one of the useful ways for removal of harmful contaminants especially dyes (Ahmad et al. 2015) using micelles, the active colloidal aggregates of surfactant molecules. Surfactant molecules are amphiphilic in nature, i.e., there are both hydrophobic and hydrophilic moieties in their molecules. Surfactant molecules may be ionic (cationic, anionic, zwitterion) or nonionic. As surfactant concentration increases in solution, its properties deviate from ideality and at the concentration where accumulation of monomers into micelles happens, a pronounced change is observed. This concentration is known as the critical micellization concentration (CMC) (Taşcıoglu 2018). Above the critical micellar concentration, monomers of surfactant will assemble and aggregate in order to form large isotropic clusters called micelles having hydrodynamic diameter greater than the pore size of membrane (Puasa et al. 2011). Organic pollutants are expelled toward interior of micelles (Luo et al. 2010) and get entrapped there. Micelles having incorporated organic dyes are retained by the membrane and a permeate stream going through the membrane has minimum amount of dyes (Ahmad et al. 2015).

Unbound surfactant monomers and dye molecules/ions go through the ultra-filter membrane to the permeate side. In the MEUF procedure, rejection efficiency and permeate flux depend upon the nature of solute and membranes as well as on working conditions (Bade and Lee 2011). MEUF has been also used for recovery of hazardous organic compound or heavy metal ions from waste water (Schwarze et al. 2009). The benefits of MEUF are low operating pressure and low energy requirement. The required pressure expected to drive MEUF ranges from 97 to 587 kPa. MEUF using mixed surfactant systems include the combination of ionic (cationic or anionic) and nonionic surfactants to improve efficiency of MEUF (Puasa et al. 2011). It is to be noticed that MEUF gives surprisingly higher filtration as compared to UF treatment (Deriszadeh et al. 2010). It has been discovered that the effectiveness of dye retention depends upon molecular weight and concentration dyes in feed solutions in addition to hydrophilicity of the membrane (Korzystka and Prochaska 2006).

11.3.2 Selection of Surfactant for Removal of Dyes

Although large number of surfactants are available but we need those having low critical micellar concentration, little or no adsorption to the surface of membrane, large micelles, low cost, greater solubilization limit, high aggregation number, and being biodegradable. Unfortunately no individual surfactant can satisfy all these conditions simultaneously. For the maximum removal of dyes from fluid, the main point has to be very good dye–surfactant interaction. The post-treatment steps, like recovery of surfactant after removal of dye should be ensured (Schwarze 2017).

11.4 Experimental Procedure

In the ultrafiltration cell a membrane of suitable pore size is placed in a compartment. A pump is used to push the aqueous solution towards that membrane. In this setup, two rotameters and a manometer were positioned for monitoring and controlling the rate of permeates flux (Huang et al. 2010).

A cylinder of nitrogen gas is used to apply pressure on membrane and permeate from the bottom of the cell is collected and analyzed using UV/Visible spectrophotometer or atomic absorption spectrophotometer (Purkait et al. 2004a). Washing of membrane is done, with distilled water, after each run. The experiment is usually performed at room temperature (Ahmad and Puasa 2007). Zeta-potential of the micelle of surfactant is controlled by dynamic light scattering and estimated by instrument known as Zetasizer (Huang et al. 2010).

11.5 Types of Membrane

MEUF requires membrane of suitable material, pore size, and hydrophilicity/hydrophobicity depending on nature of dye and other experimental conditions i.e. pH and temperature of the solution (Schwarze 2017). Usually regenerated cellulose (RC), polysulfone, cellulose acetate, polyacrylonitrile, polyethersulfone, polyvinyl difluoride (molecular weight 10,000 and permeability 6.28×10^{-11} m/Pas) PVDF, polyamide (PA), and polytetrafluoroethylene (PFTE) are used (Ahmad and Puasa 2007). Sometimes ceramic membrane is also used because it has the advantage to be used under highly basic or acidic conditions. The main problem with MEUF is membrane contamination. In laboratory scale testing, the membrane is gradually cleaned between two experiments to restore its performance and to remove any impurity. The purpose of the purification process is to eliminate adsorbents i.e. dyes or surfactant molecules. A survey revealed that membranes of hydrophobic nature are preferred as they cause less fouling. In most of cases membranes of low MWCO are used i.e. 5 or 10 kDa but in some cases membranes of 20 or 30 kDa may also be used depending on nature of dye. The washing of membrane is most often accomplished by placing it in distilled water to remove adsorbed surfactant. Sometimes dilute solution of hydrochloric acid and sodium hydroxide may also be used. In addition, combinations of water and minor amounts of alcohols such as 2-propanol are also useful.

The permeability of the membrane is checked after cleaning and the membrane is recycled as long as the deviation of the groundwater flow is less than 5–10% (Schwarze 2017). The polyethersulfone membrane showed slightly better separation than the regenerated cellulose membrane. The high molecular weight dye e.g. Direct Black was rejected almost 100% independent of the nature of membranes (Majewska-nowak 2009).

11.6 Filtration Mode

The filtration cell should not be filled till maximum volume limit. The pressure is usually applied using a cylinder of inert gas like argon or nitrogen gas. A cross-flow filtration process has been found to be much better; consequently, many researchers have used such setup. A driving force is provided by the rotational speed of the rotator in case of the centrifugal micellar enhanced ultrafiltration (Schwarze 2017).

11.7 Ultrafiltration Study

Since dye molecules are smaller than membrane pores, the retention causes the dye adsorption at surface or in the pores membrane. After micelle formation, incorporation of dye takes place leading to its solubilization and, then, held by membrane for

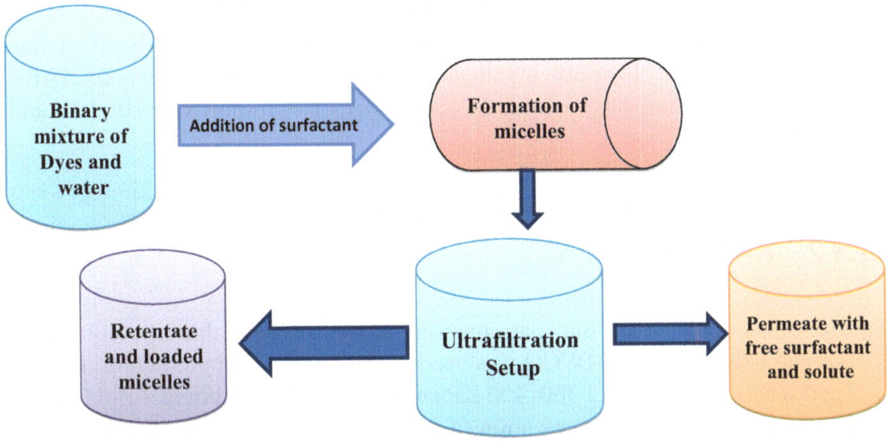

Fig. 11.1 Simple ultrafiltration setup by MEUF

ultrafiltration. The micelles, after some time, get attached to the surface of membrane, bringing about high concentration of surfactant close to the membrane (Zaghbani et al. 2007). Most of researchers have kept trans-membrane pressure (TMP) constant at 1.4 bars. The membrane is completely washed with distilled water. It has been seen that in the presence of dye, without surfactant, the flux decreases with time whereas, in the presence of surfactant reduction in the flux is observed in the beginning and then constant flux is noticed (Zaghbani et al. 2008).

A simple schematic diagram showing micellar enhanced ultrafiltration setup has been represented in Fig. 11.1.

11.8 Removal of Dyes by MEUF

Different dyes have been removed by MEUF using suitable surfactant and membranes are as follows.

11.8.1 Removal of Mordant Black 11

The removal phenomenon was performed by using three different membranes such as cellulose acetate (CQ), polysulfone (PES), and polyvinylidene fluoride (PVDF) and combination of surfactant has been made, so for this purpose cetyl trimethylammonium bromide (CTAB) and oxyethylated coconut fatty acid methyl esters (OMC-1) were used. A trans-membrane pressure of 0.2 MPa was applied. By using cellulose acetate (CQ) membrane along the CTAB, the retention of mordant black

11 was found to be 99.3% whereas for CTAB+ OMC-10, 98.6% was obtained. For polysulfone (PES) with CTAB resulting in a 99.2% for CTAB+ OMC-10 it was observed about 98.5%. For polyvinylidene fluoride (PVDF) with CTAB and CTAB+ OMC-10, the retention was found to be 82.6 and 77.1%, respectively (Bielska and Prochaska 2007).

11.8.2 Reactive Black 5

For the removal of dye by micellar enhanced ultrafiltration, a thin film composite membrane having 10,000 MWCO was used with a cetyl pyridinium chloride CPC as a surfactant. A pressure of 300–400 KPa was applied. The pH ranges 2–11 and at a temperature of 50 °C. The retention of reactive black 5 was found to be 99.7% (Ahmad and Puasa 2007; Majewska-nowak 2009). At a pressure of 1.5 bars and at concentration greater than their CMC value at pH in the range of 2–10, using 30,000 MWCO regenerated cellulose membrane, retention of 83% was observed with CPC while 98% was obtained with CTAB at room temperature of 25 °C (Khosa and Shah 2011).

11.8.3 Crystal Violet Dye

A rejection of 92% was observed using CTAB as a surfactant and a polysulfone ultrafiltration membrane was used. The variation in pressure was made in the range of 50–250 KPa (Rawal et al. 2018).

11.8.4 Removal of Mordant Black 17

Cellulose acetate (CQ), polysulfone (PES), and polyvinylidene fluoride (PVDF) membrane was used with a combination of surfactant such as cetyl trimethylammonium bromide (CTAB) and oxyethylated coconut fatty acid methyl esters (OMC-1). A trans-membrane pressure of 0.2 MPa was applied. By using cellulose acetate (CQ) membrane along the CTAB, the retention of mordant black 17 was found to be 98.9% whereas for CTAB+ OMC-10, 97.6% was obtained. For polysulfone (PES) with CTAB resulting in a 98.9% for CTAB+ OMC-10 it was observed about 97.1%. For polyvinylidene fluoride (PVDF) with CTAB and CTAB+ OMC-10, the retention was found to be 74.6 and 69.6%, respectively (Bielska and Prochaska 2007).

11.8.5 Reactive Orange 16

For the removal of dye by micellar enhanced ultrafiltration, a thin film composite membrane was used with a cetyl pyridinium chloride CPC as a surfactant. A pressure of 300–500 KPa was applied. The pH ranges 2–11 and at a temperature of 50 °C. The retention of reactive black 5 was found to be 99.6% (Ahmad and Puasa 2007); a rejection of 99.98% was also observed with same setup above (Ahmad et al. 2006).

11.8.6 Removal of Methylene Blue

By using single and a combination of surfactants i.e. sodium dodecyl sulfate (SDS) and oxyethylated coconut fatty acid methyl esters (OMC-1), the retention percentage of dye was measured by using cellulose acetate (CQ), polysulfone (PES), and polyvinylidene fluoride (PVDF) membranes. A trans-membrane pressure of 0.2 MPa was applied. By using cellulose acetate (CQ) membrane along the sodium dodecyl sulfate (SDS), the retention of methylene blue was found to be 94.3% whereas for SDS+ OMC-10, 72.3% was obtained. For polysulfone (PES) with SDS resulting in a 91.4% for SDS+ OMC-10 it was observed about 95.5%. For polyvinylidene fluoride (PVDF) with SDS and SDS+ OMC-10, the retention was found to be 88.8 and 73.6%, respectively (Bielska and Prochaska 2007). The methylene blue was also removed by using SDS and polysulfone hollow fiber membrane. The effective membrane area was noted as 0.8 m^2. The pressure was set at 0.03 MPa. The pH ranged 2–13 and a temperature of 5–45 °C. A rejection of 93.3% was observed (Huang et al. 2010). A 95–99% rejection of methylene blue was also observed with a cellulose membrane of MWCO (molecular weight cut off) 10 kDa by using SDS. The pressure was maintained at 1.4 bars. The pH was ranging in between 2 and 11 at a temperature of 30 °C (Zaghbani et al. 2007). A mixed surfactant system has been made using SDS and oxyethylated coconut fatty acid (OMC-10). A rejection of 93–94% was obtained by a hydrophilic membrane made of cellulose along a pressure of 0.2 MPa in the pH range of 2–11 (Bielska and Szymanowski 2006). Using SDS with ultrafiltration regenerates cellulose membrane 30,000 MWCO at room temperature with a pressure of 1.5 bar, a retention of 92.53% was observed at pH ranging 2–10 (Khosa et al. 2011a).

11.8.7 Methyl Orange

Twenty to Sixty percent rejection of methyl orange was found using CTAB with a polysulfone and cellulose acetate membrane at a pressure of 0.2 MPa (Korzystka and Prochaska 2006).

11.8.8 Removal of Eosin (Acid) Dye

The removal of dye by micellar enhanced ultrafiltration was done using 10,000 MWCO ultrafiltration membrane and cetyl pyridinium chloride CPC was used as a surfactant. The molecular weight of eosin dye was 691.86. A pressure of 276–414 KPa was applied and a maximum operating pressure was 518 kpa. The temperature was set at 32 °C. The retention of eosin was found to be 68–74% (Purkait et al. 2004a, b). Rejection of 74% was also observed at a pH of 6–7 and a pressure of 500 KPa (Purkait et al. 2006).

11.8.9 Removal of Safranin T

The molecular weight of Safranin T is 350.85. A rejection percentage of 99% was found by using SDS and a cellulose membrane of MWCO 10KDa at room temperature within a pH range of 2–11 and at a pressure of 1.4 bar (Zaghbani et al. 2008).

11.8.10 Direct Black

Almost a 100% rejection of direct black was found using CTAB with a polyethersulfone membrane at a pressure of 0-1-0.20 MPa (Purkait et al. 2004c).

11.8.11 Eriochrome Blue Black R (EBBR)

Eriochrome blue black R is an anionic dye and is removed by using CTAB with a cellulose membrane with 10,000 MWCO at 30 °C. The pH was maintained in the range of 2–13. Retention of 99% was observed at a pressure of 1.4 bars (Zaghbani et al. 2009).

11.8.12 Alizarin Red S

Rejection of 98.8% was obtained at pressure of 2 bars and in the pH range of 4–9 using CTAB and a regenerated cellulose ultrafiltration membrane at room temperature (Khosa et al. 2011b).

11.9 Factor Effecting MEUF

Number of factors are responsible for the removal efficiency of dyes through micellar enhanced ultrafiltration process which are discussed below:

11.9.1 Effect of pH

The effect of pH varies and is dependent on the type of surfactant and solute. For anionic surfactants, protons are attracted by micelles, at lower pH values, and there is competition between dye cations and protons for adsorption on micelle. Nature and structure of the solute get altered with the pH, and the rejection percentage depends on original structure of solute. An example is the cationic nature of crystal violet dye, having variety of structures and is pH dependent. Under the influence of acidic medium, dye has the two positive charges. Usually the increase of the flow is reported with the pH (Schwarze 2017). At the pH of 3, removal of reactive Black 5 and reactive Orange 16 dyes decreases from 90.64% and 77.50%, while 79.59% and 61.72% at pH 9, respectively (Ahmad et al. 2006). Different experiments have been performed at pH 4 and using 5 ml of Britton–Robinson buffer (Abedi and Nekouei 2011). The retention of methylene blue (MB) was found to be same, nearly 99%, using 50 mM of SDS solution, in the pH range of 2–11. Similar results were obtained in the presence of the cationic surfactant CTAB. The influence of pH on MB retention cannot be elucidated by the ion exchange model. Hydrophobic interactions appear to contribute to binding between MB and SDS micelles. The influence of the pH on the permeation flux has also been investigated. The flow increased as the pH was increased to 2–11. This may be due to deprotonation of the carboxyl groups in the dynamic membrane layer (Zaghbani et al. 2007, 2008).

11.9.2 Effect of Ionic Strength

The discharges from the textile and dyeing industries possess high salt contents that also affect the dye removal because of increase in concentration of free counter ions. In this manner decreasing repulsion force between the charged heads of micelle causes reduction in critical micellar concentration and increase in the aggregation number of micelles. Subsequently, adding little amount of the electrolyte to the ionic nature of surfactants increases the solubilization of the hydrocarbons and decreases that of polar compounds. For methylene blue, the solubilization and afterward the retention rise slightly after the addition of sodium chloride and permeate flux accordingly decreases (Zaghbani et al. 2007). The salt addition, e.g. NaCl,

Na_2SO_4, and $NaHPO_4$, is responsible for changing ionic strength in solutions. Salt addition also causes salting out of many impurities (Zaghbani et al. 2007). The effect of the addition of NaCl on the retention of Safranin T (ST) retention has been investigated, where a decrease in the ST retention has been observed after increasing concentrations of NaCl because of decreases in available space for dye in the outer region of micelle (Zaghbani et al. 2008).

11.9.3 Effect of Electrolyte

The rejection of methylene blue remains constant by increasing the concentration of NaCl i.e. at 99.3%. A reduction in the forces of repulsions between charged surfactant head groups has been observed by the addition of electrolyte in the aqueous medium, causing lower critical micellar concentration and, consequently, an increase in solubilization of hydrocarbons, to be solubilized in the inner core of micelle but a decrease in that of polar solutes that get incorporated in the outer portion of the palisade layer. The rejection capacity of surfactant (SDS) increases, with the concentration of NaCl, from 63.48% in the absence of NaCl to 96.0% at 200 mM or 300 mM of NaCl (Huang et al. 2010).

11.9.4 Effect of Trans-Membrane (TMP) or Operating Pressure

Trans-membrane pressure can be defined as the pressure needed to press the water through membrane or gradient of pressure that exists between the feed sides and the permeate side (Rawal et al. 2018). For micellar enhanced ultrafiltration, the pressure is considered as a driving force and much lower pressure (0–1.0 MPa) is required as compared to nano-filtration. High pressure may affect membrane performance due to membrane compression. In some cases, the membrane is pretreated at high pressure before the main filtration experiment (Puasa et al. 2011; Korzystka et al. 2003; Hankins et al. 2005; Schwarze 2017). It has been observed that by increasing operating pressures, the rejection of dye decreases. Dye rejections for Reactive Black 5 (RB5) and Reactive Orange 16 (RO16) decreased from 97.0% and 98.8% at 300 kPa to 91.1% and 97.4% at 400 kPa. Decreasing dye rejection may be due to fact that at higher operating pressures, micelles are compressed and, therefore, their solubilization capability is decreased. By increasing pressures, the effective driving force for the transportation of solvent is greater that will overcome the resistance of membrane. Thus, more water will pass by the membrane by increasing pressure, resulting in a high permeate flux (Ahmad and Puasa 2007). Although for nonionic surfactant, the rejection of the dyes decreases with the trans-membrane pressure but in case of anionic, the rejection increases at the start and then decreases. The reason is the increase in the thickness of the polarization and the gel layer appears by increasing trans-membrane pressure. Hence, more adsorption of pollutants will be possible on the surface of membrane and hence increase in rejection is observed (Salmani 2016). It has been noticed that MB and SDS rejection decreased a little with the increase of operating pressure, ranging from 99.62 to 62.5% at 0.01 MPa to 98.96 and 57.84% at 0.09 MPa, the permeate flux increased linearly with the operating pressure. By increasing the pressure, more solutes will move towards the surface of membrane, creating a rise in the adsorption on membrane adsorption and a concentration polarization (Huang et al. 2010). So, an increase in operating pressure leads to increase in the gel layer, and therefore rejection of CPC increases at the permeate stream (Ahmad et al. 2006).

The permeate flux decreases with feed CPC concentration at a fixed operating pressure. At constant CPC concentration, the flux increases with pressure. For example, at CPC-to-eosin ratio of 2500, (i.e., feed CPC concentration 25 kg/m^3), the flux increases from 3.5×10^{-6} to 4.25×10^{-6} m^3/m^2 s while the pressure increases from 276 to 414 kPa. With increase in the surfactant concentration, the micelle concentration in the solution increases. This leads to a stronger solubilization of eosin in CPC micelles, which increases the retention of eosin. At a fixed ratio of CPC to eosin, eosin retention increases with decreasing operating pressure. As the permeate concentration increases, the eosin retention observed at high pressure decreases. At constant pressure, the increase in eosin retention is faster at lower CPC-to-eosin concentration ratios and slows as this ratio increases. Therefore, a lower operating pressure and a ratio of CPC-to-eosin feed concentration of 2000 may be considered optimal for the maximum removal of eosin (Purkait et al. 2004b).

11.9.5 Effect of Temperature

Normally MEUF is performed at room temperature (from 20 to 40 °C). Solubility of surfactant, CMC, size, shape, and aggregation number of micelle depend on temperature. The rise in temperature causes an increase in the CMC. Viscosity is another temperature dependent factor responsible to alter the performance of ultrafiltration and it decreases with temperature and flux increases (Zaghbani et al. 2007). The increase in solubility of micellar solution with temperature results in a high extraction of dye. The results described that the effects of operating temperature on the efficiency of dye extraction increases with temperature at various concentration of the dye (20, 50, and 100 mg L^{-1}) till 75 °C then becomes constant (Abedi and Nekouei 2011).

11.9.6 Effect of the Nature of Surfactant

Rejection of methylene blue has been studied in the presence of cationic (CTAB), anionic (SDS), and nonionic (TX-100) surfactants, and the retention of 2.10, 2.82, and 99.45% was obtained, respectively. So from these results it is clear that a higher retention was observed due to anionic surfactant SDS owing to opposite charge of dye and surfactant. So, it is concluded that for retention of dyes, the electrostatic interaction between dye and surfactant plays important role (Zaghbani et al. 2007).

11.9.7 *Effect of Concentration of Surfactant*

It has been observed that, as the concentration of surfactant (CPC) increases, the dye rejection also increases. At the 1000 ppm concentration of CPC, the highest rejections of dyes (RB5 and RO16) were found to be 99.7% and 99.6%, respectively. At operating pressure of 300 kPa, a decrease in the permeate flux was observed by increasing the concentration of surfactant (CPC). Comparing to reactive black 5, the flux of reactive orange 16 was found to be very high at a same concentration of surfactant i.e. 0.500 g/L due to higher molecular weight of RB5 dye (MW 991.82) in comparison to RO16 (MW 617.54). Low molecular weight dyes can pass through the membrane quiet easily, whereas dyes having high molecular weight will remain present on the surface of membrane. Some of particles of dye were deposited on the surface of membrane, offering resistance for solvent flux and leading to a decrease in the flux of RB5 dye as compared to RO16 (Ahmad and Puasa 2007). In a suitable pressure range and in the absence of surfactants, the concentration of permeate of dye started from 5 to 6 ppm and the concentration further increases (up to 9 ppm) because of the forced convection of dyes through pores of membrane and the concentration reached 7–9 ppm. So, in both cases, with and without surfactants at low operating pressure, minimum concentration is present leading to maximum retention. When the micelles of CPC is present, the extent of the unsolubilized dyes governs the concentration of permeate (Purkait et al. 2004a). As compared to nonionic surfactant, a more permeate flux has been observed for anionic (Salmani 2016). When the concentration of surfactant (SDS) is lower than the critical micellar concentration, the obtained results were credited to three reasons: (1) the reactive precipitates of dye (MB) and surfactant (SDS), aggregated and gathered forming a cake on the surface of membrane as well as in the pores of membrane and (2) the fouling of the membrane by a MB adsorption. Whereas, above the CMC at high concentrations of SDS, the deposited layer was due to aggregates of micelle and dye molecules causing more resistance to the flux of solvent through membrane (Huang et al. 2010). Solute having low solubility in water showed a greater solubilization constant compared to solute with high water solubility in a micellar solution. Reactive orange O16 have a high tendency to be solubilized in micelles as compared to Reactive Black 5 and, thus, RO16 have a greater rejection percentage. The surfactant present in the filtrate can be precipitated and recovered and, thus, reused and recycled (Ahmad et al. 2006). It has been observed at a fixed temperature and feed dye concentration, the extraction efficiency of dye increases with surfactant concentration (Abedi and Nekouei 2011). Spectral studies revealed that hydrophobic and electrostatic forces are the reason behind the interaction between ionic dyes and surfactant and, thus, surfactant causes change in spectra of dyes. For the oppositely charged surfactants, the interaction of dyes is mainly of a Coulombic nature. Hydrophobic interaction has been able to overcome the electrostatic repulsion, if the interaction between dyes and surfactants of same charge occurs. If only the nonionic surfactant is present, just the hydrophobic interaction arises. As the hydrophobic interaction is weak as compared to electrostatic repulsion, the dye will not interact with surfactant. If no interaction is present between dye and surfactant, then it was due to the weakness of hydrophobic interaction (Zaghbani et al. 2007).

11.9.8 Effect of Concentration of Dyes

It has been noticed that by increasing dye concentration, permeate flux decreases due to adsorption of dye on membrane and, thus, formation of deposited layer. In most of cases almost 99% rejection of dye has been observed, showing that the concentration of dye in the feed has little effect on the rejection efficiency. The reason behind greater dye rejection is large scale solubilization of dye molecules in micelles. Sometimes presence of dye decreases CMC and increases aggregation number of micelles and, thus, solubilization capacity is enhanced (Huang et al. 2010). The change in dye concentrations, at constant surfactant concentration, results in almost constant retention efficiency (Purkait et al. 2004b).

11.9.9 MEUF Using Mixed Micellization

Anionic surfactants provide the most suitable and cost-effective micellar media to remove cations of metals and dyes from waste water by MEUF. Although nonionic surfactants have lower CMC than ionics but they are not able to attract ions. However, addition of small quantity of nonionic surfactant into solution of ionic surfactants can facilitate solubilization because they reduce repulsions between ionic heads and thus decrease CMC and, therefore, increase aggregation number and size of micelle. The mixed micellar solution, being formed by the combination of two or more surfactants provides better media for removal of dyes. Although presence of nonionic surfactants decrease charge density and thus cation attracting ability of micelle but they increase solubilization capacity to much greater extent (Schwarze 2017).

11.9.10 Effect of Coagulant Concentration

By increasing the dose of coagulant, the percentage removal efficiency of dye increases. For dyes, i.e. Reactive Black 5 (RB5) and Reactive Orange 16 (RO), in the dosage range of 1.0 and 4.0 g/L, respectively, a major enhancement in the removal efficiency was found. At the pH of 3 and with a concentration of coagulant between 4.5 and 9.0 g/L for RB5 and RO16, the removal percentage has been observed to be 95.61 and 79.39%, respectively. For maximum removal, the required concentration of coagulant is low for RB5 as compared to RO16. At a very high dose, Nalco coagulant represents a reversal effect due to the coagulation charge reversal phenomenon where the colloidal system becomes destabilized because coagulant charge becomes greater than total charge on colloidal system (Ahmad et al. 2006).

11.10 Summary (in Tabulated Form)

Removal of a variety of dyes by MEUF has been studied, and the effects of various factors like the type of surfactant, the effect of pH, temperature, pressure, type of membrane, etc. have been reviewed. The removal efficiency is summarized in Table 11.1.

Table 11.1 Percentage rejection of dyes by MEUF

Dyes	Membrane	Surfactant	Pressure	Rejection percentage	References
Mordant black 11	Cellulose acetate (CQ)	CTAB	0.2 MPa	99.3	Bielska and Prochaska (2007)
	Cellulose acetate (CQ)	CTAB+ OMC-10		98.6	
	Polysulfone (PES)	CTAB		99.2	
	Polysulfone (PES)	CTAB+ OMC-10		98.5	
	Polyvinylidene fluoride (PVDF)	CTAB		82.0.6	
	Polyvinylidene fluoride (PVDF)	CTAB+ OMC-10		77.1	
Reactive black 5	Thin film composite membrane having 1000 MWCO	CPC	300–400 KPa	99.75	Ahmad and Puasa (2007) and Ahmad et al. (2006)
Crystal violet dye	Polysulfone ultrafiltration membrane	CTAB	50–250 KPa	92	Rawal et al. (2018)
Mordant black 17	Cellulose acetate (CQ)	CTAB	0.2 MPa	98.9	Bielska and Prochaska (2007)
	Cellulose acetate (CQ)	CTAB+ OMC-10		97.6	
	Polysulfone (PES)	CTAB		98.9	
	Polysulfone (PES)	CTAB+ OMC-10		97.1	
	Polyvinylidene fluoride (PVDF)	CTAB		74.6	
	Polyvinylidene fluoride (PVDF)	CTAB+ OMC-10		69.6	
Reactive orange 16	Thin film composite membrane	CPC	300–500 KPa	99.6	Ahmad and Puasa (2007)

(continued)

Table 11.1 (continued)

Dyes	Membrane	Surfactant	Pressure	Rejection percentage	References
Methylene blue	Cellulose acetate (CQ)	SDS	0.2 MPa	94.3	Zaghbani et al. (2007)
	Cellulose acetate (CQ)	SDS+ OMC-10		72.3	
	Polysulfone (PES)	SDS		91.4	
	Polysulfone (PES)	SDS+ OMC-10		75.5	
	Polyvinylidene fluoride (PVDF)	SDS		88.8	
	Polyvinylidene fluoride (PVDF)	SDS+ OMC-10		73.6	
	Ultrafiltration regenerate cellulose membrane 30,000 MWCO	SDS	1.5 bar	92.53	Khosa et al. (2011a)
	Hydrophilic membrane made of cellulose	Oxyethylated coconut fatty acid (OMC-10) and SDS	0.2 MPa	93–94	Bielska and Szymanowski (2006)
	Cellulose membrane of MWCO 10 kDa	SDS	1.4 bars	95–99	Zaghbani et al. (2007)
	Polysulfone hollow fiber membrane	SDS	0.03 MPa	93.3	Huang et al. (2010)
Methyl orange	Polysulfone and cellulose acetate membrane	CTAB	0.2 MPa	20–60	Korzystka and Prochaska (2006)
Eosin (acid) dye	1000 MWCO ultrafiltration membrane	CPC	276–414 KPa	74	Purkait et al. (2004a, b, 2006)
Safranin T	Cellulose membrane of MWCO 10KDa	SDS	1.4 bar	99	Zaghbani et al. (2008)
Direct black	Polyethersulfone membrane	CTAB	0–1-0.20 MPa	100	Purkait et al. (2004c)
Eriochrome blue black R (EBBR)	Cellulose ultrafiltration membrane/10,000 MWCO	CTAB	1.4 bars	99	Zaghbani et al. (2009)
Alizarin red S	Cellulose ultrafiltration membrane	CTAB	2 bars	98.8	Khosa et al. (2011b)

11.11 Conclusion

Industrial waste having dye content are source of water pollution and cause irreparable damage to ecosystem. Conventionally, the waste water is being treated by several physical and chemical processes, extraction, adsorption, chlorination, oxidation, nanofiltration and ultrafiltration, etc. Among the said techniques, micellar enhanced ultrafiltration (MEUF) is a wonderful technique for dye removal from the aqueous stream. The key to success of MEUF is the selection of most suitable micellar media, preferably mixed micellar system, and optimum operating conditions. Generally, the surfactant with charge opposite to that of pollutant is the best choice so as to develop a strong dye–surfactant interaction. Micellar solution has proved itself to be an effective medium to capture dye molecules. In this way true solution of dye is converted to colloidal solution and, resultantly, can be filtered using ultrafiltration membrane of suitable pores size. The effects of various factors like trans-membrane pressure, pH, temperature, ionic strength, nature of surfactant, concentration of dye, concentration of surfactant, etc. have been studied. It has been concluded that low trans-membrane pressure, low pH, room temperature, high ionic strength, combination of ionic and nonionic surfactants, and surfactant concentration higher than CMC are optimum factors to accomplish maximum dye removal.

References

Abedi S, Nekouei F (2011) Removal of direct yellow 12 from water samples by cloud point extraction using triton X-100 as nonionic surfactant. Eur J Chem 8(4):1588–1595

Adeel S, Khan SG, Shahid S, Saeed M, Kiran S, Zuber M, Suleman M, Akhtar N (2018a) Sustainable dyeing of microwave treated polyester fabric using disperse yellow 211 dye. J Mex Chem Soc 62(1):1–9

Adeel S, Saeed M, Ajaz A, Rehman FUR, Salman M, Kamran M, Zuber M, Iqbal M (2018b) Microwave assisted modulation of vat dyeing of cellulosic fiber: improvement in color characteristics. J Nat Fibers 15(4):517–526

Adeel S, Shahid S, Khan SG, Rehman FUR, Muneer M, Zuber M, Akthar N (2018c) Eco-friendly disperse dyeing of ultraviolet treated polyester fabric using disperse yellow 211 dye. Pol J Environ Stud 27(5):1935–1939

Ahmad AL, Puasa SW (2007) Reactive dyes decolourization from an aqueous solution by combined coagulation/micellar-enhanced ultrafiltration process. Chem Eng J 132:257–265

Ahmad AL, Puasa SW, Zulkali MMD (2006) Micellar-enhanced ultrafiltration for removal of reactive dyes from an aqueous solution. Desalination 191:153–161

Ahmad A, Mohd-Setapar SH, Chuong CS, Khatoon A, Wani WA, Kumar R, Rafatullah M (2015) Recent advances in new generation dye removal technologies: novel search for approaches to reprocess wastewater. RSC Adv 5(39):30801–30818

Bade R, Lee S (2011) A review of studies on micellar enhanced ultrafiltration for heavy metals removal from wastewater. J Water Sustain 1:85–102

Bielska M, Prochaska K (2007) Dyes separation by means of cross-flow ultrafiltration of micellar solutions. Dyes Pigments 74:410–415

Bielska M, Szymanowski J (2006) Removal of methylene blue from waste water using micellar enhanced ultrafiltration. Water Res 40:1027–1033

Chen Y, Tsao TM, Wang M (2011) Removal of crystal violet and methylene blue from aqueous solution using soil nano-clays. In: International Conference on Environ Sci and Engg, vol 8. IACSIT Press, Singapore, pp 252–254

Das C, Maity P, DasGupta S, De S (2008) Separation of cation-anion mixture using micellar-enhanced ultrafiltration in a mixed micellar system. Chem Eng J 144(1):35–41

Dehghani MH, Jaafari J, Alghasi A, Porkar G (2011) Using medium pressure ultraviolet reactor for removing azo dyes in textile wastewater treatment plant. World Appl Sci J 12:797–802

Deriszadeh A, Husein MM, Harding TG (2010) Produced water treatment by micellar-enhanced ultrafiltration. Environ Sci Technol 44:1767–1772

Ghaffar A, Adeel S, Habib N, Jalal F, Haq AU, Munir B, Ahmad A, Jahangir M, Jamil Q (2019) Effect of microwave radiation on cotton dyeing with reactive blue 21 dye. Polish J Environ Stud 28(3):1687–1691

Hankins N, Hilal N, Ogunbiyi OO, Azzopardi B (2005) Inverted polarity micellar enhanced ultrafiltration for the treatment of heavy metal polluted wastewater. Desalination 185:185–202

Hassani H, Mirzayee R, Nasseri S, Borghei M, Gholami M, Torabifar B (2008) Nanofiltration process on dye removal from simulated textile wastewater. Int J Environ Sci Technol 5(3):401–408

Huang JH, Zhou CF, Zeng GM, Li X, Niu J, Huang HJ, He SB (2010) Micellar-enhanced ultrafiltration of methylene blue from dye wastewater via a polysulfone hollow fiber membrane. J Membr Sci 365:138–144

Jana S, Saikia A, Purkait MK, Mohanty K (2011) Chitosan based ceramic ultrafiltration membrane: preparation, characterization and application to remove Hg(II) and As(III) using polymer enhanced ultrafiltration. Chem Eng J 170(1):209–219

Khosa MA, Shah SS (2011) Micellar enhanced ultrafiltration of reactive black 5 from aqueous solution by cationic surfactants. J Dispers Sci Technol 32:1002–1007

Khosa MA, Shah SS, Nazar MF (2011a) Application of micellar enhanced ultrafiltration for the removal of methylene blue from aqueous solution. J Dispers Sci Technol 32:260–264

Khosa MA, Shah SS, Nazar MF (2011b) UV-visible spectrometric study and micellar enhanced ultrafiltration of alizarin red s dye. J Dispers Sci Technol 32:1634–1640

Kiran S, Adeel S, Rehman FUR, Gulzar T, Jannat M, Zuber M (2019) Ecofriendly dyeing of microwave treated cotton fabric using reactive violet H3R. Global NEST J 21(1):43–47

Korzystka B, Prochaska K (2006) Application of micellar ultrafiltration for separation of dyes from aqueous solutions. Ars Separatoria 11:60–62

Korzystka B, Adamczak H, Sobczyńska A, Szymanowski J (2003) Ultrafiltration characteristics of colloid solutions containing oxyethylated methyl dodecanoate, hexadecyltrimethylammonium bromide and selected phenols as pollutants. Colloids Surf A Physicochem Eng Asp 212:175–183

Luo F, Zeng GM, Huang JH, Zhang C, Fang YY, Qu YH, Zhou CF (2010) Effect of groups difference in surfactant on solubilization of aqueous phenol using MEUF. J Hazard Mater 173:455–461

Majewska-nowak K (2009) Ultrafiltration of dye solutions in the presence. Environ Prot Eng 35(4):111–120

Pozniak G, Pozniak R, Wilk KA (2009) Removal of dyes by micellar enhanced ultrafiltration. Chem Eng Trans 17:1–5

Pramauro E, Prevot AB (2007) Solubilization in micellar systems. Analytical and environmental applications. Pure Appl Chem 67(4):551–559

Puasa SW, Ruzitah MS, Sharifah ASAK (2011) An overview of micellar-enhanced ultrafiltration in wastewater treatment process. In: Proceedings of international conference on environment and industrial innovation, vol 12. IACSIT Press, Singapore, pp 167–172

Purkait MK, DasGupta S, De S (2004a) Removal of dye from wastewater using micellar-enhanced ultrafiltration and recovery of surfactant. Sep Purif Technol 37:81–92

Purkait MK, DasGupta S, De S (2004b) Resistance in series model for micellar enhanced ultrafiltration of eosin dye. J Colloid Interface Sci 270:496–506

Purkait MK, Vijay SS, DasGupta S, De S (2004c) Separation of Congo red by surfactant mediated cloud point extraction. Dyes Pigments 63:151–159

Purkait MK, DasGupta S, De S (2006) Micellar enhanced ultrafiltration of eosin dye using hexadecyl pyridinium chloride. J Hazard Mater 136:972–977

Rawal K, Kothari I, Sinha MK (2018) Micellar-enhanced ultrafiltration for removal of crystal violet dye from an aqueous solution. Proceedings of Academics World 89th International Conference 2–5

Saeed M, Abdullah A, Khan SG, Habib N, Kamran M, Zuber M (2018) Ultrasonic assisted improved dyeing of cellulosic fabric using Vat Blue 4. J Nat Fibers 17:1–8

Salmani HJ (2016) Micellar-enhanced ultrafiltration of dairy wastewater with anionic and nonionic surfactants. Desalin Water Treat 57:12956–12962

Schwarze M (2017) Micellar-enhanced ultrafiltration (MEUF)-state of the art. Environ Sci Water Res Technol 3(4):598–624

Schwarze M, Rost A, Weigel W (2009) Schomacker R selection of systems for catalyst recovery by micellar enhanced ultrafiltration. Chem Eng Process Intensif 48:356–363

Taşcıoglu S (2018) Nanofiltration mediated by surfactant micelles: micellar- enhanced ultrafiltration. In: Nanofiltration. Intech, London, pp 85–117

Zaghbani N, Hafiane A, Dhahbi M (2007) Separation of methylene blue from aqueous solution by micellar enhanced ultrafiltration. Sep Purif Technol 55:117–124

Zaghbani N, Hafiane A, Dhahbi M (2008) Removal of Safranin T from wastewater using micellar enhanced ultrafiltration. Desalination 222:348–356

Zaghbani N, Hafiane A, Dhahbi M (2009) Removal of Eriochrome blue black R from wastewater using micellar-enhanced ultrafiltration. J Hazard Mater 168:1417–1421

Chapter 12
Biofilm: An Innovative Modern Technology for Aquatic Pollution Remediation

Shakeel Ahmad Dar, Farooq Ahmad Lone, Shahid Ahmad Dar, Rouf Ahmad Bhat, Ishrat Bashir, Shafat Ahmad Mir, and Zubair Ahmad Dar

12.1 Introduction

Water is the basic necessity of life and its use is increasing globally by 1% per annum. According to Burek et al. (2016), increase in water use at this rate will lead to the total increase of 20–30% by 2050. Water bodies are seriously polluted due to increase in the rate of urbanization, industrialization, and population explosion (Qian et al. 2007). Pollution of surface water bodies mainly occurs by domestic waste effluent, surface runoff due to precipitation, and storm drainages. Besides this, industrial activities are responsible for creating tons of waste globally and its ultimate disposal into the aquatic environment (Edwards and Kjellerup 2013). World Health Organization has attributed 30% of all global diseases and 40% of global deaths to the water pollution (Kantawanichkul et al. 2009). There are different types of contaminants present in water such as detergents, pesticides, fats, and oils; microorganisms such as bacteria, viruses, protozoa, and helminthes; nutrients such as nitrogen, phosphorus, and ammonia, etc.; metals and inorganic materials such as mercury, lead, cadmium, nickel, and hydrogen sulfide, etc. (Sehar and Naz 2016). All such contaminants are responsible for the deterioration of water quality and make it unfit for use which in turn adds to the global water stress. Therefore the need of the time is to restore and remediate polluted water bodies that will ultimately help in fulfilling the demands of growing population. Different techniques have been adopted by scientists throughout the world for the remediation of pollutants from water bodies. One such technique is bioremediation of pollutants from the water. The key tools

S. A. Dar · F. A. Lone · R. A. Bhat (✉) · I. Bashir · S. A. Mir · Z. A. Dar
Sher-e-Kashmir University of Agricultural Sciences and Technology of Kashmir, Srinagar, Jammu and Kashmir, India

S. A. Dar
PG Department of Environmental Science, University of Kashmir, Srinagar, Jammu and Kashmir, India

R. A. Bhat et al. (eds.), *Bioremediation and Biotechnology, Vol 2*, https://doi.org/10.1007/978-3-030-40333-1_12

used for decontamination of water involve chemical precipitation, electrochemical deposition, evaporation, membrane process, ion exchange, and activated carbon adsorption. But the applicability of these methods is usually in question due to certain limitations such as high operational cost, inefficiency, and limited use. Therefore, the development of innovative, effective, sustainable, and low-cost technique for the remediation of the environmental pollutants has become necessary (Das et al. 2012). One such technique is the bioremediation of pollutants. "Bioremediation is the application of biological processes for the removal of hazardous pollutants present in the environment (Gianfreda and Rao 2004)." In the year 1972, bioremediation was used for the first time in Pennsylvania for the removal of gasoline leaked from the pipeline (Wang et al. 2012). Bioremediation may be natural or artificial and may find its applicability as in-situ or ex-situ. Among bioremediation techniques biofilm remediation of pollutants has recently gained much more importance particularly in remediating xenobiotic compounds from water (Das et al. 2012). The term biofilm was coined by Costerton et al. (1978). Biofilm can be defined as "An as an aggregation of bacteria, algae, fungi and protozoa enclosed in a matrix consisting of a mixture of polymeric compounds, primarily polysaccharides generally referred as extracellular polymeric substances (EPS) (Das et al. 2012)." Biofilm based bioremediation techniques of water are much better due to the sessile nature of biofilms and their ability to tolerate the extremely stressful conditions and the very nature of accumulation and stabilization of contaminants. In this chapter, emphasis has been laid on the use of aquatic biofilm community as an innovative strategy for remediation of pollutants from water.

12.2 Benefits of Using Biofilms for Remediation of Pollutants

The diverse group of microorganisms in a biofilm is surrounded by extracellular polymeric substances. They easily communicate and there is also exchange of food, materials, and energy which ultimately helps the community to tolerate variety of external pressures. This range of features demonstrates a multifaceted and strong mechanism to diverse kinds of cells working as an organ (Fux et al. 2005). Biofilms are sessile in nature and therefore continuously exposed to the contaminants in water bodies due to which these communities have developed the potential to degrade contaminants (Dar and Bhat 2020). The potential to tolerate and detoxify hazardous contaminants in water is of prime attention. While applying biofilms for bioremediation and biotransformation of contaminants, the chemicals to which biofilms are exposed may be in such concentrations that may cause severe damage to microorganisms living in isolation. Thus the accumulation of microbes in sessile or moving biofilms has the main benefit of augmented tolerance towards changing environmental conditions (Hall-Stoodley et al. 2004; Edwards and Kjellerup 2013). Other major characteristics that make biofilms suitable for bioremediation are:

1. their stress tolerance ability against toxic chemicals, changing nutrient conditions, etc. (Boles et al. 2004; Accinelli et al. 2012)
2. diverse metabolism and symbiotic relations for utilization of waste from surrounding environments (Boles et al. 2004; Chen et al. 2013)
3. their unique physical structure provided with channels of water for the transport of different nutrients, metals, and waste products (Latch et al. 2003; Wang et al. 2011; Xiao et al. 2012; Edwards and Kjellerup 2013)
4. help in treatment of xenobiotic substances such as surfactants and nutrient cycling (Rodriguez and Bishop 2008; Seo et al. 2009; Galiana et al. 2008; Verhagen et al. 2011).

12.3 Application of Biofilms for the Bioremediation of Aquatic Pollutants

12.3.1 Biofilm Based Remediation of Heavy Metals

Heavy metal remediation from aquatic environment can be achieved by immobilization of metals, biotransformation, bio-stabilization, complexation, etc. Researchers have reported the successful removal of heavy metals by using biofilms. Removal of Cr(III) by bacterial biofilm in a continuous flow biofilm reactor has been reported by Sundar et al. (2011). *Bacillus subtilis* and *Bacillus cereus* biofilms developed on coarse sand have the capacity to remediate approximately 98% of Cr(III) (Sundar et al. 2011; Das et al. 2012). A study on Cr(VI), Cd(II), Fe(III), and Ni(II) removal from the polluted water by using biofilm forming *Escherichia coli* grown on NaY zeolite proved to be an effective tool for the remediation of metal pollutants (Cristina et al. 2009a, b). The common representative groups of biofilm forming microbes on different substrates with their potential of remediation of specific types of heavy metals are presented in Table 12.1.

Table 12.1 Remediation of heavy metals by biofilm forming microbes on different substrates

Heavy metal remediated	Biofilm forming microbes	References
Cr(III)	Biofilms of "*Bacillus subtilis* and *Bacillus cereus*" on coarse sand substrate	Sundar et al. (2011)
Cr(VI), Cd(II), Fe(III), Ni(II)	"*Escherichia coli*" biofilm grown on zeolite NaY substrate	Cristina et al. (2009a, b)
Cr(VI)	Biofilm of "*Escherichia coli*" used on granulated activated carbon substrate	Rabei et al. (2009)
Zn(II)	"*Pseudomonas putida*" biofilm sorption of a Zinc metal by forming Zn-phosphoryl complex	Brandy et al. (2005)
Cu(II), Zn(II), Cd(II)	Activated sludge from a sewage	Costley and Wallis (2001)
Cd(II), Cu(II), Zn(II), Ni(II)	"*Pseudomonas* sp. NCIMB (11592)" biofilm bound on granular activated carbon (GAC) substrate by extracellular polymeric substance for metal and organic pollution removal	Scot et al. (1995)

12.3.2 Mechanism Behind Metal-Microbe Interaction

Microbes interact with metals by different ways and the classification of interaction between the two is based on type of metabolism involved for the uptake of metals. Thus on the basis of metal-microbe interaction, the mechanism of heavy metal remediation is divided into active and passive interactions (Alluri et al. 2007; Gupta and Diwan 2017). The benefit of metal-microbe interaction is that metal is changed into less toxic, immobile, and unusable form and is no longer available to cause injury. The capacity of a microbe to resist or tolerate the metal contamination in a water body may be specific such as bio-methylation of elemental mercury into more toxic methyl mercury or non-specific such as formation of cysteine rich protein by metallothioneins by *Cyanobacterium synechococcus*, *E. coli*, *Pseudomonas putida*, etc. (Rajendran et al. 2003; Gupta and Diwan 2017). Similarly in a constructive non-specific mechanism, a slime layer or exopolysaccharide substance is produced which is the main characteristic of biofilms. The metal-microbe mechanism and various ways of fixing of metal pollutants are shown in Fig. 12.1 (Gupta and Diwan 2017). Biosorption of metal by biofilms is beneficial over other mechanisms due to its ability to work passively without the help of metabolic activity and may also work with live and dead biomass. The advantage of dead cells over living cells is that dead cells are not influenced in terms of viability. Besides, dead cells are free from any hazard created by live cells.

12.3.3 Mechanism of Extracellular Polymeric Substance (EPS) to Remediate Metal Pollutants

Extracellular polymeric substances have the potential to show the phenomenon of biosorption by attracting positively charged metal ions towards its negatively charged surface. There are number of functional groups and non-carbohydrate groups such as acetamido group of chitin, structural polysaccharides of fungi, amine groups, sulfhydryl groups, carboxylic groups, proteins, phosphate, hydroxyl groups that ultimately give the negative charge to EPS (Wang et al. 2010; Gupta and Diwan 2017). In comparison to homopolysaccharides, extracellular heteropolysaccharides are frequently polyanionic due to presence of certain functional groups with polysaccharide mainstay (Pal and Paul 2010). The mechanism of sorption happens by ion exchange, complexation, precipitation, etc. (DePhilippis et al. 2011; Wang et al. 2010). As already discussed, biofilm forming microbes are potential players of remediating metal pollution from polluted ecosystems. *Herminiimonas arsenicoxydans* has the ability to induce biofilm formation and remediate arsenic through EPS up to a tolerance limit of 5 mM metal ion concentration (Marchal et al. 2010; Muller et al. 2006; Muller et al. 2007, p. 104). *Thiomonas sp.* CB2 strains remediate arsenic by using its exopolysaccharide and induce biofilm formation

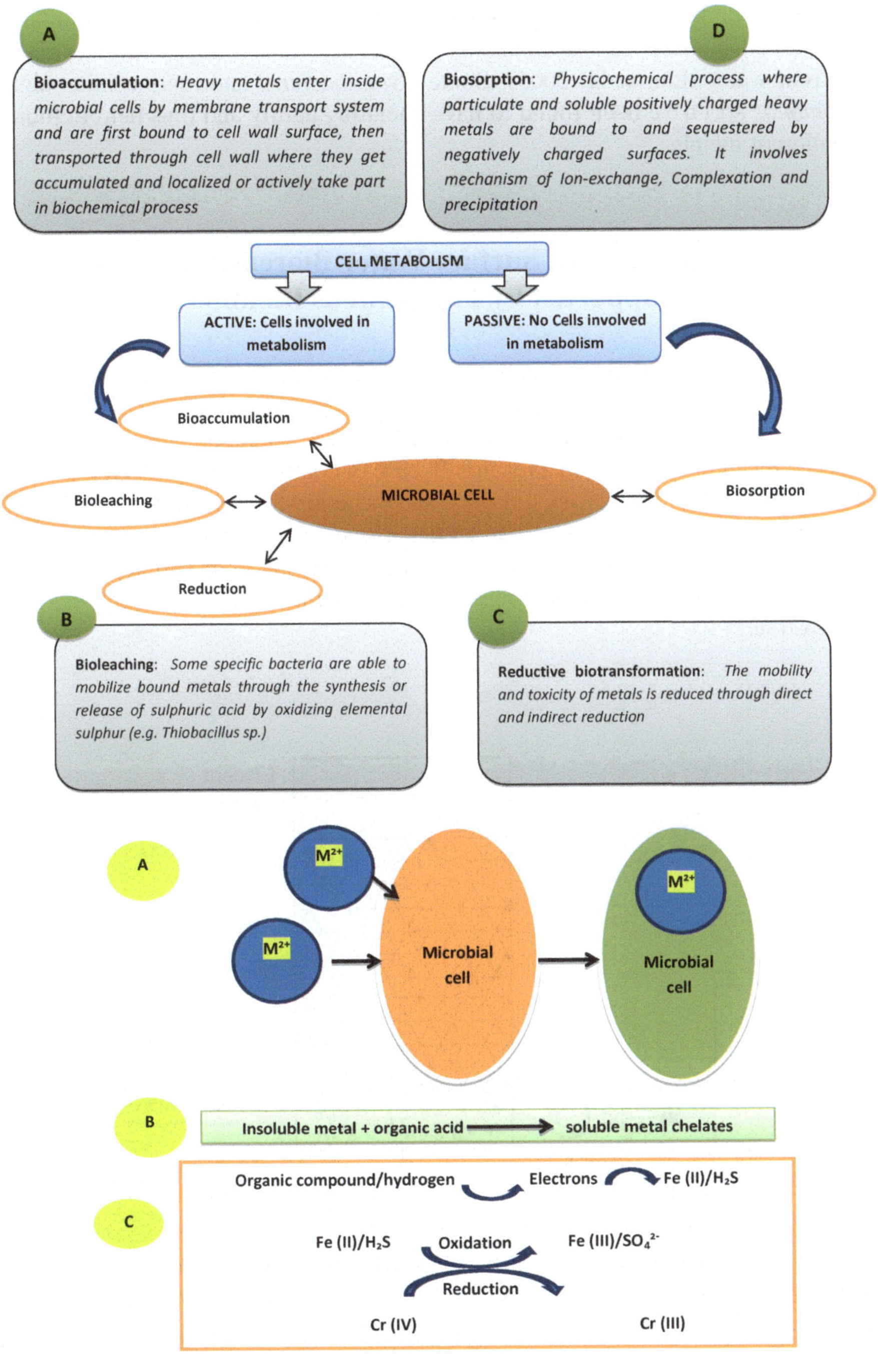

Fig. 12.1 Microbial-metal interaction mechanism (Active and Passive) showing bioaccumulation, biosorption, bioleaching, and redox arbitrated reactions for metal remediation (Gupta and Diwan 2017)

under stress conditions created by arsenic pollution (Marchal et al. 2011). Similarly, EPS and cell biomass of *Pseudomonas* isolates activated sludge scavenge metal ions from water (Lau et al. 2005). Some *Cyanobacteria* and marine microalgae (*Halomonas* sp.) have been found to have chelating ability and thus help in bioremediation of metals.

12.4 Innovative Way of Surface Water Bioremediation by Biofilms Supported by Filamentous Bamboo

Cao et al. (2012) modified batch and continuous flow bioreactor by introducing in it biofilm laden filamentous bamboo (Fig. 12.2). They reported that biofilm can remove COD, Cr, up to 66.1% in a batch hybrid reactor which can further be improved by adding glucose in a reactor. In case of continuous flow reactor, the rate of removal of COD, Mn, NH_4-N, turbidity, and bacteria was found to be 11.2–74.3%, 2.2–56.1%, 20–100%, and more than 88.6%, respectively. They further concluded that water surfaces with refractory organic pollution, low transparency, and high nitrogen pollution can easily be remediated by using biofilm on filamentous bamboo substrate.

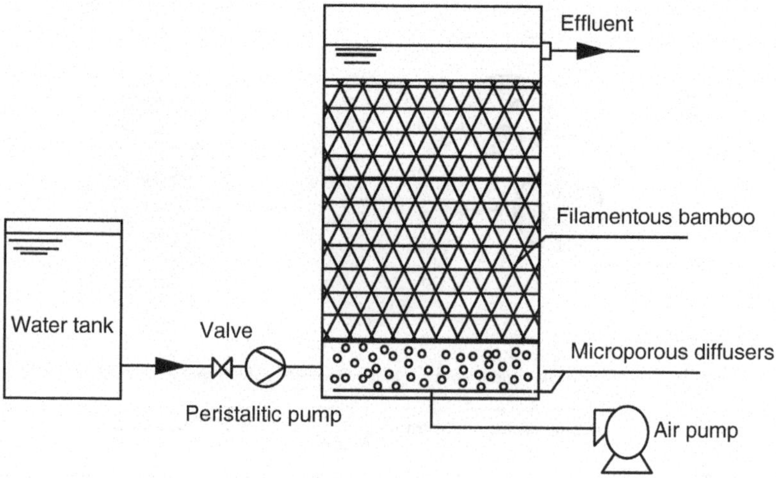

Fig. 12.2 Experimental setup for biofilm based surface water treatment plant with filamentous bamboo substrate used by Cao et al. (2012)

12.5 Biofilm Based Remediation of Recalcitrant and Toxic Chemicals

The urge for development has resulted in production of certain types of recalcitrant and toxic pollutants in water bodies that not only deteriorate the quality of water but are also dangerous for the ecosystem health. Much of the toxic chemicals like polychlorinated biphenyls, polynucleated aromatic hydrocarbons, and DDT, etc. have been banned by environmental agencies. But years after their ban, these recalcitrant compounds still exist in the environment in some form. The best way to remediate such pollutants is to use sustainable and low-cost remediation techniques. Recently yeast biofilm (*Candida tropicalis*) grown on gravel substrates has been utilized to degrade 97% of oil polluted water just in a short time span of 10 days. This biofilm can be potentially used for treating liquid waste in trickling filters on gravel bed before releasing the waste into water bodies (Chandran and Das 2011). The recent innovative technique of removal of left over organic pollutants of toxic nature after treating water has been developed by Rafida et al. (2011). The technique involved the application of PVC made vertical flow biofilter (VFB) of anaerobic nature. The upper part of this reactor was filled with small bottles of polyethylene terephthalate (PET) and lower part was filled with gravel. The biofilm of diverse microorganisms was grown on it and the efficiency of this technique proved to be 90% in terms of removal of hydrocarbon pollutants. Chlorinated ethene is usually added to drinking water to kill pathogenic microbes in it. A specific group of bacteria *Dehalococcoides* have the potential of reductive de-chlorination of chlorinated ethenes (Löffler et al. 2013). Different types of biofilm based microbes that help in the remediation of toxic chemicals from water are shown in Table 12.2.

12.6 Biofilm Based Remediation of Plastic and Micro-Plastic Wastes

The main constraint with plastic waste is its non-degradable nature in the environment. Biofilm based microbial consortia has been successfully used to degrade plastic wastes which causes several physical and chemical and biological obstructions in the aquatic ecosystem. In recent times, remediation of polyethylene wastes by the fungal biofilm formed of *Aspergillus niger* (ITCC No.6052) isolated from dump sites of plastic waste has been in use (Mathur et al. 2011). The fungi show proliferous growth when grown on medium with optimum environmental conditions. The biofilm of this fungus was able to reduce polythene by 3.44% and 61% of mass and tensile strength, respectively, in a period of 1 month. This property of *Aspergillus niger* (ITCC no. 6052) to remediate plastic waste was related to its high cell surface hydrophobicity and thermal oxidation potential. A wide variety of algal species including *Chaetophora, Cymbella, Navicula, Cocconis, Oocystis, Oedogonium,* etc. have been evaluated to show the potential of degrading plastic waste. "*Rhodococcus*

Table 12.2 Biofilm based remediation of toxic chemicals (Das et al. 2012)

Toxic compound	Biofilm forming organisms	References
Hydrocarbon compounds	*"Candida tropicalis"* biofilm grown on gravel	Rafida et al. (2011)
	"*Filamentous cyanobacteria, picoplankton, and diatom* based biofilm grown on gravel and glass plates" "*Acinetobacter calcoaceticus* and *nocardioforms* were partly found attached to filaments of cyanobacteria"	Al-Awadhi et al. (2003)
	Gravel supported biofilm of *"Cyanobacteria Dermocarpella* sp. *and Acinetobacter calcoaceticus, Acinetobacter calcoaceticus, Micrococcus* and *nocardioforms*"	Radwan and Al-Hassan (2001)
2,4-Dichlorophenol	Rotating perforated tubes biofilm reactor (RTBR) laden with *"Pseudomonas putida"* to remove 2,4-Dichlorophenol pollutant from water	Kargi and Eker (2005)
2,4,6 trichlorophenol "2,3,4,6-Tetrachlorophenol and pentachlorophenol"	*"Pseudomonas sp., Rhodococcus sp"* based fluidized bed biofilm for chlorophenol mineralization	Puhakka et al. (1995)
2 chlorophenol	Hydrogenotrophic biofilm cultivated in three silicone-tube membrane bioreactors under denitrification, sulfate-reduction, and dechlorination conditions for remediation of anaerobic sludge from a swine wastewater	Chang et al. (2004)
4 chlorophenol	Bacterial consortium from rhizosphere of *Phragmites australis*	Caldera (1999) and Das et al. (2012)
Polychlorinated biphenyl (PCB)	Anaerobic bacteria isolated from secondary sludge wastewater	Josephine et al. (2006)
4,6-Dinitro-ortho-cresol	"Biofilm based bacterial culture and fixed bed culture" biodegradation	Gisi et al. (1997)
Dinitrotolune (DNT)	Fludized biofilm reactor containing sludge from wastewater treatment plant; various aerobic and anaerobic bacteria	Lendermann and Spain (1998)

rubber strain (C208) has the capacity to degrading polythene at the rate of 0.86% per week due to its very nature of using polythene as a sole source of carbon and also it's potential to use polythene as a substrate during biofilm development (Suseela and Kiran 2007)". "Micro-plastics are defined as particles less than 5 mm in size with various chemicals and toxins attached to their surfaces (Bakir et al. 2016)". Though microplastic pollution and its impact on water bodies has been extensively studied, its remediation by using biofilms remains unexplored. The biofilm grown on the surfaces of plastics could be utilized to degrade microplastics and biotransform them into usable form. This goal can be achieved by pilot and micro-level studies by using microbial biofilms of diverse nature.

12.7 Bioremediation of Pharmaceuticals and Personal Care Products by Aquatic Biofilms

Pharmaceuticals and other personal care products are disposed into the environment as a part of effluent. These products are usually toxic in nature and prove fatal to the aquatic ecosystem. Sometimes these products may get transformed into more toxic pollutants. One such pollutant is triclosan, an antimicrobial drug which gets phototransformed into dioxin congeners in water (Latch et al. 2003; Buth et al. 2009). Removal of pharmaceuticals such as diclofenac, trimethoprim, metoprolol, etc. is efficiently removed by membrane biofilm reactors (MBRs) as compared to simple activated sludge processes (Sui et al. 2011). Onesios and Bouwer (2012) reported the in-situ bioremediation of pharmaceuticals and personal care products by using activated sewage sludge as inoculum for the formation of biofilm on sand bed column filters. In this experiment out of 14 pharmaceuticals and personal care products, 10 were completely degraded. Thus the use of biofilms for degradation of pharmaceuticals is an easy and more reliable way to get rid of these types of aquatic pollutants.

12.8 Biofilm Based Bioremediation of Synthetic Dyes from Water

Most of the synthetic dyes are carcinogenic and pose serious threat to the environment. Aquatic pollution due to synthetic dyes may also retard esthetic value of the water besides other problems by changing its color. Jiang and Bishop (1994) reported the aerobic degradation of azo dyes. Out of Acid Orange 8 (AO8), Acid Orange 10 (AO10), and Acid Red 14 (AR14), AO8 was the only dye that was decomposed aerobically. But under anaerobic conditions, all the dyes were decomposed easily. Pilot scale rotating drum biofilm reactor (ex-situ) also showed bioremediation of azo dye Everzol Turquoise Blue G by *Coriolus versicolor* activated sludge-based biofilm (Zhang et al. 1995).

12.9 Application of Periphytic Biofilm in Quantification of Denitrification Rate in Water Bodies

Periphytic biofilms are widely found in shallow water and sediment systems. It plays an important role in nitrogen cycling and helps in sustaining aquatic ecosystems and its role in denitrification by constructing the model. They divided water and sediment into three separate parts, viz. water column, sediment, and periphytic biofilm. They further used species, periphytic biofilm, and bacterial nitrifiers and denitrifiers as

Fig. 12.3 Denitrification process by periphytic biofilm (https://www.thescientificnews.com/periphyticbiofilm)

potential factors. For this research structural modeling was used. "The study pointed that periphytic biofilm increased net denitrification directly as well as indirectly through the pH, the dissolved oxygen, dissolved organic carbon, and Nconcentrations in the water column and sediment" Fig. 12.3. Further research can be helpful in utilizing the potential of periphytic biofilm in nitrogen removal from wastewater.

12.10 Conclusion

Biofilm is an important and useful tool for the remediation of potentially hazardous pollutants from water bodies. Biofilm based innovative techniques are helpful in the degradation and removal of heavy metals and recalcitrant pollutants from the polluted aquatic ecosystems. The microbial biofilm application is a sustainable way of remediating toxic chemicals, synthetic dyes, pharmaceutical contaminants, plastic waste, and hydrocarbon pollutants from the wastewater. This type of substrate on which biofilm is grown plays an important role in removal of pollutants.

References

Accinelli C, Sacca ML, Mencarelli M, Vicari A (2012) Application of bioplastic moving bed biofilm carriers for the removal of synthetic pollutants from wastewater. Bioresour Technol 120:180–186

Al-Awadhi H, Al-Hasan RH, Sorkhoh NA, Salamah S, Radwan SS (2003) Establishing oil-degrading biofilms on gravel particles and glass plates. Int Biodeteriorat Biodeg 51(3):181–185
Alluri HK, Ronda SR, Settalluri VS, Bondili JS, Suryanarayana V, Venkateshwar P (2007) Biosorption: an eco-friendly alternative for heavy metal removal. Afr J Biotechnol 6:2924–2931
Bakir A, O'Connor IA, Rowland SJ, Hendriks AJ, Thompson RC (2016) Relative importance of microplastics as a pathway for the transfer of hydrophobic organic chemicals to marine life. Environ Pollut 219:56–65
Boles BR, Thoendel M, Singh PK (2004) Self-generated diversity produces "insurance effects" in biofilm communities. PNAS 101:16630–16635
Brandy T, Alain M, Matthew AM, Dylan BM, Garrison S (2005) Zinc sorption by a bacterial biofilm. Environ Sci Technol 39:8288–8294
Burek P, Satoh Y, Fischer G, Kahil MT, Scherzer A, Tramberend S, Nava LF, Wada Y (2016) Water futures and solution- fast track initiative (Final Report). IIASA, Laxenburg. WP-16-006
Buth JM, Grandbois M, Vikesland PJ, McNeill K, Arnold WA (2009) Aquatic photochemistry of chlorinated triclosan derivatives: potential source of polychlorodibenzo-P-dioxins. Environ Toxicol Chem 28:2555–2563
Caldera M (1999) 4-Chlorophenol degradation by a bacterial consortium: development of a granular activated carbon biofilm reactor. Appl Microbiol Biotechnol 52:722–729
Cao W, Zhang H, Wang Y, Pan J (2012) Bioremediation of polluted surface water by using biofilms on filamentous bamboo. Ecol Eng 42:146–149
Chandran P, Das N (2011) Degradation of diesel oil by immobilized Candida tropicalis and biofilm formed on gravels. Biodegradation 22:1181–1189
Chang CC, Tseng SK, Chang CC, Ho CM (2004) Degradation of 2-chlorophenol via a hydrogenotrophic biofilm under different reductive conditions. Chemosphere 56:989–997
Chen X, Suwarno SR, Chong TH, McDougald D, Kjelleberg S, Cohen Y, Fane AG, Rice SA (2013) Dynamics of biofilm formation under different nutrient levels and the effect on biofouling of a reverse osmosis membrane system. Biofouling 29:319–330
Costerton JW, Geesev GG, Cheng KJ (1978) How bacteria stick. Sci Am 238:86–95
Costley SC, Wallis FM (2001) Bioremediation of heavy metals in a synthetic wastewater using a rotating biological contactor. Water Res 35:3715–3723
Cristina Q, Zelia R, Bruna S, Bruna F, Hugo F (2009a) Biosorptive performance of an *Escherichia coli* biofilm supported on zeolite naY for the removal of Cr(VI), Cd(II), Fe(III), and Ni(II). Chem Eng J 152:110–115
Cristina Q, Zelia R, Bruna S, Bruna F, Hugo F (2009b) Removal of Cd(II), Cr(VI), Fe(III) and Ni(II) from aqueous solutions by an *E. coli* biofilm supported on kaolin. Chem Eng J 149:319–324
Dar S, Bhat RA (2020) Aquatic pollution stress and role of biofilms as environment cleanup technology. In: Qadri H, Bhat RA, Dar GH, Mehmood MA (eds) Freshwater pollution dynamics and remediation. Springer Nature, Singapore, pp 293–318
Das N, Lakshmi V, Basak G, Salam JA, EvyAlice AM (2012) Application of biofilms on remediation of pollutants – an overview. J Microbiol Biotech Res 2(5):783–790
DePhilippis R, Colica G, Micheletti E (2011) Exopolysaccharide-producing cyanobacteria in heavy metal removal from water: molecular basis and practical applicability of the biosorption process. Appl Microbiol Biotechnol 92:697–708
Edwards SJ, Kjellerup BV (2013) Applications of biofilms in bioremediation and biotransformation of persistent organic pollutants, pharmaceuticals/personal care products, and heavy metals. Appl Microbiol Biotechnol 97:9909–9921. https://doi.org/10.1007/s00253-013-5216-z
Fux CA, Costerton JW, Stewart PS, Stoodley P (2005) Survival strategies of infectious biofilms. Trends Microbiol 13:34–40
Galiana E, Fourre S, Engler G (2008) Phytophthora parasitica biofilm formation: installation and organization of microcolonies on the surface of a host plant. Environ Microbiol 10:2164–2171
Gianfreda L, Rao MA (2004) Laccases: a useful group of oxido-reductive enzymes. Enzyme MicrobTechnol 35:339–354
Gisi W, Hanselman KW, Stucki G, Gisi D (1997) Biodegradation of pesticide 4,6 – dinitro- orthocresol by microorganisms in batch cultures and in fixed bed column reactor. Appl Microbiol Biotechnol 48:441–448

Gupta P, Diwan B (2017) Bacterial exopolysaccharide mediated heavy metal removal: a Review on biosynthesis, mechanism and remediation strategies. Biotechnol Rep 13:58–71

Hall-Stoodley L, Costerton JW, Stoodley P (2004) Bacterial biofilms: from the natural environment to infectious diseases. Nat Rev Microbiol 2:95–108

Jiang H, Bishop PL (1994) Aerobic biodegradation of azo dyes in biofilms. Water Sci Technol 29(10–11):525–530

Josephine A, Joseph A, Susan MG (2006) Biodegradation of polychlorinated biphenyls by activated sludge obtained from secondary sledge wastewater. AJChE 6(2006):44–52

Kantawanichkul S, Kladprasert S, Brix H (2009) Treatment of high-strength wastewater in tropical vertical flow constructed wetlands planted with *Typha angustifolia* and *Cyperus involucratus*. Ecol Eng 35(2):238–247

Kargi F, Eker S (2005) Removal of 2, 4-dichlorophenol and toxicity from synthetic wastewater in a rotating perforated tube biofilm reactor. Process Biochem 40:2105–2111

Latch DE, Packer JL, Arnold WA, McNeill K (2003) Photochemical conversion of triclosan to 2,8-dichlorodibenzo-p-dioxin in aqueous solution. J Photochem Photobiol A Chem 158:63–66

Lau T, Wu X, Chua H, Qian P, Wong P (2005) Effect of exopolysaccharides on the adsorption of metal ions by Pseudomonas sp. CU-1. Water Sci Technol 52:63–68

Lendermann U, Spain JC (1998) Simultaneous biodegradation of 2,4-dinitrotoluene and 2,6-dinitrotoluene in an aerobic fludized bed biofilm reactor. Environ Sci Technol 32:82–87

Löffler FE, Ritalahti KM, Zinder SH (2013) Dehalococcoides and reductive dechlorination of chlorinated solvents. In: Stroo HF et al (eds) Bioaugmentation for groundwater remediation, vol 5. Springer, New York, pp 39–88

Marchal R, Briandet R, Koechler S, Kammerer B, Bertin PN (2010) Effect of arsenite on swimming motility delays surface colonization in *Herminiimonas arsenicoxydans*. Microbiology 156:2336–2342

Marchal M, Briandet R, Halter D, Koechler S, DuBow MS, Lett MC, Bertin PN (2011) Subinhibitory arsenite concentrations lead to population dispersal in Thiomonas sp. PLoS One 6:e23181

Mathur G, Mathur A, Prasad R (2011) Colonization and degradation of thermally oxidized high-density polyethylene by *Aspergillus niger* (ITCC no. 6052) isolated from plastic waste dumpsite. Biorem J 15(2):69–76

Muller D, Simeonova DD, Riegel P, Mangenot S, Koechler S, Lièvremont D, Bertin PN, Lett MC (2006) *Herminiimonasarsenicoxydans* sp. nov., a metalloresistant bacterium. Int J Syst Evol Microbiol 56:1765–1769

Muller D, Médigue C, Koechler S et al (2007) A tale of two oxidation states: bacterial colonization of arsenic-rich environments. PLoS Genet 3:e53

Onesios KM, Bouwer EJ (2012) Biological removal of pharmaceuticals and personal care products during laboratory soil aquifer treatment simulation with different primary substrate concentrations. Water Res 46:2365–2375

Pal A, Paul A (2010) Microbial extracellular polymeric substances: central elements in heavy metal bioremediation. Indian J Microbiol 48:49–64

Puhakka J, Melin E, Järvinen K, Koro P, Rintala J, Hartikainen P, Shieh W, Ferguson J (1995) Fluidized-bed biofilms for chlorophenol mineralization. Water Sci Technol 31(1):227–235

Qian Y, Wen XH, Huang X (2007) Development and application of some renovated technologies for municipal wastewater bioremediation in China. Front Environ Sci Eng 1:1–2

Rabei MG, Gad Elrab SMF, Abskharon RNN, Hassan SHA, Shoreit AAM (2009) Biosorption of hexavalent chromium using biofilm of *E. coli* supported on granulated activated carbon. World J Microbiol Biotechnol 25:1695–1703. https://doi.org/10.1007/s11274-009-0063-x

Radwan SS, Al-Hassan RH (2001) Potential application of coastal biofilm-coated gravel particles for treating oily waste. Aquat Microb Ecol 23:113–117

Rafida AI, Elyousfi MA, Al-Mabrok H (2011) Removal of hydrocarbon compounds by using a reactor of biofilm in an anaerobic medium. World Acad Sci Eng Technol 73:153–156

Rajendran P, Muthukrishnan J, Gunasekaran P (2003) Microbes in heavy metal remediation. Indian J Exp Biol 41:935–944

Rodriguez S, Bishop P (2008) Enhancing the biodegradation of polycyclic aromatic hydrocarbons: effects of nonionic surfactant addition on biofilm function and structure. J Environ Eng 134:505–512

Scot JA, Karanjkar AM, Rowe DL (1995) Biofilm covered granular activated carbon for decontamination of streams containing heavy metals and organic chemicals. Miner Eng 8:221–230

Sehar S, Naz I (2016) Role of the biofilms in wastewater treatment. In: Microbial biofilms - importance and applications, vol 7. Intech Open Science, London, pp 139–144. https://doi.org/10.5772/63499

Seo Y, Lee WH, Sorial G, Bishop PL (2009) The application of a mulch biofilm barrier for surfactant enhanced polycyclic aromatic hydro- carbon bioremediation. Environ Pollut 157:95–101

Sui Q, Huang J, Deng S, ChenW YG (2011) Seasonal variation in the occurrence and removal of pharmaceuticals and personal care products in different biological wastewater treatment processes. Environ Sci Technol 45:3341–3348

Sundar K, Sadiq IM, Amitava M, Chandrasekaran N (2011) High chromium tolerant bacterial strains from Palar river basin. Hazard Mater 196:44–51

Suseela MR, Kiran T (2007) Algal biofilms on polythene and its possible degradation. Curr Sci 92:285–287

Verhagen P, De Gelder L, Hoefman S, De Vos P, Boon N (2011) Planktonic versus biofilm catabolic communities: importance of the biofilm for species selection and pesticide degradation. Appl Environ Microbiol 77:4728–4735

Wang S, Teng S, Fan M (2010) Interaction between heavy metals and aerobic granular sludge. In: Sarkar SK (ed) Environmental management. Sciyo, Rijeka, pp 173–188

Wang R, Khan BA, Cheung GY, Bach TH, Jameson-Lee M, Kong KF, Queck SY, Otto M (2011) *Staphylococcus epidermidis* surfactant peptides promote biofilm maturation and dissemination of biofilm-associated infection in mice. J Clin Investig 121:238–248

Wang J, Liu XD, Lu J (2012) Urban river pollution control and remediation. Procedia Environ Sci 13:1856–1862

Xiao J, Klein MI, Falsetta ML, Lu B, Delahunty CM, Yates JR 3rd, Heydorn A, Koo H (2012) The exopolysaccharide matrix modulates the interaction between 3D architecture and virulence of a mixed- species oral biofilm. PLoS Pathog 8:e1002623

Zhang TC, Fu YC, Bishop PL (1995) Transport and biodegradation of toxic organics in biofilms. Hazard J Mater 41:267–285

Chapter 13
Heavy Metal Soil Contamination and Bioremediation

Malik Muzafar Manzoor, Pooja Goyal, Ajai P. Gupta, and Suphla Gupta

13.1 Introduction

Heavy metals are naturally occurring inorganic chemical hazards of high atomic weight, which are usually toxic at lower concentrations. Metals like zinc (Zn), copper (Cu), mercury (Hg), arsenic (As), lead (Pb), chromium (Cr), cadmium (Cd), and nickel (Ni) appear in the WHO list of chemicals of public concern (Kirpichtchikova et al. 2006; Adriano 2001). The significant portion of heavy metals released by anthropogenic activities into the environment finds their way into the soil. Unlike organic contaminants which are oxidized to carbon (IV) by microbes, most of the metal contaminants are not degraded by microbial and chemical processes after their introduction into the soil (Adriano 2001), as a result of their total concentration persists for an extended time (Maslin and Maier 2000). However, they may change in their chemical forms (speciation) and bioavailability in due course of time. The point of concern is that the biodegradation of organic contaminants can get affected by heavy metal contaminated soils (McLaughlin et al. 2000b). Heavy metal contamination of soil has adverse influences on human health and environment through drinking of contaminated groundwater, reduction in land usability for agriculture, and also land tenure. This demands regular characterization and remediation of soil and sufficient reclamation of soil ecosystem systems contaminated by heavy metals. Contemporary enactment respecting environmental protection and general well-being over the world depends on the data that characterizes environmental phenomena based on their chemical properties, particularly that resided in our food chain (Zhao and Kaluarachchi 2002). Soil categorization would give an insight of heavy

M. M. Manzoor · P. Goyal · S. Gupta (✉)
Plant Biotechnology Division, CSIR-Indian Institute of Integrative Medicine, Canal Road, Jammu, India

A. P. Gupta
Quality Control and Quality Assurance Division, CSIR-Indian Institute of Integrative Medicine, Canal Road, Jammu, India

R. A. Bhat et al. (eds.), *Bioremediation and Biotechnology, Vol 2*,
https://doi.org/10.1007/978-3-030-40333-1_13

metal bioavailability and their speciation, remediation of soil contaminated with heavy metals would necessitate information of the type of contaminant and source of contamination, respectively. This will help in calculating the associated health risks. The scientific assessment of associated risk would facilitate decision-makers to manage contaminated sites cost-effectively while preserving the ecosystem and public health. Phytoremediation, immobilization, and soil washing strategies used for remediation of heavy metal-contaminated sites are among the best demonstrated available technologies (BDATs) for remediation purposes (Kirpichtchikova et al. 2006). In spite of being ecologically suitable and cost-effective application of these technologies has been reported only in developed nations. Due to lack of awareness for their ecological benefits and principles of operation for their field application, these technologies are not commercially available in underdeveloping nations. Due noteworthy mindfulness by the legislature and the general public of the hazardous effects of the soil contaminated with heavy metals on animal and human health, researchers have shown their keenness in remediating contaminated sites using novel technologies (Pierzynski et al. 2000). Developing nations have limited funds for environmental restoration programs thus they need cost-effective and ecologically sustainable alternatives for the restoration of contaminated soil to lessen the associated risks, enhance food security, availability of the land resource for agricultural production. This could enable scaling down the land tenure problems. In this chapter, we have compiled the literature from various sources related to the contamination and health hazardous of heavy metals and their bioremediation.

13.2 Heavy Metals: Their Origin

Heavy metals exhibit properties like conductivity, ductility, cation stability, malleability, and ligand specificity which are also characteristic properties of metals. Several heavy metals like traces elements (Co, Mn, Cu, Mo, Fe, Ni, V, and Zn) are beneficial for organisms and are required in minute quantities. These metals are found naturally in the soil environment at micro levels and are hardly toxic (Pierzynski et al. 2000; Kabata-Pendias and Pendias 2001). However, excess of these elements can be harmful to organisms. The contaminated sites are found to be predominantly rich in As, Cu, Pb, Zn, Cr, Cd, and Hg (Jiang et al. 2001) having relatively high density and atomic weight (<20) (Smejkalova et al. 2003). Other heavy metals (Hg, Cd, Pb, and As) are not beneficial to organisms and pose major threats to the ecosystem. Over-exploitation of natural resources and other anthropogenic activities causes a disturbance in the delicate balance of the ecosystem, resulting in the accumulation of different heavy metals (Fig. 13.1) above defined values jeopardizing ecosystems, plants, animals, and human health (D'amore et al. 2005). The natural balance between heavy metal formation and degradation gets disturbed essentially because of rapid generation/usage of commercially important minerals through human made techniques altering their natural concentrations. This may also influence their abundance and bioavailability in the soil environment (D'amore

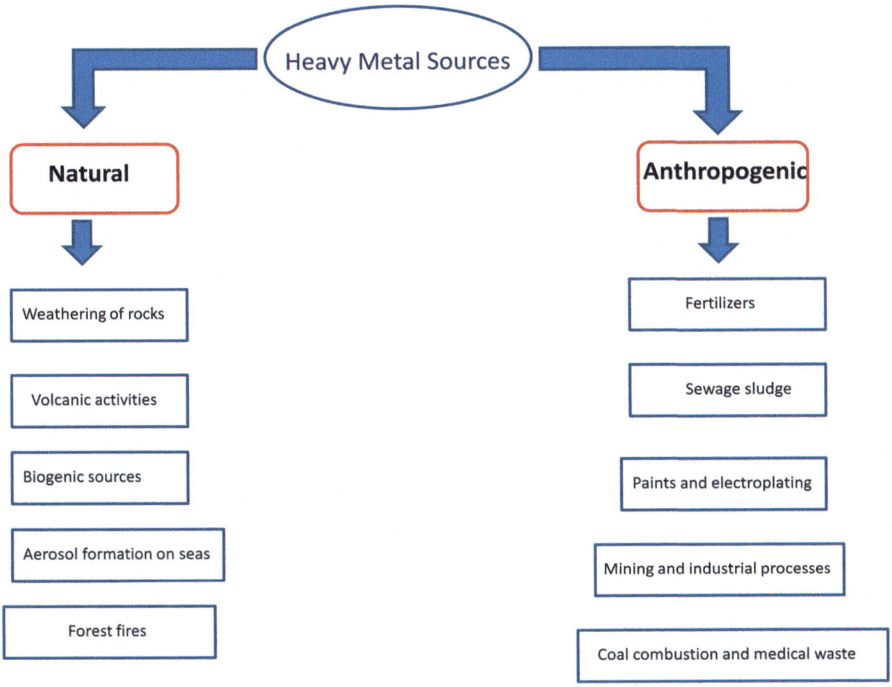

Fig. 13.1 Sources of heavy metals

et al. 2005). According to Lombi and Gerzabek (1998) the mass balance of heavy metal (M) is the sum total of parent material (p), atmospheric deposition (a), fertilizer sources (f), agrochemical sources (ag), organic waste sources (ow), and other inorganic pollutants (iP) minus activities due to crop removal (cr) and losses by leaching, volatilization (l).

$$\mathrm{M\ total} = \mathrm{Mp} + \mathrm{Ma} + \mathrm{Mf} + \mathrm{Mag} + \mathrm{Mow} + \mathrm{Mip} - (\mathrm{Mcr} + \mathrm{Ml})$$

Emission through natural fluxes of several metals into the environment is three folds lower as compared to emission caused due to anthropogenic activities (Sposito and Page 1984). Bioavailability and mobility of heavy metals present in contaminated soils, which are added through anthropogenic activities vary from pedogenic or lithogenic ones (Kuo et al. 1983; Kaasalainen and Yli-Halla 2003). Several anthropogenic sources like metal mine tailings, leaded gasoline and paints, petrochemicals, sewage sludge, fertilizer, compost, pesticides, residues from coal combustion, disposal of high metal wastes in inappropriately protected landfills are the originator of metal-bearing solids at contaminated sites (Khan et al. 2008; Basta et al. 2005; Zhang et al. 2010). To estimate environmental, health effects of heavy metals and to understand their speciation, possible remedial options, and bioavailability, it is prerequisite to have the knowledge of their basic chemistry. It is the chemical form and speciation of heavy metals which determines their fate and transport in soil.

Adsorption of heavy metals on the soil is a complex process which took years. After adsorption on soil heavy metals get reallocated into various chemical forms with changeable toxicity, bioavailability, and mobility (Shiowatana et al. 2001; Buekers 2007). There are several factors that control re-distribution which include adsorption and desorption between the metals, ion exchange, precipitation of mineral and dissolution, aqueous complexation, and plant uptake, biological immobilization and mobilization (Levy et al. 1992).

13.2.1 Effects of Fertilizer Application

Historically the first and foremost human influence on soil was in the form of agriculture. Not only macronutrients (S, Ca, Mg, N, K, and P) but also important micronutrients are acquired by plants for their complete lifestyle and growth. Heavy metals such as (Cu, Co, Fe, Ni, Mn, Mo, and Zn) which are crucial for plant growth and development are deficient in some soils (Lasat 1999), and they are supplied to crops as an addition to the existing soil nutrients. Similarly, Cu and Mn are supplied to cereal crops which are growing in deficient soils in addition to the soil composition. In intensive farming system, a large quantity of fertilizers is added to soil to provide sufficient N, P, and K for the growth of a crop. Heavy metals like cadmium and lead, in trace amounts, are generally present in these fertilizers. Their content in soil is significantly increased because of the continuous and large scale use of fertilizers (Jones et al. 1981). Though metals like Cd and Pb do not show any noted physiological activity; however, certain phosphatic fertilizers used in farming unknowingly add Cd and other toxic elements to soil together with Pb, F, and Hg. Application of fertilizer though is beneficial for short-term plant productivity but in the long term it leads to more harm if used non-judiciously.

13.2.2 Pesticides

In the past pesticides were commonly used in horticulture and agriculture containing significant concentrations of metals. About 10% recently approved chemicals for use as fungicides and insecticides in the UK contained metals like Mn, Cu, Hg, Zn, or Pb. They include copper oxychloride and Bordeaux mixture which is a copper-containing fungicidal sprays (Jones et al. 1981). In fruit orchards lead arsenate was used to control parasitic insects for several years. Similarly to control banana pests and cattle ticks arsenic-containing compounds were extensively used in New Zealand and Australia. For the preservation of wooden logs formulations of

Cu, Cr, and As (CCA) were used in various parts of the world, and there are currently several polluted sites where the concentration of these elements in the soil is greatly surpassing the background concentrations. The use of such formulations is generally localized to a particular site or job and for a specific application. However, problems arise when such contaminated locations are used for different agricultural or non-agricultural purposes (McLaughlin et al. 2000a).

13.2.3 Natural Fertilizers: Manures and Biosolids

Accumulation of heavy metals such as Cd, Pb, Hg, Mo, As, Cu, Ni, Sb, Se, Cr, Tl, Zn in the soil is primarily due to the application of various biosolids like composts, sewage sludge, and livestock manures that unintentionally leads to their buildup (Lasat 1999). Poultry, pig manures, and cattle waste are considered as valuable fertilizers and are thus applied to crops as slurries or solids (Sumner 2000). Copper (Cu) and zinc (Zn) metals are generally added to pig and poultry diet as growth promoters (Sumner 2000; Chaney and Oliver 1996). In the long run, repeated application of manure to restricted areas of land from animals on such diets containing high concentrations of Zn and Cu metal can cause a substantial buildup of these heavy metals in the soil. Potential of biosolids with heavy metals to contaminate the soil has produced abundant concern regarding their application for agricultural purposes (Canet et al. 1998). Cd, Cr, Pb, Ni, and Cu are the most commonly found heavy metals in biosolids, and it is their nature, intensity of the industrial process, and the process employed for their treatment which governs their concentration (Canet et al. 1998). Groundwater usually gets contaminated because of leaching down of metals through the soil profile (McLaren et al. 2005). In New Zealand, the recent analysis had shown that soils treated with biosolids increased Zn, Cd, and Ni levels in drainage leachates (McLaren et al. 2004; Keller et al., 2002).

13.2.4 Wastewater and Industrial Effluents

Since ages, municipal and industrial wastewater were used for watering land in several countries (Reed et al. 1995). It is estimated worldwide that wastewater is utilized to irrigate about 20 million hectares of arable land. Studies suggest that in numerous African and Asian cities, about 50% of agriculture land is based on wastewater irrigation (Bjuhr 2007). Irrigation of land for a long duration of time with industrial effluents and wastewater containing heavy metal in less concentrations ultimately results in their accumulation in the soil. Now several developed countries have invested hugely in wastewater treatment plants for its use in irrigation and other activities with minimum or no harmful effects.

13.2.5 Industrial Waste and Mining Activities

Modernization and infrastructure development have left the legacy of a wide range of heavy metal contaminants in soil, air, and water equally. Direct discharge of byproducts from the mining process (tailings) into natural or human-made reservoirs results in elevated concentrations (DeVolder et al. 2003; Mushtaq et al. 2020; Qadri and Bhat 2020). Large scale smelting and mining of Zn and Pb ores have resulted in contamination of soil that hugely affects the health of the living organism. In spite of being lengthy and expensive, many strategies used for the reclamation of these sites may not restore soil productivity with high efficiency. Heavy metals accessibility in the soil in large amounts causes environmental risk directly and indirectly to humans, directly and indirectly, both. Assimilation or entry pathways of metals include the oral bioavailability that is contaminated soil ingested directly or contaminated soil ingested by plant materials (Basta and Gradwohl 1998). Different materials are released in the environment by several industries like tanning, accidental oil spills, textile, pesticides, petrochemicals, and pharmaceutical manufacturing units and have high variability in their composition. Among these, some are beneficial for forestry or agriculture when disposed on land. But many are potentially hazardous and pose risks to ecosystem and human health because of high contents of toxic organic components and heavy metals like Pb, Cr, and Zn (Sumner 2000). Several toxic influences of heavy metal published are compiled in Table 13.1.

13.3 Bioremediation

The unlimited release of industrial waste to the environment because of the growing industrial sector, especially in the urban sector, has lead to the accumulation of heavy metals. Gradually these heavy metals are depleted through various processes like erosion, deflation, leaching, and plant uptake. Heavy metals which are released into the water and soil are least amenable to changes thus cannot be transformed into less toxic forms and consequently have long-term influences on the ecosystem. Several studies have advocated bioremediation approach to be one of the most effective solutions for the protection of the environment as well as human health (Martin and Ruby 2004; Shafi et al. 2018).

Generally, remediation processes are subjected to an assortment of regulatory necessities. However, for processes where no known standards occur or where standards are recommended, their assessment is based on ecological risks and their effects on human health. The regulatory authorities usually allow that remediation strategies that are centered on decreasing metal bioaccessibility only if it demonstrates the decreased risk and manifests long-lasting effect (Martin and Ruby 2004). The choice of suitable remediation treatment depends on the type of metal-contaminated soils and their physical and chemical form. Several techniques are

Table 13.1 Toxicity of heavy metals

Heavy metal	EPA regulatory limit (ppm)	Toxic effects	Reference
As	0.01	ATP synthesis and oxidative phosphorylation	Tripathi et al. (2007)
Ni	0.2 (WHO permissible limit)	Skin rashes, lung cancer, nose and throat infections, neurotoxic, genotoxic, affects fertility	Khan et al. (2007), Das et al. (2008), Duda-Chodak and Blaszczyk (2008)
Hg	2.0	Tremors, fatigue, temper outbursts, depression, restlessness, hair loss, vision disturbance, brain damage, failure of kidney and lung	Gulati et al. (2010), Ainza et al. (2010)
Cd	5.0	Mutagenic, carcinogenic, influences calcium regulation in biological systems, lung damage	Salem et al. (2000), Degraeve (1981)
Cr	0.1	Causes hair loss	Salem et al. (2000)
Pb	15	Cardiovascular disease, impaired development in children, short-duration memory loss, reduced intelligence, difficulties in coordination and learning debilities	Padmavathiamma and Li (2007), Salem et al. (2000)
Se	50	Influences endocrine function, hepatotoxicity, and gastrointestinal disorders, impairment of NK cells activity	Vinceti et al. (2001)
Zn	0.5	Fatigue, dizziness	Hess and Schmid (2002)
Ba	2.0	Elevated blood pressure, cause cardiac arrhythmias, muscle twitching, gastrointestinal dysfunction	Jacobs et al. (2002)
Ag	0.10	Stomach pain, exposure causes skin and tissues to turn gray or blue-gray, irritates in lung and throat, breathing problems	Abadin et al. (2007)

available which are employed to remove these heavy metals, which include oxidation or reduction, membrane technology, electrochemical treatment chemical precipitation, ion exchange, and reverse osmosis.

However, most of these techniques are expensive or ineffectual to lower concentrations of the metal (>100 mg/L) (Ahluwalia and Goyal 2007). Alternately, the attractive and the best alternate to physicochemical methods for removal of heavy metals are biological approaches such as bioaccumulation and biosorption (Kapoor and Viraraghavan 1995). To resolve and restore the natural condition of the soil usage of plants and microorganisms are the best possible options for remediation purposes. Heavy metals despite being important for microbial growth and development at relatively low concentrations, they exert considerable stress on microbial communities or desired plant community utilized for the bioremediation (Doleman 1994). Microbial makeup is mainly modified when the essential metal ions of biological molecules are displaced or modified; it is the blockage of essential functional groups that can also change the microbial makeup (Wood and Wang 1983; Li and

Tan 1994). Heavy metal contaminations trigger a complex response in microbial communities depending on their concentration, availability and kind of heavy metal, the nature of the medium, and microbial species (Coblenz and Wolf 1994).

Despite being a nondisruptive strategy for soil remediation, Bioremediation is time-consuming and is influenced by geological and climatic conditions of the remediation site (Schmöger et al. 2000). The process of bioremediation of heavy metals adopts a different approach. Here the heavy metals are not degraded but are transmuted from one organic complex or oxidation state to others. Because of the change in the oxidation state of the heavy metals, change in volatility, toxicity, bioavailability, and solubility can occur (Garbisu and Alkorta 1997, 2003). If they become more or less water soluble, they may be removed through leaching or precipitation from the environment. Since metal contamination varies from site to site and type of effluent, no single process may be used for its bioremediation. Plants, microorganisms, or the combination of both organisms can be employed to achieve the target.

13.3.1 Microbial Remediation

The microbial remediation involves a cocktail of several microorganisms like bacteria (*B. subtilis*), *Enterobacter* (*E. cloacae*), and *Pseudomonas* (*P. putida*) have been effectively used for the reduction of Cr(VI) to the less toxic Cr(III) (Garbisu et al. 1997; Garbisu et al. 1998; Ishibashi et al. 1990; Wang et al. 1989; Shafi et al. 2018). It is also reported in the literature that *B. subtilis* can reduce non-metallic elements. For example, Garbisu et al. (1998) studied that selenite has been decreased to the less toxic elemental selenium using *B. Subtilis*. Further, Ajaz et al. (2010) revealed that *B. thuringiensis* and *B. cereus* could increase the cadmium and zinc extraction from the soil rich in Cd and soil contaminated with sewage from the metal industry. Studies have revealed that heavy metals stimulated the siderophore production (Fe complexing molecules) using bacteria, thereby modulating their bioavailability (Van der Lelie et al. 1999). In a study, it was observed that *Azotobacter vinelandii* showed increased siderophore production in the presence of Zn(II) (Huyer and Page 1988). The change in bioavailability could have led to the natural extraction of these metals from the soil. Besides the change in bioavailability, precipitation by sulfate-reducing bacteria (*Desulfovibrio desulfuricans*) could also lead to bioremediation. This process is indirect and mostly ex-situ where the sulfate-reducing bacteria convert sulfate forms of heavy metal into their hydrogen sulfate, as in heavy metals Zn and Cd, to form insoluble forms of these metal sulfides (White et al. 1998). Mercury is a very toxic pollutant, and its bio-magnification has created havoc to land and aquatic life in the past. In in-situ setup soluble mercuric ions Hg(II) were converted to volatile metallic mercury Hg(0) by mercury-resistant bacteria (Hobman 1997). This microbial assisted remediation process is very crucial in controlling mercury-based contamination. The reduced form of mercury Hg(0) can be easily volatilized, subsequently decreasing atmospheric level (Lovley and Lloyd 2000).

Genetic engineering is the recent and designer technology that can be adapted to modify microbes to make microbe assisted remediation more effective. Genetically modified *Ralstonia eutropha* was used to sequester Cd-contaminated soils (Valls et al. 2000). Using this approach, the *R. eutropha* was genetically engineered and designed to produce a cysteine-rich metal binding protein (metallothionein) on its cell surface. These receptors could quench Cd available in the soil. Presence of sequestered metals in soil is less harmful because of their less bioavailability. Two major loop holes for this remediation approach are the controversies for the use of genetically modified organisms (Urgun-Demirtas et al. 2006) and the remains of sequestered metals in soil.

One of the strategies used during bioremediation of contaminated soils to make it favorable for soil microbes is biostimulation which involves the use of organic amendments and addition of nutrients to the soil. Organic matter in the form of manure serves as a carbon source for soil microbes and the growth and activities of microorganisms involved in the process of remediation increase due to added nutrients, the only limitation here is to identify the right strain for biostimulation. Typically biostimulation is used to biodegrade organic pollutants (Abioye 2011) and can also be employed for the bioremediation of soils contaminated with heavy metals. Since heavy metals are less amenable to biodegradation, biostimulation indirectly enhances their bioremediation by altering soil PH. It is the introduction of organic matter that actually reduces the soil pH (McCauley et al. 2009). Solubility and bioavailability of heavy metals get increased by low pH of the soil, which subsequently increase its extractability from the soil (Karaca 2004). Biochar is an organic material which is known for its potential in the management of soils contaminated with heavy metals. Currently, it is being used for remediation purposes in various parts of the world. Experiments conducted by Namgay et al. (2010) stated that the presence of heavy metal in the contaminated soil amended with biochar was decreased. Unlike most other organic amendments biochar because of its ability to increase soil pH (Novak et al. 2009) reduces the bioavailability of metals for plant uptake due to increased sorption. Biochar properties differ broadly depending on the feedstock used in its production, and it also depends on the method employed during its production. Thus it is essential to note the effect of different amendments of biochar will have on its effectiveness when using for soil remediation. Further, in-depth research is required to comprehend how soil microorganism are influenced by biochar and how remediation process of contaminated soils is amended because such reports are rare in literature.

13.3.2 Phytoremediation

Phytoremediation ("Phyto" means plant) is a broad term used for a group of technologies that uses plants and associated microorganisms for in-situ partial or complete removal of selected contaminants from soil, sediments, sludge, groundwater, and wastewater. Phytoremediation can be employed to remove heavy metal contaminants, organic pollutants, as well as radionuclides (Ali et al. 2013; Bhat et al. 2018). Plants because of their unique absorption, metabolic capabilities, and remarkably efficient transportation systems selectively absorb nutrients or contaminants from the growth matrix and environment. Over the years, the emphasis was on phytoremediation since plants because of this unique property can be used to remediate soils contaminated with heavy metals. Phytoremediation is a technique that includes growing plants in a contaminated matrix, to eliminate contaminants or simplify immobilization (containment/binding) or degradation (detoxification) of the pollutants. The plants can be consequently processed, harvested, and disposed of. Driven by solar energy phytoremediation is an efficient, cost-effective, and environmentally friendly in situ remediation technology. There are different processes involved in phytoremediation which contain phytoextraction, phytofiltration, phytostabilization, photo-degradation, and phytovolatilization (Alkorta et al. 2004). The preliminary step of phytoremediation is the uptake of contaminants (phytoextraction) by plant roots from soil or water and their translocation and accumulation in plant tissues (Sekara et al. 2005). Translocation of metals to the aerial tissues of a plant is an important biochemical process. During photo-filtration, the plant absorbs water along with soluble components through roots (rhizofiltration), seedlings (blast filtration), or excised plant shoots (call filtration) (Mesjasz-Przybyłowicz et al. 2004). In phytoremediation, movement of metals in underground water is restricted because of their adsorption and absorption by the plants. In addition to this, other processes like phytoimmobilization or phytostabilization also reduce metal bioavailability and movement, which ultimately prevent their migration into the food chain and groundwater (Erakhrumen 2007). There are several plant-based approaches which immobilize metals present in contaminated soils which include precipitation, sorption through roots, reduction in the rhizosphere, and complex formation or metal valence (Poschenrieder and i Coll 2003). To metabolize organic pollutants, plants employ enzymes like oxygenase and dehalogenase, which do not need rhizospheric microorganisms for their action (Vishnoi and Srivastava 2007).

However, several volatile heavy metals like Hg and Se present in the contaminated soils are made ineffective by the process of phytovolatilization. In this process, plants convert them into volatile forms and are consequently released into the atmosphere (Karami and Shamsuddin 2010). However, phytovolatilization has limitations that it does not remove metals completely from the contaminated site. Instead, they are transferred from the site of contamination to the atmosphere from which they have chances to reenter the previous medium. The list of plants used in phytoremediation is given in Table 13.2. Further research in this area to identify and experimentally validate plant species resistant to heavy metal contaminated soil would be highly valuable in saving the environment.

Table 13.2 List of selected plants reported for phytoremediation of heavy metals

Heavy metal	Plant species	References
Cd	Castor (*Ricinus communis*)	Huang et al. (2011)
Cd, Pb, Zn	Corn (*Zea mays*)	Meers et al. (2010)
Hg	*Populus deltoides*	Che et al. (2003)
Se	*Brassica juncea, Astragalus bisulcatus*	Dixit et al. (2015)
Zn	*Populus canescens*	Bittsánszky et al. (2005)
Hg	*Eichhornia crassipes*	
Cr	*Spartina argentinensis*	Nalla et al. (2012)
Cd, Cu, Ni, Pb	Jatropha (*Jatropha curcas* L.)	Abhilash et al. (2009)
As	*Pteris vittata*	Datta et al. (2017)
Cd, Cu, Pb, Zn	*Populus* spp. (*Populus deltoides, Populus nigra, Populus trichocarpa*)	Ruttens et al. (2011)
Ni	*Alyssum bertolonii, Alyssum murale*	Mengoni et al. (2012)
Hg	*Pteris vittata*	Su et al. (2008)
Cd, Cu, Pb, Zn	*Salix* spp. (*Salix viminalis, Salix fragilis*)	Pulford and Watson (2003)
Pb	*Medicago sativa*	Su et al. (2008)

13.3.3 Natural Phytoextraction

Certain plants have the inherent property of metal accumulation. Phytoextraction utilizes the natural hyperaccumulator plants which comprise more than 1% of metal in their dry biomass. Hyper-accumulator plants accumulate 100-fold higher metals compared to non-accumulator ones. Studies have shown that a hyperaccumulator plant can collect more than 10 mg kg^{-1} Hg, 100 mg kg^{-1} Cd, 1000 mg kg^{-1} Pb, Cu, Cr, and metals and up to 10,000 mg kg^{-1} of Zn and Ni (Lasat 2002; Baker and Brooks 1989). This property of high metal tolerance of hyperaccumulator plant species has been employed on the metalliferous sites for bioremediation. Literature reports around 400 plant species belonging to 45 plant families can hyperaccumulate metals (Ghosh and Singh 2005; Henry 2000). The important families include *Euphorbiaceae, Brassicaceae, Lamiaceae, Fabaceae, Scrophulariaceae, and Asteraceae* (Dushenkov 2003; Salt et al. 1998). A report on the evaluation of hyperaccumulation potential of crops has shown that alpine pennycress (*Thlaspi caerulescens*), *Astragalus racemosus,* Ipomoea alpine, *Sebertia acuminate, Haumaniastrum robertii* contain great bioaccumulation potential for Co, Se, Cu, Ni, and Zn, respectively (Lasat 1999).

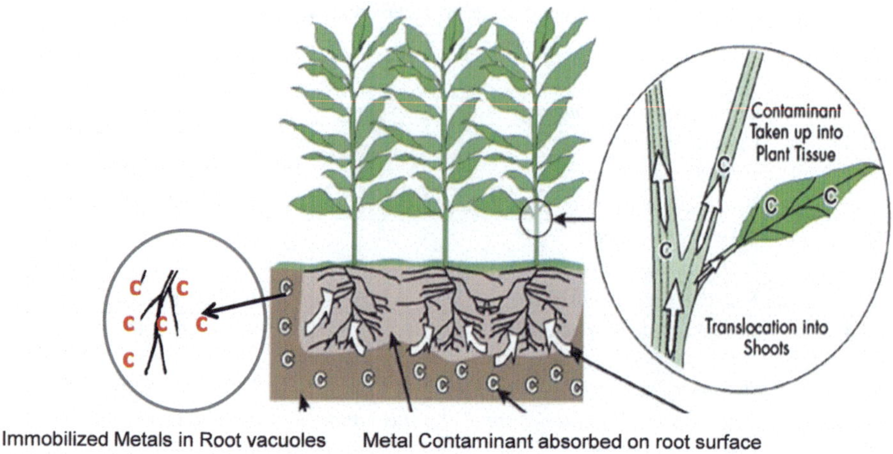

Fig. 13.2 Phytoextraction of metals from soil

Similarly, a report by Schmidt (2003) has shown that following plants have high uptake and tolerance to heavy metals: *Brassica juncea* L., *Helianthus annuus* L., *Zea mays* L., and *Salix viminalis* L. Some of the plants with hyper metal accumulation properties are listed in Table 13.2. Lasat (1999) had proposed steps involving metal absorption from the soil by the plant system. According to him, phytoextraction of metals from soil involves multiple processes (Fig. 13.2). Initially, the metal present in the proximity of the root surface gets sorbed on its surface and enters the root cells through the cellular membrane. Subsequently, some parts of the metal cross the metal cellular membranes into root vascular tissue (xylem), while the remaining part gets immobilized in the vacuole. From the vascular tissues, the metal gets translocated to stem and leaves. Most of the metals once inside the plant become insoluble and cannot move in the vascular system due to optimal cellular pH.

The insoluble metals are generally stored in extracellular (apoplastic) and intracellular (symplastic) compartments as precipitates of sulfate, carbonate, or phosphate (Raskin and Ensley 2000). Metal hyperaccumulator plants though are efficient in absorbing metals in large quantities from the soil, but they produce little biomass and are slow growing. To clean up contaminated sites using these plants, it may take years to decades. To overcome these shortfalls chemically improved phytoextraction has been employed. In this approach, chelating organic acids are used to enhance high biomass-producing crops to take up large amounts of metals (Nowack et al. 2006).

13.4 Combination Approach for the Remediation

To speed up the remediation process of contaminated soil, the combined use of microorganisms and plants has resulted in a faster and more proficient clean-up process (Weyens et al. 2009). Mycorrhizal fungi have been utilized in several heavy metal remediation processes. Studies on mycorrhizal metal uptake have shown that mycorrhiza use different approaches for the remediation of soils contaminated with heavy metals. Few studies have indicated greater accumulation of heavy metals in plants thus improved phytoextraction (Jamal et al. 2002; Weyens et al. 2009; Marques et al. 2006; Joner and Leyval 2001), others reports showed reduced metal concentration in plants and enhanced phytostabilization through metal immobilization (Janoušková et al. 2006; Heggo et al. 1990). The mycorrhizal associations with the plant are not limited to metal quenching only but it extends to water acquisition, plant growth, plant resistance to diseases, and increased nutrient content of the plant (Cardoso and Kuyper 2006; Harrier and Watson 2004; Wright et al. 2007). Many studies revealed that mycorrhizal association enhances the soil fertility, helps in the survival of plants growing in contaminated soils, and assists in the regeneration of remediated soil (Chibuike 2013). But there are also some reports that mycorrhizal association does not always help in soil remediation that is contaminated with heavy metals (Diaz et al. 1996; Joner and Leyval 1997) this is because not all the species of mycorrhizal fungi are effective in all the metals at any concentration (Chibuike 2013; Marques et al. 2009). Also, there are reports in the literature that heavy metals may inhibit the activities of mycorrhizal fungi (Del Val et al. 1999; Chao and Wang 1990). As compared to plants, some mycorrhizal fungi species *(Arbuscular mycorrhizal)* are found more sensitive to pollutants. Microorganism in association with plants is also used for bioremediation process apart from mycorrhiza. Among these are plant growth-promoting rhizobacteria (PGPR) which are present in the rhizosphere. These microbes assist plant growth by several mechanisms such as by supplying of nutrients and phytohormone production (Glick et al. 1995), siderophores production, and other chelating agents (Kamnev and Van der Lelie 2000). These PGPR also assist through specific enzyme activity, nitrogen fixation (Khan 2005), and reduction in ethylene production, which encourages root growth (Glick et al. 1998). In general, plant growth-promoting rhizobacteria (PGPR) have been used to reduce heavy metal associated plant stress in phytoremediation studies (Reed and Glick 2005). It has been observed that inoculation of *B. juncea* and *B. napus* plants with Bacillus species showed enhanced nickel and cadmium accumulation (Sheng and Xia 2006; Reed and Glick 2005; Zaidi et al. 2006). On the other hand, it was shown that inoculation of tomato plant with *Burkholderia species* and *Methylobacterium oryzae* showed improved plant growth due to reduction in Cd and Ni accumulation in the root and shoot tissues of the plant (Madhaiyan et al. 2007). This indicates that the mechanism employed in the process of phytoremediation by PGPR is a complex intertwined network and may dependent on the species of PGPR, the mechanisms employed by PGPR and plant involved in the process. Though reports on phytoremediation including both mycorrhizal fungi and PGPR are unusual, (Vivas et al. 2006) observed increased mycorrhizal efficiency due to PGPR (*Brevibacillus* sp.) which leads to decrease in metal accumulation and enhanced white clover growth on a heavy metal (Zn) contaminated soil.

13.5 Conclusion

Knowledge and possible threats of toxic heavy metals present in contaminated soils are vital for the selection of suitable remedial method. It is necessary to remediate the soil contaminated by heavy metals to make the land resource available for agricultural purposes, reduce the associated risks, scale down the land tenure problem, and enhance food security. Immobilization, soil washing, and phytoremediation are the best available technologies frequently used for cleaning up heavy metal contaminated soils. Bioremediation is the most appropriate disruptive method when used for cleanup of land used for agriculture purposes. Compared to microorganism assisted bioremediation use of plants (phytoremediation) is more common strategy to remediate the heavy metal contaminated soil. There are several methods employed by plants to remediate contaminated soil, which includes the most common method (phytoextraction). Combined use of microorganism and plants to remediate the contaminated soil will increase the efficacy of bioremediation process.

References

Abadin H, Ashizawa A, Stevens Y, Llados F, Diamond G, Sage G, Citra M, Quinones A, Bosch S, Swarts S (2007) Agency for toxic substances and disease registry (ATSDR) toxicological profiles. Toxicological profile for Lead. Agency for Toxic Substances and Disease Registry, Atlanta

Abhilash P, Jamil S, Singh N (2009) Transgenic plants for enhanced biodegradation and phytoremediation of organic xenobiotics. Biotechnol Adv 27(4):474–488

Abioye OP (2011) Biological remediation of hydrocarbon and heavy metals contaminated soil. In: Soil contamination. IntechOpen, London

Adriano D (2001) Trace elements in terrestrial environments: biogeochemistry, bioavailability, 2nd edn. Springer, New York

Ahluwalia SS, Goyal D (2007) Microbial and plant derived biomass for removal of heavy metals from wastewater. Bioresour Technol 98(12):2243–2257

Ainza C, Trevors J, Saier M (2010) Environmental mercury rising. Water Air Soil Pollut 205(1):47–48

Ajaz H, Arasuc R, Narayananb V, Zahir H (2010) Bioremediation of heavy metal contaminated soil by the exigobacterium and accumulation of Cd, Ni, Zn and Cu from soil environment. Int J Biol Technol 1(2):94–101

Ali H, Khan E, Sajad MA (2013) Phytoremediation of heavy metals—concepts and applications. Chemosphere 91(7):869–881

Alkorta I, Hernández-Allica J, Becerril J, Amezaga I, Albizu I, Garbisu C (2004) Recent findings on the phytoremediation of soils contaminated with environmentally toxic heavy metals and metalloids such as zinc, cadmium, lead, and arsenic. Rev Environ Sci Biotechnol 3(1):71–90

Baker A, Brooks R (1989) Terrestrial higher plants which hyperaccumulate metallic elements. A review of their distribution, ecology and phytochemistry. Biorecovery 1(2):81–126

Basta N, Gradwohl R (1998) Remediation of heavy metal-contaminated soil using rock phosphate. Better Crop 82(4):29–31

Basta N, Ryan J, Chaney R (2005) Trace element chemistry in residual-treated soil. J Environ Qual 34(1):49–63

Bhat RA, Dervash MA, Qadri H, Mushtaq N, Dar GH (2018) Macrophytes, the natural cleaners of toxic heavy metal (THM) pollution from aquatic ecosystems. In: Environmental contamination and remediation. Cambridge Scholars Publishing, Cambridge, pp 189–209
Bittsánszky A, Kömives T, Gullner G, Gyulai G, Kiss J, Heszky L, Radimszky L, Rennenberg H (2005) Ability of transgenic poplars with elevated glutathione content to tolerate zinc (2+) stress. Environ Int 31(2):251–254
Bjuhr J (2007) Trace metals in soils irrigated with waste water in a periurban area downstream Hanoi City, Vietnam. Sveriges lantbruksuniversitet, Uppsala
Buekers J (2007) Fixation of cadmium, copper, nickel and zinc in soil: kinetics, mechanisms and its effect on metal bioavailability. PhD, Department of Microbial and Molecular Systems, Katholieke Universiteit Leuven, Flanders, Belgium
Canet R, Pomares F, Tarazona F, Estela M (1998) Sequential fractionation and plant availability of heavy metals as affected by sewage sludge applications to soil. Commun Soil Sci Plant Anal 29(5–6):697–716
Cardoso IM, Kuyper TW (2006) Mycorrhizas and tropical soil fertility. Agric Ecosyst Environ 116(1–2):72–84
Chaney RL, Oliver DP (1996) Sources, potential adverse effects and remediation of agricultural soil contaminants. In: Contaminants and the soil environment in the Australasia-Pacific region. Springer, Dordrecht, pp 323–359
Chao C, Wang Y (1990) Effects of heavy-metals on the infection of vesicular-arbuscular mycorrhizae and the growth of maize. J Agric Assoc China 152:34–45
Che D, Meagher RB, Heaton AC, Lima A, Rugh CL, Merkle SA (2003) Expression of mercuric ion reductase in eastern cottonwood (Populus deltoides) confers mercuric ion reduction and resistance. Plant Biotechnol J 1(4):311–319
Chibuike G (2013) Use of mycorrhiza in soil remediation: a review. Sci Res Essays 8(35):679–1687
Coblenz A, Wolf K (1994) The role of glutathione biosynthesis in heavy metal resistance in the fission yeast Schizosaccharomyces pombe. FEMS Microbiol Rev 14(4):303–308
D'amore J, Al-Abed S, Scheckel K, Ryan J (2005) Methods for speciation of metals in soils. J Environ Qual 34(5):1707–1745
Das, N., R. Vimala, and P. Karthika. 2008. Biosorption of heavy metals–an overview
Datta R, Das P, Tappero R, Punamiya P, Elzinga E, Sahi S, Feng H, Kiiskila J, Sarkar D (2017) Evidence for exocellular arsenic in fronds of Pteris vittata. Sci Rep 7(1):2839
Degraeve N (1981) Carcinogenic, teratogenic and mutagenic effects of cadmium. Mutat Res 86(1):115–135
Del Val C, Barea J, Azcon-Aguilar C (1999) Diversity of arbuscular mycorrhizal fungus populations in heavy-metal-contaminated soils. Appl Environ Microbiol 65(2):718–723
DeVolder PS, Brown SL, Hesterberg D, Pandya K (2003) Metal bioavailability and speciation in a wetland tailings repository amended with biosolids compost, wood ash, and sulfate. J Environ Qual 32(3):851–864
Diaz G, Azcón-Aguilar C, Honrubia M (1996) Influence of arbuscular mycorrhizae on heavy metal (Zn and Pb) uptake and growth of Lygeum spartum and Anthyllis cytisoides. Plant Soil 180(2):241–249
Dixit R, Malaviya D, Pandiyan K, Singh U, Sahu A, Shukla R, Singh B, Rai J, Sharma P, Lade H (2015) Bioremediation of heavy metals from soil and aquatic environment: an overview of principles and criteria of fundamental processes. Sustainability 7(2):2189–2212
Doleman P (1994) Effect of heavy metals in soil on microbial diversity and activity as shown by the sensitivity-resistance, index, an ecologically relevant parameter. Biol Fertil Soils 12:5–9
Duda-Chodak A, Blaszczyk U (2008) The impact of nickel on human health. J Elem 13(4):685–693
Dushenkov S (2003) Trends in phytoremediation of radionuclides. Plant Soil 249(1):167–175
Erakhrumen AA (2007) Phytoremediation: an environmentally sound technology for pollution prevention, control and remediation in developing countries. Educ Res Rev 2(7):151–156
Garbisu C, Alkorta I (1997) Bioremediation: principles and future. *Journal of Clean Technology Environmental Toxicology and*. Occup Med 6(4):351–366

Garbisu C, Alkorta I (2003) Basic concepts on heavy metal soil bioremediation. Eur J Miner Process Environ Prot 3(1):58–66

Garbisu C, Alkorta I, Llama MJ, Serra JL (1998) Aerobic chromate reduction by Bacillus subtilis. Biodegradation 9(2):133–141

Garbisu C, Llama MJ, Serra JL (1997) Effect of heavy metals on chromate reduction by Bacillus subtilis. J Gen Appl Microbiol 43(6):369–371

Ghosh M, Singh S (2005) A review on phytoremediation of heavy metals and utilization of it's by products. Asian J Energy Environ 6(4):18

Glick BR, Karaturovic DM, Newell PC (1995) A novel procedure for rapid isolation of plant growth promoting pseudomonads. Can J Microbiol 41(6):533–536

Glick BR, Penrose DM, Li J (1998) A model for the lowering of plant ethylene concentrations by plant growth-promoting bacteria. J Theor Biol 190(1):63–68

Gulati K, Banerjee B, Lall SB, Ray A (2010) Effects of diesel exhaust, heavy metals and pesticides on various organ systems: possible mechanisms and strategies for prevention and treatment. Indian J Exp Biol 48(7):710–721

Harrier LA, Watson CA (2004) The potential role of arbuscular mycorrhizal (AM) fungi in the bioprotection of plants against soil-borne pathogens in organic and/or other sustainable farming systems. Pest Manag Sci 60(2):149–157

Heggo A, Angle J, Chaney R (1990) Effects of vesicular-arbuscular mycorrhizal fungi on heavy metal uptake by soybeans. Soil Biol Biochem 22(6):865–869

Henry JR (2000) An overview of the phytoremediation of lead and mercury. National Network of environmental management studies (NNEMS). Environmental Protection Agency, Washington, DC, pp 1–31

Hess R, Schmid B (2002) Zinc supplement overdose can have toxic effects. J Paediatr Haematol Oncol 24:582–584

Hobman J (1997) Bacterial mercury-resistance genes. Met Ions Biol Syst 34:527–568

Huang H, Yu N, Wang L, Gupta D, He Z, Wang K, Zhu Z, Yan X, Li T, Yang X-e (2011) The phytoremediation potential of bioenergy crop Ricinus communis for DDTs and cadmium co-contaminated soil. Bioresour Technol 102(23):11034–11038

Huyer M, Page WJ (1988) Zn2+ increases siderophore production in Azotobacter vinelandii. Appl Environ Microbiol 54(11):2625–2631

Ishibashi Y, Cervantes C, Silver S (1990) Chromium reduction in Pseudomonas putida. Appl Environ Microbiol 56(7):2268–2270

Jacobs IA, Taddeo J, Kelly K, Valenziano C (2002) Poisoning as a result of barium styphnate explosion. Am J Ind Med 41(4):285–288

Jamal A, Ayub N, Usman M, Khan AG (2002) Arbuscular mycorrhizal fungi enhance zinc and nickel uptake from contaminated soil by soybean and lentil. Int J Phytoremediation 4(3):205–221

Janoušková M, Pavlíková D, Vosátka M (2006) Potential contribution of arbuscular mycorrhiza to cadmium immobilisation in soil. Chemosphere 65(11):1959–1965

Jiang W, Liu D, Hou W (2001) Hyperaccumulation of cadmium by roots, bulbs and shoots of garlic (Allium sativum L.). Bioresour Technol 76(1):9–13

Joner E, Leyval C (1997) Uptake of 109Cd by roots and hyphae of a Glomus mosseae/Trifolium subterraneum mycorrhiza from soil amended with high and low concentrations of cadmium. New Phytol 135(2):353–360

Joner E, Leyval C (2001) Time-course of heavy metal uptake in maize and clover as affected by root density and different mycorrhizal inoculation regimes. Biol Fertil Soils 33(5):351–357

Jones L, Jarvis S, Green D, Hayes M (1981) The fate of heavy metals in the chemistry of soil processes. Wiley, New York

Kaasalainen M, Yli-Halla M (2003) Use of sequential extraction to assess metal partitioning in soils. Environ Pollut 126(2):225–233

Kabata-Pendias A, Pendias H (2001) Trace elements in soils and plants. CRC Press, Boca Raton

Kamnev AA, van der Lelie D (2000) Chemical and biological parameters as tools to evaluate and improve heavy metal phytoremediation. Biosci Rep 20(4):239–258

Kapoor A, Viraraghavan T (1995) Fungal biosorption—an alternative treatment option for heavy metal bearing wastewaters: a review. Bioresour Technol 53(3):195–206

Karaca A (2004) Effect of organic wastes on the extractability of cadmium, copper, nickel, and zinc in soil. Geoderma 122(2–4):297–303

Karami A, Shamsuddin ZH (2010) Phytoremediation of heavy metals with several efficiency enhancer methods. Afr J Biotechnol 9(25):3689–3698

Keller C, McGrath S, Dunham S (2002) Trace metal leaching through a soil–grassland system after sewage sludge application. J Environ Qual 31(5):1550–1560

Khan AG (2005) Role of soil microbes in the rhizospheres of plants growing on trace metal contaminated soils in phytoremediation. J Trace Elem Med Biol 18(4):355–364

Khan MA, Ahmad I, Rahman IU (2007) Effect of environmental pollution on heavy metals content of *Withania somnifera*. J Chin Chem Soc 54(2):339–343

Khan S, Cao Q, Zheng Y, Huang Y, Zhu Y (2008) Health risks of heavy metals in contaminated soils and food crops irrigated with wastewater in Beijing, China. Environ Pollut 152(3):686–692

Kirpichtchikova TA, Manceau A, Spadini L, Panfili F, Marcus MA, Jacquet T (2006) Speciation and solubility of heavy metals in contaminated soil using X-ray microfluorescence, EXAFS spectroscopy, chemical extraction, and thermodynamic modeling. Geochim Cosmochim Acta 70(9):2163–2190

Kuo S a, Heilman P, Baker A (1983) Distribution and forms of copper, zinc, cadmium, iron, and manganese in soils near a copper smelter1. Soil Sci 135(2):101–109

Lasat M (1999) Phytoextraction of metals from contaminated soil: a review of plant/soil/metal interaction and assessment of pertinent agronomic issues. J Hazard Subst Res 2(1):5

Lasat MM (2002) Phytoextraction of toxic metals. J Environ Qual 31(1):109–120

Levy D, Barbarick K, Siemer E, Sommers L (1992) Distribution and partitioning of trace metals in contaminated soils near Leadville, Colorado. J Environ Qual 21(2):185–195

Li F, Tan T (1994) Monitoring BOD in the presence of heavy metal ions using a poly (4-vinylpyridine)-coated microbial sensor. Biosens Bioelectron 9(6):445–455

Lombi E, Gerzabek MH (1998) Determination of mobile heavy metal fraction in soil: results of a pot experiment with sewage sludge. Commun Soil Sci Plant Anal 29(17–18):2545–2556

Lovley DR, Lloyd JR (2000) Microbes with a mettle for bioremediation. Nat Biotechnol 18(6):600

Madhaiyan M, Poonguzhali S, Sa T (2007) Metal tolerating methylotrophic bacteria reduces nickel and cadmium toxicity and promotes plant growth of tomato (Lycopersicon esculentum L.). Chemosphere 69(2):220–228

Marques AP, Oliveira RS, Rangel AO, Castro PM (2006) Zinc accumulation in Solanum nigrum is enhanced by different arbuscular mycorrhizal fungi. Chemosphere 65(7):1256–1263

Marques AP, Rangel AO, Castro PM (2009) Remediation of heavy metal contaminated soils: phytoremediation as a potentially promising clean-up technology. Crit Rev Environ Sci Technol 39(8):622–654

Martin TA, Ruby MV (2004) Review of in situ remediation technologies for lead, zinc, and cadmium in soil. Remediat J 14(3):35–53

Maslin P, Maier RM (2000) Rhamnolipid-enhanced mineralization of phenanthrene in organic-metal co-contaminated soils. Biorem J 4(4):295–308

McCauley A, Jones C, Jacobsen J (2009) Soil pH and organic matter. Nutr Manag Modul 8:1–12

McLaren R, Clucas L, Taylor M (2005) Leaching of macronutrients and metals from undisturbed soils treated with metal-spiked sewage sludge. 3. Distribution of residual metals. Soil Res 43(2):159–170

McLaren R, Clucas L, Taylor M, Hendry T (2004) Leaching of macronutrients and metals from undisturbed soils treated with metal-spiked sewage sludge. 2. Leaching of metals. Soil Res 42(4):459–471

McLaughlin MJ, Hamon R, McLaren R, Speir T, Rogers S (2000a) A bioavailability-based rationale for controlling metal and metalloid contamination of agricultural land in Australia and New Zealand. Soil Res 38(6):1037–1086

McLaughlin MJ, Zarcinas B, Stevens D, Cook N (2000b) Soil testing for heavy metals. Commun Soil Sci Plant Anal 31(11–14):1661–1700
Meers E, Van Slycken S, Adriaensen K, Ruttens A, Vangronsveld J, Du Laing G, Witters N, Thewys T, Tack F (2010) The use of bio-energy crops (Zea mays) for 'phytoattenuation' of heavy metals on moderately contaminated soils: a field experiment. Chemosphere 78(1):35–41
Mengoni A, Cecchi L, Gonnelli C (2012) Nickel hyperaccumulating plants and Alyssum bertolonii: model systems for studying biogeochemical interactions in serpentine soils. In: Bio-geo interactions in metal-contaminated soils. Springer, Berlin, pp 279–296
Mesjasz-Przybyłowicz J, Nakonieczny M, Migula P, Augustyniak M, Tarnawska M, Reimold W, Koeberl C, Przybyłowicz W, Głowacka E (2004) Uptake of cadmium, lead nickel and zinc from soil and water solutions by the nickel hyperaccumulator Berkheya coddii. Acta Biol Cracov Ser Bot 46:75–85
Mushtaq N, Singh DV, Bhat RA, Dervash MA, Hameed UB (2020) Freshwater contamination: sources and hazards to aquatic biota. In: Qadri H, Bhat RA, Dar GH, Mehmood MA (eds) Freshwater pollution dynamics and remediation. Springer, Singapore, pp 27–50
Nalla S, Hardaway CJ, Sneddon J (2012) Phytoextraction of selected metals by the first and second growth seasons of Spartina alterniflora. Instrum Sci Technol 40(1):17–28
Namgay T, Singh B, Singh BP (2010) Influence of biochar application to soil on the availability of As, Cd, Cu, Pb, and Zn to maize (Zea mays L.). Soil Res 48(7):638–647
Novak JM, Busscher WJ, Laird DL, Ahmedna M, Watts DW, Niandou MA (2009) Impact of biochar amendment on fertility of a southeastern coastal plain soil. Soil Sci 174(2):105–112
Nowack B, Schulin R, Robinson BH (2006) Critical assessment of chelant-enhanced metal phytoextraction. Environ Sci Technol 40(17):5225–5232
Padmavathiamma PK, Li LY (2007) Phytoremediation technology: hyper-accumulation metals in plants. Water Air Soil Pollut 184(1–4):105–126
Pierzynski G, Sims J, Vance G (2000) Soils and environmental quality. CRC Press, Boca Raton
Poschenrieder C, i Coll JB (2003) Phytoremediation: principles and perspectives. Contrib Sci 2(3):333–344
Pulford I, Watson C (2003) Phytoremediation of heavy metal-contaminated land by trees—a review. Environ Int 29(4):529–540
Qadri H and Bhat RA. (2020). The concerns for global sustainability of freshwater ecosystems. Freshwater pollution dynamics and remediation. Qadri H, Bhat RA Dar GH, Mehmood MA. Springer, Singapore: 1–15
Raskin I, Ensley BD (2000) Phytoremediation of toxic metals. Wiley, New York
Reed ML, Glick BR (2005) Growth of canola (Brassica napus) in the presence of plant growth-promoting bacteria and either copper or polycyclic aromatic hydrocarbons. Can J Microbiol 51(12):1061–1069
Reed SC, Crites RW, Middlebrooks EJ (1995) Natural systems for waste management and treatment. McGraw-Hill, New York
Ruttens A, Boulet J, Weyens N, Smeets K, Adriaensen K, Meers E, Van Slycken S, Tack F, Meiresonne L, Thewys T (2011) Short rotation coppice culture of willows and poplars as energy crops on metal contaminated agricultural soils. Int J Phytoremediation 13(sup1):194–207
Salem HM, Eweida EA, Farag A (2000) Heavy metals in drinking water and their environmental impact on human health. In: ICEHM2000, Cairo University, Egypt, pp 542–556
Salt DE, Smith R, Raskin I (1998) Phytoremediation. Annu Rev Plant Biol 49(1):643–668
Schmidt U (2003) Enhancing phytoextraction. J Environ Qual 32(6):1939–1954
Schmöger ME, Oven M, Grill E (2000) Detoxification of arsenic by phytochelatins in plants. Plant Physiol 122(3):793–802
Sekara A, Poniedzialeek M, Ciura J, Jedrszczyk E (2005) Cadmium and lead accumulation and distribution in the organs of nine crops: implications for phytoremediation. Pol J Environ Stud 14(4):509–516
Shafi S, Bhat RA, Bandh SA, Shameem N, Nisa H (2018) Microbes: key agents in the sustainable environment and cycling of nutrients. In: Environmental contamination and remediation. Cambridge Scholars Publishing, Cambridge, pp 179–188

Sheng X-F, Xia J-J (2006) Improvement of rape (Brassica napus) plant growth and cadmium uptake by cadmium-resistant bacteria. Chemosphere 64(6):1036–1042
Shiowatana J, McLaren RG, Chanmekha N, Samphao A (2001) Fractionation of arsenic in soil by a continuous-flow sequential extraction method. J Environ Qual 30(6):1940–1949
Smejkalova M, Mikanova O, Boruvka L (2003) Effects of heavy metal concentrations on biological activity of soil micro-organisms. Plant Soil Environ 49(7):321–326
Sposito G, Page AL (1984) Cycling of metal ions in the soil environment. Met Ions Biol Syst 18:287–332
Su Y, Han FX, Chen J, Sridhar BM, Monts DL (2008) Phytoextraction and accumulation of mercury in three plant species: Indian mustard (Brassica juncea), beard grass (Polypogon monospeliensis), and Chinese brake fern (Pteris vittata). Int J Phytoremediation 10(6):547–560
Sumner ME (2000) Beneficial use of effluents, wastes, and biosolids. Commun Soil Sci Plant Anal 31(11–14):1701–1715
Tripathi RD, Srivastava S, Mishra S, Singh N, Tuli R, Gupta DK, Maathuis FJ (2007) Arsenic hazards: strategies for tolerance and remediation by plants. Trends Biotechnol 25(4):158–165
Urgun-Demirtas M, Stark B, Pagilla K (2006) Use of genetically engineered microorganisms (GEMs) for the bioremediation of contaminants. Crit Rev Biotechnol 26(3):145–164
Valls M, Atrian S, de Lorenzo V, Fernández LA (2000) Engineering a mouse metallothionein on the cell surface of Ralstonia eutropha CH34 for immobilization of heavy metals in soil. Nature biotechnology, 18(6):661–665
Van der Lelie D, Corbisier P, Diels L, Gilis A, Lodewyckx C, Mergeay M, Taghavi S, Spelmans N, Vangronsveld J (1999) 15. The role of bacteria in the phytoremediation of heavy metals. In: Phytoremediation of contaminated soil and water. CRC Press, Boca Raton, p 265
Vinceti M, Wei ET, Malagoli C, Bergomi M, Vivoli G (2001) Adverse health effects of selenium in humans. Rev Environ Health 16(4):233–252
Vishnoi SR, Srivastava P (2007) Phytoremediation–green for environmental clean. Paper read at proceedings of Taal2007: the 12th world lake conference
Vivas A, Biro B, Ruiz-Lozano J, Barea J, Azcon R (2006) Two bacterial strains isolated from a Zn-polluted soil enhance plant growth and mycorrhizal efficiency under Zn-toxicity. Chemosphere 62(9):1523–1533
Wang P-C, Mori T, Komori K, Sasatsu M, Toda K, Ohtake H (1989) Isolation and characterization of an Enterobacter cloacae strain that reduces hexavalent chromium under anaerobic conditions. Appl Environ Microbiol 55(7):1665–1669
Weyens N, van der Lelie D, Taghavi S, Newman L, Vangronsveld J (2009) Exploiting plant–microbe partnerships to improve biomass production and remediation. Trends Biotechnol 27(10):591–598
White C, Shaman AK, Gadd GM (1998) An integrated microbial process for the bioremediation of soil contaminated with toxic metals. Nat Biotechnol 16(6):572
Wood JM, Wang H-K (1983) Microbial resistance to heavy metals. Environ Sci Technol 17(12):582A–590A
Wright S, Green V, Cavigelli M (2007) Glomalin in aggregate size classes from three different farming systems. Soil Tillage Res 94(2):546–549
Zaidi S, Usmani S, Singh BR, Musarrat J (2006) Significance of Bacillus subtilis strain SJ-101 as a bioinoculant for concurrent plant growth promotion and nickel accumulation in Brassica juncea. Chemosphere 64(6):991–997
Zhang M-K, Liu Z-Y, Wang H (2010) Use of single extraction methods to predict bioavailability of heavy metals in polluted soils to rice. Commun Soil Sci Plant Anal 41(7):820–831
Zhao Q, Kaluarachchi JJ (2002) Risk assessment at hazardous waste-contaminated sites with variability of population characteristics. Environ Int 28(1–2):41–53

Chapter 14
Environmental Biotechnology: For Sustainable Future

Malik Muzafar Manzoor

14.1 Introduction

Environmental sustainability is one of the biggest issues faced by humankind. Rapid and rampant industrialization and urbanization have put global land resources under severe threat. To make our planet a sustainable ecosystem and suitable habitable for future generations, we not only need to make corrections in existing methods but also need to remediate the polluted natural resources using natural ways. Recycling of biomass, development of production processes that makes optimal use of natural resources, recovering energy, and minimizing waste generation can have favourable effects on the environment. Despite escalating efforts to prevent waste accumulation and to promote recycling, the amount of environmental damage caused by the high standard of living, over-consumption, the quantities of waste generated, and the degree of unsustainable land use appear likely to continue growing. Environmental biotechnology is one of the options to prevent, arrest, and reverse environmental degradation through the appropriate use of biotechnology in combination with other technologies while supporting safety procedures as a primary component of the program. Biotechnological processes and their products can be applied with a view of long-term ecological security to protect environmental integrity (Dervash et al. 2020). While bioremediation is acknowledged as a promising technology for restoring polluted and degraded lands, its field potential is limited for various reasons. However, recent biotechnological advancements, including producing efficient microbial consortia, applying enzymes with higher degrees of specificity, and designing plants with specific microbial partners, are opening new prospects in remediation technology.

M. M. Manzoor
Plant Biotechnology Division, CSIR-Indian Institute of Integrative Medicine, Jammu, India

R. A. Bhat et al. (eds.), *Bioremediation and Biotechnology, Vol 2*,
https://doi.org/10.1007/978-3-030-40333-1_14

14.2 Recent Biotechnological Advancements for On-Site Cleaning

To speed up the remediation process of the contaminated environment, the combined use of microorganisms and plants has resulted in a faster and more proficient cleanup process (Weyens et al. 2009). With the advent of the transgenic technology, plants and microorganisms can be engineered to degrade pollutants. Most of the sites are contaminated with different pollutants; thus, onsite bioremediation needs different strains of bacteria with the diverse catabolic potential to degrade pollutants. Advanced biotechnological approaches offer opportunities to isolate and culture these useful bacterial strains for their in-depth study. Next generation sequencing technology offers clues to engineer strains so that their remediation potential can be enhanced (Su et al. 2015).

High-throughput RNA sequencing can identify viable but nonculturable (VBNC) states under unfavourable conditions of biphenyl/polychlorinated biphenyl (PCB)-degrading bacteria *Rhodococcus biphenylivorans* (Su et al. 2015, Khan et al. 2017). VBNC state identification of microorganisms which are used for remediation of the polluted environment is vital for knowing their functional state during field application, and it also assists in providing favourable conditions for enhancing their remediation potential. Microbe capturing using immunomagnetic beads and colorimetric detection of their outer membrane peroxidase by oxidizing 3,3′,5,5′-tetramethylbenzene in the presence of H_2O_2 is another cost-effective method for the rapid enumeration and sensitive detection of uranium-reducing *Shewanella oneidensis* from sediment samples (Wen et al. 2014).

Genetic engineering is the recent and designer technology that can be adapted to modify microbes to make microbe assisted remediation more effective. Genetically modified *Ralstonia eutropha* was used to sequester Cd-contaminated soils using this approach, the *R. eutropha* was genetically engineered and designed to produce a cysteine-rich metal-binding protein (metallothionein) on its cell surface. These receptors could quench Cd available in the soil. The presence of sequestered metals in soil is less harmful because of their less bioavailability. Two major loopholes for this remediation approach are the controversies for the use of genetically modified organisms and the remains of sequestered metals in soil.

14.3 Bioremediation of Heavy Metal Contaminated Soils

Expanding industries, land application of fertilizers, spillage of petrochemicals, coal combustion, animal manures, wastewater irrigation, and sewage slime are the potential sources of heavy metals which contaminate soil. Different techniques of physical, chemical to biological methods for remediating metal contaminated soils exist. Most physical and chemical methods (vapour extraction, solidification, electrokinetics, soil flushing, soil washing, and stabilization) left the soil unsuitable for plant growth. On the other hand, biological approach (Bioremediation) which is achieved through natural processes encourages the use of remediated soil for the growth of plants. Phytoremediation ("Phyto" means plant) is a broad term used for a group of technologies that uses plants and associated microorganisms for in situ partial or complete removal of selected contaminants from soil, sediments, sludge, groundwater, and wastewater. Phytoremediation can be employed to remove heavy metal contaminants, organic pollutants as well as radionuclides (Ali et al. 2013). Plants because of their unique absorption, metabolic capabilities, and remarkably efficient transportation systems selectively absorb nutrients or contaminants from the growth matrix and environment.

To speed up the remediation process of contaminated soil, the combined use of microorganisms and plants has resulted in a faster and more proficient cleanup process (Weyens et al. 2009). The microbial remediation involves a cocktail of several microorganisms like bacteria (*B. subtilis*), Enterobacter (E cloacae), and Pseudomonas (P putida) that have been effectively used for the reduction of Cr (VI) to the less toxic Cr (III) (Garbisu and Alkorta 1997, Garbisu et al. 1998, Ishibashi et al. 1990). It is also reported in the literature that B. subtilis can reduce non-metallic elements. For example, Garbisu et al. (1998) studied that selenite has been decreased to the less toxic elemental selenium using *B. Subtilis*. In a study, it was observed that *Azotobacter vinelandii* showed increased siderophore production in the presence of Zn(II) (Huyer and Page 1988). The change in bioavailability could have led to the natural extraction of these metals from the soil. Besides the change in bioavailability, precipitation by sulphate-reducing bacteria (*Desulfovibrio desulfuricans*) could also lead to bioremediation. This process is indirect and mostly ex situ where the sulphate-reducing bacteria convert sulphate forms of heavy metal into their hydrogen sulphate, as in heavy metals Zn and Cd, to form insoluble forms of these metal sulphides (White et al. 1998).

One of the strategies used during bioremediation of contaminated soils to make it favourable for soil microbes is biostimulation which involves the use of organic amendments and accumulation of nutrients to the soil. Organic matter in the form of manure serves as a carbon source for soil microbes and the growth and activities of microorganisms involved in the process of remediation increase due to added nutrients, the only limitation here is to identify the right strain for biostimulation, which can be identified by using next generation sequencing and other biotechnological approaches. Typically biostimulation is used to biodegrade organic pollutants (Abioye 2011) and can also be employed for the bioremediation of soils contaminated with heavy metals. Since heavy metals are less amenable to biodegradation, biostimulation indirectly enhances their bioremediation by altering soil pH. It is the introduction of organic matter that actually reduces the soil pH (McCauley et al. 2009). Solubility and bioavailability of heavy metals get increased by the low pH of the soil, which subsequently increase its extractability from the soil (Karaca 2004). Biochar is an organic material which is known for its potential in the management of soils contaminated with heavy metals. Currently, it is being used for remediation purposes in various parts of the world. Experiments conducted by Namgay et al. (2010) stated that the presence of heavy metal in the contaminated soil amended with biochar was decreased. Unlike most other organic amendments biochar because of its ability to increase soil PH (Novak et al. 2009) reduces the bioavailability of metals for plant uptake due to increased sorption. Biochar properties differ broadly depending on the feedstock used in its production, and it also depends on the method employed during its production. Thus it is essential to note the effect of different amendments of biochar will have on its effectiveness when using for soil remediation. Further, in-depth research is required to comprehend how soil microorganisms are influenced by biochar and how remediation process of contaminated soils is amended because such reports are rear in literature. The insoluble metals are generally stored in extracellular (apoplastic) and intracellular (symplastic) compartments as precipitates of sulphate, carbonate, or phosphate (Raskin and Ensley 2000). Metal hyperaccumulator plants though are efficient in absorbing metals in large quantities from the soil, but they produce little biomass and are slow growing. To clean up contaminated sites using these plants, it may take years to decades. To overcome these shortfalls chemically improved phytoextraction has been employed. In this approach, chelating organic acids are used to enhance high biomass-producing crops to take up large amounts of metals (Nowack et al. 2006).

14.4 Wastewater Treatment

Conventional wastewater treatment systems are energy inefficient processes and inadequate to achieve discharge standards in effluent prescribed by various agencies (Caputo and Pelagagge 2001). It involves primary and secondary processes followed by disinfection for disposal of wastewater. With the advancement of technologies nutrients and harmful chemicals are removed, which ultimately decrease the oxygen demand of the wastewater and improve the quality of the treated water so that it can be reused. To stop eutrophication and elevated oxygen demand of wastewater an additional operation unit (Tertiary treatment) is added to the conventional treatment process which is generally employed to remove nutrient load (Phosphate and nitrates) which are the major drivers of eutrophication and high oxygen demand (Schaar et al. 2010). Biological wastewater treatments gained popularity in recent years. Conventional aerobic treatments used to reduce nutrients and organic load from wastewater are not efficient enough to remove most of the contaminants and it cannot be achieved without the combined use of both that is aerobic and anaerobic treatments (Okoh et al. 2010). Algae are known to breakdown the organic material to obtain energy and nourishment for their growth and development present in wastewater. They are wonderful biological mediators that own potential to convert solar energy and carbon dioxide into valuable products such as biodiesel, bioethanol, and biobutanol. Due to non-pathogenic nature and capability of algae biomass to grown in different effluents like municipal, industrial, and agricultural effluents, they are used for bioprocessing of wastewater (Chinnasamy et al. 2010). Two stage Sludge digestion system in which organic mass is digested by bacteria in absence of oxygen. The first stage of this system involves the break down of large molecules like proteins and lipids into smaller water soluble molecule by acid forming bacteria which finally ferment these molecules into fatty acids. The second stage involves the movement of soluble molecules into the second tank where bacteria convert soluble molecules into biogas (Carbon dioxide and Methane). Methane is used to generate heat for the operation of the treatment plant and is also used for the generation of electricity (Andreoli et al. 2007). Ballasted floc reactor, integrated fixed-film activated sludge (IFAS), and membrane bioreactor process are some examples of improved treatment methods of wastewater.

14.4.1 Molecular Techniques in Wastewater Treatment

Application of molecular techniques to treat wastewater is quite new and are economical than the already established techniques. Majority of these are used to make wastewater xenobiotic free. Specific nucleic acid probes are used to detect harmful microorganism from wastewater (Khan et al. 2004; Sanz and Köchling 2007). Other techniques which proved to be most effective and interesting for detection process are fluorescent in situ hybridization (FISH), genomics library, and denaturant gradient cell electrophoresis (DGGE) (Sanz and Köchling 2007). Use of recombinant technology has also its own disadvantages and limitations (Timmis et al. 1994) such as instability of recombinant strain, complex xenobiotic degradation pathways, and accidental or deliberate release of genetically modified strains into the environment.

14.5 Microbial Enzymes in Bioremediation

14.5.1 Microbial Oxidoreductase

Various bacteria, fungi, and higher plants detoxify toxic organic compounds through oxidative coupling which is mediated by oxidoreductases (Gianfreda et al. 1999: Bollag et al. 1998). These enzymes are also involved in microbial energy extraction via energy-yielding biochemical reactions. Oxidoreductases also assist in the transfer of electrons from the reduced substrate to acceptor compound, which results in oxidization of contaminant to the harmless compound. Oxidoreductases participate in the humification of the phenolic compound resulting from the decomposition of lignin, and they also can detoxify xenobiotics by binding them to humic substances or through polymerization and copolymerization with other substrates (Park et al. 2006). Enzymes of microbial origin have been exploited in the degradation and decolourization of dyes (Vidali 2001; Williams 1977; Husain 2006). Some bacterial strains have tendency to reduce the radioactive metals from their oxidized soluble form to reduced insoluble form. Some bacteria during their energy production process takes up electrons from organic compounds and make radioactive metals as the final electron acceptor. To reduce radioactive metals some strains of bacteria use intermediate electron donor. It is the redox reactions system of metal-reducing bacteria, which finally precipitate the contaminant (Leung 2004). Effluents generated from paper and pulp industry contain chlorinated phenolic compounds which are produced from the partial degradation of lignin during the process of pulp bleaching. Many fungal species due to their extracellular oxidoreductase enzymes like manganese peroxidise, laccase, and peroxidise are considered suitable for the treatment of environment contaminated with phenolic compounds. Fungi release these enzymes from their mycelium into the environment (Rubilar et al. 2008). Phytoremediation of phenolic compounds using plants from Fabaceae, Gramineae, and Solanaceae families. They degrade contaminants using their oxidoreductases is a promising natural approach to decontaminate the environment. Organic contaminants like explosives, petroleum hydrocarbons, and chlorinated solvents are generally focused (Duran and Esposito 2000, Newman et al. 1998).

14.5.2 *Microbial Dioxygenases*

Dioxygenases are multicomponent enzyme systems that introduce molecular oxygen into their substrate Rieske nonheme iron oxygenases is one of the large family of dioxygenases which include aromatic hydrogen dioxygenase. These dioxygenases have wide application in environmental remediation because of their involvement in the oxygenation of wide range substrates (Fig. 14.1). All members of this family have one or two electron transport proteins preceding their oxygenase components. The presence of Rieske (2Fe–2S) cluster and mononuclear iron in each alpha subunit of the enzyme has been confirmed from the crystal structure of

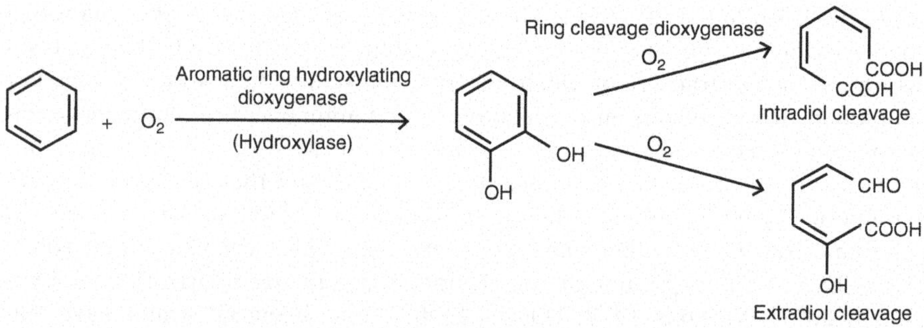

Fig. 14.1 Degradation of aromatic compound by dioxygenase

naphthalene dioxygenase (Dua et al. 2002). Catechol dioxygenases found in soil bacteria are involved in the transformation of aromatic precursors into aliphatic products.

14.5.3 Microbial Hydrolytic Enzymes

The industrialization has led to the pollution of soil and water through the discharge of effluents by the industrial units. Industrial discharge onto land contains many harmful chemicals, petroleum hydrocarbons, mineral acids, bases, and many more chemicals of unknown nature. Bioremediation provides an economical and safe alternative to commonly used chemical treatments. Organic pollutants are mostly hydrolyzed by bacterial activity. It is the extracellular enzyme activity that has the key role in degradation since bacterial cell pores can only allow molecules lower than 600 daltons to enter the cell (Vasileva-Tonkova and Galabova 2003). Disruption of major chemical bonds of toxic compounds by bacterial hydrolytic enzymes results in the reduction of their toxicity. Hydrolytic enzymes of microbial origin are effective in the biodegradation of insecticides (DDT) and oil spill (Vasileva-Tonkova and Galabova 2003; Lal 1982; Williams 1977; Shafi et al. 2018). Hydrolases are considered to be effective players in bioremediation because of their ready availability, tolerance to the addition of solvents which are water miscible and lack cofactor stereo selectivity.

14.6 Solid Waste Biotreatment

Stringent standards for the discharge of waste into the environment, as well as the cost of routine treatment options, have motivated the scientific community to look for the different treatment options (Nicell 2003; Hamer et al. 2007; Mazzanti and Zoboli 2008). Keeping the following objectives in view, these treatment processes are developed by (1) reducing costs and resource conservation, (2) minimal effluent disposal and efficient waste recycling within the facility, (3) reduction in effluent quantity, and (4) transformation of waste into valuable products. There are two ways in which waste and pollutants can be treated one is biological, and the second one is chemical. We produce biowaste in our day to day life, which includes agricultural waste, horticulture waste, wastewater treatment plants, etc. Biowaste can be categorized as: process waste, manures, and plant material. In European countries, biowaste consists of 40–60% of municipal solid waste, and it is being collected separately and used for many purposes like composting or degradation (aerobic) which improves soil texture by supplying nutrients to the soil. Biowaste is also used for the generation of biogas in many countries of the world. Biological treatment of waste involves the careful selection of organisms (Biocatalysts) to degrade the waste. It is considered as most

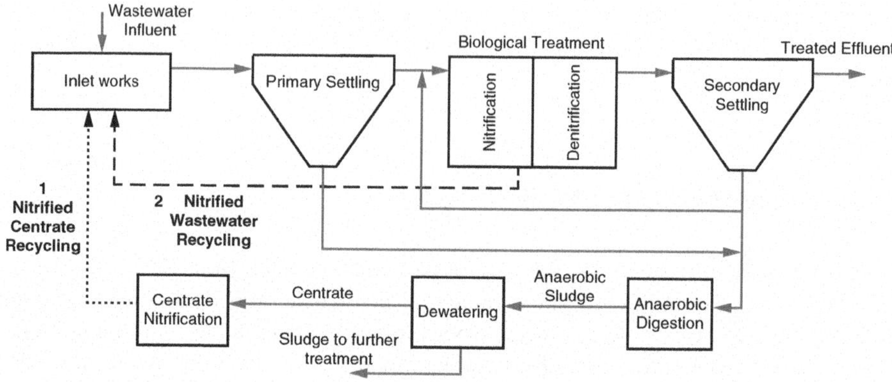

Fig. 14.2 Schematic diagram of biological wastewater treatment

favourable, stable, and bulk, reducing option which reduces the potential adverse effects on health and environment. Generating useful products and reclaiming valuable minerals for reuse are the added advantages of biotreatment. Biological treatment of solid waste can be done in situ or ex situ (Doble et al. 2004). Treatment of biological waste can be done under aerobic conditions or anaerobic conditions (Fig. 14.2). Digestion of organic waste is accelerated in anaerobic conditions under optimum conditions of temperature, pH, and moisture (Mata-Alvarez et al. 2000; Salminen and Rintala 2002), anaerobic processes are energy efficient, maintain enclosure of order, produce low sludge compared to aerobic ones.

14.7 Pollution Detection and Monitoring: Role of Environmental Biotechnology

Measuring a set of selected essential parameters on a regular basis which determine the environmental quality is environmental monitoring. There are two methods available (Biological and physiochemical) to judge the environmental health (Kaiser 2001; Lam and Gray 2003; Hagger et al. 2006). Environmental monitoring a couple of decades before was solely dependent on the measurement of physiochemical parameters and the monitoring of biological parameters were occasionally done. Physiochemical methods have limitations because of their heavy dependence on sophisticated equipment and complex analytical methodology (Gu et al. 2004).

For the protection of environment pollution detection at low concentrations is very important so that protective measures can be placed at the beginning (Durrieu et al. 2006). Monitoring of contaminants typically involves the frequent and regular measurements of various chemicals in soil, water, air, and sediments over a fixed time period. Integration of informational technology with biotechnology has made data collection, and monitoring work quit easy and accurate. It has also revolutionized the real-time analysis and data interpretation.

14.7.1 Bioindicators/Biomarkers

Environmental monitoring, apart from physiochemical measurements of contaminants, also needs to assess various biological/ecological parameters and responses. The temporal and spatial changes of selected biological responses which are used by ecologist and environmental biologists for the assessment of environmental quality (Durrieu et al. 2006; Conti 2008; Lam 2009). Organisms or communities react to changing environmental conditions by showing a measurable change in their chemical compositions/biological processes. These changes/responses are called as biomarkers/bioindicators (Kaiser 2001; Conti 2008). Thus biomarkers are used to give biological information (effect of environmental pollution). Bioindicators can be categorized into three types: those obtained on exposure, on susceptibility, and on effect.

Bioindicators that have potential to be used for biological monitoring are

- Molecular markers (Change in expression, DNA damage)
- Biochemical markers (Enzymatic, specific proteins)
- Histo-cytopathological (Cytological, histopathological)
- Physiological and behavioural

14.7.2 Biosensors

Biosensors are analytical devices consists of biomarker in the form of enzyme, antibody which is in close contact with transducer, which together relate the concentration of an analyte to a measurable electrical signal (Reiss and Hartmeier 2001; Rodriguez-Mozaz et al. 2004). Biosensing research has gained considerable attention in recent years because of the growing environmental pollution. Biosensing devices produce signals exploiting specificity of biological molecules to measure pollution levels. In broader terms biosensors are referred to any system that produces quantifiable signal on the detection of substrate using biological component. The main part of the biosensor is its biological component which can be whole cell, enzyme, antibody, genetically engineered organism. Biosensors represent the wonderful synergistic combination of microelectronics and biotechnology (Verma and Singh 2005). Biosensors can be used to detect both chemical pollutants as well as biological contaminants. Microchip biosensor with very high selective sensitivity was developed by combining biological activity with nanoelectronics (Cui et al. 2001).

Microalgal whole cell biosensors based on (Chlorophyll fluorescence or phosphatase and esterase inhibition) are used for the detection of environmental pollutions. Biosensor based on genetically modified yeast as used to detect endocrine disruptors such as oestrogen or 17-oestradiol. It was initially designed for therapeutic uses in humans. But later it was found to have potential to be used for detection of environmental pollutants (Tucker and Fields 2001). Use of biosensors has addressed many vital environmental issues which arose with other conventional methods.

14.8 Biotechnology for Cleaner Production

Use of biotechnological approach to combat environmental pollution is considered as the motor for the sustainable future. Contrary to other pollution control strategies which struggle for the tail end of the process and manage pollution once it has been generated, biotechnology works to stop pollution at its source by using different practices like (1) use of efficient raw material, (2) substituting less harmful substances for hazardous materials, and (3) eliminating hazardous toxic substances in effluents. Although the primary focus of environmental biotechnology is to develop technologies to treat waste but the basic information has been gained by people that how biotechnology can handle these wastes and the main focus is how to implement these processes as Available Technology Not Entailing Excessive Costs (BATNEEC) (Grommen and Verstraete 2002). Biotechnological processes after their successful set up are considered cost effective than conventional methods. Use of biotechnology for production processes not only contribute for the protection of environment, but also make companies economically sound and continuously improve their market image (Olguín 2000; Willke et al. 2005).

14.8.1 Pulp and Paper Industry

Biotechnological techniques like enzyme engineering, proteomics, metabolomics genetic engineering have helped in developing methods for paper production which are economically feasible, cost effective, and environmental friendly. Paper and pulp industry plays a vital role in world economy. Cellulosic pulp which is the target product is produced from pulping of wood (lignocellulosic material) by chemical or mechanical treatment for the release of fibres from the material. Most of the cellulosic pulp (90%) is produced from woody plants. Sugar cane, straw, and bamboo are the major raw material source for the pulp production (Sixta 2006).

The overall paper and pulp production process can be divided into six steps.

1. Wood preparation
2. Pulping

3. Pulp washing
4. Pulp screening
5. Bleaching
6. Paper manufacturing

From the last few decades, pulp and paper industry mostly in developed countries has been using biotechnological approaches for technological needs, which has resulted in a cleaner environment.

14.8.2 Biopulping

Use of ligninolytic enzymes or fungi cultures for the pre-treatment of raw material or wood is one of the interesting biological approaches used in the pulp and paper industry. Although there are other microorganisms which can be used for the bioprocessing of wood but fungi being the primary colonizers of wood utilizes its sugars without affecting its strength. Soft rot fungi and white fungi have high capability to degrade the wood because of their different extracellular enzymes. It is the white rot fungi in particular which are efficient ligninolytic organism producing all the necessary extracellular oxidative enzymes for the complete degradation of wood (Tekere et al. 2001). There are some fungi which are non-selective (*Trametes versicolor*) that is they simultaneously degrade hemicelluloses lignin and cellulose. While some others like *Phlebia tremellosa, Ceriporiopsis subvermispora,* and *Phellinus pini* degrade lignin preferentially (Otjen et al. 1987). The aim of biopulping is to utilize the special ability of selectively delignifying wood with cellulose decomposition of these specific fungi.

Fungal pre-treatment of wood chips before kraft pulping showed a significant increase in pulping yield (Mendonça et al. 2002) and decreased energy consumption (Young and Akhtar 1997). Fungal pre-treatment of woodchips before mechanical pulping process makes the wood soft for grinding. Thus to facilitate pulping this methodology does not require chemical agents, sophisticated equipment, and additional energy. Biotreatments enhance pulp quality, in addition to this biotreatment reduces energy consumption and environmental impact of the refining process, which make it more environmentally friendly than conventional methods used (Tian et al. 2012).

14.8.3 Bio-bleaching

Bleaching process of pulp needs chlorine, which is disastrous for the environment (Bajpai et al. 2006). Presence of lignin content is responsible for the brown colour of pulp. To improve the paper quality and printability, the removal of lignin content is required. Bio-bleaching is an alternative to conventional bleaching treatments by microorganisms such as fungi or their enzymes. Enzyme bio-bleaching is considered as an excellent environmentally friendly alternative due to its cost effectiveness, specificity, and decreased dependence on chemicals. Following are the main advantages for the use of bio-bleaching:

- Mill trials very simple and inexpensive with minimal risk
- Reducing the pollution from bleaching
- Enhance pulp physical properties
- Process can be easily combined with traditional bleaching
- It enhances fibrillation to give stronger paper
- Simplify the process and reduce the severity of treatment of wastewater.

Xylanase enzyme treatment in bio-bleaching reduces the operation cost and consumption of chlorine. Microbial thermostable xylanases are stable under alkaline conditions of pulping and are generally preferred enzymes for bio-bleaching (Raghukumar et al. 2004). It has been reported that xylanase pre-treatment of kraft pulps has reduced AOX discharge by 5–20%. Deposits of pitch are controlled by the use of lipases. Dewatering rate of pulp is improved by using cellulase enzymes (Foster 1986) that are also used for deinking and fibre modification. Catalase also has an important role it converts hydrogen peroxide to water and oxygen. Thus saves water and makes effluent ecologically harmless. Other enzymes like manganese peroxidise and laccase have also been used. Laccase delignify pulp directly thus replaces chemical stages of bleaching such as oxygen or ozone stage. For the control of pitch buildup and for deinking purposes, particularly when inks contain formulations of vegetable oil lipase has been used (Sharyo et al. 1993). It has been reported that in chip pre-treatment ascomycetes albino fungi reduce pitch and saves bleach cost up to 36% (Trivedy and Pathak 2015).

14.9 Bioplastics

The main threat to the environment and ecology because of plastic is due to its non-biodegradable nature and due to its production from synthetic polymers, which consumes the vast wealth of non-renewable resources (Stevens 2002; Reddy et al. 2003). Bioplastic production is based on natural substances (Proteins, fibres, sugar, and oils) thus avoids the use of non-renewable resources which in turn reduces the release of harmful gases into the environment. Microbes have an important role in converting vegetable and plant material into the building blocks of bioplastic (Luengo et al. 2003; Moldes et al. 2004). Bioplastic production from organic waste and enzyme assisted plastic reduction has addressed many environmental issues like the reduction of plastic waste and a decrease in consumption of non-renewable resources. OECD report (2001) assessed the wide spreading of industrial biotechnology, which was based on 21 companies from different sectors like chemical, paper, pharma, textile, etc. In this study, they found that industrial biotechnology has environmentally sound profound character because of its cleaner production.

14.10 Biotechnology and Agriculture

There is no doubt that the use of chemical herbicides, fungicides, and fertilizers has increased the yield and productivity of crops, but they have caused environmental hazards due to their low biodegradability. The use of biopesticides and genetically modified plants may considerably diminish the use of these harmful chemicals. Biopesticides are classified (Table 14.1) as

- Microbial pesticides (Containing microorganism as active ingredient)
- Plant-incorporated protectants (Genetic material incorporated into the plant produces active pesticide)

Table 14.1 Biopesticides and their control targets (MCD 2008)

S. no.	Organism	Target	Example
1	Bacteria	Insects	*Bacillus thuringiensis*
			Bacillus sphaericus
			Serratia entomophila
			Paenibacillus popilliae
2	Fungi	Insects	*Beauveria* spp.
			Metarhizium
			Lecanicillium lecanii
			Nomuraea
			Paecilomyces fumosoroseus
			Zoophthora
			Entomophaga
3	Protozoa	Insects	*Nosema*
			Vairimorpha
			Thelohania
4	Fungi	Weed control	*Colletotrichum gloeosporioides*
			Chondrostereum purpureum
			Cylindrobasidium leave
			Xanthomonas campestris
5	Fungi	Plant disease control	*Ampelomyces quisqualis*
			Candida spp.
			Clonostachys rosea
6	Competitive inoculants	Plant disease control	*Coniothyrium minitans*
			Pseudozyma flocculosa
			Trichoderma spp.
7	Composts, soil inoculants	Plant disease control	*Bacillus pumilus*
			Bacillus subtilis
			Pseudomonas spp.
			Streptomyces griseoviridis
			Burkholderia cepacia
8	Mollusc parasitic nematode	Nematicides	*Phasmarhabditis hermaphrodita*

- Biochemical pesticides (Include substances which control pests by non-toxic mechanisms)

Biopesticides decompose quickly and are often effective in small quantities (Boyetchko et al. 1999). According to Fraser et al. biopesticides have the following characteristics (Fraser 2005):

- Low risk to human health.
- Low potential to contaminate environmental components and resources.

Biopesticides and genetically engineered insect resistant plants have considerably diminished the use of harmful pesticides and chemicals. Thus, ultimately lead to less exposure to chemicals.

14.11 Chemicals and Biofuels

Excellent research and development is already in force across the globe for production of biofuels by the enzymatic conversion of different substrates (Agricultural waste, municipal waste, and vegetable oils) into bioethanol and biogas, etc. (Dale 2006; Willke et al. 2005). Different applications have conceived the idea of anaerobic digestion of wastes from agriculture and food industries for methane production (Fig. 14.3). Complex organic molecules present in the waste are broken-down by anaerobic bacteria for the production of methane rich combustible biogas having energy value between 21 and 28 MJ/m^3 (Doble et al. 2004).

The idea of chemical synthesis from natural renewable resources is still in its infancy, but the conversion of biomass feedstock into biofuels can have a major economic as well as environmental benefit (Gavrilescu and Chisti 2005; Willke et al. 2005; Chisti 2007). Several companies have started to produce their products from renewable resources (e.g., DuPont) (Willke et al. 2005). Production of fine chemicals using traditional methods has created severe environmental problems. Analysis performed by environmental organizations in order to evaluates two production processes, viz., chemical and biotechnological. It was found that it is the biotechnological process which is more economical and coefficient (e.g., Vitamin B_2 production process). It was found that riboflavin can be produced at cost reduced by 50% using biotechnological approaches (BIO–PRO 2008). Hydroxylation of steroids using microbes leads to the development of steroid drug used for arthritis (Dutta and Samanta 1997).

The key intermediate for the production of penicillins used for chemotherapy is 6-aminopenicillanic acid (6-APA). Synthesis of aminopenicillanic acid (6-APA) using biological approach was found to be 20% cheaper than its chemical synthesis.

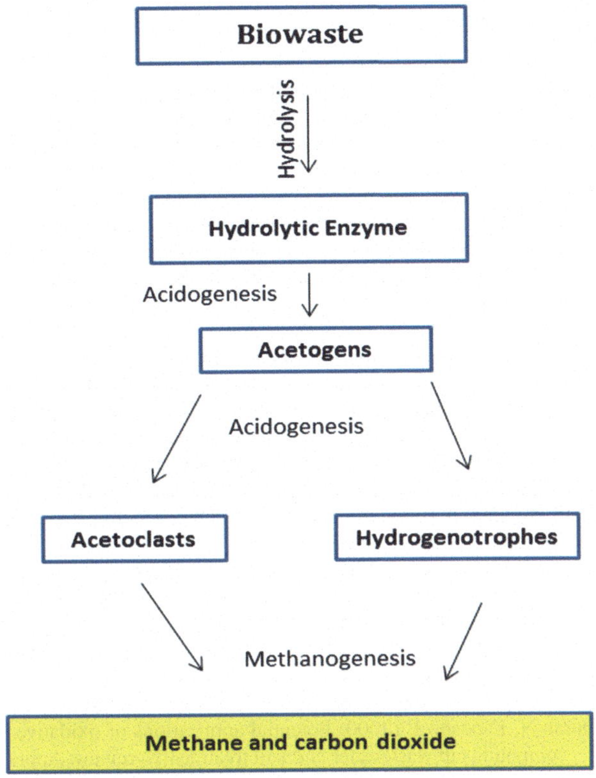

Fig. 14.3 Biowaste methanisation

14.12 Conclusion

Globally there are thousands of environmental challenges, and new challenges are evolving at a faster rate. New technologies are currently under development, and some already established ones are gaining more and more ground in practice. There is a need of a technology which can harness the potential of plants and microbes as robust and eco-efficient agents in a variety of practical situations. Along with a vast group of technologies to achieve the objectives of sustainability, biotechnology is fighting from the front and will continue to play a vital role in the field of environmental protection, food, renewable materials, and bioremediation. Environmental biotechnology is a sustainable approach to develop efficient protocols, clean processes, and products with reduced environmental impact. Environmental and economic benefits that biotechnology can offer in manufacturing, monitoring, and waste management are in balance with technical and economic problems which still need to be solved. All this is being achieved with reduced environmental impact and enhanced sustainability. An evaluation of the consequences, opportunities, and challenges of modern biotechnology is important for both policymakers and the industry.

References

Abioye OP (2011) Soil contamination. IntechOpen, London
Ali H, Khan E, Sajad MA (2013) Chemosphere 91:869–881
Andreoli CV, Von Sperling M, Fernandes F, Ronteltap M (2007) Sludge treatment and disposal. IWA Publishing, London
Bajpai P, Anand A, Sharma N, Mishra SP, Bajpai PK, Lachenal D (2006) Bioresources 1:34–44
Bollag JM, Dec J, Krishnan SB (1998) Use of plant material for the removal of pollutants by polymerization and binding to humic substances. Citeseer
Boyetchko S, Pedersen E, Punja Z, Reddy M (1999) Biopesticides: use and delivery. Springer, Totowa, pp 487–508
Caputo AC, Pelagagge PM (2001) Waste-to-energy plant for paper industry sludges disposal: technical-economic study. J Hazard Mater 81:265–283
Chinnasamy S, Bhatnagar A, Hunt RW, Das K (2010) Bioresour Technol 101:3097–3105
Chisti Y (2007) Biodiesel from microalgae. Biotechnol Adv 25:294–306
Conti ME (2008) Biological monitoring: theory & applications: bioindicators and biomarkers for environmental quality and human exposure assessment. WIT Press, Southampton
Cui Y, Wei Q, Park H, Lieber CM (2001) Science 293:1289–1292
Dale BE (2006) Biomass refining global impact—the biobased economy of the 21st century. Biorefin Ind Process Prod 1:41–66
Dervash MA, Bhat RA, Shafiq S, Singh DV, Mushtaq N (2020) Biotechnological intervention as an aquatic clean up tool. In: Qadri H, Bhat RA, Dar GH, Mehmood MA (eds) Freshwater pollution dynamics and remediation. Springer, Singapore, pp 183–196
Doble M, Kruthiventi AK, Gaikar VG (2004) Biotransformations and bioprocesses. CRC Press, Boca Raton
Dua M, Singh A, Sethunathan N, Johri A (2002) Appl Microbiol Biotechnol 59:143–152
Duran N, Esposito E (2000) Potential applications of oxidative enzymes and phenoloxidase-like compounds in wastewater and soil treatment: a review. Appl Catal B Environ 28:83–99
Durrieu C, Tran-Minh C, Chovelon JM, Barthet L, Chouteau C, Védrine C (2006) Algal biosensors for aquatic ecosystems monitoring. Eur Phys J Appl Phys 36:205–209
Dutta TK, Samanta TB (1997) Novel catalytic activity of immobilized spores under reduced water activity. Bioorg Med Chem Lett 7:629–632
Foster RN (1986) Innovation: the attacker's advantage. Summit Books, New York
Fraser H (2005) Ministry of Agriculture, Food and Rural Affairs, Ontario. http://www.omafra.gav.on.ca/english/crops/hort/news/vegnews/2005/vg1105_a5.htm
Garbisu C, Alkorta I (1997) Bioremediation: principles and future. J Clean Technol Environ Toxicol Occup Med 6:351–366
Garbisu C, Alkorta I, Llama MJ, Serra JL (1998) Aerobic chromate reduction by Bacillus subtilis. Biodegradation 9:133–141
Gavrilescu M, Chisti Y (2005) Biotechnology—a sustainable alternative for chemical industry. Biotechnol Adv 23:471–499
Gianfreda L, Xu F, Bollag JM (1999) Laccases: a useful group of oxidoreductive enzymes. Biorem J 3:1–26
Grommen R, Verstraete W (2002) Environmental biotechnology: the ongoing quest. J Biotechnol 98:113–123
Gu MB, Mitchell RJ, Kim BC (2004) Biomanufacturing. Springer, Basel, pp 269–305
Hagger JA, Jones MB, Leonard DP, Owen R, Galloway TS (2006) Integrated environmental assessment and management. Int J 2:312–329
Hamer K, Arevalo E, Deibel I, Hakstege A (2007) Assessment of treatment and disposal options. Sustain Manag Sediment Res 2:133–159
Husain Q (2006) Potential applications of the oxidoreductive enzymes in the decolorization and detoxification of textile and other synthetic dyes from polluted water: a review. Crit Rev Biotechnol 26:201–221

Huyer M, Page WJ (1988) Zn2+ increases siderophore production in Azotobacter vinelandii. Appl Environ Microbiol 54:2625–2631

Ishibashi Y, Cervantes C, Silver S (1990) Chromium reduction in Pseudomonas putida. Appl Environ Microbiol 56:2268–2270

Kaiser J (2001) Bioindicators and biomarkers of environmental pollution and risk assessment. Science Publishers, New Hampshire

Karaca A (2004) Effect of organic wastes on the extractability of cadmium, copper, nickel, and zinc in soil. Geoderma 122:297–303

Khan M, Satoh H, Katayama H, Kurisu F, Mino T (2004) Environmental biotechnology: advancement in water and wastewater application in the tropics. IWA Publishing, London, pp 349–355

Khan S, Pandotra P, Gupta AP, Salgotra RK, Malik MM, Lone SA, Gupta S (2017) Plant molecular breeding: way forward through next-generation sequencing. In: Zargar SM, Rai V (eds) Plant OMICS and crop breeding. Apple Academic Press, New York, pp 227–260

Lal R (1982) Accumulation, metabolism, and effects of organophosphorus insecticides on microorganisms. Adv Appl Microbiol 28:149–200

Lam PK (2009) Use of biomarkers in environmental monitoring. Ocean Coast Manag 52:348–354

Lam PK, Gray JS (2003) The use of biomarkers in environmental monitoring programmes. Mar Pollut Bull 46:182–186

Leung M (2004) Bioremediation: techniques for cleaning up a mess. BioTeach J 2:18–22

Luengo JM, Garcı́a B, Sandoval A, Naharro G, Olivera ER (2003) Bioplastics from microorganisms. Curr Opin Microbiol 6:251–260

Mata-Alvarez J, Mace S, Llabres P (2000) Anaerobic digestion of organic solid wastes. An overview of research achievements and perspectives. Bioresour Technol 74:3–16

Mazzanti M, Zoboli R (2008) Waste generation, waste disposal and policy effectiveness: evidence on decoupling from the European Union. Resour Conserv Recycl 52:1221–1234

McCauley A, Jones C, Jacobsen J (2009) Soil pH and organic matter. Nutr Manag Mod 8:1–12

Mendonça R, Guerra A, Ferraz A (2002) Delignification of Pinus taeda wood chips treated with Ceriporiopsis subvermispora for preparing high-yield kraft pulps. J Chem Technol Biotechnol Int Res Process Environ Clean Technol 77:411–418

Moldes C, García P, García JL, Prieto MA (2004) In vivo immobilization of fusion proteins on bioplastics by the novel tag BioF. Appl Environ Microbiol 70:3205–3212

Namgay T, Balwant S, Bhupinder PS, (2010) Influence of biochar application to soil on the availability of As, Cd, Cu, Pb, and Zn to maize (Zea mays L.). Soil Research 48(7):638–647

Newman LA, Doty SL, Gery KL, Heilman PE, Muiznieks I, Shang TQ, Siemieniec ST, Strand SE, Wang X, Wilson AM (1998) Phytoremediation of organic contaminants: a review of phytoremediation research at the University of Washington. J Soil Contam 7:531–542

Nicell JA (2003) Chemical degradation methods for wastes and pollutants. CRC Press, Boca Raton, pp 395–441

Novak JM, Busscher WJ, Laird DL, Ahmedna M, Watts DW, Niandou MA (2009) Impact of biochar amendment on fertility of a southeastern Coastal Plain soil. Soil Sci 174:105–112

Nowack B, Schulin R, Robinson BH (2006) Critical assessment of chelant-enhanced metal phytoextraction. Environ Sci Technol 40:5225–5232

Okoh AI, Sibanda T, Gusha SS (2010) Inadequately treated wastewater as a source of human enteric viruses in the environment. Int J Environ Res Public Health 7:2620–2637

Olguín EJ (2000) Environmental biotechnology and cleaner bioprocesses. Taylor and Francis, London, pp 3–17

Otjen L, Blanchette R, Effland M, Leatham G (1987) Assessment of 30 white rot basidiomycetes for selective lignin degradation. Holzforschung 41:343–349

Park JW, Park BK, Kim JE (2006) Remediation of soil contaminated with 2, 4-dichlorophenol by treatment of minced shepherd's purse roots. Arch Environ Contam Toxicol 50:191–195

Raghukumar C, Muraleedharan U, Gaud V, Mishra R (2004) Xylanases of marine fungi of potential use for biobleaching of paper pulp. J Ind Microbiol Biotechnol 31:433–441

Raskin I, Ensley BD (2000) Phytoremediation of toxic metals. Wiley, New York
Reddy C, Ghai R, Kalia VC (2003) Polyhydroxyalkanoates: an overview. Bioresour Technol 87:137–146
Reiss M, Hartmeier W (2001) Monitoring of environmental processes with biosensors. In: Biotechnology Set. Wiley-VCH, Weinheim, pp 125–139
Rodriguez-Mozaz S, Marco MP, De Alda ML, Barceló D (2004) Biosensors for environmental applications: future development trends. Pure Appl Chem 76:723–752
Rubilar O, Diez MC, Gianfreda L (2008) Transformation of chlorinated phenolic compounds by white rot fungi. Crit Rev Environ Sci Technol 38:227–268
Salminen E, Rintala J (2002) Anaerobic digestion of organic solid poultry slaughterhouse waste–a review. Bioresour Technol 83:13–26
Sanz JL, Köchling T (2007) Molecular biology techniques used in wastewater treatment: an overview. Process Biochem 42:119–133
Schaar H, Clara M, Gans O, Kreuzinger N (2010) Micropollutant removal during biological wastewater treatment and a subsequent ozonation step. Environ Pollut 158:1399–1404
Shafi S, Bhat RA, Bandh SA, Shameem N, Nisa H (2018) Microbes: key agents in the sustainable environment and cycling of nutrients. In: Environmental contamination and remediation. Cambridge Scholars Publishing, Cambridge, pp 179–188
Sharyo M, Shimoto H, Sakaguchi H, Isaji M, Fujita Y, Awaji H, Matsukura M, Hata K (1993) The recent progress and general status of the lipase pitch control technology in Japan. Jpn Tappi J 47:1223–1233
Sixta H (2006) Handbook of pulp. Wiley-VCH, Weinheim
Stevens ES (2002) Green plastics: an introduction to the new science of biodegradable plastics. Princeton University Press, Princeton
Su X, Sun F, Wang Y, Hashmi MZ, Guo L, Ding L, Shen C (2015) Identification, characterization and molecular analysis of the viable but nonculturable Rhodococcus biphenylivorans. Sci Rep 5:18590
Tekere M, Mswaka A, Zvauya R, Read J (2001) Growth, dye degradation and ligninolytic activity studies on Zimbabwean white rot fungi. Enzym Microb Technol 28:420–426
Tian XF, Fang Z, Guo F (2012) Impact and prospective of fungal pre-treatment of lignocellulosic biomass for enzymatic hydrolysis. Biofuels Bioprod Biorefin 6:335–350
Timmis KN, Steffan RJ, Unterman R (1994) Designing microorganisms for the treatment of toxic wastes. Annu Rev Microbiol 48:525–557
Trivedy R, Pathak R (2015) Role of biotechnology in pulp and paper industry. J Indus Poll Control 31:285–288
Tucker CL, Fields S (2001) A yeast sensor of ligand binding. Nat Biotechnol 19:1042
Vasileva-Tonkova E, Galabova D (2003) Hydrolytic enzymes and surfactants of bacterial isolates from lubricant-contaminated wastewater. Zeitschrift für Naturforschung C 58:87–92
Verma N, Singh M (2005) Biosensors for heavy metals. Biometals 18:121–129
Vidali M (2001) Bioremediation. An overview. Pure Appl Chem 73:1163–1172
Wen J, Zhou S, Chen J (2014) Colorimetric detection of Shewanella oneidensis based on immunomagnetic capture and bacterial intrinsic peroxidase activity. Sci Rep 4:5191
Weyens N, van der Lelie D, Taghavi S, Newman L, Vangronsveld J (2009) Exploiting plant-microbe partnerships to improve biomass production and remediation. Trends Biotechnol 27:591–598
White C, Shaman AK, Gadd GM (1998) An integrated microbial process for the bioremediation of soil contaminated with toxic metals. Nat Biotechnol 16:572
Williams PP (1977) Metabolism of synthetic organic pesticides by anaerobic microorganisms. In: Gunther FA (ed) Residue reviews, vol 66. Springer, New York
Willke T, Prüße U, Vorlop KD (2005) Biocatalytic and catalytic routes for the production of bulk and fine chemicals from renewable resources. In: Biorefineries-industrial processes and products: status quo and future directions. Wiley-VCH, Weinheim, pp 385–406
Young RA, Akhtar M (1997) Environmentally friendly technologies for the pulp and paper industry. Wiley, New York

Chapter 15
Global Environmental Regulations for Management of Pesticides

Rezwana Assad, Zafar Ahmad Reshi, and Irfan Rashid

15.1 Introduction

Excessive and improper pesticide usage results in contamination and deterioration of soil and aquatic ecosystems and disturbs ecological equilibrium (Hui et al. 2003; Morillo and Villaverde 2017; Spina et al. 2018; Sun et al. 2018; Bhat et al. 2019; Kumar et al. 2019). Pesticides are extremely deleterious due to their persistence, toxicity, bioaccumulation potential, and long range environmental transport potential (Teran et al. 2012; Pariatamby and Kee 2016; Bharat 2018). During the past decade, pesticides have drawn mass public, political, and scientific concern due to their carcinogenic, immunotoxic, mutagenic, and neurotoxic prospective (Dixit et al. 2019). Although new harmless pesticides with relatively enhanced safety profile and novel modes of action have been developed as a substitute to extremely toxic ones (Gill and Garg 2014; Dixit et al. 2019); however, due to the dearth of a comprehensible pesticides management approach, large amount of obsolete pesticide residues have already stockpiled over the time (Dasgupta et al. 2010).

Pesticide exposure occurs primarily through improper pesticide storage, pesticide mishandling and leakage, inappropriate and indiscriminate use, inadequate disposal, and by consuming pesticide contaminated food and water (Morillo and Villaverde 2017; Dixit et al. 2019). A mixture of toxic banned pesticide residues have been detected in large quantities (above WHO's maximum daily intake) in a wide variety of food items that we eat like eggs, fish, flour, fruits, grains, milk, poultry, pulses, rice, vegetables, and wheat, which cause several chronic human diseases (Pimentel et al. 1992; ICMR 2001; Rani and Dhania 2014; Pandey et al. 2018). World Health Organization (WHO) estimated that globally there occur 3,000,000 cases of pesticide poisoning which result in approximately 220,000 deaths annually (Lah 2011; Dixit et al. 2019).

R. Assad (✉) · Z. A. Reshi · I. Rashid
Department of Botany, University of Kashmir, Srinagar, Jammu and Kashmir, India

R. A. Bhat et al. (eds.), *Bioremediation and Biotechnology, Vol 2*,
https://doi.org/10.1007/978-3-030-40333-1_15

Nowadays, besides some traditional means of monitoring pesticides residues, several novel and innovative technologies are being applied (Xing-lu et al. 2019). Several developed regions have established pesticide risk assessment models which are of great significance for environmental sustainability (Xing-lu et al. 2019). These models play a crucial role in creating mass awareness and thus, facilitate involvement of all the stakeholders in successful recuperation and restoration of pesticide contaminated environs.

In addition to the application of various scientific methods for the treatment of pesticide pollutants, there is an immediate need to develop global guidelines, management policies, and standard regulations for overseeing pesticide usage and restoration of pesticide contaminated environments (Gill and Garg 2014; Varjania et al. 2019). Pesticide pollution causes various human health and environmental safety issues, which instigated varied global governmental and non-governmental organizations to frame environmental regulations for management and mitigation of pesticide pollution for global environmental safety, sustainable development, and human welfare. Pesticide pollution is one of the major drivers of ecosystem degradation. The United Nations General Assembly declared 2021–2030 as the "International Decade on Ecosystem Restoration" with one of the main focus on restoration of degraded ecosystems globally for sustainable development (Tripathi et al. 2019).

15.2 Environmental Risk Assessment of Pesticides

Risk assessment is a promising multi-disciplinary procedure inured to assess environmental and health risks propounded by chemical pollutants (Singh et al. 2009). Assessing the environmental providence of pesticide residues and their imminent exposure risks to environment and public health is essential for formulating risk-based management strategies for risk reduction (Rice et al. 2007; Singh et al. 2009); however, the development of environmental risk assessment of pesticides is extremely complex and challenging because of the disparity in nature and quantity of pesticide sprayed, ecotoxicity of pesticides, exposure pathways (direct and indirect), period and intensity of exposure, and environmental characteristics of the pesticide application site (Pandey et al. 2018; Spina et al. 2018; Xing-lu et al. 2019). Moreover, the regulatory and risk assessment approaches are concerned with the impact of the pollutant, rather than the pollutant concentration (Beesley et al. 2011).

Several developed regions like North America, European Union, and others have framed and established their own territory-specific pesticide risk assessment models (Xing-lu et al. 2019). Realistic and valid environmental risk assessment of pesticides furnishes fundamental information for management of risk and formulation of remediation decisions (Sun et al. 2017). SCI-GROW risk assessment model of pesticides was established to envisage the risk of five widely used pesticides in China (Cheng et al. 2007; Xing-lu et al. 2019). Levitan et al. (1995) subsequent to analyzing various categories of environmental risk assessment like anecdotal assessments, composite environmental impact rating systems, directory-format and tabular databases, economic assessments, holistic assessments, single and multiple-parameter assessments, and site-specific assessment tools reported that a comprehensive pesticide impact assessment system is still lacking.

Environmental impact assessment of pesticides influences characterization of previously permitted pesticides, approval of new pesticides, and establishment of remediation goals along with maintenance of environmental quality (Song et al. 2017; Dixit et al. 2019). New pesticides registrations based solely on strict environmental risk assessment procedures can help in combating the menace of hazardous pesticides (Schwarzenbach et al. 2010; Xing-lu et al. 2019). Moreover, biotoxicity tests reflect the biological impacts of toxicants, thus serve as important tools in environmental risk assessment (Prokop et al. 2016).

15.3 Pesticide Management Strategies

Pesticides incur adverse impacts on biodiversity, human health, and environment (Marican and Durán-Lara 2018; Dixit et al. 2019). So, management of pesticide pollution assumes great significance and demands immediate action. Novel pesticide management tools and techniques with greater reliability are being developed for improved safety and mitigation of adverse pesticide impacts.

Pesticide management strategies encompass assorted pest as well as pesticide management strategies (Fig. 15.1). Various pest management strategies are categorized as: biological control strategies (such as adoption of integrated pest management-IPM systems by introduction of biocontrol agents, use of biopesticides like biofungicides, bioherbicides, and bioinsecticides), cultural control strategies (such as

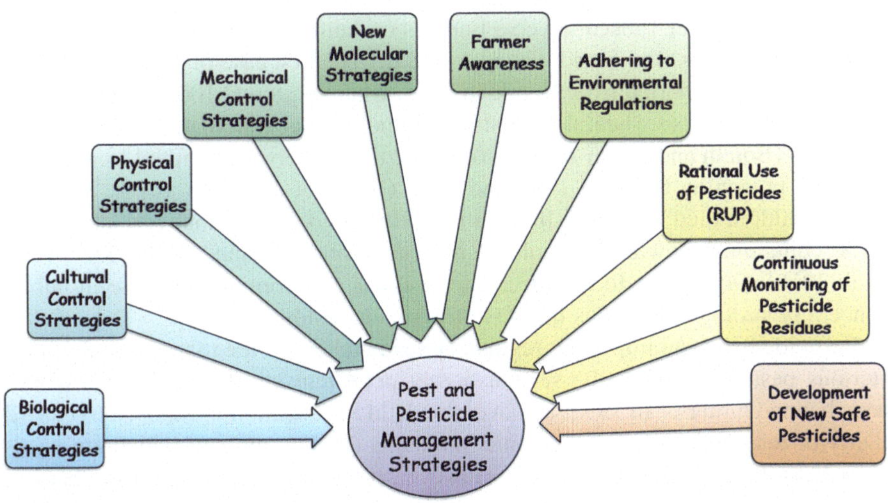

Fig. 15.1 Pest and pesticide management strategies

adoption of integrated crop management-ICM, crop rotation, intercropping, allelopathy, certified seeds, use of pest resistant varieties, farmyard manure, and organic farming), physical control strategies (like sun exposure, light traps, steaming, and moisture management), mechanical control strategies (include use of manual devices like handpicking, clipping, trapping devices, pest exclusion by barriers, and crushing/ pruning of pest infested part), new molecular strategies (for instance, development of pest resistant genotypes). Pest management by proficient alternative approaches rationalizes the pesticide consumption (Dhaliwal et al. 2006; Abhilash and Singh 2009; Gill and Garg 2014; Dixit et al. 2019).

Several effective and reliable pesticide management strategies are farmer education and awareness about basic safety guidelines in selection and application of pesticides, adhering to the environmental regulations (precautionary measures as well as safety guidelines) during all the phases of pesticide handling, rational use of pesticides-RUP (apposite pesticide selection and meticulousness in their application over space and time), continuous monitoring of pesticide residues in agricultural commodities, alteration of formerly existing pesticide formulations for secure formulations and development of new pesticides with enhanced safety profiles (Abhilash and Singh 2009; Gill and Garg 2014; Dixit et al. 2019).

15.4 Pesticide Ban

The unsustainable production and use of pesticides over past few decades resulted in colossal environmental contamination. As stated by Food and Agriculture Organization (FAO) inventory (2001), approximately 500,000 tons of obsolete pesticide residues have accumulated in several countries and are threatening the public and environmental health (FAO/UNEP/OECD/SIB 2001). Many cases of pesticide poisoning that either caused adverse impacts on human health or even death have been reported across the world. For instance, (a) The first case of pesticide poisoning in India was reported from Kerala in 1958, wherein around 100 people died due to the consumption of wheat flour that was found to be contaminated with pesticide parathion (Boudh and Singh 2019), (b) In Brazil, 123 persons suffered from severe pesticide poisoning between 1992 and 2002 (Weiss et al. 2016), (c) In 2013, 23 school children died in eastern India, after eating school lunch that was later found to be contaminated with residuals of monocrotophos (banned pesticide) (Weiss et al. 2016). During the former decade, there was an enormous increase in public concern regarding deleterious impacts of pesticide residues.

Pesticide ban is an important regulatory stride taken for the management of deleterious pesticides. Globally, several highly persistent and toxic pesticides have been banned hitherto. In 1995, a convention held in Stockholm proposed to ban 12 (Dirty dozen) toxic persistent organic pollutants (POPs), most of which are pesticides (Morillo and Villaverde 2017). However, pesticides that are currently banned in developed countries are still used or stockpiled in developing countries due to lack of awareness regarding hazardous pesticide impacts, dearth of reliable risk assessment protocols, lack of infrastructure, environmental controls and regulations in such countries (Carneiro et al. 2012). Moreover, pesticides persist even for centuries in ecosystems due to their stability and degradation resistance. The pesticides that have been banned since several decades, their residues are still today found in water and sediments (Ballesteros et al. 2014).

15.4.1 Case Study of Pesticide Dichlorodiphenyltrichloroethane (DDT)

Insecticide "Dichlorodiphenyltrichloroethane" (DDT) was discovered by a Swiss entomologist Paul Muller in 1939. Following World War II, DDT was used in large quantities. DDT was the first organic chemical to be used as a pesticide, and it protected several economical crops and eradicated malaria from numerous parts of the world (Kumar et al. 2018). Alternatively, DDT has resulted in universal environmental contamination due to its slow degradation rate that has been estimated to range from 4 to 30 years (Abhilash and Singh 2009; Pandey et al. 2018; Boudh and Singh 2019). DDT causes eggshell thinning in birds by inhibiting synthesis of prostaglandin in eggshell gland mucosa (Lundholm 1997). DDT is known to have huge bioaccumulation and biomagnification potential at higher trophic levels (Gill and Garg 2014). Thus organisms at the higher trophic levels of food chain are at more risk. DDT exposure causes cancer, endocrine hormone disruption, liver damage, neurological disorders, and reproductive abnormalities (Persson et al. 2012; Pandey et al. 2018; Dixit et al. 2019).

In 1972, production and use of DDT was entirely banned in the USA and other developed countries due to vast array of side effects like non-target toxicity, bioaccumulation, and environmental persistence (Pandey et al. 2018). However, many developing countries (including India) were exempted from DDT ban and were allowed to use DDT till March 2013 for the control of vector-borne diseases (Bharat 2018). Since the ban, large quantity of DDT have stockpiled under improper conditions in developing countries, where it pose a serious risk to human and environmental health (Weiss et al. 2016).

15.5 Global Environmental Regulations for Pesticide Management

Several national and international regulatory authorities conceded various laws and regulations regarding secure manufacture, import, export, sale, transport, proper use, and application of pesticides (FAO 1955, 1988; IPSMC 1999; Lallas 2001; Rotterdam Convention Secretariat 2004; WHO 2004; MEA 2005; FAO 2006a, b, 2007; UNEP 2006; Abhilash and Singh 2009; ECHA 2011; Teran et al. 2012; Pesticide Action Network International 2014; GlobalGAP 2016; Bharat 2018). Numerous initiatives and conventions were instigated for pesticide management during the past two decades. The initial regulations for pesticide management were introduced in the beginning of 1970 with the ban on production and use of pesticide DDT and restricted use of other harmful pesticides in the USA (Teran et al. 2012). Subsequently, several international regulations dealing with pesticide management came into force since 1980. Some of the imperative global environmental regulations for management of pesticides are:

(a) *The International Code of Conduct and Use of Pesticides (Code of Conduct)*: In 1985, the 23rd session of the main governing body of Food and Agriculture Organization (FAO) formulated and adopted the International Code of Conduct and Use of Pesticides (Code of Conduct) for providing universal principles of conduct of pesticide management for pesticide industry and national governments. This code of conduct was once amended in 1989 and then again in 2002 (FAO 2006a, b).
(b) *The Montreal Protocol*: The Montreal Protocol was adopted in 1987 with the aim to limit the production and use of substances (such as pesticide methyl bromide) that cause ozone layer depletion (Abhilash and Singh 2009). This protocol came into force in 1989.
(c) *The Rotterdam Convention*: The Rotterdam Convention was adopted in 1988 for the regulation of international trade of hazardous chemicals and pesticides (Kumer 1999; Rotterdam Convention Secretariat 2004). It later came into force in the year 2004. Its main objective is to control and ban the international trade and stockpiling of perilous pesticides. Furthermore, it promotes labelling of deleterious chemicals with safety directives and pictograms (United Nations Environment Programme, Food and Agriculture Organization 2013).
(d) *The Basel Convention*: The Basal Convention dealing with the trans-boundary passage of perilous wastes and their disposal was adopted in 1989 and came into force in 1992 (Basel Convention 2006).
(e) *The Stockholm Convention*: In 1995, a convention held in Stockholm proposed to ban 12 toxic persistent organic pollutants (POPs) (Dirty dozen), to which more chemicals were added afterwards. Later in 2001, an international legal treaty named as "Stockholm Convention" was opened for ratification and was finally signed by 179 countries, which then came into power in 2004, around after a decade after UNEP's (United Nations Environment Programme) call for global action on POPs in 1995 (UNEP 2009). This convention proposed to ban or restrict the production, release, and use of selected harmful chemicals including polychlorinated biphenyls (PCBs) and a range of organochlorine pesticides (OCPs) in all participant countries (UNEP 2009; Morillo and Villaverde 2017; Bharat 2018). Furthermore, SC amendments in 2011 engorged the list up to 22 POPs (United Nations Environment Programme 2013a).
(f) *The International POPs Elimination Network (IPEN)*: The International POPs Elimination Network (IPEN) is a non-profit organization that was established in 1998 in Sweden. IPEN deals with the sound chemical management policies and practices, as a strategy for global economic development and environmental protection (Bharat 2018).

(g) *The International Labor Organization Convention No. 184 (ILO Convention 184)*: ILO Convention 184 deals with all aspects of safety and health while chemical application in agriculture sector (ILO 2000). It was adopted in 2001 and later came to force in 2003.
(h) *The Global Harmonized System of Classification and Labelling of Chemicals (GHS)*: The GHS has been devised for the safety of human beings and the environment during the handling, use, and transport of hazardous chemicals (FAO 2006a, b).
(i) *Strategic Approach to International Chemicals Management (SAICM)*: Strategic Approach to International Chemicals Management (SAICM) is a global policy framework for proper chemical management. It was adopted by International Conference on Chemicals Management (ICCM) in 2006 (Bharat 2018). It recognizes the role of chemicals in the current scenario as well as the potential threat to sustainable development by chemical mismanagement. SAICM goals together with the support of United Nations Economic Commission for Europe (UNECE) and UNEP deal with the chemical management, environment, and sustainable development in alliance and deem in achievement of these goals by the year 2020 (Teran et al. 2012; Bharat 2018).
(j) *Registration, Evaluation and Authorisation of Chemicals (REACH)*: In 2007, a legislation called as Registration, Evaluation and Authorisation of Chemicals (REACH) was adopted and enforced by the European Union (EU) with the rationale of information gathering on chemicals entering European markets without prior knowledge vis-a-vis their detrimental impacts on environment and human health (EU 2006; EEA 2007; ECHA 2008). This is achieved through the recognition of appropriate pesticide risk assessment models for the screening, proper approval, and registration of these new pesticides being marketed in Europe, with the purpose of reducing future exposure risks (ECHA 2011). REACH encompass registration, evaluation, authorization, and restriction elements.
(k) *The Clean Sweep Program*: The Convention on Persistent Organic Pollutants along with the Food and Agriculture Organization (FAO) are setting up global efforts for decreasing the occurrence as well as abundance of detrimental pesticides. In this direction, for apposite disposal of pesticides "Clean Sweep Program" was instated in several regions within the USA (Dixit et al. 2019).
(l) *The Codex Alimentarius Commission*: The Codex Alimentarius Commission formulated several standards on the presence of pesticide residues in food (Abhilash and Singh 2009).
(m) *Global Good Agricultural Practices (GlobalGAP)*: The Global Good Agricultural Practices (GlobalGAP) is a European non-profit organization that delineates good agricultural practices like using small quantity of pesticides for sustainable production (GlobalGAP 2016).
(n) *Pesticide Action Network organization (PAN)*: The Pesticide Action Network organization (PAN) is a network of over 600 NGO's and institutions, operational in around 90 countries. Its main aim is to substitute deleterious chemical pesticides by ecologically sound approaches (Pesticide Action Network International 2014).

Some other international conventions that are of great importance are: The Mediterranean Action Plan for the Barcelona Convention (MAP) (UNEP 1978); The Arctic Monitoring and Assessment Programme (AMAP) (Arctic Council 1991); The Chemical Weapons Convention 1993; The Convention for the Protection of the marine Environment of the North–East Atlantic (OSPAR); Northern Contaminants Program (NCP) (NCP 2003); The Environmental Monitoring and Governance in the East Asian Hydrosphere-Monitoring of POPs (EMGEAH), and European Monitoring and Evaluation Programme (EMEP http://www.emep.int). Furthermore, globally, several acts were passed for management of pesticides such as The Prevention of Food Adulteration Act, 1954; The Insecticide Act, 1968; The Insecticide Rules, 1971; The Environment Protection Act, 1986; and The Federal Insecticide, Fungicide, and Rodenticide Act (FIFRA) (Frazar 2000; Abhilash and Singh 2009).

Environmental regulations play an indispensable role in effectual pesticide management along with maintenance of environment. However, these regulatory frameworks are not appropriately and effectively implemented even in developed countries (Morillo and Villaverde 2017). In developing countries, the environmental regulations governing pesticide application, storage, and disposal are scarcely implemented (United Nations Environment Programme 2013b, 2014). So, relying only on legislation for pesticide management is not a practicable solution. Globally, governments should adopt environmentally sustainable approaches for pest and pesticide management like public awareness campaigns, training programs regarding proper pesticide application, mass awareness about use of minimum-risk products, and adoption of integrated pest management (IPM) system. Nonetheless, a firm legislative and regulatory framework strengthens all of these approaches.

15.5.1 Legal Penalty for Violating Environmental Regulations

For strengthening the environmental regulations, several global organizations framed legal penalty for violating environmental regulations. For instance, depending on the location and nature of landfill site, the landfill operators are bound to make financial assurances and are charged different amounts of licence fees by Environment Protection Agency (EPA). Any kind of unauthorized waste disposal draws legal penalty up to US$60,000 per individual (Kuppusamy et al. 2016).

15.6 Conclusion

Extensive use of pesticides has become a serious ecological and human health hazard globally. Due to the dearth of a comprehensible pesticides management approach, large amount of obsolete pesticide residues have already stockpiled over the time. So, there is an immediate need to develop technique for the degradation and removal of pesticides from contaminated sites. Assessing the environmental providence of pesticide residues and their imminent exposure risks to environment and public health is essential for formulating risk-based management strategies for risk reduction. Environmental impact assessment of pesticides influences characterization of previously permitted pesticides, approval of new pesticides, and establishment of remediation goals along with maintenance of environmental quality. Several developed regions have established pesticide risk assessment models which are of great significance for environmental sustainability. Biological control strategies, cultural control strategies, physical control strategies, mechanical control strategies, new molecular strategies, Farmer education and awareness, adhering to environmental regulations, rational use of pesticides (RUP), continuous monitoring of pesticide residues, and development of new pesticides with enhanced safety profiles are some effective and reliable pesticide management strategies. Pesticide ban is an important regulatory stride taken for the management of deleterious pesticides. Globally, several highly persistent and toxic pesticides have been banned hitherto with DDT ban an exemplary one. This chapter provides an overview of global environmental regulations framed for management of pesticides. The International Code of Conduct and Use of Pesticides, The Montreal Protocol, The Rotterdam Convention, The Basel Convention, The Stockholm Convention, The International POPs Elimination Network (IPEN), The International Labor Organization Convention No. 184 (ILO Convention 184), The Global Harmonized System of Classification and Labelling of Chemicals (GHS), Strategic Approach to International Chemicals Management (SAICM), Registration, Evaluation and Authorisation of Chemicals (REACH), The Clean Sweep Program, The Codex Alimentarius Commission, Global Good Agricultural Practices (GlobalGAP) and Pesticide Action Network organizations (PAN) are some imperative global environmental regulations framed for management of pesticides.

References

Abhilash PC, Singh N (2009) Pesticide use and application: an Indian scenario. J Hazard Mater 165:1–12

Arctic Council (1991) Arctic Environmental Protection Strategy (AEPS), Rovaniemi, Finland, p 45

Ballesteros ML, Miglioranza KSB, Gonzalez M, Fillmann G, Wunderlin DA, Bistoni MA (2014) Multimatrix measurement of persistent organic pollutants in Mar Chiquita, a continental saline shallow lake. Sci Total Environ 490:73–80

Basel Convention (2006) Technical working group updated technical guidelines on environmentally sound management of waste consisting of, containing or contaminated with persistent organic pollutants (POPs). Basel Convention, Geneva

Beesley L, Moreno-Jiménez E, Gomez-Eyles JL, Harris E, Robinson B, Sizmur T (2011) A review of biochars' potential role in the remediation, revegetation and restoration of contaminated soils. Environ Pollut 159:3269–3282

Bharat GK (2018) Persistent organic pollutants in Indian environment: a wake-up call for concerted action. Policy Brief, The Energy and Resources Institute (TERI). www.teriin.org

Bhat RA, Beigh BA, Mir SA, Dar SA, Dervash MA, Rashid A, Lone R (2019) Biopesticide techniques to remediate pesticides in polluted ecosystems. IGI Global, Hershey, pp 387–407

Boudh S, Singh JS (2019) Pesticide contamination: environmental problems and remediation strategies. In: Bharagava RN, Chowdhary P (eds) Emerging and eco-friendly approaches for waste management. Springer, Singapore, pp 245–269

Carneiro FF, Pignati W, Rigotto RM, Augusto LGS, Rizollo A, Muller NM, Alexandre VP, Friedrich K, Mello MSC (2012) A warning about the impact of pesticides on health - part 1: pesticides, food safety and health. Brazilian Association of Collective Health (ABRASCO). www.abrasco.org

Cheng Y, Zhou JY, Shan ZJ, Kong DY (2007) SCI-GROW model for groundwater risk assessment of pesticides. J Ecol Rural Environ 23:78–82

Dasgupta S, Meisner C, Wheeler D (2010) Stockpiles of obsolete pesticides and cleanup priorities: a methodology and application for Tunisia. J Environ Manag 91:824–830

Dhaliwal GS, Singh R, Chhillar BS (2006) Essentials of agricultural entomology. Kalyani Publishers, New Delhi

Dixit S, Srivastava MP, Sharma YK (2019) Pesticide and human health: a rising concern of the 21st century. IGI Global, Hershey, pp 85–104

ECHA (2008) Guidance on information requirements and chemical safety assessment. ECHA, Helsinki

ECHA (2011) Guidance on information requirements and chemical safety assessment part A: introduction to the guidance document, No./ECHA/G/15EN. ECHA, Helsinki, p 46

EEA (2007) Europe's environment: the fourth assessment. State of the environment ReportNo2/2007, Luxembourg, p 28

EU (2006) Regulation ECNo.1907/2006 of the European parliament and of the council. Off J Eur Union L 396:278

FAO (1955) Pesticide registration legislation, FAO legislative study no. 51. Food and Agricultural Organization, Rome

FAO (1988) Pesticide labelling legislation, FAO legislative study no. 43. Food and Agricultural Organization, Rome

FAO (2006a) Strategic program 2006–2011 for the implementation by FAO of the revised version of the international code of conduct on the distribution and use of pesticides. Food and Agricultural Organization, Rome

FAO (2006b) The implementation of the globally harmonized system classification and labelling of chemicals-FAO's past and present activities. Food and Agricultural Organization, Rome

FAO (2007) Designing national pesticide legislation, legislative study no. 97. Food and Agricultural Organization, Rome

FAO/UNEP/OECD/SIB (2001) Baseline study on the problem of obsolete pesticides stocks. Food and Agricultural Organization, Rome

Frazar C (2000) The bioremediation and phytoremediation of pesticide-contaminated sites. U.S. Environmental Protection Agency, Office of Solid Waste and Emergency Response Technology Innovation Office, Washington, DC

Gill HK, Garg H (2014) Pesticides: environmental impacts and management strategies. In: Pesticides- toxic aspects. IntechOpen, London, pp 187–230

Hui X, Yi Q, Bu-zhuo P, Xiliu J, Xiao-mei H (2003) Environmental pesticide pollution and its counter measures in China. AMBIO J Hum Environ 32:78–80

ICMR (2001) Pesticide pollution: trends and perspectives. ICMR Bull 31(9):87–88

ILO (2000) Safety and health in agriculture, international labour conference, 88th session, report VI (I). International Labour Organization, Geneva

IPSMC (1999) Developing and strengthening national legislation and policies for the sound management of chemicals, Inter-organization program for the sound management of chemicals. IPSMC, Geneva

Kumar PS, Carolin CF, Varjani SJ (2018) Pesticides bioremediation. In: Varjani SJ et al (eds) Bioremediation: applications for environmental protection and management, energy, environment, and sustainability. Springer, Singapore, pp 197–222
Kumar S, Singh R, Behera M, Kumar V, Sweta RA, Kumar N, Bauddh K (2019) Restoration of pesticide-contaminated sites through plants. In: Phytomanagement of polluted sites. Elsevier, Amsterdam, pp 313–327
Kumer K (1999) Prior informed consent for chemicals in international trade: the 1998 Rotterdam convention. RECIEL 8:322–329
Kuppusamy S, Palanisami T, Megharaj M, Venkateswarlu K, Naidu R (2016) Ex-situ remediation technologies for environmental pollutants: a critical perspective. In: de Voogt P (ed) Reviews of environmental contamination and toxicology, vol 236. Springer, Basel, pp 117–192
Lallas P (2001) The Stockholm convention on persistent organic pollutants. Am J Int Law 95:692–708
Levitan L, Merwin I, Kovach J (1995) Assessing the relative environmental impacts of agricultural pesticides: the quest for a holistic method. Agric Ecosyst Environ 55:153–168
Lundholm CE (1997) DDE-induced eggshell thinning in birds: effects of p,p'-DDE on the calcium and prostaglandin metabolism of the eggshell gland. Comp Biochem Physiol C: Pharmacol Toxicol Endocrinol 118:113–128
Marican A, Durán-Lara EF (2018) A review on pesticide removal through different processes. Environ Sci Pollut Res Int 25(3):2051–2064
MEA (2005) Ecosystems and human well-being: current state and trends, vol I. Millennium Ecosystem Assessment, Washington, DC
Morillo E, Villaverde J (2017) Advanced technologies for the remediation of pesticide-contaminated soils. Sci Total Environ 586:576–597
NCP (2003) Canadian arctic contaminants assessment report II-physical environment. Sources, occurrence, trends and pathways in the physical environment. Indian and Northern Affairs Canada, Ottawa, p 361
Pandey C, Prabha D, Negi YK (2018) Mycoremediation of common agricultural pesticides. In: Prasad R (ed) Mycoremediation and environmental sustainability. Fungal biology. Springer, Basel, pp 155–179
Pariatamby A, Kee YL (2016) Persistent organic pollutants management and remediation. Procedia Environ Sci 31:842–848
Persson EC, Graubard BI, Evans AA, London WT, Weber JP, Leblanc A, Chen G, Lin WY, McGlynn KA (2012) Dichlorodiphenyltrichloroethane and risk of hepatocellular carcinoma. Int J Cancer 131:2078–2084
Pimentel D, Acquay H, Biltonen M, Rice P, Silva M, Nelson J, Lipner V, Giordano S, Horowitz A, D'Amore M (1992) Environmental and human costs of pesticide use. Bioscience 42:750–760
Prokop Z, Nečasova A, Klanova J, Čupr P (2016) Bioavailability and mobility of organic contaminants in soil: new three-step ecotoxicological evaluation. Environ Sci Pollut Res 23:4312–4319
Rani K, Dhania G (2014) Bioremediation and biodegradation of pesticide from contaminated soil and water—a noval approach. Int J Curr Microbiol App Sci 3:23–33
Rice PJ, Rice PJ, Arthur EL, Barefoot AC (2007) Advances in pesticide environmental fate and exposure assessments. J Agric Food Chem 55:5367–5376
Rotterdam Convention Secretariat (2004) Guideline on the development of national laws to implement the Rotterdam convention, Rome/Geneva (revised 2005)
Schwarzenbach RP, Egli T, Hofstetter TB, Von Gunten U, Wehrli B (2010) Global water pollution and human health. Annu Rev Environ Resour 35:109–136
Singh A, Kuhad RC, Ward OP (2009) Biological remediation of soil: an overview of global market and available technologies. In: Singh A et al (eds) Advances in applied bioremediation, soil biology. Springer, Berlin
Song B, Zeng G, Gong J, Liang J, Xu P, Liu Z, Zhang Y, Zhang C, Cheng M, Liu Y, Ye S, Yi H, Ren X (2017) Evaluation methods for assessing effectiveness of in situ remediation of soil and sediment contaminated with organic pollutants and heavy metals. Environ Int 105:43–55

Spina F, Cecchi G, Landinez-Torres A, Pecoraro L, Russo F, Wu B, Cai L, Liu XZ, Tosi S, Varese GC, Zotti M, Persiani AM (2018) Fungi as a toolbox for sustainable bioremediation of pesticides in soil and water. Plant Biosyst 152:474–488
Sun J, Pana L, Tsangb DCW, Zhana Y, Zhu L, Li X (2017) Organic contamination and remediation in the agricultural soils of China: a critical review. Sci Total Environ 615:724–740
Sun S, Sidhu V, Rong Y, Zheng Y (2018) Pesticide pollution in agricultural soils and sustainable remediation methods: a review. Curr Pollut Rep 4:240
Teran T, Lamon L, Marcomini A (2012) Climate change effects on POPs' environmental behaviour: a scientific perspective for future regulatory actions. Atmos Pollut Res 3:466–476
Tripathi V, Edrisi SA, Chaurasia R, Pandey KK, Dinesh D, Srivastava R, Srivastava P, Abhilash PC (2019) Restoring HCHs polluted land as one of the priority activities during the UN-international decade on ecosystem restoration (2021–2030): a call for global action. Sci Total Environ 689:1304–1315
UNEP (United Nations Environment Programme) (2009) Global monitoring report under the global monitoring plan for effectiveness evaluation. In: Proceedings of the UNEP/POPS/COP.4/33 conference of the parties of the Stockholm convention on persistent organic pollutants, fourth meeting, Geneva, Switzerland
United Nations Environment Programme (2013a) Stockholm convention on persistent organic pollutants (POPs). Text and annexes as amended in 2009, 2011 and 2013. http://chm.pops.int/TheConvention
United Nations Environment Programme (2013b) Global chemicals outlook-towards sound management of chemicals. ISBN 978-92-807-3320. www.unep.org/chemicalsandwaste/Portals
United Nations Environment Programme (2014) Basel convention on the control of transboundary movements of hazardous wastes and their disposal. UNEP/BRS/2014/3. www.basel.int/Portals/4/Basel%20Convention/docs/text/BaselConvention
United Nations Environment Programme, Food and Agriculture Organization (2013) Rotterdam convention on the prior informed consent procedure for certain hazardous chemicals and pesticides in international trade. www.pic.int/TheConvention/Overview
Varjania S, Kumarb G, Rene ER (2019) Developments in biochar application for pesticide remediation: current knowledge and future research directions. J Environ Manag 232:505–513
Weiss FT, Leuzinger M, Zurbrügg C, Eggen RIL (2016) Chemical pollution in low and middle-income countries. Swiss Federal Institute of Aquatic Science and Technology, Dübendorf
WHO (2004) WHO recommended classification of pesticides by hazard and guidelines to classification. World Health Organization, Geneva
Xing-lu P, Feng-shou D, Xiao-hu W, Jun X, Xin-gang L, Yong-quan Z (2019) Progress of the discovery, application, and control technologies of chemical pesticides in China. J Integr Agric 18:840–853

Web-Links

GlobalGAP (2016) Global Good Agricultural Practices. www.globalgap.org
Lah K (2011) Effects of pesticides on human health. Toxipedia. www.toxipedia.org
Pesticide Action Network International (2014) About. http://pan-international.org/about/
UNEP (1978) Barcelona convention: convention for the protection of the Mediterranean Sea against pollution. http://www.unep.ch/regionalseas/regions/med/t_barcel.html
UNEP (2006) Manual on compliance and enforcement of multilateral environmental agreements. http://www.enep.org

Index

R. A. Bhat et al. (eds.), *Bioremediation and Biotechnology, Vol 2*,
https://doi.org/10.1007/978-3-030-40333-1

C

N

O

The manufacturer's authorised representative in the EU is Springer Nature Customer Service Centre GmbH, Europaplatz 3, 69115 Heidelberg, Germany. If you have any concerns regarding our products, please contact ProductSafety@springernature.com

Printed and bound by CPI Group (UK) Ltd, Croydon, CR0 4YY
22/09/2026
02221057-0009